W9-DDE-512

PHYSICAL SCIENCE

PHYSICAL SCIENCE

SECOND EDITION

Jerry S. Faughn, Ph.D.
Department of Physics and Astronomy
Eastern Kentucky University

Raymond Chang, Ph.D.
Department of Chemistry
Williams College

Jon Turk, Ph.D.
Darby, Montana

BROOKS/COLE

THOMSON LEARNING

Australia • Canada • Mexico • Singapore • Spain
United Kingdom • United States

BROOKS/COLE

★

™

THOMSON LEARNING

Chemistry/Physics Editor: John Vondeling
Developmental Editor: Laura Maier
Marketing Strategist: Marjorie Waldron
Production Manager: Charlene Squibb
Project Editors: Bonnie Boehme, Laura Shur
Text Designer: Jeannette Jacobs Design
Art Director: Jennifer Dunn
Picture Developmental Editor: Lori Eby
Copy Editor: Ellen Thomas
Illustrator: J.A.K Graphics
Cover Designer: Larry Didona

Cover Printer: Lehigh Press
Compositor: Tech Books
Printer: Von Hoffmann Press
Cover Image: Lightscapes by The Stock Market
Frontispiece: NASA

Credits for focus box icons:
Physics: Timothy J. Ebersole
Chemistry: Charles D. Winters
Earth Science: USGS Photo Library
Astronomy: NASA

COPYRIGHT © 1995 Thomson Learning, Inc. Thomson Learning™ is a trademark used herein under license.

ALL RIGHTS RESERVED. No part of this work covered by the copyright hereon may be reproduced or used in any form or by any means—graphic, electronic, or mechanical, including but not limited to photocopying, recording, taping, Web distribution, information networks, or information storage and retrieval systems—without the written permission of the publisher.

Printed in the United States of America
8 9 10 11 12   05 04 03 02

For more information about our products, contact us at:
Thomson Learning Academic Resource Center
1-800-423-0563

For permission to use material from this text, contact us by:
Phone: 1-800-730-2214
Fax: 1-800-730-2215
Web: http://www.thomsonrights.com

Library of Congress Catalog Number: 94-066359
Physical Science, Second Edition
ISBN: 0-03-001112-4

Asia
Thomson Learning
60 Albert Street, #15-01
Albert Complex
Singapore 189969

Australia
Nelson Thomson Learning
102 Dodds Street
South Melbourne, Victoria 3205
Australia

Canada
Nelson Thomson Learning
1120 Birchmount Road
Toronto, Ontario M1K 5G4
Canada

Europe/Middle East/Africa
Thomson Learning
Berkshire House
168-173 High Holborn
London WC1 V7AA
United Kingdom

Latin America
Thomson Learning
Seneca, 53
Colonia Polanco
11560 Mexico D.F.
Mexico

Spain
Paraninfo Thomson Learning
Calle/Magallanes, 25
28015 Madrid, Spain

ABOUT THE AUTHORS

Jerry Faughn's primary contributions to his field have been in the role of interpreter of physics for others. To this end he has been a professor at Eastern Kentucky University since 1968, where he now serves as Chair of the Department of Physics and Astronomy. During this time he has taught courses at all levels, from the lower division to the graduate level, but his primary interest has been and still remains those students just beginning to struggle with the concepts of physics. Faughn has written two general education texts in physics for beginning students and a microprocessor interfacing text for upper-division physics students. He is co-author of the national bestseller *College Physics,* fourth edition, from Saunders College Publishing. He has been the major author of the physics and astronomy portions of *Physical Science,* second edition.

Raymond Chang was born in Hong Kong and grew up in Shanghai, China. He was educated at London University in England and received his Ph.D. in physical chemistry from Yale University. Since 1968 he has taught at Williams College. Chang is the author of many books, including a bestselling general chemistry textbook, as well as physical chemistry and industrial chemistry texts. He is also the co-author of a book on the Chinese language, a novel for juvenile readers, and a children's picture book. His hobbies include tennis, violin, gardening, and photography.

Jonathan Turk received his Sc.B. in chemistry from Brown University in 1967 and his Ph.D. in organic chemistry from the University of Colorado in 1971. Following graduation, he wrote the first environmental science textbook published in this country, and has remained active as a textbook writer for the past 25 years. In 1988, Turk began collaborating with Gray Thompson, a geologist from the University of Montana. Together, Jon and Gray have traveled to remote regions of North America, the Arctic, and Asia on field studies, and have published several successful geology and earth science texts.

With the second edition of *Physical Science,* Thomson Learning has brought together a unique author team. Jerry Faughn, Raymond Chang, and Jon Turk have each authored best-selling texts in their respective fields, and their talents have been combined here to produce a complete, up-to-date, and highly readable introduction to the physical sciences.

Preface

TO THE INSTRUCTOR

This textbook, now in its second edition, contains sufficient material for a two-semester course in physical science but can easily be used in a one-semester course by omitting certain sections or chapters according to the interests or needs of the instructor. The course for which this book is used is one usually taken by nonscience students as a general eduction course in science. The primary emphasis of the text is to help the student reader understand the basic concepts and principles of physics, chemistry, earth science, and astronomy. To this end, the book is primarily conceptual rather than mathematical in its presentation. The mathematical techniques used in the book include a small amount of algebra, but "supermarket" arithmetic will suffice for a substantial majority of the numerical problems.

Changes in the Second Edition

A number of changes and improvements have been made in preparing the second edition of this text. Most of these changes are in response to comments and suggestions offered by users of the first edition and reviewers of the manuscript. Listed below are the major changes in this edition.

1. The astronomy chapters have been substantially revised to improve the accuracy and clarity of the presentation as well as to include up-to-date information concerning new discoveries. As only one among many examples, a discussion of the results obtained by the Cosmic Background Explorer (COBE) is used to amplify our knowledge of the early moments of the Universe.
2. The chemistry chapters have also been substantially revised and expanded to ensure that they are up-to-date and complete. One feature of these changes is to include material related to applications of chemistry in the day-to-day life of the student readers. For example, a discussion of soaps and detergents teaches the students some important features of how molecules behave while at the same time showing them how a product familiar to them does its work.
3. The earth science chapters have been updated and improved. Many users asked for the inclusion of material on historical geology, and this important topic is now covered.
4. About 30 percent of the problems have been revised, with much attention devoted to adding more conceptual problems.
5. Several topics have been added to the physics sections, including the conservation of angular momentum and a discussion of quarks and the standard model of the nucleus.
6. Several new focus boxes have been added to include new topics that are usually applications of science in the world of the student reader. Examples of these everyday applications include seatbelts and rearview mirrors, but topics of a more esoteric nature, such as buckyballs, are also considered.
7. The line drawings and photographs have been updated and improved to ensure that they serve as an effective pedagogical tool.

Features

The most important factor in any successful classroom is the instructor. If the instructor is interested, concerned, and knowledgeable in the field, the students will learn regardless of the nature of the textbook. However, the task of learning any subject matter becomes considerably easier for the student if this primary source material, the textbook, is as "user friendly" as possible. With this in mind, we have retained many features from the first edition and added certain others to enhance the text's usefulness to the student and the instructor. These features are discussed in that which follows.

- *Color* is used to add substance to the presentations. The following overall scheme for color renderings should be noted.

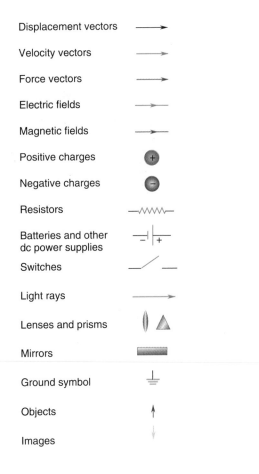

Displacement vectors	
Velocity vectors	
Force vectors	
Electric fields	
Magnetic fields	
Positive charges	
Negative charges	
Resistors	
Batteries and other dc power supplies	
Switches	
Light rays	
Lenses and prisms	
Mirrors	
Ground symbol	
Objects	
Images	

In addition to the use of color in the figures, various sections and features of the book are color coded or set off in some way for ease of identification and location.

Important equations are shaded with a light orange screen:

$$v \, (9.8 \text{ m/s}^2) \text{time.}$$

Example solutions are colored with a blue screen:

$$200 \text{ m/s}^2$$

Exercises are highlighted with a blue bar:

Exercise

Important statements are set off in italic type:

The orbit of a planet is an ellipse with the Sun at one focus. The other focus is empty.

- *Flexibility* is built into the text to accommodate the instructor's needs in course length and rigor. Some courses are completed in one semester, others in two, and naturally, the abilities and backgrounds of different groups of students differ as well. Also, there are inescapable connections between the various branches of physical science. For example, earth science can tell us what the conditions were like on the surface of our planet during the time when those chemistry experiments that eventually led to life were taking place. We make these connections at the appropriate places and times, but we do not let these connections dominate the course. That is, it is not necessary to follow the textbook in lock-step order. If you prefer to teach astronomy first, then the organization of the book is such that you are completely free to do so.

- *The use of mathematics is held to a minimum.* In fact, this textbook can be used as a so-called conceptual text in which no mathematics is used at all. Very frequently, however, the use of some mathematics will truly clarify a concept in ways that a verbal description cannot. Where mathematics is used, it is in the form of simple equations such as Ohm's law. Students need to know only basic arithmetic and simple algebra to do the computations. End-of-chapter problems that require some math manipulation are identified with a blue color so that you can skip them in your assignments if you so desire.

- *Example problems* are incorporated into the body of the text to help the student understand concepts by following the reasoning process offered in the solution. Examples also serve as a model to help students in their approach to similar end-of-chapter problems. For ease of location, these examples are set off from the body of the text with a color bar, and most examples

are given titles to describe their content. Examples that are strictly conceptual are indicated as such to facilitate incorporating the appropriate level of mathematics into the course.

- **Exercises,** which may follow examples, serve as a self-check for students' understanding of the concepts. Exercises are usually followed by a brief answer.

- **Special focus boxes** expose the student to various practical and interesting applications of physical principles. These sections are supplemental in the sense that studying them is not required to continue through the book.

- **Environmental and practical applications** are integrated throughout the text to broaden the students' understanding of theory, to retain their interest, and to relate their learning to the world in which we live.

- **Interviews** opening each section of the text are intended to provide inspiration for the student and to pique his or her interest. Interviews provide an opportunity for students to learn how leading researchers started in their chosen fields, and they enable those who have made great contributions to science to communicate what they find exciting in their areas of study.

- **A comfortable, relaxed writing style** is considered an aid for rapid comprehension. We have taken great care to ensure that the concepts of physical science are written not only

correctly but in such a manner that the student can read the explanations with understanding. Reviewers and previous users of this text indicate that we have developed a book in which concepts sometimes considered formidable—such as relativity or the study of the life and death of the stars—no longer seem so difficult, and in which learning science becomes an enjoyable undertaking.

Ancillaries

The following ancillaries are available with this text to assist the instructor teaching the course and to help the student learn the material covered.

Instructor's Manual/Test Bank contains answers to selected end-of-chapter problems, a bank of sample test questions, and a suggested reading list for each chapter.

Overhead Transparencies are available for about 100 full-color figures selected from the text.

A World View of Environmental Issues discusses environmental problems facing all of us. This supplement accompanies the text for those who wish to emphasize applications directed toward this aspect of the physical sciences.

Student's Study Guide with At Home Experiments contains chapter summaries, concept reviews, self-tests for each chapter, and experiments the students can do on their own to reinforce concepts learned in the chapter.

A Laser Disk containing a library of geology images can be accessed and shown on a video screen to enhance classroom presentations.

ACKNOWLEDGMENTS

The reviewers selected for this text were both careful and incisive with their comments. We thank them most sincerely. They are:

Stephen R. Addison, *University of Central Arkansas*

Robert J. Backes, *Pittsburg State University*

J. Bennett, *Arkansas State University*

Richard C. Brill, *Honolulu Community College*

Basil Curnutte, *Kansas State University*

Stewart Farrar, *Eastern Kentucky University*

Sandra Harpole, *Mississippi State University*

John Kalko, *Rancho Santiago College*

Paul D. Lee, *Louisiana State University*

William J. McIntosh, *Delaware State College*

James Merkel, *University of Wisconsin–Eau Claire*

John T. Netterville, retired, *David Lipscomb University*

Charles W. O'Neill, *Edison Community College*

Shanta Pal, *Moraine Valley Community College*

Jay Pasachoff, *Williams College*

Frederick R. Smith, *Memorial University of Newfoundland*

Francis M. Tam, *Frostburg State University*

Laura Thurlow, *Jackson Community College*

Karl Wetzel, *University of Portland*

Reviewers for the first edition were:

Robert Backes, *Pittsburg State University*

Basil Curnutte, *Kansas State University*

Stewart Farrar, *Eastern Kentucky University*

Fred Gamble, *Pensacola Junior College*

Joe Greever, *Delta State University*

Ted Morishige, *Central State University*

Douglas Magnus, *St. Cloud State University*

Donald E. Rickard, *Arkansas Tech University*

Oswald Schuette, *University of South Carolina*

Harry Shipman, *University of Delaware*

Ralph Thompson, *Eastern Kentucky University*

Aaron Todd, *Middle Tennessee State University*

Maurice Witten, *Fort Hays State University*

Finally, we owe a debt of gratitude to our friends at Saunders College Publishing who worked diligently on our behalf: John Vondeling, Vice President/ Publisher; Laura Maier, Developmental Editor; Bonnie Boehme and Laura Shur, Project Editors; Jennifer Dunn, Art Director; and the marketing team led by Marjorie Waldron, Vice President/Marketing.

TO THE STUDENT

Many valuable facts about this textbook are presented in the preceding section. Pay particular attention to the key on page vii for using color in the illustrations, since the use of color in this text has many functional purposes.

Study Hints

One of the most commonly asked questions of any instructor is, "How do I study for a science class?" The answer is essentially the same way that you do for any other class: Keep up on reading and do assignments on a day-to-day basis rather than try to cram the night before an exam. Set a study schedule for yourself and keep to it. There is plenty of time in your college or university life for outside activities and just plain fun. But, remember you should leave college knowing much more about the world in which you live than you did when you entered. The information discussed in science classes forms an important part of our understanding of how the world works. If approached with a positive attitude, you will find the task of understanding science can be enlightening and even fun.

It is, of course, vital that you attend class regularly and pay attention. Ask questions, both inside and outside of class, whenever you do not understand a topic being covered. Take concise notes in class and review them regularly, filling in with any additional material necessary to make the information easier to comprehend and retain. Science, perhaps more than most other courses that you will take, requires that you understand *concepts*. Memorizing every detail of Newton's laws of motion will not help a great deal if you cannot see the "big picture" of their meanings and applications. To this end, we have presented many worked-out examples to aid you in your understanding.

Consider the homework problems as a way to test your new skills. Do not lose patience if you cannot work every problem assigned or if you cannot derive a complete answer for every question asked. The successful student will *try* to answer every question. If you have already made a diligent effort to work these homework problems, you will get much more out of an instructor's review of these problems in class.

The level of mathematics used in the text has been held to simple manipulation of a few algebraic equations. If you do not know how to do these already, some useful refreshers on math techniques can be found in the appendices of this text. If you perceive yourself as having a weakness in mathematics, there is no better time than the present to correct it. Your college career is not just for you to excel in what you can already do; it is also a time for you to learn new skills and correct deficiencies.

The importance of taking good notes in class cannot be emphasized strongly enough. Many students, even good students, prefer to concentrate on what the instructor is saying as he lectures rather than trying to "write it all down" while the class is in progress. Perhaps that has been your approach, and if so, let us attempt to convince you that there is a better way. You need to develop a procedure for taking notes that will give you the best of both worlds. A simple shorthand approach of your own devising will enable you to take notes and listen at the same time, or to say it in a different way [shthnd, notes + list—same time]. No one cares how your notes look, whether they are punctuated properly, whether the spelling is correct, or even whether anyone except you can read them. The important thing is that you need something to refer to that you can understand when the time comes to study for an exam. Otherwise, your fallible memory will cause you to forget what topics were emphasized, what problems were discussed, and so forth. Your notes will not let you down.

Conclusion

Throughout your life you will make many decisions that will rely on your understanding of science. This textbook provides the fundamental principles that can aid you in making informed decisions enabling you to act rather than react. If you do not prepare yourself for such decisions, you will have to rely on the good will of others—others who may or may not have your best interests in mind. An understanding of science is valuable to you and to the world. We have made every effort to make science understandable and interesting for you. Now you must do your part and become an active participant in the learning process.

J.S. Faughn
R. Chang
J. Turk
October, 1994

Contents

6   Thermodynamics   113
7   Wave Motion: Sound   130
8   Electricity   155
9   Electromagnetism   188
10 Properties of Light   212
11 The Nature of Light   239
12 Inside the Atom   269
13 Nuclear Physics   289
14 Relativity and Elementary Particles   304

PART TWO / CHEMISTRY   333

15 Elements, Compounds, and Three States of Matter   337
16 The Periodic Table and Chemical Bonds   360
17 Principles and Applications   388

PART THREE / EARTH SCIENCE   409

18 The Atmosphere and Meteorology   415
19 Earth   460
20 Environmental Geology   507

PART FOUR / ASTRONOMY   539

21 Motion in the Heavens   544
22 The Solar System   569
23 The Life and Death of Stars   609
24 Galaxies and Time   629
    Appendices   A-1
    Glossary   G-1
    Index   I-1

1   Boundaries and Measurements in Physical Science   1

PART ONE / PHYSICS   17

2   Motion   22
3   Conservation Laws: Momentum and Energy   52
4   Newton's Law of Gravity and Some Special Kinds of Motion   75
5   Thermal Physics   95

Contents

**1 BOUNDARIES AND MEASURE-
MENTS IN PHYSICAL SCIENCE
1**

1.1   Boundaries   1
1.2   Systems of Measurement   7
1.3   Significant Figures   10
1.4   Conversion Factors   10
1.5   Scientific or Exponential Notation   11

PART ONE / PHYSICS   17
Interview: Steven Chu   18

2   MOTION   22

2.1   Early Ideas About Motion   22
2.2   Speed   23
2.3   Vectors and Scalars   24
2.4   Velocity   26
2.5   Acceleration   27
2.6   Acceleration Due to Gravity   28
2.7   Motion with Constant Acceleration   29
2.8   Forces and Motion   30
2.9   The Pythagorean Theorem   32
2.10   Newton's First Law of Motion   34
2.11   Inertia   35
2.12   Newton's Third Law of Motion   36
2.13   Newton's Second Law of Motion   39
2.14   Examples of Newton's Second Law   40
2.15   The Nature of Friction   42
2.16   Density   44
2.17   Pressure   46
Focus On
        Seatbelts and Inertia   37
        Friction Forces in the Human Body   43
        Density and Floating   45

**3   CONSERVATION LAWS: MOMEN-
TUM AND ENERGY   52**

3.1   What Is a Conservation Law?   53
3.2   Momentum   53
3.3   Conservation of Momentum   53
3.4   Work   56
3.5   Kinetic Energy   57
3.6   Gravitational Potential Energy   60
3.7   Conservation of Energy   61
3.8   Friction Forces and the
        Conservation of Energy   64
3.9   Power   65
3.10   Simple Machines   66
3.11   Conservation of Angular
        Momentum   68
3.12   Derivation of Kinetic Energy
        Equation   70
Focus On
        Transfer of Momentum in a Rocket Engine
        55
        Energy in the Human Body   64

**4   NEWTON'S LAW OF GRAVITY AND
SOME SPECIAL KINDS OF MOTION
75**

4.1   Early Ideas About Gravitation   75
4.2   The Law of Gravity   76
4.3   Weightlessness   79
4.4   Projectile Motion   80
4.5   Circular Motion   83
4.6   Centripetal Force   85
4.7   Orbiting the Earth   88
4.8   Centrifugal Forces   89
Focus On
        The Tides   76
        Artificial Gravity   86

5   THERMAL PHYSICS   95

5.1   Temperature   95
5.2   Temperature Scales   96
5.3   Early Ideas About Heat   97

5.4   Measuring Heat   98
5.5   Thermal Energy and Temperature   98
5.6   Specific Heat   100
5.7   Specific Heat at Work in Nature   102
5.8   Heat and Changes of Phase   103
5.9   Heat Transfer   105
5.10   Hindering Heat Transfer   108
5.11   The Greenhouse Effect   109
Focus On
        Evaporation and Boiling   99
        Boiling a Liquid and Freezing It at the
        Same Time   104
        The Hot Water Versus Cold Water Race
        106

6   THERMODYNAMICS   113

6.1   Work and Internal Energy   114
6.2   The First Law of Thermodynamics   115
6.3   Heat Engines   115
6.4   The Gasoline Engine   116
6.5   The Second Law of Thermodynamics   118
6.6   Alternative Forms of the Second Law   119
6.7   Consequences of the Second Law   122
6.8   The Second Law and the Generation of Electric
        Power   122
6.9   Thermal Pollution   124
Focus On
        A Flywheel-Operated Subway Train   116
        Planetary Winds   121
        The Drinking Bird   125

7   WAVE MOTION: SOUND   130

7.1   Wave Motion   130
7.2   Describing a Wave   130
7.3   Sound Waves   135
7.4   Sound: A Longitudinal Wave   136
7.5   The Speed of Sound   137
7.6   Frequency and Wavelength of Sound   138
7.7   Loudness of Sound and the Decibel Scale   139
7.8   Interference   140
7.9   Standing Waves   143

7.10 *Musical Sounds*   145 .

7.11 *The Doppler Effect*   150

Focus On

Telephone Frequencies   133

Noise   140

Resonance   144

Beats   146

Sonic Booms   148

8   ELECTRICITY   155

8.1   *Static Electricity*   155

8.2   *Measuring Charge*   158

8.3   *Coulomb's Law*   158

8.4   *Concept of Fields*   159

8.5   *Van de Graaff Generator*   163

8.6   *Voltage*   165

8.7   *Electric Circuits*   168

8.8   *Electric Currents*   169

8.9   *Resistance*   171

8.10 *Ohm's Law*   173

8.11 *Different Types of Electric Circuits*   174

8.12 *Power*   179

8.13 *Cost of Electrical Energy*   180

8.14 *Electrical Safety*   181

Focus On

How Does a Photocopying Machine Work?
157

A Black-and-White Television Receiver   166

Superconductivity   172

Microphones and Telephone Receivers
174

9   ELECTROMAGNETISM   188

9.1   *Magnets*   188

9.2   *Magnetic Fields*   189

9.3   *Magnetic Field of the Earth*   191

9.4   *The Connection Between Electricity and Magnetism*   192

9.5   *What Makes A Permanent Magnet Magnetic?*
196

9.6   *Cosmic Rays and the Earth's Magnetic Field*
198

9.7   *Electromagnetic Induction*   199

9.8   *Electric Motors*   203

9.9   *Electric Generators*   204

9.10 *Transformers and Transmission of Electricity*
206

Focus On

Magnetic Bacteria   192

Electric Guitars   201

Tape Recorders   202

10   PROPERTIES OF LIGHT   212

10.1   *Speed of Light*   213

10.2   *Rays of Light*   215

10.3   *Law of Reflection*   216

10.4   *Formation of Images*   217

10.5   *Flat Mirrors*   217

10.6   *Curved Mirrors*   219

10.7   *Refraction*   221

10.8   *Lenses*   225

10.9   *A Simple Magnifier*   226

10.10 *Telescopes*   230

10.11 *Total Internal Reflection*   235

Focus On

Rearview Mirrors   218

The Rainbow   226

Vision Correction with Lenses   228

Fiber Optics   232

11   THE NATURE OF LIGHT   239

11.1   *Is It a Wave or a Particle?*   239

11.2   *Wave Nature of Light—Young's Double-Slit Experiment*   240

11.3   *Particle Nature of Light—The Photoelectric Effect*   242

11.4   *Source of Light*   246

11.5   *Diffraction Grating*   249

11.6   *Spectral Analysis*   250

11.7   *Spectra and Astronomy*   253

11.8   *Electromagnetic Spectrum*   255

11.9   *What Is an Electromagnetic Wave?*   261

11.10 *Polarized Light*   263

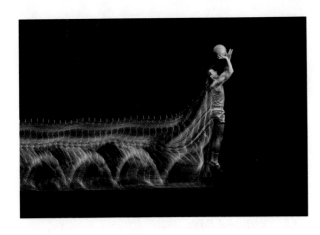

Focus On

Applications of the Photoelectric Effect 244

Color 256

Blue Skies and Red Sunsets 259

12 INSIDE THE ATOM 269

12.1 Atoms 269

12.2 The Discovery of the Electron 271

12.3 Early Models of the Atom 272

12.4 Inside the Nucleus 274

12.5 Isotopes 275

12.6 The Bohr Atom 276

12.7 Particles as Waves 277

12.8 Wave Mechanics and the Hydrogen Atom 278

12.9 Lasers 280

Focus On

Making a Hologram 283

Lasers in the Check-Out Line 285

13 NUCLEAR PHYSICS 289

13.1 Nuclear Stability 289

13.2 Radioactivity 291

13.3 The Decay Process 294

13.4 Half-Life 296

13.5 Neutrinos 296

13.6 Nuclear Reactions 298

13.7 Artificially Produced Nuclei 299

13.8 Carbon Dating 300

Focus On

Radon Pollution 291

The Geiger Counter 292

Smoke Detectors 293

14 RELATIVITY AND ELEMENTARY PARTICLES 304

14.1 Relativity Before Einstein 304

14.2 The Rest Frame of the Ether 305

14.3 The Michelson-Morley Experiment 306

14.4 Einstein's Postulates 308

14.5 Einstein and Time 309

14.6 Length Contraction 312

14.7 Relativity and Mass 314

14.8 Mass and Energy 314

14.9 Pair Production and Annihilation 315

14.10 Nuclear Fission 316

14.11 Nuclear Power Plant Considerations 318

14.12 The Design of a Nuclear Power Plant 319

14.13 Disposal of Radioactive Wastes 321

14.14 Nuclear Fusion 322

14.15 Quarks 326

Focus On

Atomic Bombs 310

PART TWO / CHEMISTRY 333

Interview: *Jacqueline K. Barton* 334

15 ELEMENTS, COMPOUNDS, AND THREE STATES OF MATTER 337

15.1 Study of Chemistry 337

15.2 Elements 338

15.3 Molecules 342

15.4 Compounds and Mixtures 342

15.5 Chemical Forces 343

15.6 States of Matter: Classification of Physical States 344

15.7 Properties of Gases 345

15.8 Boyle's Law 346

15.9 Charles' Law 347

15.10 Avogadro's Law 349

15.11 *Kinetic Theory of Gases* 350

15.12 *Liquids* 351

15.13 *Solids* 353

15.14 *Liquid Crystals* 354

Focus On

Buckyball—A New Allotrope for Carbon 339

The Pressure of Gases 349

16 THE PERIODIC TABLE AND CHEMICAL BONDS 360

16.1 *Electronic Structures of Atoms* 360

16.2 *Valence—Chemical Combining Capacity* 361

16.3 *Periodic Table* 362

16.4 *Chemical Equations and Types of Chemical Bonding* 366

16.5 *Ionic Substances* 368

16.6 *Covalent Bond* 372

16.7 *Hydrogen Bond* 375

16.8 *Organic Chemistry* 376

16.9 *The Miller-Urey Experiment and the Origin of Life* 378

16.10 *Network Covalent Substances* 380

16.11 *Metallic Substances* 382

Epilogue 384

Focus On

Bonding in DNA 379

17 PRINCIPLES AND APPLICATIONS 388

17.1 *Atomic Mass, Molecular Mass, and the Mole* 388

17.2 *Acids and Bases* 390

17.3 *The pH Scale* 391

17.4 *Oxidation and Reduction* 392

17.5 *Acid Rain* 394

17.6 *Catalysis* 397

17.7 *Ozone in the Atmosphere* 399

17.8 *Pollution by Automobile Exhaust* 401

17.9 *Soaps and Detergents* 403

Focus On

Pure Water 392

Expressions of Fractional Concentrations 400

PART THREE / EARTH SCIENCE 409

Interview: *Steven Schneider* 410

18 THE ATMOSPHERE AND METEOROLOGY 415

18.1 *Evolution and the Atmosphere* 415

18.2 *The Atmosphere Today* 418

18.3 *The Energy Balance of the Earth* 420

18.4 *Wind Systems* 425

18.5 *Condensation and Relative Humidity* 433

18.6 *Clouds* 435

18.7 *Precipitation* 436

18.8 *Oceans and World Climate* 438

18.9 *Weather Patterns* 441

18.10 *Frontal Weather Systems* 442

18.11 *Earth's Changing Climate* 448

18.12 *Natural Factors and Change of Climate* 449

18.13 Human Activities and Global Climate Change 454

18.14 Nuclear Warfare and Climate Change 456

Focus On

The Effects of the Speed of the Wind—
The Wind Chill Factor 426
Relative Humidity and Comfort 434
Common Symbols Used in Weather Maps
to Indicate Various Types of Fronts 447
Volcanoes and Weather 452

19 EARTH 460

19.1 Introduction 460

19.2 Structure of Earth 461

19.3 Rock Cycle 465

19.4 Geological Time 469

19.5 Historical Geology 470

19.6 Continents and Continental Movement—The Idea Is Born 476

19.7 Plate Tectonics—The Modern Theory 479

19.8 Plate Tectonics and Mountain-Building 482

19.9 Earthquakes and Volcanoes 486

19.10 Weathering and Erosion 495

19.11 Formation of Mineral Deposits 499

Focus On

Seismometers 464
The Richter Scale 488
The Los Angeles Earthquake of 1994: A Case History 490

20 ENVIRONMENTAL GEOLOGY 507

20.1 Soil 507

20.2 Soil Erosion 510

20.3 Loss of Farmland to Urbanization 513

20.4 The Hydrologic Cycle (Water Cycle) 514

20.5 Human Use of Water 515

20.6 Water Diversion Problems 516

20.7 Pollution of Inland Water by Nutrients 518

20.8 Industrial Wastes in Water 519

20.9 Ground Water 521

20.10 Pollution of Ground Water 523

20.11 Ocean Pollution 528

20.12 Nonrenewable Mineral Resources 529

Overview 534

Focus On

The Lipari Landfill: A Case History 525
Oil in the Ocean: The Wreck of the Exxon Valdez: A Case History 526

PART FOUR / ASTRONOMY 539

Interview: *Ben Peery* 540

21 MOTION IN THE HEAVENS 544

21.1 Constellations 545

21.2 Motion in the Heavens 546

21.3 Motion of the Moon 549

21.4 Motion of the Planets and the Sun 553

21.5 Aristotle 555

21.6 Copernicus 556

21.7 Tycho 557

21.8 Kepler 557

21.9 Galileo 560

21.10 Newton 562

22   THE SOLAR SYSTEM   569

22.1   Formation and Structure of the Solar System   569

22.2   The Sun   572

22.3   Study of the Moon and the Planets   577

22.4   The Moon   578

22.5   Mercury   581

22.6   Venus   582

22.7   Mars   586

22.8   Jupiter   589

22.9   Saturn   594

22.10   Uranus and Neptune   596

22.11   Pluto   599

22.12   Vagabonds of the Solar System   600

Focus On

How Is the Geological History of the Moon and the Planets Deduced?   577

How a Lot of Information Can Be Deduced from a Small Amount of Data   583

Extraterrestrial Life   598

The Titius-Bode Rule   601

Halley's Comet   603

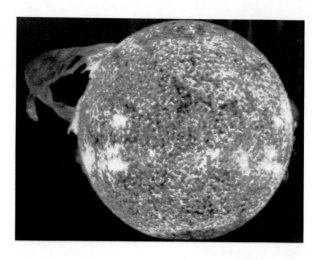

Focus On

Astronomical Distances   610

Measuring the Distance to Nearby Stars by Parallax   612

24   GALAXIES AND TIME   629

24.1   Milky Way Galaxy   629

24.2   Galaxies and Clusters of Galaxies   633

24.3   Energetic Galaxies and Quasars   635

24.4   Cosmology—A Study of the Beginning and the End of Time   637

Epilogue   641

Appendices   A-1

A.   Significant Figures   A-1

B.   Logarithms to the Base 10   A-2

C.   Direct and Inverse Proportionality   A-3

D.   Algebra   A-3

E.   Table of Atomic Weights   A-4

Glossary   G-1

Index   I-1

23   THE LIFE AND DEATH OF STARS   609

23.1   Studying the Stars   609

23.2   Brightness of Stars   614

23.3   Hertzsprung-Russell Diagram   615

23.4   Life of a Star   616

23.5   Death of a Star Like the Sun   618

23.6   Death of Massive Stars   620

23.7   Neutron Stars, Pulsars, and Black Holes   622

CHAPTER 1

BOUNDARIES AND MEASUREMENTS IN PHYSICAL SCIENCE

The subject of physical science extends from the smallest subatomic particles, to the energy in the faintest ray of light, to the primordial explosion of the entire Universe. This beautiful sunset only hints at the worlds of the incredibly small and incredibly large around us. (Courtesy Larry Ulrich, Tony Stone Worldwide)

1.1 BOUNDARIES

Physical science is the study of matter and energy. This incredibly broad subject covers a range from the smallest particle of the atom to the largest collection of stars and galaxies, from the energy in the faintest ray of light to the primordial explosion of the entire Universe. The physical scientist looks back in time to the formation of our Universe and forward to its ultimate fate. From the smallest to the largest, from the beginning to the end—these are the boundaries of this field of study. Within this realm, only life and living objects are omitted from the physical sciences. Living systems are covered by the biological sciences.

This book is an introduction to physical science. To begin, let us quickly pass from boundary to boundary, from the smallest to the largest. The remainder of this section is broken into short paragraphs, each accompanied by an illustration. Each paragraph covers an object or entity that is 10, 100, or 1000 times larger than the object in the previous paragraphs.

The unit of distance used in this section is the centimeter (cm). One centimeter is about the width of your little fingernail. Although 1 cm is easy to visualize in human terms, other numbers used in physical science are not at all easy to visualize. Within these short paragraphs, a tremendous range in size is covered from a scale of approximately 0.0000000000001 cm to 1,000,000,000,000,000,000,000 cm. Some of these quantities are so large or so small that it is often difficult to keep track of all the zeros. To make matters easier, the position of the decimal point in each number is noted in parentheses. In this system, the number 1 is taken as a reference. Now consider a large number such as 1000. The three zeros in 1000 indicate that the decimal point has been moved three places to the right of the number 1. Therefore, 1000 is followed by the reminder (+3 decimal places). In the number 0.1, the decimal point has been moved one place to the left of 1 and is followed by the reminder (−1 decimal place). Later in this chapter, an even more convenient shorthand for writing large and small numbers will be introduced.

One centimeter in the picture represents about 10 cm (+1 decimal place) in real life. Many common objects that people deal with in their everyday lives are 10 to 50 cm in length, width, or diameter. This textbook, for example, is about 20 cm by 30 cm.

(Courtesy NASA)

Our first jump brings us to objects approximately 100 times larger than your textbook. Therefore, the scale of our drawing has been changed, so that in this picture 1 cm represents 1000 cm (+3 decimal places). This first leap immediately takes us outside the range of things that can be picked up

and carried about, but the scale is definitely within the realm of ordinary human experience. A tree, an apartment building, or a large truck could all be shown here, but instead a space shuttle was chosen because it stands as a symbol of our scientific era.

Advancing our outlook by a factor of 10 moves us up to the range of the largest objects built by humans. One centimeter in the picture now represents 10,000 cm (+4 decimal places). Oil tankers and skyscrapers, both about the same size, are shown superimposed on one another to represent this boundary of human endeavor.

Another jump of 100 puts us at the one to a million scale. One centimeter in the picture represents 1,000,000 cm (+6 decimal places). The numbers are starting to get large and so are the objects. On this scale, Mt. Everest would rise a little less than 1 cm above sea level, and a group of mountains fits nicely on our 5 cm square.

(Courtesy NASA)

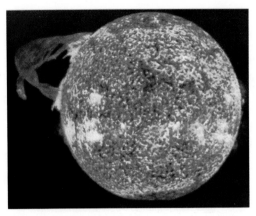

(Courtesy NASA)

One centimeter in the picture represents 1,000,000,000 (1 billion) cm (+9 decimal places). By the time our drawing reaches the one to a billion scale, the entire Earth comes into view. To a person on a ship in the middle of the ocean or on foot on one of the vast deserts, prairies, or ice caps of the planet, Earth may seem to be expansive or nearly boundless; but as our focus moves ever outward, this planet quickly recedes to a tiny speck in the cosmos.

One more jump of a factor of 10 brings the Sun into full view. One centimeter in the picture represents 100,000,000,000 cm (+11 decimal places). The diameter of the Sun is roughly 100 times the diameter of Earth and 10 times the diameter of Saturn and its magnificent rings.

(Courtesy NASA)

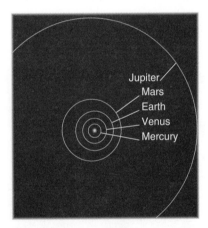

One centimeter in the picture represents 10,000,000,000 (10 billion) cm (+10 decimal places). A picture of Saturn and its most prominent rings fits nicely into our one to 10 billion scale, of the order of magnitude of ten times larger than Earth. Photographs of Saturn's rings taken by spacecraft in the early 1980s have added greatly to our knowledge of this planetary system.

One centimeter in the picture represents 10,000,000,000,000 cm (+13 decimal places). In this scale, the Sun appears as a small dot surrounded by the first four planets—Mercury, Venus, Earth, and Mars. The orbits of the planets appear as neat light blue lines in the picture, but it is important to realize that these are imaginary lines. If you were perched in a space capsule in a position to view the inner Solar System from this perspective, you would see the Sun shining in a black sky, and those planets that were visible would appear as small spots of reflected light.

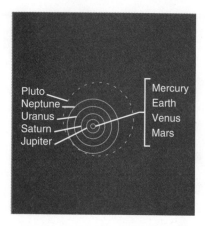

One centimeter in the picture represents about 1,000,000,000,000,000 cm (+15 decimal places). Here the entire Solar System comes into view. In this illustration, the Sun has become so small that it is impossible to draw it to scale and still see the dot. The dot is therefore larger than it should be. As you can see, in this drawing even the orbits of the inner planets are crowded together. The dominant feature now becomes the orbits of the five outer planets—Jupiter, Saturn, Uranus, Neptune, and Pluto.

Expanding our field of vision another 1000-fold brings two small spheres into the frame of vision. One of the spheres represents the Sun; the other represents the closest star, Proxima Centauri. One centimeter in the picture now represents 1,000,000,000,000,000,000 (+18 decimal places). In this vastness of outer space, it is cumbersome to express distances in centimeters. Therefore, astronomers have adopted a unit called a light-

year (ly). One light-year is the distance that a beam of light travels in a year. Since light travels at 30,000,000,000 cm/s (in a vacuum), a light-year is also proportionally large, approximately 1,000,000,000,000,000,000 (+18 decimal places), and Proxima Centauri is about 4.2 ly away from our Solar System.

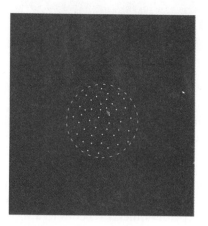

One centimeter in the picture represents about 100,000,000,000,000,000,000 cm (+20 decimal places) or 100 ly. Moving our perspective outward into space, a cluster of about 40 stars is shown. These form the local group that surrounds our Sun.

(Courtesy NOAO)

One centimeter in the picture represents about 10,000,000,000,000,000,000,000 cm (+22 decimal places) or 10,000 ly. The photograph shows a collection of some 100 billion individual stars. A collection of stars grouped together in this manner is called a **galaxy.** This picture is a galaxy like our own Milky Way galaxy. To imagine the scale of this drawing, think about the fact that a beam of light would have to travel some 100,000 years to traverse the Milky Way galaxy from edge to edge.

to scale, for in reality the spots of light would be much smaller than they are pictured here, and therefore most of the field of vision would be empty blackness. Astronomers estimate that several hundred billion galaxies exist and that the farthest one from us is about 15 billion ly distant.

One centimeter in the picture represents about 1,000,000,000,000,000,000,000,000 cm (+24 decimal places) or about 1 million ly. In a scale in which 1 cm represents 1 million ly, about a dozen separate galaxies, each one containing some 100 billion stars, are brought into view. If you were to try to draw a picture of our Sun on this scale, the required dot would be smaller than a single atom.

One centimeter in this picture represents 10 cm in real life. Before you lose yourself in the mysteries of intergalactic space, return for a moment to the ordinary, the book in your hand, in preparation for an imaginary journey into the realm of very small objects.

One centimeter in the picture represents 100,000,000,000,000,000,000,000,000 cm (+26 decimal places) or 100 million ly. In this last picture, each dot represents an entire galaxy, and many, many dots are shown. The picture is hardly drawn

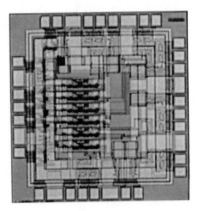

One centimeter in the picture represents about 1/10 cm or 0.1 cm (−1 decimal place). A square 0.1 cm on a side would just about fit inside the letter o in the print used in this book. As small as this seems, this is the size of an electronic circuit inside a computer chip. These chips provide the working channels and the memories of various types of computers, which have initiated the electronic revolution that is causing rapid changes in our society.

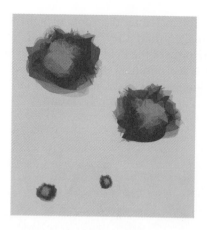

A 100-fold descent drops us into the range of objects that are barely visible. One centimeter in the picture represents 0.001 cm (−3 decimal places). Some day when you are indoors and sunlight is shining through the window, look at right angles to the sunbeam and you will see a haze of tiny pieces of dust about this size. This dust is made up of many different types of air pollutants: small particles of smoke and soot; some living organisms and pieces of dead ones; and small, oily droplets with various compounds dissolved in them or adhering to them.

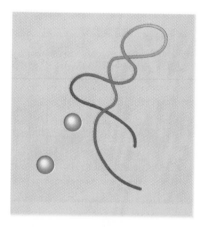

One more 100-fold descent brings the largest molecules into view. One centimeter in the picture equals 0.00001 cm (−5 decimal places). On the right side of the drawing you see a piece of a strand of DNA, the material that transmits genetic information from one generation to another. Thus, DNA is the molecular foundation for life as we know it. Next to the strand of DNA, two smaller molecules of starch are shown. Although starch molecules are still large compared with ordinary molecules such as those of water, oxygen, or sugar, they are much smaller than the strand of DNA, and less detail is seen.

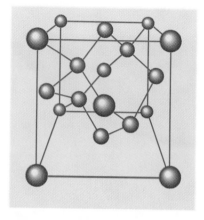

One centimeter in the picture represents 0.0000001 cm (−7 decimal places). Pictured here is a tiny segment of a diamond, which is made up of a regular array of carbon atoms. Objects of this scale are below the field of vision of even the most powerful electron microscopes; but if you could look into a world of this size, you would see individual atoms.

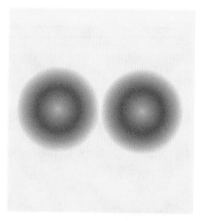

An additional ten-fold descent shows us an individual oxygen molecule floating in space. One centimeter in the picture represents 0.00000001 cm (−8 decimal places). If you wanted to draw a picture of a person on an equivalent scale, you would need a piece of paper approximately 200,000 kilometers (km) (120,000 miles) long!

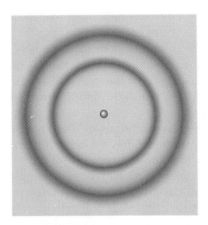

Another 100-fold drop brings us inside the atom itself. One centimeter in the picture represents 0.0000000001 cm (−10 decimal places). In this example, a carbon atom, pictured earlier as a part of the diamond, is shown. The dense central core, the nucleus of the carbon atom, is smaller than a period set in the print of this book. The rest of the space is essentially void, with an occasional electron whizzing about. Because electrons do not move in defined orbits as planets do, shading is used to picture the region in which the electrons are moving about.

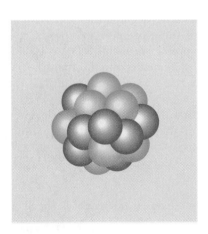

One centimeter in the picture represents 0.0000000000001 cm (−13 decimal places). In this picture, you see the nucleus of a carbon atom. Note that the scale has been magnified 10,000 times since two pictures ago when the oxygen molecule was in plain view. Thus, the nucleus, the dense central core of the atom, is much smaller than the atom itself. Yet the nucleus is made up of still smaller particles called protons and neutrons.

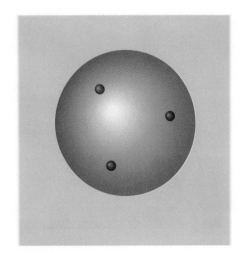

For a long time, protons and neutrons were considered to be elemental, indivisible particles, but today they are thought to be made up of even smaller particles. In this final picture, in which 1 cm represents 0.00000000000001 cm (−14 decimal places), an individual proton is shown to be composed of particles called quarks, which are surrounded by empty space. These particles are incredibly small, just as the Universe is incredibly large. No structure is shown here because no one really knows what a quark "looks like." Perhaps a quark has no structure.

1.2 SYSTEMS OF MEASUREMENT

Measurement is fundamental not only to experimental science but also to our whole technological way of life. Measuring devices are all around us. For example, clocks, thermometers, rulers, and bathroom scales are present in most homes, and automobiles are equipped with speedometers, fuel gauges, and a variety of other instruments.

A measurement must include both a quantity and a description of what is being measured. Thus, it is meaningless to say that an animal is 3. The quantity, by itself, does not give us enough information. The animal could be 3 years old, or 3 meters tall, or have a mass of 3 kilograms. Seconds (s), meters (m), and kilograms (kg) are all units. If you are asked to measure the distance between two points, your answer must include the magnitude of the length and the unit that you are using. For example, suppose you measure the length of a short pencil and find it to be 6.20 cm. The same distance

Dr. Sally Ride, mission specialist, talks to ground controllers from the flight deck of the Earth-orbiting Space Shuttle Challenger. Notice that she is surrounded by numerous gauges and measuring devices. (Courtesy NASA)

could also be expressed in other units because 6.20 cm is equal to 2.44 inches (in), or 0.0620 m, or 0.203 feet (ft).

It is much easier for different observers to communicate with one another if everyone uses the same system of measurement. The SI (for Système International), which is the present-day descendant of the metric system, is internationally recognized and is now used in nearly all the nations in the world. The use of a universal measuring "language" facilitates international communications.

As shown in the following table, the SI has seven base units from which all other units can be derived.

QUANTITY	UNIT	SYMBOL
Length	meter	m
Mass	kilogram	kg
Time	second	s
Temperature	kelvin	K
Electric current	ampere	A
Amount of substance	mole	mol
Luminous intensity	candela	cd

As we move through this text, we will encounter all of these units except the candela; it is included just to complete the set.

The SI base and derived units are used throughout this text. In some instances, references to the

British (or conventional) system of measurement are made for comparison.

The SI rules specify that the symbols or abbreviations for units are not followed by periods, nor are they changed in the plural. Therefore, scientists write, "The river is 10 m wide" (not "10 m. wide" or "10 ms wide").

Various combinations of these base units give us a large variety of **derived units.** A derived unit is made up of some combination of base units. For example, area is measured as a length times a length and can be expressed as (meter)(meter), or m^2. Similarly, volume is a length times a length times a length and can be expressed as m^3. As another example, speed is length divided by time and can be expressed as meters/second, or m/s.

The SI also includes units other than its official ones. Some of these units are accepted because they are so widely used, others are accepted only "temporarily," and still others are to be avoided. Hour (h), liter (L), and degrees Celsius (°C) are examples of accepted units. The calorie, sometimes used in the measurement of heat, is a unit that is now outside the SI.

The International System is particularly convenient because it provides a set of prefixes that express larger or smaller quantities than the standard units. The larger ones are multiples of 10, and the smaller ones are decimal fractions. The names and symbols of these prefixes are shown in Table 1.1. Thus, the prefix *kilo* means 1000, so 1 km is 1000 m. Similarly, *milli* means one thousandth (0.001), so 1 milligram (mg) is one thousandth of a gram, or 0.001 g.

The first three base units, and some units derived from them, are discussed in the following paragraphs.

TABLE 1.1	Names and symbols for SI prefixes	
PREFIX	SYMBOL	MULTIPLY BY
tera	T	10^{12}, 1 trillion
giga	G	10^9, 1 billion
mega	M	10^6, 1 million
kilo	k	10^3, 1000
deci	d	10^{-1}, one tenth
centi	c	10^{-2}, one hundredth
milli	m	10^{-3}, one thousandth
micro	μ	10^{-6}, one millionth
nano	n	10^{-9}, one billionth
pico	p	10^{-12}, one trillionth

Length

The SI unit of length is the meter. The meter originally was defined as one ten-millionth of the distance between the North Pole and the Equator. Since it is nearly impossible to calibrate ordinary measuring devices accurately against such a global standard, a more manageable meter was created by placing two marks on a platinum alloy bar (Fig. 1.1), which was called the International Prototype Meter. More recently, there has been a return to a "natural" standard for the meter, based on the speed of light. The current official definition of the meter is "the distance traveled by light in a vacuum during 1/299,792,459th of a second." Don't memorize this value; it is more important to be able to visualize the magnitude of the meter. For example, you might keep in mind that a kitchen cabinet or laboratory bench is almost a meter high, and that an average professional basketball player is about 2 m tall.

Area and volume are quantities derived from length. Area is the amount of space on a two-dimensional surface and is expressed in SI units in terms of square meters (m^2). Volume is the amount of space within a three-dimensional region and is expressed in terms of cubic meters (m^3) in SI units. For volume, the SI also recognizes the liter (L), which is $1/1000\ m^3$, or $1000\ cm^3$. A common unit derived from the liter is the milliliter (mL). One mL is $1/1000$ of a liter, or $1\ cm^3$.

FIGURE 1.1   *Length standards like the Standard Meter Bar, constructed of a platinum-iridium alloy, are ultimately compared to the standard based on the distance light travels in a vacuum in a given time.* (Courtesy National Institute of Standards and Technology, U.S. Department of Commerce)

FIGURE 1.2   *The National Standard Kilogram No. 20, an accurate copy of the International Standard Kilogram kept at Sèvres, France, is housed under a double bell jar in a vault at the National Institute of Standards and Technology.* (Courtesy National Institute of Standards and Technology, U.S. Department of Commerce)

Mass

The SI unit of mass is the kilogram. The standard kilogram is the mass of a block of platinum alloy stored at the International Bureau of Weights and Measures in Sèvres, France, and it is almost exactly equal to the mass of 1000 cubic centimeters (cm^3) of water at a temperature of 3.98°C (Fig. 1.2). You do not need to memorize this exact definition of a kilogram. We are more concerned with your ability to measure in these units and to visualize their magnitude.

Time

The SI unit of time is the second, which was originally defined as 1/86,400th of a day. The spin of the Earth varies slightly over time, however, so the length of a day is constantly changing. Therefore, a second is now defined as the duration of 9,192,631,770 vibrations of a specific type of light emitted by a specific atom (cesium-133). Once again, this is not a fact to be memorized; it is introduced merely to give an example of the type of precision that is available to and required by the scientific community of the 1990s. Other base units and various derived units are discussed in appropriate places in the text and are summarized in Appendix A.

1.3 SIGNIFICANT FIGURES

One of the problems often encountered by a beginning student of physics is that of significant figures. Just because your calculator reads to eight or more digits does not mean that you know the answer to a problem to that accuracy. For a discussion of significant figures and how to manipulate them, refer to Appendix B. Throughout the example problems and homework exercises in this text, unless otherwise specified, we shall assume that the given data are precise enough to yield an answer having three significant figures. Thus, if a problem states that a length is 5 m, it is understood that the length is actually known to be 5.00 m.

1.4 CONVERSION FACTORS

It is often necessary to change a quantity expressed in one unit to the same quantity expressed in another unit. For example, if you are traveling in France, your restaurant bill may come to 80 francs, and you might wish to know how much that is in U.S. dollars. Or if you are baking a pie, you might find it necessary to convert tablespoons to cups; in the laboratory, you might measure something in centimeters and wish to express it in meters. The mathematical manipulations do not change the cost of your dinner or the size of the object; they change only the units in which they are expressed.

To start with familiar units, consider the simple problem, "How many seconds are there in 1.5 minutes (min)?" You know the answer is 90 s, but it is instructive to study the reasoning process. First, write the equation

$$60 \text{ s} = 1 \text{ min}$$

Next, divide both sides of this equation by 1 min as

$$\frac{60 \text{ s}}{1 \text{ min}} = \frac{1 \text{ min}}{1 \text{ min}} = 1$$

or, more simply

$$\frac{60 \text{ s}}{1 \text{ min}} = 1$$

The fraction 60 s/1 min is called a "conversion factor" because it can be used to "convert" minutes to

seconds. Since this factor has a value of 1, its reciprocal

$$\frac{1 \text{ min}}{60 \text{ s}}$$

also has a value of 1 and is an equally correct conversion factor that can be used to "convert" seconds to minutes. Multiplying or dividing anything by 1 does not change its value, so it is correct to carry out such operations. How do we decide which of these two factors gives the correct answer to the problem? Let us try both to show which is right and which is wrong:

$$1.5 \text{ min} \frac{60 \text{ s}}{1 \text{ min}} = 90 \text{ s}$$

Note that the units cancel as algebraic quantities, as shown by the slash marks. If you multiply by the wrong conversion factor, you end up with a meaningless answer:

$$1.5 \text{ min} \frac{1 \text{ min}}{60 \text{ s}} = 0.025 \frac{\text{min}^2}{\text{s}}$$

This answer cannot be correct because the units are wrong. It is impossible for an incorrect use of a conversion factor to give a correct answer.

Table 1.2 gives some common conversion factors between SI and British units. A longer list is given in Appendix A and on the inside front cover of the text. Use of Tables 1.1 and 1.2 is illustrated by the examples provided.

TABLE 1.2	Handy conversion factors*	
TO CONVERT FROM	TO	MULTIPLY BY THE CONVERSION FACTOR
centimeters	inches	0.394 in/cm
feet	centimeters	30.5 cm/ft
	meters	0.305 m/ft
grams	pounds (avdp)	0.0022 lb/g
inches	centimeters	2.54 cm/in
kilograms	pounds (avdp)	2.20 lb/kg
kilometers	miles	0.621 mi/km
liters	quarts (U.S., liquid)	1.06 qt/L
meters	feet	3.28 ft/m
miles (statute)	kilometers	1.61 km/mi
pounds (avdp)	kilograms	0.454 kg/lb

*A more complete list is given in Appendix A and on the inside front cover.

EXAMPLE 1.1 GIVE THEM A THOUSAND METERS AND THEY WILL TAKE A KILOMETER

Long-distance runners run a 10,000 m race. How far is that in (**a**) km, (**b**) miles?

Solution (a) The conversion from meters to kilometers is 1 km = 1000 m or

$$\frac{1 \text{ km}}{1000 \text{ m}} = 1$$

Therefore,

$$10,000 \text{ m} \, \frac{1 \text{ km}}{1000 \text{ m}} = \boxed{10 \text{ km}}$$

(**b**) The conversion factor from Table 1.2 is

$$\frac{0.621 \text{ miles}}{\text{km}} = 1$$

Therefore,

$$10 \text{ km} \, \frac{0.621 \text{ miles}}{\text{km}} = \boxed{6.21 \text{ miles}}$$

EXAMPLE 1.2 LOOKING FOR A GAS STATION

A car can travel 500 km on a tank of fuel. How far is that in miles?

Solution The conversion factor, from Table 1.2, is

$$\frac{0.621 \text{ miles}}{\text{km}} = 1$$

Thus,

$$500 \text{ km} \, \frac{0.621 \text{ miles}}{\text{km}} = \boxed{311 \text{ miles}}$$

EXAMPLE 1.3 A RACE CAR OR A CLUNKER?

Suppose that a race car travels at an average speed of 240 km per hour. What is its speed in m/s?

Solution In this example, both units need to be changed, so two conversion factors are required. They are

$$\frac{1000 \text{ m}}{1 \text{ km}} \quad \text{and} \quad \frac{1 \text{ hour}}{3600 \text{ s}}$$

We have

$$240 \, \frac{\text{km}}{1 \text{ hour}} \, \frac{1000 \text{ m}}{1 \text{ km}} \, \frac{1 \text{ hour}}{3600 \text{ s}} = 66.7 \, \frac{\text{m}}{\text{s}}$$

EXAMPLE 1.4 GETTING A FEEL FOR SI

(**a**) A person who is 3 ft tall is probably a child. Is a person who is 3 m tall a child?

(**b**) A world-class runner can run a mile in a little under 4 minutes. Is it possible to run a kilometer in 4 minutes?

Solution These problems are designed to give you an intuitive feel for SI units. We need not solve for exact numerical answers.
(**a**) Since there are approximately 3 ft in a meter, a person who is 3 m tall is approximately 9 ft tall and is not a child, but a giant.

(**b**) A kilometer is equal to about 0.62 mile and is therefore considerably less than a mile. Thus, an athlete could easily run a kilometer in 4 minutes.

EXERCISE

Use your skills with conversion of units to determine if the conversions on the sign have been done correctly.

(Billy E. Barnes/Stock Boston)

1.5 SCIENTIFIC OR EXPONENTIAL NOTATION

During our introductory survey of the Universe from the largest objects to the smallest, very large and very small numbers were needed to express size. Recall that these numbers were cumbersome for the simple reason that it was tedious to count all the zeros. At that time, the number of decimal places in each number was noted in parentheses after the number. An even better system, called **scientific notation** or **exponential notation,** is used by scientists worldwide. This system is based on expo-

nents of 10, which are shorthand notations for repeated multiplications or divisions. The end result
is that all numbers are written as a number between
1 and 10 along with a "power of ten," which locates
the decimal point.

A positive exponent is a symbol for a number
that is to be multiplied by itself a given number of
times. Thus, the number 10^2 (read "ten squared"
or "ten to the second power") is exponential nota-
tion for $10 \times 10 = 100$. Similarly, $3^4 = 3 \times 3 \times 3 \times 3 = 81$. The reciprocals of these numbers are expressed by negative exponents. Thus, $10^{-2} = 1/10^2 = 1/(10 \times 10) = 1/100 = 0.01$.

Positive and negative powers of 10 are as follows:

$10^4 = 10 \times 10 \times 10 \times 10 = 1\underline{0000}$ (4 places to
the right)

$10^3 = 10 \times 10 \times 10 = 1\underline{000}$ (3 places)

$10^2 = 10 \times 10 = 1\underline{00}$ (2 places)

$10^1 = 10 = 1\underline{0}$ (1 place)

$10^0 = 1$ (0 places)

$10^{-1} = 1/10 = 0.1$ (1 place to the left)

$10^{-2} = 1/(10 \times 10) = 0.01$ (2 places)

$10^{-3} = 1/(10 \times 10 \times 10) = 0.001$ (3 places)

$10^{-4} = 1/(10 \times 10 \times 10 \times 10) = 0.0001$
(4 places)

Notice that to write 10^4 in longhand form you
simply start with the number 1 and move the decimal four places to the right, as 10000. Similarly, to
write 10^{-4}, you start with the number 1 and move
the decimal point four places to the left to arrive
at 0.0001.

It is just as easy to go the other way—that is, to
convert a number written in longhand form to an
exponential expression. For example, the decimal
place of the number 1,000,000 is six places to the
right of 1. Thus,

$1{,}000{,}000 = 10^6$

Similarly, the decimal place of the number
0.000001 is six places to the left of 1 and

$0.000001 = 10^{-6}$

What about a number like 3,000,000? If you
write it $3 \times 1{,}000{,}000$, the exponential expression
is simply 3×10^6. Thus, the mass of the Earth,

which, expressed in long numerical form, is
5,980,000,000,000,000,000,000,000 kg, can be written more conveniently as 5.98×10^{24} kg.

To multiply numbers in scientific notation, add
the exponents. To divide, subtract the exponents.
Thus,

$$10^4 \times 10^3 = 10^7$$

and

$$\frac{10^5}{10^3} = 10^5 \times 10^{-3} = 10^2$$

Scientific notation is used frequently with numbers in the metric system. Thus, there are 100 cm
in a meter or, in shorthand, there are 10^2 cm in a
meter. A kilometer is 1000 m, which can be written
10^3 m. Similarly, a microgram (μg) is 0.000001 g,
or 10^{-6} g. There are 1000, or 10^3, watts in a kilowatt and 1,000,000, or 10^6, watts in a megawatt.

EXAMPLE 1.5   HOW HIGH THE MOON?
(a) The distance from the center of the Earth to
the center of the Moon is approximately 3.8×10^5 km. How many meters is that? Write your answer in exponential and in longhand form.

(b) The distance between two hydrogen nuclei
in a hydrogen molecule is approximately
0.000000007 cm. Write this number in standard exponential notation.

Solution (a) There are 1000, or 10^3, m in a km.
Converting,

$$(3.8 \times 10^5 \text{ km})\left(\frac{10^3 \text{ m}}{1 \text{ km}}\right) = \boxed{3.8 \times 10^8 \text{ m}}$$

To convert from exponential to longhand
form, move the decimal point eight places to the
right:

380000000

This can be written as 380,000,000 m.

(b) As discussed in the text, the goal is to express
the number using the digit 7 (as a number between
1 and 10), followed by an exponent showing the
correct decimal place. Therefore, start with the 7
and count from right to left.

$$0.000000007 = 7 \times 10^{-9} \text{ cm}$$

FIGURE 1.3 *Since people in the United States must inevitably learn metric measurement to be compatible with the rest of the world, it is important to "think metric." Some useful rule-of-thumb relationships are shown.*

There are slightly more than 2 pounds in a kilogram

A quart is slightly less than a liter

A quart of milk has a mass of approximately 1 kilogram

An average-sized man has a mass of about 75 kilograms

A basketball player has a height of 2 meters or a little more

Mt. Everest has a height of 8850 meters

The distance between New York and Los Angeles is about 4850 kilometers

Water at 100°C is boiling

Water at 100°F is approximately body temperature

On most freeways in the United States the speed limit is 55 m/hr or approximately 88 km/hr

Water freezes at 0°C

Room temperature is about 20°C

SUMMARY

The introductory section reviews the scale of size within the Universe. The **International System of Units** (SI), an outgrowth of the metric system, is used in scientific circles throughout the world. A value in one unit can be converted to the same value expressed in another unit by multiplying by the appropriate conversion factor. Large or small numbers are conveniently expressed in terms of exponents of 10.

KEY WORDS

International System of Units (SI)
Base unit
Derived unit
Scientific or exponential notation

PROBLEMS AND CONCEPTUAL QUESTIONS

Problems requiring numerical work are identified with a blue number.

Systems of measurement

1. If something is selected to be a standard for some physical property, such as length or mass, list some properties that this standard should have in order to be as useful as possible.
2. When specifying the standard for a meter bar, the temperature is given. Why is it necessary to specify this temperature?
3. Why would a grandfather clock not be a good standard for the measurement of time?
4. How many (a) milligrams are there in a gram, (b) centimeters in a meter, (c) grams in a kilogram, (d) micrometers in a meter, (e) meters in a kilometer?
5. Discuss how your life might be affected today if at some time in the past it had been agreed that our hours would be replaced by decidays.
6. Estimate the height of a typical two-story home in meters.
7. Estimate the length or distance of the following: (a) The altitude in meters of a commercial airplane in normal flight. (b) The length of a housefly. (Use appropriate SI units.)
8. Estimate the time interval of the following: (a) The time in seconds between heart beats. (b) The average age of a college student in seconds.
9. Estimate the mass of the following: (a) The mass in kilograms of a baseball. (b) The mass of a housefly. (Use appropriate SI units.)
10. (a) A certain power plant has an output of 20,000,000 watts. Express this output in megawatts. (b) A component in a TV set is rated as 0.00000012 farad. Express this rating in microfarads.
11. The wavelength of a particular color light has a length of 0.000000580 m. Express this number in nanometers (nm).
12. A hardware store is selling chain at $1.00 per foot. What is the cost per centimeter?
13. Give two reasons why the SI is easier to use than the British system.
14. Identify an object in your everyday experience that has: (a) a mass of 1 to 2 kg, (b) a mass of 1 to 2 g, (c) a length of 1 to 2 mm, (d) a length of 1 to 2 cm, (e) a length of 1 to 2 m, (f) a length of 1 to 2 km, (g) a volume of 1 to 2 mL, (h) a volume of 1 to 2 L, (i) a volume of 1 to 2 m^3.

Conversions

15. A convenience store has several items on sale. Which are good buys and which are not? (a) Steak sells for $1.50 per kilogram. (b) Gasoline is $1.00 per liter. (c) Peanut butter is 10 cents per gram.
16. Fill in the blanks:
    2.5 in = _____ cm; 3.9 quarts = _____ L; 100 g = _____ kg; 16 miles = _____ km; 16 miles = _____ m; 80 km per hour = _____ miles per hour; 3.8 m = _____ ft; 0.1 cm = _____ in.
17. An average NFL lineman has a weight of about 280 lb. If there are about 2.2 lb in a kilogram at the surface of the Earth, what is his mass in kilograms?
18. A European dress pattern calls for 3 m of fabric. How many yards would you have to buy to make the dress?

19. The tallest building in the United States is the Sears Tower in Chicago with a height of 1455 ft. What is its height in cm?

20. A football player is 6 ft tall. How tall is he in meters? In centimeters?

21. The distance between New York and San Francisco is approximately 3000 miles. How far is that in kilometers?

22. Suppose it takes you about 2 minutes to read a page in this textbook. How many milliseconds does it take?

23. (a) A world class cyclist can pedal 300 miles in a day. Would it be possible to pedal 300 km in a day? (b) Five people would not ride comfortably in a 10 ft boat. What about five people in a 10 m boat?

24. The speed of light is approximately 3.00×10^8 m/s. Express this number in terms of cm/s, miles/s, and miles per hour.

25. The speed of sound is about 345 m/s. How fast is this in miles per hour?

26. A tablecloth is 45 cm by 60 cm. What is its area in cm^2 and in m^2?

27. Until recently, the world's land speed record was held by Colonel John P. Stapp, USAF. He rode a rocket-propelled sled that moved at a speed of 282.5 m/s. What is this speed in miles per hour?

28. A room in a house is 12 ft by 20 ft and has ceilings 8 ft high. What is the volume of this room in m^3?

29. A real estate agent is selling a lot at $10 per square yard. His competitor is selling an identical lot at $10 per square meter. Which is the better buy?

Exponential notation

30. (a) Write the following numbers in scientific notation: 1573; 0.00589; 647; 6,300,000; 0.16. (b) Write the following in longhand form: 1.5×10^4; 6.34×10^{-3}; 9.02×10^{-8}.

31. (a) Write the following numbers in longhand numerical form: The diameter of a bacterium is about 10^{-4} cm. The radius of the Sun is 6.9×10^5 km. There are approximately 10^{11} stars in the Milky Way Galaxy. (b) Write the following in exponential form: The temperature of the core of the Sun is approximately 15,000,000°C. The outer region of the Sun's atmosphere has a density of about 0.000000001 times that of the Earth's atmosphere.

32. (a) The mass of the hydrogen atom is 1.67×10^{-27} kg. Write this in longhand numerical

form. (b) The mass of the Earth is about 6×10^{24} kg. Write this in longhand numerical form.

33. (a) The distance of the Earth to the most distant quasar (a strange object that will be discussed in the section on astronomy) is about 1.5×10^{26} m. Write this out in longhand numerical form. (b) The diameter of an atomic nucleus is about 8×10^{-15} m. Write this out in longhand numerical form.

34. There are approximately 32,000 aluminum cans in a ton of scrap. How many cans are there in 10 tons? Express your answer in exponential notation.

35. An average automobile is driven about 10,000 miles per year. How many centimeters is it driven in a year? Express your answer in scientific notation.

ANSWERS TO SELECTED NUMERICAL PROBLEMS

4. (a) 1000, (b) 100, (c) 1000, (d) 1,000,000, (e) 1000

10. (a) 20 MW, (b) 0.12 microfarad

11. 580 nm

12. 3.28 cents per centimeter

15. 44300 cm

16. 2.5 in. = 6.4 cm   3.9 quarts = 3.69 L
100 g = 0.1 kg   16 miles = 25.7 km
16 miles = 25,700 m   80 km/hr = 49.8 miles/hr
3.8 m = 12.5 ft   0.1 cm = 0.039 in.

17. 127 kg

18. 3.28 yd

20. 1.83 m, 183 cm

21. 4830 km

22. 120,000 ms

24. 3×10^{10} cm/s, 186,000 miles/s, 6.70×10^8 miles per hour

25. 771 miles per hour

26. 2700 cm^2, 0.27 m^2

27. 632 miles per hour

28. 54.4 m^3

30. (a) 1.573×10^3, 5.89×10^{-3}, 6.47×10^2, 6.3×10^6, 1.6×10^{-1}; (b) 15,000, 0.00634, 0.0000000902

34. 3.2×10^5

35. 1.61×10^9 cm

PART ONE
PHYSICS

Physics will occupy the major portion of this textbook and will be the first of the physical sciences that we shall study—but why? What is there about physics that should allow it to take this premier position? To answer that question, let us briefly discuss what physics is and what it attempts to do.

Physics is the branch of science that tries to explain, often in terms of mathematical expressions, the fundamental processes that govern the world around us. These investigations include the study of mechanics, which deals with forces and motion; heat and thermodynamics; wave motion, under which we find the study of both sound and light; and electricity and magnetism. Also included are "modern" topics in physics, those explored only in the last 100 years, such as relativity, which is concerned with the behavior of objects traveling at very high speeds; and quantum mechanics, concerned with the behavior of submicroscopic particles.

The world of physics is all-inclusive. It will not be surprising, then, to find that the other areas of science make use of the fundamental laws of physics as a part of their subject matter. Let us begin our voyage toward understanding.

We must understand the laws and concepts of physics in order to understand motion in the world around us. (© David Leah, Tony Stone Worldwide)

STEVEN CHU INTERVIEW

(Courtesy of Frans Alkemade)

Steven Chu joined the Physics and Applied Physics Departments at Stanford University in 1987, where he currently serves as Chairman of the Physics Department. He earned his A.B. degree in mathematics and B.S. degree in physics from the University of Rochester in 1970 and a Ph.D. in physics from the University of California at Berkeley in 1976. He continued as a post-doctoral fellow at Berkeley until joining AT&T Bell Laboratories in 1978, where he became head of the Quantum Electronics Research Department in 1983 and remained until 1987.

In 1987 Professor Chu received the Herbert P. Broida Prize from the American Physical Society for his work on the world's first optical atom trap.

WHAT INSPIRED YOU TO BECOME INVOLVED IN A SCIENCE CAREER?

I've been very lucky my whole life, starting with my high school physics teacher, Mr. Minor, who won a national award for his teaching. I think by my senior year in high school in Garden City, L.I., I knew I wanted to be a scientist. When I was in graduate school at Berkeley, I worked side by side with my advisor, Professor Eugene Commins, sometimes eight and ten hours a day, six days a week. We shared all the tasks, even sweeping up all of the nuts and bolts that had fallen onto the floor during the week, sorting them out and putting them back in their respective drawers.

IS THAT A UNIQUE WORKING RELATIONSHIP?

Yes. Most professors don't work as closely with their students. Professor Commins truly loved the lab and liked to concentrate on one experiment at a time that he was excited about. I was very

spoiled to have worked side by side with him because the style, how you attack problems, and things of that nature rub off on you in a way you can't really define. It's like the imprint your parents leave on you even though they're not formally teaching you anything. You see how they behave, and some of it rubs off for better or worse. I had a very special relationship with my advisor.

WHAT LED YOU TO JOIN THE STANFORD FACULTY?

The whole idea of having students at a university is that you hope to imprint them with a certain way of looking at and solving problems, and also to cultivate scientific taste. You really want to teach them to try to address important problems, not necessarily problems they can solve easily, but problems that are fundamentally important.

HOW DO YOU INTERACT WITH YOUR STUDENTS?

When I first came to Stanford, I began by setting up the labs and working side by side with the students. But one of the reasons I came to Stanford was that I had so many things I really wanted to do and so little time. I am now working on four or five experiments at once, conceived of by me or by my students and me, and cannot concentrate on one experiment. In fact, the idea of graduate school is you learn by doing. It's an apprenticeship mode of teaching. My students and I interact on a technical basis and, so far, I'm at least trying to keep my hands on the nuts and bolts of each experiment, but it is difficult. For beginning exploratory experiments, where I'm not sure what's going to happen or what the rules are, I still go into the lab and play around myself to get a sense for whether something will go or not.

YOU GRADUATED FROM THE UNIVERSITY OF ROCHESTER WITH DEGREES IN BOTH MATH AND PHYSICS. WHY DID YOU CHOOSE A CAREER IN PHYSICS?

What I find so wonderful about physics is that what you're doing actually has touch with physical reality. You're trying to understand the way Nature works rather than trying to do clever things within a totally made-up world. With physics you can think about what's going to happen or what you might try to discover, and then you go into the laboratory and you try it. If there's some false logic, Nature comes right back and tells you so. It's usually a very humbling experience that doesn't allow you to go off into deep outer space. Though you must obey certain rules in math, it is not tied to a physical existence of any kind, and it is the link with Nature that excites me.

HOW DO YOU INVENT THINGS? DO YOU START WITH A MATHEMATICAL EQUATION?

"WHAT I FIND SO WONDERFUL ABOUT PHYSICS IS THAT WHAT YOU'RE DOING ACTUALLY HAS TOUCH WITH PHYSICAL REALITY AND, BECAUSE OF THAT, YOU'RE TRYING TO UNDERSTAND THE WAY NATURE IS . . . IF THERE'S SOME FALSE LOGIC, NATURE COMES RIGHT BACK AND TELLS YOU SO."

I don't invent things with mathematics. I don't sit at my desk and write a lot of equations that spawn the new idea. I have geometric pictures in my mind and visualize what something is doing—when I was working on the new theory of laser cooling, I was picturing electrons connected to springs—and the theory came from these pictures. Then the mathematics is like the paintbrush you use in order to get the idea down accurately. Whenever I invent things I seem to have to reduce it to some primitive, basic level.

HOW DO YOU TEACH THE CREATION OF NEW IDEAS?

By bringing the students in on the first glimmer of an idea, we then proceed differently' than you'd find in a textbook. Typically in a textbook you reduce things to a very clear, precise line of logic. That in itself is very beautiful, because it's the goal of physics to explain many things with a few basic principles. However, in the creation of new ideas in research it doesn't work that

way at all. When you're scrambling around for any piece of evidence that you can possibly find and you have to identify what you believe in and what you think might be wrong, you go through a much different process. What are the places you can hang your hat on? Where are the places you think it might not be o.k.? In textbooks when you learn the physics, you can hang your hat on every sentence. It almost seems as though formal physics course work is necessary to pay your dues. It becomes truly fun when you're out there on your own trying to find what new things will eventually be put in the textbooks.

WHAT TYPE OF LABORATORY EXPERIMENTS ARE YOU DRAWN TO? WHY?

It's a mixed bag. I'm fortunately drawn to many things. Because of my mathematical training, I always thought I was going to be a theoretical physicist. It seemed the most glamorous. You're taught to admire Einstein, Newton, Maxwell, and all these truly great people, and one dreams of doing that sort of thing. Especially when I was going to school during the '60s and '70s, the high point was elementary particle physics. That field was the most pure, basic area trying to get at the most fundamental questions.

WHY DO WE WANT TO ANSWER THESE FUNDAMENTAL QUESTIONS?

You want to go deeper and deeper because you're really looking for simplicity. As a matter of faith I believe that the Universe can be explained with a very small set of basic fundamental ideas and principles and that all else follows from that. As an example from ordinary physics (by ordinary physics I mean on the energy scales where things like life exist), there's a theory called Nonrelativistic Quantum Mechanics. It's felt that this theory accurately describes how chemistry works. And because it accurately describes how chemistry works, it describes how biology and life work. Over the years chemists, biologists, and physicists have learned and are still learning how to apply the understanding of very fundamental laws of quantum mechanics, which was developed to explain how electrons go around the nuclei of atoms to describe how, for example, proteins are constructed.

THE DIFFERENT AREAS OF SCIENCE THAT YOU'VE SPANNED SEEM UNRELATED. AMONG YOUR ACHIEVEMENTS IS THE FIRST LASER SPECTROSCOPY OF AN ATOM AND ITS ANTIPARTICLE, AND NOW YOUR WORK IS MOVING INTO THE BIOSCIENCE AREA. IS THIS A LOGICAL PROGRESSION?

On the surface, it looks like they're all completely unrelated, but it's a progression that resulted from following my nose. You do a piece of work led by your own and other's previous work, and that opens up a new principle that you didn't really appreciate, and you say, "Gee what can I do with this new thing?" and then you look around for the best application of this new discovery or capability and apply it to that problem. And I think the wider you look, the more likely you are able to find the best applications.

DID THIS PROCESS LEAD TO THE WORK YOU ARE DOING CURRENTLY IN BIOLOGY?

Yes. That work started with my work in cooling atoms to manipulate and trap them with lasers. A colleague of mine realized that the same types of traps that we made to hold onto atoms could be used to hold onto particles. He then discovered that he could hold onto live bacteria (that you can watch swim around under a microscope, bring in a light trap and move them around). That led me to think, "Well if you can do that, why not move around a single molecule the same way?" Then I thought, "But you have to hold onto the molecule and physics tells you that you can't readily hold onto it in the most straightforward way." So what we've done is to use biochemistry tricks to glue little handles of (commercially available) polystyrene spheres onto the ends of our molecules, and grab onto the little handles with the light traps (we call them laser tweezers) and move the molecules around. So, you see, the biology is an outgrowth of a technology we were developing for atoms for totally different reasons.

WHERE DO YOU HOPE YOUR WORK IN BIOLOGY LEADS?

From a philosophical point of view: My work and other people's work has enabled us to do experiments on single atoms. I want to be able to do similar, related experiments on individual biological molecules. Normally, chemists and biologists do experiments over millions of mole-

cules at once, and a lot of interesting features get averaged away. When you stop to think about it, some of the chemistry that's happening in the nucleus of a cell is chemistry on a single molecular basis. There's only one copy of a particular DNA in each cell. And so we want to know how a lot of these features work, because chemistry in life is working on a singular molecular basis at this level. My hope is that maybe it will lead to something that would cure an illness. What we're doing now is building a new capability to study how things work on a fundamental molecular level.

HOW DO YOU BUILD A NEW CAPABILITY?

Traditionally there is a class of physicists that basically are instrument-builders. They invent new things such as lasers, or new techniques such as magnetic resonance imaging, which in the hands of biologists and doctors have been used as important biological tools. I'm essentially trying to invent more new techniques that a large number of biologists could apply to the problem that interests them. I'm also interested in biology. There is also the hope that other biologists will take the same technique and apply it to their problems.

YOUR WORK ON THE WORLD'S FIRST OPTICAL ATOM TRAP CONTRIBUTED TO YOUR RECEIVING THE 1987 HERBERT P. BROIDA PRIZE FROM THE AMERICAN PHYSICAL SOCIETY. CAN YOU EXPLAIN SIMPLY WHY IT WAS SUCH AN IMPORTANT EXPERIMENT?

That experiment showed for the first time that you can take a neutral particle and you can hold onto it at a distance with a light beam and move it around. That means that you can move objects at a distance without getting anatomically close, and you can move them inside other things as long as the light can penetrate. To give you an example of how you might use this: It has already been shown that you can send a light beam through a living cell, pick up and hold onto an object within the cell, for example, a chromosome within its nucleus, and move it to another part of the nucleus. That means you have a new tool that's better than surgery because you don't have to cut anything open.

WHAT WAS THE KEY TO YOUR SUCCESS IN THIS EXPERIMENT?

It was a very strange situation in that all the pieces were there ready to be exploited, but you had to know that all the pieces were there. There were technical obstacles in the experiment, and you can get around them by borrowing tricks in other fields. But if you're in atomic physics, you might not know about tricks that communication engineers use.

HOW DID YOU KNOW ABOUT THIS TRICK?

At the time, I was at Bell Labs, and I went to some engineers there who were developing hardware for communications. They were making high-speed electrooptic modulators and I asked them to help me make one. And that was one of the key ingredients for trapping the atoms.

WHAT TYPE OF ENVIRONMENT CAN TODAY'S GRADUATES IN PHYSICS HOPE TO FIND?

During the time that I was working on my Ph.D. in the early '70s we were told that there were no jobs for physicists, and I knew that the chance of getting a good job in physics looked dismal. About half of my fellow students in the entering class dropped out before getting their Ph.D., in large part because they felt they couldn't get jobs. I put blinders on and said, "As long as I'm having fun, I'm going to continue as long as someone will let me do it, and we'll see what happens after that." Right now, you don't enter physics to get a good-paying job. You enter it because it's fun, and as long as people will let you do this, just do it. Because things can turn around and if they don't, you can walk away and say, "Well, I've had a lot of fun, I'm trained in a very constructive way of rigorous thinking that I can apply to any other intellectual endeavor," and you're better for that. So, you can't lose.

CAN YOU DESCRIBE THE MOST EXCITING MOMENT IN THE LABORATORY?

First, there is the conception of an experiment and that's tremendously exciting, because you say, "Aha! At last I can solve the problems that you and others have been banging their heads against the wall over, and I see a way of doing it." So the first exciting moment is when you think you can pull it off. But the most exciting moment is when you're in the lab, you've been working on an experiment for a year or several years, and you see your first sig-

nal. In the laser-cooling experiment, the first time we saw a cloud of atoms sitting there held in space by light was real ecstasy. We were on Cloud Nine for weeks.

Scientists have a very manic/depressive existence because most of the time things aren't working. As Winston Churchill said: "Progress is made by moving from failure to failure with undiminished enthusiasm." And you have to do that. You have to keep plugging away. You overcome one obstacle and there's another behind it. A lot of time it's hard work—you're plugging away, and then it works, and you're tremendously excited. That's what makes it all worthwhile.

This interview was conducted by Janice K. Mandel and first appeared in *College Physics, Third Edition* by Raymond A. Serway and Jerry S. Faughn, Saunders College Publishing, 1992. The interview appears here in abbreviated form.

CHAPTER 2

MOTION

We will begin our discussion of physics by studying motion, a subject that is a common part of our daily existence. In fact, motion is so common in our everyday lives that most of the topics covered are already somewhat familiar to you. That is, words such as speed, velocity, and acceleration are currently a part of your vocabulary. Thus, one of the primary goals of this chapter is to be sure that your intuitive ideas about motion are correct and to place some of these ideas into a slightly more quantitative context than they may be at present.

In a larger sense, this chapter begins the study of a portion of physics referred to as *mechanics*. Mechanics is the study of motion and of the forces that cause motion. Thus, this chapter is divided into two parts. The first deals with the study of motion without regard to what causes the motion. This segment of mechanics is often referred to as *kinematics*. Following this, we shall investigate the relationship of forces and motion, commonly called *dynamics*.

2.1 EARLY IDEAS ABOUT MOTION

It is somewhat surprising to find that descriptions of motion were slow in developing. In fact, great scholars worked on this subject for thousands of years before our present-day ideas became firmly established. One of the first serious students of motion was Aristotle (384–322 B.C.), who considered the movement of objects in terms of what he called "natural motion." He taught that an object moved according to the "nature" of the object. He stated that there are four elements—earth, air, fire, and water—that are the basic building blocks of the world around us. Also, according to his viewpoint, every object had a proper, or natural, place where it should be. A rock, for example, was composed of the earth element, and if it were tossed upward, it would attempt to return to the Earth. A lighter ob-

Newton's laws of motion can be used to describe the action in a football game. (Courtesy Greg Perry)

ject, such as a feather, would do the same, but because it was not as heavy as the rock, it would not try as diligently to return to its proper place, and therefore it would fall more slowly to the Earth. Smoke was composed of the fire element, and its natural place was above the air element. Therefore, smoke would drift upward from a smoldering fire.

These teachings of Aristotle, along with others that we will consider later, became a firmly entrenched part of people's outlook on nature, and, in fact, these ideas reflected everyday observations reasonably well.

The main difficulty with Aristotle's approach was that some of his theories were at odds with physical observations. Additionally, Aristotle presented his ideas without resort to experimentation. This means that his study of motion closely resembled that of a philosophical argument built on logic and debate without resorting to what Aristotle considered to be the lowly task of seeing if his laws really described events occurring in the natural world. Nevertheless, the authority of Aristotle was such that his ideas held sway for about 2000 years. Thus, his incorrect physics has been taught and believed far longer than has our present-day view of the subject. Aristotle's science was eventually challenged by Galileo Galilei (1564–1642), considered by many to be the dominant figure in leading the world of physics into the modern era. Among the changes wrought by Galileo was the introduction of the concept of time into the study of physics. For example, Aristotle held that the most important feature in determining the motion of an object was how far away it was from its proper place. Galileo, however, recognized that it was the time of fall, or the time

Galileo Galilei

Legend holds that Galileo demonstrated that all objects fall with the same acceleration by dropping various test objects from the top of the Leaning Tower of Pisa.

for the motion to occur, that was an important missing link. With the introduction of time into physics, Galileo was able to develop the concept of accelerated motion. Let us now examine some of our present-day concepts of motion.

2.2 SPEED

In some fashion, people have always been concerned with the subject of **speed.** For example, in today's world, our busy schedules cause us to glance frequently at our watches to see how fast we must travel to make it to an appointment. Likewise, prehistoric people also must have pondered speed as they sought ways to capture a rapidly moving antelope.

Everyone can already give a basic definition of speed. In fact, when you tell someone how fast you are going in a car, you are defining speed. For example, suppose you are traveling at 55 miles per hour. The units, miles per hour, are essentially a definition of speed. That is, speed is the distance an object moves divided by the time required for the object to travel this distance. In equation form this is

$$\text{speed} = \frac{\text{distance}}{\text{time}}$$

or symbolically,

$$\bar{v} = \frac{d}{t} \qquad (2.1)$$

where d is the distance traveled and t is the time required to cover this distance. The bar over the v indicates an average value.

Equation 2.1 is the first equation to be introduced in this text. Equations are simple, concise statements of the relationships between physical properties. Think of an equation as a declarative sentence, in which the verb "is" is replaced with an equal sign. For example, the declarative sentence, my height, h, is twice the length of my leg, L, can be expressed in equation form as $h = 2L$. An equation not only summarizes a given relationship, but also by omission it tells us what factors do not affect the others. For example, suppose someone asked you how the color of a vehicle affects its speed. Of course, you know that the speed of a vehicle is in no way affected by its color. This intuitive knowledge is supported by a look at the equation, for it contains no symbol for color. This example may sound trivial, but as more complex concepts are introduced, don't forget this generalization. Many students have become confused by making a problem more complex than it is and trying to include factors that are irrelevant.

As an example, suppose a car moves 5000 m in 200 s. The speed is

$$\bar{v} = \frac{d}{t} = \frac{5000 \text{ m}}{200 \text{ s}} = 25.0 \text{ m/s}$$

It is often useful to define two different types of speed, **average speed** and **instantaneous speed.** The definition presented in Eq. 2.1 is that of average speed. Average speed gives us information about the gross characteristics of a trip but omits many important details. For example, in the 5000 m trip, it may well have been that the driver traveled fast at the beginning, stopped for a traffic light, then made the last portion of the trip at a moderate speed because of a nearby police car. Nonetheless, the result was that the car traveled the 5000 m in 200 s for an average speed of 25.0 m/s.

The instantaneous speed of an object is the speed of an object at any instant of time. The speedometer on your car tells you your instantaneous speed.

EXAMPLE 2.1 STRIKE THE BUM OUT!

Sound travels at a speed of approximately 345 m/s. If you are sitting in the stands at a baseball game, how long after a player 275 m from you hits a baseball do you hear the sound of ball striking bat?

Solution The equation for speed can be solved for time as

$$\text{time} = \frac{\text{distance}}{\text{speed}}, \quad \text{or} \quad t = \frac{d}{v}$$

If you need help with such algebraic manipulations, consult Appendix E. We find

$$t = \frac{d}{v} = \frac{275 \text{ m}}{345 \text{ m/s}} = 0.797 \text{ s}$$

You should be sure to try calculations like this on your own. If you use a calculator, this device usually gives your answer out to eight or ten decimal places. However, our problems are always stated so that three-place accuracy is assumed. This means that you should consider the answer out to only three significant figures. (For a more in-depth explanation of significant figures, consult Appendix B.)

CONCEPTUAL EXERCISE

Often we follow up one of our example problems with an exercise for you to do to test your understanding. Our first one is nonmathematical.

A car travels 1000 m in 20 s. (**a**) How many average speeds does it have, and (**b**) how many instantaneous speeds does it have?

Answer (**a**) one, (**b**) an infinite number

2.3 VECTORS AND SCALARS

So far in this chapter you have encountered only three different physical quantities—distance, time, and speed—but as you continue your study, you will meet many more. All of the physical quantities that you will run into, however, can be placed in one of two categories—**vector** quantities and **scalar** quantities.

A scalar quantity is one that requires only a knowledge of its magnitude (and the units associated with the

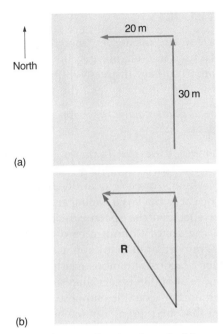

FIGURE 2.1 *The triangle method for adding vectors.*

ing sketch. Then a vector is drawn that starts at the tail of the first and ends at the tip of the second. This is vector **R** in Figure 2.1b. The length of this vector can be measured from the figure and converted to an actual displacement by using the same scale that was used initially (1 cm = 10 m). If you try this for yourself, you will find this displacement to be approximately 36 m. This general process is referred to as *adding vectors,* and the specific technique used here is called the *triangle method for vector addition.*

Force is another common physical quantity that is a vector. We have not discussed forces yet in this text, but your intuitive ideas about forces should be sufficient to enable you to understand why both a magnitude and direction must be specified for them. For example, suppose you walk up to a door that is standing open and exert a 10 pound (lb) force on it. A simple statement of the strength of the force, 10 lb, does not provide enough information to tell you what is going to happen to the door. The door moves differently if a 10 lb push is exerted on the door than it does if the force is a 10 lb pull. Thus, force is a vector quantity.

We shall indicate that a quantity is a vector with boldface notation. However, when we are concerned only with the magnitude of a vector and not its direction, we shall omit the bold notation.

magnitude) to tell you all you need to know about it. Examples of scalar quantities are the number of pages in this textbook, the amount of fluid in a container, the temperature of a room, and the speed of a car.

A vector quantity has both magnitude and direction. An example of a vector quantity is **displacement.** *A displacement is the straight-line distance and direction from where motion begins to where it ends.* To see that displacements are vectors, consider the following situation. Suppose that someone tells you that a football player made a run of 30 m, stopped momentarily while he avoided a tackler, and then ran 20 m farther. How far is the player from his initial position? The answer that may pop into your mind is 50 m. However, you have assumed that both runs, or displacements, were in the same direction, which is not necessarily the case. The 20 m run could have been in the opposite direction to the 30 m run, in which case, the distance of the player from his starting point would be 10 m. Or the 30 m run could have been north and the 20 m run toward the west. How far away from the starting point is he in this case? The steps for answering this question are shown in Figure 2.1. First, a convenient scale, such as 1 cm equals 10 m, is used to draw the displacements, maintaining their relative directions with respect to one another. Figure 2.1a shows the result-

CONCEPTUAL EXERCISE

This is a roadside "vector." Decide what vector quantity this sign is asking the driver to change. Is it displacement, velocity, acceleration, or more than one of these?

(© Pat Le Croix, The Image Bank)

The arrow on this weathervane always points in the direction the wind is blowing. What vector quantity, if any, is described by this arrow?

Answer The best answer is either displacement or velocity for the road sign.

The weathervane gives the velocity of the wind.

EXAMPLE 2.2   I'M TIRED, AND I'M WEARY, AND I WANT TO GO HOME

The triangle method for vector addition applies even if the displacements are not at right angles to one another as they were in the previous example. For example, suppose that a jogger runs 20.0 m due north and then turns to run at an angle 45° north of east for 40.0 m. How far away from her starting point is the jogger at the end of this run?

Solution Figure 2.2 shows the steps for completion of the problem. A scale is chosen to draw the vectors, and the resultant is then sketched in and its length measured. Try it for yourself to find that the resultant displacement, or vector sum, is about 55.8 m in magnitude and north–northeast in direction.

2.4 VELOCITY

In everyday life, the terms "speed" and "velocity" are often used interchangeably, but in the world of physics the two have distinctly different meanings. The basic distinction between the two is that

> *Speed is a scalar quantity, whereas velocity is a vector quantity.*

For example, if we specify that a car is moving at 20 km per hour northward, we have specified the velocity, which is a vector quantity. If all we tell you about the car is that it is traveling at 20 km per hour, we have specified the speed, which is the magnitude of the vector without regard to its direction. Speed is important in many applications, but there are other practical situations in which we need to know more about an object than simply how fast it is traveling. For example, suppose your girlfriend tells you that she is going out for her morning jog. She tells you that she is going to pace herself so that she runs at a speed of exactly 3 km per hour, that she will run in a straight-line path, and that she plans to run for exactly 1 hour. She then asks you to come pick her up in your car. Can you satisfy this request if no more information than this is given? You know that she is exactly 3 km from her starting point, but if she does not tell you the direction of her run, you will have no idea where to go to give her a ride. Velocity is a vector quantity that requires a knowledge of the magnitude and direction; thus, if she is to specify her velocity completely, she will do so by saying that she will run at, say, 3 km per hour due north on Maple Street.

FIGURE 2.2 *Example 2.2.*

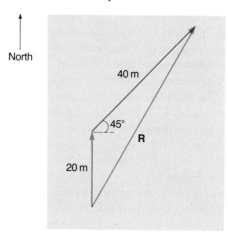

To find the *magnitude* of the velocity, you use exactly the same equation as specified earlier for average speed:

$$\text{velocity} = \frac{\text{displacement}}{\text{time}} \quad (2.2)$$

It is convenient, however, to shorten this equation such that we let $\mathbf{v}$ = velocity, $\mathbf{d}$ = displacement, and t = time. Thus, we have

$$\mathbf{v} = \frac{\mathbf{d}}{t}$$

2.5 ACCELERATION

As you travel along a highway in your car, you seldom travel at a constant velocity for long periods of time. Red lights, the flow of traffic, or a nearby state policeman will cause you to alter your velocity. **Acceleration** is defined as *the change in velocity of an object divided by the time it takes the change in velocity to occur.* In equation form, this can be expressed as

$$\text{acceleration} = \frac{\text{final velocity} - \text{initial velocity}}{\text{time}}$$

or

$$\mathbf{a} = \frac{\mathbf{v_f} - \mathbf{v_i}}{t} \quad (2.3)$$

Recall that velocity, as a vector, is a description of the speed and the direction of movement of an object. Therefore, if an object moving at a constant speed changes direction, the velocity also changes. Since acceleration is a change in velocity, a body moving at a constant speed around a curve is accelerating. This concept may seem confusing at first until you think of it in human terms. You will learn in a later section that an object will accelerate only if a net force acts on it. Think of a situation in which you are riding in a car. When the car speeds up suddenly, the seat pushes against your back. Similarly, if the car slows down suddenly while you are leaning against the dashboard, you feel the dashboard pushing against you. Now, what happens when the car turns a sharp corner? You may feel the side of the vehicle press against your shoulder as various forces cause you to change your direction and therefore your velocity. Starting, stopping, and turning are all forms of acceleration.

There are some points that need to be clarified about the equation for acceleration, and these features can best be demonstrated with an example. For that reason, go through the following example carefully; do not skip past it.

EXAMPLE 2.3   REV IT UP!

(a) A motorist traveling in his car causes his car to change from a velocity of 5.00 m/s due north to a velocity of 15.0 m/s, also north, in a time of 5.00 s. What is his acceleration?

Solution Let us call the direction of travel of the car, north, the positive direction. This means that both the velocities listed above have + signs associated with them. Thus, we have

$$\text{acceleration} = \frac{15.0 \text{ m/s} - 5.00 \text{ m/s}}{5.00 \text{ s}}$$

$$= 2.00 \, \frac{\text{m}}{\text{s}^2}$$

First note the units. An acceleration of 2.00 m/s² means that on the average the velocity of the car is increasing by 2.00 m/s every second.

(b) The motorist, now traveling at 15.0 m/s, decides that he will lower his velocity to 5.00 m/s in a time of 5.00 s. (He is still traveling due north.) What is his acceleration in this case?

Solution From the definition of acceleration, we find

$$\text{acceleration} = \frac{5.00 \text{ m/s} - 15.0 \text{ m/s}}{5.00 \text{ s}}$$

$$= -2.00 \, \frac{\text{m}}{\text{s}^2}$$

The negative sign indicates that the direction of the acceleration vector is south, and thus the car is slowing down at an average rate of 2.00 m/s every second. Such an acceleration is usually called a *deceleration*.

CONCEPTUAL EXERCISE

Airplane pilots find that it usually takes less time to fly across the United States when they move from west to east than when going east to west. How does an understanding of vectors help to explain this observation?

Answer   There is usually a prevailing wind blowing from west to east at high altitudes in the United States, and the velocity of the wind adds to the velocity of the plane as the plane moves toward the east. The vector velocity of the plane is diminished when it flies against the wind, from east to west.

CONCEPTUAL EXERCISE

If a car has zero velocity, must it simultaneously have zero acceleration?

Answer   A car can have a zero velocity at some instant and any value of acceleration at that same time. For example, a dragster starting from rest may have a very large acceleration at the start of a run. All this means is that the car's velocity is changing very rapidly with time. If a car had zero velocity and zero acceleration at the same time, the velocity of the car would not be changing, and it would remain at rest.

2.6 ACCELERATION DUE TO GRAVITY

The movement of a falling object intrigued scientists for hundreds of years. One of the first to discuss this subject was Aristotle, who taught that heavier objects fall faster than do lighter ones. To illustrate his point of view, consider dropping a baseball and a tennis ball at the same time. According to his approach, the heavy baseball would hit the ground considerably sooner than the lighter tennis ball. Before you read further, why don't you try it for yourself? If the two are allowed to fall for only a short distance, such as from eye level, you will find that they strike the ground at the same time. You should recall that Aristotle believed that natural laws could be understood by logical reasoning alone and that experimentation was largely unnecessary. In light of your simple experiment, it is surprising that Aristotle's teachings on falling bodies could have been accepted for about 2000 years.

Galileo questioned the conclusions of Aristotle and decided to test them by direct experiment. According to popular legend, Galileo took two objects, a heavy one and a light one, to the top of the Leaning Tower of Pisa and dropped them together.

They both hit at the same time! It should be obvious that if they are dropped from rest and hit at the same time, their motions are alike in all ways as they fall. Because they are accelerated, they therefore must fall with the same acceleration. It should be noted here that Galileo recognized that a fluffy object like a feather or a piece of paper falls more slowly than a denser, more compact object like a rock. He concluded correctly that air resistance is the primary factor causing the difference in the time of fall in such cases. It is to his credit that he was able to make the leap of imagination to recognize that in the absence of such resistive forces, all objects, heavy or light, would fall with the same acceleration. On August 2, 1971, a vivid demonstration of this fact took place on the Moon. Astronaut David Scott simultaneously released a hammer and a feather, and they fell with the same acceleration to the lunar surface. This demonstration would surely have pleased Galileo. *We shall define an object that falls without any resistance to its motion to be a freely falling object.*

Galileo's experiments with freely falling objects were important in that they established that objects fall with the same acceleration, but more importantly his experiments led to the idea that the process of experimentation is a necessary part of a scientific investigation. There had been experimentation before the time of Galileo. For example, some wealthy scientists had their slaves perform the menial task of doing experiments for them, and many astronomers had made and recorded observations. Galileo's work, however, was a significant milestone along the way toward the idea that experimentation is the only way to test the validity of a scientific proposition.

Galileo suggested that objects fall with the same acceleration everywhere, but actually the acceleration that a falling object experiences varies slightly at different locations. Measurements show, however, that the acceleration due to gravity is approximately 9.8 m/s^2 everywhere on Earth. This means that its speed changes at a rate of 9.8 m/s every second.

It is a simple matter to observe the speeds at various times to enable us to find a simple equation for the instantaneous speed, v, of a body that is dropped from rest. This is

velocity = (9.8 m/s^2) time

or

$$v = (9.8 \text{ m/s}^2)t \qquad (2.4)$$

You should note that this equation is just our old equation for the definition of acceleration, Eq. 2.3, in the form v = **a**t, *where we have set the initial velocity equal to zero* and have used 9.8 m/s² for the acceleration. Often, in future equations, we will use the symbol **g** to stand for the acceleration due to gravity. Thus, Eq. 2.4 could be written as **v** = **g**t.

As shown in Figure 2.3, and indicated by Eq. 2.4, an object dropped from rest off a cliff will be traveling at 9.8 m/s after 1 s, at 19.6 m/s after 2 s, and so forth.

If a ball is thrown upward, the acceleration due to gravity remains the same, but the effect of gravity is now to slow the object at a rate of 9.8 m/s every second. For example, if the ball has a speed upward of 29.4 m/s at the instant it leaves the hand, it will have decreased in speed by 9.8 m/s after 1 s. Refer to Figure 2.4. After 1 s, it will be traveling upward at a speed of 19.6 m/s. In another second, its speed will have decreased by another 9.8 m/s, so its speed will be 19.6 m/s − 9.8 m/s = 9.8 m/s. Finally, after one more second, the speed will have been reduced to zero. It then comes back down as though it had been dropped from rest and accelerates, or increases its speed, at a rate of 9.8 m/s each second.

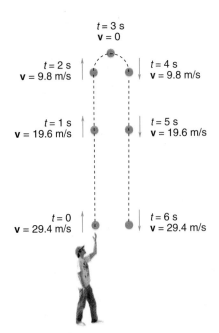

FIGURE 2.4 *As a body in free fall rises, its speed decreases at a rate of 9.8 m/s every second. While falling, its speed increases at a rate of 9.8 m/s every second.*

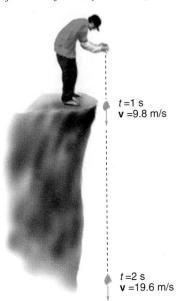

FIGURE 2.3 *The speed of an object in free fall increases at the rate of 9.8 m/s for every second it falls.*

EXAMPLE 2.4   LOOK OUT BELOW!

A student drops his textbook off a high-rise dormitory. After 1.40 s of fall, how fast is the book traveling?

Solution We use $v = (9.8 \text{ m/s}^2)\, t$ to find

$$v = (9.8 \text{ m/s}^2)(1.40 \text{ s}) = \boxed{13.7 \text{ m/s}}$$

EXERCISE

How much time must pass before the textbook reaches a speed of 26.7 m/s?

Answer 2.72 s

2.7 MOTION WITH CONSTANT ACCELERATION

The most common case of an object moving with constant acceleration is that of a freely falling body just discussed. Cars or other objects, however, often travel with constant acceleration as well (at least for short distances). As a result, let us discuss this important kind of motion more completely.

At the end of the previous section, we found that the velocity of a falling object can be found at any instant of time via the equation

$$v = (9.8 \text{ m/s}^2)t$$

We can generalize this equation to apply to objects moving with a constant acceleration, other than falling objects, by substituting a, the acceleration, for 9.8 m/s^2 in the equation. We have

$$v = at \quad [\text{if } v_i = 0] \tag{2.5}$$

You should note that this equation applies only if the object under consideration starts from rest.

Galileo was able to find a relationship between the distance covered by an object moving with constant acceleration and the elapsed time. His method was to roll balls down a slight incline and to measure the distance they moved during successive time intervals. The purpose of the incline was to slow down the movement of the ball sufficiently to enable him to take accurate measurements of time. The relationship he found was

$$d = \frac{1}{2} at^2 \tag{2.6}$$

where d is the displacement, a is the acceleration, and t is the elapsed time. You should note, again, that this equation is valid only when the object under consideration starts from rest and moves with constant acceleration.

EXAMPLE 2.5   LOOKING FOR THE CHECKERED FLAG

A jalopy starts from rest and moves in a straight line with a constant acceleration of 2.0 m/s^2 for a time of 5.0 s.

(a) Find the velocity of the car at the end of this time interval.

Solution From $v = at$, we have

$$v = (2.0 \text{ m/s}^2)(5.0 \text{ s}) = \boxed{10 \text{ m/s}}$$

(b) Find the distance covered during the 5.0 s time interval.

Solution From $d = \frac{1}{2} at^2$ we have

$$d = \frac{1}{2}(2.0 \text{ m/s}^2)(5.0 \text{ s})^2 = \boxed{25 \text{ m}}$$

EXERCISE

At the end of the 5.0 s time interval given, the car stops accelerating and begins to move at constant speed. What is its speed after another 5.0 s have passed, and what distance has it covered during this second 5.0 s interval?

Answer $v = 10$ m/s, $d = 50$ m

2.8 FORCES AND MOTION

Thus far, we have investigated some terms common to the study of motion, including "speed," "velocity," and "acceleration." However, we have not answered questions such as: What causes a car to accelerate? What causes a car to move at a constant velocity? Under what conditions does a car move at constant acceleration? The answer to these questions can be found from one of three laws of motion discovered by Isaac Newton (1642–1727). Newton was born at Woolsthorpe, England, on Christmas Day in the same year that Galileo died. He is considered by many to be the most brilliant scientist who has ever lived. During an 18-month period in his early twenties, he formulated the law of universal gravitation, invented the mathematical concepts of calculus, discovered the three laws of motion, and proposed theories on light and color. Even before this period, he had gained a measure of immortality by inventing the reflecting telescope. He was buried in Westminster Abbey with the following epitaph: "Mortals, rejoice that so great a man lived for the honor of the human race."

We shall devote the latter part of this chapter to a development of Newton's three laws of motion.

Force

Most people have an intuitive understanding of what is meant by "force"; it is usually thought of as a push or pull on an object. For our purposes, this inherent idea is sufficient, but we shall return to this topic again later in this chapter and place our ideas concerning forces on a more rigorous basis.

In the United States, the unit used to measure a force is the pound. For example, when you say that your weight is 150 lb, this means that the Earth is pulling down on you with a force of 150 lb. In scientific usage, the most commonly used unit of force is the newton. To be precise, the relation between a pound and a newton is

$$1 \text{ N} = 0.224 \text{ lb}$$

EXERCISE

Try calculating your weight in newtons. To see if you are on the right track, the weight of a 150 lb individual is about 670 N.

Forces Are Vectors

We pointed out earlier that physical quantities such as velocity and displacement are vectors. We have seen how to add vectors, such as two displacement vectors, to find the resultant displacement, and we add forces to find the resultant force by the same technique. In fact, among the most important tasks that one must perform to analyze the motion of an object is to be able to recognize all the forces that act on the object and then to find the resultant, or net, force. The **resultant force,** or **net force,** is defined as the vector sum of all the forces acting on an object. Let us examine these steps for a couple of situations.

Consider an overhead view of a safe as shown in Figure 2.5. We see two not-so-bright burglars attempting to move the safe by exerting forces on it in the directions indicated. They both push with a force of 100 N, but since the forces are in opposi-

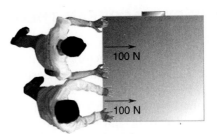

FIGURE 2.6 *Forces in the same direction add. The net force on this safe is 200 N if there is no friction.*

tion to one another, the resultant force on the safe is zero.

If the two burglars exert forces as shown in Figure 2.6, the resultant force is just the sum of the two individual pushes, and the safe moves as though a single force of 200 N were being exerted on it. (We ignore friction in all these cases.)

What about the situation shown in Figure 2.7a, however? In this case, the two burglars are exerting forces on the safe at right angles to one another. The net force can be found by the triangle method of vector addition, as shown in Figure 2.7b. You can draw the forces to scale for yourself and measure the resultant force to be about 141 N.

FIGURE 2.7 *(a) Two forces not along the same direction must be added by the triangle method. (b) The triangle method yields a resultant of 141 N.*

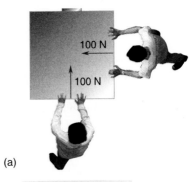

(a)

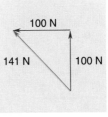

(b)

FIGURE 2.5 *Each burglar pushes on the safe with a force of 100 N, but the net force on the safe is zero.*

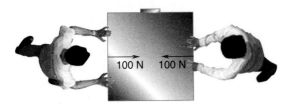

2.9 THE PYTHAGOREAN THEOREM

In our discussion of the addition of vectors, we have found the resultant of two vectors at right angles by the use of a graphical technique. There is an alternative to this approach that allows you to find the resultant by mathematical techniques alone. This approach may be traced b; k to the Greek thinker Pythagoras, who died al ` 500 B.C. According to the Pythagorean the(the lengths f the sides of a right triangle, sh igure 2.8, are related as

$$R = \sqrt{A^2 + B^2}$$

This says that the length of the longest side **R,** called the hypotenuse, is equal to the square root of the sum of the squares of the lengths of sides **A** and **B.** To see how to use this equation in a practical application, let us suppose that we have two displacements at right angles, as shown in Figure 2.9. The length of the vector forming side **A** is 4 m, and the length of the vector forming side **B** is 3 m. What is the resultant of these two vectors? To find the answer, we proceed as follows:

$$R = \sqrt{(4 \text{ m})^2 + (3 \text{ m})^2}$$
$$= \sqrt{16 \text{ m}^2 + 9 \text{ m}^2}$$
$$= \sqrt{25 \text{ m}^2}$$
$$= 5 \text{ m}$$

To verify the Pythagorean theorem result, you might check this answer with the graphical techniques discussed earlier.

FIGURE 2.8   *The Pythagorean theorem provides an alternative to the graphical process of vector addition. The resultant vector R is found by use of the equation* **R** $= \sqrt{A^2 + B^2}$.

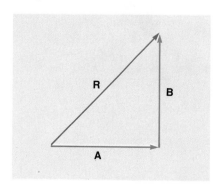

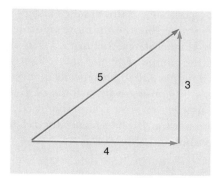

FIGURE 2.9   *Use of the Pythagorean theorem shows us that the length of the resultant is 5 units.*

Recognizing Forces

It is often necessary in solving mechanics problems to be able to find the net force on an object by use of the Pythagorean theorem. Even before one can apply the techniques of vector addition, however, one must be able to pick out all the forces that are being exerted on an object. Usually this is a trivial exercise that can be done easily, but occasionally it is a little difficult to do. Here are a few tips on how to recognize when a force exists.

As we observe the world around us, forces seem to fall into two convenient categories. These are **action-at-a-distance** forces and **contact** forces. *Action-at-a-distance forces are those that one object exerts on another even when there is no physical contact between the two.* In physics, not many of these exist. The most common is the force of gravity. As you sit in your room reading this, the Earth is pulling on you with a force that is directed toward the center of the Earth and that has a magnitude equal to your weight. If you jump off a building, this force acts on you while you are falling. This means that the force of gravity is present even in those cases in which there is no physical contact between you and the Earth. As we venture further into physics, we will find that there are other action-at-a-distance forces. Two charged objects exert electrical forces of attraction or repulsion that are of the action-at-a-distance type. Likewise, a magnet can attract a piece of iron or other magnetic material across space without the two touching. In our day-to-day experience, these are the only three action-at-a-distance forces we encounter directly. In fact, as far as the study of mechanics is concerned, the only one you need to be concerned with is the force of gravity.

Contact forces arise when there is actual physical contact between two or more objects. For example, when a batter hits a baseball, the bat exerts a force on the ball only during the time when the bat and ball are actually in contact. Similarly, a punter exerts a force on a football only while his foot is in contact with the football. As a general rule, contact forces can arise anytime there is physical contact between two objects. It should be noted that these forces, which we are calling contact forces, can be shown to be action-at-a-distance forces when we consider them on a submicroscopic level. For example, the actual origin of the contact force exerted by a bat on a ball arises from electrical interactions among atoms of the two objects. On a large-scale basis, however, these electrical forces are not obvious. Instead, we recognize their presence in terms of a gross force that we call a contact force.

Before we continue with our study, let us pause to get some practice in recognizing the presence of forces.

CONCEPTUAL EXAMPLE 2.6   A BOOK ON A TABLE

Figure 2.10a shows a simple situation in which a book rests on a table. Find all the forces that act on the book.

Solution To approach this problem, draw a sketch of the book by itself. Figure 2.10b is our sketch. Now, let us begin to find the forces that act on it. First, are there any action-at-a-distance forces present? Of course, there is always one, the force of gravity, which is manifested as the weight of the object. We have sketched this in Figure 2.10b as **w**, and we have indicated its direction toward the center of the Earth. No other action-at-a-distance forces are to be considered for our situation, so we must now attempt to recognize what, if any, contact forces are acting. To determine what may be exerting a contact force on the book, we must first ask whether there is anything touching the book. The only object that is in contact with the book is the table, and it does exert a force on the book. This force is directed perpendicular to the surface along which the book and table make contact. Often called the *normal force* this force is pictured as **N** in Figure 2.10b. Because nothing else is in physical contact with the book, there are no other contact forces present and our diagram is complete.

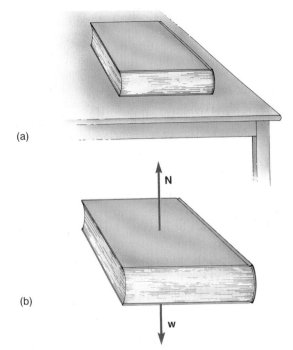

(a)

(b)

FIGURE 2.10 *Example 2.6. (a) A book at rest on a table. (b) There are two forces on the book;* **w** *its weight, and* **N** *the normal force.*

CONCEPTUAL EXAMPLE 2.7   A FALLING BOOK

Suppose the book of the preceding example falls off the table. Draw a diagram of the book that shows all forces acting on it while it is falling to the floor.

Solution The diagram is shown in Figure 2.11. The weight, the action-at-a-distance force, is always present, and this is indicated as **w** in the figure. There is nothing in contact with the book, so there can

FIGURE 2.11 *Example 2.7. An object in free fall has only one force acting on it. This force is its weight.*

be no contact forces acting. (We are ignoring forces that might arise from such factors as air resistance.) Thus, the only force acting on the book as it falls is its weight. In fact, this is the definition of a freely falling body; it is one that has only one force, the weight, acting on it. As you recall, freely falling bodies have an acceleration of 9.8 m/s^2.

2.10 NEWTON'S FIRST LAW OF MOTION

Take a look at a book lying at rest on your desk. Why is the book not moving? Imagine the Rock of Gibraltar at rest at the mouth of the Mediterranean Sea. Why does the Rock of Gibraltar not move with respect to the Earth? **Newton's first law of motion** gives us an insight into why things that are at rest stay at rest, and what must happen to them if they are to begin to move. We can state Newton's first law of motion in partial form as:

An object at rest will remain at rest unless a net force acts on it.

We saw in Example 2.6 that only two forces act on a book resting on a table—the force of gravity and the normal force. Thus, according to Newton's first law, these two forces must cancel one another to give a net force of zero if the book is to remain stationary. Figure 2.12 shows a bone being tugged by three hungry dogs. If we watch the bone and observe that it is not moving, we know that the dogs are pulling such that the vector sum of all the forces being exerted by the dogs is zero.

CONCEPTUAL EXAMPLE 2.8 STOP AND GO

A traffic light weighing 150 N hangs from a cable, as shown in Figure 2.13. Find the tension in the cable.

Solution To work this problem, we must first be able to recognize all the forces acting on the traffic light. There are two, the weight and the tension in the cable, as shown in Figure 2.14. (If you are not sure about this, re-read the section on recognizing forces.) Because Newton's first law tells us that the net force must be zero, the only way this can be possible is if the tension in the cable is also equal to 150 N.

FIGURE 2.12 *If the bone does not move, the vector sum of all forces is zero.*

Newton's first law applies as shown to objects that are at rest, but as we shall now see, it also applies under certain conditions to objects that are moving. To investigate motion via the first law, consider the following experiment. Take a book and slide it across a floor. The book slides for only a short distance and then comes to a stop. The ancient Greeks chose to concentrate on this aspect of matter. They said that the natural state of matter was for it to be stopped. When an object such as the book is set into motion, it naturally comes to a stop. But take this same book and slide it across a highly polished floor. What happens now? Again, it comes to a stop, but not nearly as soon.

Galileo looked at motion and concluded that scientists were considering it from an incorrect viewpoint. He said that the natural state of matter is to remain in motion once it is set in motion. Ob-

FIGURE 2.13 *Example 2.8. Find the tension in the cable if the traffic light weighs 150 N.*

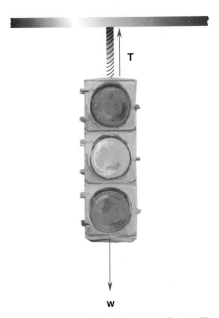

FIGURE 2.14 *The forces on the traffic light are its weight* **w** *and the tension in the cable* **T.**

FIGURE 2.15 *Example 2.9. If the jet is to fly at a constant velocity, the forward thrust by the engines* **T** *must be equal to the force of air resistance* **f.**

EXAMPLE 2.9 FREQUENT FLYER TICKETS ACCEPTED

A jet airplane is flying a straight course at a constant speed of 500 miles per hour. If the engines of the jet exert a forward thrust on the plane of 40,000 N, what is the force of air resistance acting on the plane?

Solution If the plane in Figure 2.15 is flying at a constant speed in a straight line, the net force on it must be zero. The only way this can be possible is if the force of air resistance, **f,** is equal to the forward thrust of the engines, **T.** Thus, **f** = 40,000 N.

---

jects do not "naturally" stop; they "naturally" continue to move. In our everyday world, this does not seem to make sense because every object that we see moving eventually stops. The reason that the book does stop in its slide across the floor is that there is a force of friction acting on it, and the reason that the book slides farther on a polished floor is because the force of friction is smaller than it is on a rough surface. On an air hockey table, it would slide even farther before friction brought it to rest. Now, let your imagination leap to a floor so highly polished that the force of friction could be reduced to zero. In this case, the book would slide forever because there would be no force to stop it.

This new idea of motion was formalized by Isaac newton as part of his first law. He said

An object in motion will remain in motion at the same velocity (same speed and same straight-line direction) unless a net force acts on it.

Thus, when you see a car moving down a road at a constant speed and in a straight line, according to Newton's first law, the net force on the car must be zero.

We can now form a complete statement of Newton's first law as:

An object at rest remains at rest, and an object in motion with constant velocity continues in motion with the same velocity (speed and direction) unless a net external force acts on it.

2.11 INERTIA

Newton's first law of motion is often referred to as the law of inertia. To see why it has acquired this name, consider some passengers standing in a bus. If the bus is not moving, the passengers can easily stand up, but what happens if the bus suddenly begins to accelerate forward? You know from your own experience that the passengers will fall backward in the bus. From the insights gained by our brief study of the first law, let us see if we can understand why they fall. From the viewpoint of the first law, the passengers were initially at rest, and they will remain at rest until a net force acts on them to change their motion. *The tendency for matter to remain in whatever state of motion that it is in is called* **inertia.** Thus, since the passengers were at rest, their tendency is to stay at rest, and they fall because the bus accelerates out from under them. Of course, when they regain their footing and are holding onto a strap, the strap exerts a force on them to make them accelerate along with the bus. After the bus has reached a constant velocity of 80 km per hour, the passengers again can stand comfortably without support. Now suppose the driver throws on his brakes. You know that the passengers will pitch forward. From the inertia viewpoint, the passengers when traveling at a constant speed in a straight-line path will continue to move in this fashion unless a net force causes them to change this motion. If they are not prepared to grab hold of something, their inertia will cause them to continue to move and to fall forward as the bus stops under them.

These observations were placed on a quantitative basis by Newton with the definition of a quantity called **mass.** *Mass is a term used to describe how much inertia an object has.* A very massive object requires a large force to start it into motion or to stop it once it is started. As we saw in Chapter 1, in SI units, mass is measured in kilograms.

Often, beginning students of physics confuse weight and mass, but the two are distinctly different. The weight of an object is the force of attraction that the Earth exerts on it. At the surface of the Earth, an object with a large mass also has a great weight because the Earth is pulling on it strongly. If this same object is taken thousands of kilometers into outer space, however, the pull of the Earth on the object has weakened greatly. Thus, its weight has decreased, but what about its mass? Recall that mass is a measure of how difficult it is to start an object into motion or to stop it once it is in motion. Now, consider yourself to be in outer space and a massive object comes hurtling toward you like a meteor. Even though it has little or no weight, you will still find it difficult to stop. Thus, the mass of an object does not change regardless of its physical location in space. If an object has a mass of 10 kg on Earth, it will have this same mass on the Moon or in the deepest reaches of outer space.

CONCEPTUAL EXERCISE

When you are comparing the weight of two objects, you frequently put one in each hand and shake them up and down. How does this help you find out which is more massive?

Answer   The up and down movement helps you to determine which of the two is the more difficult to move. The one that is more difficult to move is the one that has the greater mass.

2.12 NEWTON'S THIRD LAW OF MOTION

Since Newton numbered his laws as the first, the second, and the third law of motion, it seems reasonable that we should discuss them in that order. Of the three, however, the second law is a little more difficult for beginning students than are the first and the third. Therefore, the third law of motion will be examined first.

There is an age-old admonition that a parent often gives to a child while administering a spanking. The parent often says, "This hurts me as much as it does you." As we shall see, at least in principle, this description is absolutely correct from the standpoint of Newton's **third law of motion.**

Basically, what Newton discovered and presented with his statement of the third law of motion is the fact that forces always occur in pairs. When discussing these force pairs, Newton chose to call one of them the *action* and the other the *reaction*. There is no real significance in these names. He could equally well have chosen to call one of them force A and the other force B, or he could have selected designations from a multitude of other possibilities. Also, when discussing these force pairs, there is nothing that signifies that one of them is the action and the other the reaction. If you choose to call a particular one of these forces the action and the other the reaction, that is perfectly acceptable, and it is just as acceptable if someone else talking about these same forces chooses to use the opposite designations.

Newton's statement of the third law is:

For every action there is an equal and opposite reaction.

This simple statement of the third law is perfectly acceptable if one already understands all of

FOCUS ON
SEATBELTS AND INERTIA

When you pull a seatbelt in your car around you, it is easy enough to change its length so that it fits snugly. In the event of an accident, however, the obvious purpose of the seatbelt is to become taut so that you will be held firmly in place. The physical principle that underlies the action of a seatbelt is inertia, and the figure shows how one type of seatbelt mechanism works. In normal operation, the rachet turns freely to allow the belt to wind or unwind from the pulley. If an accident occurs, a rapid deceleration stops the car, but the large mass, because of its inertia, continues to slide forward along the tracks. The pin connection between the mass and the rod causes the rod to pivot about its center and to engage the rachet at the upper end. The rachet can now no longer turn, and the seatbelt maintains the length it had at the moment of impact. A slight modification of this same device enables it to serve as the mechanism to operate the airbag in a car. In this case, the movement of the mass and rod activates a valve on a cylinder that contains nitrogen under high pressure. The nitrogen rushes into an air bag on the steering column, causing it to expand rapidly and to serve as a protective cushion for the driver.

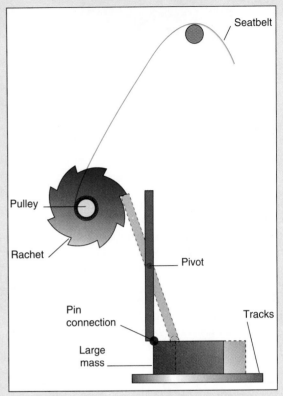

The seatbelt mechanism in an automobile.

the subtleties of the statement, but one must understand all the fine print. To recognize action-reaction pairs, first note that these pairs have three features: (1) The forces are equal in magnitude. (2) They are opposite in direction. (3) They always act on different objects.

As an example, suppose you are watching two boxers. One of them exerts a force of 10 N on the chin of his opponent. This is a force, and let us call this the action. Where is the second force, the reaction? An easy way to find the reaction is to make a simple declarative sentence concerning the action force and then invert the sentence to find the reaction. For our case, we might state:

> The action is the force that the fist exerts on the chin.

Now invert the sentence to find the reaction. The resulting sentence is:

> The reaction is the force that the chin exerts on the fist.

Thus, we see that if boxer A exerts a 10 N force on the chin of boxer B, the chin of boxer B also exerts a force of 10 N on the fist of boxer A. Now you can see that the parental statement, "it hurts me as much as it does you," is valid, at least in principle. (Of course, chins are more sensitive to forces than are fists.)

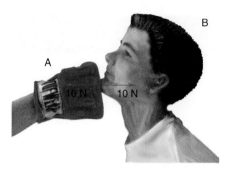

FIGURE 2.16 *The force of 10 N on boxer B and the oppositely directed 10 N force on the hand of boxer A are an action-reaction pair.*

Figure 2.16 shows these forces, and you should note that all three characteristics of action-reaction pairs are satisfied. The forces are equal in magnitude because both are 10 N forces. They act in opposite directions. This can easily be demonstrated by noting that the force exerted by the fist on the chin snaps the head of boxer B back, while the force of the chin on the fist of boxer A acts to slow down the forward motion of the hand. Finally, note that the forces act on different objects: one was on the fist, the other on the chin.

EXAMPLE 2.10   I FELT THE EARTH MOVE!

Figure 2.17 shows a 500 N student sitting on a stool. We will consider the stool to be light enough that we do not have to consider its weight. Also, we assume the student's feet do not touch the floor. This means all her weight is supported by the stool. Find all the action-reaction pairs.

Solution In a sense, this problem is also a review of the first law. The first thing we must do is to find

FIGURE 2.17 *Example 2.10. A 500 N student sits on a stool. Find all the action-reaction pairs.*

all the forces acting on the different objects—the student, the stool, and perhaps something else. We consider first the forces on the student, as shown in Figure 2.18. Force **A** in Figure 2.18a is the force that the chair exerts upward on the student. Force **B** is the weight of the student. From the first law,

FIGURE 2.18 *(a) The forces on the student are* **A,** *the force exerted on him by the stool, and* **B,** *his weight. (b) The forces on the stool are* **C,** *the downward force exerted on it by the student, and* **D,** *the normal force exerted on it by the surface supporting the stool. (c) The reaction force to the weight of the student is the gravitational force he exerts on the Earth.*

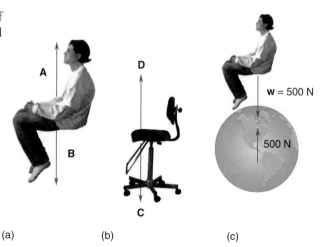

(a)   (b)   (c)

we recognize quickly that these two forces must be equal in magnitude and oppositely directed. Otherwise, the student would either sink toward the floor or levitate upward. So the forces are equal in magnitude and oppositely directed, but are they action-reaction pairs? The answer is that they cannot be because they both act on the same object, the student, and we know that action-reaction pairs are always on different objects.

Before we go further with our analysis, let us find the forces acting on the stool. These are shown in Figure 2.18b. They are **C,** the downward force exerted by the student on the stool, and **D,** the force exerted upward on the stool by the Earth. (**D** is the normal force, which we have encountered in previous problems.) These also must be equal in magnitude and opposite in direction, but they are not action-reaction pairs because both act on the chair. However, we now have enough information at hand to begin to locate some action-reaction pairs. First, consider force **A,** the force of the chair on the student. Let us call this the action and make our declarative sentence concerning this force. We have

> The action is the force that the stool exerts on the student.

This is force **A.** Now turn the sentence inside out.

> The reaction is the force that the student exerts on the stool.

This is force **C.** Thus, **A** and **C** are action-reaction pairs.

Now the situation becomes a little more difficult, but only slightly so. If we call force **B,** the weight of the student, the action, what is its reaction? Again, form the sentence.

> The weight of the student, the action, is the force exerted on him by the Earth.

Inversion gives:

> The reaction is the force exerted on the Earth by the student.

This reaction force is shown in Figure 2.18c. The force, which acts upward in the figure, is one of 500 N, the same as the weight of the student. Thus, the student is exerting an attractive force on the Earth, tending to pull it toward him, but don't expect a lot of movement on the part of the Earth because of its extremely large mass.

We will leave it to you to find the reaction to the force **D,** the force exerted by the Earth on the chair. *Hint:* You may once again have to involve the Earth in this problem.

CONCEPTUAL EXERCISE

Can you identify some of the action-reaction pairs in this tug-of-war?

(Courtesy Greg Perry)

2.13 NEWTON'S SECOND LAW OF MOTION

Newton's first law tells us what happens when the net force acting on an object is zero. The object either remains at rest if it is already at rest, or it moves with a constant velocity if it is in motion. But what happens if the net force on an object is not zero? Newton's answer to this is that the object is accelerated. The relationship between force, mass, and acceleration was summarized in his **second law of motion.**

> *The acceleration of an object is directly proportional to the net force acting on it and inversely proportional to the mass of the object.*

If the proper units are selected, Newton's second law can be stated in equation form:

$$\text{acceleration} = \frac{\text{net force on object}}{\text{mass of object}}$$

or, rearranged and condensed, we have

Force = (mass)(acceleration)

or, in symbols,

$$\mathbf{F}_{net} = m\mathbf{a} \qquad (2.7)$$

In this equation, $\mathbf{F}_{net}$ is the *net* force on the object, m is its mass, and $\mathbf{a}$ is the acceleration of the object. It should be noted here that the second law is often considered to be the most important law in mechanics because it stands as the first step in developing and understanding so many other concepts, such as the conservation of momentum and the conservation of energy, which we will discuss in the next chapter.

The Newton

Newton's second law is used as the equation by which forces are defined. For example, let us apply the second law to a 1 kg object. We ask ourselves the question, "What force must we apply to this object, in order to give it an acceleration of 1 m/s²?" To find the answer, we substitute into the second law as

$$\mathbf{F}_{net} = m\mathbf{a} = (1 \text{ kg})(1 \text{ m/s}^2) = 1 \frac{\text{kg m}}{\text{s}^2}$$

Thus, this force has a magnitude of 1. Its units, kg m/s², are lumped together and called a **newton, N.** Thus, we see that *a 1 N force is that force that will give a 1 kg object an acceleration of 1 m/s².*

The Relationship Between Weight and Mass

We demonstrated in an earlier section that weight and mass are two entirely different concepts, but there is a relationship between the two, and Newton's second law enables us to find this relationship. To do so, consider the situation shown in Figure 2.19, which shows a rock of mass m falling freely through space. The only force acting on the rock as it falls is its weight, **w,** as shown in the figure. Thus, let us substitute into Newton's second law that $\mathbf{F}_{net} = \mathbf{w}$, and $\mathbf{a} = \mathbf{g}$, where $\mathbf{g}$ is the acceleration due to gravity. We have

$$\mathbf{F}_{net} = m\mathbf{a}$$

$$\mathbf{w} = m\mathbf{g} \qquad (2.8)$$

w

FIGURE 2.19   *In the absence of air resistance, the only force on a freely-falling rock is its weight w.*

This innocent-looking relationship between **w** and m is just a special case of the second law, but it will be used often and may be considered a conversion factor between **w** and m, as long as the object is on or near the Earth's surface. For example, a typical adult has a mass of about 75.0 kg. What is the weight in newtons of this individual? From $w = mg$, we have

$$w = mg = (75.0 \text{ kg})(9.8 \text{ m/s}^2) = 735 \text{ N}$$

2.14 EXAMPLES OF NEWTON'S SECOND LAW

Newton's second law requires a little more mathematical effort than do the other two of Newton's laws, and in this section we will give you a few worked-out examples to show you its application to real-world situations. It might be helpful to read the problem first and then attempt to work it out on your own before you look at our solution.

EXAMPLE 2.11   RACE CARS AND STONES
(a) A race car with a mass of 500 kg can accelerate from 10.0 m/s to 40.0 m/s in 4.00 s. How much force is required to cause this acceleration?

(b) Imagine that you had a stone with a mass of 500 g and threw it with a net force of 100 N. What would be its acceleration while the force was acting on it?

Solution (a) We will use $F_{net} = ma$. The acceleration is calculated according to the definition of acceleration as

$$a = \frac{\text{change in velocity}}{\text{time}}$$

$$= \frac{40.0 \text{ m/s} - 10.0 \text{ m/s}}{4.00 \text{ s}} = 7.50 \text{ m/s}^2$$

Substituting into Newton's second law:

$$F_{net} = (500 \text{ kg})(7.50 \text{ m/s}^2) = \boxed{3750 \text{ N}}$$

(b) In this case, we are asked to solve for acceleration. Newton's second law is used as

$$a = \frac{F_{net}}{m} = \frac{100 \text{ N}}{0.500 \text{ kg}} = \boxed{200 \text{ m/s}^2}$$

Note that even though much less force is delivered to the stone than to the car, the acceleration of the stone is much greater. This occurs because the stone is much less massive than the car.

EXAMPLE 2.12   WATCHING YOUR WEIGHT
A sled of weight 98.0 N is pulled across an icy, frictionless lawn by a boy who pulls with a force of 5.00 N on the sled. What is the acceleration of the sled?

Solution Be careful here; you must use Newton's second law, $F_{net} = ma$, and to do so, you need to know the mass, m, of the sled. We are given the weight as 98.0 N, so we can easily find the mass from $w = mg$, so,

$$m = \frac{w}{g} = \frac{98.0 \text{ N}}{9.8 \text{ m/s}^2} = 10.0 \text{ kg}$$

Now we can find the acceleration of the sled from the second law.

$$a = \frac{F_{net}}{m} = \frac{5.00 \text{ N}}{10.0 \text{ kg}} = \boxed{0.500 \text{ m/s}^2}$$

CONCEPTUAL EXERCISE

Example 2.12 has a logical error in it, in the sense that it would be impossible for the boy to pull the sled. Why?

EXAMPLE 2.13   GET OUT OF THE WAY! I'M COMING THROUGH
How far does the sled travel in 3.00 s assuming that it starts from rest?

Solution This is not really an example of the second law, but it does point out that you cannot forget the material studied earlier in this chapter. (Physics tends to be cumulative.) If you refer back to our discussion of motion with constant acceleration, you find that we can calculate how far an object travels in a certain period of time by use of the equation $d = 1/2 \ at^2$, as long as it starts from rest and has a constant acceleration. Thus,

$$d = \frac{1}{2}(0.500 \text{ m/s}^2)(3.00 \text{ s})^2 = \boxed{2.25 \text{ m}}$$

EXAMPLE 2.14   AULD ACQUAINTANCES AND NET FORCES ARE NOT TO BE FORGOTTEN
The child continues to pull on the 10.0 kg sled of the preceding two examples with a force of 5.00 N. When he moves onto a portion of the lawn that is not quite as icy as before, however, he finds that there is a friction force of 3.00 N that begins to act on the sled. What is now the acceleration of the sled?

Solution You must recall here that $\mathbf{F}_{net}$ in $\mathbf{F}_{net} = m\mathbf{a}$ is the net force that acts on the sled. Because the friction force opposes the motion, the forces that act on the sled are shown in Figure 2.20. Thus, the net force acting on the sled is

$$F_{net} = 5.00 \text{ N} - 3.00 \text{ N} = 2.00 \text{ N}$$

and the acceleration of the sled is

$$a = \frac{F_{net}}{m} = \frac{2.00 \text{ N}}{10.0 \text{ kg}} = \boxed{0.200 \text{ m/s}^2}$$

FIGURE 2.20 *The forces on the sled are the 5 N pull and the 3 N friction force.*

2.15 THE NATURE OF FRICTION

We have mentioned the subject of friction briefly in the preceding sections, and there we relied on your understanding of forces of friction based on common experience. In this section, we will examine friction more carefully because it plays such an important role in the world of physics and technology.

Friction forces arise when one object attempts to move across another. For example, as a baseball player slides along the ground while stealing a base, there is a friction force that the ground exerts on him. *Frictional forces always act in a direction such that they oppose motion.* Thus, if the ballplayer slides toward the left, the friction force exerted on him by the ground is to the right.

Determining the direction of a force of friction is not difficult to do, but complications sometimes arise when one attempts to predict the strength of this frictional force. Factors such as the conditions of the surfaces involved and the speed of the object must be taken into consideration. Figure 2.21a shows a brick resting on a rough surface. To study the nature of frictional forces, let us begin to pull on it to the left, as shown, with only a small force of 1 N. (A force of 1 N is approximately equal to the weight of a small apple.) From your own experience, such a small force exerted on the brick will not set it into motion. Figure 2.21b shows why the

brick does not move. When motion is impending toward the left, a frictional force arises to oppose it, and this frictional force **f** is toward the right. Since the object is not moving, Newton's first law tells us that the strength of the frictional force must be equal to that of your pull, or 1 N. *Frictional forces like this, which arise even when there is no motion, are called forces of static friction.*

Now, let us pull slightly harder with a force of 2 N. Still the object does not move, so we conclude that the frictional force must have risen also and is now equal to 2 N.

Let us pull harder yet, with a force of 3 N. Still the object does not move, but now we find that if we pull only a fraction of a newton harder, the object *will* begin to move. From these observations, we see that the force of static friction between two surfaces has a maximum value that it can reach. If one pulls on an object with a force greater than the maximum value of this frictional force, the object begins to move. To see the origin of the force of static friction, consider Figure 2.22, which shows a greatly magnified view of the surface of the brick in contact with the floor. Note that little bumps on each surface have settled against one another. For the brick to slide, it must either rise over the bumps or break them off, and either action requires force. Even smooth objects have tiny, molecule-sized or atom-sized irregularities. In fact, as one object is moved across another, the atoms themselves can cling together and oppose the motion.

FIGURE 2.21   *(a) A brick resting on a rough surface. (b) When a small force of 1 N is exerted toward the left, it is opposed by a 1 N friction force to the right.*

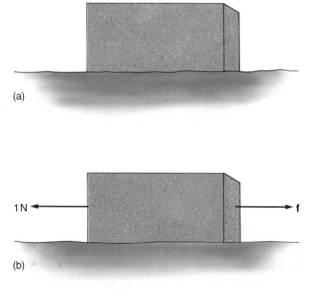

(a)

(b)

FIGURE 2.22   *Magnified view of surfaces of a floor and a brick.*

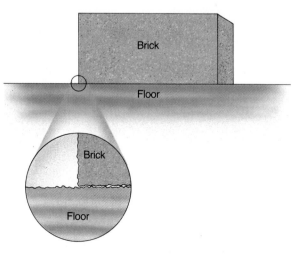

FOCUS ON
FRICTION FORCES IN THE HUMAN BODY

We take our bodies for granted . . . our muscles, joints, and organs work by the same engineering principles as the most sophisticated man-made devices. For example, nothing is taken as much for granted as the ability to stand upright or to run when we so desire. Yet if it were not for some exceptional applications of the use of frictional forces in the joints, these tasks would be impossible. Consider the simple task of standing upright. If the frictional forces within the joints were too small, the bones would tend to slip across one another, and we would fall when they moved out of alignment. At times, though, we do not want the frictional forces between the joints to be very large. An example of this would occur when you want to run or to do any exercise that requires rapid movement. If the frictional forces were too large within the joints, the bones would not be slippery enough to allow them to move freely across one another. As a result, there are times when we want the frictional forces within the joints to be large, and there are other times when we want these frictional forces to be relatively small. How does the body adjust to such varying demands?

The figure shows the construction of a typical joint in the human body. The ends of the bones are covered with cartilage, which is a sponge-like material, and the joint is encased in a capsule that contains a lubricating material called synovial fluid.

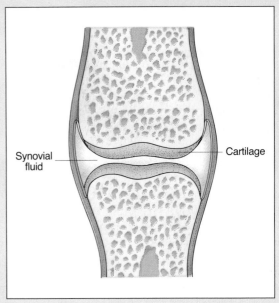

A joint in the human body.

When the bones are at rest with respect to one another, as they would be when we stand still, the cartilage absorbs much of the synovial fluid; the ends of the bones become relatively dry, and the friction force between the two increases. When the bones are in relative motion, as they would be in a knee joint when a person runs, the bones press together more firmly, and this squeezes some of the synovial fluid out of the cartilage. This released fluid lubricates the ends of the bones and allows them to slide freely across one another. Truly, this is marvelous engineering.

Once an object is set into motion, the bumps of one surface do not have time to settle fully into the holes in the other, and the friction force is less than when resting. Thus, if a force of 3 N is required to overcome the force of static friction, a smaller force, say 2 N, is large enough to keep the object moving at constant velocity once it begins to move. *The force of friction that acts on a moving object is called the force of kinetic friction.*

The preceding explanation of the origin of friction forces also explains why friction is reduced after rough objects have been rubbed back and forth against one another. This wears down the irregularities, resulting in smoother surfaces. Friction is also reduced by lubricating the surfaces with oil. The film of oil separates the surfaces so that there is no intimate contact between the irregularities of each.

This airplane is being used to test the force of friction on a specially prepared, grooved runway to see if the pavement will improve landing performance on dry, wet, flooded, and slush-covered surfaces. (Courtesy NASA)

CONCEPTUAL EXERCISE

If you drop a feather and a book, you will find that the feather falls toward the Earth more slowly than the book. However, if you place the feather on top of the book before you drop it, the feather will fall almost as fast as the book. Why?

Answer The feather falls more slowly in the first instance because of air resistance, a frictional force exerted on both objects by the atmosphere. The book is not affected as much because of its larger mass. If you place the feather on the book, the book "moves the air out of the way," allowing the feather to fall more freely.

2.16 DENSITY

There is an age-old trick question: "Which has more mass, a kilogram of feathers or a kilogram of lead?" The answer is, of course, that they both have the same mass, 1 kg. The reason that this question often catches people off guard is that there is a tendency to think of a given volume of each. For ex-

ample, you might try to imagine how much mass of feathers a small paper bag would hold, and then think of that same bag filled with lead. As a result, the off-the-cuff response would be that the lead has more mass. The point of this example is to indicate that it is frequently important to discuss not only how much mass something has, but also to indicate in some way how this mass is compressed or spread out. To this end, a quantity called **density** is defined as follows: *Density is the mass of a substance divided by the volume of the substance.* In equation form this is

$$\text{density} = \frac{\text{mass}}{\text{volume}}$$

or, in symbolic form

$$\rho = \frac{m}{V} \qquad (2.9)$$

where ρ is the symbol for density, m is the mass of the substance, and V is its volume. In the SI, density has units of kg/m^3.

One primary advantage that density has in the discussion of substances is that under the same conditions of temperature and pressure, every sample of a particular substance has its own characteristic density. That is, all iron has the same density, as does all gold, silver, and so forth. Table 2.1 gives the density for several representative substances. Note that the densities of solids are normally about 1000 times greater than densities for gases. This means that the spacing between molecules in the gaseous state is considerably greater than in the solid state.

TABLE 2.1 Densities of selected substances*

SUBSTANCE	DENSITY (KG/M^3)
Ice	0.917×10^3
Aluminum	2.70×10^3
Iron	7.86×10^3
Copper	8.92×10^3
Silver	10.5×10^3
Gold	19.3×10^3
Mercury	13.6×10^3
Water	1.00×10^3
Air	1.29
Oxygen	1.43
Hydrogen	8.99×10^{-2}

*All values are for atmospheric pressure and 0°C.

FOCUS ON
DENSITY AND FLOATING

 As noted in the discussion of density, an object will float in a liquid if its density is less than that of the liquid. An example of this in our everyday experience is at automobile service stations when we check our antifreeze or when we check the charge stored by a car battery. The first figure shows a common device used to check the state of the antifreeze in the radiator of a car. The small balls in the enclosed tube vary in density such that all of them will float when the tube is filled with pure water, none of them will float in pure antifreeze, one will float at a 5 percent mixture, two at a 10 percent mixture, and so forth. Thus, one fills the tube with the solution from the radiator, and the number of balls that float serve as a measure of

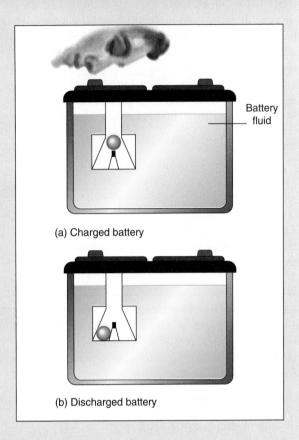

(a) Charged battery

(b) Discharged battery

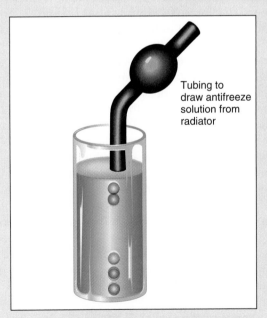

Tubing to draw antifreeze solution from radiator

The number of balls that float is a measure of the density of the antifreeze solution and, consequently, of how well it will protect your car.

the percentage of antifreeze and thus as a measure of how likely the mixture will be able to withstand the rigors of bad weather.

In a similar fashion, the state of charge of some newer car batteries can be measured with a so-called "magic dot" process built into the battery. When one looks down into a small porthole in the top of the battery, a red dot indicates that the battery is well charged, but if the dot appears black, the battery has lost its charge. The figure shows how the magic dot works. If the battery is fully charged, the battery fluid has a high density, high enough for the red ball immersed in a chamber in the battery to float. If the battery loses its charge, however, the fluid becomes less dense and the ball sinks.

EXAMPLE 2.15   ROOKING THE KING?

A king buys a block of material that a salesman says is gold. The block has a mass of 60.0 kg and measures 0.20 m by 0.20 m by 0.10 m. The king feels he may have been gypped, since the salesman operated out of the back of his car. Is the material truly gold?

Solution The volume of the gold brick is given by $V = $ (length) (width) (height). Thus, we find

$$V = (0.20 \text{ m})(0.20 \text{ m})(0.10 \text{ m}) = 0.0040 \text{ m}^3$$

The density is found to be

$$\rho = \frac{m}{V} = \frac{60.0 \text{ kg}}{0.0040 \text{ m}^3} = 15,000 \text{ kg/m}^3$$

Thus, the density of the "gold" is considerably less than the density of 19,300 kg/m³ characteristic of pure gold.

EXAMPLE 2.16   THE DENSITY OF PLANETS

The mass of Earth is 5.98×10^{24} kg, and its average radius is 6.37×10^6 m. Saturn is much more massive, 5.68×10^{26} kg, but its average radius is also considerably larger, 5.85×10^7 m. Consider each of them to be spheres ($V = \frac{4}{3} \pi r^3$) and calculate the density of each ($\pi = 3.142$).

Solution The volume of Saturn is

$$V_S = \frac{4}{3}\pi r_S^3 = \frac{4}{3}(3.142)(5.85 \times 10^7 \text{ m})^3$$

$$= 8.39 \times 10^{23} \text{ m}^3$$

and the volume of Earth is

$$V_E = \frac{4}{3}\pi r_E^3 = \frac{4}{3}(3.142)(6.37 \times 10^6 \text{ m})^3$$

$$= 1.08 \times 10^{21} \text{ m}^3$$

We are now able to calculate the density of each as

$$\rho_E = \frac{m_E}{V_E} = \frac{5.98 \times 10^{24} \text{ kg}}{1.08 \times 10^{21} \text{ m}^3}$$

$$= 5.54 \times 10^3 \text{ kg/m}^3$$

and

$$\rho_S = \frac{m_S}{V_S} = \frac{5.68 \times 10^{26} \text{ kg}}{8.39 \times 10^{23} \text{ m}^3}$$

$$= 0.677 \times 10^3 \text{ kg/m}^3$$

It is interesting to note that the density of Saturn is less than the density of water. This means that Saturn would float in a bathtub filled with water—if a large enough tub could be found for it.

2.17 PRESSURE

We saw in the last section that knowing the mass of an object is important, but that sometimes it is more important to know how that mass is distributed. To take care of this difficulty, we defined density as the mass of an object divided by the volume over which this mass is distributed. A similar problem arises with forces. Certainly they are important, but it is also important to know how they are applied to a surface. To indicate how this can become important, consider a 600 N man standing on a linoleum floor. The force he is applying downward on the linoleum is equal to his weight, but this force does no apparent damage to the surface. If he happens, however, to be wearing a pair of golf shoes with numerous metal cleats protruding from the soles, one finds that he does considerable damage to the floor by standing on it. Why? In both cases, the force applied to the floor is 600 N, yet the effect of this weight has been different in the two cases. The answer is that in the first case, the applied force is spread out over a wide area of contact between the sole of the shoe and the floor. In the second case, even though the same force is applied to the floor, the area of contact is considerably smaller. The only area in contact with the floor is the cross-sectional area of the tips of the metal cleats. Because of the importance of considering the area of contact in many practical situations, we define **pressure** as

$$\text{pressure} = \frac{\text{force}}{\text{area}}$$

or

$$P = \frac{F}{A} \tag{2.10}$$

where P is the pressure, F is the applied force, and A is the area over which this force is distributed. In the SI, the units of pressure are N/m^2, but as we shall see in Chapter 15, there are some frequently used alternatives to these units.

EXAMPLE 2.17   THE PRESSURE EXERTED BY A BRICK

A brick has a weight of 20 N and dimensions of 0.200 m by 0.080 m by 0.050 m. Find the pressure exerted by the brick on a surface (a) when it is resting with its largest face in contact with the surface and (b) when it is resting so that its smallest face is in contact.

Solution (a) The area of contact when the largest face is in contact with the surface is $A = (0.200 \text{ m}) (0.080 \text{ m}) = 0.016 \text{ m}^2$. Thus, the pressure is given by

$$P = \frac{F}{A} = \frac{20 \text{ N}}{0.016 \text{ m}^2} = \boxed{1250 \text{ N/m}^2}$$

(b) When the smallest face is in contact, the area is $A = (0.080 \text{ m}) (0.050 \text{ m}) = 0.0040 \text{ m}^2$, and the pressure exerted by the brick on the surface is

$$P = \frac{F}{A} = \frac{20 \text{ N}}{0.0040 \text{ m}^2} = \boxed{5000 \text{ N/m}^2}$$

SUMMARY

Speed is a scalar quantity and is defined as distance divided by time. A **vector quantity** is described by magnitude and direction, whereas a **scalar quantity** is specified only by its magnitude. **Velocity** is a vector quantity that describes speed and the direction of travel. **Acceleration** is change in velocity divided by time; an object is accelerating when it is speeding up, slowing down, or turning. The acceleration of a falling object near the surface of the Earth is a constant equal to 9.8 m/s^2. As a result, the velocity of a falling object is constantly increasing in magnitude as it falls and decreasing as it rises. Newton's three laws of motion are: (First) **A body at rest remains at rest, and a body in uniform motion in a straight line remains** in such motion, unless acted upon by a net force. (Second) **The net force on an object equals the mass times its acceleration ($\mathbf{F}_{net} = m\mathbf{a}$).** **Mass** is a quantitative measure of the inertia of an object and is related to the weight of the object by $w = mg$, where g is the acceleration due to gravity. (Third) **For every force there is always an equal and opposite force.** Frictional forces always arise when one object tries to slip over or pass another, and they always act in a direction to oppose such motion. The **density** of an object is its mass divided by its volume. **Pressure** is defined as the force per unit area acting on a surface.

EQUATIONS TO KNOW

$$\text{speed} = \frac{\text{distance}}{\text{time}} = \frac{d}{t}$$

$$\text{acceleration} = \frac{\text{final velocity} - \text{initial velocity}}{\text{time}}$$

$v = at$ (if the initial velocity is zero and the acceleration is constant)

$d = \dfrac{1}{2} at^2$ (if the initial velocity is zero and the acceleration is constant)

$$\mathbf{F}_{net} = m\mathbf{a}$$

$$\mathbf{w} = m\mathbf{g}$$

$$\text{density} = \frac{\text{mass}}{\text{volume}} = \frac{m}{V}$$

$$\text{pressure} = \frac{\text{force}}{\text{area}} = \frac{F_{net}}{A}$$

KEY WORDS

Average speed	Velocity	Action-at-a-distance force	Net force
Instantaneous speed	Acceleration	Contact force	Friction
Scalar quantity	Acceleration due to gravity	Inertia	Weight
Vector quantity	Force	Mass	Density
Displacement			Pressure

PROBLEMS AND CONCEPTUAL QUESTIONS

Problems requiring numerical work are identified with a blue number.

Early ideas about motion; speed

1. Can the average speed of an object ever be greater than its instantaneous speed? Can the reverse be true?

2. A speed limit sign along an interstate highway says 65 miles per hour. Is this referring to average or instantaneous speed?

3. (a) When a police officer gives someone a ticket for speeding, is the officer concerned with the driver's average speed or instantaneous speed? Explain.
   (b) On the entrance to many turnpikes, drivers are given a card that indicates the place of entry so the proper toll can be levied at the exit. The time of entry is also noted on the card. At the exit station, the toll officer can determine the driver's speed during the journey, since both the elapsed time and the distance traveled are known. Can the officer determine the driver's instantaneous speed, the average speed, or both? Explain.

4. If a car travels with a constant velocity, can there ever be a situation when the average velocity and the instantaneous velocity are different? Why or why not?

5. What is the speed of a horse that travels 20.0 km in 3.00 hours?

6. The speed of sound in air is about 345 m/s. The speed of light is so fast that at first approximation it seems to travel distances of 100 km or so instantaneously. If you see a flash of lightning and hear thunder 8 s later, how far away is the lightning storm? Express your answer in meters and in kilometers.

7. An electron in the picture tube of a TV set may travel at a speed of about 2×10^6 m/s. If so, how long does it take the electron to travel the 0.20 m length of the tube?

8. A turtle starts at one side of a room and moves toward the opposite wall 3 m away. In each second, the turtle covers half the distance to the wall. (a) How long will it take the turtle to reach the wall? (b) What is its average velocity during the first second? (c) After several moves, what value is approached for the average velocity during a movement? Why?

Vectors and scalars

9. Which of the following are vectors and which are scalars: (a) The length of one class period of the physical science class in which you are enrolled. (b) The displacement of a car from one city to another. (c) The temperature of your classroom. (d) The results of a poll taken on the popularity of the President.

10. Look around you in the room where you are reading this textbook and identify what you believe to be several physical quantities. Classify them as vectors or scalars.

11. Use the triangle method of vector addition to find the resultant displacement for a pool ball that moves 60 cm down a table, hits the bumper, and rebounds along the same path for a distance of 15 cm before stopping.

12. A shopper in a supermarket starts at the end of an aisle and pushes his cart for a distance of 20 m. He then turns at an angle of 90° and moves for a distance of 10 m. Use the triangle method of vector addition to find the resultant displacement of the shopper from his starting point. Check your result with the Pythagorean theorem.

13. An astronaut leaves Cape Kennedy in the Space Shuttle, makes eight trips around the globe, and lands safely at an Air Force base in California. Her husband watches the blastoff and then gets in a jet to meet her when she lands in California. Which of the two, the astronaut or the husband, has the largest resultant displacement vector?

14. Draw a sketch showing how three vectors can be added together to give a resultant of zero.

15. Suppose the shopper of Problem 12 made a 20° turn with respect to his initial direction of travel instead of a 90° turn. What then is his resultant displacement?

Velocity

16. Explain what is incomplete about the statement, "The velocity of the automobile was 80 km per hour."

17. A device on the dashboard of your car is called a speedometer. Why isn't it called a velocity-meter?

Acceleration

18. Can a car be increasing in speed at the same time that its acceleration is decreasing?

19. Which of the following can alter the acceleration of a car? (a) The gas pedal, (b) the brakes, (c) the steering wheel.

20. A bicyclist can accelerate from 0 km per hour to 10 km per hour in 2 s. Similarly, an automobile can accelerate from 80 km per hour to 90 km per hour in 2 s. Which vehicle is accelerating at a faster rate? Explain.

21. Find the acceleration of a car for the following circumstances: (a) The car increases its velocity from 10 km/s northward to 15 km/s northward in a time of 3 s. (b) The car changes its velocity from 10 km/s northward to 5 km/s northward in 3 s.

22. Can a car ever have an instantaneous velocity of zero and an acceleration that is not zero at the same instant?

23. Can a car ever have an instantaneous velocity of 10 km/s and a zero acceleration at the same instant?

Acceleration due to gravity

24. If a ball is dropped from the top of a tall tower, and at the same time a bullet is fired horizontally from the same tower, which object will strike the ground first?

25. You have two bricks, one heavy and one light. The heavy one is dropped, and its time of fall is measured; then the experiment is repeated for the light one. Finally, the two are glued together so that they fall as a unit, and the time of fall is measured for this combination. Discuss the expected results of this experiment from the point of view of Aristotle. Of Galileo.

26. The acceleration of gravity is stated as 9.8 m/s^2 in the text. Use conversion of units to show that this is equivalent in conventional units to 32 ft/s^2.

27. A box is dropped off a high building. At what time after release is it falling with a speed of 15 m/s?

28. A ball is thrown upward from the ground such that it reaches a height of 15 m. (a) What is its acceleration on the way up at a point 10 m above the ground; (b) on the way down at a height of 10 m above the ground; (c) at a height of 15 m above the ground?

Motion with constant acceleration

29. Imagine that you are sitting in an airplane cabin but do not have the seatbelt fastened. Describe your motion with respect to the rest of the cabin and the forces acting on you when the aircraft is (a) climbing upward, (b) flying level, (c) accelerating downward at a rate of less than 9.8 m/s^2, (d) accelerating downward at a rate equal to 9.8 m/s^2, (e) accelerating downward at a rate greater than 9.8 m/s^2.

30. A rock dropped down a deep well falls for 5.00 s before striking the water. (a) How fast was it going when it hit? (b) How deep is the well?

31. A car is moving north with a velocity of 10.0 km per hour when it begins to accelerate in the same direction at a rate of 5.0 km/h^2 and does so for a period of 2.00 s. Can the equation $d = 1/2\ at^2$ be used to find its displacement during the 2.00 s time interval? Explain.

32. A model airplane accelerates from rest with an acceleration of 5.00 m/s^2. What distance does the plane cover in a time of 3.00 s?

33. A snail starts from rest and covers a distance of 3.00 cm in 20.0 min with a constant acceleration. Find this acceleration in m/s^2.

Force

34. Two forces act on a moving object. In what circumstances would the object move at a constant velocity? Under what circumstances would the object accelerate?

35. Find the weight in newtons of a 120 lb person.

36. Find the weight in pounds of a 500 N person.

37. Can a force directed vertically on an object ever cancel a force exerted horizontally?

38. A person holds a textbook at arm's length. Identify all the forces acting.

39. A baseball is struck by a bat and flies toward an outfielder. While the ball is in flight, identify all the forces acting on it.

40. A traffic light is supported by two cables. Identify all the forces acting on the light.

41. A force of 40 N is exerted northward on an object, and a force of 30 N is directed eastward. Use the triangle method of vector addition to find the resultant force on the object. Check the magnitude by use of the Pythagorean theorem.

42. Repeat the preceding problem if the 30 N force is directed 20° east of north.

Newton's first law of motion and inertia

43. The word "inertia" is used in many nonscientific applications. Write a physical and a nonphysical

definition for "inertia" and show that the spirit of the word is similar in both cases.

44. Is an object's inertia more closely related to its mass or to its weight? Explain.

45. A woman jumps from a plane carrying a heavy briefcase. She finds that when she opens the chute, the briefcase is ripped from her hand. Explain.

46. An astronaut traveling at a steady speed along a straight path from Mars to Jupiter has to go outside the ship to do a repair. While she is working, she lets go of her wrench. Does the wrench fall far behind and get lost in space? Explain.

47. You can remove the dust from a coat by shaking it. Could you shake out the dust in free space in the absence of gravity? Explain.

48. A magic trick often used by children is to yank a tablecloth out from under a place setting of dishes. Use the concept of inertia to explain how this can be done without breaking the dishes.

49. When one prizefighter lands a blow on the chin of his opponent, the opponent is often knocked unconscious. This occurs because the force of the blow snaps the head back, but the brain, floating inside the skull, does not move because of its inertia. Loss of consciousness is caused by bruising of the brain. If the punch is to the chin, what part of the brain gets bruised? The front or the back? Explain.

50. A passenger in the rear of an airplane claims that he was injured when the plane suddenly decelerated, sending a suitcase flying toward him from the front of the plane. If you were the judge, would you let this case come to court?

51. In a rear-end collision, the passenger in the front car often suffers a whiplash injury because his neck is rapidly bent backward. Explain why the neck is snapped backward in this way.

52. It is possible for you to lie in a prone position with a concrete block resting on your chest and allow someone to break the block with a sledge hammer without causing any damage to you. Explain how this can happen.

53. A car with the emergency brake on is parked on a hill. Does gravity exert a force on the car? If so, why doesn't the car roll downhill?

Newton's third law

54. If you were to try to drive a large nail into hard wood with a light hammer, the hammer would bounce up at each blow. Explain this in terms of Newton's third law.

55. Identify the action-reaction pairs in the following situations: (a) a kangaroo jumps, (b) a pool ball glances off another ball, (c) a football player catches a punt, (d) a steady wind blows against the sail of a ship.

56. A person holds a 5 kg bag of groceries at arm's length. Identify all the forces acting and separate them into action-reaction pairs.

Newton's second law

57. An astronaut floating in free space is handed two objects and asked to determine which is more massive. Outline a simple procedure for solving the problem.

58. Consider two possible devices for measuring mass: a spring scale and a two-pan balance. In the spring scale, the object to be measured is placed on a spring, and the compression of the spring is recorded. In the two-pan balance, the object to be measured is placed on one pan, and calibrated masses are placed on the other until the two sides balance. Which device will show the same result on Earth as on the Moon? Would either device be useful in a space lab?

59. What is the mass in kilograms of an object having a weight of 20 N?

60. What is the weight in newtons of an object having a mass of 20 kg?

61. A bullet with a mass of 0.015 kg is propelled from the breech of a rifle with a force of 3000 N. What is the acceleration of the bullet while the force is acting on it?

62. A ball with a weight of 10 N is thrown forward with a force of 20 N. What is its acceleration while the force is acting?

63. In which of the following situations is the net force on the object in question equal to zero: (a) a car rolling along a road at constant velocity, (b) a car slowing down for a stoplight, (c) a bridge supporting a load of 300 automobiles, (d) a baseball leaving the pitcher's hand, (e) a person sitting in a chair, (f) a ball perched on a sea lion's nose, (g) a tree swaying in the wind? Explain.

64. It is often said that Newton's first law is just a special case of the second law. What is meant by this statement?

65. If gold were sold by weight, would you rather buy it in Denver or in Death Valley? If sold by mass, which site would be your choice?

66. A small 70 kg motorcycle carrying a 75 kg passenger has a force of 250 N acting on it. (a) What is its acceleration? (b) How far will it travel, starting from rest, in 4 s?

Friction

67. A moving company places a washing machine on a truck without tying it down, yet the machine makes the entire trip without moving on the bed of the truck. Explain carefully how this is possible. What would the driver of the truck have to do to make the machine move on the

bed? (A move like this can be dangerous. Tie down your washing machine.)

68. A small 70 kg motorcycle carrying a 75 kg passenger has a forward force of 250 N acting on it exerted by the road, while a force of air resistance of 75 N acts on the two. What is the acceleration of the unit?

69. Why can rubbing your hands together in cold weather warm them up?

70. What is wrong with this statement? A baseball player sliding toward second base was tagged out because friction pulled him toward home plate.

Density

71. (a) What volume of gold would have a mass of 10 kg? (b) What volume of air would have this same mass? (Refer to Table 2.1.)

72. Some of the gold bricks at Fort Knox have dimensions of 25 cm by 10 cm by 5 cm. What is the mass of one of the bricks?

73. If the density of an object must be less than the density of water in order to float, how can a battleship stay afloat?

74. A traveling salesman wants to sell you a solid silver block at a tremendously reduced price. The block has a mass of 0.27 kg and dimensions of 0.1 by 0.1 by 0.01 m. Is this silver? If not, identify this substance from the values listed in Table 2.1.

Pressure

75. (a) Find the maximum pressure that the gold brick of problem 72 can exert on the surface of a table by placing one of its faces flush against the surface. (b) Find the minimum pressure.

76. A king-size water bed is 250 cm long, 250 cm wide, and 30 cm deep. Calculate the mass of water in the bed and the pressure it exerts on the floor if it lies flat on the floor.

77. How does a suction cup work?

78. Indian Fakirs stretch out on a bed of nails for a nap. Explain how this is possible.

79. Why are nails made with sharp points rather than blunt ends?

ANSWERS TO SELECTED NUMERICAL PROBLEMS

5. 6.67 km per hour
6. 2760 m, 2.76 km
7. 10^{-7} s
11. 45 cm
12. 22.4 m at an angle of 26.6° with respect to the initial direction of travel
15. 29.6 m at an angle of 6.64°
20. The acceleration is 5 m/s² for both
21. (a) 1.67 m/s² north, (b) 1.67 m/s² south
27. 1.53 s
30. (a) 49 m/s, (b) 123 m
32. 22.5 m
33. 4.17×10^{-8} m/s²
35. 536 N
36. 112 lb
41. 50 N at 53.1° north of east
42. 68.9 N at 8.51° east of north
59. 2.04 kg
60. 196 N
61. 2×10^{5} m/s²
62. 19.6 m/s²
66. (a) 1.72 m/s², (b) 13.8 m
68. 1.21 m/s²
71. (a) 5.18×10^{-4} m³, 7.75 m³
75. (a) 47,200 N/m², (b) 9450 N/m²
76. 1880 kg, 2950 N/m²

CHAPTER 3

CONSERVATION LAWS: MOMENTUM AND ENERGY

This bungee jumper will experience both the conservation of energy and the conservation of momentum in his plunge. (© Pictor Uniphoto)

Newton's laws of motion, which we discussed in Chapter 2, revolutionized the world of physics and altered humans' outlook on their environment. Phenomena that had defied explanation were now understandable by people with only modest scientific knowledge. With the use of only these laws, aspects of nature as diverse as the motion of a ball rolling along the ground and the motion of a spacecraft can be explained. Additionally, they were important to physics because they led to the explanation and derivation of other laws of physics. In this chapter, we will look at two of these outgrowths of Newton's laws of motion, the conservation of momentum and the conservation of energy. Both of these concepts were known, at least to a limited extent, before the advent of Newton's work, but a true understanding and development of them were possible only following Newton.

The conservation of momentum is most frequently used in situations in which objects collide. Thus, after studying this chapter, we hope you will have a better understanding of why two pool balls move as they do when they collide on a billiard table or why a gun recoils when a shot is fired.

The concept of energy is one of the most important in science. In everyday usage, we think of energy in terms of the cost of fuel for transportation and heating, of electricity for lights and appliances, and of the foods we consume. These ideas, however, do not really define energy. They tell us only that fuels are needed to do a job and that those fuels provide us with something we call energy. We shall investigate in this chapter the many different forms in which we find energy, and we shall discover that one form of energy can be converted into another form. In fact, this transformation of energy from one form to another is an essential part of the study of physics, chemistry, biology, geology, and astronomy. Thus, a study of energy is essential to all the fields of science, but you may also be surprised at the extent to which an understanding of energy is important to your day-to-day existence, regardless of your field of study.

3.1 WHAT IS A CONSERVATION LAW?

If we are to study conservation laws, it behooves us to understand exactly what is meant when we say that something is conserved. In its simplest explanation, when something is conserved, it is never lost. If we have a certain amount of a conserved quantity at some instant of time, we find that we have exactly the same amount of that quantity at any later time. This does not mean that the quantity cannot change from one form to another during the elapsed time, but if we consider all the forms that can be taken, we will find that we always have the same amount. Let us consider the amount of money that you now have. As you are aware, your checking account balance is not a conserved quantity; it is likely to decrease during the course of a month. For the moment, however, let us assume that your money *is* conserved. This would mean that if you have a dollar in your pocket, you will always have that dollar, although it may change form. That is, one day it may be in the form of a nice crisp bill; the next day, you may have 100 pennies, and the next day, an assortment of dimes and nickels. But when you total them up, you always find exactly one dollar. It would be nice if money behaved in this way, but of course it doesn't. Momentum and energy are conserved, and the purpose of this chapter is to examine all aspects of this statement.

3.2 MOMENTUM

Frequently on the nightly news, you hear a statement such as, "Candidate Blunderbuss predicts victory based on the fact that he is gaining momentum in the polls." In such everyday usage, the term momentum expresses the idea that candidate Blunderbuss is moving forward rapidly, but let us examine the concept of momentum to see how it is defined and used in physics. **Momentum** *is defined for particles as the product of the mass of the object and its velocity.* Momentum is a vector quantity having the same direction as the velocity. In equation form, with m the mass and $\mathbf{v}$ the velocity, we have

$$\text{momentum} = m\mathbf{v} \qquad (3.1)$$

From Eq. 3.1, we see that a large object such as a locomotive can have a large momentum even if it is moving slowly. A small object such as a fly can also have a large momentum if it is moving fast enough. You should also note that because velocity is a vector quantity, so is momentum. Thus, to specify momentum completely, we also must give its direction, which is the same as that of the velocity.

EXAMPLE 3.1 To Swat or Not to Swat
A 10,000 kg locomotive is rolling leisurely along a track with a velocity of 0.25 m/s. With what speed would a 10^{-4} kg fly have to move in order to have the same momentum as the locomotive?

Solution The momentum of the locomotive is

$$\text{momentum} = (10{,}000 \text{ kg})(0.25 \text{ m/s})$$
$$= 2500 \text{ kg m/s}$$

This must also be equal to the momentum of the fly, so the speed of the fly must be

$$\mathbf{v} = \frac{\text{momentum}}{m} = \frac{2500 \text{ kg m/s}}{10^{-4} \text{ kg}}$$
$$= 2.5 \times 10^7 \text{ m/s}$$

This corresponds to a speed of 25 million m/s, about one-tenth the speed of light. In a science fiction story, our fly might have this velocity but not in real life.

3.3 CONSERVATION OF MOMENTUM

The beauty and usefulness of momentum lie with the fact that *momentum is conserved in collisions between objects.* For example, consider a hockey puck sliding across the frictionless surface of ice. The puck has mass and velocity; thus it has momentum. Now suppose that this puck collides head on with another hockey puck, at rest on the ice, and after the collision both pucks zoom away. If we measure the initial momentum of the hockey puck and measure the total momentum of the two pucks after the collision, the net momentum before is equal to the

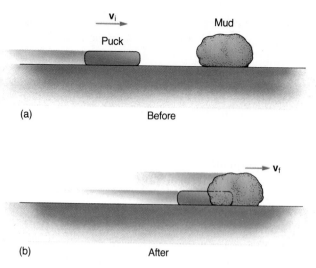

(a)   Before

(b)   After

FIGURE 3.1   (a) The hockey puck has momentum before the collision. (b) When the puck picks up the mud after collision, the mass becomes greater, so the velocity v_f must be less than the initial velocity of the puck, v_i.

net momentum afterward. In fact, regardless of how complicated the collision may seem, momentum is always conserved in collisions between two or more objects. As an example of what seems to be a slightly more complicated collision, consider the hockey puck again, but this time assume that it collides with and sticks to a piece of mud as in Figure 3.1. After the collision, the puck and mud move as a unit; nevertheless, the momentum of the puck before the collision is exactly the same as the momentum of the puck-mud combination after the collision.

In equation form, we can express the conservation of momentum in a collision or interaction as

$$\text{momentum before} = \text{momentum after} \quad (3.2)$$

EXAMPLE 3.2 BATTER UP
A 50 kg baseball pitching machine is at rest on a smooth playing surface. A 0.15 kg baseball is fired toward the right from the machine at a speed of 30 m/s. Find the recoil velocity of the machine.

Solution This is an example of conservation of momentum. In this problem, however, the initial momentum is zero because the pitching machine and the baseball are both initially at rest. After the ball is thrown, the ball has momentum toward the right, which we will call the positive direction, and the machine recoils with velocity **v** to the left. We have

$$\text{momentum before} = \text{momentum after}$$

$$0 = (50 \text{ kg})\mathbf{v} + (0.15 \text{ kg})(30 \text{ m/s})$$

Solving for **v**, we find

$$\mathbf{v} = -0.09 \text{ m/s}$$

The negative sign indicates that the pitching machine moves to the left, in the negative direction.

EXAMPLE 3.3 ROAD HOGS BEWARE
A car of mass 2000 kg traveling at a speed of 20 m/s collides with another car of the same mass that is stopped for a traffic light. The two cars become entangled and move together as a unit after the collision. What is the velocity of this wreckage?

Solution We use conservation of momentum as

$$\text{net momentum before} = \text{net momentum after}$$

The momentum before the collision is solely that of the oncoming car. Thus,

$$\begin{aligned}
\text{net momentum before} &= m\mathbf{v}_{\text{car}} \\
&= (2000 \text{ kg})(20 \text{ m/s}) \\
&= 40,000 \text{ kg m/s}
\end{aligned}$$

The momentum after is the product of the mass of the combined wreckage and the velocity **v** of this wreckage. We have

$$\text{momentum after} = (4000 \text{ kg})\mathbf{v}_w$$

Conservation of momentum yields

$$40,000 \text{ kg m/s} = (4000 \text{ kg})\mathbf{v}_w$$

From which,

$$\mathbf{v}_w = 10 \text{ m/s}$$

This velocity is in the same direction as that of the moving car before the collision.

EXAMPLE 3.4 DAREDEVIL POOL BALLS
Pool ball A of mass 0.30 kg rolls to the right across a table with a velocity of 3.00 m/s and collides head on with a 0.20 kg ball B moving toward the left with

FOCUS ON
TRANSFER OF MOMENTUM IN A ROCKET ENGINE

It is, of course, correct to say that the upward thrust of a rocket engine is the "equal and opposite reaction" to the downward thrust of the exhaust gases. But some students ask, what actually pushes the rocket skyward? The exhaust gases do not push a rocket anywhere. To understand the answer to this question, think about what happens inside the engine itself. A rocket engine is a chamber in which gases burn at a rapid but controlled rate. As the rocket fuel burns, hot (that is, fast-moving) molecules of gas shoot out in all directions. Those that hit other molecules exchange momentum with them. Those that hit the walls of the firing chamber exchange momentum with the rocket. The molecules that hit the left wall (see figure) are balanced by those that hit the right, so the rocket gains no net momentum in either horizontal direction. The molecules that hit the top of the firing chamber push the rocket upward. What about the exhaust gases, those that travel straight downward? They do not hit any part of the rocket at all, so they do not affect its motion directly. Since the rocket is forced upward, but not downward, the net momentum change of the rocket is skyward, and up it goes.

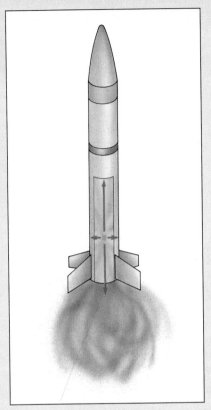

Booster rocket during liftoff. The upward momentum of the rocket and its load equals the combined downward momentum of all the tiny gas molecules ejected.

a velocity of 2.00 m/s. After the collision, the 0.20 kg ball has a velocity of 1.50 m/s toward the right. Find the velocity of ball A.

Solution This is a straightforward application of conservation of momentum. We have

$$(0.30 \text{ kg})(3.00 \text{ m/s}) + (0.20 \text{ kg})(-2.00 \text{ m/s})$$
$$= (0.30 \text{ kg})\mathbf{v}_A + (0.20 \text{ kg})(1.50 \text{ m/s})$$

Note that we have selected the positive direction for velocity to be toward the right. As a result, the velocity and momentum of the 0.20 kg ball before the collision are negative.

We solve the equation for $\mathbf{v}_A$ to find

$$\mathbf{v}_A = 0.667 \text{ m/s}$$

The positive value indicates that the velocity of the 0.30 kg ball is still to the right, in the direction we have selected as positive.

CONCEPTUAL EXAMPLE 3.5 I FELT THE EARTH MOVE AGAIN

A student hurls a piece of mud toward the right at a stationary brick wall. When the mud strikes the wall, it sticks to the wall and stops. Discuss conservation of momentum as applied to this situation.

Solution At first glance, it looks like conservation of momentum has failed us in this application. Certainly, before the collision occurred, the mud had momentum toward the right, but after the collision nothing appears to be moving. This seems to be in violation of the conservation of momentum because if something has momentum to the right before, something must have the same momentum to the right afterward. What has happened?

The answer is that something does indeed have momentum to the right after the collision. The wall is firmly affixed to the ground, and it is the entire Earth that picks up the net momentum to the right after the collision. Don't expect the Earth to change velocity much, however, because of its huge mass.

CONCEPTUAL EXERCISE

Use the conservation of momentum to explain why you are safer driving a massive car than you are one less massive.

Answer If a collision occurs, your car is going to undergo a change in its momentum, but a large mass means that there will be a smaller change in the velocity, or a smaller acceleration. The less the acceleration the driver of the car experiences, the greater is the likelihood of escaping without serious injury.

CONCEPTUAL EXERCISE

In light of the previous discussion, consider the following situation. You are standing at rest in a room, and then you begin walking toward the right. Discuss conservation of momentum in this situation.

Answer Before the walk, there was no momentum because nothing was moving, but when you took your first stride, you had momentum to the right. For the net momentum to be zero, something has to gain momentum to the left. What is it?

3.4 WORK

Imagine holding a heavy bucket of water straight out from your body at arm's length for a couple of hours. It isn't a pleasant way to spend an afternoon, but if you ever choose to do it, you will probably be quite tired at the end and you will think that you have done a lot of work. Even though your tired muscles indicate that a lot of work has been done,

we shall shortly see that according to the physics definition of work, you have not done any at all. Occasionally, we encounter a word that has a totally different meaning in physics than it does in everyday usage. Perhaps of all such cases, none differs quite as much as does the definition of work from that used in nonscientific situations.

Work is a process of energy transfer or energy transformation. That is, we shall find that when we do work on an object, we give it energy. Likewise, when an object has energy, it is capable of doing work for us. We are able to calculate a numerical value for work as the product of a force, F, and the distance, d, over which the force acts. In equation form this is

$$\text{work} = (\text{force})(\text{distance})$$

and in symbolic form

$$\text{work} = Fd \tag{3.3}$$

This equation applies only when F and d are parallel.

As an example, in Figure 3.2, a grocery cart is being pushed by a shopper with a horizontal force of 3.0 N. If the cart moves a distance of 4.0 m, how much work is done? We find

$$\text{work} = Fd = (3.0 \text{ N})(4.0 \text{ m}) = 12 \text{ N m}$$

The units of work are N m, and this combination of units is lumped together and given the name **joule** (symbol J). Thus, the work done is 12 J. In many applications, you will find that work is measured in conventional units of ft lb, called foot pounds.

FIGURE 3.2 *How much work is done by the shopper moving a grocery cart with a horizontal force of 3 N for a distance of 4 m?*

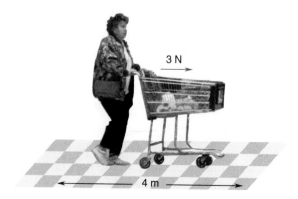

Now, return to the example with which we started this section. When you are holding the bucket of water at arm's length, you are certainly exerting a force on it, but as long as there is no movement of the bucket, d is zero and the work is also zero.

A question that has probably already occurred to you is why we have paused to discuss work. This chapter is supposed to be about momentum and energy—why this apparent digression? The reason is that energy is defined in terms of work. More specifically, *an object has **energy** if that object has the ability to do work.* We will examine this relationship between work and energy more closely in the following sections.

CONCEPTUAL EXERCISE

Two teams competing in a tug-of-war eventually have to stop because neither can move the other. Have the exhausted players done a lot of work according to the physics definition of the term?

Answer Both teams have exerted large forces but they have not moved anything. Therefore, neither team has done any work.

EXAMPLE 3.6 MORE ABOUT WORK

A farmer supports a 20 N bucket as shown in Figure 3.3 as he walks a distance of 3.0 m at constant speed. How much work is done by the farmer?

Solution It is tempting but *incorrect* to substitute into the defining equation for work to find

$$\text{Work} = Fd = (20 \text{ N})(3.0 \text{ m}) = 60 \text{ J}$$

The reason that this is incorrect lies with the definition of the distance d. Note in the definition given earlier, the distance d is the distance moved *in the direction of the force.* Granted, the person exerts a force, 20 N, and he moves a distance of 3 m, but the 3 m movement is horizontal, whereas the force is vertical. There is no movement in the direction of the force, so the work done is zero.

3.5 KINETIC ENERGY

Imagine that you have a nail driven part way into a vertical board, as shown in Figure 3.4a. You would like to drive the nail farther into the board, but no hammer is handy. How can you accomplish this task? If a rock is nearby, one way to drive the nail is to throw the rock at it, as shown in the figure. Let us examine this situation from the standpoint of work and energy. Recall that in the last section, we said that an object has energy if it has the abil-

20 N

3.0 m

FIGURE 3.3 *How much work is done by the farmer carrying the 20 N bucket at constant speed for a distance of 3 m?*

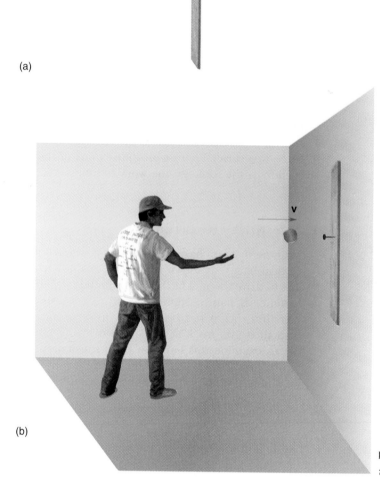

(a)

(b)

FIGURE 3.4 *(a) A nail in a board. (b) One way to drive the nail into the board.*

ity to do work. Does our moving rock have energy? It does because when it strikes the end of the nail the rock exerts a force, *F*, on the nail and the nail is moved some distance, *d*, into the plank. Thus, there is a force on the nail, a movement of the nail in the direction of the force, and work is done. As a result, we can say that the moving rock must have energy. *The energy that an object has because of its motion is called **kinetic energy**, and is a scalar quantity.*

The amount of kinetic energy that a moving object possesses can be shown to depend on two factors: the mass of the moving object and its speed. It should seem reasonable that mass is a factor be-

cause a more massive rock exerts a greater force on the nail and drives it deeper into the plank than does a less massive rock. Likewise, if one throws the rock with a high speed, the nail is driven deeper than if the rock moves slowly. In fact, the kinetic energy of an object depends on the square of the speed. The kinetic energy of an object is given by

$$KE = \frac{1}{2}mv^2 \qquad (3.4)$$

where *m* is the mass of the object and *v* is its speed. A derivation of the kinetic energy equation appears at the end of the chapter.

Energy, in joules, of some interactions and events in the Solar System.

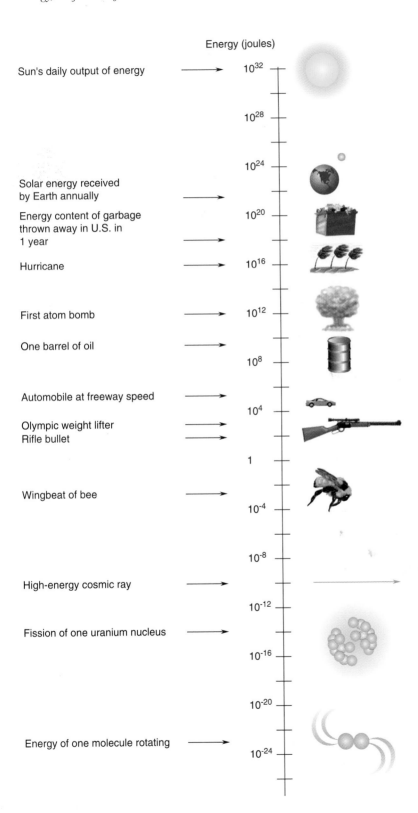

Energy (joules)

Sun's daily output of energy → 10^{32}

10^{28}

10^{24}

Solar energy received by Earth annually →

Energy content of garbage thrown away in U.S. in 1 year → 10^{20}

Hurricane → 10^{16}

First atom bomb → 10^{12}

One barrel of oil →

10^{8}

Automobile at freeway speed →

10^{4}

Olympic weight lifter →
Rifle bullet →

1

Wingbeat of bee →

10^{-4}

10^{-8}

High-energy cosmic ray →

10^{-12}

Fission of one uranium nucleus →

10^{-16}

10^{-20}

Energy of one molecule rotating →

10^{-24}

EXAMPLE 3.7 Duck!

A 0.15 kg baseball is thrown with a speed of 30 m/s. How much kinetic energy does the baseball have?

Solution This is a straightforward substitution into the defining equation for kinetic energy. We have

$$KE = \frac{1}{2}mv^2 = \frac{1}{2}(0.15 \text{ kg})(30 \text{ m/s})^2 = \boxed{67.5 \text{ J}}$$

As anyone who has ever played baseball is aware, this is enough energy to do considerable injury to someone standing in the flight path.

3.6 GRAVITATIONAL POTENTIAL ENERGY

Recall that an object is said to have energy if it has the ability to do work. We saw in the last section that a rock can have energy because of its motion, since the moving object can do work. Let us now examine a different type of energy that an object can possess, called **gravitational potential energy.** *Gravitational potential energy is the energy that an object has because of its location in space.* To examine this concept more carefully, consider once again our task of driving a nail. As shown in Figure 3.5a, the plank is now horizontal, and we are going to drive the nail by dropping our rock onto it. It should be obvious to you that the dropped rock will exert a force, *F*, on the nail and cause the nail to move a distance, *d*, into the plank. Thus, the rock can do work on the nail, and from our definition the rock must have energy. This energy arises because of the location of the rock in space above the nail.

Consider lifting an object of mass *m* a distance *h* above the surface of the Earth. To lift the object, you have to do work on it, and the amount of work you do is given by

$$W = Fd = mgh$$

where *mg* is the weight of the rock. You must exert a force at least equal to the weight of the rock in order to lift it through the height *h*. This work that you have done on the system can be returned to you. Thus, we define the gravitational potential energy, *PE*, of the rock when at the height *h* above the ground to be the same as the work you did in lifting it to that point.

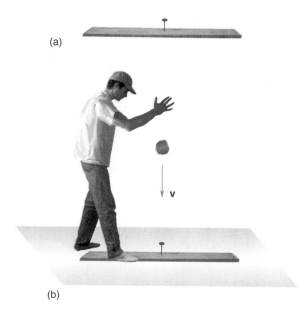

(a)

(b)

FIGURE 3.5 *(a) A nail in a horizontal plank. (b) Another way to drive the nail.*

$$PE = mgh \qquad (3.5)$$

It should seem reasonable that the potential energy depends on the mass of the object. For example, in our case of dropping the rock, it is obvious that a more massive rock does more work on the nail than a lighter rock. Also, a given object drives the nail farther into the plank, or does more work, if it is dropped from a great height than it does if it falls only a short distance.

EXAMPLE 3.8 Climbing Higher and Higher

(**a**) An employee in a highrise office building holds a 2.0 kg file folder 0.50 m above a desk, at position A in Figure 3.6. Find the gravitational potential energy with respect to the top of the desk.

Solution The potential energy can be easily found from $PE = mgh$ as

$$PE = (2.0 \text{ kg})(9.8 \text{ m/s}^2)(0.50 \text{ m}) = \boxed{9.8 \text{ J}}$$

(**b**) If he moves slightly to B so that the file folder is not above the desk, a more convenient reference level might be the surface of the floor, which is 1.0 m below the file folder. What is the potential energy with respect to the floor?

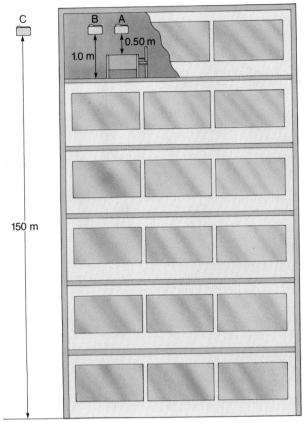

FIGURE 3.6 *The reference level for measuring gravitational potential energy is arbitrary.*

Solution Note that *h* in *PE* = *mgh* is relative to whatever level you choose. As a result, we find the *PE* as

$$PE = mgh = (2.0 \text{ kg})(9.8 \text{ m/s}^2)(1.0 \text{ m})$$

$$= 19.6 \text{ J}$$

(c) Finally, the employee moves to an open window and holds the file folder out the window at C such that it is 150 m above the street level. How much potential energy does the folder have relative to ground level?

Solution Again, we calculate the potential energy as

$$PE = mgh = (2.0 \text{ kg})(9.8 \text{ m/s}^2)(150 \text{ m})$$

$$= 2.9 \times 10^3 \text{ J}$$

In all three cases, the potential energy was measured with respect to some reference level. This reference level is completely arbitrary, and you can choose the one that you like. In a given problem, however, usually one particular reference level stands out as being the most suitable one to use. We shall have more applications later, which will help you in making judicious selections of these reference levels.

3.7 CONSERVATION OF ENERGY

The beauty and usefulness of the concept of energy arose when it was discovered that energy is conserved. We shall see that a problem that might be difficult to solve by the use of Newton's laws of motion becomes simple when approached from the viewpoint of the conservation of energy. As noted earlier, when a quantity is conserved, we find that we have as much of it at a later time as we do initially. This does not mean that the quantity cannot change form between the two times. As an example, consider the situation in which we hold a rock above the ground at position A in Figure 3.7. At this location, all of the energy of the rock is in the form of gravitational potential energy. When we release the rock, however, its gravitational potential energy decreases as the rock falls because its height above the ground becomes less. We find that its speed, and hence its kinetic energy, is increasing. Finally, just before striking the ground, all of the rock's initial potential energy is completely converted to kinetic energy. Observations such as this lead to one of the most fundamental laws of physics, the **law of conservation of mechanical energy,** which states

> *In the absence of friction, the sum of the kinetic energy and the potential energy of a system is constant.*

In equation form, this may be stated as

$$KE + PE = \text{constant} \tag{3.6}$$

EXAMPLE 3.9 GOOD TOSS FOR A ROOKIE
A baseball player throws a 0.15 kg baseball straight up into the air with a speed of 15 m/s. How high does the ball rise above the level at which it is released?

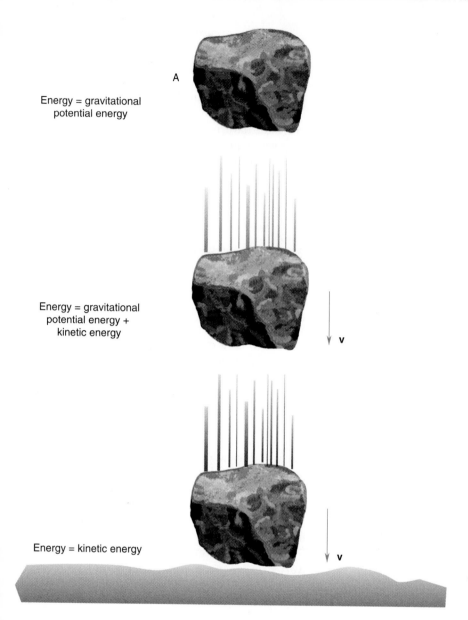

A

Energy = gravitational
potential energy

Energy = gravitational
potential energy +
kinetic energy

v

Energy = kinetic energy

v

FIGURE 3.7 *The energy of a falling object changes from gravitational potential energy to kinetic energy.*

Solution The situation is shown in Figure 3.8, in which we have called point 1 the position at which the ball is released from the thrower's hand and point 2 the highest point reached by the ball. Initially, all of the energy of the ball is in the form of kinetic energy, assuming that we set the zero level for potential energy at the release point, point 1. As the ball rises, it slows down, finally coming to a stop at the top of its flight path, point 2. At this location, all of the energy of the ball is gravitational potential energy. We find its initial total mechanical energy as

$$KE_i + PE_i = \text{constant}$$

or

$$\frac{1}{2}(0.15 \text{ kg})(15 \text{ m/s})^2 + 0 = 16.9 \text{ J}$$

The final total mechanical energy at the top is

$$KE_f + PE_f = 0 + (0.15 \text{ kg})(9.8 \text{ m/s}^2)h$$

$KE_f = 0$
$PE_f = 16.9$ J   ◯ 2

$KE_i = 16.9$ J
$PE_i = 0$ **v** ◯ 1

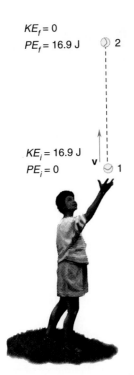

FIGURE 3.8 *Energy changes from kinetic to gravitational potential energy as a ball rises into the air.*

where h is the maximum height reached. Because mechanical energy is conserved, we have

$$(0.15 \text{ kg})(9.8 \text{ m/s}^2)\,h = 16.9 \text{ J}$$

from which we find

$$h = \boxed{11.5 \text{ m}}$$

(You should note that to cancel units in the equation, J must be changed back to its equivalent of kg m²/s².)

EXAMPLE 3.10 HOW TO BE A SWINGER
(a) An object of mass 0.50 kg is attached to a string of length 0.75 m, to make a pendulum as shown in Figure 3.9. The string is held horizontal and released at position A. How fast is the object going when it reaches the bottom of its arc, position B?

Solution Initially, all of the energy of the pendulum is in the form of gravitational potential energy, if we set the zero level for the measurement of potential energy at the bottom of the arc. Thus, at position A, the total mechanical energy is

$$mgh = (0.50 \text{ kg})(9.8 \text{ m/s}^2)(0.75 \text{ m}) = 3.68 \text{ J}$$

At B, all of this potential energy has been converted into kinetic energy, and conservation of mechanical energy gives us

$$3.68 \text{ J} = \frac{1}{2}(0.50 \text{ kg})\,v^2$$

Or, solving for v we find

$$v = \boxed{3.83 \text{ m/s}}$$

(b) At position B, the speed of the object causes it to move through the bottom of the arc and continue along its path as shown in Figure 3.9. Find how high it rises above the bottom of the arc.

Solution In this case, we find that all of the initial energy of the object at B is in the form of kinetic energy. We found in part (a) that the kinetic energy that it has is 3.68 J. When the object climbs to the same height as point A, all of this kinetic energy is converted again into gravitational potential energy. Thus, it is a simple calculation for you to show that the object will climb to a height of 0.75 m before stopping.

From this, we see that in the absence of friction, the object on the end of the string will continue to swing back and forth forever, always rising to the same height at the end of each swing and having the same speed at the bottom of the arc. You know from your own experience that this does not occur; the pendulum soon comes to a stop. In the next section, we shall examine why the motion eventually ends.

FIGURE 3.9 *How fast is the pendulum ball moving at point B? How high does it rise as it moves along path C?*

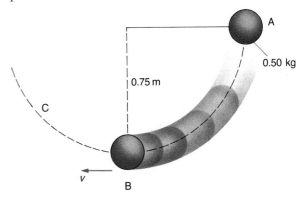

FOCUS ON
ENERGY IN THE HUMAN BODY

The ultimate source of all energy used by the human body is plant life, and because plants gain their energy from the Sun, all the energy used by us is ultimately traceable to the energy of sunlight. Before you begin to protest about the hamburger that you had for lunch, recall that the cow dined on plants, and the energy it gained from that source was ultimately transferred to you. Much of the energy supplied to your body is used just to keep the machine operational. That is, it maintains the body temperature; it keeps all the organs operating; and it allows us to walk, to move around, and to perform work.

The process by which a plant stores energy from the Sun is called photosynthesis. The steps are as follows: The foliage of a green plant stores carbon dioxide and water, and these two would forever remain separated if it were not for the presence of a substance called chlorophyll in the plants. Chlorophyll acts as an agent that allows sunlight striking the plant to cause a chemical reaction in which carbon dioxide and water are changed to a form of sugar called glucose, with some oxygen left over. The reaction is symbolized as

(carbon dioxide) + (water) + (energy)
→ (glucose) + (oxygen)

Long chains of these glucose molecules join together to form cellulose, but nevertheless this stored energy is retained and, under the proper conditions, can be reclaimed. One way to reclaim this energy is to burn a piece of wood. At a high enough temperature, the chemical reaction just discussed can proceed in the opposite direction. The glucose stored as cellulose can combine with oxygen from the air and be changed back to carbon dioxide and water. In this process, the energy originally absorbed from the sunlight is released, primarily in the form of heat. When an animal eats the plant, the energy stored in the glucose molecules is released to the animal. This release of energy takes place in mitochondria, small bodies found in cells. These mitochondria contain enzymes that break up the glucose into simpler molecules, which then react with oxygen to form carbon dioxide, water, and energy. The energy released is then used by the body to perform its various functions.

CONCEPTUAL EXERCISE

If energy is conserved, that means the Universe is always going to have the same amount. If so, how can we ever say that the Earth has an energy crisis?

Answer   Energy is indeed neither created nor destroyed, but it is converted to forms in which it is no longer useful to us. For example, when we burn a lump of coal, stored internal energy is changed to random thermal motion, a form that cannot be recovered and reused.

3.8 FRICTION FORCES AND THE CONSERVATION OF ENERGY

We pointed out in our discussion of the motion of a pendulum (Example 3.10) that a difficulty seems to have arisen with our approach to the conservation of energy. If the pendulum continues to interchange energy between kinetic and potential, it will swing forever and never lose any of its total mechanical energy. A pendulum never does this, so it is obvious that we have omitted something from our discussion—friction.

As the pendulum moves through the air, it experiences two important frictional forces. One of these arises at the point where the pendulum string is attached to the ceiling. The second is air resistance against the swinging pendulum. These frictional forces cause a decrease in the energy of the pendulum such that as the pendulum swings from an extreme position to the bottom position, the amount of kinetic energy gained is not quite equal to the amount of gravitational potential energy lost. And as the pendulum moves up toward the end of its swing, it does not end up with the same amount of potential energy at the top as the kinetic energy it had at the bottom. As a result, the pendulum does not return to the same height that it reached on its previous swing. Over a period of time, friction robs all of the mechanical energy from the system, and the pendulum ceases to swing.

The question now is, "Where does this lost energy go?" The answer is that this energy is not really lost. Instead, it appears as a third form of energy, which we will call **thermal energy.** To see how this form of energy enters the picture, imagine the effect that the swinging object has on the air molecules in its path. A molecule just in front of the object is in a position similar to a golf ball as the golf club approaches. When the pendulum hits the air molecules, it knocks them away, just as a golf club knocks away the ball. Other molecules are hit by these molecules as they fly away, and in general the billions and billions of molecules near the pendulum gain speed from the stirring effect of the pendulum's motion. Different molecules are affected differently, but the important thing to note is that, on the average, the air molecules gain speed as a result of the swinging object. In Chapter 5, we will see that this increased speed means increased temperature.

Individual molecules have mass; thus, a speed gain by the molecules means a gain in their kinetic energy. This, then, is where some of the energy of the pendulum goes. As the pendulum swings through the air, it stirs up air molecules and increases their kinetic energy. *This energy of motion of the molecules or atoms of any substance is referred to as thermal energy.* If one were able to measure the increase in energy of the molecules during one swing of the pendulum, one would find that it would account for *some but not all* of the decrease in the energy of the pendulum. The remainder of the energy loss can be accounted for by friction at the point of support. Constant rubbing at this location causes an increase in thermal energy of the support and of the string attached to it. We will examine thermal energy in more detail in Chapter 5.

3.9 POWER

Imagine a designer of heavy equipment who builds a crane that is capable of lifting a 4000 kg mass to a height of 30 m. With only this much information, it would be difficult to sell this device to a practical-minded contractor, as he wants to know not only how much the crane can lift, but also how long it takes to lift it because he may have to lift several such objects during the course of a work day at his construction site. Thus, in many practical applications, we need to know more than just how much energy output a device has. We must also consider *time,* a factor that has not appeared in our analysis of energy thus far.

> **Power** *is defined as the amount of energy converted from one form to another divided by the time required to make this conversion.*

This is

$$\text{power} = \frac{\text{energy converted}}{\text{time required}} \qquad (3.7)$$

The units of power are J/s, and this collection of units is referred to as a **watt,** W. An alternative form of measuring power is the horsepower, hp, chosen to equal 550 ft lb/s. Based on this choice, the relation between the hp and the watt is

1 hp = 746 W

Consider the definition of power in terms of a 100 W light bulb. When we say that the power of a light bulb is 100 W, we mean that energy is being converted from electrical energy to other forms at the rate of 100 J every second. These other forms are light energy and thermal, or heat, energy. (The rate at which a resting adult gives off heat is about 100 W.) Note that we are accustomed to discussing our pieces of electrical equipment in terms of power measured in watts, and we are accustomed to discussing mechanical equipment, such as the output of an automobile engine, in terms of horsepower. However, there is no fundamental reason for this; it is by convention only that this occurs. We could equally well refer to a 0.2 hp light bulb or to a 60,000 W car motor.

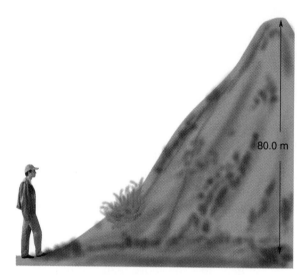

FIGURE 3.10   *Discuss the energy transformations that take place as the student climbs a hill that rises 80 m vertically.*

EXAMPLE 3.11 NOT EATING WOULD BE EASIER

A 70.0 kg student decides to exercise by climbing a hill that rises vertically by 80.0 m, as shown in Figure 3.10.

(a) Discuss the energy transformations that occur during the climb.

Solution The final gravitational potential energy of the student relative to his position before the climb is

$$mgh = (70.0 \text{ kg})(9.8 \text{ m/s}^2)(80.0 \text{ m})$$
$$= 54{,}900 \text{ J}$$

This energy is supplied by energy transformations inside the body in which stored energy from food is converted by chemical processes to gravitational potential energy.

(b) What is the power output of the student if he makes the climb in a time of 200 s?

Solution We have

$$\text{power} = \frac{54900 \text{ J}}{200 \text{ s}} = \boxed{275 \text{ W}}$$

Thus, the student is converting chemical energy to gravitational potential energy at the rate of 275 J each second. (However, the efficiency of the human "engine" is about 25 percent. Thus, the power output of the student is about 1100 W.)

3.10 SIMPLE MACHINES

In this section, we will take a look at two classifications of **simple machines,** the inclined plane and the lever. Regardless of how simple or complex a machine may be, the principle that underlies its operation is the conservation of energy. A machine is a type of tool that alters the magnitude or direction of an applied force. Thus, a machine can change a small force to a large force or, alternatively, a large force to a small force.

To understand machines, it is essential to remember the difference between force and energy. Forces are pushes and pulls, and as we will learn, it is relatively easy to amplify or reduce an applied force. Energy is different because it can be neither created nor destroyed. Thus, no machine can create energy. The difference between force and work can be emphasized by the following example. Recall that work, which is related to energy, is computed as

$$\text{work} = (\text{force})(\text{distance})$$

Imagine that 12 J of potential energy is available to you. You can design a machine to use those 12 J to exert a force of 1 N for 12 m. Another machine could be built to use the 12 J of energy to transmit a force of 12 N for 1 m. (We have assumed here that frictional forces are zero.) Of course, it would be impossible to build a machine that would use 12 J of energy to exert a force of 12 N for 12 m, or even for 1.0001 m, for either process would create energy. Thus, a given amount of energy can be used to produce either a large force for a small distance or a smaller force for a larger distance.

Inclined Planes

Suppose two people are given the task of lifting a heavy iron ball onto a platform 2 m off the ground. If the ball weighs 400 N, the work required to lift it will be

$$W = Fd = (400 \text{ N})(2 \text{ m}) = 800 \text{ J}$$

Thus, 800 J of energy are required to lift the ball—no more and no less. But there are no restrictions on how this work is to be performed. The stronger person may lift the ball straight up. In this case, the force exerted will be 400 N, and the work performed will be $W = (400 \text{ N})(2 \text{ m}) = 800 \text{ J}$ as indicated in the equation.

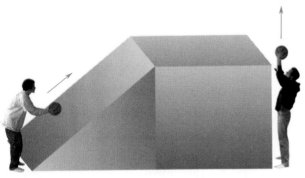

The easy way;
less force is needed

Both processes
use the same
amount of energy

The hard way;
more force is
needed

FIGURE 3.11   *Less force is needed to roll a ball up an inclined plane than to lift it straight up. However, the total energy required is the same in both instances. (Friction forces are ignored here.)*

A weaker person who is unable to lift the ball, however, might build a ramp and roll the ball up, as shown in Figure 3.11. In this example, the ramp is 10 m long. Thus, the weaker person has to exert a smaller force over a longer distance. But because the total work must be the same, the magnitude of the force is less. Because 800 J are performed over a distance of 10 m, the force needed can be calculated as

$$F = \frac{W}{d} = \frac{800 \text{ J}}{10 \text{ m}} = 80 \text{ N}$$

Thus, the weaker person is able to raise the ball to the required height by using only 80 N of force. The ramp (an **inclined plane**) reduces the force required to do a given amount of work (by increasing the distance). Don't forget that the total energy needed to perform the task is the same in both instances. An inclined plane is a simple machine. In general, if there are two ways to perform a given task,

work done by one method
= work done by any other method

or

(large force)(small distance)
= (small force)(large distance)

Levers

There are a great many other clever ways of performing work. Think of a simple seesaw, supported in the middle as shown in Figure 3.12a. If a person on one end pushes downward, the other end will naturally move upward. Because the board is supported in the middle, both ends will travel the same distance—when one end goes down 2 m, the other end goes up 2 m. If one end is pushed down with a force of 60 N, 120 J of work are performed. Because energy is conserved, it is possible to exert an equal force of 60 N on the other side for an equal distance of 2 m. Now imagine that the seesaw is positioned so that it is supported not in the middle but off to one side, as shown in Figure 3.12b. Be-

FIGURE 3.12   *A seesaw.*

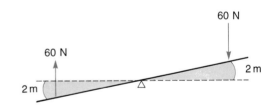

(a)   60 N × 2 m = 60 N × 2 m
          120 J = 120 J

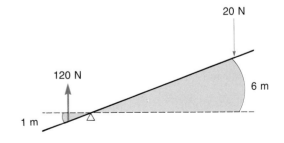

(b)   120 N × 1 m = 20 N × 6 m
          120 J = 120 J

cause one side is longer than the other, it will move upward farther. Let us say, for example, that whenever the long end travels 6 m, the short end travels only 1 m. Now what happens if the long side is pushed down with a force of 20 N, depressing it the full 6 m? The work performed is $W = (20 \text{ N})(6 \text{ m}) = 120 \text{ J}$. Since energy is conserved, 120 J of work must be performed on the other side. Because the other side travels a smaller distance (only 1 m), it must exert more force if it is to do equal work. The force, as found from $W = Fd$, is

$$F = \frac{W}{d} = \frac{120 \text{ J}}{1 \text{ m}} = 120 \text{ N}$$

Thus, once again, we see that a given amount of work can be performed by exerting a small force over a large distance or a large force over a small distance.

Any device analogous to a seesaw that pivots over a fixed point is called a **lever.** The pivot point is called the fulcrum. For convenience, the force applied to one end of the lever is called the input force, and the force developed at the other side of the lever is called the output force. For any lever, we can make use of the following relationship:

(input force)(distance from input force to fulcrum)

= (output force)(distance from output
            force to fulcrum)   **(3.8)**

It is immediately obvious how useful this relationship is. Suppose you are asked to lift a 5000 N stump, but you are not strong enough to pick up a stump that weighs more than 500 N. To do this, you can use a long rigid bar as a lever and a rock as a fulcrum. Let us suppose that you have arranged your bar so that the distance from the fulcrum to the point where you push down is 2 m and that you push down with a force of 500 N. What distance d can the weight of the stump be from the fulcrum so that you can lift it? From Eq. 3.8, we have

$$(500 \text{ N})(2 \text{ m}) = (5000 \text{ N})(d)$$

From which we find

$$d = 0.2 \text{ m}$$

EXAMPLE 3.12 Row, Row, Row Your Boat
The oars of a rowboat are simple levers that may be used as shown in either Figure 3.13a or b. In which

of these two configurations is the most force generated against the water, and in which arrangement does the oar travel the farthest distance through the water?

Solution   Figure 3.14a and b demonstrate the results of the rowing process for each situation. When the distance from the hand to the fulcrum is small, as in Figure 3.14a, a large input force is applied over a small distance, and the result is a small output force over a large distance. For the case of Figure 3.14b, a small input force applied over a long distance produces a large output force over a small distance.

Typically a rower is not concerned about applying a lot of force as much as he is about sweeping the water over a large distance. Thus, the configuration of Figure 3.14a is the more desirable rowing mode.

CONCEPTUAL EXERCISE

Is this a lever in action? If so, identify the input force, the output force, and the fulcrum.

3.11 CONSERVATION OF ANGULAR MOMENTUM

The final conservation law that we shall look at is the conservation of **angular momentum,** a conservation law applicable to spinning or revolving objects. To see this law in action, consider an observation that is a part of everyone's common experience. Tie one end of a string to an object such as a pencil and the other end to your finger; then swing the pencil around in a circular path. Once the pencil is in motion, let the string wind up

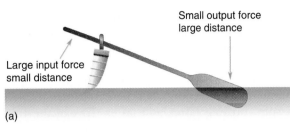

Large input force
small distance

Small output force
large distance

(a)

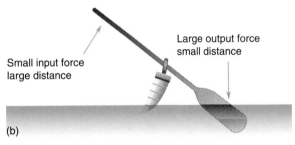

Small input force
large distance

Large output force
small distance

(b)

FIGURE 3.14   *An oar is a lever designed to change a large input force exerted over a small distance into a smaller output force exerted over a larger distance.*

(a)

(b)

FIGURE 3.13   *An oar as a lever. The oarlock is the fulcrum.*

on your finger. How does the orbital speed of the pencil change as the string gets shorter and shorter? If you don't know that the pencil will move faster and faster as the string shortens, you should try this simple experiment on your own.

The behavior of the pencil can be explained in terms of the conservation of angular momentum. The calculation of angular momentum can become a complex problem if the object that revolves has an unusual shape, but for the cases that we will consider, only a slight modification needs to be made to our expression for linear momentum considered previously. We must include a distance in our equation as:

angular momentum = mrv

The mass of the revolving object is m, the radius of the circular path is r, and the speed of the object in its path is v, as shown in Figure 3.15. In those cases in which there are no forces acting either to speed the object up in its path or to slow it down,

we find that angular momentum is conserved. That is,

$$m_i r_i v_i = m_f r_f v_f$$

Let us apply this to the pencil winding up on a finger. We can cancel the mass of the pencil from the equation because it is the same throughout the motion. We have

$$r_i v_i = r_f v_i$$

FIGURE 3.15   *The angular momentum of an object moving in a circular path equals mrv, where m is the mass of the object, $\mathbf{v}$ is its velocity, and r is the radius of the circular path.*

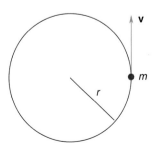

As the string shortens, the radius of the circular path becomes smaller. Thus, since r_f is less than r_i, the final velocity must be greater than the initial velocity.

You have seen the conservation of angular momentum in action when figure skaters go into a spin, turning faster and faster. To execute this maneuver, skaters start with both arms and perhaps one leg extended as they whirl in a circular path. When the limbs are pulled in toward the body, the effective radius of the path they follow is decreasing, and the result is the same as for the pencil: The final velocity increases.

On a much larger scale, the Sun rotates once on its axis approximately every 28 days, and this rotational velocity is believed to be the leftover remnant of the nebula from which it was formed. The theory of star formation says that stars begin their life as gigantic, thin clouds of dust and gas, called nebulae, stretching for light-years across space. Star formation begins when something triggers a condensation in the cloud. At first, a few molecules cluster together; then the mass grows to the size of a marble, then to a baseball, and finally after millions of years, it is at the size of a star. Any rotational motion of the original cloud would have been slow, but as the collapse of the cloud continues, the radius becomes smaller, and the rotational speed increases. It is thought also that the planets and their moons got their rotational motion in a process similar to this, as we shall discuss in Chapter 22.

CONCEPTUAL EXERCISE

An arrow shot from a bow lands with its pointed end toward the ground. A spiraling football does not land nose down, however. Why?

Answer   The nonspinning arrow is a projectile that moves such that it is always tangent to its path. This means that it will land nose down and stick up in the ground. A spinning football, however, must conserve angular momentum. This means that it must keep its axis of rotation always pointing in the same direction. Thus, it cannot land point down if thrown properly.

3.12 DERIVATION OF KINETIC ENERGY EQUATION

A moving body has energy of motion, or kinetic energy. Energy is the capacity to do work, and a moving body can do work by colliding with a stationary body and forcing it to move. Let us consider how much energy a stationary body can gain as it is accelerated at a uniform rate; this energy is its kinetic energy.

Let m = mass of body
0 = initial velocity
v = final velocity
d = distance the body travels as its velocity changes uniformly from 0 to v
t = time

Then

change in velocity = final velocity − initial velocity = $v - 0 = v$

The acceleration of the body is given by

$$a = \frac{\text{change in velocity}}{\text{time}} = \frac{v}{t}$$

and the average velocity of the body is

$$v_{av} = \frac{\text{final velocity} + \text{initial velocity}}{2} = \frac{v}{2}$$

The distance traveled by the object, d, is found by multiplying the average velocity by the time. We find

$$d = \frac{v}{2}t$$

Dividing both sides of the equation by t, we have

$$\frac{d}{t} = \frac{v}{2} \tag{a}$$

The kinetic energy of the body is the energy it gains from the work done on it. Thus,

kinetic energy = (force)(distance)

$$= (ma)(\text{distance}) = m\frac{v}{t}d = (mv)\frac{d}{t}$$

But, $d/t = v/2$ (from equation (a) above). Therefore,

$$\text{kinetic energy} = (mv)\left(\frac{v}{2}\right) = \frac{1}{2}mv^2$$

SUMMARY

Momentum is defined as the product of the mass of an object and its velocity. It is found that momentum is conserved in collisions between objects; this is the **law of conservation of momentum**. **Work** is defined as the product of a force and the distance over which the force acts. An object has **energy** if it has the ability to do work. Energy exists in many forms, two of which are **kinetic energy,** which is the energy an object has because of its motion, and **gravitational potential energy,** which is the energy of an object because of its location in space. In the absence of friction, the sum of the kinetic energy and the gravitational potential energy of an object is constant; this is the **law of conservation of mechanical energy.** Frictional forces convert kinetic energy and gravitational potential energy into **thermal energy. Power** is defined as the energy converted from one form to another divided by the time to make the conversion. A **simple machine** is a device that alters the magnitude or direction of an applied force. Two examples are the inclined plane and the lever. **Angular momentum** of an object moving in a circular path is defined as the product of its mass, velocity, and the radius of the path. Angular momentum is conserved for isolated systems.

EQUATIONS TO KNOW

Momentum $= m\mathbf{v}$

Work $=$ force $\times$ distance $= Fd$

Kinetic energy $= \dfrac{1}{2}mv^2$

Gravitational $PE = mgh$

Power $= \dfrac{\text{energy}}{\text{time}}$

For levers: (Input force)(distance from input force to fulcrum) = (output force)(distance from output force to fulcrum)

Angular momentum $= mrv$

KEY WORDS

Conservation laws	Joule	Power	Angular momentum
Momentum	Kinetic energy	Watt	Conservation of
Law of conservation	Potential energy	Law of conservation of	angular momentum
of momentum	Gravitational potential	mechanical energy	
Work	energy		
Energy			

PROBLEMS AND CONCEPTUAL QUESTIONS

Problems requiring numerical work are identified with a blue number.

Momentum and conservation of momentum

1. A piece of putty is thrown toward a wall; it has mass and is moving, so it has momentum. The putty then strikes and sticks to the wall, and all apparent motion stops. Is momentum conserved in this collision? Explain.

2. You are standing completely stationary—no momentum. You then begin to walk—now you have some momentum. Is momentum conserved in this event? Explain.

3. Consider a situation in which two guns are to fire identical shells with the same muzzle velocity. Would you prefer to withstand the kick of the heavier or the lighter of the two guns?

4. What is the momentum of a 4000 kg truck traveling at 80.0 km per hour?

5. A student is stranded in the middle of a frozen, frictionless pond with nothing except his physical science book along for comfort. How could the student use the law of conservation of momentum to get off the ice?

6. As a bullet leaves the muzzle of a gun, the gun is propelled backward, and momentum is conserved. What actually forces the gun backward? The bullet? The exploding gases? Describe in detail the forces within the gun barrel as the bullet is being fired.

7. You are standing on top of a table and drop off to collide with the Earth. Discuss how momentum is conserved in your collision.

8. A bullet with a mass of 50.0 g is fired out of a gun with a muzzle velocity of 400 m/s. If the gun has a mass of 3.50 kg, what will be its velocity after the bullet is fired, assuming that the gun is free to recoil?

9. A charging rhino moving at 2.00 m/s collides with a stationary Volkswagen. If the mass of the rhino is three times that of the Volkswagen, what is the velocity of the entangled wreckage after the collision? (Neglect friction.)

10. A puck of mass 20.0 g moves with a velocity of 1.30 m/s toward the right on an air hockey table. It collides with a puck of mass 60.0 g moving to the left with a speed of 2.00 m/s. After the collision the lower mass puck moves to the left with a speed of 0.90 m/s. What is the velocity of the heavier puck?

11. What happens to the momentum of an object if its velocity is tripled?

12. Gases are often stored in cylinders at high pressures. Signs attached to these cylinders warn that they can become dangerous projectiles if punctured. Use the law of conservation of momentum to explain this.

13. A sailor is becalmed on a lake, but he has with him a battery-powered fan. He finds that when he directs the wind from the fan against the sail he does not move. Why not? He also finds that if he turns the fan around so that it is blowing away from the sail he will move. Why?

Work

14. A gravitational force exerted on the Earth by the Sun and directed toward the Sun is the force responsible for causing the Earth to circle in its orbit. Assume that the orbit of the Earth is a perfect circle and defend the statement that the force exerted by the Sun does no work on the Earth.

15. A crane exerts a force of 40,000 N over a distance of 40 m. How much work has it performed?

16. You stand all day holding a book stationary at arm's length. At the end of the day, you may feel that you have done a lot of work, but from the standpoint of our definition of work, have you done any?

17. A truck with a mass of 5000 kg is rolling along a level road at constant velocity. If there were absolutely no friction or air resistance, how much work would be required to continue to move the truck for a distance of 1.00 km?

18. A horse is pulling a wagon down a hill. As the wagon tends to roll faster than the horse is walking, the animal must hold back the wagon. Does the horse perform work on the wagon as they move downhill? Explain.

19. (a) The engine of a model airplane exerts a forward force of 4.00 N on a plane as it moves through the air for a distance of 6.00 m. (a) How much work does the engine do? (b) If a force of air resistance equal to 2.00 N is exerted on the plane during this short flight, how much work does this force do? (*Hint:* When the force and the displacement are in opposite directions, the work is negative.) (c) Find the net work done on the plane during the flight.

Kinetic energy

20. What happens to the kinetic energy of an object if its speed is tripled?

21. A carpenter is trying out hammers of different sizes. The heaviest hammer has a mass of 1 kg but is too heavy to swing rapidly. The carpenter finds that if he uses a 0.5 kg hammer, he can swing it twice as fast. With which hammer can he generate more kinetic energy, or will it be the same with both? Explain.

22. A football player has a mass of 100 kg and can run at a rate of 7.0 m/s. (a) What is his kinetic energy? (b) If another player is lighter, having a mass of only 80 kg, how fast must he run to maintain an equal kinetic energy?

23. Can the kinetic energy of an object ever be a negative number?

24. You are sitting at rest in a car traveling down the highway. Do you have any kinetic energy?

25. A ball is thrown straight up into the air. At what point(s) is its kinetic energy a maximum? Is the momentum also a maximum at this same time?

Gravitationl potential energy

26. A roller coaster with a mass of 500 kg is sitting on top of the highest incline. The incline is 30 m above the ground and 15 m above the dip in the track below. (a) What is the potential energy of the machine with respect to the ground? (b) With respect to the track below?

27. A rock climber scaling a nearly sheer cliff reaches a ledge 500 m above the valley floor. Then he climbs 10 m above the ledge. If he falls from this point, he will land on the ledge. If he has a mass of 75 kg, what is his potential energy with respect to a fall at this point?

28. A baseball is thrown straight up into the air. At what point(s) is the gravitational potential energy a maximum?

29. A high diver weighing 700 N leaps from a 10 m tower. (a) What is his potential energy initially? (b) At 4 m above the surface of the water? (c) At the surface of the water?

Conservation of energy and friction forces: the law of conservation of energy

30. A baseball is thrown straight up into the air. At what point is its kinetic energy the greatest?

31. Tarzan drops from a tree and swings from a position where his vine is level to the bottom of the arc. At what point is his kinetic energy the greatest? His potential energy? If energy is conserved in this case, how far will he go after swinging through the bottom of the arc?

32. If energy is conserved, how can our planet ever suffer from an energy crisis?

33. A bicycle rider coasts down a hill and up another one. Use the law of conservation of energy to explain why she cannot go as far up the second hill as her initial height on the first.

34. A physics teacher constructs a pendulum by tying one end of a bowling ball to a wire and attaching the free end to the ceiling. The teacher then pulls the bowling ball aside and up against his chin. When he releases it, should he step back to keep from being hit by the ball after it swings through its complete arc?

35. Find the kinetic energy of the high diver of problem 29 at each of the positions listed.

36. A high jumper wants to clear a 2 m bar. What must be his speed on takeoff to reach this height?

37. A baseball is thrown straight up into the air and reaches a height of 18 m. What was the speed when it left the thrower's hand?

38. In the absence of air resistance, an object thrown into the air will return to ground level with the same speed as it had when thrown upward. Use the conservation of energy principle to verify this statement.

39. A diver drops off a 5 m high diving board. How fast is he going when he strikes the water?

40. (a) Water towers to supply water to a city are placed either on high hills or on large elevated supports. Why? (b) A person living on the top floor of a building often finds that the water runs very slowly out of his faucet, but a person on the ground floor finds that it comes out forcefully. Why?

41. Would we find the water cooler or warmer at the base of a waterfall than at the top?

42. A car is braked to a stop from a high speed. What happens to its kinetic energy?

Power

43. What is energy? How are the concepts of heat and power related to energy?

44. Three farmers are faced with the problem of hauling a ton of hay up a hill. The first makes 20 trips, carrying the hay himself. The second loads a wagon and has his horse pull the hay up in four trips. The third farmer drives a truck up in one load. Which process—manpower, animal power, or machine power—has performed more external work? Which device is capable of exerting more power?

45. Find the hp rating of a 100 W light bulb.

46. Find the wattage rating of a 90 hp car engine.

47. When taking an examination, you have about the same energy output as a 200 W light bulb. Find the total energy in joules you would release during a 1-hour examination.

48. Two 70 kg students are on their way to a class on the second floor of a building 2.8 m above ground level. One strolls up in a time of 60 s, while the second sprints up in 4 s. Compare the energy output and power output of each.

49. When using a simple machine, it is usually necessary to do a little more work with them than without them. If true, why do we use such machines?

Simple machines

50. Suppose you had to lift a heavy iron ball 2 m in the air. You could either lift the ball outright or roll it up a ramp. Which route would require more work? More applied force? Explain.

51. A screw can be considered to be an inclined plane wrapped around a shaft. Explain why less force is needed to turn a screw into a piece of wood than is needed to hammer a nail into an equivalent piece of wood.

52. Suppose you wish to exert a force of 2000 N against a piece of machinery to slide it across the floor. In order to do the job, you decide to use a lever and position the fulcrum 0.25 m from the load. If you are capable of exerting a force of 400 N, how long will the lever have to be to enable you to move the machine?

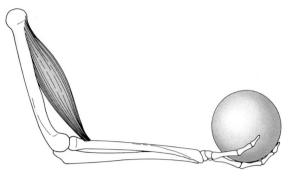

(Question 53)

53. The muscle and bone structure of your arm is shown in the figure. If your muscle exerts a force of 100 N, is more or less than 100 N delivered to a load in your hand? If your muscle moves 1 cm, does your hand move more or less than 1 cm? Explain.

Conservation of angular momentum

54. Often when a high diver wants to turn a flip in mid air, he will draw his legs up against his chest. Why does this make the diver rotate faster? What should the diver do to come out of the flip?

55. As a tether ball winds around a pole, what happens to the speed of the ball? Why?

56. When the Sun begins to die, it will increase in size to the point that it will almost reach the orbit of Mars. What will happen to its rotational velocity during this expansion?

ANSWERS TO SELECTED NUMERICAL QUESTIONS

4. 320,000 kg km per hour
8. -5.71 m/s
9. 1.50 m/s
10. -1.27 m/s
15. 1.60×10^6 J
19. (a) 24.0 J, (b) $^-$12.0 J, (c) 12.0 J
22. (a) 2450 J, (b) 7.83 m/s
26. (a) 147,000 J, (b) 73,500 J
29. (a) 7000 J, (b) 2800 J, (c) 0
35. (a) 0, (b) 4200 J, (c) 7000 J
36. 6.26 m/s
37. 18.8 m/s
39. 9.90 m/s
45. 0.134 hp
46. 67,100 W
47. 7.2×10^5 J
48. 1920 J for each, $P_{60} = 32.0$ W, $P_4 = 480$ W
52. 1.25 m

CHAPTER 4

NEWTON'S LAW OF GRAVITY AND SOME SPECIAL KINDS OF MOTION

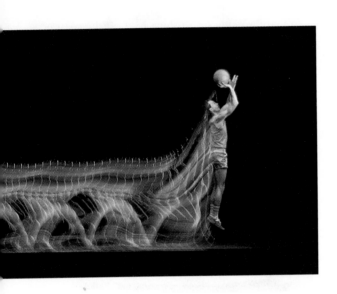

This multi-flash photograph of a basketball player shows that he follows the path of a projectile as he drives toward the basket. (© Ben Rose, The Image Bank)

Legend has it that Isaac Newton was sitting under an apple tree and was inspired to formulate the law of gravity when he saw an apple fall to Earth. It is not certain that this event actually happened (although Newton himself told the story), but it is a fact that Newton was the first to realize that objects in the heavens respond to gravitational forces. Every object, including the Moon and the Earth, feels the gravitational tug of every other object.

In this chapter, we shall examine this universal law of gravitation and some special kinds of motion. The two particular types of motion that we shall examine are that of a projectile and that of an object moving in a circular path.

4.1 EARLY IDEAS ABOUT GRAVITATION

We already examined a few of the early ideas concerning gravitation when we studied the motion of a freely falling object in Chapter 2. We saw there that the thread of thought that eventually led to our modern ideas of gravitation can be traced to the ancient Greek civilization and to the ideas of Aristotle (384–322 B.C.). For example, one of his ideas concerning falling bodies was that heavier objects fall faster than lighter ones. Aristotle's philosophy saw the Universe as heavily goal oriented. From this point of view, the "natural" place of material objects was "down," and objects sought their natural place, their goal. Heavier objects sought the Earth more than did lighter ones, so they would fall faster than lighter ones. Also, as we have seen, Galileo refuted this theory of Aristotle and opened the door to a more satisfactory approach that was eventually fully developed by Newton. We have seen that Newton's three laws of motion provide the foundation for describing and interpreting all types of motion. For example, Newton stated that an object naturally moves in a straight line with a constant speed unless caused to change this motion by a net force (the first law). But when you throw a ball toward a friend, it does not go in a straight line; it curves down toward the Earth. Aristotle would have explained that the object curves downward so that it can reach its goal, the Earth. Newton, however, concluded that there is an unbalanced force on the ball as it moves through the air that causes it to deviate from a straight-line path. The force is the object's weight, and the weight results from the phenomenon we call gravitation. In the next section, we shall present Newton's theory of gravitation, which explains not only how a ball moves while in flight, but also the movements of heavenly objects.

4.2 THE LAW OF GRAVITY

Newton stated his law of gravity as follows:

> *Every object in the Universe is attracted to, and attracts, every other object in the Universe by a force that we call the force of gravity.*

At first look, this law does not seem to relate to reality because if you hold two pencils in your hands, you do not feel the gravitational tug that one exerts on the other. Yet, according to Newton's law of gravity, they are exerting this force on each other. If you actually calculate the magnitude of this force, however, by use of the equation given here, you find that for ordinary pencils this force has a magnitude of about 2×10^{-11} N, hardly measurable. In fact, if you go to larger objects such as two automobiles, you still find that the force of gravitational attraction between the two is so small as to be unnoticeable. Thus, if you have a fender-bender with another car, don't try to use the law of gravity as your defense with the arresting officer.

In equation form, the law of gravity may be stated as

$$F = G\,\frac{m r_1 m_2}{r^2} \tag{4.1}$$

where G is the universal gravitational constant determined experimentally to have a value of 6.67×10^{-11} N m^2/kg^2 in SI units. The quantities m_1 and m_2 are the masses of the two attracting objects, and r^2 is the square of the distance between the centers of the two objects.

FOCUS ON THE TIDES

As you sit in a comfortable chair, studying a subject such as the movement of the oceans, it becomes easy to lose touch with the natural power and majesty of this phenomenon. To appreciate how people first began to think about such systems, imagine that you are working on a fishing boat off the coast of Alaska. You would notice that the level of the ocean rises and falls in a cycle of approximately 12 hours. Thus, if the water were low at noon, it would reach maximum height at about 6:00 P.M. and be low again near midnight. These vertical displacements are called **tides.** You would soon find out that low tide does not recur at the same time each day. If today's low tide occurred at 4:43 A.M., tomorrow's might not occur until 5:29 A.M., and the next day's would be still later, until eventually low tide would not occur until noon. If you were observant, you would notice that each day the tides were delayed approximately 40 to 50 minutes, or about the same amount of time that the moonrise is delayed from day to day.

To understand tides, consider the Earth-Moon system, as shown in the figure. For simplicity, imagine a situation in which there are no continents on the Earth and the surface is one giant ocean. At any one instant of time, one section of this giant ocean (marked A in the figure) lies just under the Moon, while all other regions are farther away. Because gravitational force is greater for objects that are closer together, the part of the ocean closest to the Moon is attracted with a force greater than the force on the center

---

EXAMPLE 4.1   DUELING PENCILS

A pencil of mass 0.10 kg is held at a distance of 0.20 m from a second identical pencil. Find the force of gravitational attraction between the two.

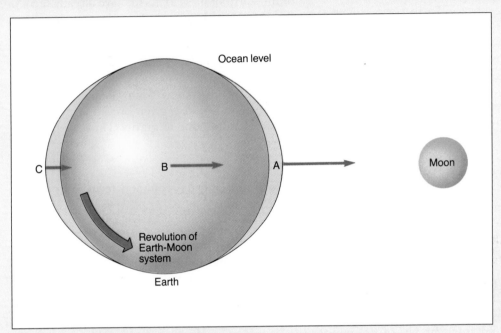

Schematic view of tide formation. (Magnitudes and sizes are exaggerated for emphasis.)

of the Earth, point B in the figure, and the force exerted on the Earth at B is greater than the force exerted on that portion of water at C. The relative magnitudes of these forces are shown by the arrows in the figure. The large force on the water at A causes the ocean to bulge outward toward the Moon, resulting in high tide. At the same time, the center of the Earth, B, is being more strongly attracted toward the Moon than is the water at C, and as a result, the Earth is "pulled" away from the water. This means that the ocean at point C is left behind a little, so a bulge is formed at C. This bulge is the high tide 180° away from the Moon. Thus, the tides rise and fall two times a day. At a given time of day,

the tide is high at a point on the Earth directly facing the Moon, and it is simultaneously high at a point exactly on the opposite side of the Earth.

Following this reasoning, it is easy to explain why the high tide appears later every day. An observer on Earth located just under the Moon at noon of one day sees that the gravitational pull of the Moon causes the ocean to bulge and the tide to rise. Twenty-four hours later, the Earth has made one revolution; but because the Moon has traveled some distance in its orbit during that time, the Earth must spin a little farther for an observer to be again directly under the Moon. Thus, the tide reaches a maximum a little later each day.

Solution The force of gravitational attraction is found by direct substitution into Eq. 4.1. We have

$$F = G\frac{m_1 m_2}{r^2}$$

$$= (6.67 \times 10^{-11} \text{ N m}^2/\text{kg}^2)\frac{(0.10 \text{ kg})(0.10 \text{ kg})}{(0.20 \text{ m})^2}$$

$$= 1.67 \times 10^{-11} \text{ N}$$

EXAMPLE 4.2   I'm Climbing Higher and Higher
(a) The typical mass for an adult is about 70 kg. Find the force of gravitational attraction exerted on a person of this mass by the Earth (mass = 5.98×10^{24} kg) at the surface of the Earth. The radius of the Earth is about 6.38×10^6 m. This force exerted on the individual by the Earth is the weight of the person.

(b) Find the weight of the person if he is in a spaceship at a height above the Earth equal to the radius of the Earth.

Solution (a) The weight of the person is found from Eq. 4.1 as

$$F_g = w = G\,\frac{m_E m_2}{r_E^2} =$$

$$(6.67 \times 10^{-11}\ \text{N m}^2/\text{kg}^2)\frac{(5.98 \times 10^{24}\ \text{kg})(70\ \text{kg})}{(6.38 \times 10^6\ \text{m})^2}$$

$$= 6.86\ \text{N} = 154\ \text{lb}$$

(b) At a height above the Earth equal to its radius, $r = 1.28 \times 10^7$ m, and from the law of gravity we find the new weight as

$$F = w = G\,\frac{m_E m_2}{r^2} =$$

$$(6.67 \times 10^{-11}\ \text{N m}^2/\text{kg}^2)\frac{(5.98 \times 10^{24}\ \text{kg})(70\ \text{kg})}{(1.28 \times 10^7\ \text{m})^2}$$

$$= 170\ \text{N} = 38\ \text{lb}$$

Notice the pattern indicated in this example. When you are twice as far away from the center of the Earth in (b) as in (a), your weight is only one-fourth as much. In a similar fashion, if you move three times farther away, your weight will be only one-ninth as much. A relationship between variables that produces this result is called an "inverse square" relationship.

EXAMPLE 4.3   Gravity and Distance
(a) Use Newton's law of gravity to find an expression for the acceleration due to gravity as a function of height above the surface of the Earth.

(b) Use your expression to find the acceleration of gravity at a height of one Earth radius above the Earth's surface.

Solution (a) At first thought, this seems to be a difficult problem, but let us assume that we have an object of mass m a distance r away from the center of the Earth of mass M. As we have already seen, in this case Eq. 4.1 gives us the weight of the object. Thus, we have

$$F = w = G\,\frac{mM}{r^2}$$

But, we also know from Chapter 2 that the weight of an object is given by $w = mg$, where g is the acceleration due to gravity. Thus, we have

$$mg = G\,\frac{mM}{r^2}$$

or, on canceling the common term m from both sides, we have

$$g = G\,\frac{M}{r^2}$$

You should note that there is nothing special about this equation that makes it apply only to the Earth. It applies equally well to any planet. Thus, we can generalize as

$$g_{\text{planet}} = G\,\frac{M_{\text{planet}}}{r^2_{\text{planet}}}$$

Thus, we see that the more massive the planet, the greater the acceleration due to gravity at its surface. Be careful, however, because in general the more massive the planet, the larger is its radius, and an increased radius tends to make the force of gravity smaller.

(b) Using the values for the mass of the Earth and the distance found in Example 4.2, we find

$$g = (6.67 \times 10^{-11}\ \text{N m}^2/\text{kg}^2)\frac{(5.98 \times 10^{24}\ \text{kg})}{(1.28 \times 10^7\ \text{m})^2}$$

$$= 2.43\ \text{m/s}^2$$

This is one-fourth the value of g at the Earth's surface, consistent with the result of Example 4.2.

EXERCISE

Find the acceleration due to gravity at the surface of Jupiter, which has a mass of 1.90×10^{27} kg and a radius of 6.99×10^7 m.

Answer 25.9 m/s^2

(a)

(b)

FIGURE 4.1 *Weightlessness in a freely falling elevator.*

4.3 WEIGHTLESSNESS

As we have seen, as an object moves higher above the surface of the Earth, it weighs a little less. Thus, you weigh less standing on the top of Mt. Everest than you do at sea level. Similarly, astronauts become lighter as they fly skyward, for they travel farther and farther from the center of the Earth. However, if their weight when they are in orbit is actually calculated, a curious fact becomes evident. The Skylab spaceship launched by the United States in 1974 orbited approximately 320 km above the surface of the Earth. Because the Earth is 6400 km in diameter, the astronauts were located about 6720 km from the center of the Earth. A person who weighed 650

N (146 lb) on Earth would have weighed 587 N (131 lb) in the orbiting Skylab. But how could that be? Television broadcasts from space showed the astronauts floating weightlessly about the capsule.

To answer this perplexing question, let us set aside the spaceship for a moment and imagine an elephant riding an express elevator down from the top of a 20-story building, as pictured in Figure 4.1a. This elevator is designed such that it goes into free-fall instantly until it reaches the fourth floor of the building, where a brake is applied, bringing it smoothly to a stop at ground level. Not expecting the sudden downward movement, the elephant drops a bag of peanuts on which it has been munching (Fig. 4.1b). There are now three objects (the elevator, the elephant, and the bag of peanuts) falling down the shaft. Remember that the acceleration due to gravity is equal for all objects. Thus, assuming that air resistance is negligible, all three accelerate downward at exactly the same rate. For every meter that the elevator falls, the peanuts and the elephant all fall 1 m. Therefore, if the bag of peanuts were one-half meter above the floor of the elevator initially, it would remain at that height relative to the elevator during the entire descent. If someone photographed the scene, the peanuts would appear to be suspended, weightless, in the air. Similarly, if the elephant had jumped 10 cm into the air just before the elevator went into free-fall, it would remain at that height, as if it too were weightless. This is only an apparent weightlessness because at any instant it would weigh just as much as if the elevator were motionless. Obviously, the elephant has to have weight, for otherwise it would not fall along with the elevator.

The apparent weightlessness of an astronaut in an orbiting spacecraft is similar to the apparent weightlessness of the elephant in the falling elevator. Rocket action carries the craft aloft until it is, say, 320 km above the Earth, then redirects the capsule and accelerates it horizontally. Finally the engines shut off. At this point, the satellite begins to fall, but as we shall see in a later section in this chapter, if the arc of the fall is the same as the arc of the Earth's surface, the satellite orbits rather than crashing into the Earth. The vehicle is falling freely all the time, however. The astronauts in the capsule are like the elephant and the peanuts in the falling elevator. They are falling, but because their enclosure is falling at the same rate, they appear to be weightless when in fact they are not.

EXERCISE

Astronaut Steven MacLean tries out gymnastics in weightless conditions on the Space Shuttle Columbia. What advantages and disadvantages would there be over similar exercises on Earth?

(Courtesy NASA)

Astronaut Tamara Jernigan uses this special device to measure her mass while in space. If her mass is 60 kg while on Earth, what value should she measure in orbit? What value should she measure for her weight?

(Courtesy NASA)

4.4 PROJECTILE MOTION

The types of motion that we have looked at thus far have been those in which the object under consideration moved along a straight-line path. For example, when we looked at freely falling objects, they were either moving vertically downward or upward along a straight-line path. We will now turn our attention in the rest of this chapter to the motion of objects moving in a plane. The first kind of motion that we shall look at is called **projectile motion.** This is the kind of motion followed by a batted baseball or a kicked soccer ball.

Before we look at this motion in detail, we ask you to repeat a mental exercise discussed in Chapter 2, that of finding all the forces that act on an object. Here is the situation. A player kicks a soccer ball: Your task is to find all the forces acting on it while it is in flight, ignoring air resistance. Before you read further, think about it a moment and remember that there are only two types of forces in nature—contact forces and action-at-a-distance forces.

If your answer is only one force, you are absolutely correct. There are no horizontal forces at all acting on the soccer ball once it is in flight. Obviously, while the ball was being kicked, there was a force acting on it exerted by the kicker as long as his foot was in contact with the ball. Once the ball has lost contact with his foot, however, there is nothing else touching the ball to exert a contact force on it. Thus, the only remaining possibilities are the action-at-a-distance forces, of which there is only one

FIGURE 4.2 *The only force on a projectile in flight is its weight* **w.**

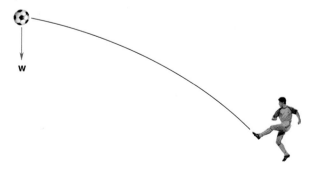

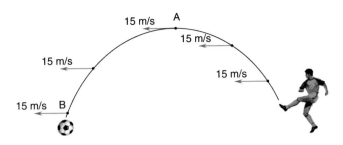

FIGURE 4.3   *The horizontal velocity of a projectile is a constant.*

that we need worry about, the force of gravity, or the weight. This force is shown in Figure 4.2.

It is important to examine the forces acting on the soccer ball because such an analysis enables us to determine some features of the motion of any object moving as a projectile. To do so, we shall consider the motion of a projectile in two parts: the horizontal motion and the vertical motion. Let us look at each of these in turn.

Horizontal Motion   There are no forces acting on a projectile along the horizontal direction. Consider this statement from the point of view of Newton's second law, $\mathbf{F} = m\mathbf{a}$. If we consider $\mathbf{F}_n$ to be the net force along the horizontal, we find

$$F_n = 0 = ma_n$$

but from this we must conclude that the acceleration a_n along the horizontal is also zero. If the acceleration along the horizontal is zero, this means that the object can neither speed up along this direction nor slow down. *Thus, the motion along the horizontal is one of constant velocity.* This aspect of the motion of a projectile is pictured in Figure 4.3. For our specific example, a player has kicked a soccer ball such that at the instant it left his foot, the ball was moving along the horizontal with a velocity of 15 m/s, and this velocity remains the same at all points. That is, it is traveling horizontally with a speed of 15 m/s at the top of its path (position A), at the end of its path just before it strikes the ground (position B), and at any of the infinite number of other intermediate points.

Vertical Motion   The motion along the vertical direction is exactly like a type of motion that we examined in Chapter 2, free fall. As we have seen, the characteristic of a freely falling body is that the only

force acting on it is its weight, and that is the case for a projectile. This means that as the object rises, its vertical velocity slows at a rate of 9.8 m/s every second, and after it passes the apex of its flight path and begins to descend, it speeds up at a rate of 9.8 m/s every second. The vertical velocity of a projectile follows a pattern like that shown in Figure 4.4. Note that initially it has a large upward velocity that gradually decreases as it rises. This is indicated by the fact that the length of the velocity vectors decreases with height. Finally, at the top of the path, its vertical velocity reaches zero for an instant. However, it continues to accelerate with the acceleration due to gravity; as a result, it now begins to increase in speed in the downward direction as it falls, as shown in Figure 4.4.

To understand some of the characteristics of projectile motion, consider the following problem. You are a punter for your college team, and you have the fantastic ability to kick the ball so that it always leaves your foot traveling at the same speed. You would like to kick the ball so that it travels the

FIGURE 4.4   *The vertical velocity of a projectile decreases as it rises and increases as it falls in the same way as that for a freely falling object.*

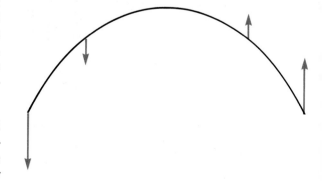

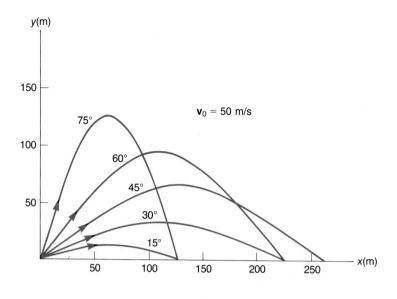

FIGURE 4.5   *The flight of a projectile with different projection angles.*

maximum distance. The question is, at what angle with the horizontal should you punt the ball? Figure 4.5 shows the result of such an analysis for balls kicked at several different angles. In the absence of air resistance, the ball travels the maximum distance when kicked so that its initial direction of travel is at an angle of 45° with respect to the horizontal. Also, note from Figure 4.6 that whatever the angle, in the absence of air resistance, the path is symmetrical about a line through the apex of the path and perpendicular to the ground. If air resistance is important, as it would be in real-world situations, the path deviates from this symmetrical pattern. The actual result is that the path is much like that shown in Figure 4.6. As you would expect, the ball will travel a shorter distance with air resistance present than it would without, as Figure 4.6 indicates.

FIGURE 4.6   *A projectile travels a shorter distance when air resistance is present than it does in its absence.*

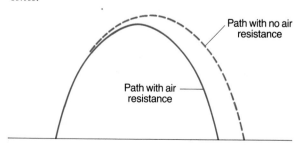

CONCEPTUAL EXAMPLE 4.4   WHICH WILL HIT FIRST?

A baseball is thrown horizontally at the same instant an identical baseball is dropped straight down, as shown in Figure 4.7. Which of the two will hit the ground first and why?

Solution They will both hit at exactly the same instant. This occurs because both balls have identical vertical motions. That is, they both fall vertically with the acceleration due to gravity. As another example of this, suppose that a baseball outfielder is sitting on a fence when a batter hits the ball directly at him. If the outfielder is to catch the ball, all he has to do is drop off the fence. He will accelerate downward at the same rate as the falling baseball, so they will both reach ground level at the same

FIGURE 4.7   *Example 4.4. Ball A is thrown horizontally at the same instant that ball B is dropped. Which will hit the ground first?*

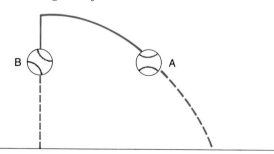

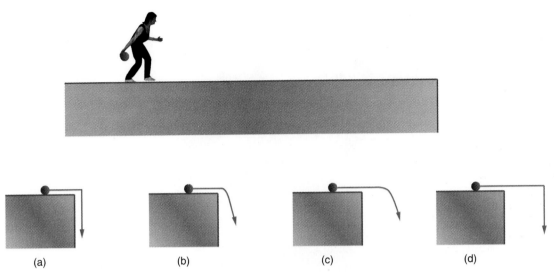

FIGURE 4.8 *Example 4.5.*

time. This assumes that the baseball has sufficient horizontal speed to reach him.

CONCEPTUAL EXAMPLE 4.5   A NEW KIND OF BOWLING ALLEY

Imagine that a flat bowling alley is built on the edge of a cliff, as shown in Figure 4.8. A bowler rolls a ball down the alley and off the cliff as shown. Which of the diagrams best describes the motion of the ball after it has rolled off the edge? (Neglect the effects of friction and air resistance.)

Solution The ball would follow path b. Initially the ball is propelled (forced to accelerate) by the bowler's arm. The ball moves in a straight line at constant speed until it reaches the edge of the cliff. What happens the instant the ball goes over the edge? There are no horizontal forces acting on the ball. Therefore, the horizontal velocity remains unchanged because there are no horizontal forces to speed it up or slow it down. There is, however, a net vertical force, its weight, which produces a net downward acceleration. The resultant motion of the ball, shown in Figure 4.8b, is the combination of these two independent components, a constant horizontal velocity and a simultaneous acceleration due to gravity. As an exercise, explain why a, c, and d are incorrect.

4.5 CIRCULAR MOTION

Here is a thought question for you to ponder before you read too deeply into this section. Figure 4.9 shows a car traveling in a circular path with a constant speed. Does the car have an acceleration? At first thought, it seems obvious that the answer should be no, but as is so often the case, what seems obvious may be incorrect. The car does indeed have an acceleration. Let us see why.

In Chapter 2, we examined the definition of acceleration and found it to be

$$\mathbf{a} = \frac{\text{change in velocity}}{\text{time for change to occur}}$$

A superficial look at this equation tells us that the change in velocity of the car is zero because the car is traveling at a constant speed. However, take a second look. This equation actually says that acceleration is equal to the change in *velocity* divided by the time over which this change occurs. This means that because velocity is a vector quantity there are two ways in which it can change. (1) *The magnitude of the velocity, the speed, can change.* This is what has been happening in all of the various types of motion that we have examined in this text thus far. (2) *The direction of the velocity can change,* and that is what is happening in the present situation. Fig-

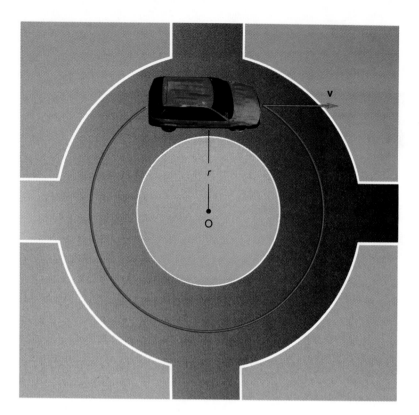

FIGURE 4.9   *Does the car moving at a constant speed have an acceleration?*

ure 4.10 shows how the direction of velocity is changing. At point A, the direction is toward the east; at point B, it is toward the south. For a point in between, such as C, it points in the direction shown.

It is a relatively straightforward derivation to find the change in velocity for the car, but we shall not do so here. Instead, it suffices to say that it can be shown that *the direction of the acceleration of an ob-*

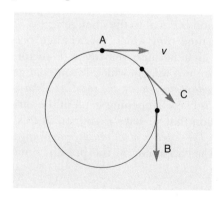

FIGURE 4.10   *The direction of the velocity of an object moving in a circular path changes as shown.*

ject moving in circular motion is always toward the center of the circular path followed and has a magnitude given by

$$a_c = \frac{v^2}{r}$$

where v is the magnitude of the velocity and r is the radius of the circular path followed by the object. The acceleration, a_c, is called the **centripetal acceleration.** The word "centripetal" means center-seeking, and this designation arises because the direction of this acceleration and the force that produces it are always toward the center of the circular path.

EXAMPLE 4.6   **EXIT RAMPS**

A particular circular exit ramp off an interstate highway is designed such that cars entering it should be traveling at a speed of 15.0 m/s. The radius of this exit ramp is 40.0 m. **(a)** Find the centripetal acceleration of a car on this ramp. **(b)** What would happen to the car on the ramp if this centripetal acceleration should vanish?

Solution **(a)** The centripetal acceleration is found as follows

$$a_c = \frac{v^2}{r} = \frac{(15.0 \text{ m/s})^2}{40.0 \text{ m}} = 5.63 \text{ m/s}^2$$

(b) The centripetal acceleration of the car arises because the direction of its velocity is changing. This means that if a car has zero centripetal acceleration, the direction of travel of the car must *not* be changing. Thus, if a_c should go to zero, the car would have to travel in a straight-line path. As we will see later, this can happen to the car when the exit ramp is icy.

EXAMPLE 4.7   CENTRIPETAL ACCELERATION ON THE EARTH

The Earth has a radius of 6.38×10^6 m at the equator and turns once on its axis in a day, 86,400 s. Find the centripetal acceleration of a bug resting at the equator as a result of the rotation of the Earth.

Solution First, we must find the speed at which the bug is turning. This is the speed at which the Earth turns, which can be found from $v = \text{distance/time}$.

The distance moved by the bug in a time of 86,400 s is the circumference of the Earth given by

$$\text{distance} = 2\pi r_E = 2(3.142)(6.38 \times 10^6 \text{ m})$$
$$= 4.01 \times 10^7 \text{ m}$$

and thus, the speed of the bug is

$$v = \frac{\text{distance}}{\text{time}} = \frac{4.01 \times 10^7 \text{ m}}{86400 \text{ s}}$$
$$= 4.64 \times 10^2 \text{ m/s}$$

From this, we are able to find the centripetal acceleration of the Earth as

$$a_c = \frac{v^2}{r} = \frac{(4.64 \times 10^2 \text{ m/s})^2}{6.38 \times 10^6 \text{ m}} = 0.0337 \text{ m/s}^2$$

This centripetal acceleration is virtually negligible in comparison to the acceleration due to gravity of 9.8 m/s².

4.6 CENTRIPETAL FORCE

In the preceding section, we found that an object that travels in a circular path always has an acceleration toward the center of the path given by $a_c = v^2/r$. Let us now examine this statement from the standpoint of Newton's second law of motion, $\mathbf{F}_{net} = m\mathbf{a}$. The second law tells us that if an object has an acceleration, $\mathbf{a}$, it also must have a net force, $\mathbf{F}_{net}$, acting on it. Thus, because an object moving in circular motion has an acceleration toward the center of the path that it follows, there must also be a resultant force toward the center of the circular path. This force is actually the force that makes the object follow the circular path and is called a **centripetal force.**

Often, when beginning students of physics encounter the term "centripetal force," they tend to think of it as some kind of mystical force that arises to keep an object in its circular path. Avoid this. Centripetal forces are just like all the other forces you have seen so far. That is, they are forces such as the tension in a string, a force of gravitational attraction, friction, or a host of other possibilities. Let us take a look at a few of these situations.

Figure 4.11a shows a rock attached to the end of a string being whirled in a vertical circle. Identify the centripetal force in this instance. The force that is acting toward the center of the circular path is a contact force, the tension in the string. To see

FIGURE 4.11 *(a) The tension T in the string causes the rock to follow its circular path. (b) If the string breaks, the ball moves away tangent to the circular path and becomes a projectile. The path followed by the projectile assumes the rock is being swung in a vertical circle.*

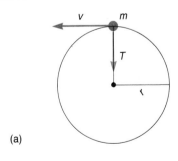

(a)

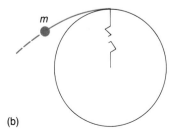

(b)

that this force does indeed cause the rock to follow its circular path, consider what would happen if the string should break, as in Figure 4.11b. At the instant the string breaks, the object is traveling toward the left, and it will fly off in a straight-line path toward the left. As it leaves, however, gravity causes it to move as a projectile. Thus, it follows the curved path shown in the figure.

Figure 4.12 shows the Moon following its almost circular path about the Earth. Identify the centripetal force acting on the Earth. Remember that centripetal forces are no different from the other forces we have run into so far. Their only particular characteristic is that they point toward the center of the circular path followed by the object. The force that acts on the Moon and causes it to move in its circular path is the force of gravitational attraction exerted on it by the Earth. Thus, the centripetal force is given by

$$F_c = F_g = G\,\frac{m_e m_m}{r^2}$$

where m_e is the mass of the Earth, m_m is the mass of the Moon, and r is the distance from the center of the Earth to the center of the Moon.

FIGURE 4.12 *Example 4.9. Identify the centripetal force acting on the Moon.*

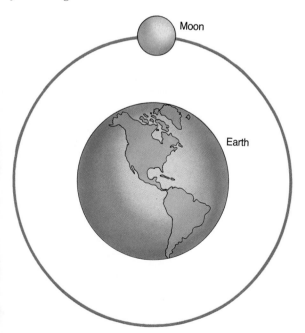

FOCUS ON ARTIFICIAL GRAVITY

There is a well-known science fiction story called "Ring World" in which an artificial world is found in space that is shaped like a huge ring that is rotating. People inside that world can move around just as though they were on the surface of a planet. Somehow the rotation of the ring produces a form of artificial gravity. Let us see how this can be done.

Consider a bug riding on the surface of a record on a turntable. If the record is turned on at a low speed such as 33 1/3 revolutions per minute, the bug may be able to turn along with the record. If it moves in a circular path, there must be a centripetal force acting on it, and this force is provided by friction between the bug's feet and the record. At a slow speed, this frictional force may be large enough to cause the bug to rotate right along with the record. If the record is moved to the next highest speed, 45 rpm, the bug may slide off the edge of the record. This occurs because the frictional force is not large enough to provide the necessary centripetal force for it to turn in the circular path at this higher speed. Suppose that the record is constructed so that it has a wall built around its rim as shown in part (a) of the figure. When the bug slams into this wall, it can now move in the circular path because the wall exerts a force on it toward the center of the circular path. This force provided by the wall is now the centripetal force. Let us now turn the speed up to 78 rpm. At this speed, the wall must exert an even larger force to provide the necessary centripetal force to keep the bug turning with the record. The wall is capable of doing so, and the bug rotates. In fact, if the bug desires, he may reorient his position so that he is standing against the wall as shown in

An artist's conception of an artificial space station. (Courtesy NASA)

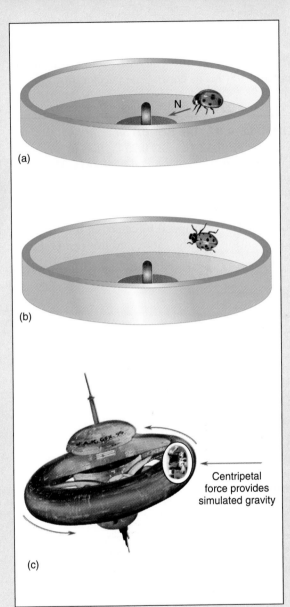

part (b). Or, if the bug is so inclined, he could take a stroll around the rim provided, walking just as though a gravitational force were acting on him. In actual practice, the bug on our turntable would stand slightly sideways because the Earth is also exerting a force on him, but in outer space this extra force would not be present.

This imaginary experience for a bug could be used in a practical way for travelers who some day may have to spend long periods of time in space capsules flying through the Solar System or even on flights to nearby stars. On such a journey, prolonged weightlessness could become inconvenient and perhaps even unhealthful. Therefore, scientists envision some sort of spinning space station as shown. In this case, the centripetal force acting on the travelers in the space vessel would act as an artificial gravity. "Ring World" is correct; such artificial worlds are possible.

(a) When the bug slams into the wall, the wall exerts a force on it, causing it to follow its circular path. (b) The bug under artificial gravity. (c) Travelers in a spinning space station experience a force like that on the bug. This force provides a type of simulated gravity. A stationary observer outside the station would see this as a centripetal force. People in the station, however, would feel that they are being pushed outward by a centrifugal force.

Note that when a problem involves only centripetal forces and centripetal accelerations, Newton's second law, $\mathbf{F}_c = m\mathbf{a}_c$, reduces to

$$F_c = ma_c = m\,\frac{v^2}{r} \qquad (4.2)$$

where F_c is the centripetal force acting on the object and v^2/r is its centripetal acceleration.

EXAMPLE 4.8   THE RED BARON IN ACTION

A model airplane of mass 0.75 kg flies in a circular path of radius 10 m at a speed of 15 m/s. The plane is held in its circular path by a cable. Find the tension in the cable.

Solution The centripetal force acting on the plane is the tension T in the cable, and this can be found from

$$F_c = m\,\frac{v^2}{r}$$

$$T = (0.75 \text{ kg})\,\frac{(15 \text{ m/s})^2}{10 \text{ m}} = \boxed{16.9 \text{ N}}$$

EXERCISE

If the cable in the exercise has a breaking strength of 20 N, what is the maximum speed at which the airplane can fly?

Answer 16.3 m/s

EXAMPLE 4.9   MOON OVER MIAMI

Find the centripetal force acting on the Moon as is shown in 4.12.

Solution The Moon is held in its circular orbit by the force of gravitation exerted on it by the Earth. This is found as

$$F = G\,\frac{m_e m_m}{r^2}$$

where m_e is the mass of the Earth, m_m is the mass of the Moon, and r is the distance from the Earth to the Moon. We have

$$F = (6.67 \times 10^{-11} \text{ N m}^2/\text{kg}^2)$$

$$\times \frac{(5.98 \times 10^{24} \text{ kg})(7.36 \times 10^{22} \text{ kg})}{14.6 \times 10^{16} \text{ m}^2}$$

$$= 2.01 \times 10^{20} \text{ N}$$

This problem re-emphasizes the idea that there is nothing special about centripetal forces. When searching for them, you should look for old friends such as tension in strings, gravitational forces, and so forth.

4.7 ORBITING THE EARTH

It is sometimes confusing to find out that an object moving in a circular path has an acceleration directed toward the center of the path. The question often asked is, if the object is accelerating toward the center of the path, why does it not fall in toward the center? To see why it does not, let us give an example used by Newton himself to explain how an object can be made to orbit the Earth. It may be argued that Newton anticipated today's satellites.

Newton considered what could be done with a powerful cannon and a mountain that extended above the Earth's atmosphere. If a cannonball were fired from the summit, its path would be somewhat as shown in Figure 4.13, path A. The cannonball

FIGURE 4.13   *Newton's idea for placing a cannonball in orbit.*

behaves exactly like you expect in that it accelerates toward the Earth and strikes the surface. If, now, one uses a greater charge of gunpowder to shoot the cannonball at a greater speed, the ball would follow path B. Even more powder yet, and you get path C. With careful adjustment (and a powerful cannon), you should be able to shoot the ball such that as it falls toward the surface of the Earth, the natural curvature of the Earth causes its surface to curve out from under the ball as it falls. The result is a cannonball in orbit. Speeding cannonball and speeding Moon are equivalent.

4.8 CENTRIFUGAL FORCES

Of all the concepts of physics, perhaps the one that is most frequently misunderstood by students is that of **centrifugal** forces. The word "centrifugal" is derived from Latin roots meaning center-fleeing. Therefore, the word relates to forces that are directed away from the center of the circular path that an object follows. Because of the subtle nature of centrifugal forces, we shall have to spend some time talking about what they are and what they are not.

Consider a car turning a corner as shown in Figure 4.14. If the car is to make the circular turn on the exit ramp, there must be a centripetal force acting on it. This centripetal force is the frictional force exerted on the car by the roadway. On an icy road where friction is minimal, this frictional force might be so small that the car would not be able to negotiate a sharp curve. According to Newton's first law, the natural tendency of the car is to maintain its motion along a straight-line path unless some external force, such as the frictional force, causes it to deviate from this straight line. Thus, on an icy road where the necessary centripetal force is not available, the car does not move in the circular path; it continues on its straight-line course and goes off the road. Now, consider a passenger in the car. If he is to deviate from the straight-line path he is following before the turn, a centripetal force *must* act on him also. What is the origin of this force? This force arises from the frictional force exerted on the person by the car seat. If the turn is sharp enough, however, or made at a high speed, this frictional force may not be large enough to deviate him from his straight-line path. As a result, he slides across the seat until he strikes the door. The

FIGURE 4.14 *All moving objects have an inertial tendency to travel in a straight line. If a person left a box sitting on the roof of a car, and the car made a sharp turn, the box would be likely to continue to travel in a fairly straight line and fall off the car. The reason is that there is no force of sufficient strength to cause the box to follow a curved path.*

door then provides a push on him toward the center of the circular path. Thus, in this case the centripetal force is the force that the door of the car exerts on him. In Figure 4.14, we see what would happen to a box left on the smooth roof of a car as it rounds a corner quickly. Frictional forces would not be large enough to produce the required centripetal force; there is not a barrier for the box to slide into to produce the necessary centripetal force, and as a result, the box would continue to travel in a straight-line path and fly off the roof of the car. Thus, the box slides off the roof of the car because of the *absence* of a centripetal force. If you were in the position of the box on the roof of the car, however, you would describe your experience in totally different terms from those presented here. Let us examine the most common explanation for what happens and see why it is incorrect.

If you were in the position of the box on the roof of the car when it begins to turn under you, you would suddenly find yourself flying off into space. Why? A person on a rooftop watching the event would say that you are obeying Newton's first law of motion and continuing on in your straight-line path as you should. But your point of view would be completely different. As you went flying off into space, you might glance back at the car and see it turning in its circular path. Glancing quickly toward the center of the circular path followed by the car, you would notice yourself getting farther and farther away from this center point. Flipping through your memory bank of past experiences, you would ask yourself what has caused such occurrences in the past. You could explain your motion by saying that some force must be acting on your body to pull you away from the center of the circular path, and you could give this force a name, a centrifugal force. This is a fictitious force created only to explain your motion. It obviously cannot be a real force because there is nothing to create it. There are no action-at-a-distance forces that would act in this way, and as you fly through the air, there is nothing touching your body that could produce a contact force. Thus, you move as you do, not because of a centrifugal force, but because of the absence of a centripetal force to cause you to make the turn with the car.

Let us consider another example to show how people often invent centrifugal forces to describe common observations. When you wash clothes, at the end of the cycle the drum of the washing machine goes into a spin cycle during which it rotates rapidly to throw excess water off the clothing. Why is the water thrown off? The common answer is that a centrifugal force is exerted on the water droplets pulling them off the fabric. As you probably might guess, however, this fictitious force is not present. Let us consider the clothes as they rotate at a low rate of speed. A particular drop of water adheres to the fabric by molecular forces, and at a low turning rate, these forces are large enough to provide the centripetal force to allow the drop to turn with the clothes. As the rotational rate increases, there comes a time when the molecular forces are not great enough to provide the required centripetal force. When this occurs, the drop flies off in a straight-line path until it hits the outside of the drum and disappears down the drain. If an intelligent amoeba were in the drop of water, it might explain what is happening to the drop by saying that it was pulled off the clothing by some centrifugal force. A person standing beside the wash-

FIGURE 4.15   *The centripetal force exerted on the can by the string causes it to travel in a circular path.*

ing machine would explain the phenomenon correctly by saying that the drop containing the amoeba was separated from the clothing simply because there was not a large enough centripetal force acting on it to cause it to turn along with the drum and clothing.

Now that we have seen what centrifugal forces are not, let us see what they are. Centrifugal and centripetal forces are action-reaction pairs. To explain this with an example, consider a situation in which a child attaches a string to a can and swings it around her head in a circular path, as indicated in Figure 4.15. The force that is causing the can to

FIGURE 4.16   *The centripetal force acts on the bucket; the centrifugal force is on the string.*

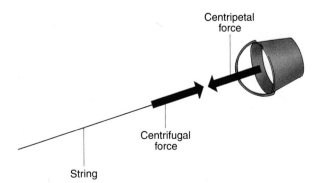

move in its circular orbit is the tension in the string. This is the centripetal force. To find out the reaction to this action, let us formulate a sentence as follows: The action is the force exerted on the can by the string. To find the reaction, we invert the sentence as follows: The reaction is the force exerted on the string by the can. This is the centrifugal force. Thus, the action-reaction pair is as indicated in Figure 4.16. Note that the centrifugal force is not on the can; it is on the string.

FIGURE 4.17 *Example 4.10. What is wrong with this figure?*

CONCEPTUAL EXAMPLE 4.10 WHY DOESN'T IT FALL?

Back in the early days of the space program, government officials were often confronted with the question of how a satellite can stay out there. Why doesn't it fall to Earth? In searching for a simple answer, they devised an explanation that uses centripetal and centrifugal forces as shown in Figure 4.17. They said that there was a force on the satellite directed toward the center of the Earth, its weight, and this was the centripetal force. If this were the only force acting, they said incorrectly, the satellite would fall. They then said that there was another force, a centrifugal force, which acted on the ship in a direction away from the center of the Earth. This force is shown as a dashed force in Figure 4.17 to emphasize that it really isn't there. The explanation went on to say that the reason the satellite did not fall was because the two forces were action-reaction pairs, and because they are equal in magnitude and opposite in direction, they would cancel, and the object would sail happily along in its orbit. From what you know about Newton's laws of motion, you should be able to verify that this explanation violates all three laws. Try it. Explain why this violates (**a**) the third law, (**b**) the first law, and (**c**) the second law, and finally, explain (**d**) why the object doesn't fall.

Solution (**a**) Recall that action-reaction pairs have three fundamental properties. They are equal in magnitude, are opposite in direction, and act on different objects. These two proposed forces are equal in magnitude and opposite in direction according to space officials, but they both act on the satellite. Thus, they cannot be an action-reaction pair.

(**b**) If these two forces are equal in magnitude and opposite in direction, they would cancel one another and produce a net force on the satellite equal to zero. According to Newton's first law, any object that has no net force acting on it will either not move at all or it will move at a constant velocity. The satellite *is* moving, and because it follows a curved path, the changing direction means it *cannot* have a constant velocity. (Don't forget that velocity is a vector quantity.) The satellite is not behaving in the way that an object should when it has no net force on it. Thus, the first law would be violated.

(**c**) We know that any object that has the direction of its velocity changing has an acceleration, which we have called the centripetal acceleration. Our satellite therefore has a centripetal acceleration, but, if so, it also must have a net force acting toward the center of its circular path. As noted in part (**b**), their explanation depends on there being no net force acting on the satellite. Thus, in $\mathbf{F} = m\mathbf{a}$, you would have an acceleration without a net force, and so the second law is also violated.

(**d**) The object does fall. As pointed out on several occasions in this chapter, the satellite *is falling* toward the Earth in an arc such that the curvature of the Earth causes its surface to bend away from the falling satellite to exactly the same extent that the satellite is bending down to try to meet the Earth.

SUMMARY

Newton's law of universal gravitation states that every object in the Universe exerts a force of gravitational attraction on every other object. The apparent weightlessness of objects in orbit occurs because they are all "falling" toward the Earth with the same acceleration.

Projectile motion is characterized by two separate types of motion. Motion along the surface of the Earth takes place at a constant velocity, whereas the vertical motion is one of constant acceleration.

Centripetal force is an inward force that causes an object to move in a curved path. Every object moving in a curved path has an acceleration toward the center of that path called a **centripetal acceleration.** If the centripetal force is the action, the reaction is an outwardly directed **centrifugal force.**

EQUATIONS TO KNOW

$F = G \dfrac{m_1 m_2}{r^2}$ (Newton's law of universal gravitation)

$a = \dfrac{v^2}{r}$ (centripetal acceleration)

$F = m \dfrac{v^2}{r}$ (centripetal force)

KEY WORDS

Gravity

Newton's law of universal gravitation

Apparent weightlessness

Projectile motion

Centripetal acceleration

Centripetal force

Centrifugal force

PROBLEMS AND CONCEPTUAL QUESTIONS

Problems requiring numerical work are identified with a blue number.

The law of gravity

1. Is there a gravitational attraction between you and this book? If so, why don't you feel the pull?
2. In a desperate effort to lose weight, you decide to move either to Denver or to Death Valley. Which would be your best choice and why?
3. According to the law of universal gravitation, how far from the Earth would you have to get before its gravitational pull on you dropped to zero?
4. In this text, we frequently refer to the mass of a planet but never to its weight. Why not?
5. A man driving his car (total mass 1500 kg) crashed into a telephone pole (mass 500 kg). He used the defense in court that the gravitational pull on his car by the pole when he was 10 m from it was so great that he was unable to control the car. Calculate the force and rule on the case.

6. Find the acceleration due to gravity at the surface of (a) Mercury (mass = 3.18×10^{23} kg and radius = 2.43×10^6 m) and (b) Saturn (mass = 5.68×10^{26} kg and radius = 5.85×10^7 m).
7. A baseball is dropped from a height of 10.0 m on the surface of Mercury. (a) Use the results of problem 6 to find the speed of the ball just before it strikes the ground. (b) Repeat your calculation for Saturn.
8. Some experimental evidence seems to indicate that the value of G in the law of universal gravitation is decreasing with time. What consequences could this have?
9. A bowling ball is held above the surface of the Earth at a height of 1 m, and the force of gravitational attraction exerted on it by the Earth is calculated. A tennis ball is then held at the same height and the calculation repeated. (a) In which case is the gravitational force the greatest? (b) In light of your answer to (a), why doesn't the bowling ball fall faster than the tennis ball?

Weightlessness

10. Would it be possible for a person to appear to be weightless 2 km above the surface of the Earth? Explain.

11. Cyrano de Bergerac describes the following method by which he could reach the Moon:

> "Finally—seated on an iron plate,
> To hurl a magnet in the air—the iron
> Follows—I catch the magnet—throw
> again—And so proceed indefinitely."

Do you think this plan could work? Why or why not?

12. Science fiction writers imagined flights to the Moon long before human flight was a reality. Edgar Allan Poe wrote of a journey to outer space by balloon, Jules Verne had his characters fired to the Moon out of a mammoth cannon, and H.G. Wells used an antigravity machine. Which of these devices (if any) could, in theory, send a spacecraft to the Moon? If any are theoretically possible, why have they not been used in modern space programs?

13. If you jump off the top of a building, are you weightless on the way down?

14. If you punch a hole in the bottom of a container filled with water, the water drains out. However, if you drop the container while the water is coming out, the water ceases to flow out while the container is falling. Why?

Projectile motion

15. One of the greatest cannons of all time was "Big Bertha" used by the Germans to shell Paris from 70 miles away. A simple physics demonstration can show that a football will have the highest range when shot at an angle of 45°. For Bertha, however, it was found that the range was almost twice as great when shot at a higher elevation. Why? *Hint:* The density of air decreases with altitude.

16. Three children stand on the edge of a tall building. One drops a rock from rest over the side of the building, one throws an identical rock downward at 10 m/s, and the third throws an identical rock horizontally at 10 m/s. Which has the greatest acceleration downward?

17. Someone is angry with you, and you hide in a tree to escape a confrontation. Nevertheless, you are seen, and the other person throws a rock such that it is heading directly toward you. To avoid being hit, should you remain seated on the limb or drop out of the tree?

18. A person standing at the side of a road drops a coin, and at that same instant a person driving by in a car drops an identical coin. (a) Which of the coins has the greater acceleration? (b) Which of the coins will strike the ground first? (c) Which of the coins will have the greater speed when it strikes?

19. An airplane moving horizontally at a constant velocity drops a load of supplies to some stranded explorers. In the absence of air resistance, will the supplies strike ahead of the plane, below it, or behind it?

20. A baseball is struck so that it leaves the bat with a vertical velocity of 10 m/s and a horizontal velocity of 8 m/s. (a) What are the values of the vertical and horizontal velocity at the maximum height reached by the ball? (b) What are the values of the vertical and horizontal velocities at the end of the flight just before it strikes the ground?

21. A jet pilot in horizontal flight fires a rocket directly forward. He then goes into an evasive dive. Describe what he would have to do, if anything, to shoot himself down.

22. You are riding in a car when you flip a coin straight up into the air as seen by you. Describe the motion of the coin as seen by an observer standing by the side of the road.

Circular motion; centripetal force; centrifugal force

23. Is it possible to go around a curve with zero acceleration?

24. Why are exit ramps on expressways often banked?

25. A coin is placed on the surface of a record. When the turntable is brought up to speed, the coin stays in place for a while, but then as the speed increases the coin flies off. Explain this observation.

26. The gravitational pull of the Sun holds the planets in orbit. Is that pull acting as a centripetal or as a centrifugal force? Explain.

27. If the gravitational field of the Sun disappeared magically and instantly, what would happen to the planets? If the gravitational field of the Earth disappeared magically and instantly, what would happen to you? What would happen to the Moon?

28. A person puts a stone inside an open tin can, ties a string to it, and spins the can around vertically. Why doesn't the stone fall out of the can?

29. Show that the expression for centripetal acceleration v^2/r has units of acceleration.

30. A car is moving around a circular racetrack when the centripetal acceleration suddenly goes to zero. What happens to the car?

31. A car is moving around a circular racetrack at a speed v. The speed is quickly increased to $2v$.

What must happen to the centripetal force to keep the car on the track at the same radius? Where does this centripetal force come from?

32. A person swings a 2.00 kg object attached to the end of a 1.50 m long string around his head in a horizontal circular path at a speed of 3.00 m/s. What is the centripetal force acting on the object, and what produces this force?

33. Why might a jet pilot tend to black out when pulling out of a steep dive?

34. A centrifuge is a device used in hospitals to separate heavy chemical substances from lighter ones. In these devices, a solution of the materials is placed in a test tube and whirled at high speeds in a circular path. The heavy materials migrate to the bottom of the test tube, and the lighter ones appear at the top of the solution. Why?

35. If the breaking strength of the string in problem 32 is 33 N, how fast can the object move?

ANSWERS TO SELECTED NUMERICAL PROBLEMS

5. 5.00×10^{-7} N
6. (a) 3.59 m/s^2, (b) 11.1 m/s^2
7. (a) 8.47 m/s, (b) 14.9 m/s
32. 12.0 N
35. 4.97 m/s

CHAPTER 5

THERMAL PHYSICS

Our study thus far has dealt with a brief introduction to that portion of physics known as mechanics. We have examined such concepts as motion, Newton's laws, momentum, and energy. We now move to a second subdivision: thermal physics. Here we shall be concerned with such topics as temperature and heat.

5.1 TEMPERATURE

One of the first sensations of childhood is that of the relative hotness or coldness of objects. But why does one object feel hot while another may be cold to the touch? We will withhold the answer to this question for a few pages because scientists had learned how to measure temperature long before they understood the internal difference between two bodies at different temperatures. The key experimental observation that enabled people to measure temperature is the fact that most materials expand when heated and contract when cooled. For example, if you blow up a rubber balloon and place it in a refrigerator, you will find that the size of the balloon diminishes as the temperature of the air inside the balloon decreases. In general, a change of size with a change of temperature is a property of all materials, regardless of whether they are in the form of solid, liquid, or gas. This change of size is, for most materials, an expansion as the temperature is increased. This property forms the basis for the most common of all temperature-measuring devices: the mercury (or alcohol)-in-glass thermometer.

A common mercury thermometer is illustrated in Figure 5.1. It consists of a glass tube with a small inside diameter from which all the air has been removed. At one end, the tube bulges out to become a small bulb filled with mercury. When placed in contact with a hot object, both the mercury and the glass expand, but the increase in volume of the glass

The Kuwait oil fires of 1991 are visible from space. The total number of burning wellheads in this area was between 550 and 600. In some cases, the fires at the wellhead sources are discernible. This thermal insult to Kuwait caused much sickness and death in the region because of pollution fallout. (Courtesy NASA)

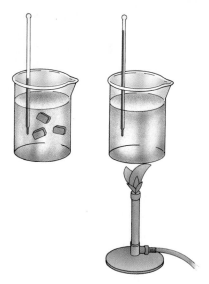

FIGURE 5.1 *As the temperature rises, the mercury expands more than does the glass, and the level of the mercury is used as a measure of the temperature.*

is not as great as the increase in volume of the mercury. The result is that the mercury gradually creeps up inside the stem. The temperature is measured by how high the mercury rises in the stem, as indicated by a scale etched or printed on the glass.

5.2 TEMPERATURE SCALES

A standard and convenient way to calibrate a thermometer is by first inserting it into a mixture of ice and water, noting the height of the mercury, and then inserting it in boiling water while again noting the height of the mercury. A scale can then be engraved on the glass by use of these reference points. For example, in the United States, the most commonly used temperature scale is the **Fahrenheit** scale, in which the freezing point of water is chosen to be 32°F and the boiling point to be 212°F. This particular scale was originally set up to relate roughly to the lowest temperatures reached on our planet during the winter and to the hottest temperatures reached during the summer months. That is, the originator of this scale wanted the coldest days to have a temperature of about 0°F and the hottest about 100°F.

Although the Fahrenheit scale remains the one most often used in everyday life in the United States, in the scientific community and throughout most of the world an alternate scale called the **Celsius** scale is the choice. This scale sets the temperature of the ice point to be 0°C and the temperature of boiling water to be 100°C.

Many weather forecasts give the temperature in both Celsius and Fahrenheit units. If you would like to check their accuracy, the equation that relates these two temperature scales is

$$T_C = \frac{5}{9}(T_F - 32) \tag{5.1}$$

where T_C is the Celsius temperature and T_F is the Fahrenheit temperature. Use this equation to verify that a Fahrenheit room temperature of 72°F corresponds to a Celsius temperature of approximately 22°C.

How cold can a substance be cooled? As any material is cooled, the molecules that are its building blocks gradually lose kinetic energy, and when this energy has decreased as much as possible, the temperature can drop no lower. This theoretical lower limit of temperature is reached at approximately −273°C (−459°F). In our day-to-day activities, we never approach this lowest limit of temperature. Even if we should travel to the planet Pluto, the planet farthest from the Sun in our Solar System, we would find that the temperature there is −233°C. In fact, theory predicts that it is impossible to reach this lowest limit of temperature, but experiments on Earth have come close. In general, these experiments are based on the principle that when a substance is magnetized it heats up, and when it is demagnetized it cools. In one experiment, a compound of copper was magnetized and then placed in liquid helium, thus dropping its temperature to about −272°C. When the copper was demagnetized while in the helium, its temperature dropped to a point only 0.000001 degree above the lowest theoretical temperature.

In view of the fact that there exists an absolute minimum of temperature, another scale, called the **Kelvin** scale, has been developed that sets its zero point at this minimum temperature. (This lowest theoretical temperature is thus often referred to as **absolute zero**.) On the Kelvin scale, the freezing point of water occurs at 273 K and the steam point of water at 373 K. (The actual value of the freezing point is 273.15 K, but we will approximate it as

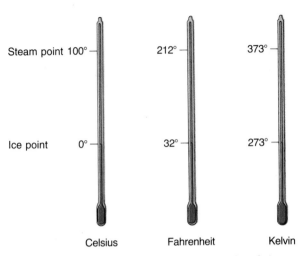

Steam point 100° — ... 212° — ... 373° —

Ice point   0° — ... 32° — ... 273° —

Celsius   Fahrenheit   Kelvin

FIGURE 5.2   *Comparison of the Celsius, Fahrenheit, and Kelvin temperature scales.*

273 K.) The difference between the two temperatures is the same on this scale as on the Celsius scale. The Celsius, Fahrenheit, and Kelvin temperature scales are compared in Figure 5.2.

The relationship between the Kelvin scale and the Celsius scale is given by

$$T_K = T_C + 273 \qquad (5.2)$$

EXAMPLE 5.1   A UNIQUE TEMPERATURE
(a) Convert the temperature $-40°C$ to the Fahrenheit scale.

(b) Repeat for the Kelvin scale.

Solution (a) Equation 5.1 can be solved for the Fahrenheit temperature to give

$$T_F = \frac{9}{5} T_C + 32$$

Thus, if we substitute -40 for T_C, we find

$$T_F = -40°F$$

The temperature -40 is unique in that it is the same on both the Celsius and the Fahrenheit scales.

(b) The temperature of -40 on the Celsius scale can be converted to the Kelvin scale by use of Eq. 5.2.

$$T_K = T_C + 273$$

or

$$T_K = -40 + 273 = 233 \text{ K}$$

This answer is read as 233 kelvins.

5.3   EARLY IDEAS ABOUT HEAT

When a hot object is placed in contact with a cold object, the two eventually reach a common temperature intermediate between the two initial temperatures. When such processes occur, we say that heat is transferred from the object at the higher temperature to the one at the lower temperature. But what is it that is being transferred? Early investigators believed that heat was an invisible, colorless, weightless material to which they gave the name **caloric** and that when two objects at different temperatures were placed in contact, caloric was transferred from one to the other.

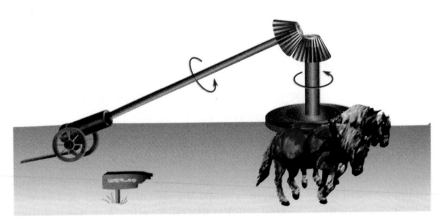

While boring cannons, Count Rumford wondered, "Where does all the work go after the hole is drilled?"

The first experimental observation suggesting that caloric does not actually exist was made by Benjamin Thompson at the end of the eighteenth century. Thompson, an American-born scientist, emigrated to Europe during the Revolutionary War because of his Tory sympathies. Following his appointment as director of the Bavarian arsenal, he was given the title Count Rumford. While supervising the boring of artillery cannon in Munich, Thompson noticed the great amount of heat generated by the boring tool, indicated by the fact that the water used to cool the tool had to be replaced continually as it boiled away. Many of the details of the caloric theory had been worked out, and Thompson found quickly that one of the predictions was not holding. The prediction that failed said that as metal shavings were produced in the boring process, they would lose their caloric to the water. In fact, the smaller the shavings, said the theory, the less able were they to contain their caloric. So, if the boring tool produced fine shavings, as it would do when it was sharp, the water would heat up quickly. To his surprise, Thompson discovered that when the boring tool became so blunt as to produce almost no shavings at all, the amount of water being boiled away actually increased. As a result, he rejected the caloric idea that said that the number and size of the shavings had something to do with the boiling of the water. Instead, he concluded that what was important was the amount of work being done by the boring tool on the cannon. He suggested that heat is not a substance but rather some form of motion that is transferred from the boring tool to the cooling water. In another experiment, he showed that the heat generated was equal to the mechanical work done by the boring tool.

Although Thompson's observations provided evidence that brought the caloric theory of heat into question, it was not until the middle of the nineteenth century that the modern model of heat was developed. In this view, heat is treated as just another form of energy, one that can be transformed into mechanical energy.

5.4 MEASURING HEAT

If heat is just another form of energy, it seems obvious that we should measure it in joules. Until recently, however, this has not been the case. In most present-day applications, heat *is* indeed measured in joules, but there is an alternative method that harkens back to the early misunderstandings of the nature of heat, as exemplified by the caloric theory. Before a correct understanding of heat came about, units in which to measure it had already been developed. These units were so widely used and had become so ingrained in the world of science that they are still often used today.

One of these units is the **caloric (cal)**, defined as *the amount of heat necessary to raise the temperature of 1 g of water from 14.5°C to 15.5°C.*

As an example of the use of this unit in practical situations, you will find that dieters often think in terms of counting calories. But be careful here because the energy unit used in describing the energy equivalent of food is a Calorie, spelled with a capital C. The Calorie is actually 1000 calories.

An alternative old way of measuring heat units is the **British thermal unit, Btu.** *This is defined as the heat required to raise the temperature of 1 lb of water from 63°F to 64°F.* Heating and air conditioning systems often have their capabilities specified in Btu. For example, a 20,000 Btu air conditioning unit is capable of removing 20,000 Btu of heat energy from a room every hour of operation.

Even though these units are still frequently used, most of our work in future chapters will specify heat as measured in joules. The relationship between the calorie and the joule was first established by James Prescott Joule and is given by

$$1 \text{ cal} = 4.186 \text{ J}$$

This relationship is called the **mechanical equivalent of heat.** In his experiment to establish this relation, Joule turned paddle wheels in a container of water and noted the temperature change of the water that was being stirred. He found that he had to do 4.186 J of mechanical work on the paddles for every gram of water that he raised in temperature by 1°C.

5.5 THERMAL ENERGY AND TEMPERATURE

Atoms and molecules are always moving, even in a block of ice or in a drop of liquid nitrogen. The combined energy of motion of all the particles in a sample is called the thermal energy. If a beaker of cold water is placed on a block of hot iron, the

FOCUS ON
EVAPORATION AND BOILING

We found in our discussion of thermal energy and temperature that temperature is a measure of the average kinetic energy of the atoms or molecules of a substance. By random processes, some of the atoms of a substance end up with kinetic energies that are higher than the average, and some have energies lower than average. With this thought in mind, let us consider a common experience and see why it occurs. If you set a beaker of water out on a table, after a few hours some of the water will have evaporated. To see why this occurs, consider the molecules near the surface of the liquid. From time to time, one of them receives a higher than average kinetic energy via collisions and interchanges of energy between it and other molecules in the liquid. In fact, it often receives a large enough kinetic energy to escape from the surface of the liquid. Thus, the molecule has gone from a condition in which it is considered a molecule of a liquid to one in which it is considered to be in vapor form.

Let us take a look at this same process from a slightly different point of view. After swimming, you surely have noticed that your wet body feels cooler than it did when it was dry. The reason for this cannot be related to the temperature of the liquid because the same sensation is felt even after a hot bath. A deeper look at the process of evaporation provides an explanation. Note that we have said that the molecules that escape a liquid by evaporation have a greater-than-average kinetic energy. But if the molecules leaving have a lot of energy, the ones left behind have, on the average, less energy than before. Thus, the remaining liquid is cooler. Therefore, evaporation is a cooling process, and the evaporation cools the swimmer.

If one adds heat to a liquid, the average kinetic energy of the molecules increases. Finally, at some temperature, the molecules attain enough energy to break free of one another even below the surface of the liquid. When this happens, bubbles of vapor form in the liquid and rise to the surface. We call this phenomenon **boiling.**

temperature of the water rises. Energy from the iron has been transferred to the water. Energy transfers of this type occur all around us all the time. *Heat is defined as the energy that is transferred from one system to another when the two systems at different temperatures are in contact.*

The relationship between heat and the energy of atoms and molecules can be understood by considering the process of heating up a liquid. When thermal energy is transferred to a liquid, the atoms or molecules speed up. Now think about one molecule. To speed it up, a force must be applied to it, just as a force must be applied to a bowling ball or a dump truck to speed it up. When a force is applied over a distance, work is done. On a microscopic level, *heat is a measure of the work required to change the speeds of a collection of atoms or molecules.*

To see that heat energy and temperature are different, consider the following experiment. Suppose you add a measured amount of heat to a 200 L container of water. You will find that the temperature of the water will change by a certain amount. Add this same amount of heat to a small cup of water, and its temperature may rise by several more degrees. The same amount of heat is transferred in both instances, but in the first case, an individual molecule receives, on the average, only a small portion of this energy, whereas in the second case, an individual molecule receives a comparatively large amount of this energy. The end result is that the molecules in the smaller sample of water have large speeds and consequently large kinetic energies, whereas the molecules of the larger sample have much smaller speeds and kinetic energies. *Temper-

ature is a measure of the average kinetic energy of the atoms or molecules of a substance.

5.6 SPECIFIC HEAT

A series of thought experiments will enable us to discover a few facts relative to what happens to the temperature of an object as we add heat to it. These are referred to here as thought experiments simply because most of these observations should be familiar to you. If they are not, try them as actual experiments.

1. Imagine two equal sources of heat energy, such as identical hot plates plugged into the same source of current (Fig. 5.3). We will place a beaker with 1 kg of water on both of the plates, measure the temperature of the water, and turn the hot plates on for the same period of time so that the *same* amount of heat energy is added to *both* beakers. At the end of this interval, we will again measure the temperature of both beakers. It should not come as a surprise to you that the temperatures of both containers of water have risen by the same amount. Now let us turn on the hot plates once again and watch the temperature of each for another period of time. Again, we will find that at the end of this interval both have risen in temperature by the same amount. The result of this simple observation is that the amount of temperature increase ΔT is proportional to the amount of heat Q added. That is,

$$\Delta T \propto Q$$

The notation used here needs some explanation. First, a change in a quantity is represented by the symbol Δ (Greek delta) before the quantity. That is, ΔT means a change in T, the temperature; Δv would refer to a change in velocity, and so forth. The symbol $\propto$ is used to designate a proportionality between two quantities.

Our experiment will work the same way if we reverse the process by extracting heat from the system. That is, for a given amount of a substance, the temperature decrease is directly proportional to the amount of heat removed.

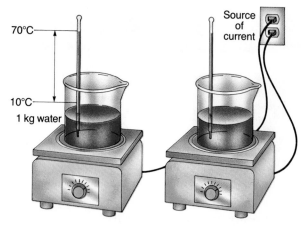

FIGURE 5.3 *If two beakers are placed on identical hot plates at the same time, and both beakers contain the same mass of water, the temperature of the two will increase identically. Thus, the temperature change is proportional to the amount of heat added.*

2. We alter our experiment somewhat by placing twice as much water in one beaker as in the other (Fig. 5.4). With 2 kg in one beaker and 1 kg in the other, we again place our beakers on the identical hot plates and add energy for a given period of time. In this case, we will find that the temperature increase of the 2 kg beaker will be only half that of the 1 kg beaker. (As previously, this process is reversible. The temperature decrease of the 2 kg beaker would be half that of the 1 kg beaker if we remove equal amounts of heat energy.) The result of this experiment indicates that the temperature increase is inversely proportional to the quantity of matter in the sample. To be a little more specific, the temperature increase is inversely proportional to the *mass* of the substance. Thus, we can write

$$\Delta T \propto \frac{1}{m}$$

3. Finally, we add one last observation to our series of investigations. We place a beaker with 1 kg of water on one hot plate and a beaker with 1 kg of ethyl alcohol on the other, as shown in Figure 5.5. If the same amount of heat energy is added to each beaker, the temperature of the ethyl alcohol

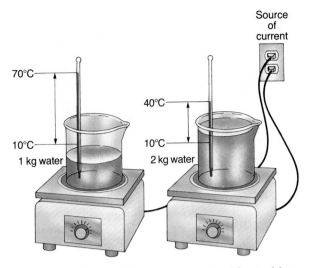

FIGURE 5.4 *If two beakers are placed on identical hot plates, and one beaker has twice as much water as the other, the temperature of the larger volume of water will rise half as much as the temperature of the smaller volume of water.*

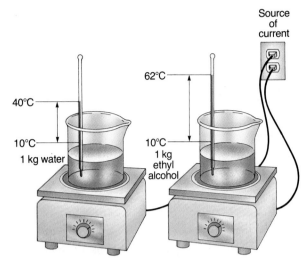

FIGURE 5.5 *If two beakers are placed on identical hot plates, and both contain equal quantities of different substances, generally the rise in temperature of the two liquids will be different.*

will rise about twice as fast as the temperature of the water. Thus, the same amount of heat energy added to equal masses of different substances will produce a greater temperature change for one than it does for another. We conclude that the temperature change depends on the substance to which we add the heat. We include this in our proportionality observations by use of a constant called *c*, defined as the **specific heat** for a particular material. As a proportionality, we express this observation as

$$\Delta T \propto \frac{1}{c}$$

Thus, the larger is *c*, the smaller is the temperature increase.

Let us collect all of our observations into a single relationship. We also shall choose the value of *c* such that we have an equation rather than a proportionality. We have

$$\Delta T = \frac{Q}{mc} \quad \text{(5.3)}$$

or

$$Q = mc\,\Delta T$$

Table 5.1 lists the specific heats for a variety of common materials in units of J/kg°C.

Note that a substance has a different specific heat in its different phases. Specifically, we see that water in its liquid phase has specific heat of 4186 J/kg°C; in its solid phase, ice, the specific heat is 2100 J/kg°C; and in the vapor phase, steam, 2010 J/kg°C.

EXAMPLE 5.2   POUR ON THE HEAT
(a) Calculate the heat energy required to raise the temperature of 0.50 kg of water by 15°C.

TABLE 5.1	Specific heats of common materials
MATERIAL	**SPECIFIC HEAT (J/KG°C)**
Aluminum	901
Copper	387
Glass	838
Ice	2100
Iron	448
Silicon	704
Lead	128
Mercury	138
Steam	2010
Water	4186

(b) Repeat for 0.50 kg of sand, which has a specific heat of about 1500 J/kg°C.

Solution (a) This is a straightforward application of $Q = mc\Delta T$. We find for water

$$Q = (0.50 \text{ kg})(4186 \text{ J/kg°C})(15°C) = \boxed{31{,}400 \text{ J}}$$

(b) For sand, we have

$$Q = (0.50 \text{ kg})(1500 \text{ J/kg°C})(15°C) = \boxed{11{,}300 \text{ J}}$$

EXERCISE

Let us slightly reverse the previous problem. We found that adding 31,400 J of heat energy to 0.50 kg of water would increase its temperature by 15°C. Now suppose that you add 31,400 J of energy to 0.50 kg of sand. What would be its temperature increase?

Answer 42°C

5.7 SPECIFIC HEAT AT WORK IN NATURE

In the previous exercise, you were asked to find what happens when you add a specific amount of heat energy to equal amounts of water and sand. We found that the temperature of the water would increase by a small amount, 15°C, whereas the temperature of the sand would rise considerably more, by about 42°C. Undoubtedly, you have observed this effect on a trip to the beach. You have found that you can walk comfortably in the water, but if you try to walk across the sand-covered beach, you may find that it is almost hot enough to burn your feet. Why? In light of our exercise, we see that the Sun has been pouring down equal amounts of heat energy on both water and sand, but these equal amounts of heat added to each warms the water only slightly, whereas the sandy beach soars in temperature. (You should note that there actually are other factors involved in this practical application besides the differences in the specific heats of the substances. Water mixes while sand does not, and as a result, the Sun is actually heating a larger mass of water than of sand.)

HEAT

(a)   (b)

FIGURE 5.6   *(a) On a hot day, the air above the warmer land becomes hot and rises as cooler air above the water moves in to replace it. The breeze is toward the shore. (b) At night, the warmer air is over the water, and the breeze blows away from the land.*

While we are at the beach, let us take a look at how the high specific heat of water compared to that of land is responsible for the pattern of air flow at the seaside. During the daylight hours, the Sun adds roughly equal amounts of heat to both the beach and the water, but the lower specific heat of sand causes the beach to reach a higher temperature than the water, as already noted. Because of this, the air above the land reaches a higher temperature than that over the water; as a result, the air near the surface of the sand decreases in density and rises as cooler, denser air over the water rushes in to displace it, as shown in Figure 5.6a. This movement of air produces a breeze from sea to land during the day. The heated air gradually cools as it rises and thus sinks to set up a circulating pattern, as shown in the figure. During the night, the land cools more quickly than the water; thus, the circulating pattern of air flow reverses because the hotter air is now over the water, as shown in Figure 5.6b. You should not expect to observe this pattern every time you go to the beach because prevailing winds caused by other factors often obscure this effect.

An effect similar to that just described produces rising layers of air, called thermals, that can help eagles to soar higher and hang gliders to stay aloft longer. Thermals are created when certain portions of the Earth get heated to hotter temperatures than nearby regions. This often occurs over plowed fields, which are heated by the Sun to higher temperatures than are vegetation-covered fields nearby. Cooler, denser air from the fields pushes the expanding air over the plowed ground upward, and thermals are formed.

The high specific heat of water is also responsible for moderating the temperatures of regions near large bodies of water. As the temperature of a body of water decreases during the winter season, it gives off heat to the air, which carries it landward when prevailing winds are favorable. For example, the prevailing winds off the west coast of the United States are toward the land, and the heat liberated by the Pacific Ocean as it cools keeps coastal areas much warmer than they would be otherwise.

5.8 HEAT AND CHANGES OF PHASE

Let us suppose that we have a 1 kg ice cube at $-10°C$. If we would like to have 1 kg of steam at a temperature of $110°C$, we can make the required transition with the addition of some heat. Let's examine the changes involved, one at a time; as we do, some surprising features will appear. Figure 5.7 shows a graph of the temperature of the material versus the amount of heat added under conditions of standard atmospheric pressure. Let us examine each portion of the curve separately.

Part A   During this portion of the curve, we are changing the temperature of the ice from $-10°C$ to $0°C$. Since the specific heat of ice is $2100\,J/kg°C$, we can calculate the amount of heat added as follows:

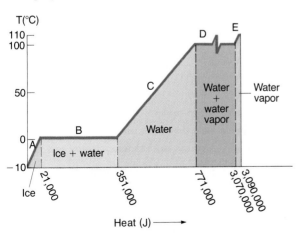

FIGURE 5.7   *A graph of the temperature change of 1 kg of water initially at $-10°C$ (ice) as heat is added, changing it to steam at $110°C$.*

$$Q = mc\Delta T = (1.0\text{ kg})(2100\text{ J/kg°C})(10°C)$$
$$= 21{,}000\text{ J} = 2.1 \times 10^4\text{ J}$$

Part B   When the ice reaches $0°C$, it remains at this temperature—even though heat is being added—until all the ice melts. The change in a substance from a solid to a liquid form or from a liquid to a gaseous form is called a **change of phase,** and the amount of heat needed to accomplish a phase change from solid to liquid is given by

$$Q = mL_f$$

where L_f is the so-called **latent heat of fusion** of the substance. The value of L_f depends on the particular material. For example, if ice is changing from ice to water at $0°C$, $L_f = 3.34 \times 10^5$ J/kg. This means that it takes 3.34×10^5 J of heat to transform 1 kg of ice at $0°C$ to water at $0°C$, or 6.68×10^5 J to change 2 kg of ice at $0°C$ to 2 kg of water at the same temperature, and so forth. This can be expressed and used in the form of an equation as $Q = mL_f$.

$$Q = mL_f = (1.0\text{ kg})(3.34 \times 10^5\text{ J}) = 3.3 \times 10^5\text{ J}$$

This can go in either direction. That is, if one removes 3.3×10^5 J of heat from 1 kg water at $0°C$, the water is transformed to ice at $0°C$.

Part C   Between $0°C$ and $100°C$, there are no phase changes, and all the heat added to the water is being used to increase its temperature. We can find the heat required as

$$Q = mc\Delta T = (1.0\text{ kg})(4186\text{ J/kg°C})(100°C)$$
$$= 4.2 \times 10^5\text{ J}$$

Part D   At $100°C$, we have another phase change occurring as the water at $100°C$ changes to steam at $100°C$. The heat required for this transition is found from

$$Q = mL_v$$

where L_v is called the **heat of vaporization.** For water, $L_v = 2.26 \times 10^6$ J/kg. We have

$$Q = mL_v = (1.0\text{ kg})(2.26 \times 10^6\text{ J})$$
$$= 2.3 \times 10^6\text{ J}$$

Part E   Finally, we must change the steam at $100°C$ to steam at $110°C$. The amount of heat required for this is

$$Q = mc\Delta T = (1.0\text{ kg})(2010\text{ J/kg°C})(10°C)$$
$$= 2.0 \times 10^4\text{ J}$$

Thus, the total amount of heat to change our ice at $-10°C$ to steam at $110°C$ is

$$Q = (2.1 \times 10^4 \text{ J}) + (3.3 \times 10^5 \text{ J})$$
$$+ \ (4.2 \times 10^5 \text{ J}) + (2.3 \times 10^6 \text{ J})$$
$$+ \ (2.0 \times 10^4 \text{ J}) = 3.1 \times 10^6 \text{ J}$$

It should be noted that this process is reversible. That is, if we remove 3.1×10^6 J of heat from 1.0 kg of steam at $110°C$, we will end up with 1.0 kg of ice at $-10°C$.

One feature that you surely noted is that it takes a tremendous amount of heat to accomplish a phase transition. For example, we saw that to transform 1 kg of ice at $0°C$ to water at $0°C$, 3.3×10^5 J of heat energy had to be added. To understand why so much heat energy is necessary, we must realize that a phase transition produces a rearrangement of the molecules of the substance. Consider the liquid-to-gas transition. The molecules of a liquid are close together, and the forces between them are stronger than they are in a gas, in which the molecules are far apart. As a result, to move the molecules farther apart by some distance d, we must exert a force. This means that work must be done on the liquid to overcome these attractive molecular forces. The amount of heat required to vaporize a substance is equal to the amount of work that must be added to the fluid to accomplish this separation. A similar explanation applies to the large amount of heat required to change a solid to liquid. Because the average distance between molecules in the gas phase is much larger than in either the liquid or the solid phase, however, we might expect that more work is required to vaporize a given amount of a substance than to melt it. Thus, the heat to vaporize, the heat of vaporization, is much larger than the heat required to melt a given substance, the heat of fusion. The heat of vaporization and the heat of fusion of several common substances are given in Table 5.2.

We shall investigate forces between molecules in more detail in our study of chemistry.

CONCEPTUAL EXAMPLE 5.3   PHYSICS GOTHIC

On nights when the temperature is expected to fall below freezing, farmers often protect fruits and vegetables stored in a cellar by placing large vats of water in the cellar with the produce. Explain how this works.

FOCUS ON
BOILING A LIQUID AND FREEZING IT AT THE SAME TIME

The temperature at which a liquid boils is affected by the pressure exerted on it. In general, as the pressure on a liquid is increased, the boiling point also increases. This is reasonable to expect from the molecular interpretation of boiling given in the box on evaporation and boiling. An increase in pressure on the surface of a liquid makes it more difficult for molecules to escape the liquid; hence, the boiling point increases.

The opposite is also true; a reduction of pressure reduces the temperature at which a liquid boils. Let us consider an interesting experiment that produces some unique results. Imagine that water is placed inside a container from which the air can be gradually pumped out. We start with the water at room temperature—hence, far below its boiling point. As the air is evacuated, the pressure on the water is gradually reduced and so is the boiling point of the liquid. Finally, when the pressure has been reduced by about a factor of 95 percent, the boiling point is reduced to room temperature, and the water boils.

We pointed out in the box on evaporation and boiling that the boiling process is actually a case of rapid evaporation, which is a cooling process. Hence, as the water in the container boils, it is also being cooled. If you reduce the pressure just a little bit more, the water will boil a little more vigorously, and its temperature will decrease a little bit more. Finally, the temperature of the liquid will reach the freezing point (even though it is boiling). *Thus, you can simultaneously boil water and freeze it.*

TABLE 5.2	**Heat of fusion and vaporization**			
SUBSTANCE	HEAT OF FUSION	HEAT OF VAPORIZATION	MELTING POINT	BOILING POINT
Ethyl alcohol	1.04×10^5 J/kg	8.54×10^5 J/kg	$-114°C$	$78°C$
Water	3.34×10^5 J/kg	$22.6 \ \times 10^5$ J/kg	$0.00°C$	$100.00°C$
Lead	0.25×10^5 J/kg	8.71×10^5 J/kg	$327.2°C$	$1750°C$
Silver	0.88×10^5 J/kg	$23.4 \ \times 10^5$ J/kg	$960.8°C$	$2193°C$
Gold	0.65×10^5 J/kg	$15.8 \ \times 10^5$ J/kg	$1063°C$	$2660°C$

Solution As the water freezes at 0°C, each kilogram of ice that forms releases about 3.3×10^5 J of heat to the surroundings. This helps to keep the cellar temperature high enough to prevent damage to the stored food.

5.9 HEAT TRANSFER

In this section, we shall examine three ways in which heat can be transferred from one place to another: conduction, convection, and radiation. Regardless of the process, there can be no heat transfer from one substance to another unless the two objects are at different temperatures.

Conduction

Each of the three processes of heat transfer can be examined by considering the various ways you can warm your hands over an open fire. For example, if you stick a poker into a roaring campfire, you will soon find that heat is transferred from the flame along the poker to your hand. In this instance, the heat reaches your hand through the process of **conduction.** There are two factors that are responsible for this heat transfer. Let us look at these in turn.

1. Before the poker is inserted into the flame, the atoms of the metal are vibrating back and forth about their equilibrium positions. As the flame heats the rod, the vibration of those atoms near the flame becomes larger and larger. Because they are swinging farther and farther away from their equilibrium positions, they collide with neighboring atoms and transfer some of their energy to them. This causes these neighbors to vibrate with larger amplitudes also, and they, in turn, collide with atoms farther up along the

poker. This increase in the amplitude of vibration of the atoms gradually works its way along the rod until the end being held is reached. We measure this increase in the amplitude of vibration (increase of kinetic energy) as a temperature increase of the rod, and thus our hand is warmed as our molecules vibrate with a larger amplitude.

2. The explanation of the transfer of heat by conduction, as given here, is only part of the story. Some heat is indeed transferred by the vibration of the atoms of the substance. But all substances contain atoms, so why are all substances not equally good conductors of heat? It is well known that some materials like copper conduct heat extremely well, whereas others like asbestos do not. Those materials that allow heat to travel through them easily are called good **conductors,** and those that inhibit the flow of heat are called **insulators.** Metals are good conductors of heat and electricity because they have a large number of free electrons that can aid in transferring heat energy throughout the material. As the end of the poker is heated, the electrons in the hot end begin to move more rapidly. This increase in kinetic energy is transferred by collision to electrons farther up the rod, and we perceive this increase in energy of the electrons as an increase in temperature of the rod. Thus, heat energy is transferred throughout a substance by the increase in the amplitude of vibration of the atoms and by the increase in kinetic energy of individual electrons of the material.

It is quite easy to recognize the difference between a good conductor of heat and a poor conductor. Suppose you remove an ice tray and a package of frozen vegetables

FOCUS ON
THE HOT WATER VERSUS COLD WATER RACE

One of the most commonly asked questions about heat transfer problems is, "Does hot water freeze faster than cold water?" The answer to the question is simple: No, it does not. It is not as simple as one might believe to prove the assertion, and in fact, many common experiences tend to show just the opposite.

Suppose you have two pipes against the wall of your home, and the temperature drops below freezing. Invariably, it seems as though it is the hot water pipe that you have to thaw out. If cold water freezes faster than hot, why did the hot water pipe freeze before the cold? The answer has nothing really to do with physics; instead it relies simply on the fact that you use more cold water than you do hot. The water in both pipes decreases in temperature when against the cold wall, but before the cold water gets to the freezing point, someone flushes a commode, washes his hands, and so forth. As a result, the cold water moves on to be replaced by another batch of water to be cooled. Because you have not used any hot water, however, the water in that pipe stays in place against the cold wall and continues to drop in temperature. Thus, there is an alternative explanation for this commonly observed phenomenon.

To see why the cold water freezes faster than the hot, think of a situation in which you are removing heat from a pan of water at 50°C and from another pan at 20°C. The hot water may cool faster at first because it is at a greater temperature difference with its surroundings and thus will radiate more heat to the atmosphere. When it cools to 20°C, however, it is in the same state as the pan of cool water and will cool down in the same way as it did until it reaches 0°C.

It would seem to be a simple matter to test the statement that the cold water freezes faster than the hot by placing two pans in the freezing compartment of a refrigerator. There are other factors, though, that influence the results of your experiment. First, if you place the hot pan and the cold pan in the compartment at the same time, there is an interchange of heat between the two such that the cool warms up a little and the hot cools down. Thus, they affect one another and the results of your experiment. Another factor that you have to watch for is the contact of the pans with the walls of the freezing compartment. If there is a frost buildup on the walls, the hot pan will melt through this frost, allowing it to make intimate contact with the colder walls of the compartment so that heat can be conducted away from it faster than from the cold pan, which remains somewhat insulated from the cold walls by the layer of frost.

from a freezer so that they are both at the same temperature. You will easily notice that the ice tray feels considerably colder than does the cardboard package. This can be explained by noting that the metal ice tray is a good conductor, whereas the cardboard is a fairly good insulator. The metal conducts heat from your hand better than does the cardboard package; as a result, the tray feels colder than the carton. By use of a similar argument, you should be able to explain why

a tile floor feels colder to your bare feet than does a carpeted floor.

The ability of a given material to conduct heat depends on its cross-sectional area and its length. Figure 5.8 shows a bar placed between two objects at different temperatures. The rate at which heat energy flows through the bar increases as the cross-sectional area increases and decreases as the length of the bar increases. It also depends on the type of material used.

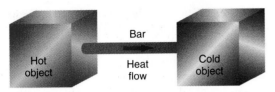

FIGURE 5.8 *The heat flow will increase as the cross-sectional area of the bar increases and will decrease as the length of the bar increases.*

CONCEPTUAL EXAMPLE 5.4   BOILING WATER IN A PAPER CUP

If you hold a paper cup filled with water over an open flame as in Figure 5.9, you can bring the water to a boil before the temperature of the cup becomes high enough to burn. Explain this phenomenon by use of the fact that as the length of a material decreases, the rate at which it is able to transfer heat increases. (*Hint:* If you try this, make sure that your cup does not have a rim around the bottom or is waxed or made of plastic. If so, it will burn. An alternative suggestion is to fill a paper bag with water rather than a cup.)

Solution The length to be considered is the thickness of the bottom of the paper cup. Because this thickness is so small, the heat from the flame is rapidly transferred through it to the water.

Convection

In the process of heat transfer by conduction, thermal energy is moved from one location to another

FIGURE 5.9 *Example 5.4. Boiling water in a paper cup.*

FIGURE 5.10 *Warming your hand by convection. Rising convection currents of air pass over your hands and warm them.*

via the vibration of molecules inside a substance (and, in the case of metals, by the transfer of kinetic energy between electrons). When energy is transferred by way of **convection,** *there is an actual movement of the material from one location to another.* In practice, these materials are usually air or water. To illustrate this, imagine warming your hands over an open flame as shown in Figure 5.10. In this situation, the air directly above the flame becomes heated, expands, and becomes less dense than nearby, cooler air. As a result, the heated air rises, and the cooler air moves in to replace it. Your hand is warmed as the rising hot air moves past it.

You should note that we have encountered convection in nature in an earlier section when we were describing the air flow patterns at a beach. This same air flow pattern takes place in a room heated by a radiator, as shown in Figure 5.11.

FIGURE 5.11 *Warm air above the radiator rises and is displaced by cool air drawn in from the room.*

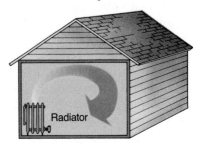

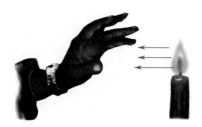

FIGURE 5.12   *Warming your hands by radiation.*

Radiation

A third method of transferring thermal energy from one place to another is via the process of **radiation.** To illustrate this effect, consider Figure 5.12. In this case, someone is warming his hand by holding it in front of an open fire. Because there is nothing in contact with the flame and the hand, there is no way that heat can reach the hand via conduction. (There is, of course, air between the hand and flame, but because air is a poor conductor of heat it will not allow much heat transfer by conduction.) Also, because the hand is not in the path of the rising hot air above the flame, heat is not transferred by convection. The process that remains is that of heat transfer by radiation.

We shall find in a later chapter that energy can be transmitted by way of a type of wave called an **electromagnetic wave.** Light is one example of electromagnetic radiation, radio waves are another, and X-rays are yet another. For the moment, it suffices to say that there is another type of radiation called **infrared radiation,** which is associated with the transfer of heat. Through electromagnetic radiation, approximately 1340 J of heat energy strikes each square meter of the top of the Earth's atmosphere every second. Some of this energy is reflected back into space and some is absorbed by the atmosphere, but enough reaches the surface of the Earth each day to supply all of our energy needs on this planet hundreds of times over—if it could be captured and used efficiently. The growth of the number of solar homes in the United States is one example of an attempt to use this free energy.

All warm objects emit infrared radiation. You should note that "warm" is a relative term because every object that is at a temperature other than absolute zero emits infrared radiation, and the hotter the object, the more radiation it emits.

If your home is heated by a fireplace, the principal way that heat enters the room is by radiation. The chimney provides an outlet for convection currents to escape; thus, the room does not receive an appreciable amount of heat by this mechanism. In fact, cooler air drawn into the room to replenish the air supply exhausted through the chimney competes with the heat supplied to the room via radiation. As a result, the temperature of a drafty room may actually decline when a fireplace is used as the primary source of heat.

As an example of a common experience with infrared radiation, consider what happens to the atmospheric temperature on a winter night. If there is a cloud cover above the ground, the water vapor in the clouds reflects back a portion of the infrared radiation emitted by the ground; as a result, the temperature remains at moderate levels. In the absence of a cloud cover, this radiation escapes into space, and the temperature drops lower than it would if it were cloudy.

As another example from common experience, you know that you are more comfortable in white clothing in the heat of the summer than you are in black clothing. Black material acts as a good absorber of the incoming sunlight, and it also acts as a good emitter of this absorbed energy. Half of this absorbed energy goes toward the body, and you get uncomfortably warm. The lighter clothing reflects away the incoming energy.

5.10 HINDERING HEAT TRANSFER

A common device to keep cool liquids cool or hot liquids hot is the thermos bottle. Let us examine its design, as shown in Figure 5.13, to see how it is able to perform this task. The walls of the bottle are made of glass and are silvered on the inner surface. The highly reflecting silvered surfaces reflect radiation, hindering heat transfer by radiation either into or out of the bottle. To prevent heat transfer by conduction or convection, the bottle has a double-walled construction with the air removed in the space between the two walls. A true vacuum conducts no heat at all; also, if there is no material medium there, convection currents cannot form in this enclosure. Thus, if all the air could be exhausted in this space, no heat would be transferred in or out by these two mechanisms. In practice, it is impossible to evacuate this interior space com-

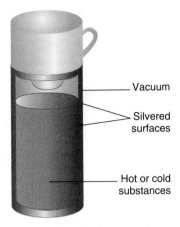

FIGURE 5.13 *Design of a thermos bottle.*

Vacuum

Silvered surfaces

Hot or cold substances

pletely, so some heat is transferred by these two means.

Another situation in which one would like to minimize heat transfer occurs in the insulation of buildings. To prevent heat transfer between the interior of a home and its surroundings, insulating materials are placed in the walls and above the ceilings. These materials, such as fiberglass, are poor conductors of heat, so little heat transfer occurs by this process. Additionally the insulating material is loosely packed so that pockets of air are trapped in cavities within the insulation. Air does not readily conduct heat, and when air is trapped, convection currents cannot be set up. This same basic principle is used to keep the body warm with wool sweaters or down jackets. Both of these trap air near the body and hence reduce heat loss from the body by both convection and conduction.

5.11 THE GREENHOUSE EFFECT

The **greenhouse effect,** as its name implies, would certainly be of interest to a florist or a gardener, but the term has appeared in the popular press as an example of a catastrophe waiting to happen to the world. (We shall return to this problem in our discussion of Earth science.) The greenhouse effect provides a good example of the principles of heat transfer and of ways to prevent heat transfer, but it is even more important in that it has global significance for life on this planet.

Let us examine the greenhouse effect first from the point of view of the glass enclosure so familiar to you at florists or plant nurseries. The underlying principle behind these buildings is that *glass allows visible light to pass through, but it does not allow infrared radiation to be transmitted as readily.* During the day, sunlight passes through the windows of a greenhouse and is absorbed by the earth, walls, and vegetation inside the structure. These objects then re-radiate this energy in the form of infrared radiation, but this radiation is now trapped because it cannot escape through the glass. As a result, the temperature of the interior rises.

An additional factor that causes the temperature to rise is the fact that convection currents are also inhibited in a greenhouse. This means that the heated air cannot circulate rapidly past those surfaces of the building that are exposed to the colder outside air. This prevents heat loss to the surroundings by this mechanism. In fact, many experts consider this to be an even more important effect in a greenhouse than the effect of the trapped infrared radiation.

Now let us turn to a phenomenon that occurs in the Earth's atmosphere that plays a role in determining the temperature of our planet. The primary constituent of the atmosphere that we need to consider is carbon dioxide. Carbon dioxide acts somewhat like the glass in a greenhouse in that it readily allows incoming visible light from the Sun to pass through to the surface of the Earth, but it does not allow infrared radiation to pass as readily. Thus, incoming visible light is absorbed at the surface of the Earth and is re-radiated in the form of infrared radiation, but this infrared radiation is absorbed and trapped by the carbon dioxide. This trapped heat energy causes the temperature of the surface of the Earth to be warmer than it would be if this infrared radiation could escape. This overall effect by which the temperature of the surface rises because of the trapping of infrared radiation by carbon dioxide is called the greenhouse effect. An example of a planet on which the greenhouse effect has run wild is Venus. Venus has an atmosphere rich in carbon dioxide, and because of the amount of trapped heat energy in this atmosphere, Venus is our warmest planet, approximately 850°F, even though it is not the closest planet to the Sun. We shall discuss this more thoroughly in the astronomy portion of this book.

As fossil fuels (oil, coal, and natural gas) are burned on Earth, large amounts of carbon dioxide are released into our atmosphere. This, of course, causes the atmosphere to retain more heat by virtue of the greenhouse effect. Many scientists are convinced that the 10 percent increase in the amount of atmospheric carbon dioxide in the last 30 years could lead to drastic changes in world climate. The drought of summer 1988 led many scientists to speculate that these worldwide temperature increases are already on the way. It has been estimated that if the average global temperature should rise by only 2°C, this would be sufficient to melt the polar ice caps, thus causing flooding and the destruction of many coastal areas, an increase in droughts, and a reduction of already low crop yields in tropical and subtropical countries. Present-day agricultural areas such as the wheat belt in the Midwest would move northward into Canada. The jury is still out as to whether or not a runaway greenhouse effect is indeed in control of this planet. It is an important problem, however, that all nations must address.

SUMMARY

Three common temperature scales are the **Celsius,** the **Fahrenheit,** and the **Kelvin** or absolute scale.

A **calorie** is the amount of heat required to raise the temperature of 1 g of water from 14.5°C to 15.5°C.

Heat is a measure of the work required to change the speeds in a collection of atoms or molecules. **Temperature** is a measure of the average kinetic energy of the atoms or molecules of a substance.

The **specific heat** of a substance is the amount of heat added to a substance divided by the mass and temperature change of the substance.

The change in form of a substance from solid to liquid, from liquid to gas, or vice versa, is called a **change of phase.** The amount of heat to change a unit mass of a substance from solid to liquid at the melting point is called the **heat of fusion.** The amount of heat to change a unit mass of a substance from liquid to vapor at the boiling point is called the **heat of vaporization.**

The three mechanisms of heat transfer are **conduction, convection, and radiation.** In the conduction process, heat is transferred by the motion of the atoms and electrons within the substance. Heat transfers by convection involve the movement of a heated body of liquid or gas. Heat transfer by radiation involves the absorption or emission of infrared radiation.

EQUATIONS TO KNOW

$T_c = \dfrac{5}{9}(T_F - 32)$ (converting between Fahrenheit and Celsius temperatures)

$T_K = T_c + 273$ (converting between Celsius and Kelvin)

$Q = mc\Delta T$ (heat transferred in the absence of a phase change)

$Q = mL_f$ (heat transferred during solid-liquid phase change)

$Q = mL_v$ (heat transferred during liquid-vapor change)

KEY WORDS

Temperature	Btu	Heat of fusion	Radiation
Celsius	Heat	Heat of vaporization	Conductor
Fahrenheit	Specific heat	Conduction	Insulator
Kelvin	Change of phase	Convection	Greenhouse effect
Calorie			

PROBLEMS AND CONCEPTUAL QUESTIONS

Problems requiring numerical work are identified with a blue number.

Temperature and temperature scales

1. If you place a thermometer in a container of hot water, you will often note that the level of the mercury falls slightly before it begins to rise. Explain why this happens.

2. Could a thermometer be made to work if the glass expanded more than the liquid trapped inside?

3. Galileo reputedly constructed one of the first thermometers by allowing water to rise and fall in a column of glass. Why do you suppose this type of thermometer did not catch on?

4. A typical warning on a medicine bottle label says that the product should be stored at a temperature lower than 86°F. Why did they pick that particular value? *Hint:* Convert to Celsius.

5. Based on the definition of temperature as given in this chapter, can one assign a temperature to a vacuum?

6. If the weatherman reports that the temperature is going to rise by 10° but does not refer to the type of scale he is using, in which of the following cases will the day end up the hottest? If he meant (a) 10 C°, (b) 10 F°, or (c) 10 K.

7. The temperature may reach −10°F on a cold day and 100°F on a hot day. Convert these temperatures to Celsius and to Kelvins.

8. You are using a Celsius thermometer to read your body temperature. It says 38°C. Do you have a fever?

9. Aluminum expands more on heating than does copper. What would happen if you heated a bar that consisted of a sheet of aluminum firmly bonded to a sheet of copper? How could such a bar be used as a thermometer? Can you think of any other use for such a device?

Measuring heat and specific heat

10. (a) How many calories are needed to change the temperature of 1000 g of water by 40°C? How many calories are needed to warm 10,000 g of water by 4°C?

11. Distinguish carefully between temperature and heat.

12. Repeat problem 10 except determine your answers in joules.

13. (a) How many joules are needed to heat 1.00 kg of copper 40.0°C? (b) How many joules are needed to heat 1.00 kg of iron 40.0°C?

14. How much energy in J will be released if 50.0 g of mercury is cooled by 25.0°C?

15. 500 cal of heat energy are added to a 5.00 kg mass of iron. What is the temperature increase of the iron?

Specific heat at work in nature and heat and changes of state

16. In general, cities have a higher average temperature than the surrounding countryside. There are many factors involved in this, but one of them is the difference in specific heat of soil and concrete. Which do you think has the higher specific heat?

17. Why does smoke rise from a burning candle?

18. Steam at 100°C will cause a more severe burn than water at 100°C. Why?

19. It takes longer to heat a cold room full of furniture than it does to heat an identical empty room. Explain. If the furnace is turned off when both rooms are warm, which one will cool off more quickly?

20. If a house is built with many large windows facing south, the sunlight passing through the glass will heat the interior, even on cold winter days. Imagine that two houses are identical except that one is constructed of wood and the other of much more massive stone and concrete. Both have many south-facing windows. If the furnace is turned off, (a) which one will be warmer during the sunlight hours and (b) which will be warmer during the evening? Explain.

21. Lead has a melting point of 327°C and a boiling point of 1750°C. Find the amount of heat in cal and J required to convert 1.00 kg of solid lead at 327°C to a vapor at 1750°C.

22. In winter, pioneers often stored barrels of water in their food storage cellars. Why?

23. In general, a city is hotter than the countryside. Would you expect breezes to blow toward or away from the city?

24. State whether your body would gain or lose heat from each of the following changes of phase. (a) An ice cube melts in your mouth. (b) Molten candle wax falls on your finger and solidifies.

25. Which of the following requires more energy: (a) melting 1 g of ice (at its freezing point) or boiling 1 g of water (at its boiling point); (b) vaporizing 1 kg of water or vaporizing 1 kg of alcohol; (c) vaporizing 1 g of water or vaporizing

5 g of mercury (both are at their boiling point)? ($L_v = 2.72 \times 10^5$ J/kg.)

26. How much heat in joules is needed to (a) melt 20.0 g of solid mercury; (b) vaporize 10.0 g of liquid mercury? ($L_f = 0.12 \times 10^5$ J/kg, $L_v = 2.72 \times 10^5$ J/kg.)

Heat transfer and hindering heat transfer

27. A piece of paper is wrapped around a rod made of wood, while another piece of paper is wrapped around a rod made of copper. When held over a flame, the paper wrapped around the wood burns first. Why?

28. An electric fan actually slightly warms the air that it circulates. Explain why this is the case.

29. Florida orange growers often protect their crops during a cold snap by spraying the trees with water. How does this help?

30. Stainless steel pots and pans often have a bottom made out of copper. Why do you suppose the manufacturers do this?

31. A piece of metal feels colder than a piece of wood when they are both at the same temperature. Why?

32. Often you see highway signs that warn that bridges freeze faster than roadways. Why does this occur?

33. You need to boil some water quickly and you have a choice of two vessels in which to perform the task. One is a pan with a thin bottom, and the second is of the same material but has a thick bottom. Which would you choose and why?

34. Windowpanes in some houses actually consist of two pieces of glass with an air space separating them. Why are these much more efficient at holding heat in the house than single pane glass?

35. Heat radiation will not pass through glass. Why then should you not have more wall area of your house covered with glass to keep the inside warm?

36. A potato can be baked more quickly by inserting a nail through it. Why?

37. Builders insulate ceilings well but the floors not so well. Why?

38. You will find that outside temperatures often drop to their lowest on nights when there is no cloud cover. Which of the three mechanisms of heat transfer do the clouds protect against?

39. Of the three possible mechanisms for heat transfer, which is the most important for maintaining the temperature balance of the Earth?

40. When your body gets cold, the blood vessels in your skin contract. Why?

41. Why are dark-colored clothes more suitable for winter wear than white clothes?

ANSWERS TO SELECTED NUMERICAL QUESTIONS

7. $-23.3°C$, 249.7 K; $37.8°C$, 310.8 K
8. $= 100.4°F$, yes
10. (a) 40,000 cal, (b) 40,000 cal
12. 1.67×10^5 J
13. (a) 1.55×10^4 J, (b) 1.79×10^4 J
14. 173 J
15. $0.935°C$
21. 257,000 cal, 1.078×10^6 J
26. (a) 2.40×10^2 J, (b) 2.72×10^3 J

CHAPTER 6

THERMODYNAMICS

Winds that sweep across the surface of the Earth are part of a great heat engine driven by the Sun. We can capture the energy of these winds to produce electrical power, to irrigate fields, and to do other useful tasks.
(© Glen Allison, Tony Stone Worldwide)

Thermodynamics is a part of physics that falls loosely within the category of heat-related concepts. Basically, ***thermodynamics*** *deals with those processes related to the use of heat in practical applications.* Let us briefly look at some of the details that will concern us in this chapter.

Consider any mechanical device that is made of metal or other materials. After years of use or neglect, the device may wear out, become obsolete, break down, or rust. Even though the device may have no inherent value in its present form, however, the atoms of the original metal are still there. If one has need of these raw materials, the bent and broken machinery can be remelted, and the materials that are reclaimed can be recycled. Likewise, the rust can be chemically reconverted to iron, which can then be recast to make some new device. Therefore, at least in principle, the world's supply of metals will never be depleted. They can be used again and again forever. But does this also apply to energy? Does the Earth have a stockpile of energy that will always be available, or is it possible that the world is slowly running down in regard to usable energy?

Coal is a storehouse of energy that can be released to do work for us. But what becomes of the energy when used? If we can find used metals and reuse them, can we also find used energy and recycle it in some way so that it can be used repeatedly? A second question that we must consider is: Is it possible that we could somehow add some energy to our stockpile "for nothing"? After all, energy is not a material substance like a metal. We have defined energy as the ability to do work, and as such, energy is not something that you can touch and feel or store in a warehouse in the same sense that you can store several bars of metal. Is it possible that we could somehow do a conversion process on energy that would actually give us more in a usable form than we had at the beginning? These two questions plagued early scientists for a long time, and the search for answers led to the study of heat-motion, or thermodynamics.

6.1 WORK AND INTERNAL ENERGY

To consider the first law of thermodynamics, we must re-examine a topic that we have already covered—work. Additionally, we must investigate a new concept—internal energy.

Work

Often the materials that we must consider when discussing applications of thermodynamics are fluids, such as a gas or a liquid. As a result, we must briefly consider exactly what is meant when we say that we do work on a fluid, or what is meant when we say that a fluid does work for us. To aid us in our understanding, consider a cylinder filled with gas and fitted with a movable piston, as shown in Figure 6.1. (This is like a bicycle pump.) When the system is in equilibrium, the gas holds the piston at some stationary height above the bottom of the container. However, we can change this height in a variety of ways. For example, suppose we play a flame over the container such that the enclosed gas is heated. As the gas warms, it expands, and this causes the piston to rise in the container. The expanding gas thus exerts a force on the piston, and this force causes the piston to move through some distance. Thus, the gas does work on the piston. In turn, a shaft leading from the piston could be used to lift a heavy weight for us. As a result, work can be done by the gas *for us*.

> *When a system is doing work for us, we say that the work done is positive.*

We could do work on the gas enclosed in the container by exerting a force on the piston and compressing the gas.

> *When we do work on a system, we say that the work is negative.*

Note that the volume of the gas must change when it does work or when work is done on it. When the system does work, the gas expands; when work is done on the system, it is compressed. When the volume remains constant, no work is done on or by the system.

FIGURE 6.1 *A gas in a cylinder fitted with a movable piston.*

Internal Energy

If we could see the atoms of a gas, we would find them moving rapidly in random directions, colliding with the walls of the container and with one another. Thus, because of their motion, the atoms of the gas have kinetic energy. *The **internal energy** of a substance is the total energy (both kinetic and potential) of all the atoms or molecules that make up the substance.*

The simplest kind of gas is one composed of the so-called noble gases, such as argon, helium, and neon. As we will see in our study of chemistry, these gases are composed of atoms that are not bound to other atoms. As a result, the only kind of energy that they can have is the kinetic energy of their random motion. In a gas made of more complicated molecules, however, the rotational and vibrational motions of the molecules also contribute to the internal energy of the gas (Fig. 6.2). The atoms of the gas are connected to other atoms by

FIGURE 6.2 *The rotational and vibrational motions of the molecules add to the internal energy of some gases.*

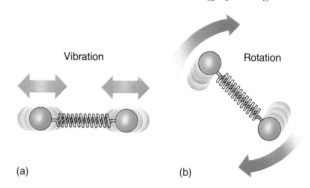

Vibration

Rotation

(a)　　　　　　　　　　(b)

forces that act as though the atoms were connected by small "springs" attached between the particles. The energy associated with the stretching or compressing of these small springs is a potential energy, and this potential energy contributes to the total internal energy of the gas (Fig. 6.2a).

Also, for these more complicated molecules, the group can rotate as shown in Figure 6.2b. The kinetic energy associated with this rotation is another part of the total internal energy of the gas.

6.2 THE FIRST LAW OF THERMODYNAMICS

The **first law of thermodynamics** is an extension of the law of conservation of energy to include possible changes in the internal energy of a system. The law is valid in all circumstances, and it can be applied in a wide variety of situations. In equation form, the first law is stated as

$$\Delta U = Q - W \qquad (6.1)$$

In this equation, Q is the heat added to or subtracted from a system. We will use the convention that Q is *positive* if heat is *added* to a system and *negative* if the system *loses* heat. W is the work done on or by the system. Work is *positive* if the system *does work* and *negative* if work is *done on* the system. The quantity ΔU is the change in internal energy of the system.

Thus, the first law of thermodynamics relates heat, work, and internal energy. The relationship expressed by Eq. 6.1 has been verified experimentally in a variety of situations.

CONCEPTUAL EXAMPLE 6.1   A CONSTANT PRESSURE PROCESS

A gas is enclosed in a container, like a bicycle pump, fitted with a piston. The piston is allowed to move freely such that the pressure of the gas always stays at the same value. (A constant pressure process is often called an isobaric process.) Heat is slowly added to the gas; as a result, the piston is pushed upward. Discuss this process from the point of view of the first law of thermodynamics.

Solution We are adding heat to the gas, so Q is a positive quantity. Likewise, the gas is doing work on its surroundings because it is pushing the piston upward. As a result, the sign of the work is positive.

The sign of ΔU is found as

$$\Delta U = Q - W = (+) - (+)$$

Thus, the sign of ΔU could be either positive or negative depending on which is larger, Q or W. In most processes of this nature, Q is greater than W. Thus, the sign of ΔU would be positive, indicating that the internal energy of the gas has increased.

CONCEPTUAL EXAMPLE 6.2 A CONSTANT VOLUME PROCESS

Heat is added to the container of gas in Example 6.1, but the piston is clamped in position so that it cannot move. Under such conditions, the pressure of the gas will not remain constant. Discuss the change in internal energy of the system for this process.

Solution Because the piston is clamped, it cannot move, and as a result, no work can be done either on or by the gas. (A force is exerted on the piston, but there is no movement.) Because heat is added to the container, the sign of Q is positive, and the first law becomes:

$$\Delta U = Q - W = (+) - (0)$$

or

$$\Delta U = Q$$

Thus, the sign of ΔU is positive, the same as that of Q, indicating that all the heat added to the system goes into increasing the internal energy of the gas.

6.3 HEAT ENGINES

One of the oldest applications of thermodynamics, and one that is still important today, is the study of heat engines. A heat engine is any device that converts heat energy to other useful forms of energy, usually mechanical or electrical energy. A heat engine is found to follow a three-step process during one of its cycles: (1) Heat is absorbed from some source at a high temperature, (2) work is done by the engine, and (3) heat is expelled from the engine at a lower temperature. For example, the internal combustion engine in an automobile extracts heat from fuel burning inside the engine and converts a fraction of this energy to mechanical en-

FOCUS ON
A FLYWHEEL-OPERATED SUBWAY TRAIN

The first law says that energy can never be found for free. It is possible, however, to build machines that operate efficiently. Many modern devices are highly inefficient and could do the same job using less fuel if clever engineering practices were employed. For example, a subway train must accelerate away from a station and brake to a halt within a few minutes and then repeat the process many times during the operating day. Acceleration requires large amounts of energy. The brakes of a subway train operate by friction; they slow the train by rubbing a brake pad against a wheel. When the brakes are applied, the energy of the motion of the train is converted to heat generated in the braking system. Thus, most of the energy used to accelerate the train is dissipated as heat. In the summertime, the braking mechanism raises the

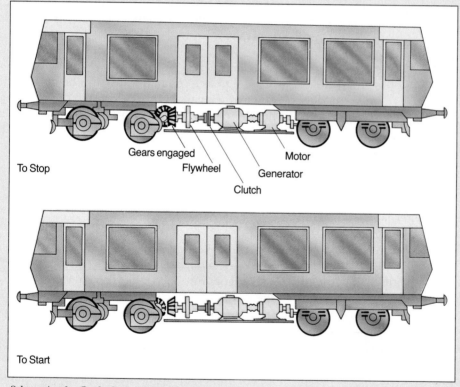

Schematic of a flywheel-operated subway train.

ergy. Some heat is then expelled from the system through the exhaust to the atmosphere. In the next section, we shall examine the gasoline engine more completely.

6.4 THE GASOLINE ENGINE

Our civilization has reached a point where engines are in constant use around us. Hardly anyone walks

temperature of the cars and creates a need for increased air conditioning. Because the air conditioners also use energy, the whole system is highly inefficient.

An alternative subway system has been proposed, as shown in the figure. This design employs a large, heavy flywheel mounted under each car. This flywheel is designed so that it can be connected to the wheels of the train or to an electric generator, also mounted under the chassis of each car. Instead of applying conventional friction brakes, the train engineer would pull a lever connecting the wheels of each car to a flywheel. A great deal of energy would be required to set all the flywheels spinning. This energy would come from the kinetic energy of the train. In other words, instead of dissipating the kinetic energy of the train as heat, the engineer could convert the kinetic energy of the train to rotational energy of the flywheels and thereby stop the train. While the subway train rests at the station to pick up and discharge passengers, the flywheels would be spinning rapidly. When power is needed to accelerate the train, the shafts of the flywheels could be connected to their electric generators, electricity could be produced, this electricity would then be used to power the train, and the train would speed up while the flywheels would slow down again. The train could produce some of its own electric power and in doing so would diminish the need for air conditioning as well.

Naturally, there would be some heat losses in such a system, and outside power would be needed but much less than is needed at present.

anymore. Most people usually move about with the aid of a motor, or engine, in an automobile, bus, or train. Similarly, most homes rumble quietly with the vibrations of furnace motors and blowers, re-

frigerators, freezers, and perhaps washing machines, dryers, and numerous other household devices. The design and development of the engines that have become so much a part of our lives provided much of the impetus for the study of thermodynamics. For this reason, let us examine a conventional type of engine.

The most common type of engine used in our automobiles is the four-stroke internal combustion engine. The principal parts of the four-stroke engine are shown in Figure 6.3. The piston, P, is a piece of metal that can slide freely up and down inside a cylindrical chamber. V_1 and V_2 are valves that allow fuel to enter the cylinder and burned gases to leave. S is the spark plug, and C is the crankshaft. An up-and-down motion of the piston causes the crankshaft to rotate, and this rotation is transmitted by gears to the wheels of the car. The steps in a single cycle of the engine are best understood by examining what happens during each of the four strokes of the engine.

1. **The intake stroke:** The first stroke of the cycle is called the intake stroke. As Figure 6.3 shows, the piston moves down and, as it does so, valve V_1 opens. The gradually increasing space inside the cylinder causes the pressure there to drop. This drop in pressure draws a mixture of gasoline vapor and air through the open valve and into the cylinder.

2. **The compression stroke:** At the end of the intake stroke, the piston begins to move up in the cylinder. This starts the compression stroke, as shown in Figure 6.3. In this stroke, both valves are closed, and the mixture of air and gasoline vapor is compressed as the piston rises.

3. **The power stroke:** The third stroke begins when the piston reaches the end of the compression stroke and the gas mixture is fully compressed. At this instant, the spark plug fires. As its name indicates, this device causes an electric spark inside the chamber, which ignites the fuel. The small explosion produced inside the chamber has sufficient force to push the piston downward, as shown in Figure 6.3. This stroke, the power stroke, generates the driving force that causes the crankshaft to rotate. This is the only one of

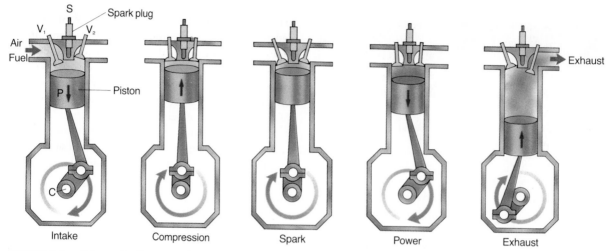

FIGURE 6.3   *Schematic diagram of a gasoline internal combustion four-stroke cycle.*

the four strokes in which energy is transferred from fuel to crankshaft. The linear motion of the pistons is eventually converted to rotational motion of the wheels.

4. **The exhaust stroke:** At the end of the power stroke, when the piston is at the bottom of its travel, valve V_2 opens. The upward motion of the piston forces the burned fuel mixture out through the exhaust system, as shown in Figure 6.3. At the end of this exhaust stroke, the cycle is repeated.

Since the power stroke took only one-fourth of the total time of the cycle, the crankshaft would have to coast through the other three strokes if only one piston and one cylinder were in the automobile. As a result, the engine would not run smoothly. In practice, most cars have four to eight cylinders and pistons to smooth out the drive of the engine and to produce significantly more power.

6.5 THE SECOND LAW OF THERMODYNAMICS

To formulate our statement of the **second law of thermodynamics,** we must examine a heat engine in terms of the first law. To do so, it will be convenient to represent the engine schematically, as shown in Figure 6.4. The engine, represented by the circle, absorbs a quantity of heat Q_h from the heat reservoir at a temperature T_h. (For the gasoline engine, Q_h is supplied at a high temperature

when the fuel is burned.) The engine then does work W. (This work is done during the power stroke of the gasoline engine.) Finally, the engine gives up heat Q_c at a temperature T_c. (For the gasoline engine, Q_c is exhausted to the atmosphere at the temperature of the atmosphere, T_c.)

FIGURE 6.4   *A representation of a heat engine. The engine (the circle) absorbs heat Q_h from a source of energy and rejects heat Q_c to the environment. In the process, an amount of work W is done.*

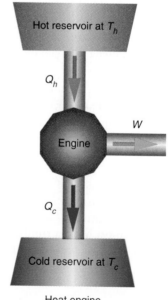

Heat engine

You should take particular note of the fact that the internal energy change of the gas is zero for the complete cycle. This occurs because we started with a compressed gas ready to absorb heat during the burning process, and at the end of the four strokes, we have been through a cycle in which we have a compressed gas again ready to absorb heat. Thus, following a complete cycle, the gas is in the same state as it was when it entered the cycle. As a result, the change in internal energy, ΔU, is zero. Now let us apply the first law to the complete cycle. Work W is done by the gas; hence, W is a positive quantity. The net heat transferred is $Q_h - Q_c$. We have used a positive sign for Q_h because this is heat absorbed by the gas, and a negative sign is used for Q_c because this is heat lost from the gas. Thus, we have

$$\Delta U = Q - W$$
$$0 = Q_h - Q_c - (+W)$$

or

$$W = Q_h - Q_c \qquad \textbf{(6.2)}$$

There are a variety of ways of stating the second law of thermodynamics, but one of the more common is based on Eq. 6.2. This form of the second law is useful in understanding the operation of heat engines.

It is impossible to construct a heat engine that absorbs an amount of heat from a reservoir and performs an equivalent amount of work.

In light of Eq. 6.2, the second law says that it is impossible for W to equal Q_h. Another way of looking at this is to note that the second law says that some heat must always be exhausted to the environment. An instructive assessment of our circumstances is that the first law says we cannot get more work out of a process than the amount of energy we put in. The second law goes further and says that we cannot break even.

Let us examine the statement of the second law a little more carefully. Suppose a mass of coal is burned in a steam locomotive. The heat is converted to useful work, and the engine travels from Paris to Amsterdam. When the engine arrives in Amsterdam, the coal is gone. What happened to the energy? Could you somehow find it, save it, and use it to drive the train back to Paris? The answer is no. The heat from the coal was spread out into the environment. The air between Paris and Am-sterdam was warmed slightly, but the locomotive cannot extract enough energy from the warm air to drive back to Paris. Thus, the energy from the coal cannot be recycled. This observation is a general one and explains why energy, once used, cannot be reused to perform work efficiently. In brief, materials can be recycled but energy cannot. The energy that was dissipated into the environment has not been destroyed—it still exists somewhere, but it is lost in the sense that it is no longer available to do work. Ingenious scientists have tried to invent heat engines that can convert all of the energy of a fuel into work, but they have always failed. It was found, instead, that a heat engine could be made to work only by the sets of processes indicated by our statement of the second law as given earlier. (1) Heat must be absorbed by the working parts from some hot source. The hot source is generally provided when some substance such as water or air is heated by the energy obtained from a fuel, such as wood, coal, oil, or uranium. (2) Waste heat must be ejected to an external reservoir at a lower temperature.

6.6 ALTERNATIVE FORMS OF THE SECOND LAW

The formulation of the second law, like that of the first, arose out of a long series of observations. If a hot iron bar is placed on a cold one, the hot bar always cools, while the cold one becomes warmer, until both pieces of metal are at the same temperature. No one has ever observed any other behavior. Similarly, if a small quantity of black ink is dropped into a glass of water, the ink disperses until the solution becomes uniformly black, as shown in Figure 6.5. A reverse process, such as one in which a jar of uniformly black ink spontaneously becomes black on one side of the jar and colorless on the other side, is never observed.

Thus, there appears to be a natural drive toward sameness, or random disorder. If there are two blocks of iron at different temperatures in a system, or a drop of ink in a glass of water, there is a differentiation of physical properties. Such differentiation results from some kind of orderly arrangement among the individual parts of the system. This is a subtle but important point. In your experience, how is order different from disorder? The answer is that order is characterized by repeated separations. Your room is orderly if all the books are sep-

(a)   (b)

(c)   (d)

FIGURE 6.5 *The second law of thermodynamics states: Any undisturbed system will naturally tend toward maximum disorder. If a drop of ink is placed in a glass of water, the ink will always disperse until it is evenly distributed.*

arated from your socks—books on the shelves, socks in the drawer. It is disorderly if books and socks are all mixed up in both places. Similarly, if a small spot of black ink is separated from colorless water, the system is orderly. Which system is more natural (that is, more probable)? That's easy—if you neglect your room, does it naturally become more orderly or more disorderly? Disorderly, of course! The reason is that there are always more ways to be disorderly than to be orderly (or there are more ways to break rules than to follow them). Therefore, any system, if left alone, tends toward disorder.

Entropy is a thermodynamic measure of disorder. It is possible to state the second law of thermodynamics in terms of entropy as follows: It has been observed that *the entropy of an isolated system always increases during any spontaneous process;* that is, the degree of disorder always increases. Thus, if you drop a spot of ink in water, the ink spreads out evenly throughout the liquid. It becomes disorderly.

If you don't clean your room regularly, it becomes messy. Similarly, a boat engine designed to extract heat from warm lake water as a source of energy and to release cool water as exhaust would cause a separation of hot and cold. This is a more orderly arrangement. Any separation leads to a decrease of entropy of the system, and this is impossible to do without adding energy from some outside source.

Let us return now from the impossible to heat engines that use fuel, where the situation continues to be discouraging. To understand this concept, let us examine a simple heat engine. Imagine that you have a box, and one side of the box is filled with hot air and the other side is filled with cold air. If left undisturbed, this system spontaneously becomes more disorderly; that is, heat is exchanged until the temperature is uniform throughout the box. But now imagine that energy is continuously added to the system. Suppose a flame is placed under one side of the box, as shown in Figure 6.6. The flame heats the air above it. In turn, the hot air rises and cold air blows along the bottom of the box, creating a convection current, as shown in the diagram. This moving air has kinetic energy and has the capacity to perform work. If a well-balanced toy windmill is placed in the box, it rotates. Thus, the heat of the flame is converted to work.

FIGURE 6.6 *A simple heat engine. The flame sustains a convection current, and the moving air turns the blades of the toy windmill. The air will stop moving when all of it is at the same temperature.*

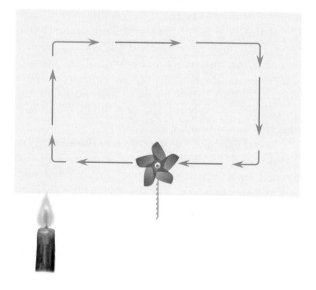

FOCUS ON PLANETARY WINDS

Planetary wind systems are another type of heat engine. For example, the Sun warms the Earth at the equator more than at higher latitudes. Thus, the air rises vertically upward over the equator, as is shown schematically in the figure. This rising air is replaced by cooler air moving along the surface. The air that moves across the surface is simply the wind and can indeed push ships across the ocean or cause windmills to rotate. Obviously a convection current is greatest if the difference between the hot part of the system and the cold portion is large. Thus, the energy created by the

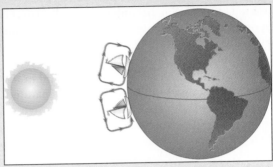

The global wind systems are heat engines powered by the Sun.

wind is related to the temperature difference of the two parcels of air.

The amount of work performed on the windmill is related to the velocity of the moving air. In turn, the air moves most rapidly if the temperature difference between the hot air and the cold air is greatest. Think of it this way: If you have hot air and cold air, the difference in density between the two types of air is great, and the hot air rises rapidly. In turn, this rapidly rising air initiates a rapid convection current, and the windmill is forced to rotate rapidly. Conversely, if the hot air is only slightly warmer than the cold air, it rises slowly, and the convection current moves slowly. The weak convection current transmits little force, and the work performed is small. This reasoning leads to another alternative statement of the second law as: *The quantity of work performed by a heat engine depends on the temperature difference between the hot reservoir and the cold reservoir.* Heat energy can be converted efficiently into work only if large differences in temperature are available.

CONCEPTUAL EXAMPLE 6.3   HOT OUTSIDE, COOL INSIDE

A refrigerator causes a separation of hot and cold; cold air is maintained inside the unit, while the kitchen remains warm. Is this separation a violation of the second law?

Solution No, the second law is not violated. The second law states that the entropy of an isolated or undisturbed system always increases during any spontaneous process. A refrigerator operating in a room is neither undisturbed nor spontaneous. An outside source of electrical energy is added to this system to cause the separation of hot from cold.

CONCEPTUAL EXAMPLE 6.4   TELLING GRAPES FROM ORANGES

Imagine a closed system consisting of a mixture of wet oranges and grapes suspended by a shelf made of paper. Below the shelf there is a second shelf made of metal screen. The mesh size of the screen is large enough to permit grapes to pass through, but small enough to hold the oranges. The box is sealed. After a while the paper shelf becomes soggy, loses its strength, and breaks. The oranges and grapes fall and are separated into two ordered collections, grapes on the bottom and oranges on the metal screen. Has the total order of the system increased? Has the entropy decreased spontaneously? Has the second law been violated? Explain.

Solution The second law has not been violated. In a sense, the separation of oranges and grapes in this system is similar to the separation of hot from cold in a refrigerator. In a refrigerator, the potential energy in some fuel is used to generate electricity, and the electricity creates a separation of hot from cold. When the total system—fuel plus refrigerator—is examined, the system becomes more disordered than ordered, and the entropy increases. The same is true with this system. At the start, the fruit has potential energy of position. When the shelf breaks, the gravitational potential energy is lost, and in the process, the oranges and grapes are separated. Thus, energy is exchanged for order, just as in a refrigerator. The system is not an isolated or undisturbed one because gravity is doing work on it; therefore, the second law is not violated.

6.7 CONSEQUENCES OF THE SECOND LAW

There are three important practical consequences of the second law. The first is that because the quantity of work performed by any heat engine depends on the temperature difference between the hot working parts and the cooler surroundings, engineers must design an engine with not only a hot working substance, but also an adequate cooling system. The second consequence is that whatever fuel is used to operate a heat engine, some waste heat is discharged into the environment, and it is therefore impossible to convert all of the potential energy of the fuel to useful work. Finally, as stated earlier, a fuel can be used only once; after it has been burned and work performed, the energy cannot be completely recycled.

6.8 THE SECOND LAW AND THE GENERATION OF ELECTRIC POWER

Most of the electricity generated in the world today is produced in steam-driven power plants. Here water is boiled using the heat available from coal, oil, gas, or nuclear fuel. The hot steam is allowed to expand against the blades of a turbine. A **turbine** is a device that spins when air or water is forced against it. You can think of it as a kind of enclosed windmill. The hot, expanding steam forces the turbine to spin, and the spinning turbine then operates a generator, which produces electricity. After the steam has passed through the turbine, it is cooled, liquefied, and returned to the boiler to be re-used,

as shown in Figure 6.7. Normally the steam is cooled with river, lake, or ocean water. The cooling action of the condenser is essential to the whole generating process.

To understand this important concept, let us return to the discussion of the second law of thermodynamics. Recall that a windmill in a box of air moves when the air on one side of the box has been heated. Also, remember that the greatest amount of work is performed if the temperature difference between the hot air and the cold air is maximized. A steam turbine operates on the same general principle. Water is boiled, the steam is heated by some fuel, and the hot steam expands against the turbine blades. If the exhaust gases are cooled, the temperature difference between the hot gas and the exhaust is increased, and more work can be performed. Therefore, some provision must be made to cool the steam at one end of the engine.

In practice, cooling is accomplished by circulating water around the condenser. It is obvious that maximum efficiency is reached with hot steam and a cold condenser. In practice, the nature of the metals in the turbine limits the temperature to a maximum of about 540°C. The low temperature is limited by the most readily available coolant, which is

The solar panels on this satellite provide the electrical energy to drive its motors and transmitters. Solar power may in the future provide a substantial fraction of U.S. energy needs. (Courtesy NASA)

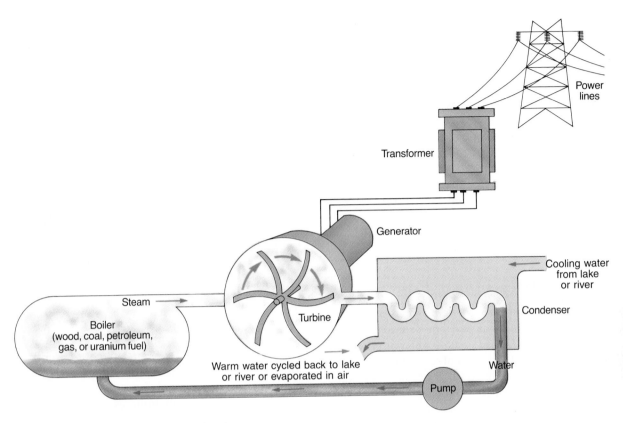

FIGURE 6.7 *Schematic view of an electric generator.*

generally water. Within the constraints of these two limits, an efficiency of 60 percent is theoretically possible, but uncontrollable variations in steam temperature and miscellaneous heat losses reduce efficiency to about 40 percent, even in the best installations. Considering the power plant as a whole, this level means that for every 100 units of potential energy derived from the fuel, 40 units of electrical energy are available as useful work and 60 units of energy are dissipated to the surroundings as heat.

Although a modern electric power plant is only 40 percent efficient, it is more efficient than other common heat engines. A large diesel is 38 percent efficient, the gasoline engine in an automobile is 25 percent efficient, and a steam locomotive is only 10 percent efficient. Therefore, use of electricity is an efficient way to perform work. Less fuel is needed to operate an electric lawn mower or car than a gas-powered machine. But the generation of electricity to be used for heating is fundamentally inefficient despite the fact that advertisements encourage homeowners to "live better with clean, efficient electric heat." True, an electric heater is 100 percent efficient *in the home,* but that is only part of the system. At the power plant, heat must first be converted to work, and 60 percent of the available energy is lost during the work-to-heat conversion. (An additional 7 percent of the original energy is lost during transmission, so that a total of 67 percent of the fuel energy is lost before the electricity is delivered to your home.) A direct fuel heater, such as a gas stove, delivers the heat directly where it is needed, and there is little waste.

CONCEPTUAL EXAMPLE 6.5 EVALUATE THIS ENGINE

In the tropics, the surface of the ocean is some 20°C warmer than the deep layers. Imagine an electric generator that operates on this temperature difference. The engine of such a device would use the warmer surface water to heat a gas, causing it to expand against the blades of a turbine. The exhaust gases would then be cooled with the subsurface water. Would such an engine violate the first law? The second law?

Solution It violates neither. Energy is derived from warm water and not "created" magically, so the first law is obeyed. The system operates on the temperature difference between a hot working substance and a cooler exhaust, so the second law is also obeyed. However, such a device would not be efficient because the temperature difference is small. In act, an experimental power plant of this kind has been built and tested near Hawaii. Although electric power has been produced, the cost of building and operating such a plant is so high that it is not economical to run, even though the energy is free.

6.9 THERMAL POLLUTION

Large amounts of heat are released from modern electric generating stations. For example, a 1000 megawatt facility, running at 40 percent efficiency, would heat 60 million liters of water by 8.5°C every hour. In many systems, lake, river, or ocean water is used as a coolant. It is not surprising that such large amounts of heat, added to aquatic systems, cause ecological disruptions. The term **thermal pollution** has been used to describe these heat effects.

What happens when the outflow from a large generating station raises the water temperature of a river or lake? Fish are cold-blooded animals, which means that their body temperature increases or decreases with the temperature of the water. In natural systems, marine life is adapted to the prevailing water temperatures and their seasonal changes. When the temperature is raised above its normal level, all the body processes of a fish (its metabolism) speed up. As a result, the animal needs more oxygen, just as you need to breathe harder when you speed up your metabolism by running. But hot water holds less dissolved oxygen than cold water. Therefore, cold-water fish may suffocate in warm water. In addition, warm water can kill fish by disrupting their nervous systems. In general, not only fish, but also entire aquatic ecosystems are rather sensitively affected by temperature changes. For example, many animals lay their eggs in the spring when the water naturally becomes warm. If a power plant heats the water so that some organism starts reproducing at the wrong time and the eggs are hatched in midwinter, the young may not find the food needed to survive.

The second law of thermodynamics assures us that it is impossible to invent a process to avoid the production of waste heat in steam-fired turbines. It is possible, however, to dispose of the waste heat with minimal disruption to the environment or, better yet, to put it to good use. Some techniques will now be described.

How to Use the Atmosphere as a Heat Sink

One approach to the problem of thermal insult to our waterways is to dispose of heat into the air. Air has much less capacity per unit volume for absorbing heat than does water, so the direct action of air as the cooling medium in the condenser is not economically feasible. For this reason, power plants must still be located near a source of water, the only other available coolant. The water can be made to lose some of its heat to the atmosphere, however, and then can be recycled into the condenser. Various devices are available that can effect such a transfer.

The two cheapest techniques are based on the fact that evaporation of water is a cooling process. Many power plants simply maintain their own shallow lakes, called **cooling ponds.** Hot water is pumped into the pond, where evaporation as well as direct contact with the air cools it, and the cooled

FIGURE 6.8 *A cooling tower at a nuclear reactor site in Oregon.* (Courtesy U.S. Department of Energy)

FOCUS ON
THE DRINKING BIRD

When you think of an engine, you probably think in terms of the kind used in your car, as discussed in the text. However, any device that is capable of transforming heat into work meets the criterion of being an engine. One of the most common of these is the drinking bird toy that you have undoubtedly seen. The figure shows how one of these toys works. The inside of the bird is constructed with two cavities, A and B, separated by a fluid, usually ether. To start the bird into motion, a piece of fabric that covers the beak of the bird is wet with water, as shown in (a). This water begins to evaporate from the beak, and since evaporation is a cooling process, the ether vapor inside the head

of the bird also begins to cool down. As this gas cools, the pressure inside chamber A becomes less than that in chamber B, and the liquid begins to creep up inside the tube, as shown in (b). At some height, the weight of the bird's head becomes greater than the weight of its body, and the bird tips forward, as shown in (c). In this way, its beak becomes wet once again. Also, in this position, chambers A and B are no longer separated, so the pressures in the two come to the same value, and the fluid can also drain back into the body of the bird. Thus, the bird now tips to the upright position, as shown in (d), and the cycle begins anew. This continuous motion of the bird could be used to drive a (small) piston and to do some work. Thus, the drinking bird qualifies as an engine.

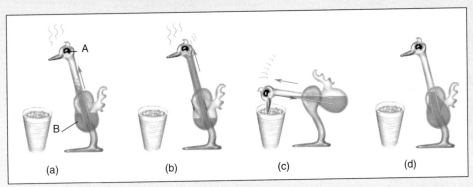

(a)   (b)   (c)   (d)

A complete cycle of a "drinking bird" engine.

water is drawn into the condenser from some point distant from the discharge pipe. Water from outside sources must be added periodically to replenish evaporative losses. Cooling ponds are practical where land is cheap, but a 1000 megawatt plant needs 1000 to 2000 acres of surface, and the land costs can be prohibitive.

A cooling tower, which can serve as a substitute for a cooling pond, is a large structure, often about 180 m in diameter at the base and 150 m high (Fig. 6.8). Hot water is pumped into the tower near the top and sprayed downward onto a mesh. Air is pulled into the tower by either large fans or con-

vection currents and flows through the water mist. Evaporative cooling occurs, and the cool water is collected at the bottom, as shown in Figure 6.9. No hot water is introduced into aquatic ecosystems, but a large cooling tower loses over 4 million liters of water per day to evaporation. Thus, fogs and mists are more common in the vicinity of these units than they are in the surrounding countryside. These fogs and mists are less ecologically disruptive than the thermal pollution of waterways, but they do affect the quality of the environment. Many new electrical generating stations are using cooling towers to dispose of waste heat.

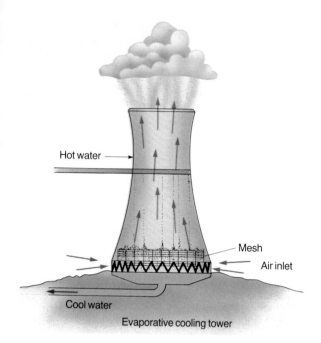

FIGURE 6.9   *Schematic view of a wet cooling tower.*

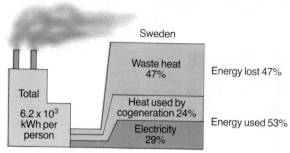

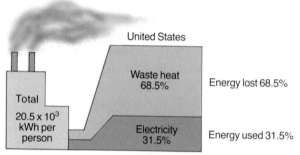

FIGURE 6.10   *Use of a fuel to produce electricity in Sweden and the United States. (kWh is an abbreviation for kilowatt-hour, a unit of energy equal to 3,600,000 J, to be discussed in Chapter 8.)*

Use of Waste Steam (Cogeneration)

No matter how well they work, the cooling systems described here are still elaborate means of throwing energy away. Some of this discarded energy, however, can still be useful. Waste steam is too cool to produce work efficiently, but it is hot enough for many other industrial processes. For example, the steam can be used to cook food in a cannery, to heat wood pulp in a paper mill, or to process petroleum in an oil refinery. In the United States, most of the waste steam for electric generation is discarded, leading to thermal pollution, but in Europe much of the waste steam is sold to other industries. Energy is used efficiently, and pollution problems are reduced (Fig. 6.10). Some companies in the United States do use their excess heat efficiently. One example is the relationship between the Baywood, New Jersey, refinery of Humble Oil

and Refining Company and the Linden, New Jersey, generating station. The Linden power plant is capable of producing electricity at 39 percent efficiency. This efficiency is lowered by less than optimum cooling of the condenser, and some of the waste heat is sold as steam to Humble. If the two-plant operation is considered as a single energy unit, the overall efficiency of power production is raised to a level of 54 percent. The process is beneficial to many; the companies save money, fuel reserves are conserved, and thermal pollution of waterways is reduced.

SUMMARY

When a system does work on its environment, the work is defined as a positive quantity; when the environment does work on a system, the work is defined as negative. Heat is considered a positive quantity when it is added to a substance and a negative quantity when it is removed. The **internal energy** of a substance is the total energy of all the atoms or molecules that compose the substance. The **first law of thermodynamics** is an extension of the law of conservation of mechanical energy to include internal energy.

An **internal combustion engine** is a four-stroke engine—the intake stroke, the compression stroke, the power stroke, and the exhaust stroke.

The **second law of thermodynamics** can be stated in a variety of ways: (1) It is impossible to construct a heat engine that absorbs an amount of heat from a reservoir and performs an equivalent amount of work. (2) The **entropy** of an isolated system always increases during any spontaneous process. (3) The quantity of work performed by a heat engine depends on the temperature difference between the hot reservoir and the cold reservoir.

EQUATIONS TO KNOW

$\Delta U = Q - W$ (the first law of thermodynamics)

KEY WORDS

Work
Internal energy

The first law of
    thermodynamics
Engine

The second law of
    thermodynamics
Entropy

Thermal pollution
Cogeneration

PROBLEMS AND CONCEPTUAL QUESTIONS

Problems requiring numerical work are identified with a blue number.

Work and internal energy

1. A gas is contained in a cylinder fitted with a piston. The gas is then taken through the following processes. In each case, state whether work is done on or by the gas and whether heat is added to or removed from the gas. What is the sign of the work and the heat transferred in each case? (a) The piston is clamped firmly in position, and a flame is played across the base of the container. (b) The cylinder is covered with a material that will not conduct heat, and the gas is allowed to expand. (c) The gas is compressed while the nonconducting wall is still in place.

2. Discuss how the internal energy changes of the following gases might manifest themselves. That is, does an increase of the internal energy produce a change solely in the random kinetic energy of the atoms, do internal molecular rotations or vibrations play a role, and so forth? (a)

A monatomic gas (composed of individual atoms), (b) a diatomic gas (composed of two atoms bound together).

First law of thermodynamics

3. An engineer designed and built the roller coaster shown in the figure. He was fired. Why?

(Question 3)

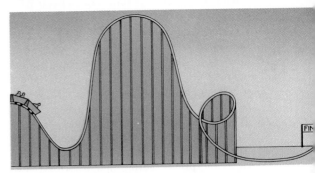

4. Two bicycles are coasting down one big hill and up another. Both bicycles are identical except that one cyclist has a small electric generator attached to the rear wheel to power a portable cassette tape player. Which cyclist will coast farther uphill? Explain.

5. Imagine an isolated system that consists of a battery and an electric motor connected through pulleys to a set of weights. When the battery is connected to the motor with properly insulated wires, the motor can do 50,000 J of work lifting the weights before the battery runs down. If shoddy wires are used and some partial short circuits develop, the system will spark and sputter, and a fully charged battery will do only 25,000 J work. Does this observation disprove the first law of thermodynamics? Explain.

6. A car designer has the idea to position large flywheels inside a conventional automobile. He reasons that if the flywheels can be connected to the wheels while the car is coasting downhill, energy can be conserved for the next uphill pull. Do you think that this is a good idea? Explain.

7. A gas is compressed so that 10,000 J of work are done on it; at the same time, 5000 J of heat are released to the environment. What is the internal energy change of the gas?

8. During an expansion process of a gas, 3000 J of heat are added to the gas, and it does 2000 J of work. What is the internal energy change of the gas?

9. In what way is the statement, "You can't get something for nothing," related to the first law of thermodynamics?

10. The first law of thermodynamics is a modification of a conservation law encountered earlier in your study in this text. What is that conservation law and how are they related?

Heat engines and the gasoline engine

11. Why is an automobile engine called an "internal combustion" engine?

12. What is a cylinder? A piston? A spark plug? What is the function of each?

13. It can be said that the human body is an engine. Discuss the comparison.

14. One of the world's most awesome engines is the hurricane. Look up hurricanes in a reference source and discuss them in terms of their relationship to an engine.

15. Explain what happens inside an automobile engine during the power, exhaust, intake, and compression strokes. During which parts of the cycle is the piston moving upward? Downward? At what point in the cycle do the valves open? When do they close? When does the spark plug

fire? How many valves are there in a four-cylinder engine?

16. What would happen if a small hole were burned in the face of one intake valve so that gases could leak through even when the valve was closed? Explain.

17. What would happen if a small hole were burned in the face of one exhaust valve so that gases could leak through even when the valve was closed? Explain.

18. The piece of metal connecting the piston to the crankshaft (which ultimately drives the wheels) is called the connecting rod. What would happen to a car if a connecting rod broke?

19. The piston of an internal combustion engine is surrounded by a ring that presses tightly against the piston and the wall of the cylinder. Explain why a car with worn piston rings loses power.

20. In the 1970s, the Wankel engine gained a measure of notoriety as a possible serious competitor to the four-stroke gasoline engine. Do some research to find out how the Wankel engine works and compare the steps in one of its cycles to those of the four-stroke engine.

Second law of thermodynamics and alternative forms of the second law

21. In what way is the statement, "You can't even break even," related to the second law of thermodynamics?

22. Write three statements of the second law of thermodynamics.

23. Does a salad have more entropy before or after it is tossed?

24. A certain refrigerator says that the heat it adds to a room is only that which it has removed from the food. Can you use the laws of thermodynamics to refute this statement?

25. It is found that some process occurs such that the entropy decreases by one unit. You can bet that there is another entropy change somewhere that has one or more of these values. Which one(s) are possible? −1 unit, 0 units, 1 unit, 2 units?

26. If a room is the same temperature as the surrounding air, and an air conditioner is turned on in the house, the room will get cooler while the outside is heated. Is this a violation of the second law of thermodynamics? Explain.

27. Suppose a refrigerator is placed in the middle of a closed room, turned on, and the refrigerator door left open. Will the temperature of the room rise, fall, or remain the same? Explain.

28. Mountain ranges slowly erode and crumble, ultimately weathering down to flat land. Is entropy increasing or decreasing during this process? Explain.

29. (a) Automotive engineers are experimenting with building engine blocks out of ceramic materials instead of metals. Since ceramics can withstand greater thermal stress without failing, ceramic engines operate at higher temperatures than ordinary metal ones do. All other factors being equal, will a ceramic engine be more, or less, efficient than a metal engine? Explain. (b) All other factors being equal, will a car give better gas mileage in the summer or in the winter? Explain.

30. One of the early pioneers in the science of thermodynamics, Julius Mayer, started thinking about work and heat while he was a physician on board a trading ship. Dr. Mayer noticed that the sailors ate less when they were in the tropics than they did when they were in colder regions, yet they performed the same amount of work regardless of location. Was this an observation of the first law or of the second law? Explain.

31. Does water have more entropy when frozen or when liquid?

32. When rolling dice, it is far more probable that one will roll a seven than a two. Defend the statement that the entropy of a system of dice is higher for seven than it is for two.

The second law and the generation of electric power

33. Consider two power plants, one that uses 500°C steam as its working substance and one that uses 150°C steam. In all other respects, both are identical. Both use cooling water at 30°C and produce the same amount of electricity. Which power plant discharges more thermal energy to the environment? Which uses less fuel? Explain.

34. Why is electric heat less efficient than a small propane or oil furnace that is installed in a house? Would it be possible to generate electric heat as efficiently as it is to produce heat by burning fuel directly in the home?

35. In a hydroelectric generating system, a turbine is driven by a stream of falling water. The water itself may be cool, and no auxiliary cooling system is required, yet these facilities are nearly 99 percent efficient. No fossil fuels are used. Do hydroelectric power plants violate the laws of thermodynamics? Review the entire cycle, answering the questions: What is the ultimate source of energy of the system? How is the second law obeyed?

36. Would it be practical to increase the efficiency of a power plant by cooling the condenser with a giant refrigerating unit? Explain in terms of the second law.

Thermal pollution

37. Define thermal pollution. How does it differ in principle from air or water pollution?

38. Explain why nuclear-fueled power plants require more cooling water than fossil-fueled plants.

39. Since marine life is abundant in warm tropical waters, why should the warming of waters in temperate zones pose any threat to the environment?

40. Describe some difficulties with the use of waste steam for home heating. Describe the potential benefits.

ANSWERS TO SELECTED NUMERICAL QUESTIONS

7. 5000 J

8. 1000 J

CHAPTER 7

WAVE MOTION: SOUND

The source of all sound is a vibrating object. For this tympani, sound waves are produced by the vibrating drumhead. What is it that vibrates to produce the sound of a clarinet, a violin, a bird's whistle, or a jackhammer? (Courtesy The Image Bank)

7.1 WAVE MOTION

When most people think of a wave, they visualize water waves traveling across the surface of a lake or crashing against the seashore. A water wave is made up of a collection of moving particles. Wave motion, however, is different from the other types of motion that we have studied so far, in that the wave itself takes on an identity separate from that of the moving particles of which it is composed. In fact, the wave pattern has its own special properties and behavior.

There are many different types of waves; three common ones are water, sound, and light waves. In one sense, these three are so different that they hardly seem to resemble each other at all. There are certain features of waves, however, that are common to all the various types that one may encounter in nature. In this chapter, we shall attempt to determine some characteristics of waves that are common to all the various types, and in doing so, we shall focus much of our attention on one particular type of wave, the sound wave.

7.2 DESCRIBING A WAVE

Longitudinal and Transverse Waves

All of the different types of waves that we will discuss in this text can be placed into one of two categories. They are either **longitudinal** waves or **transverse** waves. Let's look at the difference between these two categories.

A wave can be set up on a rope by tying one end to a wall and moving the other up and down to form a pattern moving along the rope. In Figure 7.1, a bump has been created on the rope by one upward motion of the hand. The bump moves away from the hand and down the rope toward the right. Let us now examine the motion of the rope a little more carefully as the bump passes by a particular location. We shall do so by painting a small section of the rope and focusing our attention on the movement of the painted segment. We find that the motion of the segment is up and down while the bump travels to the right. *Waves of this kind, in which the moving particles vibrate at right angles to the direction in which the wave travels, are referred to as transverse waves.* Waves traveling along the surface of a pool of water are basically transverse waves. This can be demonstrated by considering the motion of a cork placed in the water. As the wave passes by, the cork bobs up and down. (For a water wave, there is a small back-and-forth motion of the cork, but the primary motion is up and down.) Individual water drops follow a path in the water like that followed by the cork. We shall see in a later chapter that light waves are transverse waves.

A different kind of wave pattern can be set up by attaching one end of a spring to a wall and moving the other end in a back-and-forth motion along the direction of the spring, as shown in Figure 7.2. As the hand moves to the right in Figure 7.2b, the coils of the spring near the hand are compressed, and this compression begins to travel along the spring. When the hand moves to the left, as shown in Figure 7.2c, the coils near the hand move farther apart, and this expansion then begins to travel to the right. If this to-and-fro motion of the hand is continued, as in Figure 7.2d, a series of compressions followed by expansions travels along the spring toward the right, forming a wave. Note that the individual coils of the spring do not travel along the spring. It is the wave—the disturbance—that travels. Also, note that if you focus your attention on one particular coil, its motion is a to-and-fro motion that basically follows the motion of the hand. *When the individual particles that constitute the wave move to-and-fro along the same direction as the wave travels, the wave is called a longitudinal wave or a compressional wave.* We shall see in Section 7.2 that a sound wave is a longitudinal wave.

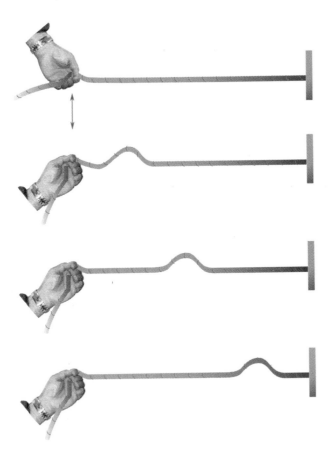

FIGURE 7.1 *A transverse wave. Any given point on the string moves up and down, but the wave travels horizontally.*

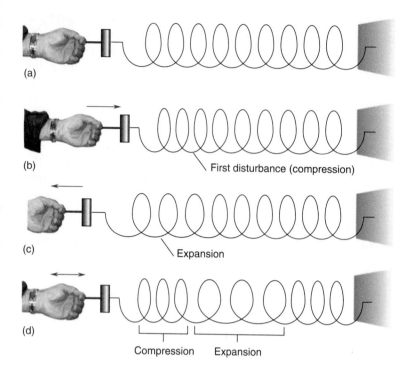

(a)

(b)

First disturbance (compression)

(c)

Expansion

(d)

Compression   Expansion

FIGURE 7.2 *(a) The spring is held at rest. (b) Moving the hand toward the right compresses the coils near the hand. (c) Moving the hand to the left creates an expansion in the coils. (d) To-and-fro motion of the hand sets up a series of compressions and expansions.*

CONCEPTUAL EXERCISE

In our discussion of earthquake waves in the earth science section of this book, we will find that there can be both longitudinal and transverse "sound" waves emanating from the site of the quake. Explain why this is possible.

Answer The force between molecules is very weak in a gas, so only longitudinal waves can be sustained in air, which is our focus of attention in this chapter. However, in a solid, where forces between neighboring molecules are much stronger, a slight push in *any* direction will be felt over a much longer distance. Thus, the waves can produce a transverse disturbance as well as one that is longitudinal.

FIGURE 7.3 *A wave can be represented graphically by a sine curve.*

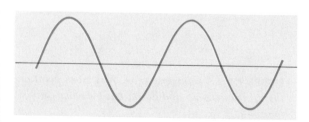

Graphical Representation of a Wave

A general pattern used to represent a wave is shown in Figure 7.3. (This type of pattern is called a sine curve.) It should be noted that this representation is similar in appearance to an actual transverse wave on a rope. This same figure can be used to represent a longitudinal wave. To see this, refer to Figure 7.4. We note that at those points where the coils of the spring are compressed in part a of the fig-

FIGURE 7.4 *Representing a longitudinal wave by a sine curve.*

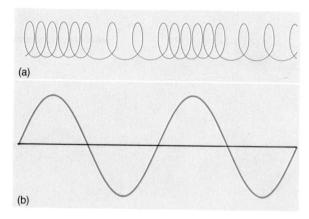

(a)

(b)

FOCUS ON
TELEPHONE FREQUENCIES

Have you ever wondered how the equipment at the telephone office knows that you have pushed, say, the 6 button on your telephone when you are making a call? The information is passed along by pure tones generated when you press the button, as shown in the figure. For example when you press the number 6, two pure tones are generated; one has a frequency of 770 Hz, the other 1477 Hz. These tones are transmitted along connecting cables to the telephone company, where they activate switches to complete your call.

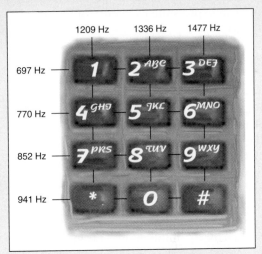

When you press a button on your touch-tone phone, two frequencies are transmitted.

ure, the sine curve of part b is at a crest. Likewise, at those points where the coils are far apart in part a, the curve in part b is at a trough. Thus, although a transverse and a longitudinal wave are distinctly different, the same kind of graph can be used to visualize both.

Speed of a Wave

Imagine yourself resting on a beach and watching the waves come rolling in. If you focus your attention on one particular crest of a wave, you will see it come moving in toward the shore with a speed of perhaps a few feet per second. Thus, a wave has a velocity in that it travels with some speed and in some given direction.

As we saw in Chapter 2, the speed of a car is relatively easy to measure. All that you have to do is measure the distance the car travels and divide by the time it takes to travel this distance. The speed of a wave can be measured by exactly the same technique. In this case, you would focus your attention on, say, a crest of a wave, watch how far it travels in a given time interval, and divide this distance by the time interval.

Frequency of a Wave

While watching a wave at the beach, you might decide to focus your attention on a fisherman's cork bobbing up and down in the water. An examination of the motion of this cork would enable you to find yet another important feature of a wave, its **frequency.** *The frequency of a wave is the number of complete vibrations that the wave makes each second.* Since the cork is following the vibrations of the water, its frequency of vibration is the same as that of the water. As another example, suppose you attach one end of a rope to a wall and vibrate the other end up and down with your hand. A wave travels down the rope, with each segment moving such that it emulates the movement of the hand. That is, if your hand makes three up-and-down cycles each second, each point of the rope will also make three up-and-down vibrations each second. In this case, the frequency of the wave is

$$f = 3 \,\frac{\text{vibrations}}{\text{s}} = 3 \,\frac{\text{cycles}}{\text{s}}$$

Note the different ways in which frequency can be expressed. Units often used are vibrations per second or cycles per second. In calculations, only the "per second" is used, as we will see in our example problems yet to come. The official name given to the unit is the **hertz** (abbreviated Hz). Thus, we can express the frequency of the wave on the rope as

$$f = 3 \text{ Hz}$$

EXAMPLE 7.1   TOCK TICK GOES THE ERRATIC CLOCK
A clock repairman is attempting to repair an erratic grandfather clock. He notes that the pendulum of the clock makes 12 complete vibrations every 4.0 s. What is the frequency of motion of the pendulum?

Solution This is not a problem dealing with waves, but every object that undergoes oscillatory motion, such as the clock pendulum, has a frequency of vibration. We can calculate this frequency by noting that the definition of frequency is the number of complete vibrations made every second. Thus,

$$f = \frac{\text{vibrations}}{\text{s}} = \frac{12 \text{ vibrations}}{4.0 \text{ s}} = 3.0 \text{ Hz}$$

Wavelength of a Wave

Figure 7.5 shows a graph of a wave on a string. Another important characteristic of a wave, called the **wavelength,** is shown in the figure by the horizontal arrows. (The symbol used to represent wavelength is the Greek letter lambda, λ.) *The wavelength of a wave is defined as the distance between two consecutive points on a wave that are behaving identically.* Thus, a wavelength could be the distance between two consecutive crests on a wave; between two consecutive troughs; or between any two identical, consecutive points.

Relationship Between Speed, Wavelength, and Frequency

We will develop a relationship between the speed of a wave, its wavelength, and its frequency by use of a thought experiment. Let us consider waves moving under a fishing dock. Suppose fishing is slow, and a fisherman starts wondering about the waves he sees rolling in toward him. He estimates that the distance between crests is about 3 m, and

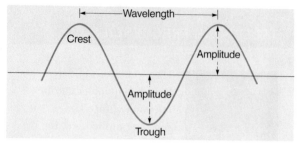

FIGURE 7.5   *Graph of a wave on a string showing the terminology used to describe waves.*

he notes that a crest hits the dock once every second. What, he wonders, is the speed of these waves? If each wave is 3 m long and one hits the dock every second, the waves must be moving at a speed of 3 m per second.

Now let us examine these observations in a slightly different light. The fisherman was actually able to determine two of the fundamental characteristics of waves just by watching the waves come in. He noted that the distance between consecutive crests was 3 m; thus, he found the wavelength ($\lambda = 3$ m). He noted that the waves were hitting the dock at a rate of one each second; thus, he found the frequency ($f = 1$ Hz). From his observations, he was able to find the speed by multiplying the wavelength times the frequency as

$$\text{speed} = (\text{wavelength})(\text{frequency})$$

or symbolically

$$v = \lambda f \tag{7.1}$$

where we use v for the speed of the wave.

EXAMPLE 7.2   SOUND WAVES
A tuning fork produces a 256 Hz note, corresponding to middle C on a piano. It is found that this fork produces a sound wave having a wavelength of 1.30 m. From this information, determine the speed of sound.

Solution Equation 7.1 is valid regardless of the type of wave under consideration. Thus, we can find the speed of sound as

$$V = \lambda f = (1.30 \text{ m})(256 \text{ Hz}) = 333 \text{ m/s}$$

As we shall see, the speed of sound depends on temperature, but this is a reasonably accurate value for the speed of sound on a winter day when the outside temperature is 0°C.

EXAMPLE 7.3   LIGHT WAVES

Light travels at a speed of 3×10^8 m/s in a vacuum and, for all practical purposes, through air. The wavelength of a particular shade of yellow light is 580 nm (1 nm $= 10^{-9}$ m). Find the frequency of this light.

Solution Equation 7.1 can be used to find the frequency as

$$f = \frac{v}{\lambda} = \frac{3 \times 10^8 \text{ m/s}}{580 \times 10^{-9} \text{ m}} = 5.17 \times 10^{14} \text{ s}^{-1}$$

$$= 5.17 \times 10^{14} \text{ Hz}$$

CONCEPTUAL EXERCISE

Is it possible to increase the frequency of a sound wave and lengthen its wavelength at the same time?

Answer The product of the wavelength and frequency of a sound wave is always a constant, and equal to the velocity of the wave, as the wave moves through a given material. That is, $v = \lambda f$. From this equation, we see that increasing λ means that f must decrease and vice versa. Thus, if you double the wavelength, you must decrease the frequency by a factor of 2.

Amplitude of a Wave

Shown in Figure 7.5 is one last term that is characteristic of a wave, the **amplitude.** Note the horizontal line in the figure, which we have called the reference line. If the wave under consideration were a wave on a string, this center line would be the straight-line shape taken by the string in the absence of a wave. If the wave were, instead, a water wave, the reference line would be the undisturbed surface of the water in the absence of the wave. Thus, from the figure we see that *the amplitude of a wave is the maximum distance that the wave has been raised (or lowered) from the undisturbed position.*

You might try a simple experiment on your own to discover the relationship between the speed of a wave and its amplitude. Run a bathtub full of wa-ter and wait until the surface becomes reasonably smooth. Then, wiggle your finger in the center of the pool to set up a small amplitude wave on the surface. Watch the wave and try to estimate its speed. Now repeat the process except this time hit the water more violently so that a large amplitude wave is set up. Again, try to estimate the speed of this wave. Admittedly, it is a little difficult to do this experiment with any accuracy, but we hope that you will see that both waves seem to have the same speed. If you were to do this more accurately, you would find that *the speed of a wave and the amplitude of the wave are not related.*

The amount of energy carried by a wave determines the amplitude of the wave. This last statement should be obvious to you from your experiences at a beach. On a day when the water is relatively calm, you are able to play in the surf and allow the small amplitude waves to break against you with no difficulty. If the wave amplitudes become large because of a storm, however, you may find that you are unable to stay on your feet when they surge into you. The difference occurs because the larger amplitude waves are carrying more energy. In fact, the energy carried by a wave is proportional to the square of the amplitude.

7.3   SOUND WAVES

So far in our discussion of waves, we have focused our attention on terms and concepts that are common to all types of waves. From now on in this chapter, we will narrow our point of view and concentrate exclusively on one important type of wave, the sound wave. We shall begin our discussion by considering how a sound wave is produced. Regardless of whether you are listening to a heavy-metal rock star's guitar or the melodious sound of a violin concerto, the sounds have the same basic source. Whether you are listening to the sweet trill of a robin or to the harsh sound of an angry shout, the sound waves originate in the same basic way. *The source of any sound is a vibrating object.* When the rock star strums his guitar, the source of sound is the vibrating string; likewise, the sound from a violin comes from a vibrating string. When the robin trills, its vibrating vocal cords are the source of the sound, just as they are for the angry shouter. You pick your own sound; whether it is pleasant or irritating, the source of the sound can always be traced to a vibrating object.

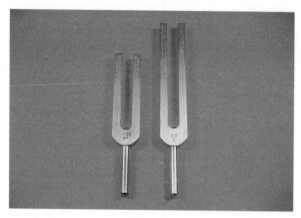

FIGURE 7.6   *Tuning forks.*

When an object vibrates, it disturbs the air near it, and this disturbance moving through the air is what is referred to as a sound wave. Let us examine exactly how a sound wave is produced by considering how a tuning fork produces a sound. A tuning fork like that shown in Figure 7.6 consists of two metal prongs that, when struck, vibrate back and forth. Consider Figure 7.7 to see how the sound wave is produced. When one prong of the tuning fork moves to the right, as in Figure 7.7a, air molecules near the fork are forced more closely together than normal. We call these regions **compressions** or **condensations.** When the prong swings to the left, as in Figure 7.7b, the air molecules to the right of the prong have room into which they can move; hence, the air molecules are not squeezed together as closely as normal. Such a region is called a **rarefaction.**

As the tuning fork continues to vibrate, additional compressions and rarefactions are produced, and these spread out through the air like ripples in a pool of water. After a short period of time, the air might look as shown in Figure 7.8, if one were able to see the molecules. We are able to represent the sound wave by our familiar curve, as shown by a comparison between parts a and b of Figure 7.8. We see that where the sound wave has a compression in part a, the curve in part b is shown to be at a crest. Likewise, at points where the sound wave has a rarefaction, the curve is seen to be at a trough. To simplify the drawings in Figures 7.7 and 7.8, we have shown the wave spreading out only from one tine and moving along a horizontal path from left to right. In an actual situation, the wave spreads out from both tines and in all directions—up, down, into the page, out of the page, and so forth.

7.4 SOUND: A LONGITUDINAL WAVE

We can determine whether sound is a longitudinal wave or a transverse wave by once again considering what a tuning fork does to the air near it. In Figure 7.9, we will focus our attention on one particular air molecule as the prong of a tuning fork moves to the right and then back to the left. As we see in Figure 7.9a, as the prong swings to the right, the air molecule is also forced to the right, and in Figure 7.9b, as the prong swings back to the left, the air molecule follows its motion and also moves

FIGURE 7.7   *(a) When the tuning fork prong moves to the right, air molecules in front of it are pushed close together. (b) A rarefaction is produced when the prong swings back to the left.*

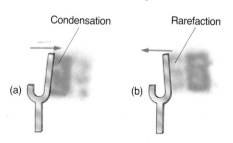

FIGURE 7.8   *(a) After several vibrations, the air in front of the tuning fork is filled with compressions (condensations) and rarefactions moving away from the fork. (b) A graphical representation of the sound wave. Note that a compression is represented by a crest in the sine curve and a rarefaction by a trough.*

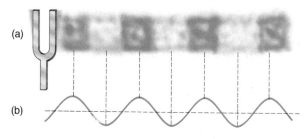

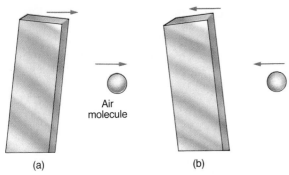

FIGURE 7.9 *(a) As the prong of the fork swings to the right, the air molecule shown moves to the right. (b) When the prong moves to the left, the air molecule also moves to the left.*

to the left. We know that the sound wave is spreading out from the tuning fork, toward the right in our figure. Thus, we see that as the sound wave moves outward toward the right, an individual air molecule vibrates to and fro from right to left. If you go back and check the definition of a longitudinal wave, you will see that such a wave is characterized by the fact that the vibrations of a material are along the same direction as the wave travels. Thus, sound is an example of a longitudinal (or compressional) wave.

7.5 THE SPEED OF SOUND

The speed of sound in air at 0°C is about 331 m/s (1090 ft/s, or 741 miles/hour). This is a fast speed, but it is small when one considers that the speed with which light travels through air is 3.00×10^8 m/s. Everyone has at one time or another observed vivid demonstrations of the enormous differences between these two speeds. For one example, imagine yourself in the centerfield bleachers at a baseball game, some 300 m from home plate. When a batter hits a baseball, the sound of bat striking ball will take about 1 second to reach you, but the speed of light is so fast that you observe the batter make contact almost instantaneously. Thus, you see the batter hit the ball and a brief moment later you hear the contact. As another example, when you are in the stands at a track meet, at a large distance from the starter, you will notice a delay between the appearance of smoke from his pistol and the sound of the shot.

The speed of sound also depends on the type of material through which the sound wave is traveling. In general, sound travels fastest through a solid, next fastest through a liquid, and slowest through a gas. Sound has to have a material to travel through; it does not move through a vacuum at all. This can be demonstrated by setting an electric bell ringing inside a vacuum chamber, as shown in Figure 7.10. When air is in the container as in Figure 7.10a, you can hear the bell ringing quite well, but note what happens when a vacuum pump is attached to the container and the air is pumped out. As the air is evacuated, the sound of the bell gradually diminishes, until when the air is almost all gone, the sound of the bell disappears completely (Fig. 7.10b). *Because light does travel through a vacuum, you can still see the bell ringing inside the container—you just can't hear it.*

The reason that sound travels differently through different types of materials depends on a property of a material called its **elasticity.** An elastic material quickly springs back to its original shape when something distorts it. In such a material, the individual atoms act as though they were attached to their neighbors by springs. This makes it easy for

FIGURE 7.10 *(a) With air in the container, we can hear the bell, but (b) when the air is removed, sound cannot travel through a vacuum, and we cease to hear the bell.*

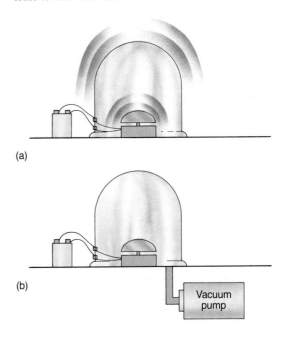

(a)

(b)

Vacuum pump

a disturbance of one molecule to be transferred to its neighbors. Thus, when one atom is caused to swing to and fro by a sound wave, this motion is quickly transmitted to nearby atoms, and the sound wave spreads rapidly through the material. Solids are more elastic than liquids, which in turn are more elastic than gases. Thus, sound travels with decreasing speed through solid, liquid, and then gas.

A final factor affecting the speed of sound in air is the temperature of the material. The reason for this is that as the temperature of a substance increases, the random motion of the individual atoms in the material increases. As a result, the atoms can collide more often; thus, a disturbance of one atom is transmitted more quickly to its neighbors. For a sound wave in air, it is found that the speed of sound increases with temperature according to the following equation

$$v = v_0 + (0.61)\,T \qquad\qquad (7.2)$$

where v_0 is the speed of sound at $0°C$, 331 m/s, and T is the temperature in degrees Celsius.

EXAMPLE 7.4   THE HAPPY WANDERER

A hiker glances at his Acme wrist thermometer and notes that the temperature is $21°C$. What is the speed of sound through air at his location?

Solution The speed can be found from Eq. 7.2. We find

$$v = v_0 + (0.61)\,T$$
$$= 331 \text{ m/s} + (0.61)(21°C) = \boxed{344 \text{ m/s}}$$

EXAMPLE 7.5   THE HAPPY YODELER

The hiker of Example 7.4 yells out and finds that an echo returns to him in a time of 2.00 s. How far away is the mountain that reflected the sound?

Solution From the definition of speed, we find that the distance traveled by the sound wave is

$$d = vt = (344 \text{ m/s})(2.00 \text{ s}) = 688 \text{ m}$$

However, the sound wave had to travel to the mountain and back. Thus, the actual distance to the mountain is $688/2 = \boxed{344 \text{ m}}$.

CONCEPTUAL EXERCISE

It is said that when you see a bolt of lightning, you can tell how far away it is by counting to 5 over and over until you hear the sound of thunder. Each time that you count to 5 adds a mile to the distance the storm is away from you. Explain why this is true. (*Hint:* The speed of sound is about 1100 ft/s.)

Answer If sound travels at 1100 ft/s, then in 5 s it will have traveled 5500 ft, about 1 mile (1 mile = 5280 ft). So, if it takes you 5 s to count to 5, each time you make the count means the sound has had time to travel another mile.

7.6 FREQUENCY AND WAVELENGTH OF SOUND

A longitudinal disturbance traveling through the air is called a sound wave only if its frequency is between 20 and 20,000 Hz because it is only within this range that the normal human ear is sensitive. Waves having a frequency greater than 20,000 Hz are called **ultrasonic** waves and can be heard by some animals, including dogs.

Ultrasonic waves have important medical applications. For example, these waves are often used to clean objects. The object to be cleaned is placed in a liquid through which an ultrasonic beam is then passed. These waves set the contaminants on the surface of the object into a rapid vibration that shakes them free. A second important medical application of ultrasonic waves is as an imaging tool to examine internal organs or to observe a fetus. In the latter application, an instrument that is both a source and a detector of ultrasonic waves is passed across the mother's abdomen. An ultrasonic wave emitted by the instrument penetrates the body of the mother and is reflected to the receiver by the fetus. The reflected waves are converted to an electrical signal that is then used to produce an image on a phosphorescent screen. Certain birth defects such as spina bifida are easily detected by this technique.

A relatively new medical application of ultrasonics is the cavitron ultrasonic surgical aspirator (**CUSA**). These devices have made it possible to surgically remove brain tumors that were previously inoperable. The CUSA is a long needle that emits very high frequency ultrasonic waves, about 23 kHz, at its tip. When the tip touches a tumor, the part of the tumor near the needle is shattered, and the residue can be sucked up (aspirated) through the hollow needle.

Another interesting application of ultrasonic techniques is the ultrasonic ranging unit designed by the Polaroid Corporation. This device is used to determine the distance from the camera to the object to be photographed. A burst of ultrasonic waves is emitted from the camera and travels toward the subject, which reflects part of it back to the camera, where it is picked up by a detector. The time interval between emission and the detection of the echo can be used to find the distance to the object to be photographed, using the same technique demonstrated in Example 7.5.

In everyday life, the terms "frequency" and "pitch" are often used interchangeably. For example, the flute is an instrument that characteristically emits sounds having a high frequency, whereas some horns emit sounds of low frequency. When speaking of these, it is common to refer to the flute as having a high pitch and to the horn as having a low pitch.

CONCEPTUAL EXAMPLE 7.6 SPEED AND FREQUENCY

Imagine yourself at a symphony performance and use your experience with what you hear to convince yourself that the speed of sound does not depend on the frequency of the sound.

Solution If you are to enjoy a symphony, the musical notes played must reach you at the precise instant they should, relative to other notes. Suppose that speed did depend on frequency, such that high frequencies travel faster than low frequencies. If that were the case (and it isn't), the notes from a high-frequency instrument would reach you in the audience more quickly than would the notes from a bass instrument. Instead of being in perfect timing and coordination, the sound would end up being a jumble and not pleasing at all. Thus, the speed of sound is completely independent of the frequency of the sound. This seems somewhat surprising in view of the equation $v = \lambda f$ because this equation seems to indicate that as the frequency f

goes up, so should the speed v. This is not the case, however; as one moves to higher frequencies, the wavelength decreases such that the product of λ and f always remains equal to the speed of sound.

7.7 LOUDNESS OF SOUND AND THE DECIBEL SCALE

We noted earlier that the energy carried by a wave depends on the amplitude of the wave. Thus, a loud sound causes large-amplitude vibrations of the molecules in the air, whereas a soft sound produces small vibrations. Be careful here, however, because your definition of a large vibration may not be in agreement with what actually happens in a sound wave. The faintest sound that the human ear can detect, at a particular frequency, is called the **threshold of hearing,** and the loudest sound that the ear can tolerate is called the **threshold of pain.** If one could examine an individual molecule of air when a sound equal to the threshold of hearing is passing by, one would find that the molecule moves with an amplitude of vibration of only about 10^{-11} m. This is an incredibly small number, so we see that the ear is an extremely sensitive detector of sound. Suppose that a sound at the threshold of pain passes by. In this case, the amplitude of vibration of the molecule increases to about 10^{-5} m. This is still a small amplitude of vibration, but it is extremely large in comparison to the vibration amplitude at the threshold of hearing.

To discuss the loudness of sounds, the **decibel** scale was devised. This scale is constructed so that the lowest perceptible sound for the normal ear is assigned a 0 decibel level. A sound carrying 10 times as much energy as the 0 decibel sound is said to have a decibel level of 10. A reasonable guess is that a 20 decibel sound carries 20 times as much energy as the 0 decibel scale, but that is incorrect. The scale is devised so that the 20 decibel level sound carries 100 times the energy of the 0 decibel sound and 10 times the energy of the 10 decibel sound. Thus, an increase of 10 decibels means that the energy content of the sound increases by 10 times. For example, a 50 decibel sound carries 10 times more energy than a 40 decibel sound. On this scale, a sound at the threshold of pain has a decibel level of 120. Table 7.1 indicates the decibel level of several common sounds.

TABLE 7.1   Decibel level of common sounds

NOISE LEVEL	SOURCE OF SOUND
0	Threshold of hearing
10	Rustle of leaves
30	Soft whisper
40	Mosquito buzzing
50	Average home
60	Ordinary conversation
70	Busy street traffic
100	Power mower
120	Threshold of pain
130	Rock concert
150	Jet engine (at 30 m)
180	Rocket engine (at 30 m)

CONCEPTUAL EXERCISE

If a tree falls in a region where there are no creatures, human or otherwise, to hear the sound, is there a sound?

Answer You have heard this question raised since you were a small child, and the answer to it depends on which definition of sound you are choosing to use at the moment. In this book, we are emphasizing the physical disturbance that a sound wave makes in air, and the tree will create this disturbance with or without the presence of a listener. Thus, from this point of view there is a sound. On the other hand, if you choose to concentrate on the physiological effect that the wave produces when it strikes an eardrum, there will be no sound. Take your pick.

7.8 INTERFERENCE

A characteristic of all types of waves is that they can interact with one another. To get an idea of how this phenomenon works, let us consider two rivers, as in Figure 7.11, coming together and melding into a single river. We are going to assume that these are perfect rivers in that we don't have to worry about sand bars, turbulence, or any other effect that would disturb a wave moving along the surface. With all these restrictions placed on our example, it should be obvious to you that you are not going to be able to observe these effects on a real river. The phenomenon *can* be demonstrated with water waves, but it takes a more ideal setting, such as a laboratory, to produce it.

FOCUS ON NOISE

An occasional noise interferes with hearing, but a person's hearing recovers when quiet is restored. If the exposure to loud noise is protracted, however, some hearing loss becomes permanent. The general level of city noise, for example, is high enough to deafen people gradually as they grow older. In the absence of such noise, hearing ability need not deteriorate with advancing age. Thus, inhabitants of some quiet regions hear as well in their seventies as New Yorkers do in their twenties.

It is important to understand that most instances of loss of hearing that result from environmental noise are not traumatic; in fact, the victim is often unaware of the effect. Let us picture a worker who completes his first day in a noisy factory. Of course, he recognizes the noisiness and may even feel the effect as a "ringing in the ears." He has suffered a temporary hearing loss that is localized in the frequency range around 4000 Hz, as shown in the figure. Curve A indicates that at 4000 Hz, the sound has to be about 15 decibels louder than normal to reach the threshold of hearing. The worker does not hear moderately high frequencies well, but low-frequency and very high frequency sources are unaffected. As he walks out of the factory, then, most sounds seem softer. His car seems to be better insulated because he does not hear the rattles and squeaks so well. He judges people's voices to be just as loud as usual, but they seem to be speaking through a blanket. He also feels rather tired. By morning he will be rested, the ringing in his ears will have stopped, and his hearing will be partly but not completely restored. The factory will therefore not seem quite so noisy as it was on the first day. As the months go by, he will become more accustomed to his condition, but his condition will be getting worse. Can he recover if he is removed from his noisy environment? That will depend on how noisy it has been

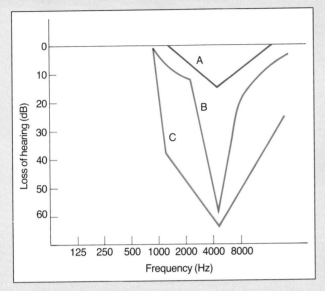

Patterns of hearing loss from exposure to industrial noise. (A) Temporary loss of hearing. (B) After 20 years. (C) After 35 years.

and how long he has been exposed. In many cases, his chances for almost complete recovery will be fairly good for about a year or so. If the exposure continues, hearing loss becomes irreversible, and eventually he will become partially deaf. Look at Curves B and C in the figure to see a typical downward progression caused by prolonged exposure to industrial noise.

In general, noise levels of about 80 decibels or higher can produce permanent hearing loss, although, of course, the effect is faster for louder noises, and it is somewhat dependent on the frequency. At a 2000 Hz frequency, for example, it is estimated that occupational exposure to 95 decibel noise (about as loud as a power lawn mower) will depress one's hearing ability by about 15 decibels in 10 years. Occupational noise, such as that produced by bulldozers, jackhammers, diesel trucks, and aircraft, is deafening many millions of workers.

What about extremely loud noises? Concern over exposure of people to rock music stems from the fact that such music is often played very loudly. Sound levels of 124 decibels have been recorded in some nightclubs and concerts. Such noise is at the edge of pain and is unquestionably

deafening. Noise levels as high as 135 decibels should never be experienced, even for a brief period, because the effects can be instantaneously damaging. Such an acoustic trauma might occur, for example, as the result of an explosion. If the noise level exceeds about 150 or 160 decibels, the eardrum may be ruptured beyond repair.

A living organism, such as human being, is a complicated system, and the effects of a stress or disturbance follow intricate pathways that may be difficult to understand. Many investigators believe that loss of hearing is not the most serious consequence of excess noise. The first effects are anxiety and stress reactions or, in extreme cases, fright. These reactions produce body changes, such as increased rate of heart beat, constriction of blood vessels, digestive spasms, and dilation of the pupils of the eyes. The long-term effects of such overstimulation are difficult to assess, but scientists do know that in animals it damages the heart, brain, and liver and produces emotional disturbances. The emotional effects on people are, of course, also difficult to measure. One known effect is that work efficiency goes down when noise goes up.

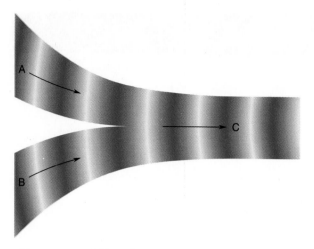

FIGURE 7.11   *When the waves in rivers A and B meet in river C, what does the resulting wave look like?*

Let us consider identical waves moving along branches A and B of the river in Figure 7.11. These waves have the same frequency, wavelength, and amplitude. The question is: What happens to these waves when they come together in river C? The answer is that the resulting disturbance in C depends on the relationship between the waves in A and B. For example, suppose at some instant of time, the wave in river A and the wave in river B match up as shown in Figure 7.12, such that crest matches up with crest and trough matches up with trough. When these waves come together in river C, the resultant disturbance of the water looks like that shown in the figure. The two waves, A and B, are said to have undergone **constructive interference.** The resultant wave, C, has the same frequency and wavelength as A and B, but its amplitude is the sum of the amplitudes of A and B.

FIGURE 7.12   *When two waves interfere such that crest meets crest and trough meets trough, constructive interference occurs.*

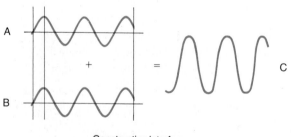

Constructive interference

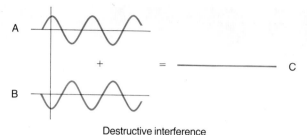

Destructive interference

FIGURE 7.13   *Destructive interference occurs when crest meets trough.*

If the relationship is like that shown in Figure 7.13, the crest of one matches up with the trough of the other. When these two waves come together in river C, they cancel one another and the surface of the water remains calm. This occurs because when one of the waves is trying to produce a crest by pulling the water upward, the other wave is attempting to produce a trough by pulling the water downward. The result is no motion of the water at all. In this case, we say that the waves are undergoing **destructive interference.**

Interference effects are important in understanding many day-to-day occurrences, particularly in our understanding of most musical instruments. It should be noted that if the two waves interfering have different frequencies, the situation becomes more complicated, for they interfere constructively part of the time and cancel part of the time (Fig. 7.14).

EXAMPLE 7.7   UNEQUAL AMPLITUDES
Two waves interfere constructively. One of the waves has an amplitude of 2 ft and the other has an amplitude of 1 ft. What is the amplitude of the resultant wave?

FIGURE 7.14   *Interference of waves with different frequencies.*

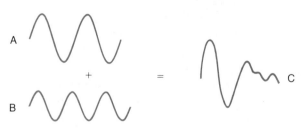

Solution To find the resultant amplitude when constructive interference is occurring, we must add the two amplitudes together. As a result, the resultant wave has an amplitude of 3 ft.

EXERCISE

Repeat Example 7.7 assuming the waves undergo destructive interference.

Answer Amplitude = 1 ft.

7.9 STANDING WAVES

In the preceding section, we examined what happens when two waves of the same frequency traveling in the *same* direction come together. Many of the interference effects important to musical instruments occur when two waves traveling in *opposite* directions interfere. One way in which such a set of conditions can occur arises when a wave is sent down a string that is attached to something at the end toward which the wave is traveling. For example, the string could be tied to a wall, or it could be attached to one of the standards on a guitar. When the wave hits the fixed end, it is reflected back, and the reflected wave interferes with the oncoming wave.

Before we can truly understand the kind of interference pattern that results on the string, we must consider what happens to a wave when it is reflected. You can find out on your own by way of a simple experiment. Tie one end of a rope to a wall and send a pulse (a small portion of a wave) down the rope by shaking the other end briefly. You might end up with a situation like that shown in Figure

FIGURE 7.15   *(a) A bump sent down a rope that is tied to a wall is (b) reflected so that it turns over on the rope.*

7.15a, with the bump traveling on top of the rope. The bump travels along the rope, hits the wall, and is reflected. You would find that upon reflection, the bump turns over on the rope such that it returns on the bottom of the rope (Fig. 7.15b).

While you have the rope tied to the wall, try another experiment to see what happens when you shake the rope at different frequencies. In this case, you are causing a wave to move toward the wall; likewise, a wave reflects off the wall and returns toward you. If you are careful and shake the rope at just the right frequency, you will find that you can cause ongoing crests to meet reflected troughs at the same point on the rope, resulting in a cancellation of motion at these points. The various patterns that you could set up on the rope are shown in Figure 7.16. The frequency that produces the pattern shown in Figure 7.16a is called the **fundamental frequency** (or first harmonic). A higher

A time-blurred photograph of a standing wave pattern on a string. This corresponds to the pattern in Figure 7.16c.

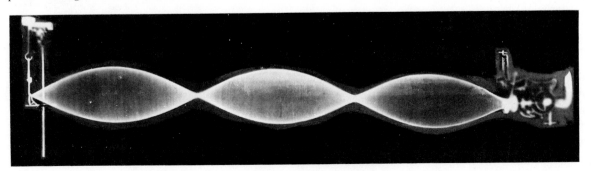

FOCUS ON RESONANCE

Every object that is free to vibrate has a particular set of frequencies at which it "prefers" to vibrate. For example, a guitar string prefers to vibrate at its fundamental or one of its overtones. When we pluck or push an object in such a way that our pushes match one of the object's preferred frequencies, we say that **resonance** exists. Under resonance conditions, the amplitude of vibration of the object can become extremely large. Opera singers have demonstrated this vividly by breaking crystal goblets with their powerful voices. In this case, resonance occurs when the wavelength of the sound wave emitted by the singer is the same length as

The destruction of the Tacoma Narrows Bridge, caused by resonant vibrations activated by the wind. (United Press International Photo)

shaking frequency would produce a pattern like that in Figure 7.16b. This is called the **first overtone** (or second harmonic). The pattern in Figure 7.16c corresponds to the second overtone, and so forth. All of these patterns are characterized by the fact that the wave does not appear to move along the string at all. There are large amplitude vibrations, called antinodes, at points like A in Figure 7.16b, whereas other points, called nodes, like those labeled N have no vibration amplitude. The over-the distance around the rim of the glass. If the singer is able to sustain the note, the amplitude of vibration can increase to the point where the glass shatters.

Our vocal cords disturb small masses of air, but the vibrating cords cause resonant vibrations within various cavities in a person's neck and head. If you ever have a bad cold and some of these cavities are filled with fluids, the quality of your voice changes because the fluids alter your natural resonant frequencies.

Resonance can occur in mechanical situations as well. Any structure, such as a building or a bridge, naturally sways in the wind. If such a structure is made to vibrate at its natural frequency, the vibrations increase in amplitude. In 1940, the Tacoma Narrows Bridge was set into a resonant vibration by a wind blowing down the canyon that it spanned. The normal oscillation increased to the point that the bridge collapsed, as shown.

A more recent example of the destruction that structural resonance can cause occurred during the Loma Prieta earthquake near Oakland, California, in 1989. The Nimitz Freeway was constructed as a double-decker road, and in one section, almost a mile in length, the upper deck collapsed onto the lower deck, causing the loss of several lives. The reason that this particular section of roadway collapsed, while other sections escaped serious damage, has been traced to the fact that the earthquake waves had a frequency of approximately 1.5 Hz. This was a close match to the natural resonant frequency of the section of roadway that collapsed.

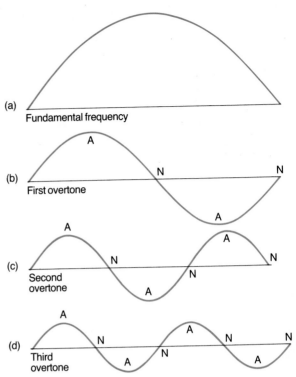

FIGURE 7.16 *Fundamental frequency and overtones. The points labeled A have large amplitude vibrations, whereas at the points labeled N, there is no motion of the string at all.*

100 Hz to produce the fundamental frequency, you will have to cause it to vibrate at 200 Hz to produce the first overtone, at 300 Hz to produce the second overtone, and so forth. If f_f is the frequency of the fundamental, the overtone frequencies are given by

$$f_0 = nf_f \ (n = 2, 3, 4, 5 \ldots)$$

This says that the frequency of any overtone is some integral multiple of the fundamental frequency.

7.10 MUSICAL SOUNDS

When a guitar string is plucked, the waves produced on the string reflect back and forth between the end supports, and thus standing wave patterns are set up. Similarly, standing waves are set up on other stringed instruments such as the violin or cello when they are bowed. Surprisingly, perhaps, standing waves also produce the musical sounds when a wind instrument, such as an organ, is played. In this case, sound waves reflect back and forth between

all wave pattern, however, does not move along the rope at all. As a result, this kind of wave pattern is referred to as a **standing wave.**

The relationship between these patterns is that if you have to vibrate the string at a frequency of

FOCUS ON
BEATS

Many interference effects occur when two waves of the same frequency travel through the same region of space in opposite directions. This is how standing waves on a guitar string are formed. But what happens if the two waves moving through the same region of space travel in the same direction and have slightly different frequencies? The answer is that the waves move into constructive interference at some instant of time, and a brief instant later they undergo destructive interference. Consider part (a) of the figure. There we see two waves that we can consider as having been emitted by two tuning forks having slightly different frequencies. Part (b) shows the interference pattern formed by the addition of these two waves. At some particular time indicated by t_a, the two waves interfere destructively. This occurs because one of the forks is emitting a compression at the same instant that the other is emitting a rarefaction. At some later time, t_b, however, the vibrations move into step with one another, and constructive interference occurs. This means that at this instant, the two forks are simultaneously

emitting compressions and rarefactions. As time continues, the two forks continually move into and out of step because of their differing frequencies. Consequently a listener hears an alternation in loudness, known as **beats.**

The number of beats heard per second is known as the beat frequency. The beat frequency is found to be equal to the difference in frequency between the two sound sources. One can use beats to tune a stringed instrument, such as a piano, by listening to the beats produced between a string and a tuning fork. The string can be loosened or tightened to change its frequency and bring it into tune with the accurate fork. For example, suppose a particular string on a piano is supposed to emit a frequency of 440 Hz but is not doing so. To bring the string into its desired niche, a tuning fork of frequency 440 Hz is sounded. Let us suppose that two beats per second means that the string and the tuning fork differ by 2 Hz in frequency. Thus, the string could have a frequency of either 438 Hz or 442 Hz. The string has probably worked loose, which would cause it to have a lower than normal frequency. Thus, its frequency is probably 438 Hz.

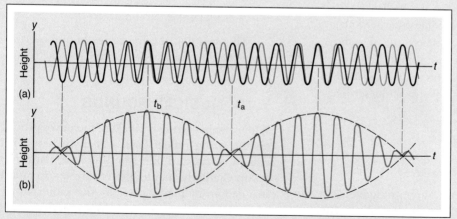

(a) Two sound waves of slightly different frequencies are shown traveling in the same direction. (b) The interference pattern produced by these two waves. Note that the amplitude of the resultant wave oscillates in time, producing alternating loud and soft sounds.

the ends of a pipe, and these reflecting waves interfere to produce standing waves. However, the sounds produced by musical instruments are quite complex, as we shall see in what follows.

If you are careful, you can cause a guitar string, or the string of any other musical instrument, to produce a single frequency sound. For example, if you distort a guitar string to a shape that looks like any of those pictured in Figure 7.16, you will find that it vibrates only at the corresponding frequency. Such a pure tone, however, seldom happens in practice. When a string on a guitar is plucked, the sound that you hear is basically that of the fundamental, but the string does not vibrate with a simple pure frequency. Instead, the string vibrates such that it emits not only the fundamental, but also several overtones. This means that if we try to cause a string to vibrate at a fundamental of 220 Hz, the resultant sound will really be a 220 Hz sound with a little 440 Hz mixed in, a little 660 Hz, and so forth. All of these frequencies add together to produce a complicated sound.

As another example, if you try to play a particular note on a flute, the resultant sound is composed of a certain amount of the fundamental, a lot of the first overtone, not much of the second overtone, and so on. Figure 7.17 displays the char-

FIGURE 7.17   *The characteristic mixture of overtones for a flute.*

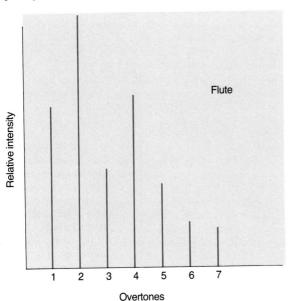

FIGURE 7.18   *The characteristic mixture of overtones for a clarinet.*

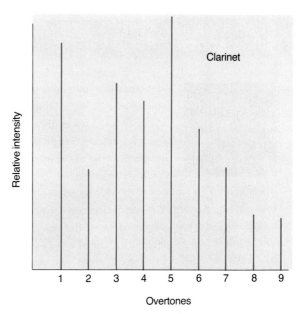

acteristic mixture of overtones for a flute. If you try to play the same note on a clarinet, the amount of each overtone mixed in with the fundamental is different. As Figure 7.18 shows, you get some of the first overtone, not much of the second overtone, a little more of the third overtone, and so forth. Because all different kinds of musical instruments produce their own characteristic mix of overtones with the fundamental, it is easy to tell the difference between, say, a flute and a clarinet even when both are attempting to play the same fundamental frequency. A note from any instrument contains a mixture of the fundamental and its overtones. In a quality instrument, this mixture is more complex and therefore fuller than in an inferior one. Thus, it is easy to distinguish a cheap piano from a quality grand piano.

The musical scale now in use is an example of survival of the fittest among many competing patterns. The present arrangement seems to be the one that produces the most satisfying tones to the ear for people accustomed to "Western" music. The scale of a typical piano keyboard is shown for reference purposes in Figure 7.19 (p. 150). The notes on the scale are labeled from A to G with 12 keys, both black and white, separating these two extremes. Let us start with middle C to follow the pattern of frequencies. Middle C has a basic frequency *(text continued on page 150)*

FQCUS ON
SONIC BOOMS

To visualize a sonic boom, think of something moving rapidly through water. If the object is traveling faster than the speed of the waves it creates, it therefore leaves its waves behind (see photo of the ducks). Moreover, the wave energy is being continuously reinforced by the forward movement of the object. The result is a high-energy wave, called a wake, that trails the object in the shape of a V and that slaps hard against other vessels or against the shoreline. The sonic boom is a high-energy air wave of the same type. The tip of the wake moves forward with the airplane, while the sound itself moves out from the wake at the speed of sound

in air. The faster the airplane, the more slender is the wake.

To understand the geometry of the wake, consider the next figure. A stationary object (A) remains in the center of the circular waves it generates. The waves from a moving object crowd each other in the direction of the object's motion (B). The object is, in effect, chasing its own waves. Recall that sound travels in air at sea level

Notice the V-shaped bow wave behind the ducks. This is similar to the shock wave behind a supersonic airplane except the shock wave is three-dimensional, in the shape of a cone.

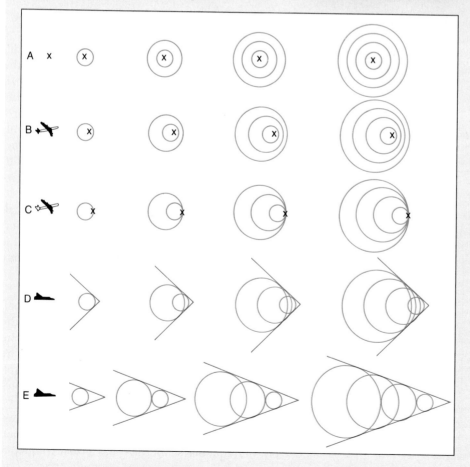

Wave patterns of subsonic and supersonic speeds: (A) Stationary; (B) subsonic; (C) sonic, Mach 1; (D) supersonic, Mach 1.5; (E) supersonic, Mach 3. From left to right, the waves are shown at equal time intervals as they expand.

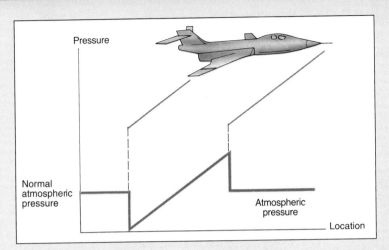

Sonic boom. Shock waves originate from both the front and rear of a supersonic plane.

at a speed of about 345 m/s. When the speed of the object equals the speed of the wave, the object will not see any waves before it; it will just be keeping up with them (C). Such high speeds are usually expressed in Mach numbers after Ernst Mach (1838–1916), a physicist who made important discoveries in sound. The Mach number is defined as

$$\text{Mach number} = \frac{\text{speed of object}}{\text{speed of sound}}$$

If an object is traveling at the speed of sound, then the numerator and the denominator of the equation are the same, and the Mach number equals 1. Mach 2 is twice the speed of sound, Mach 3 is three times the speed of sound, and so forth.

Figures (D) and (E) show the wave pattern at supersonic speeds. Note that the object is always ahead of its waves. A passenger in a supersonic transport would therefore not hear the sound of its motion. Instead, the waves crowd each other, and the effect is a significant elevation of pressure at the advancing boundary of the overlapping wave fronts.

Of course, an airplane travels within its medium and not, like a boat, on the surface of its medium. Therefore, the outlines shown in (D) and (E) are two-dimensional projections of what are really conical shapes.. Furthermore, an airplane is more than a point in space; therefore, a whole series of such cones is generated. It is sufficient to consider only the nose and the tail of the plane and to represent the entire space between the forward and rear cones as the volume of the disturbance, as shown in the next figure. As the conical shock wave strikes the ground, people within the volume of the shock wave hear the sonic boom.

To be struck unexpectedly by a sonic boom can be quite unnerving. It sounds like a loud, close thunderclap, which can seem quite eerie when it comes from a cloudless sky. Depending on the power it generates, the sonic boom can rattle windows or even shatter them.

It is important to avoid the misconception that the sonic boom occurs only when the aircraft "breaks the sound barrier," that is, passes from subsonic to supersonic speeds. On the contrary, the sonic boom is continuous and, like the wake of a speedboat, trails the aircraft during the time that its speed is supersonic. Furthermore, the power of the sonic boom increases as the supersonic speed of the aircraft increases.

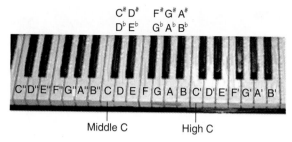

FIGURE 7.19   *A typical piano keyboard.*

of 261.6 Hz. Each key upward on the scale has a frequency of 1.0595 times the frequency of the one that precedes it. Thus, the next key is the black key, labeled either C sharp (C$^\#$) or D flat (D$^\flat$), and the frequency is given by (261.6 Hz)(1.0595) = 277.1 Hz. The next key is D, and its frequency is 1.0595 times the frequency of the black key, (277.1 Hz)(1.0595) = 293.6 Hz. If you continue this process on up the scale, you find that at the end of the basic 12-key segment, you reach high C, which has a frequency of exactly twice that of middle C, 2 × 261.6 Hz = 523.2 Hz. The basic 12 keys are called an octave, and as the numbers indicate, the same notes in two adjacent octaves have frequencies such that the higher frequency is twice that of the lower. You can check this by multiplying 1.0595 by itself 12 times. This will give exactly the number 2.

EXAMPLE 7.8   GIVE ME A C
(a) Use the fact that each key on a piano scale is 1.0595 times the frequency of the one that precedes it and the fact that the frequency of middle C is 261.6 Hz to find the frequency of A.

Solution We will not go through all the arithmetic for you, but start with middle C with its given frequency of 261.6 Hz and work your way up the scale by multiplying by 1.0595 for each key until you reach A on the scale. You will find that A has a frequency of 440.1 Hz.

(b) What is the frequency of C below middle C?

Solution This is one octave below middle C, and since the corresponding frequencies in consecutive octaves have a ratio of 2 to 1, the frequency of low C is 261.6/2 = 130.8 Hz.

7.11 THE DOPPLER EFFECT

Have you ever stood beside a train track and noticed the change in frequency of a train whistle as the train approached and then passed you? The siren of an ambulance approaching and then receding from you is another situation in which you may have encountered this phenomenon. This change in frequency was first explained by Christian Doppler (1803–1853) and bears his name: the **Doppler effect.** He tested his mathematical explanation by placing trumpeters on a flatcar and having it pulled repeatedly past listeners chosen because of their ability to estimate frequencies accurately.

To understand the cause of this frequency change, consider a sound source, such as a tuning fork, emitting a constant-frequency sound. When the tuning fork is at rest, as in Figure 7.20, waves spread out from it uniformly in all directions. We picture the crests of the sound waves by the dark blue lines in the figure. Notice that the distance between consecutive crests is the same on each side of the fork, and as a result, observers A and B hear the same frequency. If the fork is set into motion toward the right, as in Figure 7.21, the situation changes dramatically because, as the figure shows, the crests are spread further apart behind the fork and are closer together in front of it. To see that this is reasonable, consider crests 1 and 2 behind the fork. Between the times when 1 and 2 are emitted by the fork, the fork itself has moved toward the right. As a result of this movement, crests 2 and 1 are farther apart than they would be if the source were still. The distance between consecutive crests, such as 3 and 4, in front of the moving fork is de-

FIGURE 7.20   *The observers at A and B hear the same frequency when the sound source is at rest.*

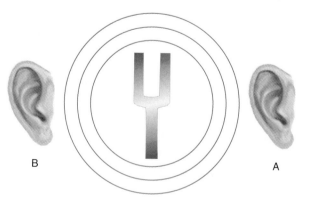

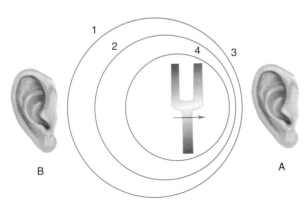

FIGURE 7.21   *When the sound source is in motion, the observer in front of the source at A hears a higher frequency than he did when the source was at rest, and the observer behind the source at B hears a lower frequency.*

creased because in the interval between emission of the two, the fork has moved to the right. This means that the motion of the fork helps 4 to keep up with 3. A listener to the right finds more crests reaching him per second than when the fork was at rest, and as a result, his ear correctly interprets the sound as having a higher frequency. Observer B behind the fork (to the left) detects fewer crests per second reaching him and interprets this as a decrease in sound frequency.

Doppler correctly predicted that a similar effect should be observed for light waves, but he was unable to demonstrate it because of the difficulty in his time of working experimentally with the high frequencies of light waves. The present-day science of astronomy uses Doppler effects to determine many important facts about distant astronomical objects.

As we shall see in Chapter 11, the gaseous material in stars emits certain frequencies that are characteristic of the elements present, and this pattern of frequencies can be used to identify the elements present in a star. For example, mercury vapor emits four prominent, specific frequencies in the visible spectrum. When we find these frequencies in starlight, we know that mercury atoms are producing these particular lines. When we examine the starlight more closely, however, we find that all of the frequencies of the pattern have been shifted slightly. For example, if we expect to find a frequency of 5.17×10^{14} Hz, we may actually discover the frequency to be 5.16×10^{14} Hz, and all the other frequencies in the pattern have also been shifted downward by the same amount. From the Doppler effect, we know that the lowering of frequency means that the source is moving away from the Earth. In the visible portion of the spectrum, red light has the lowest frequency; thus, because the light from the star has been shifted toward the lower, or red, end of the visible spectrum, we refer to the light as having been red-shifted. The fact that the light from distant galaxies is red-shifted is the rule rather than the exception. On this observation rests the now-accepted theory that the Universe is expanding.

CONCEPTUAL EXAMPLE 7.9   DOPPLER EFFECT AND THE SUN

When one observes the Sun, it is found that the light from one side of the Sun is red-shifted and the light from the other side of the sun is blue-shifted. What does this piece of information tell us about the Sun?

Solution The side of the Sun that emits the red-shifted light is moving away from us, while the side that emits the blue-shifted light is moving toward us. The only way that this could happen is if the Sun were in rotation, with one side rotating away and the other side rotating toward the Earth.

SUMMARY

A wave in which the vibrations are at right angles to the direction of travel of the wave is a **transverse wave.** If the vibrations are in the same direction as the direction of travel of the wave, the wave is a **longitudinal wave.**

The **wavelength** of a wave is the distance between successive **crests** or **troughs.** The speed of a wave is the rate at which the disturbance moves, and the **frequency** is the number of repetitions, or cycles, per second. The **amplitude** is the intensity of the disturbance.

A sound wave is a **longitudinal** wave and is manifested as a succession of compressions and rarefactions of a medium, such as air. The speed of a sound wave is about 331 m/s at 0°C, and audible frequencies for a normal human ear range from 20 to 20,000 Hz.

The **decibel** scale used to measure the energy content, or loudness, of a sound wave (a) starts at zero, which represents the softest audible sound, and (b) represents each tenfold increase in sound intensity as an additional 10 dB.

Waves exhibit **interference.** When they come together such that crest meets crest, and trough meets trough, **constructive interference** is produced. When crest meets trough, **destructive interference** occurs.

Standing waves are produced when two waves of the same frequency attempt to travel in opposite directions through the same region of space. Musical tones contain a fundamental frequency and some combinations of overtones.

When an observer moves toward a source of sound, or if the source moves toward the observer, the frequency of the sound increases. If the relative motion is such as to separate the source and listener, the frequency of the sound decreases. This is the **Doppler effect.**

EQUATIONS TO KNOW

$v = \lambda f$ (speed = wavelength for all types of waves times frequency

$v = v_0 + (0.61)\, T$ (the speed of sound as a function of temperature)

KEY WORDS

Longitudinal wave	Wavelength	Rarefaction	Standing waves
Transverse wave	Speed	Ultrasonic waves	Overtones
Sine curve	Frequency	Decibel	Resonance
Crest	Amplitude	Interference	Doppler effect
Trough	Compression	Fundamental frequency	

PROBLEMS AND CONCEPTUAL QUESTIONS

Problems requiring numerical work are identified with a blue number.

Describing a wave
Unless otherwise specified, use 340 m/s as the speed of sound in air.

1. Give some examples from nature to prove that energy can be carried by a wave.
2. Theater goers are lined up at a ticket booth. How could you set up a longitudinal wave in this line?
3. When fans do the "wave" at a football game, is their wave more nearly a longitudinal wave or a transverse wave?
4. The speed of light is 3.00×10^8 m/s. Calculate (a) the wavelength of a gamma ray with a frequency of 10^{22} Hz; (b) the frequency of a radio wave with a wavelength of 30 m (gamma rays and radio waves travel at the speed of light); (c) the wavelength of a sound wave with a frequency of 1000 Hz.
5. Can you change the wavelength of a wave without simultaneously changing its frequency?
6. While observing waves at a beach, you note that the distance from the crest of a wave to the next consecutive trough is 3 m. What is the wavelength of this wave?
7. You hold one end of a spring in your hand with the other end attached to a wall. How would you move your hand to set up (a) a longitudinal wave in the spring? (b) A transverse wave?
8. For the sketch shown in the figure, what is (a) the amplitude of the wave? (b) The wave-

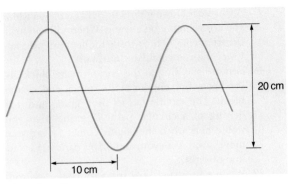

20 cm

10 cm

(Question 8)

length of the wave? (c) If its speed is 30 m/s, what is its frequency?

9. A metronome is set so that it makes 20 complete vibrations in 25 s. What is the frequency of the metronome?

Sound waves

10. In many science fiction movies, one starship will explode and the sound of the explosion can be clearly heard in a nearby ship. Is this possible? Why or why not?

11. Sound waves of all frequencies have the same speed. Discuss how you could prove this by listening to a marching band.

12. Will the sound of an echo return to you more quickly on a hot day or a cool day? Defend your answer.

13. When someone says that a sound wave can be represented by a sine curve, which of the following statements is implied? (a) The molecules move along in a wavy motion like water waves. (b) A graph of air density versus distance will be a sine curve. (c) A graph of the degree of increase or decrease of air pressure above or below atmospheric pressure versus distance will be a sine curve. (d) A graph of pressure versus time at any one point will be a sine curve. (e) The molecules move back and forth like the inking on a piece of paper used to draw the sine curve. (f) Since the air does not ripple, there is no real sine function, but rather we are using a figure of speech to describe a sound wave.

14. The distance between the compression of a sound wave and the next consecutive rarefaction is found to be 1 m. Can this wave be heard as an audible sound wave?

15. If sound travels at 340 m/s, what is the range of wavelengths to which a human ear can respond if the person can hear all sounds with frequencies between 20 Hz and 20,000 Hz?

16. What happens to the frequency of a sound wave if the wavelength is tripled?

17. What happens to the speed of a sound wave if the wavelength is tripled?

18. What happens to the speed of a sound wave if the amplitude is tripled?

19. Devise an experiment to prove that the speed of sound does not depend on the frequency of the sound.

20. Devise an experiment that would enable you to measure the speed of sound in (a) water, (b) aluminum.

21. A tugboat lost in the fog in a harbor sounds its whistle, and the captain hears an echo returned from the surface of an oil tanker 6 s later. How far away is the tanker? Don't forget the sound has to travel to the tanker and return.

22. Sketch a sine curve to represent two waves described as follows: They have equal amplitudes but one has twice the frequency of the other.

23. The sound waves used in normal conversations are usually between 500 Hz and 5000 Hz. What are the wavelengths of these sounds?

24. Find the speed of sound in air at (a) 0°C, (b) 22°C, and (c) 50°C.

25. Find the speed of sound in air at (a) 0°F, (b) 22°F, and (c) 100°F.

26. Some animals can hear sounds as high as 100,000 Hz. What is the wavelength of this wave?

Loudness of sound and the decibel scale

27. A person hears a cry in the woods that is 1000 times the intensity of the faintest audible sound. What is the sound level in decibels (dB)?

28. (a) How much more sound energy is carried by a 50 dB sound than a 0 dB sound? (b) How much more sound energy does the 50 dB sound carry than a 30 dB sound?

29. Can there be a negative decibel level?

30. A dog can hear sounds inaudible to a human. Suppose a dog could just hear a sound whose intensity is 100 times less than the faintest audible sound for humans. What is the minimum decibel level to which the dog is sensitive?

31. Estimate the decibel level of the following sounds: (a) a train passing by while you are near the tracks, (b) a pin dropping on a wooden table, (c) the background noise in a church.

32. The sound intensity of a motorcycle at a distance of 8 m is 90 dB. How many times greater is the energy level of this sound than the faintest audible sound?

33. (a) The sound intensity of a garbage disposal unit is 80 dB. How many times greater is the energy content of this sound than the faintest audible sound? (b) How many times greater is the energy content in a sound of a 120 dB thunderclap than the sound of a garbage disposal unit?

Interference

34. Imagine that two equal pure tones are generated by speakers mounted on a wall and situated 5 m apart. What will you observe if you stand in the center of the room and move a sensitive decibel meter slowly back and forth in front of them? Explain.

35. Two speakers emitting identical sound signals are placed side by side. You start to move one of them back and a listener in front of the speakers listens for an interference effect to be produced. (a) At what minimum distance would you move the speaker to hear destructive interference? (b) At what minimum distance would you have moved the speaker to hear constructive interference?

36. Can a transverse wave and a longitudinal wave interfere?

37. Two water waves are interfering to produce constructive interference with a wave of amplitude 2.2 m. One of the individual waves has an amplitude of 1.6 m. What is the amplitude of the other?

38. If the two waves of problem 31 were moved into destructive interference, what would be the amplitude of the wave?

Standing waves and musical sounds

39. Sketch the wave pattern for a string vibrating in its fourth overtone.

40. The fundamental frequency of a string is 125 Hz. What is the frequency of the fifth overtone?

41. A string driven by a tuning fork is found to vibrate in its third overtone when the frequency of the fork is 400 Hz. What is the frequency of the fundamental?

42. If the frequency of middle C is 262.6 Hz, find the frequency associated with the key labeled G.

43. What is the wavelength emitted when middle C is sounded?

44. What is the frequency of the first overtone of middle C?

45. If a standing wave pattern is set up in a small tank of water, will a cork placed in the water move at all? Explain.

Doppler effect

46. Certain types of stars pulsate in size. (a) If you observe the light from such a star while it is growing, would you see the light shifted toward higher or lower frequencies? (b) Repeat part (a) for the case in which the light is observed while the star is diminishing in size.

47. A whistle is blowing directly in front of you while you are swinging on a swing. When will you hear the frequency of the whistle higher than normal, lower than normal, and normal?

48. You are standing at the end of a city block observing a fire engine at the other end of the block. The engine has its siren going and it is driving in a circular path. Describe what you would hear when the engine comes toward you, moves away from you, and drives parallel to your line of sight.

49. When a fire truck is at rest with respect to you, you hear a sound from its siren having a frequency of 500 Hz. Late at night you hear the same engine emitting a frequency of 502 Hz. Is the engine coming toward you or away from you?

50. A wind is blowing toward you in the same direction as a fire engine moves with its siren sounding. In what way would the wind affect the sound you hear?

51. In some movies, a spaceship accelerates to near the speed of light and suddenly the stars disappear. How could the Doppler effect account for this?

52. How could the Doppler effect be used to prove that our Sun is rotating?

53. A sound source and an observer are both traveling in the same direction at the same speed. Will a Doppler shift in the sound be heard?

54. A binary star system consists of two stars in revolution about one another. Describe how the Doppler effect will change the light reaching us from these stars.

ANSWERS TO SELECTED NUMERICAL PROBLEMS

4. (a) 3×10^{-14} m, (b) 10^7 Hz, (c) 0.34 m
6. 6 m
9. 0.8 Hz
14. yes, $f = 170$ Hz
15. 0.017 m to 17 m
21. 1020 m
23. 0.68 m and 0.068 m
24. (a) 331 m/s, (b) 344 m/s, (c) 362 m/s
25. (a) 320 m/s, (b) 328 m/s, (c) 369 m/s
26. 3.4×10^{-3} m
27. 30 dB
28. (a) 10^5, (b) 100
30. -20 dB
32. 10^9
33. (a) 10^8, (b) 10^4
37. 0.6 m
38. 1 m
43. 1.29 m

CHAPTER 8

ELECTRICITY

Multiple lightning bolts flicker in the Arizona sky above the observatory on Kitt Peak. (Courtesy Gary Ladd)

Many people reading this book have had elderly people, such as grandparents, tell them about the way things were in their childhood. They can remember when there were no radios, TVs, stereos, and a whole host of other electronic devices. Technology has changed a lot during the lives of these people, but there have also been many changes in electrical equipment that have occurred in your own memory. Many of you can recall when an electronic appliance was a large, bulky item. Likewise, such accepted modern-day devices as the hand-held calculator, the computer, the VCR, the compact disc player, and so forth were either not available or available only at a prohibitive cost. Thus, you too will have stories to tell your grandchildren.

Much of the impetus for these devices arose with the advent of microminiaturization of electronic components. This led to the production of circuits filled with transistors, diodes, resistors, and capacitors on a chip so small that its individual parts can be seen only with a microscope. Although these devices are quite fantastic, they all obey the basic laws of physics. The trail that has led to these modern conveniences started with the Greeks about 500 B.C. The interest of these early Greek thinkers was in static electricity. Their results were modest, and it is safe to say that they could never in their wildest dreams have envisioned where their investigations would lead.

8.1 STATIC ELECTRICITY

The first observations of static electricity occurred when someone noticed that a waxlike substance called amber would attract small objects after it had been rubbed with wool. We now know that this phenomenon is not restricted to amber and wool but that a similar effect can be observed (to some extent) when almost any two nonmetallic substances are rubbed together. To describe this change in the

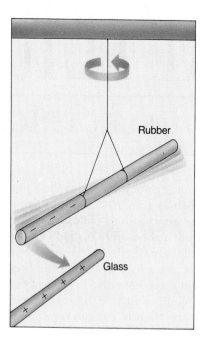

FIGURE 8.1 *A negatively charged rubber rod is attracted by a positively charged glass rod.*

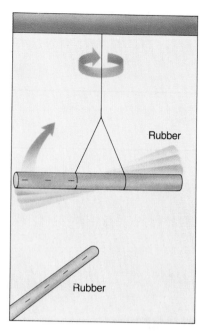

FIGURE 8.2 *Two rods having the same type of charge repel one another.*

physical properties of these objects, scientists said that they had been given an **electric charge.** For example, when hard rubber is rubbed with wool, both the rubber and wool become charged. Similarly, rubbing a glass rod with silk charges both objects. It is simple to show that there are only two different kinds of charge, which were given the names **positive** and **negative** by Benjamin Franklin (1706–1790). Figure 8.1 shows one experimental technique that can be used to demonstrate the two different kinds of charge and to show some additional properties of charged objects. In the figure, a rubber rod that has been charged is suspended by a string so that it can swing freely, and a charged glass rod is brought nearby. It is found that the rubber rod swings toward the glass rod, indicating that the charges on the two rods are attracting one another. It has become customary to say that the glass rod has a positive charge and the rubber rod has a negative charge. Based on our experiment, we can state that *unlike charges attract one another.* A similar experiment, illustrated in Figure 8.2, shows that a negatively charged rubber rod is repelled by another negatively charged rubber rod. This is also true for two positively charged glass rods. Thus, we can state that *like charges repel one another.*

To understand what is happening when an object becomes charged, let us digress briefly to examine a model of the atom, often called the planetary model because of its similarity to our Solar System. (It should be noted here that this planetary model of the atom has now been replaced by a more accurate quantum mechanical model, which will be discussed in Chapters 12 and 13.) In the Solar System, the planets orbit the Sun, bound into their paths by the gravitational attraction exerted on them by the Sun. Likewise, in the planetary model of the atom, small, light, negatively charged particles called **electrons** circle the relatively massive central core of the atom, called its **nucleus.** The nucleus has two fundamentally different types of particles in it. One of these is called a **neutron** and is so named because it carries no electric charge. The second type is called a **proton,** and the *proton is nature's most readily available carrier of positive charge. The planetlike electrons are nature's most readily available carriers of negative charge.* Under normal circumstances, there is an electron circling a nucleus for every proton in the nucleus. The magnitude of the electric charge on an electron is identical to the magnitude of the charge on the proton. Thus, an atom is electrically neutral. Similarly, ordinary material made of neutral atoms also has no net charge.

Thus, even though a block of copper may contain literally trillions of charges, it exhibits no electrostatic effects because the net charge is zero.

When atoms are assembled such that they form a solid object such as a rubber rod, it is found that the atoms are basically fixed in a given location. Individual atoms may vibrate back and forth about some equilibrium position, but they do not migrate freely throughout the solid. Some of the electrons of the material can get loose from the nucleus and become relatively free to wander throughout the

body of the substance. Thus, when two objects are charged by friction, the basic mechanism that occurs is that electrons have been transferred from one of the objects to the other. *It is always the electrons that move around in a solid and never the protons.* For example, when a rubber rod is rubbed vigorously with wool, electrons move from the wool to the rubber. Because the rubber now has an excess of electrons, it has a net negative charge, and because the wool has lost electrons, it now has a net positive charge. All materials have their own char-

FOCUS ON
HOW DOES A PHOTOCOPYING MACHINE WORK?

In past generations, if you wanted copies of a document, you could either write it over and over with, perhaps, a quill pen, or, after vast technological developments occurred, you could use carbon paper to do the task. However, the job is now done quickly and effortlessly by a photocopier. The basic process for producing photocopies was discovered by Chester Carlson in 1940, and in 1947 the Xerox corporation used his methods to produce the now familiar office machines.

The figure illustrates the various steps taking place in a machine when you press the "copy" button. In part (a), a drum coated with the element selenium or a selenium compound is given a positive charge in the dark. The page to be

copied is then projected on the surface of the drum as in part (b). Selenium is a substance that is photoconductive. This means that when light strikes it, the lit portion becomes a conductor of electric charge. Thus, those positive charges on the selenium surface are neutralized when struck by light, and only the dark portions, the writing, that light could not penetrate retain a charge. This means that a copy of the original is formed on the drum by an arrangement of positive charges that duplicate the print on the original document. Next a negatively charged powder, called a toner, is dusted onto the drum, as shown in part (c). The toner sticks to the drum only at the places that still have a positive charge. In this way, the writing becomes visible. The image is then transferred by heat to a sheet of paper placed in contact with the drum.

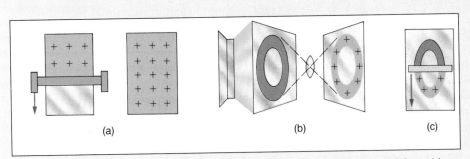

(a) A positive charge is deposited on a selenium drum. (b) Light neutralizes some of the positive charge. (c) A toner adheres to the positively charged image.

acteristic ability to hold onto their electrons; for the case here, the rubber rod has the greater ability to hold onto its electrons than does the wool. As an exercise, explain to yourself what happens when a glass rod is rubbed with silk.

Note that in the process of charging by friction, no charge is ever created or destroyed; it can only be transferred. Thus, we can say that *electric charge is conserved.* An object becomes charged negatively by gaining electrons from some other object; electrons are not created out of thin air. Also, an object becomes charged positively by losing electrons; positive charges are not being created nor are negative charges being destroyed.

8.2 MEASURING CHARGE

When we discussed the measurement of length and mass, we decided that some fundamental unit of comparison was needed for the measurement process. Thus, we adopted the basic unit of length to be the meter and the basic unit of mass to be the kilogram. It is also necessary to be able to measure charge, and the basic unit by which charge is measured is called the **coulomb,** C. This unit is more complex for beginning students to understand because a coulomb of charge is not something that you talk about in everyday life. To give you an example of how this unit is used, it is found that within an atom an electron has a charge of -1.6×10^{-19} C and a proton has a charge of 1.6×10^{-19} C.

You cannot give an object any charge that you like. This is considerably different from measuring the length of an object. For example, you may measure the length of one object to be 1.39869 m long, whereas the length of another can be 1.39868 m long. A length can be subdivided in an infinite number of ways. This is not the case for electric charge. The smallest negative charge that an object can have occurs when that object gains a single electron. In that case, its charge would be -1.6×10^{-19} C. The next smallest negative charge would arise if the object gains two electrons, leaving it with a charge of -3.2×10^{-19} C. Never will you find an object that has a charge of, say, -2.0×10^{-19} C.

As a point of interest, in modern-day theoretical study of matter, it has been proposed that there should exist particles that are even more fundamental than the neutron and proton. These particles are called **quarks,** and they are predicted to have charges that are either one-third or two-thirds the magnitude of the charge of the electron or proton. Neutrons and protons are actually bound units of three quarks of different types. We shall discuss quarks in our study of modern physics.

8.3 COULOMB'S LAW

We have seen that charged objects exert forces of either attraction or repulsion on other charged objects. In 1789, Charles Coulomb investigated these forces in an attempt to find an equation that would predict their strength. He found that the force is given by an equation that is quite similar to an equation you have already encountered, the force of gravitational attraction. This force law is referred to as **Coulomb's law,** and is given in equation form by

$$\mathbf{F} = k\,\frac{q_1 q_2}{r^2} \qquad (8.1)$$

where q_1 is the charge on one of the objects and q_2 is the charge on the other. The quantity r is the distance of separation of the charges, assuming that they are small enough to be considered to be located at a point, or if the charges are spread out over the surface of a sphere, it is the distance between the centers of the spheres. The constant k is found experimentally to have the value 8.99×10^9 N m^2/C^2, but we shall round this off to 9.0×10^9 N m^2/C^2.

As we have noted, there are similarities between Newton's law of universal gravitation and the Coulomb force law, Eq. 8.1. However, there are also differences. For example, Newton's law predicts that the force between objects having mass is always one of attraction, whereas the force between charged objects, given by Coulomb's law, can be either attractive or repulsive. Also, as we shall see in the following example, electrical forces are considerably stronger than gravitational forces.

EXAMPLE 8.1   COMPARING GRAVITATIONAL AND ELECTRICAL FORCES

In a simple hydrogen atom, a single electron circles a single proton such that their distance of separation is 0.53×10^{-10} m. **(a)** Find the magnitude of the gravitational force of attraction between the two, and **(b)** find the magnitude of the electrical force of attraction.

Solution (a) Figure 8.3 gives us the masses of the electron and proton. The gravitational force of attraction is given by Newton's law of universal gravitation as

$$F_g = G\,\frac{m_1 m_2}{r^2}$$

$$= (6.67 \times 10^{-11}\ \text{N m}^2/\text{C}^2)$$

$$\times\ \frac{(1.67 \times 10^{-27}\ \text{kg})(9.11 \times 10^{-31}\ \text{kg})}{(0.53 \times 10^{-10}\ \text{m})^2}$$

$$=\ 3.61 \times 10^{-47}\ \text{N}$$

(b) We use the charge of the electron and proton to find the force of attraction between the two from Coulomb's law as

$$F_e = k\,\frac{q_1 q_2}{r^2}$$

$$= (9.0 \times 10^{9}\ \text{N m}^2/\text{C}^2)\,\frac{(1.6 \times 10^{-19}\ \text{kg})^2}{(0.53 \times 10^{-10}\ \text{m})^2}$$

$$=\ 8.20 \times 10^{-8}\ \text{N}$$

Thus, we can see that electrical forces between electrons and protons in an atom are considerably stronger than gravitational forces between the two.

We did not include the negative sign for the electron in our calculation because we are concerned only with finding the magnitude of the force. Usually, positive and negative signs associated with vector quantities, such as forces, are used to determine the direction of the force, and we wanted only the magnitude.

CONCEPTUAL EXERCISE

Imagine yourself as Coulomb investigating electrical forces. A skeptic says that you are wasting your time because electrical forces are really gravitational forces and not a different phenomenon at all. Give a simple explanation that would prove him wrong.

Answer Gravitational forces are always attractive; otherwise we could have antigravity devices such as magic carpets to carry us around. Electrical forces, on the other hand, can be either attractive or repulsive.

8.4 CONCEPT OF FIELDS

In our study of mechanics, we ran across a type of force called an action-at-a-distance force. This was the force of gravitational attraction between two objects that have mass. These were in distinction to the more common contact forces. To refresh your memory on these forces, contact forces arise when there is actual physical contact between the objects. For example, when a fighter punches his opponent, there is actual physical contact between his fist and the chin of the challenger. The Earth exerts a force of gravitational attraction on the Moon even though there are some 240,000 miles of empty space between the two. There is no physical contact between the two at all, and forces of this kind are called action-at-a-distance forces. We have now encountered our second action-at-a-distance force, the electrical force between two charged objects. These forces are somewhat unusual and rare; in fact, we will meet only one more example of this class of forces in our study—magnetic forces. There are many ways to discuss this classification of forces, but a method developed by Michael Faraday (1791–1867) is of practical importance.

Gravitational Fields

Consider bricks held above the surface of the Earth as in Figure 8.3. What happens to the bricks when they are released? The answer, of course, is that they fall. But why do they fall? A complete answer to the question is difficult and is still a topic of theoretical investigation. A partial answer might be that they fall because the Earth exerts a gravitational force on them which pulls them downward. An al-

FIGURE 8.3 *Bricks in the gravitational field of the Earth.*

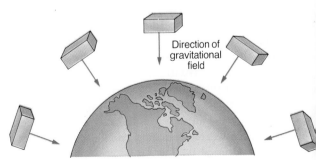

Direction of gravitational field

FIGURE 8.4   *A representation of the gravitational field of the Earth.*

ternative to this answer is provided by the field approach. In this method, it is said that the Earth somehow alters the space around it such that a field is set up in this space; we call this field a gravitational field. When any object that has mass, such as a brick, moves into this gravitational field, it finds a force exerted on it. In this case, the force is one of attraction, which pulls the brick toward the Earth. This alternative approach does not produce much new insight into gravitational forces; therefore, we did not introduce the concept of gravitational fields into our earlier discussions of gravity. Because of the importance of the field concept in electricity and magnetism, however, let us continue with the idea of gravitational fields. A field is defined to be a vector quantity. Thus, it must have magnitude and direction.

The direction of a gravitational field is defined to be in the direction of the gravitational force on an object placed in the field.

Thus, in Figure 8.3, we see that each of the bricks has a force on it directed toward the center of the Earth when the brick is in the gravitational field of the Earth. Thus, at the point where the brick is located, the gravitational field is in the direction indicated in the figure. Regardless of where we go above the surface of the Earth, however, the force of gravitational attraction is always toward the center of the Earth, so we could use lines to represent this gravitational field as shown in Figure 8.4.

The strength of the field is defined as the magnitude of the gravitational force of attraction on the brick divided by the mass of the brick. In equation form,

$$\text{gravitational field} = \frac{F_g}{m_b} \qquad (8.2)$$

where F_g is the force of attraction exerted on the brick by the Earth and m_b is the mass of the brick.

Electric Field

Let us now parallel our discussion of gravitational fields as presented here in an attempt to find an alternative way of looking at electrical forces. When a small positively charged object is near an object having a considerably larger positive charge, what happens? Figure 8.5 shows us that the small object is repelled from the larger object. But why? One answer is that the larger object exerts a force on it given by Coulomb's law; thus, the object is pushed away by this force. Our objective here, however, is to examine the field concept. In this alternative viewpoint, the large charged object somehow alters the space around it such that an **electric field** is set up in this space. According to this outlook, when another charged object is in this electric field, it finds an electrical force acting on it and moves accordingly.

As for gravitational fields, we must define a direction for electric fields:

*The direction of an electric field at any point in space is in the direction of the force that would be exerted on a small **positively** charged object if it were placed at that point.*

Let us consider this definition for a moment. In Figure 8.6a, a large positively charged object is shown. The question is: What is the direction of the

FIGURE 8.5   *The small charge is repelled by the larger charge.*

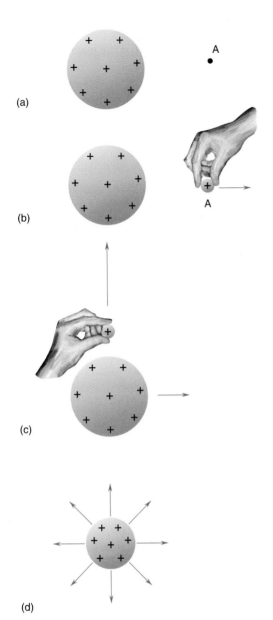

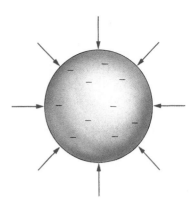

FIGURE 8.7 *The electric field around a negatively charged object.*

FIGURE 8.6 *(a) To find the electric field at point A, we (b) place a tiny positive charge at that point. The direction in which an electrical force acts on that charge is the direction of the field. (c) The electric field can be found at other points to show that the field (d) radiates away from the center of the large, spherical charged object.*

electric field at the point A? To find this direction, in our imagination we place a tiny positive charge at this location and ask in what direction would there be an electrical force acting on it? In Figure

8.6b, we have placed this charge at A, and since positively charged objects repel one another, the electrical force on the test charge is toward the right. Thus, we say that the direction of the field at A is to the right, or it is away from the center of the larger object that sets up the field. If we move all around the large object as in Figure 8.6c, determining the direction of the field at each point, we find that we can represent the field as shown in Figure 8.6d. The electric field of a positively charged object is represented by lines radiating away from the center of the object.

Can you use the definition of the direction of electric field lines to show that the electric field around a large negatively charged object looks like that shown in Figure 8.7?

By analogy with our discussion of gravitational fields, the magnitude of the strength of an electric field is given by the strength of the electrical force exerted on a small object with a charge q divided by the strength of the charge. Thus,

$$\mathbf{E} = \frac{\mathbf{F}_e}{q} \tag{8.3}$$

CONCEPTUAL EXAMPLE 8.2 **WHICH WAY DOES THE CHARGE MOVE?**

(a) You walk into a room carrying a small positive charge. When you release it, you discover that it moves from east to west because of an electrical force acting on it. What is the direction of the electric field in the room?

Solution Recall that the direction of an electric field is the direction in which an electrical force is exerted on a small positive charge. Thus, the direction of the field is from east to west. A conclusion that will be important to us later is that *a positive charge always moves in the direction of an electric field.*

(b) You walk into the room of part a and find that all you have with you is a small negative charge. When you release it, which way will it move?

Solution When you released a positive charge in a, you noticed that it moved from east to west. This must have occurred because of a force of repulsion exerted on the positive charge by the charged object setting up the field. A little thought should convince you, however, that a negative charge should behave exactly opposite a positive charge. Thus, the negative charge will move from west to east. The conclusion that we can draw is *that a negative charge always moves in a direction opposite to that of an electric field.*

We shall discover later that a battery is capable of setting up electric fields inside wires attached to it. Thus, if there are free positive charges in the wire, they move in the direction of this field, and if there are free negative charges in the wire, they move in a direction opposite to that of the electric field.

EXAMPLE 8.3   SPEEDING UP A PROTON
A proton is released at rest in a region where a constant electric field of magnitude 500 N/C exists.

(a) Find the magnitude of the force on the proton.

Solution The force on the proton (charge = 1.6×10^{-19} C) can be found from Eq. 8.3 as

$$F = qE = (1.6 \times 10^{-19} \text{ C})(500 \text{ N/C})$$

$$= 8.0 \times 10^{-17} \text{ N}$$

This is a constant but extremely small force. Remember that it is acting on an extremely light particle, so it can produce large accelerations and large speeds in a short time. Let us investigate this.

(b) What will be the speed of the proton after it has traveled a distance of 1.00 cm? Assume that the proton starts from rest.

Solution The force is constant. Thus, from Newton's second law, we see that the acceleration is also a constant. The acceleration is

$$a = \frac{F_{net}}{m} = \frac{8.0 \times 10^{-17} \text{ N}}{1.67 \times 10^{-27} \text{ kg}}$$

$$= 4.79 \times 10^{10} \text{ m/s}^2$$

To find the speed, we can use the equations of motion with constant acceleration. First, let us find the time it takes for the proton to travel 1 cm from $d = 1/2 \ at^2$. We have

$$0.01 \text{ m} = \frac{1}{2}(4.79 \times 10^{10} \text{ m/s}^2)t^2$$

from which

$$t = 6.46 \times 10^{-7} \text{ s}$$

We can now find the speed from $v = at$. We have

$$v = at = (4.79 \times 10^{10} \text{ m/s}^2)(6.46 \times 10^{-7} \text{ s})$$

$$= 3.09 \times 10^4 \text{ m/s}$$

This is an extremely high speed, achieved in a small interval of time.

The concept of electric fields is important in the study of electricity, and one important value is indicated by the way that sketches of electric fields are drawn. When a small positive charge is placed close to a positively charged sphere, the small charge will be pushed away with a great force. This force of repulsion gradually diminishes as the small charge moves farther and farther away. If the force $\mathbf{F}_e$ is diminishing in strength, however, Eq. 8.3 shows us that the strength of the electric field is also diminishing. Thus, the electric field is strong close to the large charged sphere and becomes weaker as we move away from it. One advantage of the electric field concept is that we can represent the strength of the field by the way in which we sketch the electric field lines. For example, in Figure 8.8a, we see that the lines representing the field are closely spaced near the large positively charged object and become farther apart as they move away. Thus, the spacing of the lines is used to convey the strength of the field. The more closely spaced the lines, the stronger the field.

Figure 8.8b illustrates the field between a positive charge and a negative charge.

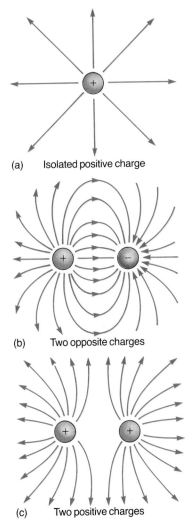

(a)   Isolated positive charge

(b)   Two opposite charges

(c)   Two positive charges

FIGURE 8.8 *Lines of force surrounding electrical charges.*

CONCEPTUAL EXERCISE

What can you say about the strength of the field between the two charges? Is it relatively large or small?

Answer Because the lines are relatively closely spaced in this region, the field is strong.

CONCEPTUAL EXERCISE

Figure 8.8c shows the field lines between two positively charged objects. Describe the field in the region between the two charges.

Answer Because the lines are not close together in this region, the field must be relatively weak.

8.5 VAN DE GRAAFF GENERATOR

Many of the concepts dealing with electric charges, electrical forces, and electric fields can be demonstrated with a device designed and built in 1931 by Robert J. Van de Graaff. His device, called a **Van de Graaff generator,** serves as an excellent teaching tool when the device is constructed on a small scale, but larger models are still in use in nuclear physics research laboratories, where they are used to investigate the nucleus. A classroom-size demonstration unit like that shown in Figure 8.9 may be about 1 m tall, but those in use in nuclear research may be as tall as a two-story building. Regardless of the size, the basic principles of operation remain the same.

Figure 8.10 shows the schematic details of such a generator. A motor in the base drives a pulley, which causes a rubber belt to turn as indicated by the arrows in the figure. In a small-scale device, the rubber belt rubs against a material that produces a negative charge on the belt by a process that is much like that of charging by friction, discussed earlier in this chapter. This negative charge is carried upward by the belt, and at the top of the de-

FIGURE 8.9 *A small-scale Van de Graaff generator that can be used for classroom demonstrations.*

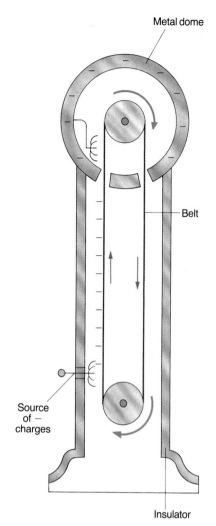

Metal dome

Belt

Source
of −
charges

Insulator

FIGURE 8.10 *Cut-away view of a Van de Graaff generator.*

FIGURE 8.11 *A person touching the charged dome also becomes charged.*

it is being charged, the person becomes a part of the Van de Graaff dome, and charge spreads out all over the body (Fig. 8.11.) (The purpose of the insulating stand is to prevent charge from leaking off the person through the feet to the floor.) The fact that the person's body is covered with a large charge can easily be seen if the person has straight,

FIGURE 8.12 *Protons are repelled by the positively charged dome. When they strike the target, nuclear reactions may be produced.*

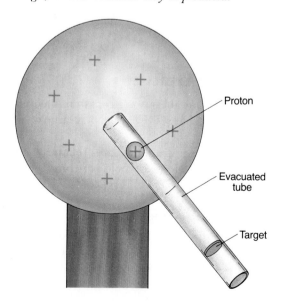

Proton

Evacuated
tube

Target

vice, a comblike row of points is suspended near the belt. Because the negative charges are repelled by one another, they move as far away from one another as possible, so they move from the belt and onto the metal contacts. From there they spread out all over the surface of the metal dome. In this fashion, it is possible to give the dome quite a large negative charge. (In larger machines, the charge is sprayed onto the rubber belt by devices called ion sources instead of being produced by friction. These are capable of giving the belt either a negative or a positive charge; hence, the dome can end up with a charge of either sign.)

If a person stands on an insulating stool and places his or her hand on the top of the dome while

fine hair. The charges spread out all over the hair, and the repulsion between these charges causes the hairs to stand out away from the head as they repel one another.

In a more important application, these devices are used in physics laboratories to produce nuclear reactions. For our example, let us suppose that the dome has been given a positive charge, as shown in Figure 8.12. A tube from which the air has been removed is attached to the dome, and protons are released inside the tube at a location near the dome. The repulsion between the positively charged dome and the positively charged protons causes the latter to accelerate away from the dome through the evacuated tube. At the end of the tube, the protons strike a target made of the nuclei that the investigator wishes to study. Nuclear reactions will be discussed in detail in Chapter 13.

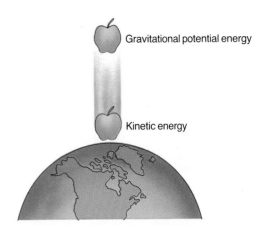

FIGURE 8.13   *The gravitational potential energy of the apple is converted to kinetic energy as it falls toward the Earth. The falling apple is capable of doing work.*

CONCEPTUAL EXERCISE

Explain why touching the generator causes this person's hair to stand on end.

8.6 VOLTAGE

The study of practical electric circuits can be made easier by referring to certain analogous topics covered when we examined mechanics. Think of an apple near the surface of the Earth. If we grasp the apple and lift it into the air, we do work on it, and as we saw in our study of energy, the potential energy of the apple is increased. If the apple is then dropped, the potential energy is converted into kinetic energy, as shown in Figure 8.13. In turn, the moving apple can perform work for us. The same situation arises for electric charges. Suppose you have a large positively charged object (analogous to the Earth) and a small negatively charged object (analogous to the apple) (Fig. 8.14). To pull the negatively charged object away from the positively charged one, you must do work. When the two are separated, the small charge has an energy due to its position, and we call this kind of energy **electrical potential energy,** by analogy to our use of the term "gravitational potential energy." If the negative charge is released, it accelerates toward the positive charge, and the potential energy is converted into kinetic energy. Once again, work can be done for us.

FIGURE 8.14   *Charged objects can have electrical potential energy.*

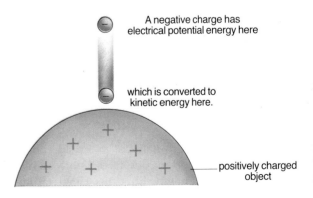

FOCUS ON
A BLACK-AND-WHITE TELEVISION RECEIVER

The essential pieces of a black-and-white picture tube are shown in part (a) of the figure. (It should be noted here that a common piece of electronic equipment to view electrical signals is the oscilloscope, and the method of operation of the tube in one of these devices is similar to that of a picture tube in a television receiver.) In the neck of the picture tube, a metal cylinder A, called an electron gun, is heated by an electric current until the temperature of the cylinder becomes hot enough to boil off electrons. Located a short distance down the tube is a plate B that has been given a strong positive charge by electrical circuits not shown in the figure. The negatively charged electrons are attracted toward this

plate and gain a high speed just before they reach it. Most of the electrons strike this plate, but a few pass through a hole in its center and continue on in a straight-line path until they strike the front surface of the picture tube. The front of the tube is painted with small phosphorescent dots that emit white light when struck by electrons. Thus, an observer watching the screen would see a bright spot of light when electrons strike it.

Watching a single spot of light like this does not make for stimulating programming, so to enhance the excitement, two sets of plates are installed in the neck of the tube to make the electron beam, and hence the spot of light, move around. These plates are the horizontal and vertical deflection plates. To understand how these deflection plates work, let's consider them sepa-

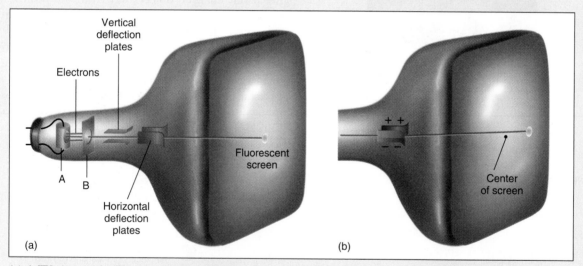

(a) A TV picture tube. Electrons from A are accelerated toward plate B and then move toward the fluorescent screen. The charged plates deflect the beam from point to point on the screen. Gradually increasing the charge on the plates as shown would cause the electron beam to move to the right side of the screen.

The concept of gravitational potential energy is an extremely important one in the study of mechanics. Thus, one might think that electrical potential energy would be just as valuable in our study of electricity. Granted, electrical potential energy is important to us now, but it turns out that another

rately. (As we shall see in Chapter 9, the beam is actually moved by magnets placed along the neck of the tube.)

Initially, let's assume that the electron beam travels directly down the center of the tube and forms a spot of light at the center of the screen. External circuits can change the amount and the sign of the charge present on the horizontal plates. Assume that positive charge is gradually placed on one of the plates and negative charge on the other. This charge produces an electric field on the space between the plates, which causes the electron beam to be gradually deflected from its straight-line flight path. You can follow the movement of the electron beam by watching the movement of the spot of light on the face of the screen. For the situation shown in part (b) of the figure, you would see the spot gradually move to the right side of the screen. What would happen if the following steps were taken in sequence? (1) The horizontal plates are charged so that the beam starts at the extreme left side of the screen. (2) Now the charge on the plates is gradually changed so that the beam slowly moves to the right side of the screen. (3) The electron beam is turned off when the spot of light reaches the right side of the screen, and (4) the charge on the plates is changed so that the beam is again focused at the extreme left position when it is turned back on. (5) The steps are continually repeated in the order given.

The answer is that an observer would see a spot of light continually sweeping from left to right on the screen face. Still this is nothing to get excited about, so let's include the vertical plates in our discussion. Combined action of the horizontal and vertical plates causes the electron

beam to sweep across the tube face to form a line at the top of the tube. As the beam sweeps across the top, the number of electrons leaving the electron gun is controlled by varying the heating current. Thus, the line is neither all dark nor all light, but shadings of intensity of the light are observed. The plates are then charged so that the beam moves downward slightly and sweeps out another line across the screen. In a television set, 30 complete scans across the *total* face of the tube are made each second. By varying the strength of the current with a control grid, the varying shades of dark and light can produce a picture on the screen. The rapid scanning of the beam across the screen produces a series of pictures so rapidly that the eye cannot follow the production of an individual one. Thus, a series of pictures is flashed before the eye, each slightly different from the preceding one, and the perception of motion is produced. The strength of the electron beam current is controlled by a signal sent out by a TV transmitting station and received by the antenna.

The "picture tube" in a test instrument called an oscilloscope is much like that in a television receiver.

related concept is of even greater importance. We shall find that in the study of electric circuits, a quantity called **potential difference,** with units of volts, hence, in colloquial terms, **voltage,** is more frequently used. *The difference in potential between two points is defined as the work performed by a charge q as* *it moves between these points divided by the magnitude q of the charge.* An alternative form of this same definition is that *the potential difference between two points is equal to the energy that a charge q transforms from electrical energy to other forms of energy as it moves between the points divided by the magnitude of the charge q.* In equation form this is expressed as

$$\text{potential difference} = \frac{\text{electrical energy transformed}}{\text{charge}} \quad (8.4)$$

or symbolically

$$V = \frac{\Delta(\text{energy})}{q} \quad (8.5)$$

where V is the change in potential difference, $\Delta(\text{energy})$ is the energy transformed, and q is the charge in coulombs. As Eq. 8.5 indicates, the units of potential difference are joule/coulomb,

$$V = \frac{1\,\text{J}}{1\,\text{C}} = 1\text{V}$$

where the units of joule per coulomb, J/C, have been defined as the **volt**, V.

If you are a typical student beginning the study of physics, the concepts of potential difference and voltage seem somewhat obscure. To get a better feel for how voltage can be of importance, we will begin our consideration of electric circuits in the next section.

8.7 ELECTRIC CIRCUITS

Figure 8.15 shows a simple electric circuit consisting of a battery connected to a light bulb and a switch. Under normal operating conditions, charges move through the wires and the bulb, and the bulb lights as shown (Fig. 8.15a). If the switch is opened, leaving a gap in the circuit, the charges cannot cross the air gap, and the bulb does not glow (Fig. 8.15b). For the charges to move, there must be a closed conducting path for them to move through. For convenience in sketching our circuits, we shall use certain symbols to represent the various pieces that can go into a circuit. For example, a battery is represented by the symbol |ı. *A device*

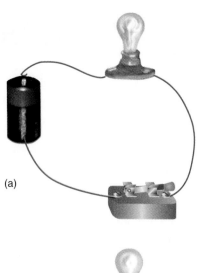

(a)

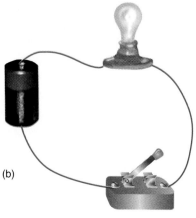

(b)

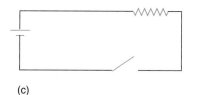

(c)

FIGURE 8.15 *(a) The bulb glows if there is a complete circuit. (b) The bulb does not glow if the switch is open. (c) Schematic diagram of the circuit in part (b).*

that hinders the motion of charge through a circuit is called a resistor and is represented by the symbol ⌐⌐. A switch is represented as ⌐⌐. These symbols enable us to draw the circuit of Figure 8.15b as shown in Figure 8.15c.

Let us consider the battery in our circuit shown in Figure 8.16, and let us assume that it is a 12 V

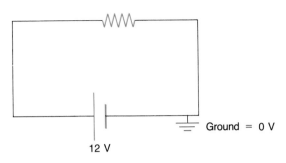

FIGURE 8.16 *A simple circuit.*

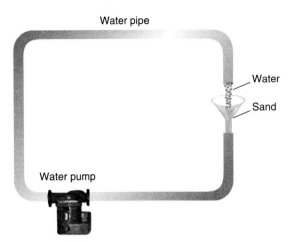

FIGURE 8.17 *A water pump analogy for the electric circuit of Figure 8.16.*

battery. When we say that a battery has a voltage of 12 V, we mean that chemical reactions occurring inside the battery maintain one of its terminals at a potential 12 V higher than that of the other. The long line drawn in the symbol for a battery is used to indicate which of the two terminals is being held at the higher potential. This high-voltage terminal is often called the positive terminal, whereas the other is often called the negative terminal. For convenience in our discussion, let us assume that the negative terminal is held at a voltage of 0 V. Any point in a circuit that is held at a voltage of 0 V is said to be grounded, and a ground location in a circuit is indicated by the symbol shown in Figure 8.16.

Now let us consider the motion of charges in our circuit, and also for convenience in our discussion, let us assume that positive charges move through the circuit. (We shall see in a later section that actually negative charges move in a circuit.) We have already noted that our battery is a 12 V, or a 12 J/C, battery. This means that every 1 C of charge that leaves the positive terminal carries with it 12 J of electrical energy. This energy is furnished to the charges by the chemical reactions occurring inside the battery. As our 1 C of charge moves around the circuit, it gives up this electrical energy to the pieces of equipment that make up the circuit. For our example, the only piece of equipment that we have included is the light bulb, indicated as a resistor. In the resistor, the 12 J of electrical energy supplied by the 1 C of charge is converted to other forms of energy, light and heat. When the 1 C of charge returns to the negative terminal, it returns with zero electrical energy. It has given up all of its energy to the devices it encounters. The charge then moves through the battery, where chemical reactions replenish its electrical energy to a 12 J level once

again, and the charge is ready to begin another transit of the circuit. In this way, charges moving through a circuit continually supply power to light bulbs, stereos, radios, or whatever else the battery is connected to, until the chemicals inside the battery are used up and the battery "runs down."

It is often helpful to compare a circuit like that of Figure 8.16 to a mechanical system like that of Figure 8.17. In Figure 8.17, water is caused to circulate through a pipe by means of a water pump. The pump is analogous to a battery, and the motion of the water is analogous to the motion of charges in the electric circuit. (Loosely speaking, a battery is a charge pump.) The water then drops into a funnel filled with loosely packed sand. The sand, which inhibits the flow of water, is analogous to the resistance of an electric circuit.

8.8 ELECTRIC CURRENTS

In the previous section, we discussed the motion of charges in an electric circuit as though positive charges do the moving. This is not the case; it is actually the electrons that move. When electric circuits were first studied, however, it was not known whether it was the positive charges or the negative charges that were in motion. As a result, early investigators discussed circuits in terms of the motion of positive charges and defined the sense of a current in terms of positive charge movement. In Fig-

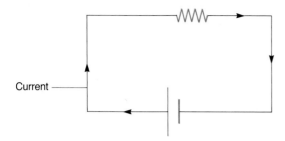

FIGURE 8.18   *The sense of an electric current is from the positive terminal of a battery toward the negative terminal.*

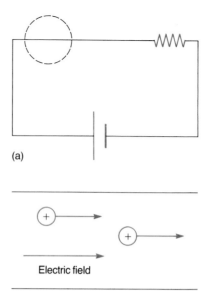

(a)

Electric field

(b)

FIGURE 8.19   *(a) The dashed circle encloses the part of the circuit we will investigate. (b) In this portion of the circuit, there is an electric field directed as shown. This field would cause positive charges to move from left to right.*

ure 8.18, we show the direction selected for the sense of the electric current. We assume that the charge exits the positive terminal, moves around the circuit, and re-enters the battery at the negative terminal. Thus, *we define the sense of a current to be in the direction of motion of positive charges.* We know that this cannot be the true state of affairs because, as we noted in our discussion of charging by friction, the positive charges are rigidly fixed in place in a wire, and it is the electrons that are free to move around. What actually happens is that electrons leave the negative terminal, move around the circuit, and re-enter the positive terminal to have their energy replenished.

To see that it makes no difference whether we consider positive charges in motion or negative charges in motion, consider the following situation. Suppose you have five positive charges stacked up against the left wall of a room and five negative charges stacked up against the wall on the right side of a room. Now, move one positive charge from its location to the side of the room where the negative charges are located. After the movement, there are four positive charges against the left wall and a net of four negative charges against the right wall. The net number of negative charges has been decreased by one because the positive charge cancelled one of the negative charges. If we had moved the negative charge from the right wall to the left wall, we would be left with four negative charges against the right wall and four positive charges against the left wall. Again the negative charge would have cancelled out the effect of one of the positive charges. So, it makes no difference which charge actually moves; the end result is the same in both cases.

Now imagine that you shrink to a size small enough to fit into a wire and that you are carrying

with you a device to "count charge" and a watch. The segment of the wire is outlined in Figure 8.19a. The first thing that you notice when you enter the wire is that there is an electric field present that has the direction shown in part (b) of the figure. A battery sets up an electric field in a wire that originates on the positive terminal of the battery and stops on the negative terminal. Thus, in the magnified segment of the wire shown in Figure 8.19b, you see an electric field directed from left to right. Positive charges tend to move in the direction of an electric field, so you see positive charges moving past you from left to right. Thus, the sense of the current in the wire is from left to right in Figure 8.19b.

The quantity of charge moving through a cross-sectional area of the wire divided by the time required for the charge to move through this area is defined as the magnitude of the current in the wire.

This is

$$I = \frac{Q}{t} \qquad (8.6)$$

where I is the symbol used to represent current, Q is the charge moving through a cross-sectional area, and t is the time required for the charge to move through the area. The units of I are coulombs per second, C/s, and these units are grouped together and referred to as an **ampere** (amp), symbol A.

$$I = \frac{1 \text{ C}}{1 \text{ s}} = 1 \text{ A}$$

Let us consider one more detail concerning the motion of charges through a wire. If one could determine the actual speed with which charges move through a wire, one would find it to be surprisingly small. Usually the speed is such that an individual charge moves only a small fraction of a centimeter in 1 s. If this is the case, why is it that when we turn on a wall switch, a light bulb comes on almost instantaneously? If the electrons are moving slowly, it seems that it might take several hours for one to leave the wall socket and finally reach the light. To understand what happens, consider the analogy of dominoes. When stood on end and lined up in close proximity to each other, a push at the end of the line will rapidly knock down each domino in succession. This same effect happens to the charges in a wire. As a charge enters the wire from the wall outlet, it pushes on those in front of it in the wire, and these pushes cause a charge to leave the wire at the other end. Thus, the movement of charges in the wire is established almost instantaneously. As we noted, it is the presence of an electric field in the wire that is causing the charges to move, and when you turn on a switch in a circuit, the electric field inside the wire is set up at the speed of light, 3.00×10^8 m/s. The electric field moves through the wire far more rapidly than do the individual charges.

EXAMPLE 8.4   GETTING A FEEL FOR CURRENTS

In a circuit, 3.0×10^{-3} C of charge moves past a cross-sectional area every 2.0 s. Find the magnitude of the current in the wire.

Solution The current can be found from Eq. 8.6 as

$$I = \frac{Q}{t} = \frac{3.0 \times 10^{-3} \text{ C}}{2.0 \text{ s}} = \boxed{1.5 \times 10^{-3} \text{ A}}$$

$$= \boxed{1.5 \text{ mA}}$$

where the symbol m = milli = 10^{-3}

EXAMPLE 8.5   COUNTING CHARGES

The current in a circuit is 3.0 A. How many charges, each having a charge of 1.6×10^{-19} C, pass through a cross-sectional area of the wire every second?

Solution Let us first find the amount of charge that flows past the cross-sectional area each second. We do this by use of the definition of current, Eq. 8.6.

$$Q = It = (3.0 \text{ A})(1 \text{ s}) = 3.0 \text{ C}$$

But if each charge has a magnitude of 1.6×10^{-19} C, the total number of positive charges is

$$\text{number} = \frac{3.0 \text{ C}}{1.6 \times 10^{-19} \text{ C/charge}}$$

$$= \boxed{1.9 \times 10^{19} \text{ charges}}$$

8.9 RESISTANCE

We have said that certain devices placed in an electric circuit impede the motion of charges through the circuit. This hindrance to the motion of charges is called **resistance.** To understand what produces the resistance in a circuit, let us consider the motion of a single charge as it moves through a wire that has resistance. When the charge leaves the positive terminal of the battery, it has a high electrical potential energy, and it is moving toward the negative terminal of the battery, where it will have zero electrical potential energy. Thus, from the conservation of energy, we see that the kinetic energy of the charge will increase as it moves. Periodically, however, the charge collides with the atoms of the wire, and when one of these collisions occurs, the moving charge is slowed. The kinetic energy of the moving charge is given to the struck atom, which causes the atom to vibrate about its fixed location with a larger amplitude of vibration than it had before the collision occurred. After the collision, the moving charge begins to speed up once again, until it collides with another atom. Thus, it is the collisions with the fixed particles of the material that hinder the motion of a charge through the wire, and this hindrance is called the resistance of the wire.

The collisions inside a wire carrying a current cause the atoms of the wire to vibrate with larger than normal amplitudes, and we recognize this fact by a temperature increase of the wire. Thus, inside a material that has resistance, kinetic energy of the moving charges is being transformed into heat energy. If the material gets hot enough it will glow, first with a dull red light and then, as it gets hotter, with a white light. Inside a light bulb, the charges move through a piece of tungsten wire, called the filament of the bulb, which has a high resistance to the motion of charges. A current of about 1 A through the filament causes its temperature to reach about 2000°C. At this temperature, the filament becomes white-hot and emits light.

Materials that have a high resistance to the motion of charges are called **insulators.** Examples include asbestos, rubber, and glass. Other materials, such as the metals gold, silver, and aluminum, allow charges to move through them freely. These materials are called **conductors.** It is convenient for us that there exists a wide range of resistances in the materials in our environment. Thus, it is easier for an electric current to pass through thousands of kilometers of low-resistance wire in an overhead transmission line than through the few centimeters

FOCUS ON
SUPERCONDUCTIVITY

In the last few years, a lot of renewed attention has been directed toward new advances in the study of a class of materials called **superconductors.** Superconductors are materials that behave basically like ordinary substances above a temperature called the critical temperature. As the temperature is decreased to and below the critical temperature, they behave as though they have zero resistance. This phenomenon was discovered by Kamerlingh-Ohnes in 1911 when he was working with mercury. He found that when the material was cooled below 4.2 K, its resistance dropped essentially to zero. Since that time, thousands of other materials have been found to exhibit superconducting properties. Aluminum, tin, lead, zinc, and indium are some common examples. The feature of superconductors that has kept them from having many practical applications and more widespread use is that the critical temperature is extremely low for all of them. In 1986, an oxide of barium, lanthanum, and copper was discovered to be superconducting at 30 K by George Bednorz and K. Alex Muller, working at the IBM Zurich Research Laboratory in Switzerland. The

two were awarded the Nobel Prize in 1987 for their research. In 1987, scientists at the University of Alabama in Huntsville and the University of Houston reported a critical temperature of 92 K for an oxide of yttrium, barium, and copper. Later in 1987, the critical temperature went to 105 K for an oxide of bismuth, strontium, calcium, and copper, and in 1988 an oxide containing thallium was discovered to be superconducting at 125 K. In December 1993, two French laboratories reported superconductivity at 250 K. Thus, the rise in critical temperature toward room temperature is increasing, and it is not out of the question that such a goal could be reached in a few years.

If room temperature superconductors could be developed, the electronics industry would be revolutionized. Superfast computers could be developed, and all electronic items would feel the impact of the new technology. One of the main problems with the distribution of electrical energy across country is that there are unavoidable power losses in the transmission lines caused by the resistance of the wires. If these wires could be constructed of zero-resistance materials, this problem would be solved, and the cost of electrical energy could drop drastically.

of insulating matter that stands between the wire and its supporting tower.

8.10 OHM'S LAW

It is debatable which of the laws that we studied during our investigation of mechanics stands as the most important, but many scientists would pick Newton's second law. Similarly, if one had to choose the most important law in the study of electricity, Ohm's law would finish high on the list. The law, discovered by George Simon Ohm (1787–1854), relates the voltage supplied to a circuit to the current in the circuit and the resistance of the circuit. In equation form, **Ohm's law** is

$$V = IR \qquad (8.7)$$

where V is the applied voltage, I is the current in the circuit, and R is the resistance. From this equation, we see that the units of resistance are V/A. These units are grouped together and called an **ohm.**

$$R = \frac{V}{I} = \frac{1\ V}{1\ A} = 1\ \Omega$$

Resistances are thus stated in terms of ohms, where the symbol Ω (Greek upper case omega) is used in place of the word ohm. Thus, a good insulator might have a resistance of $10^8\ \Omega$, and a good conductor might have a resistance of $10^{-3}\ \Omega$.

To see that Ohm's law agrees with common sense, let us consider a simple circuit consisting of only a battery and a resistor (Fig. 8.20). Solving Ohm's law for the current, I, we find

$$I = \frac{V}{R}$$

What can we do to alter the current in this circuit? To increase the current, it seems reasonable that we could do one or more of the following: We could increase the applied voltage, that is, use a higher-voltage battery; we could decrease the resistance of the circuit; or we could do both. Isn't this what Ohm's law says we should do? Similarly, we could decrease the current by doing the opposite of these suggestions.

EXAMPLE 8.6   MAKING TOAST

The heating element inside a toaster has a resistance of 12 Ω. In use, a toaster is connected to a wall outlet, which supplies 120 V. (This voltage is an alternating voltage, but for our example, the direction in which the current moves is of no significance and we can still use Ohm's law.) Find the current in the heating element of the toaster.

Solution From Ohm's law we find the current to be

$$I = \frac{V}{R} = \frac{120\ V}{12\ \Omega} = 10\ A$$

CONCEPTUAL EXERCISE

Find the resistance of a light bulb that carries 0.50 A when connected to a 120 V source.

Answer 240 Ω

CONCEPTUAL EXERCISE

You are often warned against operating electrical devices while in a shower or a bathtub. Use Figure 8.20 to explain why this is good advice.

Answer An electrical device connected to a wall outlet can provide a voltage like the battery shown in the figure. The water, which is a good conductor, and your body act like the connecting wires and resistor in the circuit. Finally, the circuit is completed when the current has a place to go, and that place is through the metal pipes to the earth. A large current can go through the body, sufficient to cause death by electrocution.

FIGURE 8.20   *What could be done to this circuit to change the current, I?*

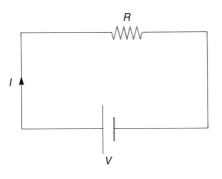

FOCUS ON
MICROPHONES AND TELEPHONE RECEIVERS

There are many different types of microphones in use today, but we shall concentrate here on the type that is found in the mouthpiece of a telephone, the carbon microphone. The figure shows the design of such a microphone and the details of how it is connected to a circuit. The primary resistance in the circuit is the carbon granules packed into a box inside the mouthpiece. When a sound wave strikes the steel diaphragm on one side of the box, it flexes in and out following the compressions and rarefactions of the sound wave. When the carbon granules are compressed tightly together, they make intimate contact with each other, and the resistance of the circuit is low. When the diaphragm flexes outward because of a rarefaction, the granules become more loosely packed, and the resistance in the circuit is high. The current in the circuit, produced by the battery, follows these changes in resistance, fluctuating between a large value and a small value. These variations in current are sent through a transformer and into the transmission lines of the telephone company. (Transformers will be discussed in Chapter 9.) These changing currents are converted by the listener's earpiece back into sound waves.

Carbon microphones are fine for telephone transmission because they reproduce sound frequencies quite well up to 4000 Hz, and most of the frequencies of normal conversation are below this value. At higher frequencies, they are not efficient. Thus, if someone tries to play a flute solo over your telephone, the quality of the sound may not be satisfactory.

Diagram of a carbon microphone.

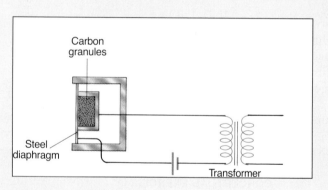

8.11 DIFFERENT TYPES OF ELECTRIC CIRCUITS

If several appliances are to be wired into a single circuit, they can be connected either in **series** (Fig. 8.21a) or in **parallel** (Fig. 8.21b). The characteristics of parallel and series circuits are different from one another. One obvious difference can be noted by a casual glance at Figure 8.21a and b. In a series circuit, connections are made such that there is only one path for charges to take as they move around the circuit. In a parallel circuit, there are branches in the circuit that allow some charges to follow one branch while other charges follow another pathway. Let us study each of these classifications of circuits in turn.

Series Circuits

Let us build a series circuit by starting with one that contains only a single 100 Ω resistor, say a light bulb, as in Figure 8.22a. Now let us add more resistors such as a stereo and a toaster, as in Figure 8.22b.

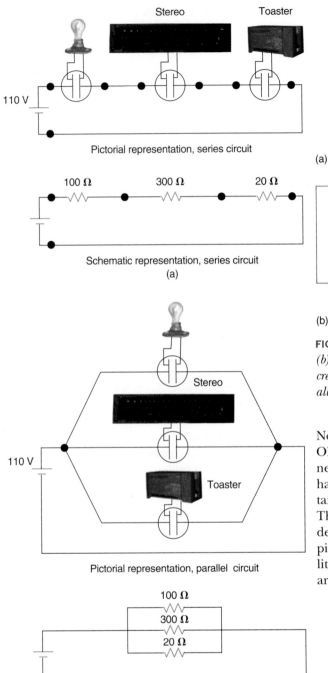

Pictorial representation, series circuit

100 Ω   300 Ω   20 Ω

Schematic representation, series circuit
(a)

Pictorial representation, parallel circuit

100 Ω
300 Ω
20 Ω

Schematic representation, parallel circuit
(b)

FIGURE 8.21 *(a) A series circuit. (b) A parallel circuit.*

As these appliances are added, the total resistance connected to the battery increases. (This is like adding more and more sand inside a water pipe.)

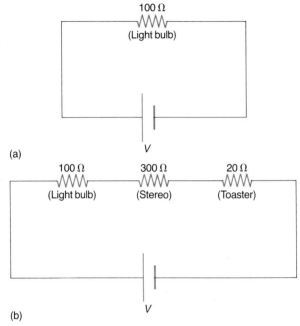

(a)

100 Ω   300 Ω   20 Ω
(Light bulb)   (Stereo)   (Toaster)

(b)

FIGURE 8.22 *(a) A series circuit with one resistor. (b) As more resistors are added, the current must decrease, but the current will still have the same value at all points in the circuit.*

Now let us examine what we have done by use of Ohm's law, written as $I = V/R$. The voltage V connected to the circuit has not changed because we have not changed the battery; however, the resistance has increased as more appliances are added. Thus, we see that the current in the circuit must decrease. Series circuits are integral parts of every piece of electronic equipment that you own, but a little thought shows that for household wiring, this arrangement is unsatisfactory for several reasons.

1. Each new resistor that you add decreases the current in the circuit. As a result, if you add a new lamp to a room, the new light bulb decreases the current supplied to that room. This would cause the light bulbs already present to glow a little less brightly, your toast would be light rather than dark, and so forth.

2. In a series circuit, there is only one pathway for the charges to move along. This means that the current through all parts of the circuit has the same value. For example, in Figure 8.22b, if the current through the light bulb is 0.5 A, the current in the stereo is also

0.5 A, and the toaster likewise has the same current. This has to be true because the charge motion through the circuit is like that of water flowing through a pipe that does not leak. If 12 gallons per minute are flowing through a pipe at one location in a series plumbing connection, there must be 12 gallons per minute flowing at all other locations in the piping. If not, either water would be escaping from the pipes or it would be building up at some location. Such a set of circumstances does not occur for water, and it certainly does not occur for charges flowing through an electric circuit. We have already discussed the fact that charge is conserved, and if the current were different at different parts of a series circuit, this would imply that charges are being either created or destroyed at some location in the circuit. Household circuits would not be effective at all if they were designed such that all devices connected in the line had to have the same current in them. Some devices such as an electric heater require large currents, whereas others such as a stereo may require only a small fraction of an amp. Also, certain devices such as a computer are designed for current at a constant, specified value. Think what would happen if you were working on your computer and you added a lamp to your series wiring arrangement. The addition of the light would decrease the current in the entire circuit.

3. A series circuit has yet another disadvantage that is perhaps even more important than the others mentioned. Imagine that the light bulb in Figure 8.22b burns out. When a light bulb burns out, the filament in the bulb actually breaks. Therefore, the circuit is interrupted as shown in Figure 8.23, and no current at all can exist because there is not a complete, uninterrupted pathway through which the charges can move. This means that if your house were wired in series and one light bulb burned out, all the lights would go out. In addition, your toaster, stereo, and any other appliances would also stop. Of course, you wouldn't know which light bulb was defective originally, so you would have to grope around to find the culprit that broke the circuit. Some Christmas tree lights are wired in series, so perhaps you

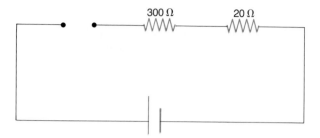

FIGURE 8.23   *Series circuit with light bulb burned out.*

already know from practical experience the frustration of trying to find the bad bulb in a long chain.

Parallel Circuits

We can construct a parallel circuit beginning as shown in Figure 8.24a, where we have connected a light bulb to our battery. As we progress from Figure 8.24b to c and beyond, we are connecting other devices to the battery. As each new device is added, entirely new pathways are formed for charges to move through. For example, in Figure 8.24c, charges leaving the battery have a choice of pathways when they reach point A. Some charges move through the light bulb to point B and then back to the negative terminal of the battery. Simultaneously, other charges follow the pathway from A through the stereo to B and back to the battery, and still others move from A through the toaster to B. Thus, each new device that we add forms an independent route from the battery through the circuit. An arrangement of this nature is ideal for a household electric circuit because all the difficulties associated with a series arrangement discussed in the previous section are avoided.

1. When more appliances are added to a parallel circuit, the current through each one is independent of the current through the others. The circuit of Figure 8.24c is not affected when, say, the light bulb is turned off because there is still a pathway for charges to move through all the other devices.
2. Because each appliance is independent of the others, the current carried by each is dictated solely by the resistance of the appliance.
3. If one appliance in a parallel circuit burns out or is unplugged, as shown in Figure 8.25,

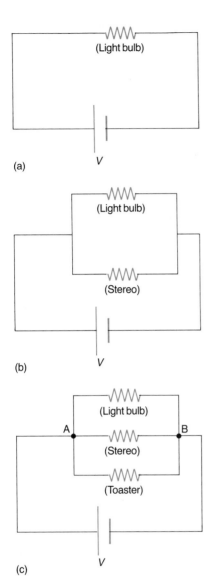

(a)

(b)

(c)

FIGURE 8.24 *A parallel circuit provides alternative pathways for currents to flow.*

the other pathways are unaffected, and the appliances in the other branches operate normally. Thus, if one light bulb burns out, the others work just as well, and it is easy to find the dark one.

CONCEPTUAL EXERCISE

In the fuse shown in Figure 8.27, the current through it causes the fuse strip to melt when the current becomes great enough. Should this device be connected in series or parallel with the equipment it is to protect?

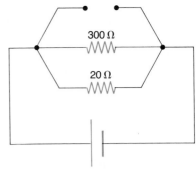

Parallel circuit with light bulb burnt out

FIGURE 8.25 *If one appliance in a parallel circuit burns out, the other pathways still operate normally.*

Answer If it were in parallel, the melting of the fuse strip would ensure that no current would flow through the fuse, but there would still be a pathway through the equipment for current to move. Thus, no protection would be provided. If connected in series, once the fuse strip melts, there is no pathway for current to move, so everything shuts down. Fuses are always connected in series.

EXAMPLE 8.7   RESISTORS IN SERIES
A certain light bulb has a resistance of 144 Ω.
(a) Find the current through this light bulb when it is connected to 120 V.

Solution Ohm's law provides the answer as

$$I = \frac{V}{R} = \frac{120 \text{ V}}{144 \text{ } \Omega} = \boxed{0.833 \text{ A}}$$

(b) If two of these light bulbs are connected in a series circuit, find the current through each.

Solution The total resistance connected to the voltage source will now be 288 Ω. Thus, the current will be

$$I = \frac{V}{R} = \frac{120 \text{ V}}{288 \text{ } \Omega} = \boxed{0.417 \text{ A}}$$

Note that the supplied voltage does not change. Thus, the current supplied to each will be one-half the value that is supplied to one alone. As a result, the bulbs will dim as more are added.

EXAMPLE 8.8   PARALLEL CIRCUITS AND THE NEED FOR FUSES

The two light bulbs of the previous example are connected in parallel as shown in Figure 8.26a.

(a) Find the current through each and the total current drawn from the voltage supply.

Solution Each of the light bulbs is connected directly across the 120 V voltage supply. Thus, the current through each is

$$I = \frac{V}{R} = \frac{120 \text{ V}}{144 \text{ }\Omega} = \boxed{0.833 \text{ A}}$$

Thus, we see from Figure 8.26b that the total current drawn from the voltage supply is 1.67 A.

(b) An electric heater is now added to the circuit as in Figure 8.26c. The resistance of this heater is such that it will draw a current of 19.00 A when turned on. Find the current now drawn from the supply voltage and discuss the implications of what is happening.

Solution The total current drawn from the voltage supply is

$$I_T = 1.67 \text{ A} + 19.00 \text{ A} = 20.67 \text{ A}$$

The supply lines running through the house have a low resistance, but they still do have *some* resistance. As a result, these wires get hotter as the current through them increases. 20.67 A is a dangerously high level for them to carry because there is a danger that they can overheat and cause a fire. To ensure that the current does not exceed safe levels, fuses are inserted in the supply line, as shown in Figure 8.26d. Fuses are protective devices that allow some maximum current, say 20 A, in the circuit before they "blow," causing all current to stop. A typical fuse is shown in Figure 8.27. It consists of a metal strip through which the current passes. This strip is made of a metal that melts when the current reaches a certain level. In modern homes, alternative devices called circuit breakers are used instead of fuses. Most of these devices are designed to use electromagnets, but the basic purpose remains the same—to cut off the current before it reaches a dangerous level.

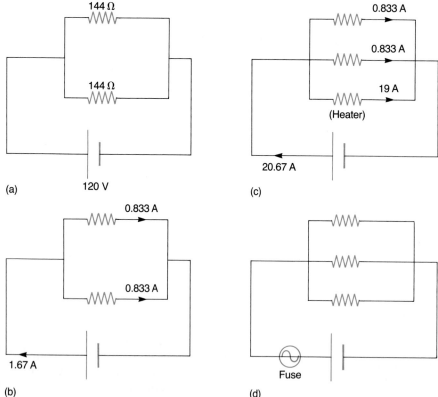

FIGURE 8.26   *Example 8.8.*

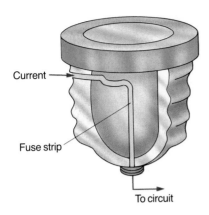

FIGURE 8.27   *A simple fuse.*

Current
Fuse strip
To circuit

$$\times \left(\frac{\text{energy given to device}}{\text{charge passing through device}} \right)$$

$$= \frac{\text{energy given to device}}{\text{time}} = \text{power} \qquad \textbf{(8.9)}$$

Thus, we can find the power supplied to a device by finding the product of the current drawn by the device and the voltage supplied to it.

The equation power = *IV* can be used regardless of the kind of circuit element you are considering. That is, you could find the power supplied to a circuit by a battery or the power converted to heat by a resistor. There is an alternative equation that applies *only* to resistive elements in a circuit, such as a light bulb. This alternative form is found by substituting $V = IR$ for V in Eq. 8.9. We find

$$\text{power} = IV = I(IR) = I^2R \qquad \textbf{(8.10)}$$

CONCEPTUAL EXERCISE

When starting your car with the headlights on, you will notice that the lights dim. Why?

Answer When starting the car, the starter motor must be activated by the battery. This requires a large amount of current from the battery, so much, in fact, that it can decrease the voltage of the battery momentarily. If *V* is decreased, the lights will be supplied with a lower current, causing them to dim.

CONCEPTUAL EXERCISE

Which will require the most power to run, a large-screen television set or a small electric heater?

Answer There is no way to answer this question from the information provided. The size of an electrical device has nothing to do with the amount of power it dissipates. To answer the question, one would have to know how much current each draws when in operation.

8.12 POWER

In our study of mechanics, we defined power as the rate at which work is done or, alternatively, as the rate at which energy is transformed from one form to another. The equation used is

$$\text{power} = \frac{\text{energy transformed}}{\text{time for the transformation}} \qquad \textbf{(8.8)}$$

and the units are power = J/s = W, where W is the symbol for a watt, as we saw in Chapter 3.

For electric circuits, power is easily calculated if one knows the voltage and the current. Current tells us how much charge passes through a device each second. Voltage tells us how much energy each of these charges gives up to the device. Let us examine the units for the product of current times voltage, *IV.*

$$IV = \left(\frac{\text{charge passing through device}}{\text{time}} \right)$$

EXAMPLE 8.9   **FINDING THE POWER**
(a) In the circuit of Figure 8.28, find the current drawn by the two resistors.

Solution The total resistance of the circuit is 2 Ω + 4 Ω = 6 Ω, and the total voltage supplied to the cir-

FIGURE 8.28   *Example 8.9.*

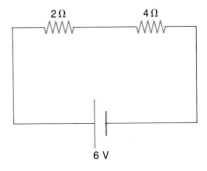

2 Ω   4 Ω

6 V

cuit is that of the 6 V battery. Thus, the current is

$$I = \frac{V}{R} = \frac{6\text{ V}}{6\text{ }\Omega} = \boxed{1\text{ A}}$$

(b) Find the power supplied to the circuit by the battery.

Solution The power supplied is found from power = IV. We have

$$\text{power} = IV = (1\text{ A})(6\text{ V}) = \boxed{6\text{ W}}$$

(c) Find how much power is consumed by each resistor.

Solution The power consumer by a resistor is found from power = I^2R. The current through both resistors is the same because this is a series circuit. Thus, the power consumed by the 2 Ω resistor is

$$\text{power} = I^2R = (1\text{ A})^2(2\text{ }\Omega) = \boxed{2\text{ W}}$$

and the power consumed by the 4 Ω resistor is

$$\text{power} = I^2R = (1\text{ A})^2(4\text{ }\Omega) = \boxed{4\text{ W}}$$

Thus, we see that the device that has the higher resistance converts more electrical energy into heat than does the lower resistance device, when the circuit is a series circuit.

EXERCISE

If houses were wired in series, similar to the circuit of Figure 8.28, tasks such as cooking breakfast could take a long time. That is, as you turned on the toaster, the coffee pot, and other appliances, the cooking time would get longer and longer. Why?

8.13 COST OF ELECTRICAL ENERGY

Each month, an electric utility company presents you with a bill for the amount of electrical energy that you have converted from electrical form to other forms, such as heat, light, or sound. Let us examine the procedure that an electric company uses to determine your bill. The amount of energy that you have converted to your needs can be determined from the total power supplied to your home and the time that it has been supplied. The definition of power shows us how to do this calculation:

energy converted

$$= (\text{power})(\text{time for conversion})$$

The electric company computes the power in units of kilowatts rather than watts, however, and they measure the time in hours rather than seconds. Thus, the units for the energy converted in your home are

$$\text{energy converted} = (\text{kW})(\text{h})$$
$$= \text{kWh (called a kilowatt-hour)} \quad \textbf{(8.11)}$$

(One kWh = 3.60×10^6 J.)

To see how your bill is determined, let us consider an example problem.

EXAMPLE 8.10 **TURN OFF THAT LIGHT BULB**
Suppose you accidentally leave a 100 W light bulb burning in your basement for a 24 hour period. How much money will the utility company charge you for the energy consumed by the bulb if electricity costs 8 cents per kWh?

Solution A 100 W light bulb is a 0.100 kW bulb. Thus, the energy consumed in kilowatt-hours is

$$\text{energy converted} = (0.100\text{ kW})(24\text{ hours})$$
$$= 2.4\text{ kWh}$$

and if electricity costs 8 cents per kWh, your oversight in leaving the bulb on will cost you

$$(2.4\text{ kWh})(8.0\text{ cents/kWh}) = \boxed{19\text{ cents}}$$

EXAMPLE 8.11 **READ THE FINE PRINT**
All electrical devices are required by federal law to have certain pieces of information supplied with them, usually on a metal plate on the back of the device. This plate gives you enough information to calculate how much the device will cost to operate. For example, a plate on a black-and-white TV set says that the TV will draw 0.75 A of current when connected to a voltage source of 120 V. Find the cost of watching 6.0 hours of soap operas on this TV in a location where electricity costs 9.0 cents per kWh.

Solution The power consumed by the device is found as

$$\text{power} = IV = (0.75\text{ A})(120\text{ V}) = 90\text{ W}$$

Thus, the device is a 0.090 kilowatt TV set, and the energy consumed in kilowatt-hours is

$$\text{energy consumed} = (0.090 \text{ kW})(6.0 \text{ hours})$$
$$= 0.54 \text{ kWh}$$

At 9.0 cents per kWh, the total cost is

$$(0.54 \text{ kWh})(9.0 \text{ cents/kWh}) = \boxed{4.9 \text{ cents}}$$

8.14 ELECTRICAL SAFETY

Almost everyone who has ever been around electricity has at one time or another shocked themselves. Most of the time a shock produces no effect other than an unpleasant tingling sensation, but if the electrical contact is made in the wrong way, death can occur.

Electricity causes damage to the body in two ways: It can cause the muscles of a vital organ to contract and stop its life-sustaining function, or it can burn the tissues of the body. The primary consideration in the first instance is the amount of current that passes through the muscles of the vital organ. It is difficult to generalize concerning the amount of current that will cause various effects on the body because factors such as the physical condition of the person involved and the exact current path are important. Generally a current of about 0.001 A will be felt as an electric shock, and a current of 0.01 A *can* be fatal if it passes through a vital organ such as the heart. Currents of 0.02 A can paralyze the respiratory muscles and result in suffocation. A current of the order of 0.1 A passing near the heart is almost certainly fatal because it can cause the heart muscles to contract irregularly.

At first thought, it seems that these are extremely small currents to produce such dramatic effects. For example, consider a 3 V flashlight battery connected in a circuit to a 6 Ω bulb. A simple Ohm's law calculation shows that the bulb draws a current of $I = 3 \text{ V}/6\Omega = 0.5$ A. Is it, then, dangerous to touch the circuit or to deal with it in any way? The answer is obviously no because you have touched such circuits many times without even feeling a hint of a shock. Why not? The reason is that the human body has a high resistance. For example, dry skin has a normal resistance of perhaps 200,000 Ω or greater, depending on where the contact is made. Thus, if you connect yourself directly across the battery, the current through your body is only about

$I = 3 \text{ V}/200{,}000 \ \Omega = 0.000015$ A, well within the safe range. Frequently, when people are killed by electricity, the reason is because their skin is wet, a factor that may reduce the resistance of the body to as low as a few hundred ohms. Thus, we see why it is especially dangerous to handle electrical equipment while you are in a bathtub or while you are standing barefoot on wet ground.

To help protect consumers, electrical equipment manufacturers now use electrical cords that have a third wire, called a case ground. To see how this works, let us consider the drill in Figure 8.29.

FIGURE 8.29   *(a) Normal path of current is from the hot wire through the motor, to ground and back to the power supply. (b) If the hot wire shorts to the case of the drill, the current path is from the hot wire, to case, through the body, and to ground. (c) If a case-ground is provided, the current goes from the hot wire, to case, through the case-ground, and back to the power supply.*

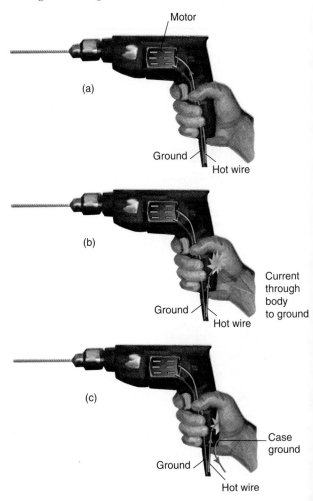

A two-wire device has one wire, called the hot wire, that is connected to the high-voltage (120 V) side of the input power line to the home, and the second wire is connected to ground, to 0 V. Under normal operating conditions, the path of the current through the drill is as shown in Figure 8.29a. The cord inside the device, however, can become frayed, and a "short circuit" can occur if the high-voltage wire comes into contact with the case of the drill, as shown in Figure 8.29b. In this circumstance, the pathway for the current is from the high-voltage wire through the consumer and to the earth—a pathway that can lead to death for the user. Protection is provided by the third wire, which is connected to the case of the drill, as shown in Figure 8.29c. In this situation, if a short occurs, the easiest path for the current is from the high-voltage wire through the case and back to ground through this third wire. The high current produced can blow a fuse or trip a circuit breaker before the consumer is injured.

CONCEPTUAL EXAMPLE 8.12   BIRD ON A WIRE

Explain why a bird can perch on a 10,000 V power line without being electrocuted.

Solution If you cannot answer this question, try a simpler one: Why doesn't a person get hurt by stepping off a curb in Denver, Colorado, at elevation 1.6 km? The answer is simple—even though Denver lies far above sea level, there is no way to fall that distance, and a person stepping off the curb falls only a few centimeters, which is hardly lethal. Just as height per se is not dangerous, but only differences in height are, so high voltage is not dangerous, only differences in voltage are. What is of concern to the bird is not the potential of the line, but the potential difference between its two feet, that is, the voltage directly across its body. If the bird stands with its feet a few centimeters apart, the resistance of the copper wire separating his feet is only about 10^{-6} Ω, and if the wire carries a current of 1 A, the potential difference between the feet of the bird is

$$V = IR = (1 \text{ A})(10^{-6} \text{ Ω}) = 10^{-6} \text{ V}$$

The bird is safe because there is so little voltage applied directly across his body.

If the bird straddled two wires, one at 10,000 V and the other at 0 V, the potential difference between its feet would be the full 10,000 V, and the bird would die instantly. Therefore, if the bird, or

10,000 V

0 V

Bird is on high-tension wire. Voltage difference between feet is small. Bird is happy.

10^5 Ω

Equivalent circuit.

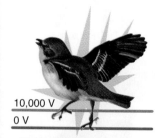

10,000 V

0 V

Bird is straddling two high-tension wires. Voltage difference between feet is large. Bird is electrocuted.

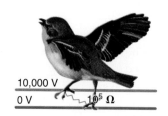

10,000 V

0 V 10^5 Ω

Equivalent circuit.

FIGURE 8.30   *A bird on a high-voltage wire.*

a human for that matter, touched one foot to the high-voltage line and the other foot to the ground, the resulting shock would be lethal (Fig. 8.30).

CONCEPTUAL EXERCISE

When caught out in an electrical storm, you are advised to kneel down rather than lie flat on the ground. Why? (*Hint:* Consider the bird in the last exercise.)

Answer If lightning strikes nearby, the current moves away from the strike-point through the earth. Consider what would happen if you were lying on the ground with your head pointing toward the strike-point. There would be a large current through the high-resistance earth and, therefore, a large voltage between your head and feet. This voltage can easily be high enough to produce a current through your heart sufficiently great to cause death.

SUMMARY

If two electric charges are the same ($++$) or ($--$), they repel each other; opposite charges attract. Ordinary matter is made up of **atoms.** Every atom has one or more **electrons,** which are negatively charged, and **protons,** which are positively charged, along with **neutrons,** which have no charge. The magnitude of the charge on a proton and an electron is 1.6×10^{-19} C **(coulombs).** The magnitude of the force between charged particles can be found from **Coulomb's law.**

An **electric field** is said to exist in a region of space if a force of electrical origin is exerted on a charged particle brought into that region. The direction of an electric field is defined to be the direction that a positive charge would move under the action of the field.

Voltage is the difference in potential energy between two points, divided by the magnitude of the charge. An **electric current** is any concerted nonrandom movement of electric charge. A current of 1 **ampere** is said to exist in a circuit if 1 coulomb per second of charge passes a cross-sectional area of the circuit.

Ohm's law states that the voltage drop across a resistor is equal to the current times the resistance. In a **series circuit,** the resistors are aligned so that there is a single pathway for charges to move along; in a **parallel circuit,** the pathway is branched, and charges move through all the branches simultaneously. Electric **power** is equal to current times the voltage and is measured in **watts.**

EQUATIONS TO KNOW

$F = k \dfrac{q_1 q_2}{r^2}$ (Coulomb's law)

$E = \dfrac{F_e}{q}$ (definition of electric field)

$V = \dfrac{\Delta(\text{energy})}{q}$ (definition of potential difference)

$I = \dfrac{Q}{t}$ (definition of electric current)

$V = IR$ (Ohm's law)

$P = IV = I^2 R$ (power)

KEY WORDS

Electric charge	Electric field	Ampere	Parallel circuit
Electron	Van de Graaff generator	Resistance	Series circuit
Proton	Voltage	Electric circuit	Electric power
Neutron	Potential difference	Ohm	Watt
Coulomb's law	Direct current	Ohm's law	Kilowatt-hour
Coulomb	Electric current		

PROBLEMS AND CONCEPTUAL QUESTIONS

Problems requiring numerical work are identified with a blue number.

Static electricity and measuring charge

1. An object is given a net positive charge via friction. Does the mass of the object increase, decrease, or stay the same?

2. A thin metal ball cannot be given a static charge by rubbing it against another material. Why not?

3. Often when you try to charge something by friction on a humid day, your experiment proves unsuccessful. Why?

4. How many electrons would be required to produce a charge of -1 C?

5. If you rub a coin briskly with a piece of either silk or fur, it will not become charged. Why?

6. When you walk across the floor to turn off the TV, you often find that a spark jumps from your hand to the metal knob. Explain how your body got charged.

7. If the people who first designated the electron as the basic carrier of negative electricity and the proton as the basic carrier of positive electricity had reversed their decision, how would our world have been affected?

8. A thin piece of aluminum foil is draped over a piece of wood. When a rod with a positive charge is touched to the foil, the leaves stand apart. Why do they repel one another?

9. One object is given a positive charge and another a negative charge. What happens when these two objects are placed in contact with one another?

Coulomb's law

10. Compare and contrast Coulomb's law and the law of universal gravitation.

11. Suppose someone tells you that he is hypothesizing that the force that holds the Moon in its orbit is a force of electrical attraction. How could you prove him wrong? Could he be proved wrong?

12. In the text, it is said that gravitational forces are weaker than electrical forces. Yet, when you try to pick up something heavy, you recognize gravitational force instantly, but you seldom are affected by electrical forces. Why not?

13. Two protons are placed 1 m apart. Could another proton be placed at some point in space so that it would feel no net electrical force from these two? Where would this point be, and why?

14. Isolated charges as large as 1 C are seldom encountered in nature. (a) Find the electrical force of repulsion between two 1 C charges separated by 1 m. (b) Charges actually encountered are usually of the order of 10^{-9} C. Find the force of repulsion between two of these charges separated by 1 m.

15. An electrical force of 10^{-11} N is exerted on an electron. (a) Find its acceleration. (b) Repeat for a proton.

The concept of fields

16. A proton is released in a room where an electric field is present, and you discover that electrical forces move it upward. What is the direction of the field in the room?

17. Do positive charges tend to move in the direction of an electric field or in the opposite direction? Explain. What if the charges are negative?

18. If the charge released had been an electron in problem 16, would your answer for the direction of the field still be the same?

19. An electron is placed 1 m away from an object having a charge of 10^{-9} C. (a) What is the acceleration of the electron? (b) What is the strength of the electric field set up by the object having the 10^{-9} C charge?

20. A proton is released from rest in an electric field of strength 100 N/C. (a) What is the acceleration of the proton in magnitude and direction? (b) How fast will the proton be moving after 10^{-3} s?

21. Electric field lines never cross one another. Why not? (*Hint:* What would happen to an electron if it was moving along a field line and reached the point of intersection?)

Van de Graaff generator

22. (a) When a metal object is given a charge, all the charge resides on the outside surface of the metal. Why? (b) If the object is hollow, the electric field inside the metal is zero. Why?

23. In light of problem 22, why would it be safe to be on the inside of a car if it is struck by lightning?

24. Would it be safe to be inside the dome of a Van de Graaff generator when it is given an extremely large charge?

25. Suppose you were wearing clothes affected by static cling. How would the clothing be affected when you come near a Van de Graaff generator?

26. When a demonstration is done in which a person's hair is made to stand on end by having him touch a Van de Graaff generator, the person is asked to stand on an insulating stool. Why?

Voltage and electric circuits

27. Do positive charges tend to move from points of high potential to low potential or vice versa? What about negative charges?

28. Distinguish carefully between potential and potential energy.

29. A high-voltage transmission wire is at a potential of 120,000 V, local power lines are at a potential of 2200 V, and household wiring is at 110 V. Assume these voltages can be treated as though they were direct current voltages and find the potential difference between (a) the high-voltage line and the local power line; (b) the local power line and the ground; (c) the high-voltage line and the household wiring.

30. Some devices operate directly off 110 V, whereas others operate off a small transformer that converts the voltage to, say, 5 V before it is applied to the device. Based on the definition of potential difference, what do you suppose would happen to the 5 V device if it were connected directly to 110 V?

31. Two identical electrical devices are connected one after the other and to a 12 V battery. Discuss the energy transformations that occur in this circuit.

Electric current

32. We have defined the sense of an electric current to be in the direction of motion of positive charges rather than in the direction of motion of electrons, which are the charges that really move in a circuit. This is often defended by saying that "an electron moving right to left has the same effect as an equal positive charge moving left to right." Defend this position.

33. A current of 10^{-3} A exists in a circuit. How long a time would have to pass before 1 C of charge moved past a cross-sectional area in the circuit?

34. A current of 10^{-3} A exists in a circuit. How many electrons move past a given cross-sectional area of the circuit in 10 minutes?

Resistance and Ohm's law

35. (a) Calculate the resistance of a simple circuit in which a 40 V power supply produces a current of 4 A. (b) Calculate the voltage of a battery if it can push 1.5 A through an 8 Ω resistor.

36. Based on what you know about the cause for resistance in a circuit, explain why the resistance of a light bulb should increase as the temperature of the filament goes up.

37. A light bulb usually burns out when it is first turned on. Why?

38. You would like to use an electric chain saw to trim some branches at the back of your lawn. The saw works fine when connected to one extension cord, but to reach your location, two cords must be used. You find that the saw does not work properly. Why not?

39. It is found that the resistance of a material increases with temperature. On the basis of what is happening inside a wire as charges move through it, explain why you would expect this increase in resistance.

40. Suppose you insert batteries into a flashlight such that they oppose one another. On the basis of Ohm's law, explain why you would not expect to get a current in the circuit.

41. An air conditioning unit has a compressor that is connected to 120 V and draws 90 A when first turned on. What is the effective resistance of the compressor at this instant?

42. The resistance of 10 cm of a certain wire is about 0.006 Ω. How many meters of this wire would you have to connect to a 12 V battery to limit the current to 3 A?

Different types of electric circuits and power

43. (a) Explain why there must always be two wires leading to an electric light bulb if it is to draw current and light up. (b) In an automobile, the wire from one terminal of the battery (the "ground") generally is connected directly to the metal chassis. The other terminal is said to be "hot." Therefore, any electrical device, such as a light bulb, can be energized by first connecting one of its wires to the chassis and then running the other wire to one of the terminals of the battery. Explain. From which terminal, the ground terminal or the "hot" terminal, should the wire be run? Which terminal, the positive or the negative, should be grounded, or does it matter?

44. The statement has been made that it is not the current or the voltage that kills you, but the product of the two. Discuss this statement.

45. New energy-efficient light bulbs have stamped on the end the words, 100 W at 125 V. These bulbs will never operate at 125 V. What is the purpose of this information?

46. Calculate the resistances of (a) a 25 W light bulb, (b) a 50 W light bulb, (c) a 50 W stereo amplifier, (d) a 100 W light bulb, (e) a 250 W toaster, and (f) an 800 W space heater, assuming all are designed to operate at 110 V.

47. During a "brown-out" in a big city, fuel shortages force a power company to supply 90 V potential instead of the usual 110 V potential. If a 250 W toaster is plugged in during such a brown-out, how much power will it actually deliver?

48. If two different heaters are compared, the one with the lower resistance uses more power. Since pure copper wire has a lower resistance than heater elements, why doesn't the wiring in your house or the wire in an extension cord get hot when it is drawing current?

49. A customer brought her new portable radio into the repair shop, complaining that although the radio operated well, a set of batteries lasted only a short time before running down. Speculate in a general way on the nature of the problem. Would you think the radio was drawing excess voltage or excess current? Discuss the reasons for your answer.

50. Explain why Christmas tree lights are generally wired in parallel. If they were strung together in series and one bulb blew out, would it be hard or easy to find that bulb? Explain.

51. Are the lights in your automobile wired in series or parallel? Defend your answer.

52. Calculate the current in the wire and the power delivered to each resistor in each of the diagrams in the figure below.

53. Assume that you have a 12 V battery and a small appliance with a resistance of 6 Ω. The appliance manufacturer states that the appliance will burn out if more than 15 W are drawn. Can the appliance be safely connected to the battery? If not, design a circuit using the 12 V battery that will operate in such a manner as to supply 15 W to the appliance.

54. If a household fuse burns out, the circuit can be re-established by inserting a penny into the socket. Explain why this works but why this is a dangerous practice that should never be done.

55. All automobile circuits are provided with fuses. What would happen if an automobile were wired without fuses and a short circuit developed?

56. A certain light bulb has a resistance of 200 Ω. (a) What is the current in this bulb when connected to a 12 V battery? (b) If three of these light bulbs are connected in series to this same battery, what is the current?

57. The three light bulbs of the preceding problem are connected in parallel to the 12 V battery. What is the current in each bulb, and what is the potential difference across each?

58. Two 100 Ω resistors are connected in series to a 30 V battery. (a) What is the power supplied to each? (b) What is the power supplied to each if the two are connected in parallel?

59. Two resistors in parallel, R_1 and R_2, can be replaced by an equivalent resistance R given by the equation

$$\frac{1}{R} = \frac{1}{R_1} + \frac{1}{R_2}$$

Use this equation to find the resistor that will replace two resistors of 6 Ω and 3 Ω in parallel.

60. Use the equation in the preceding problem to find the equivalent resistance of the resistors in Figure 8.25(b).

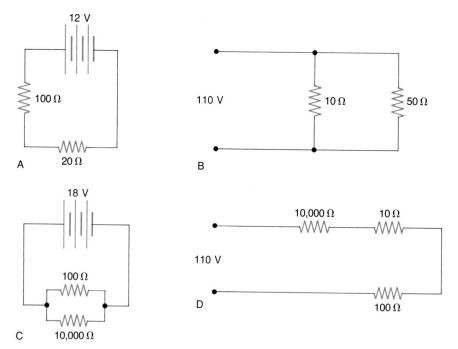

A   12 V   100 Ω   20 Ω

B   110 V   10 Ω   50 Ω

C   18 V   100 Ω   10,000 Ω

D   110 V   10,000 Ω   10 Ω   100 Ω

(Problem 52)

Cost of electrical energy

61. A black-and-white TV has a power rating of about 90 W. If electricity costs 8.0 cents per kWh, how much will it cost you to watch TV 4 hours a day for a year?

62. Repeat problem 61 for a 300 W color set.

63. Suppose you have a job that requires you to load 600 N bags of fertilizer on a platform that is 1 m above ground level. You lift one bag every 30 s. (a) How much work do you do in an 8 hour day? (b) What is your power output in kilowatts? (c) If you are paid at the rate of 10 cents per kWh, how much money do you make in a day?

64. A certain electric heater draws 10 A of current when operated at a voltage of 110 V. If electricity costs 8.0 cents per kWh, how much does it cost to run this heater for an 8 hour day?

Electrical safety

65. If it is the current that passes through your body that does the damage, why do you see warning signs that say, Danger: High Voltage?

66. Why is cutting off the third prong on a plug so that it can be used in a two-prong outlet a dangerous thing to do?

67. You leap off the top of a tall building and grab a high-voltage wire on the way down. (a) Are you in danger while you are swinging from the wire? (b) Suppose the wire breaks. Are you in danger while you are falling while holding onto the wire? (c) While still holding onto the wire, you touch the ground. Are you injured?

68. A power line falls on your car while you are in it. Are you in danger while you are in the car? Are you in danger if you open the door and step out of the car?

69. Why is it suicidal to use a hair dryer while in the bathtub?

ANSWERS TO SELECTED NUMERICAL PROBLEMS

4. 6.25×10^{18} electrons
14. (a) 9.0×10^9 N, (b) 9.0×10^{-9} N
15. (a) 1.10×10^{19} m/s^2, (b) 5.99×10^{15} m/s^2
19. (a) 1.58×10^{12} m/s^2, (b) 9×10^{-9} N/C
20. (a) 9.58×10^9 m/s^2, (b) 9.58×10^6 m/s
33. 1000 s
34. 3.75×10^{18} electrons
35. (a) 10 Ω, (b) 12 V
41. 1.33 Ω
42. 66.7 m
46. (a) 484 Ω, (b) 242 Ω, (c) 242 Ω, (d) 121 Ω, (e) 48.4 Ω, (f) 15.1 Ω
47. 167 W
52. A. $I_{100} = I_{20} = 0.10$ A, $P_{20} = 0.2$ W, $P_{100} = 1.0$ W
    B. $I_{10} = 11.0$ A, $I_{50} = 2.2$ A, $P_{10} = 1210$ W, $P_{50} = 242$ W
    C. $I_{100} = 0.18$ A, $I_{10000} = 0.0018$ A, $P_{100} = 3.24$ W, $P_{10000} = 0.0324$ W
    D. $I = 0.109$ A, $P_{10000} = 1.19$ W, $P_{10} = 0.0012$ W, $P_{100} = 0.0119$ W
53. No, the resistor would have to dissipate 24 W. Connect a 1.60 Ω resistor in series with the 6 Ω resistor.
56. (a) 0.06 A, (b) 0.02 A
57. 0.06 A, 12 V
58. (a) 2.25 W, (b) 9 W
59. 2 Ω
60. 15.8 Ω
61. $10.51
62. $35.04
63. (a) 5.76×10^5 J, (b) 2×10^{-2} kW, (c) 1.6 cents
64. 70.4 cents

CHAPTER 9

ELECTROMAGNETISM

The electrical power lines dotting the skyline carry electrical power produced at Hoover Dam. The production and distribution of this energy are possible because of a connection between electricity and magnetism.

The first magnetic materials discovered were in the form of mineral deposits found in Magnesia (now Iran). The mineral was given the name lodestone or magnetite. The term "lodestone" is derived from the Saxon word *laedan,* meaning "to lead." The association between leading and the mineral arises because ancient navigators learned that if a bar of this mineral was mounted so that it could rotate freely, one end would always point north. Scientists now know that these materials are permanent magnets and that they align themselves with the magnetic field of the Earth.

The study of magnetism began with the investigation of the properties of these naturally occurring permanent magnets. The use of magnets as compasses to find direction opened up new trade and exploration routes and led to alterations in the life of the people of that time. The impact of magnetism on society, however, became even more profound when it was discovered that an electric current can produce magnetic effects in the space around the wire. Because of the connection between electricity and magnetism, we shall embrace all of the phenomena discussed in this chapter under the single heading of electromagnetism.

9.1 MAGNETS

Experiments show that every magnet, regardless of its shape, has two poles, called **north** and **south.** The names north and south pole arose from the use of magnets as direction finders. The end of a magnet that points toward the north of the Earth is called the "north seeking" end, or more simply the north pole of the magnet. The interaction between the poles of two magnets is similar to that observed between charged objects. For example, like electric charges repel one another, and so do like magnetic poles. Also, as unlike charges attract, so do unlike poles. Because of this correspondence between electric forces and magnetic forces, it is tempting to carry the analogy a step farther and say that magnetic effects are produced in a material by tiny isolated north and south poles, somewhat like isolated positive and negative charges. If true, then perhaps what is happening when a material becomes magnetized is that there is a separation of these tiny poles, with the north species accumulating at one end of the magnet and the south species at the other.

If such tiny magnetic entities do exist, it seems that it would be easy to separate them. One way would be to start with a bar magnet and break it in half, leaving one end with an excess of tiny north poles and the other with an excess of tiny south poles. If you attempt this experiment, however, you find that cutting a magnet in half produces two smaller magnets, *each* with its own north and south pole. Slice again, and you get four magnets, and so on. If your instruments were fine enough, you could keep cutting until you had an enormous collection of atomic-sized magnets, but each would still

have a north pole and a south pole (Fig. 9.1). It should be noted here that attempts to detect an isolated magnetic monopole (an isolated south or north pole) have been unsuccessful, but the search continues. If such isolated monopoles are indeed found someday, our basic understanding of ordinary magnetic effects will not be drastically altered. As we shall see, an understanding of what makes a magnet a magnet is explained by the fact that the ultimate source of magnetic effects is the motion of charged particles at the atomic or the subatomic level. We shall return to this topic after we have investigated the relationship of electricity and magnetism a little more carefully.

An unmagnetized piece of iron can become magnetized by stroking it with a bar magnet. After stroking, the iron is found to be magnetic, with a north pole and a south pole. There are other ways in which an unmagnetized piece of iron can be magnetized. One way is to place the iron near a strong magnet and either hammer the iron or heat it. The hammering and the heating accelerate the process, but neither is necessary. Hammering or heating will weaken a magnet if no other strong magnet is nearby. The iron will also become magnetized if it is left alone near a strong magnet; however, the time period for the magnetism to occur will be longer. For example, iron fences left standing for many years often become magnetized because of the presence of the magnetic field of the Earth. Also, the naturally occurring magnetic material, lodestone or magnetite, became magnetized because of the field of the Earth.

FIGURE 9.1 *If a permanent magnet is cut in half, two smaller magnets are produced. Is it impossible to isolate separate north and south magnetic poles by repeated cutting?*

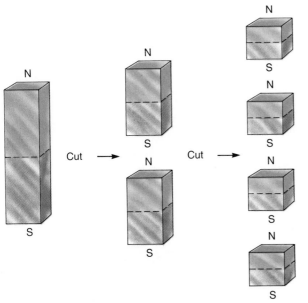

9.2 MAGNETIC FIELDS

The concept of electric fields was introduced earlier to provide a way to discuss the forces acting on charged particles. It is also convenient to discuss forces that are magnetic in origin in terms of a field concept. One of the tasks that must be considered when discussing a field is to define a process by which we determine the direction of the field. For a magnetic field, this is done by defining the *direction of a magnetic field to be in the direction that the north pole of a compass needle points when placed in the field.* Figure 9.2a shows how a magnetic field line of a bar magnet can be traced with a small compass. The

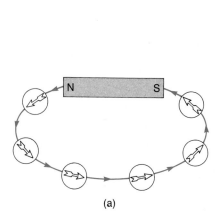

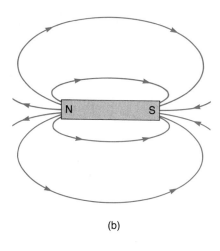

(a)

(b)

FIGURE 9.2 *(a) Tracing the magnetic field of a bar magnet. (b) Several field lines of a bar magnet.*

compass is first placed near the north pole of the magnet, and the direction in which the compass needle points is noted. The compass is then displaced slightly in the direction indicated by the needle, and the direction in which the needle points is once again noted. Figure 9.2b shows several magnetic field lines around a bar magnet.

One of the characteristics that we observed for electric fields is that an electric field originates on a positive charge and terminates on a negative charge. Superficially, it appears that we might be able to make a similar statement for magnetic fields because in Figure 9.2b it appears that magnetic field lines originate on north poles and terminate on south poles. This is not the case, however. *Magnetic fields have no starting or stopping points. Magnetic fields always close on themselves.* For example, if we could trace the magnetic field of the bar magnet through the magnet itself, we would find it continues as shown in Figure 9.3.

CONCEPTUAL EXAMPLE 9.1   THE SPACING OF MAGNETIC FIELD LINES

When we studied electric fields, we found that sketches of the fields convey information about the strength of the field by the way in which the lines are spaced. At those locations where the lines are drawn closely packed, the field is strong, and where the lines are far apart, the field is weak. This same technique is used to represent magnetic fields. With this in mind, **(a)** discuss the strength of the magnetic field in the region of space between two bar magnets aligned as shown in Figure 9.4a. **(b)** Repeat (a) for two bar magnets aligned as shown in Figure 9.4b.

FIGURE 9.4   *(a) The field lines between two unlike poles, and (b) the field lines between two like poles.*

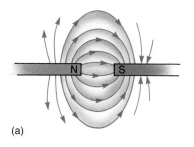

(a)

FIGURE 9.3   *Magnetic field lines never start or stop.*

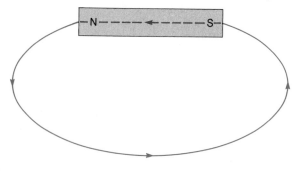

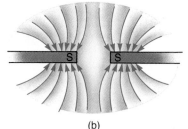

(b)

Solution (a) The magnetic field lines are drawn closely spaced in the region of space between the two magnets, indicating that the magnetic field is strong in this region.

(b) The magnetic field lines are not closely spaced in the region between the two south poles, indicating that the field is weak there.

CONCEPTUAL EXERCISE

Another characteristic common to electric and magnetic field lines is that they never cross. Explain why magnetic field lines cannot intersect.

9.3 MAGNETIC FIELD OF THE EARTH

As mentioned previously, a compass is merely a magnetic needle that is suspended so that it can rotate freely. Because navigation relies on the fact that a compass aligns itself at least roughly in the north-south direction of the Earth, it is obvious that the Earth itself must possess some sort of magnetic field. In fact, *the magnetic field of the Earth is similar to the field that would be produced if a giant bar magnet were situated inside the planet with its south magnetic pole near the north geographic pole* (Fig. 9.5). In everyday conversation, people frequently refer to the magnetic pole near the north geographic pole as the "north magnetic pole." You should bear in mind, however, that it is really a south magnetic pole. Several independent observations convince us that the magnetic field of the Earth cannot be due to large masses of permanently magnetized material buried beneath the Earth's surface. For one thing, geologists believe that sections of the interior of our planet are hot and liquid, and a solid magnet would either melt or lose its magnetism under such conditions. Second, the position of the Earth's north magnetic pole moves measurably from year to year. In fact, geologists have determined that the field has reversed *direction* from time to time in the past. Evidence for this is obtained by examination of basalt, a type of igneous rock that contains iron, which is spewed forth by volcanic activity on the ocean floor. As the lava cools, it becomes magnetized and leaves a permanent record

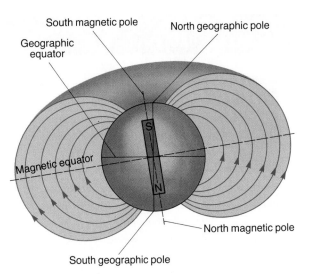

FIGURE 9.5 *A schematic view of the magnetism of the Earth. The north-seeking pole of a compass is attracted toward the north geographic end of the Earth, which is a magnetic south pole. Imagine a bar magnet positioned as shown, but, as explained in the text, there is no bar magnet in the center of the Earth. Note that the magnetic south pole does not coincide exactly with the geographic North Pole.*

of the Earth's magnetic field at the time it solidified. These rocks can be dated by their radioactivity to provide the evidence for these periodic reversals of the Earth's field. No theory can explain how a solid magnet inside the Earth could produce this effect.

It is believed that the source of the Earth's field is charge-carrying convection currents in the molten core of the Earth. As noted earlier, charges in motion can produce a magnetic field, and this field would be similar to that set up by a bar magnet. No one knows with certainty what processes are at work inside the Earth to produce its field because no one knows exactly what is happening at the Earth's core. The core is too far beneath the surface to be observed directly, and any laboratory or computer models of this region represent only approximations of reality.

Whatever the exact mechanisms, planetary magnetic fields are quite common. The magnetic field of the Earth is hardly unique. Several other planets and moons in our Solar System, our Sun, and most of the other stars all produce magnetic fields. There is some evidence to indicate that the

FOCUS ON
MAGNETIC BACTERIA

It has long been speculated that some animals may use the magnetic field of the Earth to find their way around on the surface of the planet. Do birds follow their migratory patterns by following the magnetic field lines of the Earth? Do bees use this field to find their way back to their hives after a foraging trip for food? There is one specific case in the animal kingdom in which the answer is a conclusive yes. The creature is a type of anaerobic bacteria that lives in swamps. These tiny creatures have a magnetized chain of magnetite as part of their internal structure. This chain acts as a compass that enables them to align themselves with the field of the Earth. The term "anaerobic"

means that they survive without oxygen; in fact, oxygen is toxic to them. As a result, if they find themselves lifted out of the mud and slime on the bottom of a swamp, they need to return to its oxygen-free environment as rapidly as possible. In the Northern Hemisphere, the field of the Earth has a downward component, and it is believed that their magnetic sensing ability is used to follow this downward directed field back to the ooze. This hypothesis is verified by observing similar bacteria living in the Southern Hemisphere, where the component of the Earth's field perpendicular to the surface is directed upward. The internal magnetite chain of these bacteria has a magnetic field opposite to that of their Northern Hemisphere counterparts.

rate of rotation of a planet has an effect on the magnetic field set up by that planet. For example, Jupiter rotates faster than Earth, and space probes indicate that Jupiter's magnetic field is stronger than ours. Venus rotates more slowly than the Earth, and its magnetic field is found to be weaker.

If a compass needle is allowed to swing freely in both the horizontal and the vertical directions, it is found that the needle is horizontal with respect to the Earth's surface only when it is near the equator (Fig. 9.5). As the compass moves northward, it points more and more toward the surface of the Earth, until at a point just north of Hudson Bay in Canada, the needle points directly downward. This location was first found in 1832, and it is considered to be the site of the south magnetic pole of the Earth. This site is approximately 1300 miles from the Earth's geographic North Pole. Similarly, the Earth's north magnetic pole is about 1200 miles away from the Earth's geographic South Pole. This means that a compass needle points only roughly toward the geographic poles of the Earth. For example, along a line through Florida and the Great Lakes, a compass indicates true north, but in Washington State, the needle aligns 25° east of true north.

9.4 THE CONNECTION BETWEEN ELECTRICITY AND MAGNETISM

Early scientists did not recognize the connection between electricity and magnetism. Magnetic compasses were used by navigators probably more than 2000 years ago, and scientists have experimented with electric currents for several hundred years, but the first investigation of the interrelationship between the two phenomena was conducted by Hans Christian Oersted in 1820. Oersted's simple experiment was to hold a magnetic compass needle under a wire carrying a direct current, as shown in Figure 9.6a. He noticed that the needle of the compass was always affected in the same way. When the compass was placed below the wire as shown in Figure 9.6a, the needle lined up at right angles to the wire with its north end pointing as shown. When the compass was placed above the wire, the needle again aligned so that it was perpendicular to the wire, but the needle reversed direction, as shown in Figure 9.6b.

In a second experiment, Oersted had a flexible wire running through the poles of a horseshoe magnet, as shown in Figure 9.7a. When a current passed

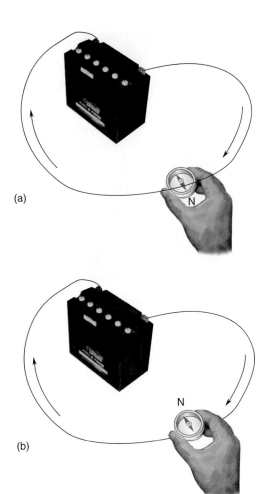

(a)

(b)

FIGURE 9.6 *(a) A compass needle points as shown when held below the wire, and (b) it reverses direction when placed above the wire.*

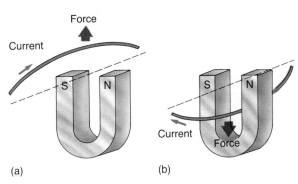

(a)   (b)

FIGURE 9.7 *(a) A wire carrying a current is deflected upward when the current direction is as shown. (b) It is deflected downward when the current is reversed.*

through the wire, the wire bent upward or downward, depending on the direction of the current, as shown in Figure 9.7b. When the wire was held in line with the direction of the magnetic field lines of the magnet, the wire was not affected. The sections that follow will explain why these effects were observed.

Magnetic Field Set Up by a Current-Carrying Wire

Whenever an electric current exists in a wire, a magnetic field is set up in the space around the wire. Thus, if a compass is brought into the field, it is affected in accordance with Oersted's observations. It is now known that a wire carrying a current produces a magnetic field such that the lines of the field form concentric circles around the wire (Fig. 9.8).

A large horseshoe magnet, equivalent to a bar magnet bent to this shape.

FIGURE 9.8 *A magnetic field around a current-carrying wire.*

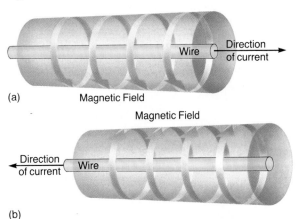

(a)   Magnetic Field

Magnetic Field

(b)

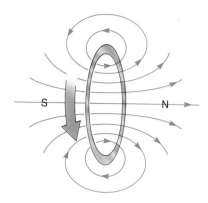

FIGURE 9.9   *The magnetic field of a loop of wire carrying a current.*

A large coil of wire used to demonstrate magnetic electric currents.

The direction of the field depends on the direction of the current, as shown in Figure 9.8a and b.

If the wire carrying a current is bent into the shape of a loop, the magnetic field pattern set up by it is like that shown in Figure 9.9. Finally, if several loops of wire are wound around a frame, the magnetic field set up is as shown in Figure 9.10. Several turns of wire can produce a magnetic field that is quite strong. Coils of wire used to produce magnetic fields are called **electromagnets,** and they have many practical applications.

Magnetic Forces and Moving Charges

If a magnet is held near a wire that is carrying a current, a force is exerted on the wire, which can cause the wire to move as Oersted demonstrated. Oersted's discovery showed that electrical energy of the moving charges can be converted into mechanical energy, which causes the wire to move. This result ultimately led to the development of the electric motor. Subsequent investigation of these forces by Oersted and others showed that the moving charges do not have to be confined to a wire.

A charge moving in a magnetic field can have a magnetic force acting on it.

Figure 9.11 shows a positive charge moving upward on the page in a region in which a magnetic field exists. The crosses are used to indicate that the direction of the magnetic field is into the paper. (It is easy to remember the meaning of the crosses if you think of the magnetic field as being represented by arrows with feathers at the back of the arrow. Thus, the crosses indicate the arrow moving away from you, and a dot indicates the tip of the arrow coming toward you.) When a charge moves through a magnetic field in this manner, a

FIGURE 9.10   *Magnetic field set up by a coil of wire.*

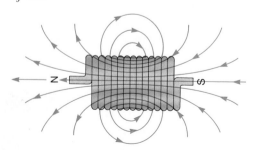

FIGURE 9.11   *A charged particle moving perpendicularly to a magnetic field has a force on it as shown.*

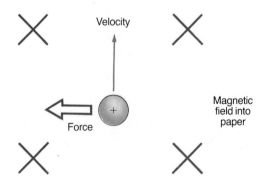

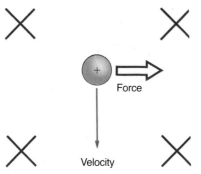

FIGURE 9.12   *Reversing the direction of the velocity of the charge reverses the direction of the force.*

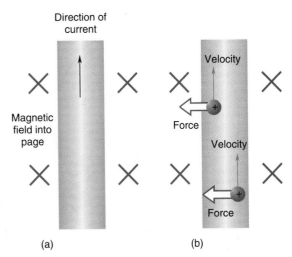

FIGURE 9.14   (a) *A wire carrying a current upward in a magnetic field directed into the page.* (b) *Each moving charge has a force on it to the left. This produces a net force on the total wire, which is to the left.*

force is exerted on the charge by the magnetic field. The force, however, is not in the direction of the field, nor is it in the direction of the motion of the charge. In fact, the direction of the force is perpendicular to both the field and the velocity of the charge. For the situation shown in Figure 9.11, the direction of the force is toward the left. If the motion of the charge is reversed as shown in Figure 9.12, the direction of the force is also reversed. If the motion of the charge is along the direction of the magnetic field lines as in Figure 9.13a, or in the direction opposite to that of the field lines as in Figure 9.13b, there is no force acting on the charges.

Moving charges in a magnetic field have a force acting on them whether they are in air or in a material. Because an electric current is a motion of charges through a wire, it is expected that there would be a force on the wire. For example, in Figure 9.14a, we see a portion of a wire carrying a current upward on the page and in a magnetic field directed into the paper. As Figure 9.14b shows, each charge has a force acting on it toward the left, and this force is communicated to the atoms of the wire

and to the "sides" of the wire, producing a net force toward the left on the wire.

Experimentally, it is found that *the strength of the force acting on a current-carrying wire is dependent on the magnitude of the current and on the strength of the field.* The larger the current and the magnetic field, the stronger the force.

CONCEPTUAL EXAMPLE 9.2   WHICH WAY DOES THE COMPASS POINT?

(a) Figure 9.15 shows a wire carrying a current coming out of the page and the magnetic field pattern

FIGURE 9.15   *Example 9.2a. Which way will a compass point at locations A and B?*

FIGURE 9.13   *If the motion of the charge is (a) in the same direction as the magnetic field or (b) in the opposite direction, there is no magnetic force on the charge.*

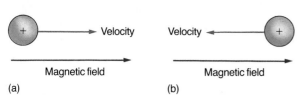

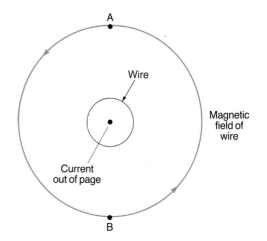

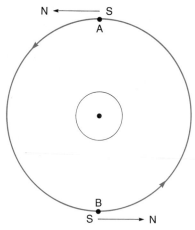

FIGURE 9.16 *Example 9.2a. A compass will align as shown at locations A and B.*

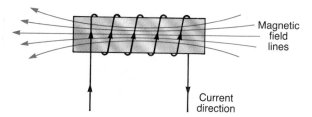

FIGURE 9.17 *Example 9.2b. Which end of the electromagnet is the north end?*

set up by this current. If a compass needle is placed at A and then at B in the figure, which way will the needle point at each location?

Solution A compass needle will align in the direction of the magnetic field. Thus, it will point as shown in Figure 9.16.

(b) Figure 9.17 shows the magnetic field set up by an electromagnet. Which end of the electromagnet is the north end? Also, discuss the strength of the field inside the coil, assuming that the field lines inside the electromagnet are drawn correctly.

Solution Magnetic field lines always appear to exit from the north end of a magnet. Thus, the left end of the electromagnet is the north pole. Since the field lines are drawn evenly spaced throughout the interior of the coil, the spacing indicates that the magnetic field has the same value at all points in the coil. The fact that a coil can set up a uniform field in its interior is an important feature of the device in many applications.

CONCEPTUAL EXAMPLE 9.3 STEERING CHARGED PARTICLES

In a "Focus On" box in Chapter 8, we discussed how a black-and-white television set works. We saw that the charges in the tube could be moved from point to point on the screen by use of electrical forces produced by horizontal and vertical deflection plates. Discuss how this steering could also be done with electromagnets. In fact, in practice, the steering of the moving charges *is* done with electromagnets.

Solution Figure 9.18 shows how the electromagnets are arranged at the neck of the tube. When a current is sent through the coils, a magnetic field is produced that exerts a force on the electrons when they pass through the field. This force deflects the electrons from their path, and by varying the current in the electromagnets and hence the strength of the field, the beam can be moved to any location on the face of the screen.

CONCEPTUAL EXERCISE

Explain why electromagnets rather than charged plates are used to steer the beam in a TV tube. Consider such factors as cost and ease of adjustment.

9.5 WHAT MAKES A PERMANENT MAGNET MAGNETIC?

Although it is possible to isolate electrically positive particles and to separate them from negative ones, no one has ever isolated particles with only a north or a south magnetic pole. As we have seen, any object that is magnetic has both a north and a south pole. A clue to the ultimate cause of magnetism is given by the kind of magnetic field that is produced by a loop of wire as shown in Figure 9.19a. We see that the field produced is quite similar to that produced by a bar magnet as shown in Figure 9.19b.

The explanation of why magnetic materials are magnetic lies in the structure of iron. This magnetism arises in two ways: (1) An individual atom

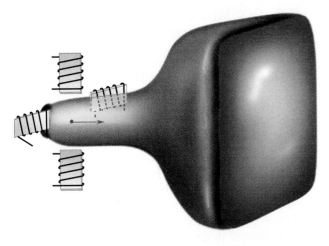

FIGURE 9.18 *Electromagnets used to steer electrons in a TV picture tube.*

can be magnetic. (2) The individual electrons of the atom possess an intrinsic magnetism. To see why an atom is magnetic, again consider the fact that a coil of wire produces a magnetic field. If a coil can produce a magnetic field, then any charge moving in a circular path should also produce a magnetic field. That is precisely what is happening in an atom as an electron circles the nucleus of the atom. Each negatively charged electron circles about the nucleus, thus constituting a motion of charge, or an electric current. From this we see that a net magnetism for the total iron bar could be produced if we could somehow cause several of these atomic magnets to align inside the material. This does occur, but only to a limited extent. The reason that this is not a more important factor in magnetic materials is that the magnetic field produced by one electron in an atom is often cancelled by the magnetic field produced by another nearby revolving electron in the same atom. The net result is that the magnetic field produced by the electrons orbiting the nucleus is either zero or very small for most materials.

The most important reason for the magnetic effects exemplified by materials such as iron lies in an intrinsic property of the individual electrons. To understand this property, let us think about the Earth orbiting the Sun. As it revolves around the Sun once each year, the Earth is also rotating on its axis once each day. In like manner, as an electron circles the nucleus, it is also spinning on its axis. These electrons also produce a magnetic field as they spin. In four materials (iron, cobalt, nickel, and gadolinium), the magnetic fields of one or more electrons do not cancel and produce a net magnetic field for an individual atom. In turn, several of these atoms act in unison to produce a small cluster of atoms with a net magnetic field for the

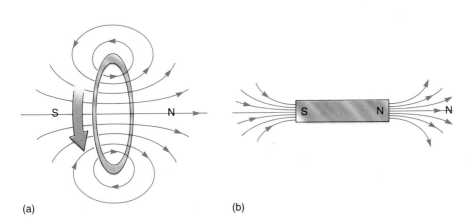

(a)   (b)

FIGURE 9.19 *The magnetic fields set up (a) by a loop of wire and (b) by a bar magnet are similar.*

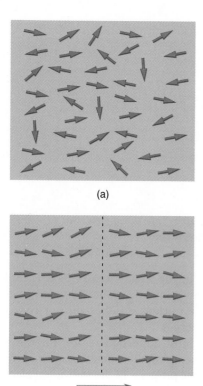

(a)

(b)

FIGURE 9.20   *(a) The domains in an unmagnetized substance point in random directions. (b) When the substance is magnetized, the domains tend to align.*

cluster. These groups of cooperating atoms are called **magnetic domains.** In an unmagnetized piece of material, we can picture these domains as pointing in random directions (Fig. 9.20a). When the material becomes magnetized, these domains have been caused to align (Fig. 9.20b). The alignment of domains explains why the strength of an electromagnet is increased dramatically by the insertion of an iron core into the magnet's center. The magnetic field produced by the current in the loops of wire causes the domains to align and thus to produce a strong field. Figure 9.20b shows why one ends up with two identical magnets when a magnet is broken in half. Breaking the magnet along the dashed line in the figure results in two pieces that still have north and south poles at each end. There is no way to break a magnet into small enough pieces to isolate a free north pole and a free south pole.

9.6 COSMIC RAYS AND THE EARTH'S MAGNETIC FIELD

The Earth's magnetic field provides a protective shield around our planet. The Earth is constantly subjected to bombardment by many different types of particles and radiation from outer space. The most energetic particles, known as **cosmic rays,** originate from interstellar space and are believed to be ejected when dying stars explode. Other high-speed particles originate from huge magnetic storms on the surface of the Sun; these will be discussed in the astronomy section. Energetic particles from either source can produce disruptions in our lives and environment that range from the extremes of producing damage in living cells and altering genetic information carried in the reproductive organs of all living things to the more minor disruptions of television and radio communication.

When charged particles approach the Earth, they come under the influence of its magnetic field. Because these field lines are roughly parallel to the surface of the Earth over much of its area, an incoming particle is likely to approach the Earth in a direction that is perpendicular to this field. Now, recall that when the velocity of a particle and a magnetic field are at right angles, a magnetic force is exerted on the particle, which in this case deflects the particle sideways, as shown in Figure 9.21. Thus,

FIGURE 9.21   *A charged particle heading toward the surface of the Earth is deflected by the Earth's magnetic field as shown.*

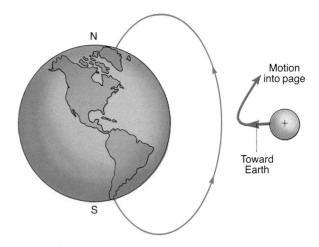

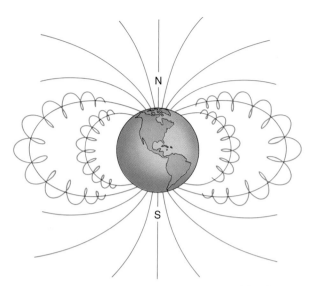

FIGURE 9.22 *A charged particle is trapped by the Earth's field and spirals along one of the field lines.*

This view of the Aurora Australis or Southern Lights shows a band of airglow above the limb of the Earth as seen from Space Shuttle Discovery. (Courtesy NASA)

most of these particles are deflected away from the surface of the Earth. Of course, some do make it through, and these constitute a part of the background radiation on Earth. If the Earth's magnetic field were to disappear, all life would be subjected to a much higher number of incoming particles. It is likely that more mutations would occur, with results that cannot be predicted. It is interesting to note that evolution has not been steady during geological time. Instead, there have been periods of rapid change and other periods when life forms have changed little. Some scientists believe that the periods of rapid evolutionary change may coincide with times when the Earth's magnetic field was weakened or nonexistent.

With the advent of orbiting satellites, a most astounding discovery was made in 1957 by James Van Allen of the University of Iowa. Instruments carried aloft in spacecraft found two belts high in the atmosphere in which many energetic charged particles were trapped. (The discovery of these belts is considered to be the first major scientific discovery of the Space Age.) These belts are filled by particles (mostly protons) coming in from outer space such that their direction of travel would bring them in almost parallel with the Earth's magnetic field. In such a case, it can be shown that a particle would be trapped by the field and would spiral around a

magnetic field line (Fig. 9.22). As the particle approaches one of the Earth's poles, the field gets stronger, and the spiral winds in on itself. Finally, near the pole the particle reverses its direction of travel and spirals back toward the other pole. A proton trapped in this way may travel from pole to pole once every few seconds and may remain trapped for several hundred years.

There are two belts that seem to favor this trapping process. One is about 2000 miles above the Earth's surface, and the second is about 10,000 miles above the Earth. For the most part, these trapped particles go unnoticed by us in our day-to-day lives. The particles do occasionally build up to such an extent that large numbers of them spill out of their to-and-fro orbits above one of the poles. These spilled particles lose energy in collisions with air molecules and produce the beautiful **auroras** that sometimes light the northern and southern skies.

9.7 ELECTROMAGNETIC INDUCTION

Shortly after Oersted's experiments, people began to speculate that since a current produces a magnetic field, perhaps a moving magnet would produce an electric current. Eleven years after Oersted's experiments, Joseph Henry of the United

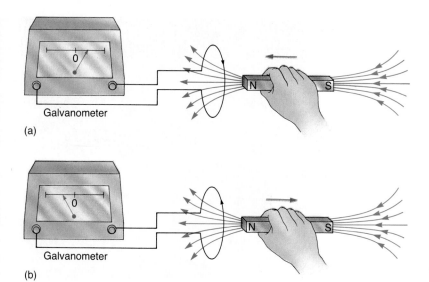

(a)

(b)

FIGURE 9.23 *Relative motion between a magnet and a coil of wire produces an electric current in the coil.*

States and Michael Faraday of England independently tried to produce an electric current by moving a magnet up and down through a coil of wire. The experiments performed by these men were relatively simple, yet on these modest grounds lie the basic principles used to generate the enormous amounts of electrical energy produced throughout the world today. Let us examine what they did.

In Figure 9.23a, a bar magnet is placed close to a coil of wire such that some of the magnetic field lines from the magnet thread through the coil. Also, a sensitive current detector is connected to the coil. Now, if you move the magnet toward the coil (Fig. 9.23a) so that more magnetic field lines thread through the coil, you will note that while the motion of the magnet takes place, the current detector deflects toward the right, indicating the presence of a current in the coil. If you move the magnet away from the coil (Fig. 9.23b) so that fewer field lines thread through, you will note that the needle deflects in the opposite direction. This result is astonishing. We have seen many examples in our study of electricity in which a current has been set up in a wire. In all those cases, however, there has *always* been a power supply, such as a battery, present to produce the current. In this case, there is a current with no apparent source. In these circumstances, we say that *the effect of the magnetic field has been to induce a current in the coil.* It is as though the *changing* field lines passing through the coil somehow produce a phantom battery in the coil. For example, if the current is counterclockwise, as shown in Figure 9.24a, this battery would have to

be connected, as shown in Figure 9.24b, to produce the current. This voltage that is produced is referred to as an **induced voltage.**

FIGURE 9.24   *(a) The current produced by the coil is in the same direction as it would be if (b) a battery were in the circuit. The voltage produced is called an induced voltage.*

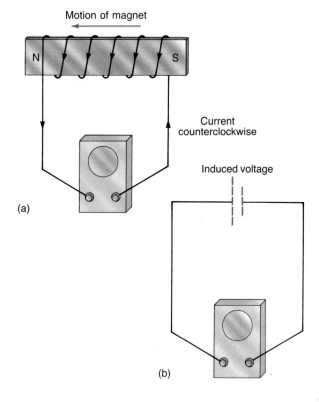

FOCUS ON
ELECTRIC GUITARS

Induced voltages are at the heart of the production of sound in an electric guitar. Let us suppose that a performer plucks a guitar string such that it vibrates with a frequency of 440 Hz. This vibration is detected and amplified as shown in the figure. A permanent magnet inside a pickup coil magnetizes a portion of a guitar string. When this guitar string is set into vibration, the magnetized segment of string produces a changing magnetic field through the pickup coil. The induced voltage in the coil is then fed to an amplifier.

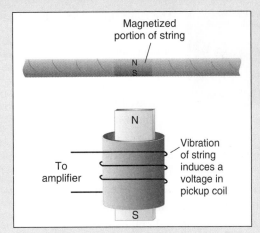

The pickup coil in an electric guitar.

In performing this experiment yourself, you could easily demonstrate several facts:

1. There is an induced current in the coil only when there is relative motion between the coil and the magnet. That is, a current is induced when the magnet is moved closer to the stationary coil or pulled farther away, or a current is induced when the coil is moved closer to or farther away from a stationary magnet.

2. The faster the relative movement, the greater the current.

FIGURE 9.25 *Example 9.4. When someone closes the switch, is there a current in the detector?*

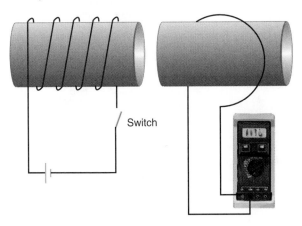

Switch

3. The number of turns of coil also affects the current. If the coil has several turns, the magnet affects each individually; thus, twice as many turns result in twice as much current.

Think of the simplicity of these experiments of Oersted, Henry, and Faraday and of their significance in world history. Oersted discovered that electrical energy can be converted to mechanical energy. Then Henry and Faraday discovered that mechanical energy can induce an electric current. The way was paved for the development of electric motors, generators, and the entire electric age.

CONCEPTUAL EXAMPLE 9.4   IS THERE MOTION OR NOT?

Figure 9.25 shows an electromagnet connected to a battery and an open switch. Nearby is a coil of wire with a current detector. When someone closes the switch, is there an induced current in the coil? Why or why not?

Solution At first thought, the answer might be no because there does not seem to be any relative motion between the two coils. There *is* a deflection of the current detector, however, when the switch is closed, indicating the presence of a current. Let us see why this occurs.

FOCUS ON
TAPE RECORDERS

The common reel-to-reel tape recorder that is a part of many music enthusiasts' collections provides some interesting examples of practical uses of the principles we have discussed thus far. The principles of a typical recorder are shown in part (a) of the figure. Tape moves from one side of a cassette reel past a recording head and a playback head to a take-up reel in the same cassette. The tape itself is a plastic ribbon coated with a metallic oxide that can become magnetized in the presence of a magnetic field.

The steps in recording are illustrated in part (b) of the figure. A sound wave is translated by a microphone into an electric current, which is then amplified and passed through a wire wound around a doughnut-shaped piece of iron. The iron ring and the wire form an electromagnet that constitutes the recording head of the instrument. The varying current in the wire produces a varying magnetic field in the iron, and this field is completely inside the iron except at the point where a slot is cut in the ring. At this location,

the magnetic field fringes out of the iron and magnetizes the small pieces of metallic oxide on the tape as the tape is pulled past the head by a motor. The pattern of magnetization preserved on the tape simulates the frequency and loudness of the sound signal originally used to produce the magnetization. Thus, *the recording process uses the fact that a current passing through an electromagnet produces a magnetic field.*

To reconstruct the sound signal, the tape is allowed to pass through the recorder again, this time with the playback head in operation. Part (c) of the figure shows that this head is similar to the recording head in that it consists of a doughnut-shaped piece of iron with a wire coil wound around it. When the magnetized tape is pulled past this head, the varying magnetic field pattern present on the tape produces changing fields through the wire coil on the playback head. These changing lines induce a current in the coil that corresponds to the current in the recording head that produced the tape originally. This changing current can be amplified and used to drive a speaker. *The playback process is thus an example of induction of a current by a moving magnet.*

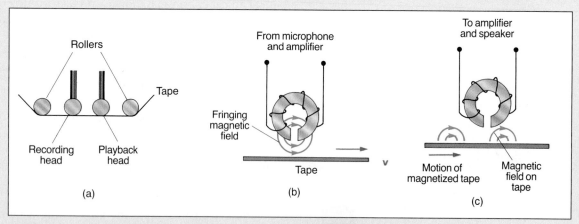

(a) The essential elements of a tape recorder. (b) As the tape is drawn past the recording head, the fringing magnetic field of the head magnetizes the tape. (c) Playback of a magnetic tape in a tape recorder.

With the switch open, there is no current in the electromagnet, and hence it produces no magnetic field. As a result, the single coil finds no magnetic field lines threading through it. When the switch is closed, there is a current; hence, the electromagnet produces a magnet field. Now the coil does have a magnetic field threading through it. Thus, even though there has been no relative motion, the net effect is as though the coil were moved from far away where the field is zero to close to the electromagnet where the field is strong. It is as though one had "tricked" the coil into believing that either it or the electromagnet had moved. Thus, there *is* an induced current.

9.8 ELECTRIC MOTORS

An **electric motor** is a device that converts electrical energy into mechanical energy. The first conversion of this kind was performed by Oersted when he held his compass near an electric current. The magnetic field produced by the wire caused the compass to turn. Thus, electrical energy was con-

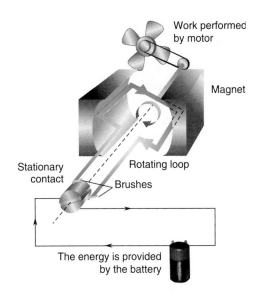

FIGURE 9.27   *A simple d.c. motor.*

verted to mechanical energy of motion. This was not a useful or efficient motor, however, for once the needle turned so that it was along the direction of the magnetic field of the current-carrying wire, the needle stopped moving. If an electric drill were constructed in this manner, it would turn only half a revolution and then stop. To keep a motor running, some way must be found to alternate the current flow so that the shaft of the motor is forced to turn continuously. To understand this concept, refer back to Oersted's original experiment, redrawn in Figure 9.26. When the current is in one direction as shown in Figure 9.26a, the compass needle moves such that it points in a direction perpendicular to the wire, and there it stops. But if the wires connected to the battery are reversed at this precise instant, the current changes direction as in Figure 9.26b, the magnetic field produced by the wire also changes direction, and the needle rotates 180° as shown. If the wires are once again reversed, the needle rotates again to the position of Figure 9.26a, *ad infinitum.* Thus, the needle rotates continuously if some way can be found to switch the wires rapidly and in synchronization with the compass needle.

The essential parts of a slightly more realistic motor are shown in Figure 9.27. Here a coil of wire is mounted so that it is free to rotate, and a shaft connected to it turns as the coil turns. Connected to the coil are two semicircular pieces of metal called **brushes.** To see how the motor works, refer to Fig-

FIGURE 9.26   *Demonstration of the principle behind an electric motor. (a) The compass needle aligns as shown when the current is right to left, and (b) it reverses direction when the current direction reverses.*

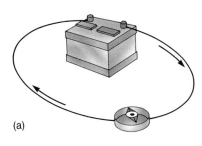

(a)

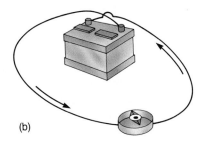

(b)

FIGURE 9.28 *Several stages in the motion of the coil of a d.c. motor.*

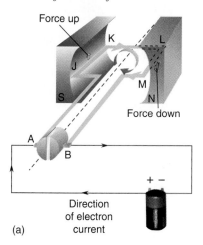

(a)

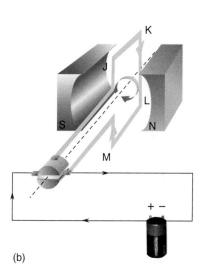

(b)

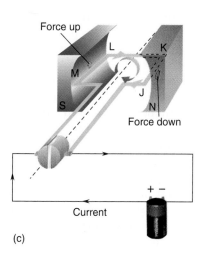

(c)

ure 9.28, where we show a battery as the source of current in the coil. In Figure 9.28a, current leaves the positive terminal of the battery, moves through brush A around the coil in the direction indicated, through brush B, and back to the battery. The side of the coil labeled JK is in a magnetic field, and it is carrying a current. Thus, there is a magnetic force acting on it. The direction of this force is upward, as shown. Likewise, the wire labeled LM has a force directed downward acting on it. These forces acting on the coil cause the coil to rotate until it reaches the position shown in Figure 9.28b. At this position, the brushes momentarily become disconnected from the rings and, as a result, no current passes through the coil. Therefore, no forces are acting on the sides JK and LM. The motion of the coil, however, causes it to continue to rotate under its own momentum until it reaches the position shown in Figure 9.28c. Sides JK and LM have now changed position relative to that shown in Figure 9.28a, and the forces on these wires are now in the direction indicated. Thus, the coil continues to rotate, always in the same direction.

The motor discussed here is not likely to be very powerful. Recall that the magnitude of the force on the segments JK and LM depends on the strength of the current in the wires and on the strength of the magnetic field that these wires are in. A practical motor differs from our simple picture in two respects. First, one loop of wire as shown in our diagram carries only a small amount of current, so the core of any useful electric motor must have many loops of wire rather than one. Second, we have produced a magnetic field by use of a permanent magnet. These magnets are not suitable for a rugged, long-lasting motor because they are extremely heavy and may easily become demagnetized. As a result, the magnetic field is provided by an electromagnet.

9.9 ELECTRIC GENERATORS

An electric motor operates on the principle that a current-carrying wire in a magnetic field has a force acting on it. A **generator** utilizes the principle that a current is induced in a coil of wire when the number of magnetic field lines passing through the cross-sectional area of the coil changes with time. A generator looks much like a simple motor. The difference between the two is often expressed by

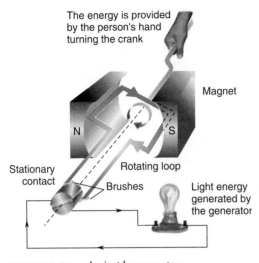

The energy is provided by the person's hand turning the crank

Magnet

N   S

Stationary contact

Rotating loop

Brushes

Light energy generated by the generator

FIGURE 9.29   *A simple generator.*

saying that a motor uses a current to produce rotary motion, whereas a generator uses rotary motion to produce a current, as indicated in Figure 9.29. A cycle of the motion of a simple generator is depicted in Figure 9.30. The basic parts of a generator consist of a coil of wire, a means to make the coil rotate, and two rings that rub (or brush) against stationary contacts. The brushes allow for connections to be made to some external circuit, which could be an entire city but in our figure is a single light bulb. Let us follow what happens as the coil is caused to rotate.

When the coil is horizontal, as in Figure 9.30a, no magnetic field lines pass through it, but a small rotation in the direction indicated by the arrows causes several lines to pass through its cross-sectional area. The rapid change in the field through

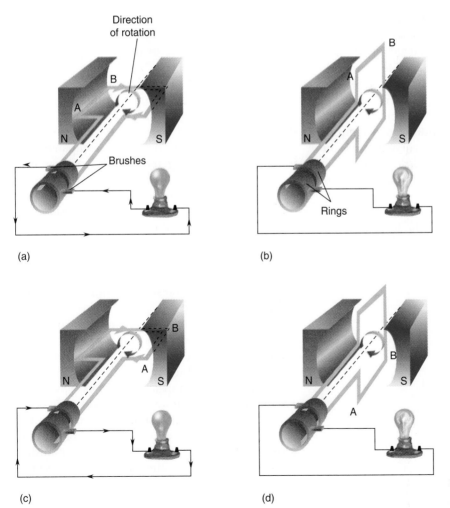

Direction of rotation

B
A
N   S

Brushes

(a)

B
A
N   S

Rings

(b)

B
A
N   S

(c)

B
A
N   S

(d)

FIGURE 9.30   *A cycle of a generator.*

the coil induces a current in the coil. This current flows around the coil in the direction indicated, through the brushes and rings to the light bulb. As the coil moves from the horizontal position to a vertical one, as shown in Figure 9.30b, the current decreases. This occurs because when the coil is almost vertical, a slight rotation of it does not change appreciably the number of field lines threading through it. In fact, the current drops to zero when the coil is exactly vertical. In Figure 9.30c, the coil has again rotated through the vertical and back to a horizontal position. During this rotation, the current becomes large again, but a careful analysis shows that the current through the light bulb has reversed direction. A continued rotation of the coil again takes it to the vertical position, as in Figure 9.30d, and again the current drops to zero.

Note an important result brought forth in our analysis. The current through the light bulb is considerably different from that which would be produced by an ordinary battery. A battery produces a constant current that always moves in the same direction through the bulb. The generator produces a current that has a large value in one direction, decreases to zero, increases to a large value in the opposite direction, and again decreases to zero, as shown in Figure 9.31. A current of this nature, one that changes direction and magnitude regularly, is called an **alternating current** and is the type provided to your home by a utility company. The alternating current produced by generators in the United States varies at 60 Hz. That is, the current goes through a complete cycle 60 times per second. In much of the rest of the world, the frequency is 50 Hz. (Converters are necessary to use American appliances overseas, but these are normally used to change voltages—not frequencies.)

The problem of causing the coil to rotate is solved in a variety of ways; a common method is to direct falling water against the blades of a device called a turbine. The water striking the blades produces rotational motion of a shaft on which the coil of the generator is mounted. Coal, oil, and nuclear power plants use turbines in a similar manner. In these cases, however, the heat generated by the burning fuel is used to convert water to steam, which is then directed at the blades of the turbine, causing the rotation.

9.10 TRANSFORMERS AND TRANSMISSION OF ELECTRICITY

Most of the electricity used today is generated in large, centralized power plants, and in North America, there is an increasing trend toward situating these plants near sources of fuel, especially coal mines, rather than near urban centers, where most of the electricity is needed. The reason for this choice is that it is cheaper to transmit the electricity than it is to transport the coal. As a result, long-distance transmission lines have become common. An important problem that utility companies face is that of transmitting the electrical energy over large distances to the consumer without losing much of the energy in the process.

Electrical energy is lost in the transmission lines primarily because some of the electrical energy is converted to heat in the lines. These heat losses occur because the wires have resistance, and the equation power = I^2R (Eq. 8.10) tells us the rate at which energy is lost through resistive heating. Utility companies have the choice of transmitting their electrical energy either at higher voltages and lower current or at higher current and lower voltage. As power = I^2R shows, less energy is lost if the electricity is transmitted at low current. As a result, modern long-distance transmission lines operate at 120,000 V or more. However, it would be disastrous to supply electricity into household circuits at this potential; appliances would spark uncontrollably, wires would overheat, and even a slight mishap would be fatal. Thus, the high voltages used to

FIGURE 9.31   *Alternating current.*

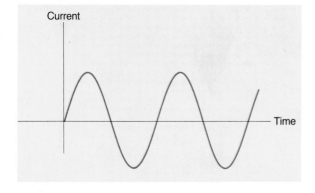

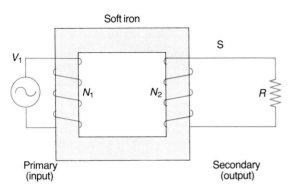

FIGURE 9.32 *A simple transformer.*

(a)

(b)

(c)

transmit the electricity from the source to the site of consumption must be "stepped down" to the normal voltages encountered in a home, approximately 120 V. A device that accomplishes this conversion is called a **transformer.** A simple transformer is diagrammed in Figure 9.32. It consists of a number of turns of wire wound around a piece of iron. One set of windings, called the **primary** coil, is connected to a source of alternating voltage that is to be changed. A second coil of wire, called the **secondary,** is also wrapped on the same iron core. As the direction and magnitude of the current in the primary change, the magnetic field set up by the primary also changes simultaneously. The iron core provides a pathway for the magnetic field lines produced by the primary to be led through the secondary coil. Thus, the secondary coil finds a changing magnetic field threading through its cross-sectional area. As we have seen, a changing magnetic field through the secondary produces an induced current and voltage in this coil. To see how such a device can be helpful to us, let us assume that there is one loop of wire in the primary and two loops in the secondary. A changing magnetic field in the primary induces a voltage in each loop of the secondary. Thus, if the input signal to the primary has an amplitude of 1 V, the output from the secondary will be 2 V (Fig. 9.33). When a transformer is used in this way to step up the voltage, it is appropriately called a *step-up transformer.* If the coil is constructed such that the secondary has one loop while the primary has two loops, the voltage at the secondary will be decreased by a factor of two, and we have a *step-down transformer.* The relationship between the number of turns and the voltages is given by

(a) A small demonstration transformer that shows the primary and secondary windings and the magnetic core that links them. (b) A step-down transformer used near a home to reduce the voltage supplied to the house from about 2200 V to 120 V. (c) A step-down transformer used near a city to reduce the voltage from the utility company down from about 200,000 V to about 2200 V.

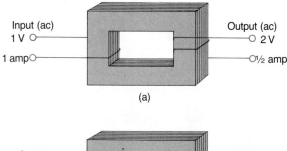

(a)

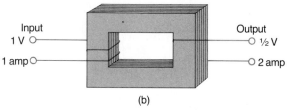

(b)

FIGURE 9.33   *(a) A 1 V potential in one loop of the primary coil produces 1 V in each loop of the secondary. Therefore, the potential in the secondary coil is amplified to 2 V, but the current is reduced by half. (b) In this situation, the voltage is cut in half, and the current is doubled.*

$$\frac{V_p}{N_p} = \frac{V_s}{N_s} \qquad (9.1)$$

where N_p and N_s refer to the number of turns on the primary and secondary, respectively, and V_p and V_s are the voltages at the primary and secondary.

If the voltage at the secondary is greater than that at the primary for a step-up transformer, does this mean that we are getting something for nothing? Since we know that energy is conserved, obviously we are not. Because the primary and secondary are operating together for the same amount of time, the power supplied to the primary also appears at the secondary. Since $P = IV$, this equation tells us that if the voltage at the secondary is greater than that at the primary by a factor of, say, two, the current at the secondary is stepped down by a factor of two. Thus, an increase of voltage at the secondary is accompanied by a simultaneous decrease in the current by the same factor. The relationship between the power at the primary and at the secondary is given by

$$(IV)_{secondary} = (IV)_{primary} \qquad (9.2)$$

This equation is for an ideal situation. In reality, some of the power is lost inside the transformer, resulting in a heating of the transformer.

Large transformer centers near every big city step down the 120,000 V arriving from the generating plant to about 2200 V for transmission along conventional "telephone pole" routes to residential and commercial buildings. The 2200 V must be stepped down again before electricity is brought into a house or a business. Therefore, another transformer, one that steps down the voltage to about 120 V, is placed near these establishments.

EXAMPLE 9.5   STEPPING DOWN

Imagine that you wish to step down a high-voltage line at 120,000 V to 2200 V for local transmission. **(a)** If there are 100,000 coils of wire in the primary circuit of the transformer, how many coils will be needed in the secondary?

(b) If the high-voltage line carries 100 A, how many amps will be present in the secondary circuit?

Solution (a) In this situation, a step-down transformer is needed. The number of turns on the secondary can be found from Equation 9.1 as

$$N_s = N_p \frac{V_s}{V_p} = (100{,}000 \text{ turns}) \frac{2200 \text{ V}}{120{,}000} \text{ V}$$

$$= \boxed{1830 \text{ turns}}$$

(b) To find the current in the secondary, let us use the fact that power is conserved between the primary and the secondary as

$$(IV)_{secondary} = (IV)_{primary}$$

or

$$I_s(2200 \text{ V}) = (100 \text{ A})(120{,}000 \text{ V})$$

from which we find

$$\boxed{I_s = 5450 \text{ A}}$$

SUMMARY

Like magnetic poles repel and unlike poles attract. The direction of a magnetic field is determined by the direction in which the north pole of a compass needle points. Magnetic field lines have no stopping or starting points.

The magnetic field of the Earth is created by some sort of electric generator in the Earth's fluid interior.

When an electric current travels perpendicular to a magnetic field, a sideways force is generated. When a magnet is forced to move perpendicularly to an electric wire, or when a wire or any charged particles are forced to move through a magnetic field, a current is generated.

Magnetism is produced by individual domains, and a permanent magnet is created when these domains are aligned. Whenever the number of magnetic field lines passing through a coil is changed, an electric current is induced in the coil.

In an **electric motor,** a current is forced to travel perpendicularly to a magnet, and mechanical work is produced. A **generator** operates on the reverse principle; a coil of wire is mechanically turned within a magnetic field, and electricity is generated. **Transformers** can be used to alter the voltage and current (but not the energy) in an alternating current circuit.

EQUATIONS TO KNOW

$$\frac{V_p}{N_p} = \frac{V_s}{N_s}$$ (voltage-turn relationship for a transformer)

$(IV)_{secondary} = (IV)_{primary}$ (power relationship for a transformer)

KEY WORDS

North pole (of a magnet)	Magnetic domain	Electromagnet	Electric generator
South pole (of a magnet)	Alternating current	Electric motor	Transformer
Magnetic field	Electromagnetic induction		Plasma

PROBLEMS AND CONCEPTUAL QUESTIONS

Problems requiring numerical work are identified with a blue number.

Magnets and magnetic fields

1. Are magnetic fields more similar to electric fields or to gravitational fields? Defend your answer.
2. You have two pieces of metal; one is magnetized and one is not. By using *nothing* else except the two pieces of metal, how could you determine which is the one that is magnetized?
3. You have two equivalent iron magnets and keep one in the refrigerator and the other in a hot oven for the same length of time. Which one do you think will preserve its magnetism more completely? Defend your answer.
4. You have a magnet with unlabeled poles. How could you find which is the north end of the magnet? What equipment would you need?
5. Some iron fences that have been in place for a long time are found to be magnetized. What has caused this to happen?

6. Magnetic field lines never cross one another. Explain why this has to be true.
7. Based on what you know about the magnetic field of the Earth, draw a sketch that represents this field. Your pattern of lines should also continue through the interior of the Earth.

Magnetic field of the Earth

8. What is the source of energy for the magnetic field of the Earth?
9. The Moon has virtually no magnetic field. Does that information tell us anything about the structure of the interior of the Moon? Defend your answer.
10. Sketch a diagram of the Earth showing the direction a compass would point at (a) the South Pole, (b) the North Pole, and (c) the equator.
11. Airline pilots have to reset their compasses to correct for irregular variations in the magnetic field of the Earth. What could cause these variations?

12. During the history of our planet, its magnetic field has undergone periodic reversals. Suppose one of these reversals should take place now. How would our lives be affected?

Connection between electricity and magnetism

13. Modern ships and airplanes use both conventional compasses and sophisticated electronic equipment for navigation. Would it be possible for the electronic equipment to damage the effectiveness of the compass? Do you think that precautions would be needed to shield the compass from the ship's electronics? Explain.

14. A positive charge is not deflected as it passes from one location to another in a room. Can we be sure that there is no magnetic field present in the room?

15. You charge a balloon by friction and then throw it across the room. Does the moving balloon set up a magnetic field?

16. A positive charge hangs between the poles of a horseshoe magnet. Is there a magnetic force acting on it?

17. If you rub a balloon on your hair to charge it and then place the charged balloon near a compass, will the compass needle deviate from pointing northward? What will happen if you wave the charged balloon back and forth across the top of the compass? Explain.

18. Imagine that you are a space explorer who discovers a rain of high-energy, fast-moving, glowing particles in the atmosphere of some faraway planet. Design a simple experiment to determine whether or not these particles are electrically charged.

19. Will a magnet held near the screen of a TV tube affect the picture? Why?

20. A beam of protons is moving across the room when the north pole of a magnet is pointed at the beam such that the magnetic field and the motion of the particles are perpendicular. Will the beam be deflected toward the magnet, away from it, sideways, or not at all?

21. How could a magnet be used to determine whether a beam of charged particles is positive, negative, or neutral?

22. Explain why an electromagnet made with ten loops of wire will be stronger than one with five loops of wire. How much stronger will it be?

23. If you were building an electromagnet to lift scrap iron, would you use alternating current or direct current, or would it make any difference? Explain.

What makes a permanent magnet magnetic?

24. An iron nail will be attracted to either end of a bar magnet. Why?

25. The magnetic pattern around a magnet such as a horseshoe magnet can be traced out by placing a piece of plastic over the magnet and sprinkling iron filings on the plastic. The filings align along the field lines. Why?

26. A magnet is capable of supporting long chains of paper clips, each clip dangling from one above it. Explain on the basis of magnetic domains what is happening in each clip.

Cosmic rays and the Earth's magnetic field

27. Would the magnetic field of the Earth deflect high-speed neutrons from outer space? Explain.

28. As mentioned in problem 12, the Earth has undergone periodic reversals in its magnetic field. If the field should drop to zero, what effect would it have on (a) the Van Allen belts? (b) Auroras? (c) The number of charged particles striking the Earth from outer space?

Electromagnetic induction

29. If a powerful electromagnet is suspended by a long cable and held stationary over a coil of wire, will a current be induced in the wire? If a wind comes up and the magnet sways in the breeze, will this induce a current in the wire? Explain.

30. Explain why lightning bolts produce large induced voltages in electronic equipment.

31. A popular classroom demonstration consists of dropping a light, but strong, magnet down the length of a piece of copper pipe. Instead of falling rapidly through the pipe, the magnet floats down the pipe. Explain how this could occur.

32. A coil of wire is placed close to the north pole of a magnet, and the flexible coil is rapidly crushed to zero cross-sectional area. Is there a current in the coil while it is being crushed? Explain.

33. A bar magnet is placed on the turntable of a record player and set into rotation at 45 rpm. A coil of wire near the turntable is found to have a current induced in it. Describe how this current would vary with time.

Electric motors and generators

34. Electric motors can be constructed to operate from a direct or an alternating power source, but inside the motor there must always be alternating current. Explain.

35. As the price of copper wire continues to increase, do you think that small portable electric motors might someday be built with permanent magnets rather than electromagnets? Defend your answer.

36. The motor described in the text and illustrated in Figure 9.27 has a loop of wire spinning in a magnetic field. Would it be possible to build a motor in which a permanent magnet rotates in an alternating electric field? Explain.

37. When you start an electric motor, there is an initial rapid surge of current, followed by a rapid approach to constant current flow. Explain why this current surge occurs. Can the surge be harmful to the motor itself? If so, how?

38. Does a generator create energy? Explain.

39. Would a generator work if the coil were held stationary and the magnet rotated? Explain.

Transformers and transmission of electricity

40. If the primary coil of a transformer has 100 loops of wire carrying 10 A at 2000 V, how many loops are needed in the secondary coil to produce 4000 V? What about 200 V? 2000 V? What would the current be in each case?

41. A transformer to be used with a toy is designed to be plugged into 120 V at the primary and to deliver 6 V at the secondary. What would happen if the connections were accidentally reversed?

42. A transformer can be built to step up the potential of a line from 50 V to 500 V. Is this a violation of the conservation of energy? Explain.

43. Slot cars and electric trains operate at about 10 to 15 V. The 110-V household current is stepped down to lower voltage in a transformer. The speed of the toys can be changed by varying the voltage applied to them. Draw a labeled diagram of a transformer that can supply a variable output voltage.

44. Explain why direct current is not generally supplied in household circuits.

45. Why is a transformer constructed such that an iron core passes through both coils?

46. The primary of a transformer is connected to a battery, and a direct current flows in the coil. If a switch in the primary circuit is opened, will there be a voltage induced in the secondary?

47. A transformer is constructed with 100 turns in the primary and 400 turns in the secondary. If a voltage of 20 V is applied at the primary and 2 A of current exists in the primary, find (a) the current and (b) voltage in the secondary.

48. According to the text, power losses from generating stations are reduced when the voltage is stepped up to large values. Why doesn't the company then step the voltage up to even higher values?

ANSWERS TO SELECTED NUMERICAL PROBLEMS

40. (a) 200 turns, (b) 10 turns, (c) 100 turns, (d) 5 A, 100 A, 10 A

47. (a) 0.5 A, (b) 80 V

CHAPTER 10

PROPERTIES OF LIGHT

Reflections in a peaceful stream.

Throughout this book, there has been a consistent attempt to relate the concepts of physics to everyday experiences. This was reasonably easy to do in our study of mechanics because the motion of objects and the forces that cause this motion are readily apparent to our senses. The task became more complex when we examined electricity and magnetism. We can understand how the principles of electricity and magnetism are applied on a large scale to provide us with electrical energy to heat our homes and do a myriad of other wonderful things for us. However, electricity per se is elusive. To understand all of these large-scale phenomena, we must search for answers in the invisible world of electrons and atoms.

As we begin our study of light, we will find that it becomes even more difficult to comprehend the concept of light. Although we will be able to explain many everyday occurrences related to light, the question of what light is will be quite difficult to answer. The nature and behavior of light are quite different from anything that can be explained in terms of experience with large-scale natural phenomena. It is no wonder that many great physicists of the past have struggled to understand the nature of light.

Newton, for example, considered light to be a stream of particles. According to his approach, a source of light emitted tiny particles, something like a stream of BBs, and we are able to see objects when these little bullets bounce off them and enter the eye. Other scientists found that many of the properties of light could be explained only by considering it to be a form of wave motion. Thus, when we see an object, we do so because light from it breaks like a wave on the retina of our eye. Most of the concepts discussed in our study of light will be explained by the wave model of light. Keep in mind, however, that sometimes the behavior of light can be explained only from the Newtonian viewpoint of it being a stream of particles. This so-called dual nature of light, in that it sometimes behaves as a wave and sometimes as a particle, is unique among the topics and concepts that we have studied, but there are other fascinating characteristics of light.

10.1 SPEED OF LIGHT

We pointed out that light has many unusual characteristics. One of them is its enormous speed, which we now know to be 3.00×10^8 m/s, or 186,000 miles/s. Think of the enormity of this value. If you turn on a flashlight and its light could spread out through space without being absorbed or diminished, an observer 186,000 miles distant would detect the light beam 1 s later. It is no wonder that if you turn on an electric light bulb in your room or throw a switch that turns on a light across the road or even on a distant hilltop, the rays seem to appear almost instantly. How is something as fast as the speed of light measured?

The first recorded experiment in which an attempt was made to determine the speed of light was conducted about A.D. 1600 by Galileo Galilei. He attempted to measure the speed of light by positioning two people on hilltops separated by a distance of about 1.5 km (Fig. 10.1). Both experimenters carried shuttered lanterns. One person quickly removed the cover from his lantern and began to record the time. As soon as his partner on the neighboring hill saw the light beam reach him, he uncovered his lantern. When the first observer saw that the second lantern had been uncovered,

he stopped his measurement of time. From these observations, Galileo hoped to be able to calculate the speed of light from $v = $ distance/time. As you might guess, however, Galileo found that human reaction time was so slow compared with the speed of light that all he was really able to measure was the time required to uncover the lanterns. As far as he could tell, the light made the round trip instantaneously, or at least with an unmeasurably fast speed.

Several techniques have been used to measure the speed of light. Because it is impossible to turn switches on and off fast enough to time the speed of a light ray as one would time the speed of, say, a human runner, all earthbound measurements have been performed with the aid of some continuously moving device. The most famous experiment of this sort was performed by Albert Michelson in 1880.

His apparatus, shown in Figure 10.2, consisted of a rotating eight-sided mirror, an intense source of light, a flat mirror approximately 35 km from the rotating one, and an observer located as shown. With the mirror originally at rest, the light source and observer were aligned so that the light would shine on face A of the eight-sided mirror, reflect to the distant mirror, bounce back to side B, and enter the eye of the observer, as shown in Figure 10.2a. The mirror could be rotated at different speeds, and it was found that light was reflected into the eye only at certain rates of rotation. If the mirror is rotating just right, the light will reflect off side A and return at the instant side C has moved into place to reflect the beam into the observer's eye. If the rotation rate is either too slow or too fast, however, C will not be in the proper position to reflect

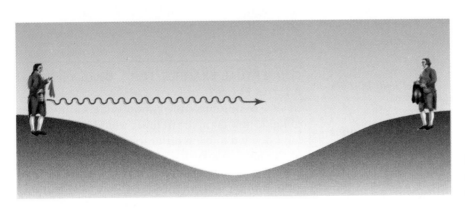

FIGURE 10.1 *Galileo's attempt to measure the speed of light.*

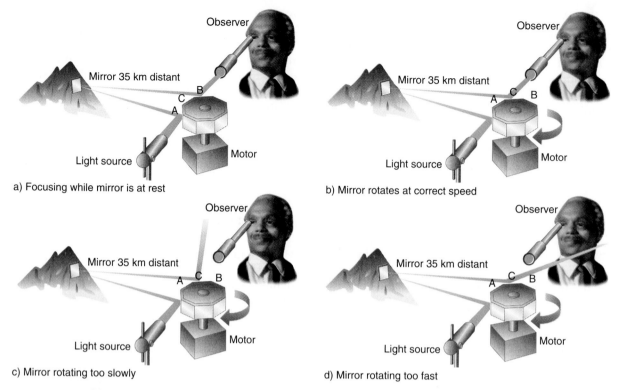

a) Focusing while mirror is at rest

b) Mirror rotates at correct speed

c) Mirror rotating too slowly

d) Mirror rotating too fast

FIGURE 10.2 *Measurement of the speed of light by Michelson's method.*

light into the eye, and the light will bounce off in some random direction, as indicated by Figure 10.2c and 10.2d. Thus, by accurately adjusting the speed of rotation of the mirror until the light was reflected to the observer, Michelson was able to calculate the speed of light. His experimental value was 2.9992×10^8 m/s.

Various other methods have been used to measure the speed of light since this attempt by Michelson. The currently accepted value for the speed of light is 2.997924574×10^8 m/s. For convenience in calculation, we shall round off this value to 3.00×10^8 m/s.

Scientists now know that the speed of light in a vacuum is a universal constant.

At any time and any place anyone chooses to measure the speed of light in a vacuum, it will travel at 3.00×10^8 m/s, never faster and never slower.

If a beam of light travels in an uninterrupted path for an entire year, it will cover a distance of 9.5 trillion km (9.5×10^{12} km). This distance is called a **light-year.** But numbers can be misleading: On an astronomical scale, the speed of light no

longer seems so great. For example, the closest star to our Solar System is about 40×10^{12} km (40 trillion km) away. A light ray leaving that star for Earth must travel for about 4 years before it reaches us. If our focus of interest is expanded even farther out into space, even a distance of 4 light-years can seem small. The most distant galaxies now known are about 10 billion light-years from Earth. This means that if one of those galaxies had exploded 9 billion years ago, we still wouldn't know about it for another billion years to come.

CONCEPTUAL EXAMPLE 10.1 COULD YOU WORK A LITTLE FASTER?

In view of our currently known value for the speed of light, how far apart in miles would Galileo's hills have needed to be in order for the round-trip flight of the light to take 1 s?

Solution In 1 s, the light would travel a distance of 186,000 miles. Thus, the hills would have to have been 93,000 miles apart for 1 s to elapse for the round trip.

EXAMPLE 10.2 WHAT HAPPENED YESTERDAY?

(a) Explain why astronomers often say that when they are looking through a telescope, they are looking backward in time.

(b) If the Sun is 93,000,000 miles away, how far back in time are you looking when you glance at the Sun?

Solution (a) When we look at a distant object, we see it not as it is now, but as it was when the light left it. For example, when we look at a star 10 light-years away, we see that star as it was 10 years ago because it has taken 10 years for that light to reach us. Thus, if an observer 70,000,000 light-years away should at this instant focus an absurdly powerful telescope on the Earth, he would not see us; instead, he would see dinosaurs peacefully grazing on a now ancient meadow.

(b) The time for light to reach us from the Sun is found from the definition of speed as

$$t = \frac{d}{v} = \frac{93000000 \text{ miles}}{186000 \text{ miles}} = 500 \text{ s} = \boxed{8.33 \text{ min}}$$

Thus, we see the Sun not as it is now, but as it was approximately 8 minutes ago.

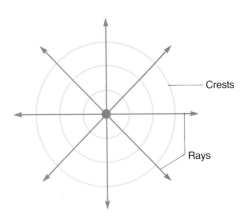

● = Source of light

FIGURE 10.3 *A small source emitting rays of light.*

10.2 RAYS OF LIGHT

When an object emits light, the waves spread out uniformly in all directions. We will discuss the motion of these waves in terms of rays. *A ray is a line drawn in the direction in which waves are traveling.* For example, a sunbeam passing through a darkened room traces out the path of a ray. Figure 10.3 shows a small object emitting light waves. We have drawn the figure such that the circles indicate crests of the

(a) The mountainous terrain is clearly reflected in a still, calm lake. The smooth water is an effective mirror. (b) When the lake surface is disturbed by waves, reflected light is scattered, and the image is blurred. Rough surfaces may reflect light but do not make good mirrors.

wave. The same type of picture could be used to represent ripples spreading out from a vibrating object placed in a pool of water. Several rays of light are indicated in the figure. Observations indicate that *these rays of light always travel in straight lines, as long as the light does not move from one material into another.* As we shall see later, when light moves from one material into another, it changes direction.

10.3 LAW OF REFLECTION

Most of the objects around us are not emitters of light but instead act as reflectors of the light that falls on them. For example, if light moving through air strikes a mirror, the wave bounces back from the mirror. When waves bounce back from an object, they are said to be reflected. Light is reflected from the surface of wood, paper, metal, soil, and anything else that is described as opaque (nontransparent). Transparent substances such as glass or water allow most of the light to pass through them, although they do reflect some of the light.

To discover the law of reflection, refer to Figure 10.4, where we see an incoming beam of light reflecting off a surface. At the point where the incoming beam of light strikes the surface, point A, we sketch in a reference line drawn perpendicularly to the surface. The angle between the incident beam of light and the perpendicular is called the

FIGURE 10.4   *When light is reflected from the surface, the angle of incidence is equal to the angle of reflection.*

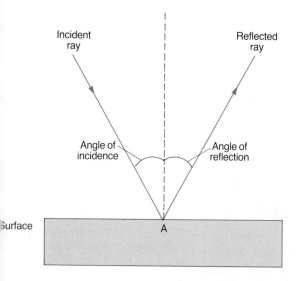

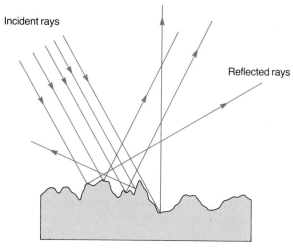

FIGURE 10.5   *Scattered reflections from a rough surface produce diffused light.*

angle of incidence. When the light ray is reflected, it bounces off on the other side of the perpendicular line as shown. The angle between the perpendicular and the reflected ray is called the **angle of reflection.** A careful measurement in an actual experiment would show that *the angle of reflection equals the angle of incidence.* Additionally, it is found that *the incident ray, the reflected ray, and the perpendicular line all lie in the same plane.* These pieces of information are referred to as the **law of reflection**. In Figure 10.4, we have sketched the reflecting surface as perfectly flat and smooth, but what happens if the surface is rough, as shown in Figure 10.5? As a group of rays approach the surface, different ones strike at different angles of incidence, yet all obey the law of reflection. The result is that the reflected rays are scattered in all directions as shown. Light rays reflecting in this manner are said to be diffusely reflected, and such light is called diffuse light.

CONCEPTUAL EXAMPLE 10.3   STAY HOME WHEN IT RAINS

It is common experience that it is more difficult to drive at night after a heavy rain than it is when the roadway is dry. Use the law of reflection to explain why this is so.

Solution When the roadway is dry, the light from an oncoming car is scattered in random directions off the rough pavement. Thus, light reaches your eye from all directions. After a rain, however, microscopic irregularities in the pavement surface are filled in by water, and the roadway is covered by a smooth reflecting surface. Thus, light from the headlights of an oncoming car reflected off the smooth water surface comes to your eye from only one direction, making it more difficult to see a wide expanse of the road, which reduces the ability to drive safely.

10.4 FORMATION OF IMAGES

The fact that light forms images is useful to us, and it provides us with recreational opportunities. One example of a useful image is that formed in a flat mirror. A glance at your image tells you whether you look as good as you would like before meeting the public. The image formed on a movie screen provides us with entertainment. The image in a mirror and the image on a theater screen are both examples of images, but the two types are distinctly different. Before we look at some optical devices that form images, we will briefly discuss how images are formed.

Images are formed in two ways: *(1) An image is formed when light rays actually intersect, and (2) an image is formed when light rays appear to intersect.* We shall look shortly at the actual processes of forming images by these two methods. For now, remember two more facts about these images. An image formed when light rays actually intersect is called a **real image.** A characteristic of real images is that they can be caught on a screen. Thus, the image formed by a movie projector is a real image. An image formed when light rays only appear to intersect is called a **virtual image**. A virtual image cannot be caught on a screen. This means that if you hold a sheet of paper at the location where the image appears to be, no image is formed on the paper. When you look at your image in a flat mirror, the image appears to be formed some distance behind the mirror. However, if you move a sheet of paper around behind the mirror, no image is ever formed on the paper.

10.5 FLAT MIRRORS

The kind of mirror that you probably have in your bathroom is a flat mirror, which means that the glass surface of the mirror is not curved. To see how such a mirror forms an image of an object placed before it, we will have to follow at least two different rays of light that leave the same point of the object.

In Figure 10.6, we follow two rays leaving the top of an apple placed before the mirror. One of these rays, labeled A, leaves the apple's top and strikes the mirror head-on. In this case, the angle of incidence is zero, and the ray of light is reflected such that the angle of reflection is also zero. This means that it reflects back directly on itself. The next ray, labeled B, leaves the top of the apple at any arbitrary angle and strikes the mirror. This ray, too, obeys the law of reflection and follows the path labeled C after reflecting. When these rays enter the eye of an observer, they are traced back along a straight-line path to the point from which they appear to originate. Thus, as far as the eye is concerned, the reflected ray A appears to have come from some point along the line labeled D in the figure, and the ray C appears to have come from some point along the path labeled E. The dashed lines drawn for D and E show that these lines intersect at point G; thus, this is the point from which the eye believes the rays originated. At point G, where the rays appear to intersect, is the location of the image. If you refer back to the last section, you will find that the image formed is like that dis-

FIGURE 10.6 *Tracing rays of light to find the location of an image formed by a flat mirror.*

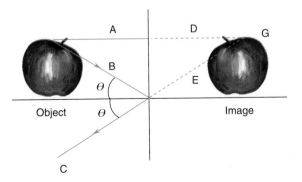

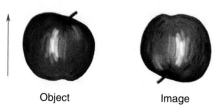

Object   Image

FIGURE 10.7   *If the image is turned over with respect to the object, it is said to be an inverted image.*

cussed as case two, and the image is a virtual image. You could follow rays of light from all points on the apple to see where the images of these points are found, and you would end up with a complete image formed at the position G indicated in Figure 10.6.

You could draw Figure 10.6 to scale and measure all the angles and distances carefully to discover some facts about the images formed by flat mirrors. Some of the features that you would find are:

1. The image is as far behind the mirror as the object is in front of it. This means that if a book is held 20 cm in front of a flat mirror, the image will be formed 20 cm behind the mirror.

FIGURE 10.8   *Reflection from a curved surface. The angle of incidence equals the angle of reflection with respect to a line drawn perpendicular to the tangent to the surface, as shown.*

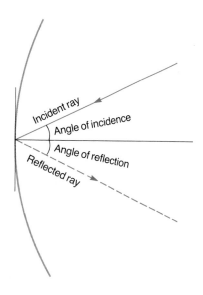

Incident ray
Angle of incidence
Angle of reflection
Reflected ray

FOCUS ON REARVIEW MIRRORS

Most rearview mirrors in cars have a day setting and a night setting. In the latter case, the image produced is greatly diminished in intensity so that lights from trailing cars will not blind the driver. To see how these mirrors use the laws of reflection and refraction, refer to the figure. These mirrors are actually a wedge of glass with the reflecting side of the mirror on the back side as shown. The day setting for the mirror is shown in part (a) of the figure. In this case, light from an object behind the car strikes the mirror at point 1. Most of the light enters the wedge, is refracted, and reflects from the back of the mirror to return to the front surface, where it is refracted again as it reenters the air as ray B (for bright). At the surface of any transparent object, some light is reflected and some is transmitted. Thus, as shown, when the light first struck the wedge, a small portion was reflected as D (for dim). This dim light becomes important when we move to the night setting as in part (b). Here the wedge has been rotated slightly, and the path now followed by most of the light, path B, does not lead to the eye. Instead, the dim light from the front surface is reflected to the eye, and the intensity of the bright lights of the car behind does not produce a hazard.

2. The image is unmagnified. This means that if the object is 2 m tall, the image will also be 2 m tall.
3. The image is erect. This means that if an object in the shape of an arrow is placed in front of a flat mirror such that the tip of the arrow points toward the ceiling, the image formed of the arrow will also point toward the ceiling. An image is said to be inverted if it points in a direction 180° away from that pointed by the object (Fig. 10.7).

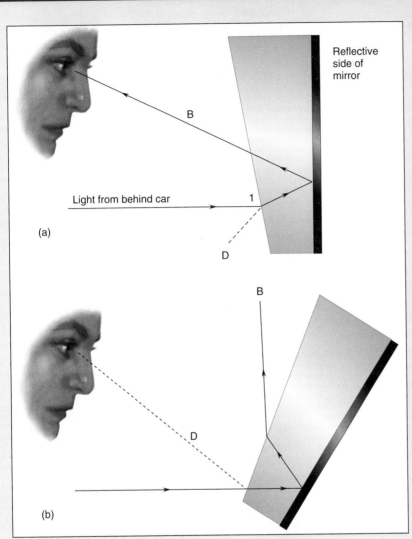

A rearview mirror.

4. The image has right-left reversal. This means that if you raise your right hand while standing in front of a flat mirror, your image will seem to be raising its left hand.

10.6 CURVED MIRRORS

Curved mirrors are used in many practical applications. If light strikes a curved surface, it is reflected so that the angle of incidence equals the angle of reflection, just as with a flat surface (Fig. 10.8). There are many different types of curved mirrors, but for our present discussion, let us choose a curved mirror that is a segment of a sphere. For example, suppose you cut out a segment of a round Christmas tree ornament and use it as a mirror. You would find that you could use it in two ways. If you allowed the light to reflect off the inward curving side, as in Figure 10.9a, the type of mirror that you have is called a **concave mirror.** If you are interested in the light reflected from the other side of the seg-

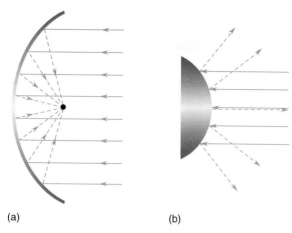

FIGURE 10.9 *(a) A spherical concave mirror and (b) a spherical convex mirror.*

ment, as in Figure 10.9b, you have a **convex mirror.** For convenience, let us examine only one type of mirror, the concave mirror, and see what kind of image(s) it can form.

Figure 10.10 follows two rays of light that leave an object in the shape of an apple placed at point O. Ray A strikes the mirror at the center of the curved segment and reflects, obeying the law of reflection, to follow path B. The second ray of light leaving the object follows path C, hits the mirror head-on, and reflects back on itself. These two rays *actually* intersect at point I. Thus, the image formed at I is a real image. The figure has been drawn to scale, so other pertinent facts about the image can be determined by inspection: (1) The image is smaller than the object, and (2) it is inverted.

FIGURE 10.10 *Formation of an image by a concave mirror.*

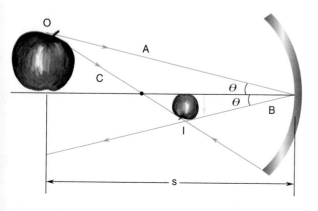

Figure 10.11 follows two rays from an object at O that reflect off a convex mirror. We will leave it for you to follow the path of the rays leaving the object and show that an image is formed at point I. Note the characteristics of the image formed: (1) The image is erect, and (2) it is smaller than the object.

When describing two apparently similar types of mirrors, a distinguishing characteristic called the **focal length** is often used. *The focal length of a mirror is the distance of the image from the mirror when the object is placed an infinite distance away.* For example, when you use a mirror to form an image of the Sun, for all practical purposes the image is formed at the focal point of the mirror. This is true because the Sun is so far away that it can be considered to be at infinity. Figure 10.12 diagrams this situation. Because the object is so far away, by the time the light rays reach the mirror, they are traveling parallel to one another as shown. After reflection, the light rays come to a focus at I, and the distance from the image to the mirror is called the focal length.

Curved mirrors are often used in telescopes except that the shape of the mirror is usually not a part of a segment of a sphere. In this application, the shape is usually chosen to be a **parabola.** Parabolic shapes are better than spherical shapes because the light reflected by a parabola is much more sharply focused at the image than it would be by a spherical segment, which smears the light out somewhat at the image position. Figure 10.13 shows a small parabolic reflector that can concentrate enough sunlight to boil water, cook a hamburger, or bake a chicken. In France, there is a multistory parabolic mirror that can concentrate proportionally larger amounts of solar energy. This solar tower can produce enough steam by heating water to power a small electric generating station. Similarly, parabolic antennas that reflect radio waves are used to collect signals from objects in outer space that emit radio waves. Many important astronomical discoveries have been made with these radio telescopes.

CONCEPTUAL EXAMPLE 10.4   OBJECTS MAY BE CLOSER THAN THEY SEEM
The rearview mirrors in most modern automobiles are designed to use convex spherical mirrors. These mirrors have a cautionary warning printed on them that reads, "Objects may be closer than they seem." Explain why this message is necessary.

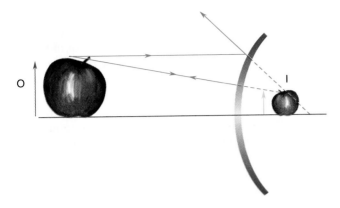

FIGURE 10.11 *Example 10.4. Image formation by a convex mirror.*

Solution Note the characteristics of the image formed by a convex mirror, as shown in Figure 10.11. The key feature here is that *the image is always demagnified.* Thus, even if a car is right on your rear bumper, the small image formed could mislead you into believing that the car is farther away than it really is. Thus, a dangerous driving maneuver might be undertaken.

10.7 REFRACTION

A light ray travels in a straight-line path as long as it does not move from one material into another. If the ray moves from one transparent material into

FIGURE 10.12 *The focal point of a concave mirror.*

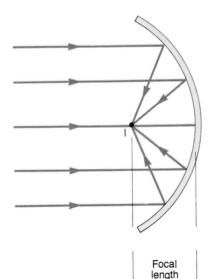

Focal
length

another, however, the ray undergoes a change in direction. This bending is called **refraction.** The reason that light bends is because it has a different speed in the two materials. Let's see why this causes bending.

Light travels at a speed of 3.00×10^8 m/s in a vacuum or, for all practical purposes, in air. When light moves into another material, it slows down. For example, the speed of light in glass is about 2.00×10^8 m/s, and the speed in water is about 2.25×10^8 m/s. To see why this change in speed causes bending, consider an analogy from common experience. Suppose you are sledding down a snowy hill and you hit a roadway head-on, perpendicular to the roadway. The sled will not move as fast on the roadway, but because you hit it head-on, you will not change direction. If you hit the road

FIGURE 10.13 *A reflecting parabola can concentrate enough light energy to cook food.* (Courtesy Dan Halacy, University of Wisconsin)

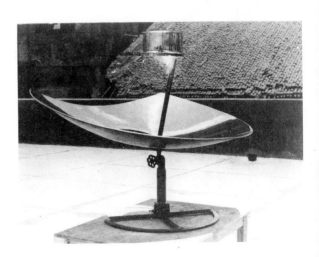

Note that the image formed in this convex mirror is diminished and erect. (Courtesy Bill Schulz)

at an angle, one runner will reach the pavement before the other, as shown in Figure 10.14a. The runner that hits first slows down, while the other runner, which is still on the snow, continues to travel at a rapid speed. The result is that the sled turns abruptly and follows the path shown in the figure. Light acts in a comparable manner. If light traveling in air enters a transparent liquid or solid head-on, it slows down, but its direction of travel does not change. If it moves at an angle from one material into another in which it travels more slowly, however, it turns (undergoes refraction) just as the sled did when it hit the road (Fig. 10.14b).

To describe the refraction process, consider the situation shown in Figure 10.15, where a ray of light in air is incident on the surface of a pool of water. At the point at which the incident ray strikes the water, we construct a reference line perpendicular, or normal, to the surface. The angle between the incident ray and the normal is called the angle of incidence, and the angle between the ongoing ray and the normal is called the angle of refraction. Figure 10.15a has been drawn in a manner that demonstrates a characteristic of the refraction process for this situation. *As light moves into a material in which it travels more slowly, its direction of travel bends toward the perpendicular.* The reverse of the situation also holds. For example, suppose the source of light was beneath the surface of the pool of water. A ray directed toward the surface at an angle would follow the path indicated in Figure 10.15b. Note the identification of the angles of incidence and refraction for this situation. The general rule is that *when light moves from a material into one in which it travels faster, the ray is bent away from the perpendicular.*

The refraction of light explains many common optical phenomena. For example, when you look at a fish in a pool of water, the fish appears to be closer to the surface than it really is because of refraction. Figure 10.16 shows why. We follow two rays leaving the fish, passing through the surface, and bending. When these rays enter the eye of an ob-

(a) Sled analogy

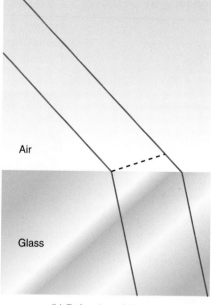

(b) Refraction of light

FIGURE 10.14 *(a) A sled turns upon striking a paved roadway at an angle because one runner hits the roadway and slows down before the other does. (b) A ray of light behaves in the same manner.*

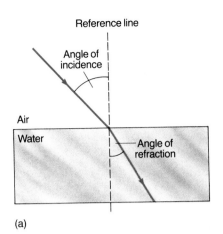

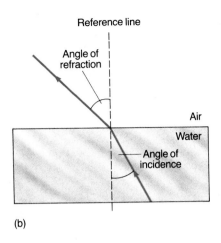

(a)

(b)

FIGURE 10.15   *(a) Light moving into a material in which it travels more slowly is bent toward the perpendicular reference line. (b) When light moves into a material in which it travels faster, it is bent away from the perpendicular reference line.*

server, the eye traces the rays back to the point where they appear to have originated. This point is point I, where an image of the fish is formed. Thus, the fish appears to be lifted toward the surface of the water. The apparent bending of a pencil when inserted in water is also explained in the same way. Each point of the pencil that is beneath the surface appears to be lifted toward the surface, just as was the fish, and the pencil appears to be bent (Fig. 10.17).

A mirage in the desert is a well-known optical illusion caused by refraction. Travelers in the desert have reported "seeing" water only to find out that no such pool ever existed. On a hot day in the desert, the earth is extremely warm, and thus the air next to the earth is also warm. Warm air is less dense than cool air, so a ray of light headed toward

FIGURE 10.17   *(a) The pencil appears to be bent because of refraction. (b) Solid lines show actual position of pencil and actual path of light ray. Dashed lines show observed position of pencil and the mentally projected light ray.*

(a)

FIGURE 10.16   *A fish appears to be closer to the surface than it actually is because of refraction.*

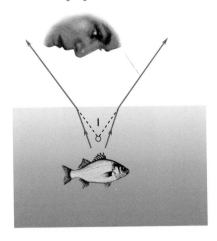

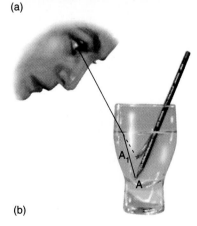

(b)

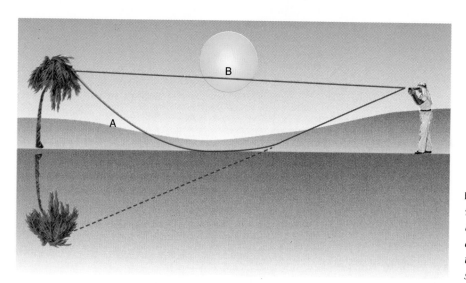

FIGURE 10.18 *A desert mirage occurs when light rays coming from a distant object are bent as they pass through air of different densities.*

the desert surface, such as ray A in Figure 10.18, is moving through a material that continuously becomes less dense as the ray approaches the ground. As the density of the air becomes less, the speed of the ray also changes, and this leads to a continual refraction that causes the ray to follow the curved path indicated in the figure. An observer thus sees rays of light from the tree reaching him from two different directions. One is the direct path B, and the other is the curved path A. The eye of the observer traces these rays back to the points where they appear to originate. The brain is fooled and assumes the rays coming from the ground originate below the surface, while the direct rays produce the expected view of the tree. Thus, the viewer sees the tree and an inverted image of the tree. Based on prior experience, the observer would interpret the inverted image of the tree to have been produced by reflection in a pool of water in front of the real tree.

FIGURE 10.19 *Example 10.5. Trace a possible path for the ray.*

CONCEPTUAL EXAMPLE 10.5   FOLLOW THE BENDING LIGHT

A ray of light in air strikes the surface of a glass block, as shown in Figure 10.19. The ray is then observed to exit the block at some point along the bottom surface. Trace a possible path for the ray through the block, based on refraction.

Solution At the top surface, we construct a perpendicular line and identify the angles of incidence, i, and refraction, r, as shown in Figure 10.20. Because the speed of light is slower in glass than in air, the light is shown to have been bent toward the

FIGURE 10.20 *Example 10.5. Following the ray of light.*

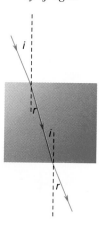

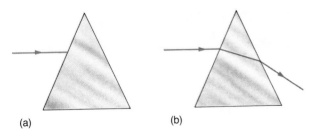

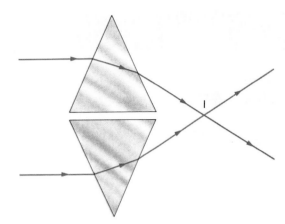

FIGURE 10.21 *You follow the ray through the prism.*

perpendicular. While in the glass, the light follows a straight-line path, but it will refract again when it exits into air. This occurs at the lower surface, where we have again constructed a perpendicular line and identified the angles of incidence and refraction for this surface. In this case, the light will bend away from the perpendicular as it moves into the air.

FIGURE 10.22 *Two prisms would produce an image at I.*

10.8 LENSES

To see how lenses work, let us consider another problem similar to that of Example 10.5. In this case, we will trace a ray of light through the prism shown in Figure 10.21a. The ray is incident on the left surface and is observed to exit the glass at the right surface. By now, you should be able to identify the angles of incidence and refraction at each surface and, by use of the law of refraction, convince yourself that a possible path for the ray of light is like that shown in Figure 10.21b.

Being able to trace a ray of light through a prism is important because one useful type of lens, called a **converging lens,** can be made by stacking two prisms base to base, as shown in Figure 10.22. In this case, incoming rays of light follow the path shown and intersect at I. Because rays are actually intersecting at I, a real image is formed at this location. Also note what effect our two stacked prisms have had on the incoming parallel beams of light. The rays have been deviated from their path such that they come together, or converge—thus, the name converging lens. In actual practice, a converging lens is produced by grinding the shape out of a single piece of glass and rounding off all the sharp corners to produce a shape like one of those shown in Figure 10.23. *Any lens that is thicker at its center than it is at the rim will act as a converging lens.*

The transformation shown in Figure 10.24a will produce another important type of lens, called a **diverging lens.** Note that incoming rays of light are diverged, or spread apart, after having passed through the lens. Figure 10.24b shows a variety of different types of diverging lenses, all of which have the characteristic that they are *thinner at the center than at their rim.*

An important descriptive feature of a mirror is its focal length, and this same terminology is also used to describe lenses. Figure 10.25 shows incoming parallel rays of light from a distant object such as the Sun. The lens converges these rays such that an image is formed at I. *The point at which parallel rays are converged to an image by a converging lens is called the focal point of the lens, and the distance of this point from the lens is called the focal length.* These concepts are illustrated in Figure 10.25.

FIGURE 10.23 *Several types of converging lenses.*

FOCUS ON
THE RAINBOW

To see a rainbow, an observer must be between the Sun and a rain shower. The concept of how a rainbow is formed can be understood by means of the laws of refraction. To see how this works, you must first understand that white light from the Sun is really a composite of all the colors from red through violet. (The subject of color spectra will be discussed more thoroughly in the next chapter.) Consider a ray of light from the Sun passing over your head and striking a raindrop, as shown in the figure. The white light passing into the drop is bent out of its original direction of travel, as we know it will be from our study of refraction. We also find, however, that some colors are bent more than others. Specifically, violet light is bent the most and red the least. These red and violet rays, and all the ones in between, continue until they hit the back side of the drop, where much of the light is reflected back toward the front surface. When these colors strike the front of the drop, they are bent a little more, and they end up separated as shown.

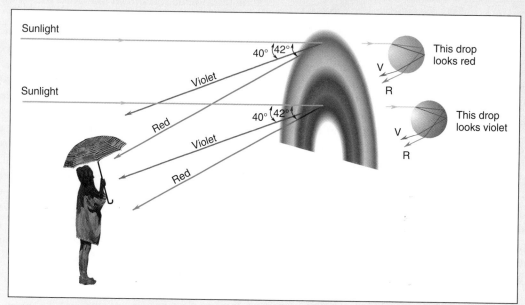

The formation of a rainbow by light refracted in water drops.

If incoming parallel rays of light strike a diverging lens, they are bent as shown in Figure 10.26. *The point from which the rays appear to originate after bending by a diverging lens is called the focal point, and the distance of this point to the lens is called the **focal length**.*

10.9 A SIMPLE MAGNIFIER

As you are no doubt aware, an important use of a converging lens is as a simple magnifier. To see how this magnification occurs, consider Figure 10.27. Here we see an object placed at O, and we follow

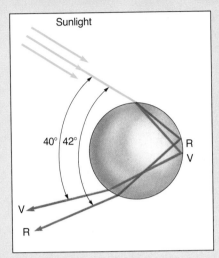

Sunlight

40° 42°   R
V

V

R

Light refracted by a drop of water is separated into its component colors, thus forming a rainbow.

To see how this separation of the colors of light leads to the formation of a rainbow, consider the man shown in the figure. If he observes a raindrop high in the sky, the red light from it reaches his eye, whereas the violet light passes over his head. The observer would interpret the color of this drop to be red. A drop lower in the sky would direct violet light toward his eye and would be seen as having that color. The red light from this drop would strike the ground and not be seen. Colors between the red and violet extremes would be sent to the eye by drops located between these two, and thus the rainbow would be red on the outside and violet on the inside.

two rays of light that leave the object, move through the converging lens, and pass into the eye of an observer. In this case, unlike any case that we have discussed so far, the object to be viewed is placed close to the lens. When the object is close to the lens, the lens converges the rays of light as usual, but it is

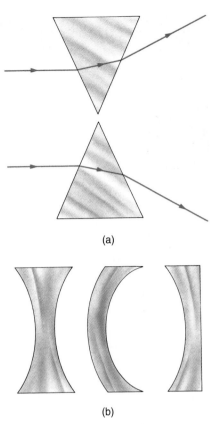

(a)

(b)

FIGURE 10.24 *(a) Another way to stack prisms. (b) Several types of diverging lenses.*

not able to converge them enough that they intersect. Thus, a real image is not formed. As shown in Figure 10.27, the rays A and B *have* been converged somewhat by the lens, but not enough to cause them to cross. When these rays enter the eye of the observer, the eye follows these two rays back to the

FIGURE 10.25 *The focal point and focal length for a converging lens.*

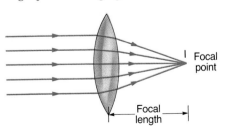

Focal point

Focal length

FOCUS ON
VISION CORRECTION WITH LENSES

One of the most common applications of lenses is in the correction of imperfect eyesight. To follow how this is done, consider the figure, which shows the important parts of the eye. The front of the eye is covered by a transparent membrane called the cornea. This is followed by a clear liquid region (the aqueous humor), an aperture (the iris and pupil), and the lens. When light enters the eye, it is bent slightly by the cornea, but most of the bending takes place as it passes through the lens. The cornea and the lens bring light rays together such that an image is formed on the back surface of the eye at the retina. The surface of the retina contains millions of light-sensitive receptors, called rods and cones, which generate electrical impulses when struck by light. These impulses are transmitted via the optic nerve to the brain, where the image is interpreted. It should be obvious from this brief discussion that distinct vision is possible only when the image is formed directly on the retina.

In the defect commonly called nearsightedness (myopia), the eye is too deep or the lens too strong; as a result, the image is brought to a focus in front of the retina (see part (a) of figure). The distinguishing feature of this imperfection is that distant objects are not seen clearly. To correct this problem, an ophthalmologist or an optometrist fits the eye with a diverging lens, which prevents the light from coming to a focus until it reaches the retina, as shown in part (b) of the figure.

A second common defect of the eye is called far-sightedness (hyperopia). Here distant objects are seen clearly, but near objects are indistinct. The lens of the eye is unable to bring diverging rays of light to a focus on the retina and instead tries to form an image behind it. Part (a) of the figure shows the problem, and part (b) shows how a converging lens can correct the difficulty.

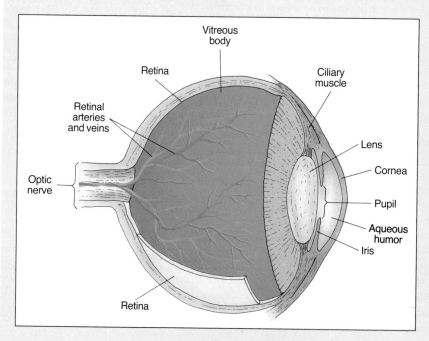

The primary parts of a human eye.

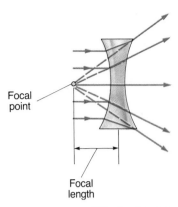

FIGURE 10.26 *The focal point and focal length for a diverging lens.*

point where they appear to have originated; this is I in the figure. Note that the image at I is much larger than the object. Thus, it is magnified. Is the image real or virtual?

CONCEPTUAL EXAMPLE 10.6   A COMPOUND MICROSCOPE

A simple magnifier provides only limited assistance in inspecting the minute details of an object. Greater magnification can be achieved by combining two lenses in a device called a compound microscope (Fig. 10.28). Figure 10.28 traces rays of light that leave the object O and move through the lens system. Use your knowledge of refraction and image formation to explain qualitatively how this device works.

Solution We will follow two rays of light from the specimen at O through both lenses to see how magnification occurs. The rays first pass through a lens called the objective lens, which converges the rays and causes them to intersect at point I_1. Examina-

FIGURE 10.27 *Using a converging lens as a simple magnifier.*

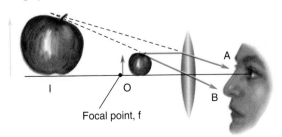

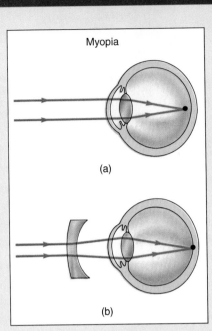

(a) A nearsighted eye. (b) Its correction with a diverging lens.

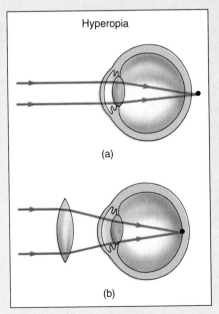

(a) A farsighted eye. (b) Its correction with a converging lens.

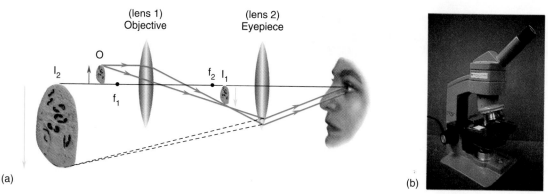

FIGURE 10.28   *(a) Image formation by a compound microscope. (b) A compound microscope.* (Courtesy Bill Schulz)

tion of I_1 shows that this image is real, inverted, and slightly enlarged. The second lens in the system is called the eyepiece, and it acts as a simple magnifier because its object (the image from lens 1) is close to the lens. This lens is used to examine the image I_1 and to magnify it. Thus, a final image is formed by the eyepiece at I_2. Because both lenses have produced magnification, the final image is much larger than it would be if only a single lens were used.

10.10 TELESCOPES

There are two fundamentally different kinds of telescopes, both of which have the same basic purpose: to aid us in viewing distant objects such as the planets in our Solar System. The two classifications are (1) the **refracting telescope,** which uses a combination of lenses to form an image, and (2) the **reflecting telescope,** which uses a curved mirror and a lens to form an image. Let us examine these in turn.

Refracting Telescope

The first type of telescope to be constructed was the refracting telescope, and it was first used in a systematic way for observation of the heavens by Galileo. With it, Galileo observed mountains on the Moon, the phases of Venus, the stars in the Milky Way, and the moons of Jupiter. These observations were important to the development of astronomy because, as we shall see later, when these observa-

tions were made, there were two competing views of our Universe. One of these, called the **geocentric model,** held that the Universe was Earth-centered. This meant that the Earth was considered to be at the center of the Universe, and all planets, moons, stars, and so forth were thought to revolve about the Earth.

In the 1500s, a competing view, called the **heliocentric model,** was developed by Copernicus, a Polish astronomer. Copernicus was not the first to hold the belief that the Sun was at the center, but he was the first to put this model on a firm scientific footing. In 1543, he published a book detailing his observations and beliefs while on his death bed; however, the book did not receive wide circulation and probably would have faded into obscurity for many years had it not been for the work of Galileo. Many of Galileo's observations with the telescope could be explained only by the heliocentric model, and as a result, he became an outspoken advocate of this view of the heavens. His ability to communicate caused the heliocentric viewpoint to become the accepted belief by many educated people during his lifetime. Let us now examine the refracting telescope to see how it is used to bring the heavens closer to Earth.

Two lenses are used in a fashion similar to that used in a compound microscope. Figure 10.29 shows the construction details and the image formation processes. Light from a distant object enters a large-diameter lens, called the **objective lens** (or just the objective), as parallel rays, and the rays are converged to form an image at point I_1. The second lens, called the **eyepiece,** is then used as a mag-

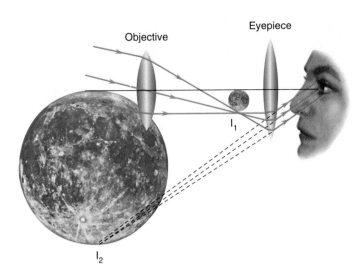

Objective

Eyepiece

I₁

I₂

FIGURE 10.29   *Diagram of a refracting tele-scope.*

nifier to form an image at I_2. We shall not derive this result here, but the magnification, M, of either a refracting or a reflecting telescope is given by

$$M = \frac{f_o}{f_e} \qquad (10.1)$$

where f_o is the focal length of the objective and f_e is the focal length of the eyepiece. The largest refracting telescope in the world, located at Yerkes Observatory in Williams Bay, Wisconsin, is designed such that the diameter of the objective lens is 1 m.

Surprisingly enough, when a telescope is used as a tool in astronomy, the magnification is seldom of importance because stars are so far away that regardless of how much magnification you attempt to use, they always appear as simple points of light. No telescope can magnify a star enough to enable one to see any detail on its surface. The primary uses for an astronomical telescope are stated in terms of three types of powers: (1) the **light-gathering power,** (2) the **resolving power,** and (3) the **magnifying power.**

The light-gathering power of a telescope is important because the objects that are of astronomical interest are distant; hence the light reaching Earth is faint. Thus, it is important to collect as much light as possible from them to form as distinct an image as possible. Imagine a rainstorm occurring, and someone sets you the task of going outside in the rain and collecting as much falling water as possible. Won't you achieve the most success by going outside with a container having as large a diameter as you can find? Likewise, the light from a distant astronomical object is falling on the Earth like rain, and the job for an astronomer is to collect as much of this light as possible. The obvious solution is to use a telescope with a large-diameter objective. Thus, *the light-gathering power of a telescope depends on the diameter of the objective.*

The resolving power of a telescope refers to the ability of a telescope to distinguish between objects that are close to one another. For example, suppose you use a telescope that does not have a good resolving power to view some distant object in space. You see something that looks like Figure 10.30a.

(a)

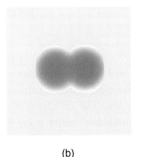

(b)

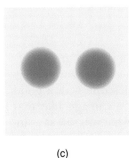

(c)

FIGURE 10.30   *An image seen through a telescope with (a) a poor resolving power, (b) better resolving power, and (c) the best resolving power. What appeared to be a single object in (a) is found to be two close-together objects under better resolution.*

FOCUS ON FIBER OPTICS

An important application of the principle of total internal reflection lies in the burgeoning field of **fiber optics.** To understand what is happening here, consider light entering one end of a small, plastic fiber, as shown in the figure. If the fiber is not bent at sharp angles, the light will always strike the surface of the plastic such that it undergoes total internal reflection. Thus, the fiber can transmit the light from one location to another as if it were a "wire" for light.

The analogy between a wire and a light-transmitting fiber can be taken one step farther. An unvarying direct current in a wire carries no information and cannot, say, drive a speaker or transmit data from one computer to another. But if the current is turned on and off, information can be conveyed. The simplest form of electronic communication is Morse code, in which letters and numbers are carried by a series of dots, which are merely short bursts of current, and dashes, which are longer bursts of current. Obviously a light-carrying fiber can carry information in this manner.

Of course, modern communications have gone far beyond the use of Morse code. Voice and other forms of information are carried by rapid oscillations in an electrical signal. (See AM and FM transmission in Chapter 11.) The same effect can be achieved by using fiber optics. There are many advantages in transmitting information via light because the density of information that can be im-

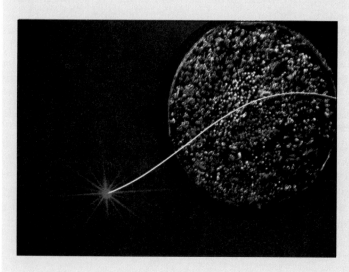

Light is seen emerging from the end of a single optical fiber. The cable behind the fiber contains hundreds of these fibers, which are used for transmission of telephone signals in place of copper wires. (Courtesy Corning Incorporated)

You then use a telescope with a slightly greater resolving power, and the image looks as shown in Figure 10.30b. You then use a telescope with a good resolving power, and you see an image like that shown in Figure 10.30c. Thus, what you saw as a single object with the poorly resolving telescope is found really to be two close-together objects when observed with a telescope with a good resolving power. *The characteristic of a telescope that determines its resolving power is the diameter of the objective.* Thus,

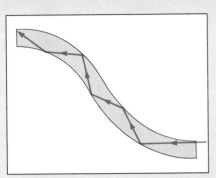

Reflection of light inside a transparent medium such as a piece of glass or plastic. Note how light that travels in straight-line paths can turn a corner by repeated total internal reflections.

pressed on light waves is considerably higher than that on conventional electrical distribution systems. A typical fiber optics system, for example, can transmit 300,000 telephone conversations simultaneously through a single glass fiber. By comparison, a single copper wire can carry only 24 different voice channels. In addition, the light rays are not affected by static from lightning or other electrical signals. Glass fibers are not only better than copper wires for carrying telephone conversations, but also are cheaper. As a result, it seems inevitable that fiber optics will be used in the future both for conventional telephones and for more advanced communication systems. The first major fiber optics transmission system became operational in February, 1983, for a telephone link between Washington, D.C., and New York City. At present, many large cities have changed completely to fiber optics systems, and all new installations are of this type.

the larger the diameter, the greater is the ability of the telescope to distinguish the fine details in an image.

The final power, the magnifying power of a telescope, is given by Eq. 10.1. From this equation, we see that it is the focal lengths of the objective and the eyepiece that determine the magnifying power of the instrument. As noted earlier, magnification is not important when viewing objects beyond our Solar System, but is can be important when viewing nearby objects such as planets.

CONCEPTUAL EXAMPLE 10.7   BUYING THE PROPER TELESCOPE

A shopper for a telescope is unable to decide between two models. The two telescopes are described below.

TELE-SCOPE	FOCAL LENGTH OF EYEPIECE	FOCAL LENGTH OF OBJECTIVE	DIAMETER OF OBJECTIVE
A	2.4 cm	150 cm	12 cm
B	0.06 cm	100 cm	8 cm

(a) If the main interest of the shopper were in light-gathering power, which would you advise him to buy?

Solution Light-gathering power is determined by the diameter of the objective lens. Thus, he should buy telescope A.

(b) If the main interest of the shopper were in resolving power, which would you advise him to buy?

Solution Resolving power also depends on the diameter of the objective. Thus, telescope A is once again the choice.

(c) If the main interest of the shopper were in magnification, which would you advise buying?

Solution We will have to calculate the magnification of each to answer this question. This is done for A as

$$M_A = \frac{f_{oA}}{f_{eA}} = \frac{150 \text{ cm}}{2.4 \text{ cm}} = 62.5$$

and the magnification of B is

$$M_B = \frac{f_{oB}}{f_{eB}} = \frac{100 \text{ cm}}{0.06 \text{ cm}} = 1670$$

Thus, telescope B is the scope of choice.

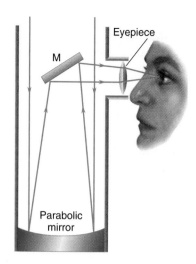

FIGURE 10.31 *Image formation by a reflecting telescope. The parabolic mirror collects light over a large area and focuses it toward the center of the tube. Before the image is formed, a small flat mirror redirects the image to the eyepiece.*

Reflecting Telescopes

Even before Isaac Newton gained immortality via his discovery of the three laws of motion and the law of universal gravitation, he had achieved some fame in scientific circles because he invented the reflecting telescope. The heart of the reflecting telescope is a parabolic mirror mounted at the base of a tube. The mirror collects light from a large area and focuses it, as shown in Figure 10.31. Incoming light from an astronomical object converges toward a focus, but before the image can be formed, a small flat mirror reflects the light toward

the side of the telescope barrel. A real image is formed when this light converges, and an eyepiece is then used to observe this image. To illustrate the potential power of such a collection system, the pupil of your eye collects light available in a circular area a few millimeters across, but the light-gathering mirror in the Hale telescope at Mt. Palomar, California, is about 5 m in diameter.

All new astronomical telescopes at observatories around the world are reflectors. This occurs because there are certain features in the design and construction of large refractors that are difficult or impossible to overcome. Let us look at some of these difficulties and see how they are either not present in a reflector or can be easily overcome with a reflector.

1. If one is to collect as much light as possible, it is necessary to make the objective lens in a refractor large. The grinding and polishing required for the front and back surfaces of this lens are quite difficult and expensive. A reflector has only one surface to be ground, since light bounces off the objective mirror rather than passing through it.

2. The already faint light from a distant object is dimmed even more as it passes through the objective lens of a refractor because the glass absorbs a percentage of the light passing through. Additionally, bubbles or imperfections in the interior of the lens will scatter some of the light, thus removing it from the portion eventually collected. Both of these factors are unimportant in a reflector because the light does not pass through any glass on its way to the eyepiece.

3. It is difficult to support the objective lens rigidly in a refracting telescope. Because

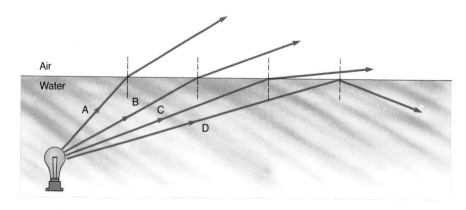

FIGURE 10.32 *A ray such as D that strikes the surface at a large angle of incidence is totally internally reflected.*

light must pass through the lens, the only way that it can be held in place is by supports around the rim of the lens. This leads to sagging in the heavy glass and can cause distortions in the final image. The mirror in a reflector can rest on a support at the base of the telescope.

10.11 TOTAL INTERNAL REFLECTION

The phenomenon of total internal reflection is one that embodies facts related to both reflection and refraction. Let us examine how the principle works and then investigate some of its practical applications. To understand what happens, consider Figure 10.32. Shown are four rays of light leaving a source that is beneath the surface of a pool of water. Ray A strikes the surface and bends away from the perpendicular, as it must. Ray B strikes the surface, and it also is bent away from the perpendicular, but because its angle of incidence is greater than that for A, it is bent away from the perpendicular much more. Finally, consider ray C, which strikes the surface at an angle of incidence such that when it passes out of the water, it skims right along the surface. Any ray of light, such as D, striking the surface farther out simply cannot get out of the water at all. It is reflected from the surface just as though the surface were a perfectly reflecting mirror. Such rays are said to have undergone

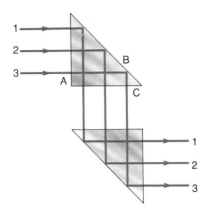

FIGURE 10.33   *A periscope.*

total internal reflection. *Total internal reflection can occur only when light rays are attempting to pass from a material in which they travel slower to one in which they travel faster.*

Figure 10.33 shows a way in which this phenomenon can be used in a practical situation. Shown are rays of light incident on the face of a prism having angles of 90°, 45°, and 45°. The rays strike side A and pass into the glass without being bent. They then strike side B, where they undergo total internal reflection and are reflected toward side C, where they exit the prism. A combination of two such prisms arranged as indicated in the figure will produce a change in direction of the light as shown. This combination could be placed in a tube and used as a periscope.

SUMMARY

The speed of light is 3.00×10^8 m/s and was originally measured with an apparatus using rotating mirrors.

When a ray of light is **reflected,** the angle of incidence equals the angle of reflection.

Images are formed when light rays actually intersect or appear to intersect. When the rays actually intersect, the image is called a **real** image; when they only appear to intersect, the image is **virtual.** An image that is turned over relative to the original object is said to be **inverted**; otherwise, it is **erect.** The image formed by a flat mirror is as far behind the mirror as the object is in front of it, unmagnified, erect, and left-right reversed. Curved mirrors may be either **concave** or **convex,** In either case, the position of an image for a distance object is called the **focal length** of the mirror. Curved mirrors may be used to form **reflecting** telescopes.

Refraction occurs when a light ray bends as it passes at an angle from one medium to another.

A **lens** is a piece of glass shaped such that it uses refraction of light to form images. Lenses are of two types: **converging** and **diverging.** In either case, the distance from the lens to the point where the image of a distant object is formed is called the **focal length** of the lens. Lenses may be used as simple magnifiers or, in combination, to form a **refracting** telescope.

When light attempts to pass from a material in which it travels at a certain speed into one in which it travels faster, **total internal reflection** can occur. In this situation, no light escapes, and the surface acts like a perfectly reflecting mirror.

EQUATIONS TO KNOW

$M = \dfrac{f_o}{f_e}$ (magnification of a telescope)

KEY WORDS

Light-year	Real image	Focal length	Light-gathering power
Rays of light	Erect image	Lenses	Resolving power
Reflection	Inverted image	Converging lens	Magnifying power
Refraction	Flat mirror	Diverging lens	Total internal reflection
Image	Convex mirror	Refracting telescope	
Virtual image	Concave mirror	Reflecting telescope	

PROBLEMS AND CONCEPTUAL QUESTIONS

Problems requiring numerical work are identified with a blue number.

Speed of light

1. In another version of Galileo's experiment, he is said to have quickly uncovered a lantern and tried to measure the time required for light to travel to a mirror 1.5 km away and back. How long does it take for light to travel that distance? Explain why this attempt also failed.

2. If Michelson's mirror had had six sides instead of eight, would he have had to spin it faster or slower to measure the speed of light? Explain.

3. Find the length of a light-year in miles.

4. Our Sun is approximately 93 million miles away. (a) Find the distance to this star in light-seconds. (b) Find the distance to the Moon, about 240,000 miles distant, in light-seconds.

Law of reflection

5. Draw a diagram showing light reflecting from a mirror when the angle of incidence is 0°.

6. Why does a highway seem darker when wet than when it is dry?

7. Some store windows are slanted inward so that reflections from the sun, streetlights, and so forth do not hinder someone trying to see inside. Draw a sketch showing how this would work to improve the view.

8. Aristotle said that the Moon is smooth and polished as a looking glass. If the Moon truly were this way, what would it look like to the naked eye?

9. A person stands at the center of a floral clock with a mirror. A beam of light comes in toward him from the 12 o'clock position, and he wants to use his mirror to deflect the beam so that it passes over the 3 o'clock position. What will the angle of incidence and reflection have to be in order to do this? What if he wants to deflect it over the 4 o'clock position?

10. A mirror, A, is placed flat on the surface of a table, and a second mirror, B, is placed against it, but vertical. A ray of light strikes A at an angle of incidence of 60° and bounces off it and finally off of B. Draw a sketch of this situation, and find the angle of incidence and reflection at mirror B.

11. Repeat problem 10 for the case in which the mirrors are placed at an angle of 120° with respect to one another, rather than 90°.

Formation of images and flat mirrors

12. The word AMBULANCE is written in a very strange way on the front of such a vehicle. It is written so that when you look in your rearview mirror, you are able to read the word. Write the word ambulance as it is usually printed on the front of the vehicle. Use a mirror to look at what you have written to see if you did it correctly.

13. A person 2 m tall stands in front of a flat mirror. What is the minimum height of the mirror so that the person can see all of himself, but no more? Be careful, the answer is not 2 m.

14. If you are standing 3 m in front of a mirror, you see yourself essentially as another person would if he were standing how far from you?

15. Often, when you look out a window at night, you see two images formed by the glass. How are these images produced?

16. You have a flat mirror, but you don't know whether it is silvered on the front or the back. How could you use a beam of light bounced off the mirror to determine how it is silvered?

Curved mirrors

17. Curved mirrors in amusement parks can make a person appear to be fat, skinny, or even wiggly. Using a diagram, explain briefly how a curved mirror distorts an image.

18. When you look at yourself in the back of a teaspoon, what kind of mirror are you using? When you look at yourself in the bowl portion of the spoon, what kind of mirror are you using?

19. Television crews at sports events often try to pick up the call of the quarterback by directing a concave reflecting mirror at him. The soundwaves are reflected by the mirror toward a microphone. Where should this microphone be located?

20. Convex mirrors are often placed in the corners of stores so that observers can watch for shoplifters. Why do you think convex mirrors are selected?

21. A small light bulb is placed 10 cm from a concave mirror, and a real image is formed 20 cm from the mirror. Is the focal length of the mirror 20 cm? Why or why not?

22. A dish antenna to receive television signals from satellites is made in the shape of a concave mirror. Why do you think this shape is used rather than the shape of a convex mirror?

23. All the properties of concave mirrors were not discussed in the text. A dentist uses a small concave mirror to examine teeth. The mirror must be held closer to the tooth than the focal length of the mirror. Under these circumstances do you believe the image is (a) real or virtual, (b) erect or inverted, (c) magnified or unmagnified?

Refraction

24. If you place a pencil straight down into a glass of water and look at it from above, it does not appear to be bent. But if you place it in the water at an angle and eye it obliquely, it does appear bent. Explain.

25. When you look at a fish in an aquarium, is it closer to you than it appears or farther away?

26. A ray of light bends as it enters the body of a lens and bends again when it emerges. Explain.

27. The Earth's atmosphere is denser near the surface than at high altitudes. If a scientist is trying to locate the exact position of a weather balloon flying in the upper atmosphere, is the density gradient of the atmosphere a factor to consider? Explain.

28. Mirages are common in the arctic as well as in the desert. Explain why an arctic mirage might appear and how it is different from a desert mirage.

29. Light travels faster in water than in does in glass. (a) Sketch the path followed by a ray of light moving obliquely from water into glass. (b) Repeat for the ray moving from glass into water.

30. We are able to see the Sun for a short period after it has actually sunk below the horizon. Explain why refraction in the Earth's atmosphere could produce this effect.

Lenses and a simple magnifier

31. Explain why either a converging lens or a concave mirror could be used to build a solar cooker. Which would be more effective? Defend your answer.

32. You want to use a lens to start a fire with sunlight. Should you use a converging or a diverging lens? Where should the object to be burned be placed with respect to the lens?

33. Lenses used in eyeglasses may be either converging or diverging, depending on the defect of vision. One of the problems of lenses used in eyeglasses is that they must be ground such that the lash of the eye does not rub against them as the wearer blinks. (a) Show the general shape that a converging lens might have to avoid this problem. (b) Repeat for a diverging lens.

34. A small light bulb is placed 10 cm from a converging lens and an image is formed 20 cm from the lens. Is the focal length of the lens 20 cm? Why or why not?

35. (a) If you were trying to use the light from the Sun and a lens to set a fire, what kind of image would you be forming, a real image or a virtual image? (b) When you use a lens as a simple magnifier, is the image real or virtual?

Telescopes

36. Sometimes distant galaxies are so faint that they cannot be seen even with a powerful reflecting telescope. If, however, a galaxy is followed for several hours while a piece of photographic film is placed at the eyepiece of the telescope, the galaxy can be detected. Explain how the film can detect an object in space whereas the eye cannot, even though film, per se, is no more sensitive than a person's eye.

37. Two telescopes have the properties listed below.

TELE-SCOPE	FOCAL LENGTH OF EYEPIECE	FOCAL LENGTH OF OBJECTIVE	DIAMETER OF OBJECTIVE
A	25 mm	1250 mm	5 cm
B	6 mm	500 mm	8 cm

(a) Which has the greater resolving power?
(b) Which has the greater light-gathering power?
(c) Which has the greater magnification?

38. List the advantages of reflecting telescopes over refracting telescopes.

39. Do some research in the library to find out how a Cassegrain telescope works. What advantage does it have over the Newtonian focus scope described in the text?

Total internal reflection

40. Light travels faster in water than in glass. (a) Can total internal reflection occur when light traveling in water is incident on glass? (b) Can it occur when light traveling in glass is incident on water?

41. Medical devices use fiber optics to view locations such as the colon and the stomach without having to resort to invasive surgery. Describe how these devices might work.

42. Rearrange the prisms used in the periscope example given in the text so that you could see where you have been rather than where you are going.

43. Recall that the path followed by a light ray is reversible, and use the principles discussed in the section on total internal reflection to describe how a fish would view the world outside the water.

ANSWERS TO SELECTED NUMERICAL PROBLEMS

1. 10^{-5} s

3. 5.87×10^{12} miles

4. (a) 500 light-sec, (b) 1.29 light-sec

9. 3 o'clock = 45°
    4 o'clock = 60°

10. 30°

11. 60°

13. 1 m

14. 6 m

CHAPTER 11

THE NATURE OF LIGHT

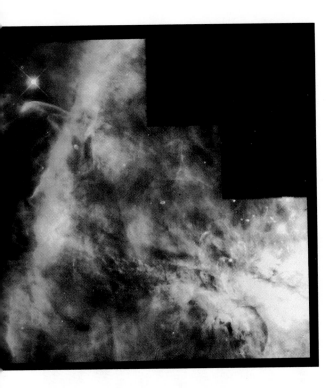

All that we know about the beautiful Orion nebula, the background stars, or any other object outside our Solar System has been read from the messages carried from them by light. The nebula is a giant gas cloud illuminated by surrounding young hot stars. This photo of Orion Nebula was taken with The Wide Field Planetary Camera of the Hubble Space Telescope. (Courtesy NASA)

The beginning of Chapter 10 explained that light sometimes acts like a wave and sometimes like a particle. Thus far, we have been able to explain optical phenomena by adhering strictly to a picture derived from its wave nature. This chapter, however, will look a little more closely at the dual nature of light. We shall examine some of the experiments that have been done to illustrate its wave nature, and we shall also look at some that require a particle explanation. Along the way, we shall attempt to answer the question: Where does light come from? We shall find that light is only one kind of wave from a spectrum of other similar types of radiation collectively referred to as electromagnetic waves.

11.1 IS IT A WAVE OR A PARTICLE?

What exactly is the nature of light? Two conflicting answers to this question were proposed in the late 1600s and early 1700s. Robert Hooke and Christian Huygens argued that light travels in waves, whereas Isaac Newton postulated that light rays consist of streams of particles—so-called packets of light. As we have already noted, this disagreement was ultimately resolved. Hooke and Huygens were correct—light exhibits wave behavior—and Newton was also right—light acts as if it is composed of a stream of particles. But how can light be two things at the same time, both a wave and a particle?

In a sense, this is an unfair question to ask. In the world around us, it is easy to distinguish between waves and particles. A thrown baseball unquestionably behaves like a particle, and a breaker crashing in on a beach unquestionably acts like a wave. In the submicroscopic world of electrons, protons, and light waves, however, the distinction be-

239

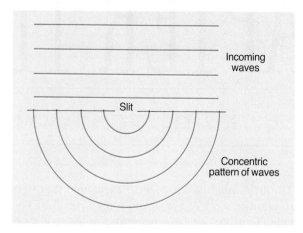

FIGURE 11.1 *Incoming waves, from above in the figure, spread out in a concentric circular pattern after passing through a slit.*

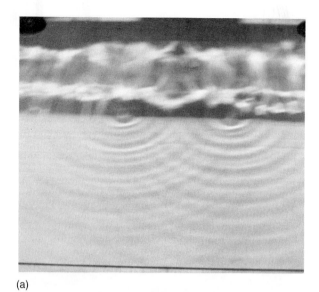

(a)

(b)

FIGURE 11.2 *(a) Two point-source waves will interfere with each other, producing alternating patterns of wavy and quiet water. (b) An artist's graphical representation of the alternating pattern produced when two point-source waves interfere.*

tween waves and particles is not as sharply drawn. We must be content with saying that light is light, and it happens to exhibit properties of both waves and particles. There is no fundamental reason why it should be like ocean waves or speeding bullets. This blurring between wave-like and particle-like behavior is not unique to light. In fact, we shall soon see that electrons, protons, and other elementary particles also exhibit a dual nature in their behavior.

11.2 WAVE NATURE OF LIGHT— YOUNG'S DOUBLE-SLIT EXPERIMENT

For many years, scientists had attempted to establish whether or not light travels in waves. Then, in 1801, Thomas Young, a British physicist, decided that the best way to resolve this question would be to determine whether or not light could exhibit interference. If so, its behavior would resemble that of known waves, such as water waves. Young knew that if a wave such as a water wave passed through a narrow opening, the opening would act as a source of new waves. Figure 11.1 is a sketch of what would happen. Incoming waves from above would spread out after passing through the slit and form a concentric pattern of circular waves below. If two narrow openings are placed side by side, two independent sets of waves would be produced. As these waves met, at some locations they would be in phase,

and constructive interference would occur, whereas at other locations they would be 180° out of phase, and destructive interference would be produced. As a result, alternating patterns of quiet water and moving water would be produced (Fig. 11.2). This works for water, but would it work for light? Young's experiment provided the answer.

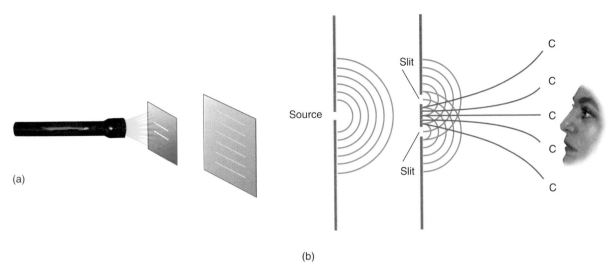

(a)

(b)

FIGURE 11.3 *(a) Young's experiment was simple, yet it became a milestone of physics. He passed a beam of light through two narrow slits. When alternate patterns of light and dark appeared on a viewing screen, he deduced that light must exhibit wave behavior. (b) Constructive interference occurs along lines labeled C, and destructive interference occurs along lines midway between these.*

The details of **Young's double-slit experiment** are indicated in Figure 11.3. He scratched two fine lines on a painted piece of glass and allowed light from a small source to pass through the openings (Fig. 11.3a). The light spread out from each slit, and interference between the two sources occurred. Along certain lines, labeled C in Figure 11.3b, the waves of light emerging from the two slits interfered constructively, whereas along other lines, midway between these, destructive interference occurred. This means that if the eye is placed in the position shown, alternating bright and dark lines are seen (Fig. 11.4).

This demonstration of interference gave the wave model of light a strong boost. It was inconceivable that particles of light coming through these slits could cancel each other in a way that would explain the regions of darkness. Today, we still use the phenomenon of interference to distinguish wave-like behavior in any observation.

Although we will not discuss the details of the calculations, it is of historical interest to note that Young's experiment was the first to provide a way of measuring the wavelength of light. White light is composed of all the colors and hues ranging from red to violet. Measurements on these individual components of white light showed that the wavelength of red light is approximately 750 nm, and

violet light has a wavelength of about 400 nm ($1 \text{ nm} = 10^{-9}$ m).

FIGURE 11.4 *The alternating light and dark bands produced in Young's double-slit experiment.*

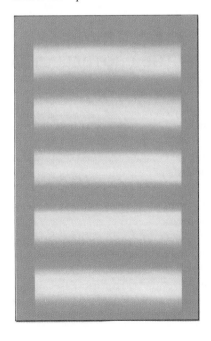

EXAMPLE 11.1   VISIBLE LIGHT FREQUENCIES

If the wavelength of red light is 750 nm and that of violet light is 400 nm, find the frequencies of these two extremes of visible light.

Solution In our study of wave behavior in Chapter 7, we found that all waves obey the equation $c = \lambda f$, where the symbol c represents the speed of light, 3.00×10^8 m/s, λ is the wavelength of the light, and f is its frequency. Thus, the frequency of red light is

$$f = \frac{c}{\lambda} = \frac{3.00 \times 10^8 \text{ m/s}}{750 \times 10^{-9} \text{ m}} = 4.00 \times 10^{14} \text{ Hz}$$

and that of violet light is

$$f = \frac{c}{\lambda} = \frac{3.00 \times 10^8 \text{ m/s}}{400 \times 10^{-9} \text{ m}} = 7.50 \times 10^{14} \text{ Hz}$$

CONCEPTUAL EXERCISE

Use what you know about the behavior of particles to describe the kind of image formed on a screen in Young's experiment if light had been composed of particles instead of being a wave.

Answer Particles do not bend around corners, so they would have passed through the slits and formed two images on the screen directly behind the slits.

11.3 PARTICLE NATURE OF LIGHT—THE PHOTOELECTRIC EFFECT

Approximately 100 years after Young's experiment demonstrated the wave nature of light, a series of experiments was performed that showed that light sometimes acts as though it is composed of a stream of particles. There are many processes in nature in which light exhibits its particle-like behavior; the one we consider here is the **photoelectric effect.** The photoelectric effect is a process in which electrons are observed to be emitted by certain metals when light is shined on the metal (Fig. 11.5).

Wave theory can explain how light can knock electrons off a metal surface. After all, water waves can dislodge pebbles from a sandy beach, and a loud noise can knock down a delicate house of

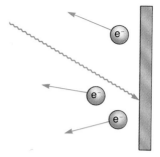

FIGURE 11.5   *Incoming radiation dislodges electrons from certain metallic surfaces.*

cards, so why shouldn't light waves dislodge electrons from a metal surface? Further analysis of the photoelectric effect, however, led to some results that could not be explained satisfactorily by assuming that light acts as a wave. Let us look at the key observations and show how the wave theory failed.

1. *Electrons are released by the metal when high-frequency light strikes the metal but not when low-frequency light strikes.* According to the wave model of light, both high-frequency and low-frequency light waves carry energy. Therefore, it should make no difference what frequency we use; electrons should be ejected for all types.

2. *Dim high-frequency light can produce the effect, but even very bright low-frequency light cannot.* Presumably, dim light would carry less energy than bright light. As a result, bright low-frequency light should eject electrons much more easily than dim high-frequency light.

3. *The electrons begin to leave the surface as soon as the light strikes the surface—there is no time lag at all, even for very dim light.* Calculations can easily be done to find out how much energy the incoming light carries and how much energy is required to eject electrons from the metal. These calculations show that light would have to shine on the metal for several minutes before individual electrons could gain sufficient energy to leave the metal.

4. *The higher the frequency of the incoming light, the greater are the speed and kinetic energy of the ejected electrons.* Again, it would be expected that the energy of the ejected electrons would depend on the brightness of the light but not on its frequency.

To indicate the extremely unusual behavior of the electrons emitted by the photoelectric effect, let us consider an analogy with water waves striking a beach and dislodging rocks from hard-packed sand. The amplitude of the water waves is analogous to the brightness of the incident light, and, of course, the frequency of the water waves is analogous to the frequency of the light. Consider only one of the observations listed, number two, to see just how different the photoelectric effect is from our experiences with water waves. Observation two says that dim high-frequency light produces the photoelectric effect, but very bright low-frequency light does not. This is the same as saying that large-amplitude water waves crashing into a beach will not release any pebbles if the frequency of the waves is low. A gentle, small-amplitude wave lapping against the shore, however, will scatter rocks away from the shore easily if the waves have a high frequency. We know that water waves do not behave this way; large-amplitude waves dislodge rocks easily and push them away from the shore with large kinetic energies. As an exercise, use this water wave analogy for the other experimental observations for the photoelectric effect to see that they too are not in tune with common sense.

The results of the experiments on the photoelectric effect were most puzzling, but in 1905 Albert Einstein published a theory that explained the observations. Five years earlier, another physicist, Max Planck, had postulated that light is emitted discontinuously from a source of light. By this he meant that the light coming from any source is not in the form of a continuous wave. Instead the light is emitted in tiny bundles or packets. The original name for one of these discrete bundles of energy was the **quantum** (plural quanta), but the present-day accepted name is the **photon.** The amount of energy carried by an individual photon can have only a certain value and no others. The energy carried by a photon was determined by Planck to have a value given by

energy = h(frequency of the light)

or symbolically

$$E = hf \qquad (11.1)$$

where h is Planck's constant = 6.626×10^{-34} J s. Note that Eq. 11.1 says that the higher is the energy car-

ried by a photon. Thus, photons of violet light carry more energy than do photons of red light.

With this photon picture of light in mind, let us see how Einstein was able to explain the photoelectric effect. Figure 11.6 indicates the model used by Einstein. The incoming light acts like a stream of bullets, with each bullet being an individual photon (Fig. 11.6a). An electron in the metal interacts with only one of these photons by absorbing the photon and taking on all of its energy; in this process, the photon disappears (Fig. 11.6b). The electron now has an abundance of energy, which it uses in two ways. Part of it is used simply to get out of the metal. In other words, the surface of the metal acts like a fence that the electron must jump over to be freed from its "corral," as shown in Figure 11.6c. Any excess energy that the electron has after it uses some to escape from the metal now appears as kinetic energy, causing the electron to move away at a high speed if it has a lot of energy remaining or at a low speed if it has used most of it to escape (Fig. 11.6d).

FIGURE 11.6 *(a) A stream of incoming photons strikes a metal. (b) The photon at A in (a) is absorbed by an electron and disappears. (c) The electron uses part of the photon's energy to escape the metal, and (d) the remainder of the photon's energy appears as kinetic energy of the electron.*

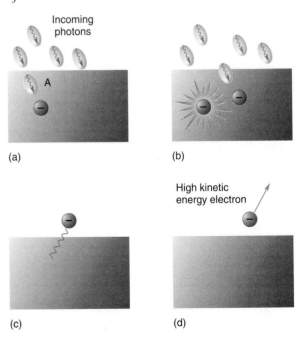

FOCUS ON
APPLICATIONS OF THE PHOTOELECTRIC EFFECT

Practical applications of the photoelectric effect usually make use of an electronic component called a photo-tube, shown in the figure. A phototube acts much like a switch in an electric circuit in that no current can flow through it when it is in the dark, but a substantial current can flow if the tube is exposed to light. The curved plate inside the tube of the figure is made of a photoelectric material that will emit electrons when exposed to light.

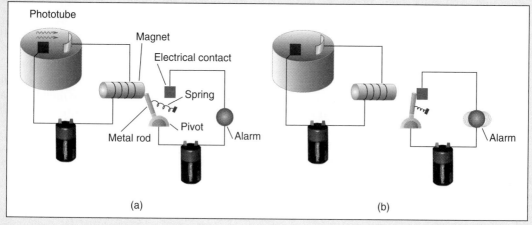

A burglar alarm.

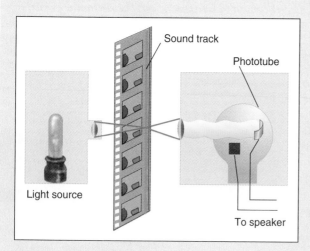

The sound track on a movie film.

With this picture in mind, let us examine all the experimental observations once again.

1. High-frequency photons have more energy than low-frequency photons. Thus, apparently, low-energy photons may not have enough energy to allow electrons to "jump the fence" and escape from the metal.

2. A bright beam of light would have a large

Thus, when light shines on this surface, the electrons emitted move through the vacuum in the tube to the collector (shown in black); these electrons constitute a current in the external circuit. One application of such a phototube as a burglar alarm is shown. When light shines on the phototube in part (a) of the figure, there is a current present in the external circuit, and this current energizes the electromagnet. The magnet then attracts the pole of a switch, and no current is allowed to flow in the portion of the circuit that contains the alarm. If a burglar interrupts the light beam, however, the current to the electromagnet is cut off, and the switch pivots to the right, as shown in part (b). The switch in the circuit containing the alarm is closed, and the alarm sounds. Obviously, these devices work better if ultraviolet light is used to activate the phototube so that its pathway will not be obvious to the burglar.

The second figure shows how the photoelectric effect is used to produce the sound information on a movie film. The sound track is impressed on the film as an alternating pattern of dark and light lines along the side of the film. When light falls on this track from a light source, it can penetrate at some locations and activate the phototube shown. Likewise, at points where the sound track is dark, no light penetrates and the phototube is not illuminated. These changes in the amount of light reaching the phototube produce a fluctuating current in its circuit. This fluctuating current can be used to recreate the original sound signal by driving a speaker connected as shown.

ergy to cause the electrons to be ejected. Dim, high-frequency light has few photons, but each of them carries enough energy to free an electron from the metal.

3. Time does not enter the picture at all. As soon as a photon is absorbed by an electron, the electron instantly either has enough energy to escape the metal or it does not. There is no need to wait for the electron to accumulate energy gradually. Either a single photon does it, or it isn't going to be done.

4. When a photon is absorbed by an electron, only a certain amount of the energy is needed by the electron to escape the metal. Any excess energy left over after the escape appears as kinetic energy of the freed electron. Thus, when an electron absorbs a high-frequency (or high-energy) photon, it has more energy left over than it does if it absorbs a low-frequency (low-energy) photon.

So what is light anyway? Is it a wave traveling through space, or is it a series of particle-like photons? This is a question that cannot readily be answered. Instead, all we can do is describe how light acts. It does exhibit wave behavior such as interference. Furthermore, as with other waves, there is a frequency associated with light. But it also acts like a series of particles. Because these particles can strike objects, be absorbed, and carry energy, when they strike an object, they can cause the object to behave as though it had been struck by a particle. Thus, light acts as though it had a split personality—sometimes it acts like a wave, sometimes like a particle. Luckily for us, it never acts like both in the same experiment.

CONCEPTUAL EXERCISE

Explain why blue light might produce a photoelectric effect in a certain metal, but red light will not.

Answer Photons of blue light have a higher frequency than red light and, therefore, a higher energy. Apparently, the blue light photons are able to give enough energy to an electron to enable it to escape the metal, while the red light photons do not have enough energy to allow the electron to escape.

number of photons in it, while a dim light would carry far fewer photons. Bright, low-frequency light would have many photons, but none of them would carry enough en-

EXAMPLE 11.2   THE ENERGY OF A RED AND A VIOLET PHOTON

(a) Find the energy carried by a "red" photon having a frequency of 4.00×10^{14} Hz.

Solution The energy is given by Eq. 11.1 as

$$E = hf = (6.626 \times 10^{-34} \text{ J s})(4.00 \times 10^{14} \text{ Hz})$$
$$= 2.65 \times 10^{-19} \text{ J}$$

(b) Repeat part (a) for a violet photon of frequency 7.50×10^{14} Hz.

Solution

$$E = hf = (6.626 \times 10^{-34} \text{ J s})(7.50 \times 10^{14} \text{ Hz})$$
$$= 4.97 \times 10^{-19} \text{ J}$$

EXAMPLE 11.3   ANALYZING THE PHOTOELECTRIC EFFECT FOR A PARTICULAR METAL

It is found that for an electron to escape from a certain metal, the electron must gain 3.00×10^{-19} J just to get out.

(a) Will a beam of either the red photons or the violet photons in Example 11.2 have enough energy to cause the photoelectric effect to occur?

Solution The red photons have an energy of 2.65×10^{-19} J, but 3.00×10^{-19} J is required to eject electrons from the metal. So if the beam is red light of this frequency, the photoelectric effect will not oc-

cur. The violet photons, with an energy of 4.97×10^{-19} J, have enough energy to eject the electrons, so the photoelectric effect will occur.

(b) What is the kinetic energy of the ejected electrons when violet light shines on the metal?

Solution The energy of a photon is 4.97×10^{-19} J, and when this energy is absorbed by an electron, 3.00×10^{-19} J is used to escape from the metal. The remaining energy, 1.97×10^{-19} J, appears as kinetic energy of the electron.

11.4 SOURCE OF LIGHT

We have seen that a beam of light can be considered to be composed of literally trillions of small particle-like packets of energy called photons. But where do these photons come from? In this section, we shall find that photons are produced when electrons inside an atom are jostled around. To explain exactly what is happening, we shall consider the simplest atom, the hydrogen atom, which has one proton in its nucleus and one electron orbiting about the nucleus (Fig. 11.7). Under normal circumstances, the electron always orbits the hydrogen atom at a specific average distance from the nucleus and with a specific amount of energy. The electron is said to be in its **ground state** under these conditions. Now suppose that you want to move an electron farther away from the proton. Because the

FIGURE 11.8   *The ground state and a few excited states for the hydrogen atom.*

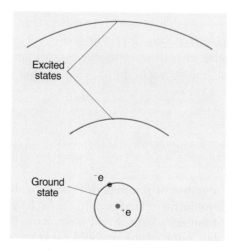

FIGURE 11.7   *The hydrogen atom consists of a single negatively charged electron circling a single positively charged proton.*

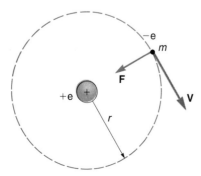

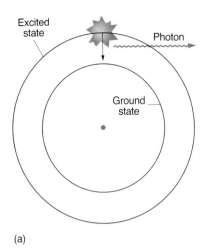

(a)

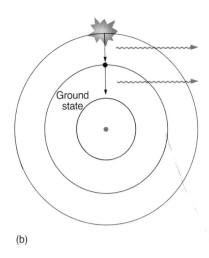

(b)

FIGURE 11.9 *(a) When an electron jumps from an excited state to the ground state, it emits a photon. (b) In a high excited state, the electron can "stair-step" down to the ground state, emitting photons with each step.*

two are being held together by electrical forces, it should seem reasonable to you that you will have to do work on the electron to move it—that is, you will have to add energy to lift the electron farther out into space. There are some complications, however, that we must address.

In 1913, Niels Bohr developed a model of the atom that was successful in explaining, among other things, how light is emitted by an atom. One of the assumptions made by Bohr was that there are only certain orbits in an atom in which an electron could orbit, and an electron will never be in any other orbit except one of these. The electron has its lowest energy in the ground state, and it has higher energy as it moves to orbits farther from the nucleus (Fig. 11.8). When an electron is in one of these higher orbits, the atom is said to be in an **excited state.**

This idea of only certain allowed orbits is a strange concept. If you get into your car and decide to drive in a circular path, there is nothing to prevent you from driving in a circle having any radius that you choose; there are no forbidden circular paths in between. But for the electrons in an atom, there are. Thus, if you want to move an electron from its ground state to an excited state, you must give the electron just enough energy to cause it to reach the excited level. (There are a variety of ways by which you could add energy to the atoms. You could heat a material made up of the atoms, pass an electric current through the material, or shine light on the material.)

If an electron has been caused to reach an excited level, it will not stay in that level for a long pe-

riod of time. Instead, it will return quickly to the ground state. To fall back to the ground state, however, the electron must release its excess energy. It does so by emitting a photon, as shown in Figure 11.9a. The energy carried away by this photon is exactly equal to the energy difference between the excited state and the ground state. You should also note that the electron does not have to move to the ground state in one single jump. Instead, if it is in a high excited state, it can stair-step down to the ground state, emitting photons in each jump, as shown in Figure 11.9b. The energy carried by a photon is given by $E = hf$; thus, if the energy is such that the frequency of the photon is in the visible range, we can see these emitted photons.

Therefore, according to the Bohr model of the atom, light is produced as electrons tumble down to the ground state from excited states. Specifically, in the hydrogen atom, it is found that visible light is produced when electrons fall from higher excited states down to the first excited state (Fig. 11.10). Photons are released in other transitions also, such as from the first excited state to the ground state, or from the fourth excited state to the third excited state, but these photons do not have a frequency that places them in the visible portion of the spectrum. As we shall see later, this radiation can be ultraviolet (a radiation having a frequency greater than visible light) or infrared radiation (which has a frequency lower than visible light).

The origin of light has been discussed here in terms of the Bohr model of the atom, which pictures the atom as being much like our Solar System, with electrons replacing planets and the nu-

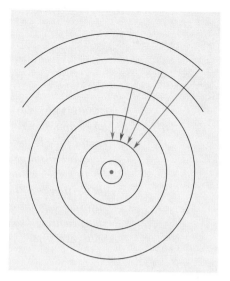

FIGURE 11.10   *Visible light is produced from hydrogen by jumps that start on highly excited energy levels and end on the first excited state.*

cleus of the atom replacing the Sun. This model is effective in helping one to gain a mental picture of the atom, but this simple model has been replaced by a more accurate quantum mechanical viewpoint, to be discussed in Chapters 12 and 13. We will find that the model of an atom with perfectly defined orbits and energy levels is not really the way that nature behaves.

CONCEPTUAL EXERCISE

An electron jumps from a highly excited state down to the ground state and emits a photon of light. A second electron jumps from a lower excited state to the ground state. In which case is the frequency of the emitted photon lowest? Explain.

Answer The atoms get rid of excess energy by emitting photons as electrons jump downward in their allowed orbits. The electron has more energy in a highly excited state than it does in a lower excited state; so the photon it emits carries more energy. The higher the energy of a photon, the higher is its frequency. Thus, the high-frequency photon comes from the photon moving out of the highly excited state.

EXAMPLE 11.4   LIGHT AND HYDROGEN ATOM

When the electron in a hydrogen atom is in the first excited state, it has a certain amount of energy. When the electron is in the second excited state, it has 3.02×10^{-19} J more energy than when in the first, and when it is in the third excited state, it has 4.08×10^{-19} J more than it does when in the first. Find the energy and the frequency of all photons that could be emitted as the electron moves from the third excited state to the first excited state.

Solution If the electron jumps directly to the first excited state, it must get rid of its excess energy by emitting a photon with an energy equal to the difference in energy between these states. This energy is 4.08×10^{-19} J, and the frequency corresponding to this energy is

$$f = \frac{E}{h} = \frac{4.08 \times 10^{-19} \text{ J}}{6.626 \times 10^{-34} \text{ J s}}$$

$$= 6.16 \times 10^{14} \text{ Hz}$$

Note that this frequency is between 4.0×10^{14} Hz and 7.5×10^{14} Hz, the limits of the visible spectrum, and thus this photon will be seen as violet visible light.

The electron, however, does not have to jump directly down to the first excited state from the third; instead, it can stair-step down by first jumping to the second excited state and then from there to the first. In the jump from the third excited state to the second, the energy given off is

$$E = 4.08 \times 10^{-19} \text{ J} - 3.02 \times 10^{-19} \text{ J}$$

$$= 1.06 \times 10^{-19} \text{ J}$$

corresponding to a frequency of

$$f = \frac{E}{h} = \frac{1.06 \times 10^{-19} \text{ J}}{6.626 \times 10^{-34} \text{ J s}}$$

$$= 1.60 \times 10^{14} \text{ Hz}$$

This frequency is not within the frequency limits of the visible spectrum. In fact, this frequency is lower than that of red light, and this radiation is referred to as infrared radiation.

Finally, the electron will jump from the second excited state to the first excited state. The energy emitted in this transition will be

$$E = 3.02 \times 10^{-19} \text{ J}$$

corresponding to a frequency of

$$f = \frac{E}{h} = \frac{3.02 \times 10^{-19}\,\text{J}}{6.626 \times 10^{-34}\,\text{J s}}$$

$$= 4.56 \times 10^{14}\,\text{Hz}$$

This frequency corresponds to that of red light.

11.5 DIFFRACTION GRATING

We have stated at a number of different points throughout this book that white light is actually made up of all colors ranging from red through violet with all hues and shades appearing. As a child, you probably verified this statement for yourself by using a prism to separate sunlight into its component colors. A device that is even better for breaking up light into its components is the **diffraction grating.** To understand how this device works, consider our discussion of Young's interference experiment. There we considered what happened when light waves interfered after having passed through *two* slits. The diffraction grating allows light to pass through not just two but a number of slits and to undergo interference.

Gratings are made by engraving closely spaced, parallel grooves on a piece of flat glass. A typical grating has about 6000 of these grooves per centimeter. As a result, the machine work is detailed and precise. Figure 11.11 shows what happens when light is incident on a grating. Light passes through the glass unobstructed at locations between the grooves, but the light that strikes the grooves is ei-

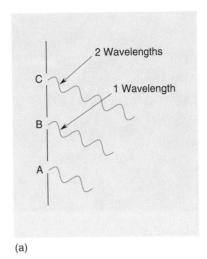

(a)

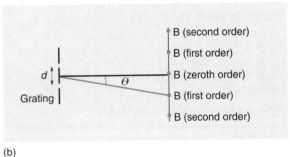

(b)

FIGURE 11.12 *(a) Three rays from a diffraction grating that will undergo constructive interference. (b) The angle θ is measured above or below the central axis to a bright line B on the screen. The distance d is the distance between successive openings on the grating.*

FIGURE 11.11 *A side view of a diffraction grating.*

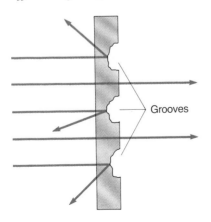

ther reflected or refracted to the side and is no longer a part of the beam.

Let us consider what happens to the unobstructed light when it falls on a screen. Figure 11.12a shows three rays of light with the same wavelength passing through adjacent openings and moving toward the same location on a screen located far to the right. For convenience, let us refer to the openings as slits. The light from slit A and the light from slit B move toward the screen, but we have selected a location on the screen such that the light from B will arrive exactly one wavelength behind the light from A. Thus, because of the extra distance B has to travel, it lags behind, but still when the two combine at the screen, crest will overlap crest and trough will overlap trough. This means that the light from these two slits will undergo constructive interference. The light from slit C also

FIGURE 11.13   *In this simple spectrometer, light passes through a set of focusing lenses so that the rays are parallel before striking the diffraction grating on the platform at the center. The resultant pattern is observed through the telescope.* (Courtesy Bill Schulz)

heads toward this location on the screen, but this light is one wavelength behind that from B and two wavelengths behind that from A.

This analysis can be continued for all the openings of the grating, but the result will be the same. Constructive interference will be occurring for all the waves headed toward this particular location on the screen. If we look at the screen at this position, we will see a bright line of light. Another bright line will be seen on the screen at a location such that the light from B has fallen two wavelengths behind that from A, and the light from C is two wavelengths behind B, or four wavelengths behind that from A. Thus, we will see alternating patterns of bright and dark on the screen.

A bright line occurs at a point along the central axis, as shown in Figure 11.12b. This line is called the **zeroth order** line. The first bright line for a particular wavelength on each side of this central bright line is called the **first order** line, the second bright line for a particular wavelength is the second order, and so on. These lines can be observed easily by looking through a device called a spectrometer (Fig. 11.13).

You should take particular note here of the fact that for constructive interference to occur at some location, the light coming through an opening must fall an integral number of wavelengths behind the wave passing through a slit directly above it. Each of the colors that constitute white light has a different wavelength; therefore, a red bright line

will not be formed at the same location on a screen as will, say, a green bright line. Thus, white light fans out into a spectrum of all the colors after passing through a grating.

As we shall see in a later section, the diffraction grating is an extremely important tool in the study of astronomy. No one has ever visited a star and sampled the material of which it is made. Yet astronomers have a good idea of the elements present in a star, the abundance of each element, the star's temperature, and so forth. This information is derived by studying the light from that star with a diffraction grating. In addition, an analysis of the diffracted light from a star or a galaxy can tell us whether that star is orbiting about another object or how fast the object is moving away from us or toward us. In short, the diffraction grating spectacularly enhances the ability of the telescope to give us information about astronomical objects.

11.6 SPECTRAL ANALYSIS

The colors falling on a screen after white light passes through a diffraction grating is called a spectrum. This complete rainbow of colors is pretty and nice to look at, and it has some usefulness in the world of physics, but there are other kinds of spectra, and these have affected the world of physics and astronomy in a more far-reaching way than has that produced by white light spectra. In this section, we will look at the three types of spectra that can be produced. They are the **continuous spectrum,** the **emission (or line) spectrum,** and the **absorption spectrum.**

Continuous Spectrum

The general approach to observing spectra is shown in Figure 11.14. The light from a heated source is allowed to pass through a diffraction grating, which fans out its various colors and allows them to fall on a screen. For careful analysis, the screen is usually replaced with a sheet of photographic film so that the preserved images can be studied. That portion of Figure 11.14 consisting of the diffraction grating and screen or film is called a **spectroscope.** The type of pattern that appears on the film depends on the type of source that is used. If the object is a hot solid or a hot, high-pressure gas, as shown in Figure 11.15, the type of spectrum that

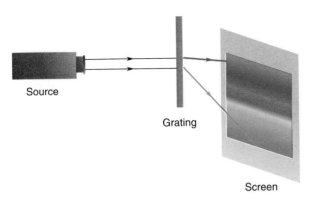

FIGURE 11.14   *Observing spectra.*

appears on the film is one in which all the colors of the rainbow are present. A spectrum in which all the hues and shades appear is called a **continuous spectrum.**

In view of our discussion of how light is produced by the hydrogen atom, it should seem surprising that a heated solid produces all the colors in the rainbow. In an isolated gas atom, such as the hydrogen atom, photons are produced when an electron jumps from an excited state toward the ground state. Since there are only a few pathways for jumps, it seems that photons of only a few different energies or frequencies would be emitted. Since frequency is the determining factor in the colors that we see, it would seem reasonable that only a few jumps would produce only a few distinct colors—not all the colors of the rainbow. The key to understanding why this vast array of colors does appear is to note that our energy level theory of the atom basically applies only to widely separated atoms, such as those in a gas. In solids or high-pressure gases, the atoms are packed so tightly together

that neighboring atoms can influence one another. This alters the energy level pattern that they would have if they were isolated. Some of these closely packed atoms may have their first excited state quite far above the ground state, whereas another atom will have its first excited state quite close to the ground state, and other atoms will have all the possibilities in between these two extremes. As a result, there are possible jumps in the atoms that will produce a photon of virtually any energy within the visible spectrum and consequently virtually any color. (Also included in the spectrum [but invisible] will be frequencies below red—the infrared—and frequencies above the violet—the ultraviolet.)

Emission Spectra

A second type of spectrum, called an **emission spectrum,** can be produced by using a hot, low-pressure vapor as the source of light (Fig. 11.16). In this case, the spectrum that falls on the photographic film contains only a few distinct frequencies or colors instead of the rainbow found for a continuous spectrum. For example, if the heated vapor is sodium, only one color appears on the screen, a bright yellow. (Actually the spectrum consists of two closely spaced wavelengths, but, since the two are separated by only 0.6 nm, we shall refer to the two collectively as a single line.) The yellow light often used to illuminate streets is produced by sodium vapor lamps.

We find that sodium is characterized by emitting only one frequency of light, and its emission

FIGURE 11.16   *Production of an emission, or line, spectrum.*

FIGURE 11.15   *Production of a continuous spectrum.*

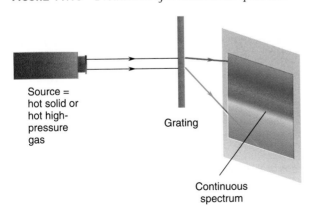

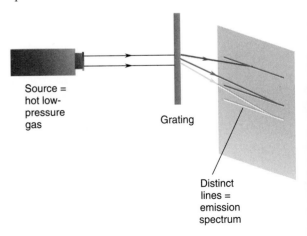

spectrum contains only one line. If the heated vapor is mercury, it is found that four distinct colors, or frequencies, are observed. These are a particular frequency of yellow light, one of green, one blue, and one purple. (Actually, mercury contains several more lines that are faint, but these would be seen only if long exposure times were used for the film.) In fact, it has been found that every chemical element produces its own characteristic pattern of spectral lines. In a sense, the lines produced by a heated vapor serve as fingerprints to identify the type of vapor that produced the lines.

The procedure of identifying materials according to the spectral lines they emit has been an important part of chemical analysis since it was first discovered by the German chemists Robert Bunsen and Gustav Kirchhoff in 1859, and the technique is still used in many laboratories, such as crime labs, to identify unknown substances. When Bunsen and Kirchhoff developed this technique, they immediately began to catalog the spectra of all the known elements, and soon some surprising results turned up. For instance, when they were investigating the emission spectrum of mineral water, they found that an impurity in the liquid was producing an emission spectrum that had not been previously observed. These new lines were primarily in the blue region of the spectrum, and they named the element responsible for producing these lines cesium (from the Latin *caesium* for gray-blue). Later, while observing the vapor from a vaporized mineral sample, they found another pattern of previously unobserved lines. These lines were at the red end of the spectrum, and they named the element responsible for the production of these lines rubidium (from the Latin *rubidum* for red).

The explanation for why the elements emit the particular colors that are found in their emission spectra follows the basic interpretation used to explain why the hydrogen atom emits its particular frequencies, or colors. In a vapor, the atoms are so far apart that they do not influence one another as they do in a solid. Thus, only a small number of jumps are possible as electrons move from excited states toward the ground state. A few possible jumps mean that only a few possible energies exist for the emitted photons; so only a few possible frequencies are observed.

Absorption Spectra

An absorption spectrum is produced by the technique shown in Figure 11.17. The light from a hot solid or high-pressure gas is first passed through a *cool gas*, and then the light is passed through a diffraction grating. The type of spectrum that is observed on the film appears at first glance to be a continuous spectrum. This is what you might expect because, as we saw earlier, the light produced by a heated solid or a high-pressure gas is a continuous spectrum. A more careful look, however, shows that the spectrum is really not continuous at all. Instead, it is found that there are black lines scattered through the spectrum. Apparently, some particular frequencies, or colors, have been removed or absorbed out of the continuous background spectrum.

A clue to finding out what is causing these absorption lines is found by noting the effect that the cool gas has on the absorption spectrum. For example, let us assume that the cool gas is sodium. We saw in the section on emission spectra that heated sodium vapor produces a single yellow line. If sodium vapor is used as the cool gas in Figure 11.17, it is found that all the colors of the rainbow are present in the resulting spectrum except one, and this missing color is identical to the one that would be produced by heated sodium vapor. That is, the missing color is the characteristic yellow line of sodium. Likewise, if we use mercury as our cool gas, we find that there are four lines missing in the absorption spectrum, and they are the characteristic yellow, green, blue, and purple lines of mercury.

FIGURE 11.17   *Production of an absorption spectrum.*

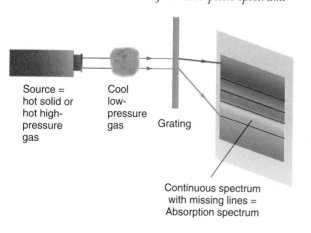

Source = hot solid or hot high-pressure gas

Cool low-pressure gas

Grating

Continuous spectrum with missing lines = Absorption spectrum

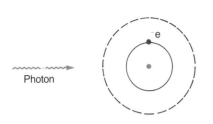

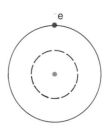

Photon

(a)   (b)   (c)

FIGURE 11.18 *(a) A photon strikes an electron and (b) is absorbed. (c) The electron is raised to an excited state.*

To understand why the absorption spectrum looks as it does, we must examine the atomic processes that are responsible for it. An emission spectrum is produced as the electrons in excited atoms return to lower energy levels, emitting photons of light. As we have noted, there are many ways to excite atoms. Two ways are by heating the gas of atoms or by passing an electric current through the gas. There is yet another way, which is important to us now, and this process is somewhat like the photoelectric effect. Consider Figure 11.18. Shown in Figure 11.18a is a photon heading toward an atom. If this photon has just the right energy, it can be absorbed by an electron in the atom, and this excess energy of the electron raises it to an excited state, as shown in Figure 11.18b and c.

Let us now return to our discussion of absorption spectra. In the white light coming from the heated solid, there are many yellow light photons. When these pass through the cool sodium vapor, these yellow photons are absorbed by the sodium atoms because these photons have the precise amount of energy to lift an electron to an excited state. Thus, the cool vapor is removing from the beam precisely those photons that it would emit if it were already excited. These excited electrons will soon return to the ground state, emitting yellow photons as they go. These "new" photons, however, can be sent out in any direction. As a result, few photons of this frequency or color pass through the diffraction grating. This means that this color will be missing from the continuous spectrum when it is observed.

11.7 SPECTRA AND ASTRONOMY

No earthly being or human-made object has ever visited even one single star in the sky and brought back a sample for us to analyze here on Earth. Yet the store of known information about the stars is overwhelming. Astronomers can tell us the chemical composition of the stars, how fast they are moving toward us or away from us, whether they are orbiting other stars, and so forth. Certainly, this fund of knowledge would be far less were it not for the telescope. But even with the telescope, additional tools are necessary, and one of the most important, if not *the* most important, is the **spectroscope.** Most of our knowledge about the stars and other objects outside our Solar System comes from the study of the spectra produced by these objects.

As a brief quiz to see if you were paying attention in our discussion of spectra, what type of spectrum would be observed by looking at the light from a star such as our Sun? At first thought, it might seem that we should see a continuous spec-

A continuous spectrum and line spectrum for sodium and mercury. (Courtesy Welch Scientific)

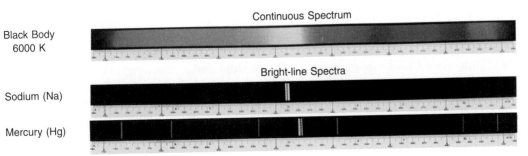

Continuous Spectrum

Black Body 6000 K

Visible Spectrum

Bright-line Spectra

Sodium (Na)

Mercury (Hg)

trum because in this case the source is a hot, high-pressure gas. However, that is not the case. The light produced initially by the star *is* a continuous spectrum, but before this light reaches an observer on Earth, it must pass through two cool layers of gas. One of these is a relatively cool layer of gas surrounding the star like a halo, and the second is the atmosphere of the Earth. In both instances, certain frequencies are removed, so the spectrum observed is an absorption spectrum. The fact that the spectrum from our own star, the Sun, is an absorption spectrum was first noted in 1814 by Joseph von Fraunhofer. He found that the solar spectrum contained literally hundreds of dark lines missing from the background continuous spectrum. These missing lines, now called the Fraunhofer lines, provide the key that enables astronomers to determine the elements present on the Sun as well as on any star.

For example, if the solar spectrum is compared to an absorption spectrum of sodium made in a laboratory here on Earth, it is noted that there are matching absorption lines in the Sun's spectrum. This tells us that sodium did the absorbing, but it actually does not tell us whether the sodium was present in the solar atmosphere or in our own because the light must pass through both. We can make use of the Doppler effect to determine which atmosphere produced the absorption lines.

When we discussed the Doppler effect as applied to sound waves, we found that a sound has a slightly different frequency when it is moving toward us or away from us than it does if it is stationary with respect to us. This applies to all types of waves, including light waves. If we look at the edge of the Sun that is rotating away from us, we find that the light reaching us has been lowered in frequency. Let us now prepare a film of the absorption spectrum of sodium made in a laboratory here on Earth. The result would look like Figure 11.19a. Now let us look at the line corresponding to the absorption spectrum of sodium from light reaching us from the Sun. The pattern is identical, as shown in Figure 11.19b; except if the light is from the side of the Sun that is moving away from us, the lines are shifted slightly toward lower frequencies, toward the red end of the spectrum. These shifted lines must originate in the atmosphere of the Sun because the atmosphere of the Earth is not moving with respect to us. By the same means, we are able

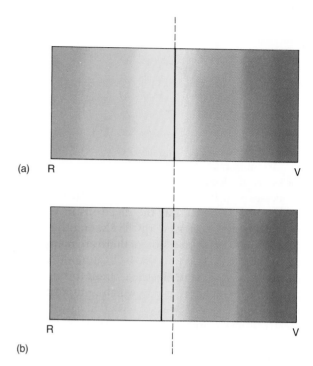

(a)   R   V

(b)   R   V

FIGURE 11.19 *(a) Absorption spectrum of sodium made in a laboratory on Earth. (b) The spectrum line is shifted toward the red if the source is a star moving away from Earth.*

to examine the absorption spectra from stars and other astronomical objects and to determine what elements they contain. Our Sun is found to be composed of approximately 84 percent hydrogen, with most of the rest being helium. In fact, about two-thirds of the elements known on Earth have been found in trace amounts in the Sun. The remainder are probably also there, but in amounts so small that they cannot be detected by this procedure.

It is of interest to note that in 1868, astronomers found a pattern of absorption lines in the Sun that had never been seen in an element here on Earth. The conjecture at the time was that a new element unique to the Sun had been discovered. This element was named helium from the Greek word *helios*, meaning Sun. Of course, you know that this element has now been found on the Earth. It was only discovered, however, in 1895 on the Earth when it was found as a gas emitted from uranium-containing minerals.

11.8 ELECTROMAGNETIC SPECTRUM

In our study of light, we have been concerned only with those waves that have frequencies detectable by the eye. Actually, this region of visible light constitutes only a small portion of the **electromagnetic spectrum**. Electromagnetic radiation spans a wide range of frequencies (Fig. 11.20). Because each frequency has a different amount of energy per photon, waves of different frequencies affect us and our environment in different ways. Some warm our bodies, others can kill germs, others destroy living tissue, and so forth. There are many sources of radiation in our environment. Sunlight, the ultimate source of most of the Earth's energy, contains a broad range of different frequencies. Some of this energy passes through the Earth's atmosphere and strikes the Earth, some is absorbed in the upper atmosphere and re-emitted at different frequencies, and some is reflected off the top of the atmosphere back into space. In addition to these natural radiations, there are technological sources, such as communication systems, radar, light bulbs, and X-ray machines, to name a few. To appreciate the significance of electromagnetic energy in our environment and to understand how photons of different frequencies affect matter in different ways, we will take a brief look at all the different categories of waves that make up the electromagnetic spectrum.

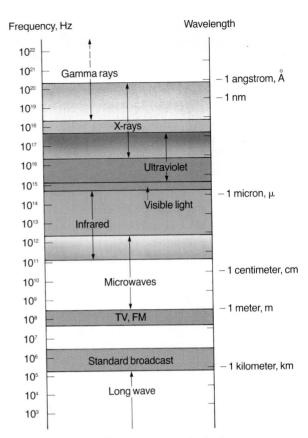

FIGURE 11.20 *The electromagnetic spectrum.*

Radio Waves

Our discussion of the production of light by an atom revealed that visible light is produced when electrons in an atom jump from a state of high energy to a state of lower energy. There is an alternative way by which certain types of electromagnetic radiation can be produced. It is found that any time a charge is accelerated, the charge will release some of its energy in the form of electromagnetic waves. This is what happens in a radio antenna here on Earth. Charges in the antenna are caused to surge back and forth along a metal rod, and as they do so, electromagnetic waves of low frequency and long wavelength are produced. An AM (for *amplitude modulated*) radio station designs its antenna and associated electronic equipment such

that the wave emitted by the station has a particular frequency between 530 and 1605 kHz, whereas the antenna of an FM (for *frequency modulated*) station emits a wave with a particular frequency between 88 and 108 MHz.

The frequency of the wave that a station is allowed to transmit is called the carrier wave. If the station did nothing more than emit that wave, however, it would not be an exciting programming achievement. Some means must be provided to allow this wave to carry information. Let us examine this process for an AM station. Figure 11.21a is a representation of the carrier wave emitted by the station. The sound signal to be sent out by the station at some instant of time is represented by the wave pattern shown in Figure 11.21b. Electronic equipment converts the sound signal to an electrical signal, and this signal is superimposed on the carrier wave, as shown in Figure 11.21c. Thus, the effect of the superimposed sound signal is to modify, or to modulate, the amplitude of the carrier

FOCUS ON COLOR

We have discussed in the text that white light consists of all the colors of the visible spectrum ranging from red to violet. When all of these colors are mixed together, the eye does not distinguish these individual components, but blends them together into what we refer to as white light. The eye also has the characteristic that all the colors of the spectrum can be produced by blending together only the three colors red, blue, and green. These colors are often called the **primary additive colors.** The color wheel, shown in the figure, demonstrates how this mixing works. In fact, rather than take our word for it, you can do this on your own. Take three flashlights and cover the front of one of them with red cellophane, the face of another with blue cellophane, and the face of the last with green cellophane. If you now shine them on the wall in the pattern indicated by the figure, you will find that various colors are produced when they overlap. For example, in the region where the red and blue overlap, the color magenta appears, and where the red and green overlap, yellow is produced. Note that in the center of the color wheel, white is produced in the region where all colors come together.

This technique of producing colors by mixing light beams is used to produce the image on the face of a color television set. If you examine the screen of a color TV closely, you will see that it is covered by a multitude of small dots painted on the glass. These dots are arranged in groups of three, where one of the three emits red light when struck by electrons, the second emits blue,

and the third emits green light. Thus, if you are watching a western movie, at the location on the screen where the white hat of the hero is to appear, all the dots will be turned on by electron beams such that they glow with the same intensity. The three colors will mix together, and at a distance of a few feet from the screen, you will be unable to tell that the light is coming from closely spaced dots. Instead, you will see only the combined effect of white light. At the location where the black-hatted villain appears, no dots are turned on at the location of his hat. The yellow scarf of the school marm is produced by turning on only the red and green dots with equal intensity at the location of the scarf. How would you produce the pink hat of a hero of questionable character?

To understand why you see the colors of nature as you do, let us, for convenience, assume that white light consists of only the primary additive colors, red, blue, and green. The colors of objects around us are produced primarily by the colors that they reflect. For example, a red shirt appears red because it reflects the red light and

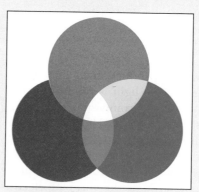

The color wheel.

wave. This modulated carrier wave travels through space until it is intercepted by an antenna in your home receiver. The electronic circuits in your radio work somewhat in reverse of those at the radio station. They separate the incoming wave (Fig. 11.21c) into two parts, the carrier wave and an electrical sig-

absorbs all the other colors, assumed here to be blue and green. A yellow banana appears yellow because it will reflect the two primary colors red and green while absorbing blue light. These two colors mix together as yellow according to the color wheel. What color will an American flag take on in a room where we have only a green light bulb? To answer this question, let us assume the flag consists of a blue square in one corner and of alternating red and white stripes over the rest of the flag. Because there is only green light in the room, this color will be absorbed by the blue square and by the red stripes. Thus, since they are reflecting no color, they will appear black. Black is not really a color; it is the absence of color. The white stripes are capable of reflecting any color that falls on them, so since the only color present for them to reflect is green, these stripes will take on this color. Here is one more for you to try. A red rose with a green stem is brought into a room where a red light is present. What color does the rose appear to have? Try it to convince yourself that the rose petals will appear red, but the stem will be black.

Artists and others who study and work with color often state that the primary colors are not red, blue, and green as we have stated. Instead, they say that they are yellow, magenta (bluish red), and cyan (turquoise green). These colors are often called the **subtractive primaries** because all the colors can be produced by selectively mixing pigments of these colors. The absence of these pigments will produce white paint, and the presence of all these pigments in equal proportions will produce black. To see how these pigments work, consider placing a drop of cyan paint on a piece of paper, as shown in the second figure. When white light falls on the cyan paint, it reflects the colors blue and green while absorbing red (see the color wheel in the first figure). The two colors blue and green are blended together by the eye, and the drop is interpreted as having its characteristic cyan color. What color will be produced when equal proportions of magenta paint and yellow paint are mixed? To answer this, recall that magenta will reflect only red and blue, while yellow will reflect only red and green. Thus, the combination will absorb every color except red, and the mixture appears red. Predict what color will be produced when magenta and cyan paint are mixed in equal amounts. The answer, for you to work out for yourself, is blue.

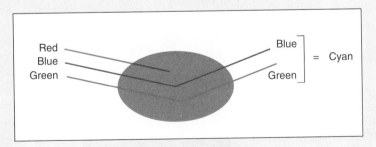

Cyan pigment reflects blue and green light and absorbs red.

nal that carries the audio information. The carrier wave portion is tossed aside, and the electrical signal carrying the information is sent to a speaker. This signal causes the cone of the speaker to vibrate to and fro, following the pattern shown in Figure 11.21b, thus reproducing the original sound wave.

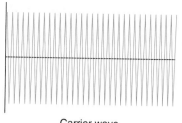

Carrier wave
(a)

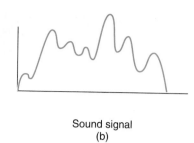

Sound signal
(b)

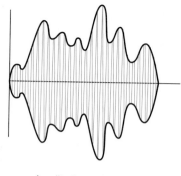

Amplitude-modulated wave
(c)

FIGURE 11.21 *Amplitude modulation.*

EXAMPLE 11.5 TUNE IN YOUR BOOM-BOX

(a) Find the wavelength of the carrier wave broadcast by an AM radio station assigned a frequency of 1430 kHz.

(b) Repeat for an FM station assigned a carrier wave frequency of 92.9 MHz.

Solution (a) Regardless of the type of wave used, the equation that relates the speed of the wave c, the wavelength λ, and the frequency f is given by $m = \lambda f$. We also must note that the speed for all the various types of electromagnetic waves in air is 3.00×10^8 m/s. Thus,

$$\lambda = \frac{c}{f} = \frac{3.00 \times 10^8 \text{ m/s}}{1430 \times 10^3 \text{ Hz}} = \boxed{209 \text{ m}}$$

(b) For the FM wave, we find

$$\lambda = \frac{c}{f} = \frac{3.00 \times 10^8 \text{ m/s}}{92.9 \times 10^6 \text{ Hz}} = \boxed{3.23 \text{ m}}$$

Microwaves

Microwaves are short-wavelength radio waves that have wavelengths between about 1 mm and 30 cm. One important use of these waves is in communication. For example, some cable television signals are transmitted as microwaves and received by antennas (Fig. 11.22). In an application such as this, the frequencies used are selected so that the wave will not be absorbed appreciably by water molecules or other molecules in the atmosphere.

Another common use of microwave radiation is in the cooking of food. Water and oil molecules have a natural frequency of vibration that is in the microwave range. Thus, when you direct a beam of microwave radiation at the food, much of the radiant energy is absorbed by the water and oil molecules. This absorption is noted by the fact that these molecules begin to vibrate with larger amplitude, which is a signal of an increase in temperature of the substance. In fact, the increase in temperature is sufficient to cook the food. If there are no water or oil molecules in the substance, the microwaves either reflect off or pass through without absorption. This explains why glass or ceramic dishes do not become even lukewarm when they are placed alone in a microwave oven.

FIGURE 11.22 *These antennas are used to receive cable television signals transmitted as microwaves.*

FOCUS ON
BLUE SKIES AND RED SUNSETS

The sky on Earth takes on its characteristic blue color because of a resonance effect between the frequencies of the colors in the visible spectrum and that of molecular oscillators in the atmosphere of the sky. In our discussion of sound resonance, we found that any object that can vibrate prefers to vibrate at certain specific frequencies. If pushed at one of these frequencies, its amplitude becomes larger than if pushed at any other frequency. In the sky, the molecular oscillators have a preferred frequency of vibration that is the same as the frequency of blue light. Thus, when white light passes through the atmosphere, it is absorbed by the molecular oscillators, which in turn re-radiate it at the same frequency in all directions, as shown in the figure. This phenomenon is called scattering. Thus, when we look in any direction in space, we see light coming to us from these molecules, and this light has the characteristic blue color of the sky. On the Moon, where there is no atmosphere, there are no oscillators to re-radiate skylight to an observer. Thus, a sky watcher on the Moon would see a black sky above him.

At sunrise or sunset, the light reaching an observer on Earth has to travel a long distance through the atmosphere. Thus, before the light reaches the observer, the blue light is scattered out, but so also are most of the other colors from blue downward. The last color to be affected is that of the red light, so it is still present in the light when it arrives. Thus, sunsets and sunrises are red. The beauty of these effects is enhanced by the fact that the red light is reflected off clouds. The water droplets in clouds are capable of reflecting all colors that strike them, and as a result, the clouds take on a pinkish hue as they reflect the red light from the Sun.

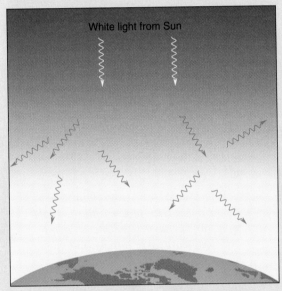

White light from Sun

White light from the Sun has the blue portion of it scattered in all directions in the atmosphere of the Earth.

Microwave radiation has given astronomers some insight into the question of how the Universe came into being about 20 billion years ago. The most widely accepted theory is that at the time of formation, the Universe was compressed into a ball of pure energy. Then an explosion occurred, which gives this theory its name as the **Big Bang** theory. Enormous amounts of radiation were produced at the time of the explosion, and from the remnants of the explosion came our stars, galaxies, planets, and all other parts of our Universe. Astronomers have predicted that if this scenario is correct, there should remain evidence of this initial radiation still around after all these years. The prediction is that the radiation, called **cosmic background radiation,** has changed its character such that now it is in the microwave region of frequencies. In 1965, radiation of the predicted frequency was detected striking Earth uniformly from all directions in the sky. This discovery helped to substantiate the Big Bang theory.

Infrared Radiation

Infrared radiation is characterized by wavelengths between approximately 1 mm and those of the red end of the visible spectrum. Infrared radiation is commonly called *heat radiation* because this radiation is the type emitted by warm objects, and when it is absorbed by materials, it increases the kinetic energy of the molecules of the absorber, thus increasing its temperature.

Humans cannot see infrared radiation, but it is of interest to note that a rattlesnake is able to sense these radiations. Special sensors above the snake's eyes absorb infrared, and thus the snake "sees" warm animals even in the dark. Instruments using infrared radiation can help us to "see" through the darkness, also. Because trees are generally at a different temperature from that of the earth, and nonliving objects such as houses or automobiles are at other temperatures, all these objects emit different frequencies of infrared radiation. Photographic film with an emulsion sensitive to infrared waves can detect these waves (Fig. 11.23). Such photos are important in many applications. For example, the leaves of diseased trees are slightly cooler than the leaves of healthy ones, so infrared photographs can be used to pinpoint centers of disease in a forest before the unhealthy plants start to die. Similarly, if you take an infrared picture of a house on a cold winter night, you can map the insulating qualities of the structure; more heat escapes through poorly insulated regions than through well-insulated ones, and these differences in temperature show up as variations in brightness on the photo. Slightly revised versions of this process enable doctors to diagnose cancerous growths in the human body. An infrared scan may reveal hot spots in the body, which are characteristic of malignancies.

Ultraviolet Light

Just below the visible portion of the electromagnetic spectrum lies a region referred to as ultraviolet waves. These waves have wavelengths ranging from about 60 to 380 nm. As we moved through our discussion of the electromagnetic spectrum, we have moved from low-frequency, low-energy radiation toward higher frequencies and consequently higher-energy photons. Ultraviolet light, which is more energetic than visible light, can excite elec-

(a)

(b)

FIGURE 11.23   *(a) A mountain scene photographed with film sensitive to visible light. (b) The same scene photographed with film sensitive to infrared radiation. Notice that while the black-and-white photograph of the tree and the tepee are significantly different, the infrared radiation emitted by the two is more similar.*

trons in most molecules and cause a multitude of chemical reactions. For example, the ultraviolet rays filtering through the Earth's atmosphere are largely responsible for several reactions on the surface of the skin. For example, they can cause a sunburn. An intense red or infrared light does not affect a person's skin, but even a moderate dose of ultraviolet light can produce a burn. Inhabitants of the Earth should be thankful that most ultraviolet radiation is absorbed in the upper atmosphere, or stratosphere, because larger doses reaching Earth could be harmful to humans. As we shall see later in our discussion of astronomy, there is fear that this protective covering is being depleted. A so-called ozone hole over the south polar regions is allowing ultraviolet radiation to penetrate to a much greater extent than in the past.

Ultraviolet light initiates many reactions in the atmosphere. For example, sunlight reacts with air pollutants to produce photochemical smog, and it is primarily the ultraviolet rays that produce these reactions.

Ordinary glass provides an effective shield against ultraviolet rays. As a result, you cannot get a suntan through a glass window. Some light sources such as mercury vapor lamps inherently emit a lot of ultraviolet. As a result, these lamps are usually enclosed in a glass housing to protect the user.

X-rays

X-rays have wavelengths in the range from about 10 to 10^{-4} nm. Thus, they have a high frequency and high-energy photons. In an X-ray generator, such as the one used in a hospital, electrons are boiled off a heated metal wire by a current passing through the wire, accelerated through a voltage of about 50,000 V, and directed toward a metal target, as shown in Figure 11.24. When the electrons strike the target, they slow down, and their kinetic energy is converted to electromagnetic energy in the form of X-rays.

Because of their high energy, X-rays have great penetrating power. As we all know, they can easily

FIGURE 11.24 *Diagram of an X-ray tube.*

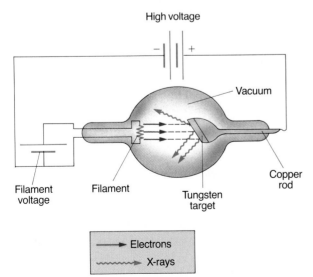

penetrate skin and flesh, but they cannot penetrate bone. As a result, if the X-rays are directed toward a photographic plate with an injured limb between the source and the film, the X-rays penetrate the flesh and expose the film, but the portion of the film covered by bone is not exposed. Thus, the presence of broken or injured bones is revealed when the film is developed. Because of their use as a diagnostic tool, X-rays have provided many medical benefits. However, X-rays can also seriously damage living cells. They can ionize molecules or even break bonds between atoms, and this damage can seriously disrupt the normal function of the cells. Studies show that high doses of X-rays lead to cancer and birth defects in animals. We shall discuss X-rays in more detail in a later chapter.

Gamma Rays

Gamma rays have wavelengths ranging from about 0.1 nm to less than 0.00001 nm. Thus, gamma ray photons have very high frequencies and energies. Gamma rays are emitted by nuclei during some nuclear transformations. Under controlled usage, gamma rays are effective in combating cancer cells in the body, but in uncontrolled releases, they can severely damage the body and produce cancerous growths. Gamma rays will be discussed more completely later in our study of nuclear physics.

11.9 WHAT IS AN ELECTROMAGNETIC WAVE?

Earlier in this chapter, we discussed how light waves are produced when an electron in an atom jumps from an excited state to a lower energy level. We also pointed out in our discussion of radio waves that these kind of waves can be produced by accelerated charges within an antenna. Regardless of the type of electromagnetic wave or how it is produced, its fundamental makeup is the same. For that reason, let us take a look at the production of an electromagnetic wave by an antenna to see exactly what constitutes such a wave.

To understand the fundamental principles of the production of a radio wave, consider Figure 11.25, which shows an alternating current generator connected to two metal rods. (The metal rods are the antenna.) The generator causes charges to

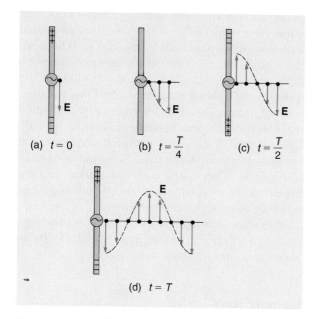

(a) $t = 0$

(b) $t = \dfrac{T}{4}$

(c) $t = \dfrac{T}{2}$

(d) $t = T$

FIGURE 11.25   *The electric field set up by oscillating charges in an antenna.*

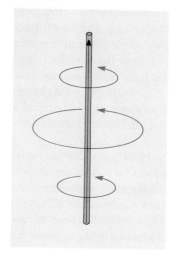

FIGURE 11.26   *The magnetic field lines around an antenna carrying a changing current.*

move back and forth between the two rods. For example, at some particular instant of time, as shown in Figure 11.25a, the generator has forced an excess of positive charges into the top rod and an excess of negative charges into the bottom rod. This separation of charge sets up an electric field in the space surrounding the rod. As you recall from our discussion of electric fields in Chapter 8, this electric field is pointing downward, as shown in Figure 11.25a. Like a ripple in a smooth pond, this downward pointing electric field begins to move away from the antenna. As this field spreads out into space, however, the generator is changing the amount of charge on each of the rods, such that the downward-pointing field is decreasing in strength. This effect is pictured in Figure 11.25b. At some later time, the distribution of charge on the rods is such that the lower rod is positive, and the upper one is negative. An upward-directed field is produced near the rod, as shown in Figure 11.25c. As the oscillations in the rods continue, the electric fields continue to spread away from the antenna, as shown in Figure 11.25d. The speed of this varying field through space is the speed of light. The name "electromagnetic" used to describe these waves now begins to make sense. As the "electro" portion of the name indicates, an electromagnetic wave is partially composed of a vibrating electric field.

The charges moving back and forth between the two rods constitute an electric current, and as we saw in our discussion of magnetism in Chapter 9, a current in a wire produces a magnetic field around it. Figure 11.26 shows the magnetic field pattern around the antenna during the portion of the cycle when the current is upward in the rods. Thus, at the same time that an electric field is being produced around the wire, a magnetic field is also being produced, and just as the electric field spread out away from the antenna like ripples on a pond, so does the magnetic field. Note that the direction of the magnetic field is always perpendicular to the electric field. In the diagrams in Figure 11.25, the vibration of the electric field is up and down in the plane of the page, while the vibration of the electric field is into and out of the page. Thus, we see that *an electromagnetic wave consists of an electric field and a magnetic field vibrating at right angles to one another.* We also see that *the direction of travel of the wave is at right angles to the direction of vibration of both the electric field and the magnetic field.*

This discussion of how an antenna produces an electromagnetic wave is useful in that it provides a convenient way to show the makeup of these waves. The method discussed, however, is only partially correct because the fields produced in this manner would become extremely weak at short distances away from the antenna. The waves that actually reach us from an antenna use a principle that can be understood based on Lenz's law. Lenz's law says that a changing magnetic field can produce a changing electric field, which, in turn, can produce a current in a wire. Maxwell found that a changing electric field can induce a changing magnetic field. At large distances from the antenna, these effects predominate. Thus, the changing electric field and magnetic fields regenerate one another, and an electromagnetic wave is produced. Regardless of the method of production of a wave, however, the end result is the same: An electromagnetic wave is made up of a vibrating electric field and a vibrating magnetic field, and the two vibrations are perpendicular to the wave's direction of travel.

11.10 POLARIZED LIGHT

In our discussion of the production of an electromagnetic wave in the preceding section, we found that an electromagnetic wave meets the characteristics of a transverse wave—that which vibrates, the electric field and the magnetic field, moves at right

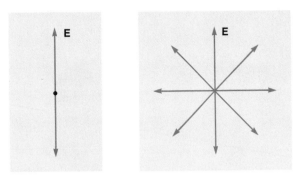

FIGURE 11.28   *(a) A polarized electromagnetic wave. (b) An unpolarized wave.*

angles to the direction of travel. Figure 11.27 shows a representation of an electromagnetic wave moving away from an antenna. Imagine yourself standing at the position shown in Figure 11.27, such that the wave is coming directly toward you. If you could actually see the electric field portion of the wave, it would consist of an up-and-down vibration as shown in Figure 11.28a. Such a wave in which the vibrations of the electric field vector are back and forth along a particular direction is called a **polarized wave.** The vibrations emitted by a light source, such as an ordinary light bulb, are not of this nature, and the light is said to be **unpolarized.** To see why this is true, consider each atom in the filament of a light bulb to be a tiny antenna that is sending a wave toward you. The end result is a collection of trillions of waves, oscillating in all conceivable transverse directions. Such a mixture of waves is typical of all light sources, from the Sun down to the smallest light bulb. We picture an unpolarized light wave as shown in Figure 11.28b, which indicates that the direction of vibration of the electric field could be in any direction as it comes toward you.

There are a variety of ways in which an unpolarized beam can be polarized, but we shall investigate only one of these. The process that we shall examine is that of selective absorption, which today is the most common method used. In 1938, E. H. Land discovered a material, which he called **polaroid,** that allows vibrations only along a particular direction to pass through. This material is produced in sheets consisting of long-chain organic molecules, such as polyvinyl alcohol. During the

FIGURE 11.27   *A representation of an electromagnetic wave leaving an antenna. The observer would see that the wave is polarized.*

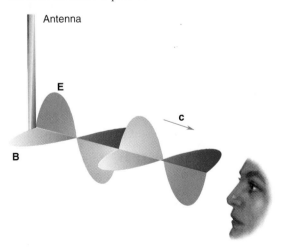

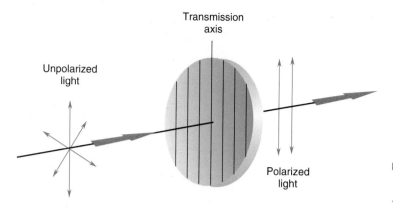

FIGURE 11.29   *Unpolarized light becomes polarized after passing through a sheet of polaroid.*

manufacturing process, these sheets are stretched, which causes the long-chain molecules to align like pickets in a fence. The sheets are then dipped into a solution containing iodine, which causes the long-chain molecules to become conductors of electricity along the lengths of the chains. When light falls on these sheets, that portion of the light that has its electric field vibrations along the lengths of these chains is absorbed, while those vibrations at right angles to the chains are transmitted with little loss in intensity. Thus, the picket fence analogy is almost correct except the light can pass through when the vibrations are perpendicular to the "pickets" but not when the vibrations are aligned with them.

The direction in the material that is perpendicular to the alignment of the long-chain molecules is called the transmission axis. Figure 11.29 shows a beam of unpolarized light incident on a piece of polaroid material with its transmission axis aligned as shown. The light that passes through is polarized.

Polaroid lenses are popular in sunglasses to cut down on the glare from reflected light. To see how these glasses work, consider Figure 11.30, which shows unpolarized light from the Sun reflecting off a pool of water. When light is reflected off a surface, it becomes partly polarized, and if the reflecting surface is horizontal, the resulting polarization is also horizontal. (Before the advent of polaroid, the primary way of producing polarized light was by this reflection process.) As Figure 11.30 shows, the lenses of the polaroid sunglasses are oriented such that they will not let this horizontal vibration pass through.

An important use of polarized light involves the use of certain materials that are **optically active.** An

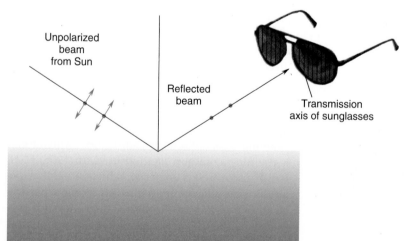

FIGURE 11.30   *Unpolarized light can become polarized after reflection from a surface, such as that of water. Sunglasses, with transmission axis properly aligned, can cut down on the reflected light.*

(a)   (b)

(a) When the polaroids are uncrossed, light is easily transmitted, but (b) when one is rotated by 90°, they become crossed, and no light passes through. (Courtesy Bill Schulz)

optically active material rotates the direction of polarization of a beam of light passing through the material. Figure 11.31a shows a beam of unpolarized light incident on a piece of polaroid, which is called the polarizer. After passing through the polarizer, the light is polarized in the direction shown. The light now falls on a second piece of polaroid, one that has its transmission axis perpendicular to the vibration of the light. All of the light is absorbed by the analyzer, so no light reaches the observer shown. If a piece of optically active material is placed between the polarizer and analyzer, however, as shown in Figure 11.31b, the direction of vibration of the light is rotated slightly, and some light

is able to move through the analyzer. This light can be blocked out once again by rotating the analyzer through the same angle as that through which the light beam has been rotated. It is found that the angle of rotation depends on the length of the sample and on the concentration if the substance is in solution. One optically active material is a solution of the common sugar, dextrose. A standard method for determining the concentration of the sugar solution is to measure the rotation produced by a fixed length of the solution.

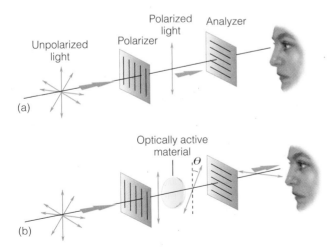

FIGURE 11.31 *(a) The polarizer polarizes the light, which the analyzer then absorbs. (b) An optically active material rotates the polarized light so that some can pass through the analyzer.*

SUMMARY

Light has a dual nature in that in some experiments it behaves like a wave and in others it behaves like a particle. It manifests itself as a wave in **Young's double-slit** experiment in that it undergoes interference, producing a series of alternating bright and dark lines.

Light manifests itself as a particle in the **photoelectric effect.** This effect is explained by assuming that light energy is carried in small packets called **photons** that are absorbed by electrons in certain metals, and the energy picked up by these electrons is sufficient to kick them out of the material.

Light is produced when electrons in **excited states** of atoms release their energy as they jump back to lower level states.

Light can be separated into its component frequencies by a **diffraction grating.** Such gratings are useful for observing the three different types of spectra, the **continuous,** the **emission,** and the **absorption.** The observation and analysis of spectra are important techniques in astronomy.

The **electromagnetic spectrum** consists of a vast array of waves, all of which are produced either by electronic transitions within atoms or by the acceleration of electric charges. The spectrum ranges from low frequency to high frequency as **radio waves, microwaves, infrared, visible, ultraviolet, X-rays,** and **gamma rays.**

EQUATIONS TO KNOW

$E = hf$ (energy carried by a photon)

$c = \lambda f$ (basic wave equation)

KEY WORDS

Wave-particle duality
Young's double-slit experiment
Photoelectric effect
Photon
Ground state

Excited state
Diffraction grating
Continuous spectrum
Emission spectrum

Absorption spectrum
Electromagnetic spectrum
Radio waves
Microwaves

Infrared
Ultraviolet
X-rays
Gamma rays

PROBLEMS AND CONCEPTUAL QUESTIONS

Problems requiring numerical work are identified with a blue number.

Young's double-slit experiment

1. Would the bright lines of Young's double-slit experiment be farther apart for blue light than for red light? Defend your position.
2. If the incident light in Young's double-slit experiment were white light, what would the interference pattern look like?
3. When light waves interfere destructively and the light is destroyed, is energy also destroyed? Likewise, when light waves interfere constructively, the resultant pattern seems more intense than the simple sum of the two. Is energy being created?
4. Often when watching TV, you get a diminution of intensity on the screen caused by two beams reaching your antenna simultaneously. One is the direct beam; the other is from, say, an airplane flying overhead. Explain why the signal is reduced.

Photoelectric effect

5. Which of the following statements are true and which are false? Defend your answers. (a) A dim, high-frequency light will eject electrons with more kinetic energy than a bright, low-frequency light. (b) You can dislodge many high-energy electrons from a metal surface with a bright, low-frequency light if you shine the light on the metal for a long time. (c) Light waves dislodging electrons from a metal behave like water waves dislodging rocks from a sandstone beach.

6. A beam of light falls on one metal, and photoelectrons are emitted. This same beam is now directed toward another metal, and no photoelectrons come off. Explain what is different between these two metals.

7. When the photosensitive cell of a light meter is exposed to light, a small electric current is produced that moves a needle to register the amount of light present. Could you obtain a true light meter reading near a powerful radio transmitter, or would the radio waves be likely to affect the instrument? Explain.

8. Sound can knock down a delicate card tower. If you are trying to destroy card towers with sound, would you be better advised to turn up the frequency or the amplitude? Explain. Light can knock electrons off the surface of a metal. If you are trying to dislodge electrons with light, would you be better advised to turn up the frequency or the brightness? Explain.

9. Would you expect the temperature of a metal to have any effect on the ease of removing electrons from it via the photoelectric effect?

10. If both visible light and ultraviolet light will eject electrons from a metal, for which of the two would you expect the electrons to have the greater kinetic energy?

11. In a certain metal, photons of red light having an energy of 2.65×10^{-19} J will eject electrons with a maximum kinetic energy of 1.43×10^{-19} J. What is the amount of energy required to allow an electron to escape?

12. An electron in a certain metal has to have an energy of 2.35×10^{-19} J just to escape from the metal. What is the minimum frequency of the photon that will eject an electron? (*Hint:* What will be the kinetic energy of the ejected electron?)

Where does light come from?

13. What is an excited state? How does an electron acquire the energy to be promoted into an excited state? What happens to the potential energy of an electron when it falls from an excited state to the ground state?

14. If a person does not have the strength to carry a rock up a hill, he may be able to roll it up slowly in stages and eventually accomplish the task. If one photon does not have the energy to raise an electron from one quantum level to the next, can many photons of the same energy move it up little by little? Explain.

15. The light emitted by hot sodium vapor consists of two colors close together in frequency. What

might the excited states responsible for this result look like?

16. According to the Bohr theory of the hydrogen atom, which has more energy, a photon emitted when an electron drops (a) from the second excited state to ground or (b) from the third excited state to the second?

17. Is it possible for there to exist a single photon of white light?

18. The wavelength of the light emitted in a certain transition from a higher excited state to the first excited state of hydrogen is 656.3 nm. What is the energy difference between these two excited states?

19. An electron is in the fourth excited state of hydrogen. How many different energy photons could be produced as the electron returns to the ground state?

Diffraction grating

20. Discuss the similarities and differences between the interference produced in Young's double-slit experiment and that produced by a diffraction grating.

21. Sodium emits two different wavelengths in the visible spectrum that are separated by 0.6 nm. That is, they are extremely close together in wavelength. Some diffraction gratings will show these as two separate lines, whereas others will not. That is, some will resolve them and some will not. What do you believe is the difference between a grating that will and one that will not?

Spectral analysis

22. If a woman said that she could create electrons using only flint, steel, and a little dry tinder, would you believe her? If she said she could create photons using the same materials, would you believe her? Explain.

23. Describe the spectrum seen for the following situations. (a) The light from a hot tungsten filament is observed. (b) Light from a heated solid is passed through cool sodium vapor. (c) Hydrogen gas is heated until it emits light, which is then observed.

24. What kind of spectrum would be observed if hydrogen vapor is heated until it emits light, and then this light is passed through cool hydrogen vapor before being observed?

25. How might the spectrum from a star reveal the following properties about that star? (a) The composition of the star. (b) The speed of the star toward us or away from us. (c) The fact that the star is a binary star.

26. How could the spectrum from the Sun prove that it is rotating?

Electromagnetic spectrum

27. Which of the following objects would emit electromagnetic radiation? (a) A balloon with a slight negative charge on it falling to the earth. (b) An electron moving in a battery-operated circuit. (c) An electron oscillating to and fro in a wire as an alternating current. (d) A baseball leaving a pitcher's hand.

28. Why is it that a study of the heavens by a telescope can only tell us how the heavens were and not how they are?

29. If you are told that it is the frequency of light reflected from an object that gives it its color, devise an experiment that would prove that frequency does not change when light moves from one substance into another.

30. Every time a wave moves, something oscillates. If it is a water wave, it is the water in oscillation. What is it that oscillates in an electromagnetic wave?

31. Do some research to find out the status of the ozone hole over the south pole at the present time.

32. What is the electromagnetic spectrum? Is there a sharp distinction between radio and microwave radiation? Microwave and infrared radiation? Infrared and visible radiation? Explain.

33. List in order of increasing energy: ultraviolet, visible, X-ray, microwave, radio, infrared.

34. When the infrared picture shown in Figure 11.23 was taken, the photographer wished to filter out some of the more energetic visible light entering the camera. Do you think he used a filter that removed mostly blue light or mostly red light? Explain.

35. Photographers who work in dark rooms with photosensitive paper generally use a low-intensity red light to see. Would a dim blue light work just as well? Explain.

36. The flame of a welding torch is blue at the inner tip, white in the middle, and red on the outside. Which part of the flame is the hottest? Explain.

37. A fire inside a pot-bellied stove emits visible light. However, the stove itself does not change color when it gets hot. What has happened to the energy of the visible-frequency photons?

38. Imagine that you have two beams of light with the same number of photons in each beam, but one consists of ultraviolet frequencies and the other of visible light. Which beam carries more energy, or are they the same? Which requires more energy to generate, or are they the same? Explain.

39. Explain why X-rays are more apt to damage living tissue than are ultraviolet rays.

40. Defend the statement that an electromagnetic wave is a transverse wave.

Polarized light

41. List some advantages and disadvantages of polarizing sunglasses.

42. What would happen if you tried to polarize a sound wave?

43. A sunbather wearing polarizing glasses notices that the glare from a pool of water is greatly diminished with the glasses in place. What does she observe about the glare if she lies down on her side on a blanket?

ANSWERS TO SELECTED NUMERICAL PROBLEMS
11. 1.22×10^{-19} J
12. 3.55×10^{14} Hz
18. 3.03×10^{-19} J

CHAPTER 12

INSIDE THE ATOM

A physicist uses a lens to spread out the beam from a helium-neon laser. (Courtesy Greg Perry)

When science is presented in a textbook, the various fields are neatly separated and packaged into units, chapters, and sections. In this book, for example, motion, energy, electricity, and magnetism have all been discussed as discrete topics. The history of the development of scientific thought is not nearly so orderly and compartmentalized. In many cases, seemingly different branches of science progress independently, and at first it seems as though the separate discoveries bear little relevance to each other. Then a key insight is brought forth that shows that the different components are really interconnected. Sometimes experimental technique is not refined enough to answer a crucial question until some apparently unrelated event occurs that enables a forward leap to be taken.

The study of the atom is one such agglomeration of many different types of research in different fields that were brought together to form one complete picture. During the eighteenth and early part of the nineteenth centuries, various aspects of physics, such as electricity, magnetism, and light, were considered to be different from each other, and the field of chemistry was thought to be pretty much unrelated to physics. Then in a relatively short period of time around the beginning of the twentieth century, scientists discovered that these seemingly different fields of inquiry are all intimately interrelated. These discoveries marked an exciting time in the history of science, for they changed our understanding of the physical world. In turn, this increased understanding created the foundation for the vast changes in human society that have occurred during the past few generations.

12.1 ATOMS

Consider a gold bar that has a mass of 1 kg. The volume of such a bar would be about 52 cm^3. Does this mean that there is nothing in this volume but wall-to-wall gold with no empty space? If the bar is cut in half, the two resulting pieces still retain their chemical identity as solid gold. Imagine that the bar is cut again and again, indefinitely. Will the smaller and smaller pieces always be the same substance, gold? Two ancient Greek philosophers (Leucippus and Democritus) could not accept the idea that such cutting could go on forever. Ultimately, they speculated, the process must end when it produces a particle that no longer can be cut. In Greek, *a tomos* means "not cuttable," and from this we get the word **atom.** Thus, the word atom is used to describe the smallest, ultimate particle of matter.

Many centuries later, in 1803, John Dalton moved the atomic theory from a speculative basis to a scientific foundation. Dalton was a chemist interested in **chemical changes,** Chemistry is the subject of Chapters 15 through 17, but for now let us define a chemical change as the transformation of one material (or materials) to another. For example, consider hydrogen and oxygen, which are gases at room temperature. If the two gases are mixed and ignited, they combine and are transformed into another material—water. Dalton knew that hydrogen and oxygen always combine in fixed, constant proportions. For example, he found that 8 g of oxygen would always combine with 1 g of hydrogen, no more and no less. If an excess of oxygen is added to the container such that the 8:1 ratio of oxygen to hydrogen is exceeded, the excess oxygen will be left over after the chemical reaction. Similarly, if excess hydrogen is added, that excess will not enter into he reaction. Think how different this is from baking a cake. You can bake a cake with one egg or two, a lot of sugar or no sugar, and the result will still be a cake. Different cakes may taste different, but they will still be recognizable as cake, as long as you basically follow the recipe. Water found anywhere on Earth, however, is always found, within experimental error, to contain the same ratio of elements.

Dalton recognized that atoms are the fundamental units of matter. A collection of atoms of a single type forms a special class of substances known as **elements.** Gold, silver, hydrogen, and oxygen are all elements because each of these substances is made up of only one kind of atom. Water is not an element because it is made up of two different kinds of atoms, hydrogen and oxygen. In water, the dif-ferent types of atoms are bound together as though they were connected by springs, and the combination of atoms is called a **molecule.** Any substance, such as water, made up of a fixed proportion of *different atoms* is called a **compound.**

The major contribution that Dalton made was that he recognized that all atoms of the same element have the same mass. To explain the transformation of hydrogen and oxygen into water, Dalton would have reasoned as follows. A specific mass of oxygen always combines with a specific mass of hydrogen because the atoms of oxygen (with their fixed mass) are combining with the atoms of hydrogen (with their fixed mass). The resultant compound, water, made up of billions of these atoms, always has a definite percentage, by mass, of these two atoms. If matter were continuous, Dalton reasoned, this definite ratio of mass would not always be found.

During a chemical reaction, atoms are not created or destroyed, nor are they divided into parts; instead, the reactions merely change the way the atoms are bonded to one another. During chemical change, bonds between atoms are made or broken so that new compounds are formed or previously existing compounds are broken up. The concept of bonding will be investigated more thoroughly in our study of chemistry.

Dalton placed the atomic theory on a sound footing with his experimentation, but he never attempted to study the composition of an individual atom. Like the ancient Greeks, he still considered atoms to be ultimate, indivisible particles. Their makeup, or internal structure (if any), was not even considered.

CONCEPTUAL EXERCISE

Explain the atomic theory of matter to someone using only a piece of cheese and a sharp knife.

Answer You could begin to cut the cheese into smaller and smaller pieces. Soon, it becomes easy to imagine that there could be a smallest piece that either could not be cut, or if it were cut, it would leave you with something other than cheese. If cheese were an element, this smallest piece would be an atom, but since cheese is a compound, the smallest piece is a molecule.

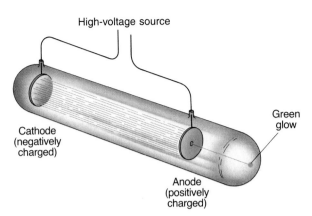

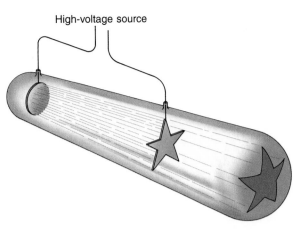

FIGURE 12.1   *A simple cathode ray tube. If a high voltage is imposed between two electrodes in an evacuated tube, a current will pass between them. If a small slit is placed in the anode, some of the rays will pass through the slit and strike the glass. When they hit the glass, a greenish glow appears. Thus, the position of the rays can be located precisely.*

FIGURE 12.2   *If the anode is cut in some specific shape, a shadow is observed behind the anode. This result shows that the cathode rays travel from the cathode to the anode. It also implies that the rays consist of a stream of particles that move in straight lines.*

12.2 THE DISCOVERY OF THE ELECTRON

The concept of a structureless atom began to change as a result of experiments with electricity. In 1752, Benjamin Franklin learned that lightning was a burst of electrical energy and therefore that electricity could travel through a gas. Other scientists found that the phenomenon of electrical discharge could be duplicated in the laboratory with a device called a **cathode ray tube.** This device is constructed by sealing two pieces of metal into a glass tube and then pumping most of the air out of the tube. The pieces of metal are then connected to a high-voltage source, as shown in Figure 12.1. The positive plate is called the **anode,** and the negative plate is the **cathode.** When the high-voltage source is connected to these plates, an unusual phenomenon occurs. With a hole drilled in the anode, as shown in Figure 12.1, a green glow is seen on the phosphorescent coating on the end of the glass tube. The obvious question was: What is causing this glow? A series of experiments was devised in an attempt to answer this and other questions, and the results were as follows:

1. Scientists observed that if the anode is placed in the center of the tube, as shown in Figure 12.2, a shadow will be cast on the glass behind it. The existence of this shadow shows that something is being emitted by the cathode and is traveling toward the anode. As a result of this observation, the radiation became known as **cathode rays.**

2. Two charged plates were mounted inside the tube. When the ray passed between the plates, it was bent, or attracted, toward the positive plate and repelled by the negative one (Fig. 12.3). This experiment proved that the beam was negatively charged.

3. Similarly, the cathode rays could be bent by a magnetic field. Thus, the cathode rays behaved more like a stream of charged particles than a beam of light. By measuring how much the cathode rays were deflected by a given electric and magnetic field, physicist J. J. Thomson calculated the ratio of the charge to mass for the particles. He found this ratio to be 1.76×10^{11} C/kg.

4. The tubes were constructed using a variety of materials for the cathode and a variety of residual gases inside the tube. (Although most of the gas inside the tube is pumped out, a small amount does remain.) Experiments always gave the same value for the charge-to-mass ratio.

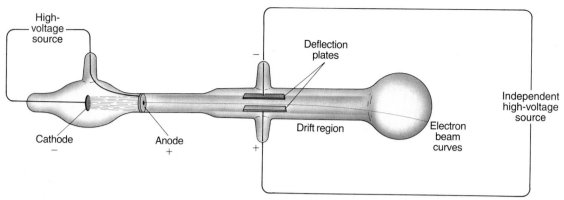

FIGURE 12.3   *Thomson's cathode ray apparatus. J. J. Thomson's experiment was performed in the following manner: The cathode and the anode were placed in one end of a long evacuated tube, and a slit was cut in the anode. Some of the cathode rays would pass through the anode and drift through the tube. If no other fields were present, they would strike the center of the end of the tube and could be located by the green glow that appeared on the glass. Thomson then created an electric field perpendicular to the rays by placing two charged plates inside the tube and applying a high voltage between the plates. When this was done, the cathode rays were deflected toward the positive plate. Cathode rays were also deflected by a magnetic field. From these experiments, Thomson was able to measure the mass/charge ratio of the cathode ray particles.*

Thomson's interpretation of the data was bold. Since the cathode rays contained a fixed and constant ratio of charge to mass, he reasoned that the ray must consist of discrete particles. Today, these particles are called **electrons.**

At the time, Thomson had no way to prove that an electron was more or less massive than an atom. He assumed, without any definite proof, that electrons must be much less massive than atoms. Therefore, it followed that atoms must be made up of smaller, more fundamental particles. Thus, he concluded that atoms are not the smallest indivisible units of matter. Following this bold proposition, there arose an immediate interest in measuring the absolute value for either the mass or the charge of a single electron. This was accomplished in 1911. The most accurate modern measurements give the following values:

charge of electron $= 1.60219 \times 10^{-19}$ C

mass of electron $\;\;= 9.10953 \times 10^{-31}$ kg

By contrast, the mass of the lightest atom, hydrogen, is 1.67×10^{-27} kg, so the mass of an electron is only about 1/2000 that of the lightest atom. Electrons are indeed a small subdivision of the atoms, as Thomson had predicted.

Ordinarily, matter is electrically neutral. Because electrons are negatively charged, it is obvious that matter must also contain positive electrical charge. What is the nature of this positive charge?

Electrons can be pulled out of atoms relatively easily. Thus, electrons can be removed from a piece of rubber by rubbing it with fur, and they can be pulled out of a cathode by a high voltage. This ease of removal makes it possible to perform experiments such as Thomson's cathode ray studies that identified the electron and measured some of its properties. The positively charged portion of the atom, however, is much harder to study because it is difficult to separate and to isolate the positive charge from ordinary matter.

12.3 EARLY MODELS OF THE ATOM

The earliest model of the atom was that of a tiny, hard, indestructible sphere. When J. J. Thomson began to unravel the electrical nature of the atom, however, new models arose. One of these was the so-called plum pudding model suggested by Thomson. As shown in Figure 12.4, this model pictured

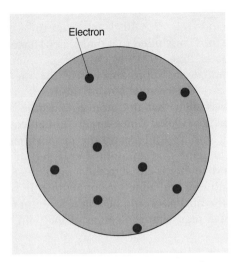

FIGURE 12.4 *Thomson's plum pudding model of the atom.*

the atom as a volume of positive charge with electrons embedded in it like plums in a plum pudding.

In 1911, an important experiment was conducted by Geiger and Marsden under the guidance of Lord Ernest Rutherford (1871–1937) that proved that the plum pudding model could not be correct. Before discussing the details of the experiment, let us ask, in general, how scientists can study something as small and seemingly unfathomable as the interior of an atom. To understand the approach, consider the following analogy. Imagine that you are a warrior in one of those fantasy novels where dragons, spells, and wizards are commonplace. As you wander across the mysterious countryside, you come upon an open valley. In the middle of the valley, you see a dense cloud hanging low over the ground. You cannot see into the cloud, and some magic spell prevents you from walking into it. As a guardian of your people, however, you feel that it is important to find out what lies inside the cloud. How would you go about your task?

Perhaps the first thing you would do is sit at the edge of the cloud and observe it. If an occasional burst of fire, steam, and smoke came forth, you might deduce that a dragon is hidden inside. If volleys of arrows flew out at you, you might suspect that hostile people lurk within.

A second and more active procedure would be to shoot something into the cloud. If you fired 100 arrows randomly into the cloud and all passed through and emerged on the other side, you might

guess that there wasn't much solid matter hidden inside it. If 50 arrows passed through and 50 did not, then you would guess that there was a mixture of solid objects and open space. Other possibilities exist. The arrows might be absorbed, but new objects, say dragon scales, might fly back at you. Each observation would tell you something about the nature of the matter inside the cloud.

The procedure of shooting something into the cloud and observing what happens is analogous to what Rutherford and his colleagues did. As we shall see later, some atoms are radioactive, which means that they spontaneously emit certain kinds of radiation. One type of this radiation is a positively charged particle called an **alpha particle**, which is now known to be the nucleus of a helium atom. The experiment that Geiger, Marsden, and Rutherford conducted used alpha particles from radioactive atoms as arrows to bombard atoms in a thin metal foil (Fig. 12.5). The results of the investigation were quite astounding to Rutherford and his colleagues. They found that some of the alpha particles were deflected through large angles, as shown

FIGURE 12.5 *Rutherford's experiment with alpha particles. When some of the massive alpha particles were scattered at large angles, Rutherford deduced that the nucleus must be a small, dense, central core.*

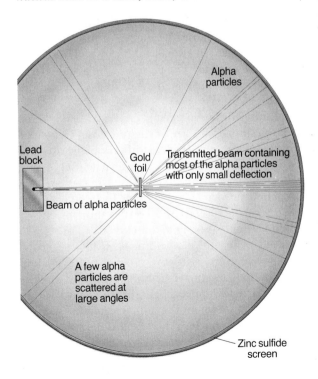

FIGURE 12.6 *Head-on collisions of alpha particles with nuclei enabled Rutherford to estimate the size of a target nucleus.*

in Figure 12.5; in fact, a few of them were deflected back along their initial direction of travel. To use Rutherford's words, "It was quite the most incredible event that had ever happened to me in my life. It was almost as incredible as if you fired a 15-inch shell at a piece of tissue paper and it came back and hit you."

If the plum pudding model were correct, large-angle scatterings would have been impossible. In such an atom, an alpha particle passing through the atom would be equally attracted and repelled in all directions, so the different forces would cancel and the alpha particle would pass right through the foil with little or no deflection.

To explain his observations, Rutherford theorized that the atom is mostly empty space, which accounts for the fact that most of the alpha particle arrows pass on through the foil without striking anything. To account for the occasional deflections, however, Rutherford also concluded that the positive charge and most of the mass of the atom must be concentrated in a small volume, which he called the **atomic nucleus.** The electrons, he assumed, circled about the nucleus in orbits like planets about the Sun.

Rutherford's scattering experiments also led him to estimate the size of the nucleus. Figure 12.6 shows how he approached this problem. Rutherford considered only those alpha particles that were reflected back on themselves when they approached the nucleus. He knew the initial kinetic energy of the alphas, and conservation of energy enabled him to calculate how close the alpha could approach the nucleus before the Coulomb repulsion force exerted on the alpha by the nucleus would stop it and turn it around. He found that the alpha particles approached the nuclei to within 3.2×10^{-14} m when the foil was made of gold. Thus, the gold nucleus must be less than this value. For silver atoms, the distance of closest approach

was found to be 2×10^{-14} m. From these results, Rutherford concluded that the nucleus has a radius no greater than 10^{-14} m. Other experiments had determined an approximate size for the entire atom, and the relative size of the nucleus to the atom was such that the atom as a whole was approximately 10,000 times larger in diameter than the nucleus. To understand the proportions in an atom, imagine that an atom is enlarged in size so that its diameter is equal to the length of a football field and one end zone (about 100 m). In such an atom, the nucleus would be a sphere about the size of a small marble at the center of the field.

12.4 INSIDE THE NUCLEUS

Following Rutherford's discovery that the atom consists of a central nucleus with electrons swirling around it, the question arose as to whether or not the nucleus has structure. That is, is the nucleus a solid unit of charge, or does it consist of a collection of individual particles? The exact composition of the nucleus has not been defined completely even today, but by the early 1930s, a model of the nucleus had evolved that is useful in helping us to understand much about how it behaves. By that time, physicists had determined that the charges of the nuclei of the different elements were all whole number multiples of the charge on the electron. For example, a neutral iron atom has 26 electrons, and a neutral copper atom has 29 electrons. Consequently the nucleus of an iron atom must have a charge equal but opposite to the charge of 26 electrons, the nucleus of a copper atom must have a charge equal but opposite to 29 electrons, and so on. This observation led to the obvious conclusion that there must be particles inside the nucleus and that each particle bears a positive charge equal but opposite to that of the electron. These particles are, of course, protons, the existence of which was verified by Rutherford in 1920.

According to this model, the simplest atom, hydrogen, would consist of a single proton with a single electron swirling around it (Fig. 12.7). From this model, we see that the mass of a proton must be equal to the mass of the hydrogen atom less the mass of the electron. This gives us a proton mass of 1.67×10^{-27} kg. Complications arise as one moves to the next simplest atom, helium. Neutral

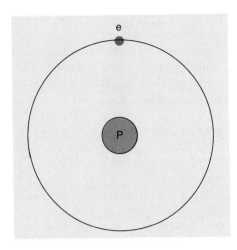

FIGURE 12.7 *The hydrogen atom.*

TABLE 12.1		
PARTICLE	**RELATIVE CHARGE**	**RELATIVE MASS**
Electron	−1	0
Proton	+1	1
Neutron	0	1

helium must have in its nucleus two protons, yet the mass of a helium atom is about twice the mass of two protons. This led to the conclusion that there is another constituent part of the nucleus, a particle that must have no charge and that must have a mass about equal to that of the proton. This particle is the **neutron**, and its existence was verified conclusively by nuclear reaction experiments in 1932 by James Chadwick (1891–1974).

In our future discussions of the nucleus, we shall make use of the following quantities:

1. The **atomic number** of an atom is the number of protons, or units of positive charge, in the nucleus. It is also the number of electrons in the neutral atom. Thus, a hydrogen atom with its single proton has an atomic number of 1. An atom of carbon has an atomic number of 6, indicating that it contains 6 protons and, when neutral, 6 electrons. (We shall often refer to the charges on the electron and proton in terms of relative charges. For example, a proton has a relative charge of $+1$ and an electron has a relative charge of -1. This is just a shorthand way of saying that the proton has a charge of 1.6×10^{-19} C and an electron has a charge of -1.6×10^{-19} C.)

2. The **mass number** of an atom is the total number of protons and neutrons in the nucleus of the atom. For example, the most common form of carbon atoms has a mass number of 12 and an atomic number of 6.

This means that since there are 6 protons in the nucleus, there must be 12 minus 6, or 6, neutrons. We shall also often use a relative scale for discussing masses of protons, neutrons, and electrons. Because the proton and the neutron have almost equal masses, we shall set their relative mass equal to 1, and because the electron is so much lighter than either of these two, we shall set its relative mass equal to zero. Table 12.1 indicates the relative charges and masses of the components of the atom.

It is convenient to have a symbolic way of representing nuclei that will show how many protons and neutrons are present in the nucleus. For the element iron, this notation is $^{56}_{26}$Fe. Fe is the chemical symbol for iron; 56 is the mass number, and 26 is the atomic number. Thus, there are 56 total particles (protons and neutrons) in the iron nucleus, and 26 of these are protons. So, there are 30 neutrons.

12.5 ISOTOPES

All atoms of a particular element have the same number of protons in their nuclei. Thus, the nucleus of an element identified chemically as carbon always contains six protons. The nuclei of a given element, however, may contain different numbers of neutrons. An alternative way of expressing this is to say that all atoms of a particular element have the same atomic number, but they may have different mass numbers.

*Atoms that have the same atomic number but different mass numbers are said to be **isotopes**.*

For example, in a sample of carbon, 98.6 percent of the carbon nuclei are the isotope $^{12}_{6}$C, and about 1.1 percent are of the form $^{13}_{6}$C. Found in even smaller percentages are the isotopes $^{11}_{6}$C and $^{14}_{6}$C. Even the simple hydrogen atom has three isotopes.

They are $^1_1$H, ordinary hydrogen, $^2_1$H, deuterium, and $^3_1$H, tritium.

EXAMPLE 12.1   COUNTING PROTONS AND NEUTRONS
(a) How many neutrons are in the nucleus of ^{37}Cl?

Solution Note that we have given you the mass number of chlorine, but we have not given the atomic number. All chlorine nuclei have the same atomic number, and you can find this in the table of elements (see Appendix E). Find chlorine in this table and verify that it has an atomic number of 17. The number of neutrons is now easily found as

number of neutrons
= mass number − atomic number
= 37 − 17
= 20

(b) What are the symbol and the mass number of an isotope of an element containing 19 protons and 22 neutrons?

Solution If the nucleus contains 19 protons, its atomic number is 19. Search for atomic number 19 in the table of elements and prove for yourself that this nucleus is potassium, symbol K.

The mass number
= number of neutrons + number of protons
= 22 + 19
= 41

Thus, this isotope is represented as $^{41}_{19}$K, or more simply as ^{41}K.

12.6 THE BOHR ATOM

Taken together, the experiments performed by Thomson and Rutherford showed that an atom consists of a small, comparatively massive, positively charged nucleus surrounded by light, negatively charged electrons swirling in orbits around the nucleus. This picture, by itself, is a start, but there were some perplexing problems that this simple model could not explain.

1. It was known that if an electric charge is accelerated, it emits energy in the form of visible light or other forms of electromagnetic radiation. This presents a problem because an electron circling the nucleus has a centripetal acceleration. As a result, the electron continuously radiates energy, but if it loses energy, it should quickly spiral into the nucleus. In effect, the atom should collapse, and the predicted collapse time is quite short, of the order of 10^{-8} s after its formation. Thus, the question arises: How have atoms remained stable for the life of the Universe?

2. As we saw in Chapter 11, atoms of a particular element emit only certain specific frequencies of light that are characteristic of the particular element. Why? Also, we have seen that an atom will absorb only those frequencies of light that it will emit. Why? As noted in (1), the accelerated electron in an atom should emit light, but calculations indicate that the light emitted should consist of *all* visible frequencies, not distinct frequencies.

The first significant attempt to answer these questions was offered by Niels Bohr in 1913. We

Niels Bohr (1885–1962). (Courtesy AIP Niels Bohr Library, Meggers Collection)

have already taken a brief look at the Bohr model of the atom in our discussion of spectra in Chapter 11. Because of its importance in the world of physics, however, let us re-examine it here and explore its assumptions in greater detail. Bohr based his model of the atom on two postulates. They are:

1. Electrons orbit the nucleus in distinct paths, or orbits, similar to the orbits of the planets around the Sun. Out of all the infinite number of possible orbits, however, only those with certain radii actually occur. Bohr did not understand why this was so; in fact, this postulate contradicted the known laws of physics. In the gravitational analogy, a spacecraft can orbit the Earth at any radius, not just a few "allowed" radii. As part of this assumption, Bohr also assumed that the allowed orbits are characterized by the fact that they are *nonradiating* orbits. This means that the electron can revolve about the nucleus indefinitely in these orbits without radiating away any of its energy. This gets us out of the predicament of having the electron continuously radiating energy and thereby spiraling into the nucleus.

2. If an electron is in a high (excited) orbit, it will fall back to its ground state, getting rid of its excess energy by emitting a photon. The photon carries an energy that is related to its frequency by $E = hf$. As we have seen, this explains why an atom emits only certain frequencies of light and no others. Conversely, if a photon of the precise frequency or energy is incident on an atom, the photon can be absorbed by an electron, causing it to move to an excited state. Thus, the atom will absorb only those photon energies that it will emit.

The validity of Bohr's assumptions about how the atom works could be determined only by how successful they were in agreeing with experimental observations. From his assumptions, Bohr was able to derive equations from which the frequencies of light that would be emitted by hydrogen could be calculated. The radius of the hydrogen atom could be calculated from his work, and the energy of the electron in each of its possible orbits could be found. These calculated values agreed with experimental findings. Thus, his work was enormously successful to scientists of the time. Perhaps of even more importance, however, was the fact that the Bohr theory gave us a model of what the atom looks like and how it behaves. Once a basic model is constructed, refinements and modifications can be made to enlarge upon the concept and to explain finer details. His model did have limitations, and the picture of an atom with perfectly defined orbits and energies has been largely superseded today. We shall look at some of these innovations in the next section.

CONCEPTUAL EXERCISE

(a) An electron drops to the ground state in an atom from a very high excited state and emits a photon of energy A. Another electron in the same atom drops from a lower excited state to the ground state and emits a photon of energy B. Which photon has the greater energy, A or B?

Answer An electron has more energy in a high excited state than it does in a low excited state. Since an electron sheds its energy by emitting a photon, the photon emitted in a transition from a high excited state must carry away more energy from the atom. Thus, photon A has the greater energy.

(b) Which photon has the higher frequency?

Answer The energy of a photon is related to the frequency by $E = hf$, so the higher energy photon also has the higher frequency. Photon A is the high-frequency photon.

(c) If one of the photons is a red light photon and the other a photon of blue light, which is which?

Answer We know from our study of light that blue light is of a higher frequency than is red light. Thus, photon A must be the blue light photon and photon B the red light photon.

12.7 PARTICLES AS WAVES

We have examined the peculiar dual nature of light on several occasions throughout our study of physics. Beginning students of physics often find it troubling that light sometimes behaves like a wave and at other times like a particle. If you are one of these students, don't be concerned; many eminent physicists have shared your concern. In 1924, however, Louis de Broglie proposed the idea that *particles of matter also have wave properties*. Thus, the circle was complete—photons sometimes act like waves and sometimes like particles; now electrons, protons, and even baseballs share this dual nature in that they may at times act like waves.

De Broglie proposed that the wavelength of a particle could be found by the equation

$$\lambda = \frac{h}{mv} \tag{12.1}$$

where h is Planck's constant, $h = 6.63 \times 10^{-34}$ J s, and the product mv is the momentum of the particle, the mass times its velocity.

If de Broglie's theory is correct, it should be possible to detect the wave-like behavior of electrons by passing a beam of them through something like a diffraction grating. After light passes through a grating, alternating patterns of light and dark are seen on a screen. Could a similar pattern be observed for electrons? The experiment was difficult to perform because for a diffraction grating to work for light, the distance between the slits must be of the order of the wavelength of the radiation used. As we shall show in an example problem that follows this section, the wavelength of electrons is too short to enable an ordinary diffraction grating to work. However, crystals are made up of orderly arrays of atoms, and as it turns out, the atoms themselves are spaced at just the right intervals to provide a diffraction grating for electrons. Therefore, if moving electrons pass through certain crystals, they will be fanned out into alternating bands, provided that they have a wave nature. In 1927, C. J. Davisson and L. Germer succeeded in measuring the wavelength of an electron by this process and thereby verified the wave nature of matter.

EXAMPLE 12.2   THE WAVELENGTH OF AN ELECTRON AND A BASEBALL

(a) Calculate the wavelength of an electron ($m = 9.11 \times 10^{-31}$ kg) moving with a speed of 10^7 m/s.

Solution Eq. 12.1 gives

$$\lambda = \frac{h}{mv} = \frac{6.63 \times 10^{-34} \text{ J s}}{(9.11 \times 10^{-31} \text{ kg})(10^7 \text{ m/s})}$$

$$= 7.28 \times 10^{-11} \text{ m}$$

This wavelength is in the range of X-rays in the electromagnetic spectrum.

(b) A baseball pitcher throws a 0.150-kg baseball with a speed of 30.0 m/s. What is the wavelength of the baseball?

Solution Again, we use Eq. 12.1 to find

$$\lambda = \frac{h}{mv} = \frac{6.63 \times 10^{-34} \text{ J s}}{(0.150 \text{ kg})(30.0 \text{ m/s})}$$

$$= 1.47 \times 10^{-34} \text{ m}$$

This wavelength is so small that one could never find a diffraction grating with slits this small through which the baseball could pass. Thus, we could never observe interference effects with a large-scale object like a baseball.

12.8 WAVE MECHANICS AND THE HYDROGEN ATOM

One of the first successes of de Broglie's wave theory of particles was in explaining the arrangement of electrons in an atom. To understand de Broglie's model, we will have to recall some facts that we learned in our study of sound. There we found that we could set up standing waves on a string by tying one end of it to a wall and shaking the other end at just the right frequency. When the end of the string is vibrated, a wave is sent down the string and reflected off the wall. The two waves, traveling in opposite directions, interfere, and standing waves are produced. A standing wave on a string will have an integral multiple of half-wavelengths present on the string, as shown in Figure 12.8.

In de Broglie's application of wave theory to the hydrogen atom, he pictured an electron in orbit as having the properties of a circular standing wave. To understand his model, consider Figure 12.9. In Figure 12.9a, we have drawn three standing wave patterns: One contains a full wavelength, the second contains two full wavelengths, and the third has three full wavelengths. Now imagine those wave patterns to be bent into a circular pattern, as shown in Figure 12.9b. De Broglie's viewpoint was that *one of Bohr's nonradiating orbits was formed when the circumference of the orbit was equal to an integral*

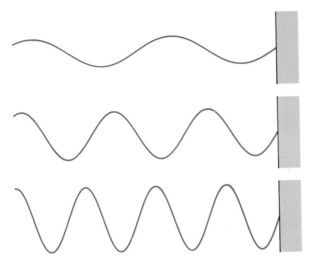

FIGURE 12.8   *Several standing wave patterns on a string.*

number of wavelengths. The ground state orbit had a circumference of exactly one wavelength, the first excited state had a circumference of two wavelengths, and so forth. Figure 12.9c shows an excited state composed of five complete electron wave-

lengths. An orbit similar to that in Figure 12.9d would never be formed because it does not have an integral number of wavelengths. A fractional wavelength orbit would annihilate itself because of destructive interference.

This initial effort by de Broglie was important to our understanding of the atom in that it provided a mechanism by which wave theory could be applied in determining the details of atomic structure. Real success with wave mechanics, however, occurred 2 years later when an Austrian-German physicist, Erwin Schrödinger, developed a wave equation that described how particle waves change with position and time. This equation demonstrated convincingly that the wave nature of particles was a necessary feature in understanding the subatomic world. Schrödinger's equation has been successfully applied to the hydrogen atom and to many other submicroscopic systems.

Schrödinger's equation was mathematically very elegant. We shall describe it here by saying that what one is attempting to determine by solution of the equation is a quantity represented by the Greek letter ψ (psi), called the **wave function.** The closest analogy between ψ and ordinary attributes of a wave is that ψ is most closely related to the amplitude of the wave. Its importance in wave mechanics lies in the fact that all that one can know about the behavior of a particle of matter can be determined from a knowledge of its wave function. For example, if ψ is a wave function for a single particle, the value of ψ^2 at some location at a given time is proportional to the probability of finding the particle at that location at that time. As a result of wave mechanics, our viewpoint of the electron arrangements in atoms has now been modified in the following ways.

FIGURE 12.9   *The de Broglie model of the atom can be visualized (a) by drawing 1, 2, 3, or more electron wavelengths on a flat piece of paper and (b) then wrapping the drawing around a nucleus. (c) An integral number of wavelengths is allowed. (d) If an orbital has a fractional number of wavelengths, the wave interferes with itself and will be destroyed. Therefore, fractional orbitals do not exist.*

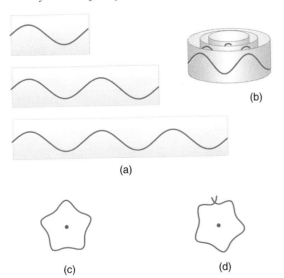

1. The electron is no longer required to remain at certain specific distances from the nucleus as in the Bohr theory. For example, the Bohr model predicts that the electron will always be found at a distance of 0.53×10^{-10} m from the nucleus when in the ground state. Wave mechanics says that it is most probable that the electron will be at this distance, but it also says that there is a probability of finding the electron at other distances from the nucleus.

2. The Bohr theory says that the electrons orbit the nucleus in a plane similar to the plan-

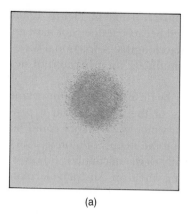
(a)

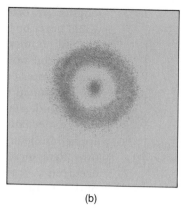
(b)

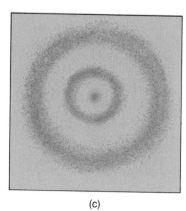
(c)

FIGURE 12.10   *Three diagrams representing the shape of the electron cloud (a) in the ground state and (b) and (c) in two excited states of hydrogen. The densely shaded regions indicate places where an electron is likely to be found; the less densely shaded zones indicate regions where the probability of finding an electron is low. Note that these are not photographs of electrons, but rather models based on a mathematical theory of the atom.*

ets revolving around the Sun. Wave mechanics says that there is a probability of finding electrons anywhere in a spherical region around the nucleus.

As a result of these modifications, the currently accepted sketch of an atom in its ground state would be pictured as an electron cloud surrounding the nucleus, as shown in Figure 12.10a. Regions where the cloud is pictured as dense represent those locations where the electron is most likely to be found. Figure 12.10b and c represent electron cloud pictures of the orbits for excited states of hydrogen.

CONCEPTUAL EXERCISE

If racecars behaved like electrons, describe a few of the features of a race.

Answer According to the Bohr theory, the racetracks could have only certain definite radii because a car could travel in only certain orbits around the center of the track. According to the wave theory of the atom, there would be some probability of locating the smeared-out car at any point on the track at the same time. In the Bohr theory, cars would never pass because they all would have the same energy (and speed) in their allowed orbits. There are other features; try thinking of some.

12.9 LASERS

The information and understanding that we have gained so far enable us to understand one of the most fantastic devices of the twentieth century, the **laser.** The word laser is an acronym for **l**ight **a**mplification by the **s**timulated **e**mission of **r**adiation. In principle, the acronym does a pretty good job of explaining how a laser works, provided that one understands the meaning of all the words.

Stimulated emission is a process that we have not considered previously, so let's examine how it works. We have seen in our earlier discussion that the electrons in atoms can be promoted to excited

FIGURE 12.11   *(a) An electron in an excited state. (b) When the electron drops to a lower level, a photon is emitted.*

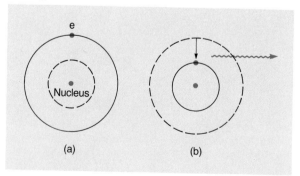
(a)   (b)

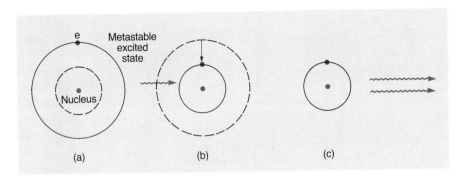

FIGURE 12.12 *(a) An electron in a metastable state. (b) An incoming photon stimulates the atom to de-excite. (c) The photon emitted joins with the photon in part b to move away in step.*

states by various processes. When the electron returns to its ground state, however, it gets rid of its excess energy by releasing a photon. This process is pictured in Figure 12.11. An electron typically does not spend much time in an excited state before it de-excites; 10^{-8} s is a typical length of time. Some atoms, however, have excited states in which an electron seems to get stuck for periods of the order 10^{-3} s to 10^{-2} s. Such states are called almost stable, or *metastable*, states. The electron will eventually return to the ground state from a metastable state and emit a photon. The process of stimulated emission can induce the electron to de-excite earlier than usual. Figure 12.12 pictures the stimulated emission process. In Figure 12.12a, we see an atom in an excited metastable state. In Figure 12.12b, another photon comes along *that has exactly the same frequency as the photon that would be emitted by the excited atom if it were to de-excite.* This incoming photon "stimulates" the excited atom to de-excite, and the result is shown in Figure 12.12c. The initial incoming photon continues on along in its initial direction, but now it is accompanied by a photon from the de-excited atom. Note that both of these photons have the *same frequency.*

Figure 12.12c shows another feature of the two photons. The two leave the atom with the *same direction of travel, and they are in phase.* Thus, when one is at a crest, so is the other. If these two photons come upon another atom that is in a metastable state, they can stimulate it to emit its radiation, and another photon joins the march, all photons in step and traveling in the same direction. Light composed of photons in step and moving in the same direction is said to be **coherent.** Coherent light is different from that emitted by an ordinary light bulb. The atoms in a light source operate independently of one another, in that once excited, they return to the ground state at random intervals and emit light of random frequencies, in random directions, and with no relationship between crests of one photon and those of another. The light of an ordinary source is often described as resembling waves created in a fountain. Little waves and ripples are created everywhere with no particular order or pattern between them.

Let us now apply these pieces of information to see how the light of a helium-neon laser is produced. The orbital arrangement of neon gas is shown in Figure 12.13. One of its levels is a metastable level from which an electron can descend to a lower orbit and in the process emit a photon with a wavelength of 632.8 nm. A mixture of helium and neon is confined to a glass tube sealed at the ends with mirrors, as shown in Figure 12.14a. One of the mirrors is completely silvered so that it reflects all light that strikes it, whereas the other is only partially silvered so that it reflects some light and transmits some. A high-frequency power supply, not shown in the figure, is then connected to the tube such that it causes free electrons in the gas to sweep back and forth. These moving elec-

FIGURE 12.13   *The metastable state of neon.*

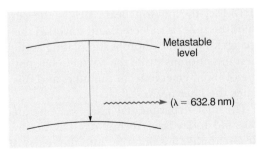

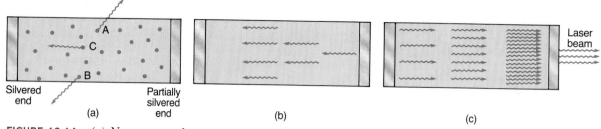

FIGURE 12.14   *(a) Neon atoms, shown as dots, are excited to a metastable level. These atoms eventually return to the ground state and emit photons. Most of these photons are lost through the sides of the tube (atoms A and B), but some move along the axis of the tube (atom C). (b) A photon moving along the axis of the tube encounters excited neon atoms and causes them to emit a photon by stimulated emission. This results in an avalanche of photons all traveling along the axis of the tube. (c) The avalanche of photons is reflected off the completely silvered mirror and continues to cause stimulated emission as it moves toward the partially silvered end. Those photons that emerge from the right end of the tube constitute the laser beam.*

trons collide with neon and helium atoms and excite them. Some of the neon atoms are excited to the metastable state. Helium is present in the tube because it has the property of being able to transfer its energy efficiently to a neon atom in a collision between the two. As a result, we end up with a huge number of neon atoms in an excited metastable state. The details of what happens next are pictured in Figure 12.14a. Some of the excited neon atoms soon de-excite and emit photons. As shown, some of these (A and B in the figure) escape from the side of the tube and do not enter into the lasing action. Other photons, however, are emitted along the axis of the tube (C in the figure). One of these photons moving along the axis of the tube encounters an excited neon atom and causes it to emit a photon by the stimulated emission process. These two photons then stimulate other excited atoms to emit photons, and thus the avalanche grows (Fig. 12.14b). Light amplification is produced, starting with a single photon. In Figure 12.14c, the photon beam strikes the silvered end of the tube, is reflected, and traverses the tube once again, with the photons stimulating as they go. When the photons strike the partially silvered end, some of them emerge from the tube, and this constitutes the laser beam.

The first experimental laser was operated in 1960, and today laser technology has become so widespread that it is entering into the realm of the commonplace. One obvious feature of the laser is that its beam is *highly directional*. Space scientists

have been able to send laser light beams to the Moon from the Earth and reflect them off mirrors left there by astronauts. The reflected beam was detected on Earth, and by this means the distance to the Moon has been determined to within a few centimeters. If conventional light sources had been used, the beam would have spread out to such an extent that its reflection would have been impossible to detect. In a similar experiment, scientists from two different continents have simultaneously bounced laser beams off the Moon. By accurately measuring the angle of the emitted beam with respect to the Earth and the time required for the round trip, and by using some simple trigonometry, the distance between continents has been measured with extreme accuracy. In more mundane applications, the highly directional feature of a laser beam has enabled engineers to dig tunnels precisely along a straight line, align bridges perfectly, and so forth.

The output of a laser is a narrow beam that carries a lot of energy, and this energy can be concentrated into an even smaller area by lenses. As a result, when the beam is focused on a material, high temperatures can be achieved quite rapidly. A temperature of 6000°C can be reached in about half a millisecond with a powerful laser. Thus, a selected portion of metal can be melted and welded without destroying nearby areas. Eye surgeons are also using lasers to "weld" detached retinas back into position without disturbing the rest of the eye. Machinists have used the beams to melt tiny holes in

FOCUS ON
MAKING A HOLOGRAM

The figure shows an arrangement that can be used to produce a hologram. Light coming from a laser is split into two parts by a half-silvered mirror H. One beam of light goes through the mirror and then through lens D, which diverges the beam. This spread-out beam then bounces off the subject and strikes a piece of photographic film. The second half of the beam is sent through lens A, which spreads it out and then bounces it off two mirrors, M1 and M2, and to the photographic film. The two beams of light then undergo interference at the location of the film, and it is this interference pattern that is really captured by the film. The pattern is so intricate that no vibration of the parts can be allowed while the hologram is being made. A movement or vibration as small as a fraction of a wavelength of the light used can ruin the interference pattern. The interference pattern produced relies on the fact that there is a constant phase relationship between the two beams during the period of time the film is being exposed. This constant phase relationship is possible only if the light is the coherent light from a laser.

When the photographic negative is produced, swirling patterns of lines are captured that cannot be interpreted by the eye. However, if laser light now is passed through the negative while one looks in the direction from which the light is coming, a three-dimensional view of the subject is seen hanging in space.

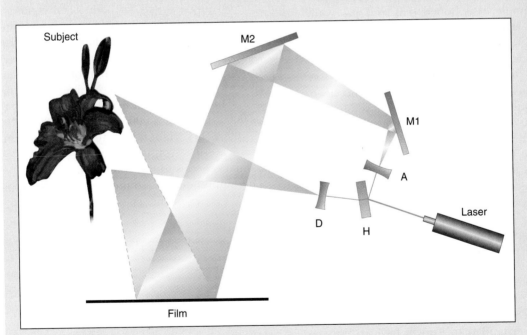

Making a hologram.

small jewels used as bearings in watches or other precision machinery. Holes can even be drilled through a diamond, the hardest substance known.

Unfortunately, many devices that can be used for the good of the human race can be used for destruction as well. A beam of energy that can create intense heat, that can be focused precisely, and that travels in a straight-line path at the speed of light has obvious uses in warfare.

The output of a helium-neon laser is at a single wavelength, 632.8 nm. More advanced lasers are tunable, however, in that they can have an output beam at several different frequencies. One application of this is in determining the kinds of pollutants present in the atmosphere. Suppose you want to determine if a specific pollutant is present. If light energy of a specific wavelength is absorbed only by that pollutant molecule and not by normal air molecules, then the absorption of light from a laser beam operating at this wavelength becomes a method of detecting the presence of the pollutant. By measuring the amount of light absorbed, the concentration of that particular air pollutant can be determined.

The fact that the light is coherent means that precise interference patterns can be produced with this device. Three-dimensional displays called **holograms** can be produced in this way. The applications of holography promise to be many and varied. For example, it is hoped that someday your television set will be replaced by one using the hologram photographic process. Instead of being flat, the picture you view will be in three dimensions.

One of the more recent applications of the laser that has found its way into the life of college students is the compact disk (CD) player that is used in place of phonographs or tape players. Figure 12.15 shows the important pieces of one of these devices. The light from a laser is sent through lens A, then toward a partially reflecting mirror. Part of the beam passes through this mirror and is redirected by another mirror through lens B and toward a CD. The light reflected off this disk retraces the original path until it reaches the partially reflecting mirror, where it is bounced toward a receiver. The receiver then uses the output it receives to drive a loudspeaker and to recreate the information that has been picked up off the CD. But how does the light retrieve information from the disk? To understand that, we will have to understand, to a limited extent, the **binary number system.** In ordinary arithmetic, we express the size of a quantity by representing its size on our ordinary decimal number system. Thus, a person 6 feet tall is taller than a 5 foot tall person. An alternative numbering system is the binary system. This system is the only one that can be understood by a com-

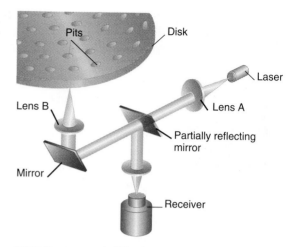

FIGURE 12.15   *A CD player.*

puter because it is a series of *on* and *off* signals. These on and off signals are segmented into groups of eight individual signals, called a **byte.** Thus, in the binary system, the number one is represented as 00000001; this could be interpreted by a computer by turning a switch "off" for seven equal intervals of time then "on" for one of those time intervals. The number two in binary is 00000010. Turn your switch off for six equal intervals of time, on for one of these intervals, then off for one interval, and you have told the computer the number is 2. The decimal number three is 00000011, four is 00000100, five is 00000101, six is 00000110, seven is 00000111, eight is 00001000, ad infinitum. (We will leave it to you to determine how to relay these binary signals to the computer.)

The amplitude of a sound signal can be converted to a series of binary numbers (Fig. 12.16). The amplitude is sampled at regular intervals, and

FIGURE 12.16   *Converting the amplitude of a sound to binary.*

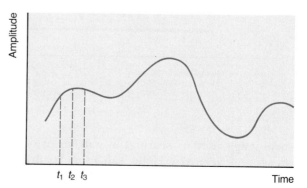

FOCUS ON
LASERS IN THE CHECK-OUT LINE

Lasers are becoming more and more a part of everyday life, as a visit to your local supermarket will verify. In most larger supermarkets, the cashier moves your grocery items across the beam from a helium-neon laser that senses information on the Universal Product Code, or bar code, label on a package. These bar codes are printed such that the spacing and darkness of successive lines can be interpreted by a computer as representing specific numbers. One of these bar codes is pictured in the figure. The two thin lines at the far left are used to indicate to the computer that a new bar code is on the way to be interpreted. The next two lines, a heavy one and a thin one, are the basic code for the number zero. The number zero is interpreted by the computer to indicate that the bar code for a grocery item is to follow. The next several lines are interpreted by the computer as the number 21140, which is the number that identifies the manufacturer. Two more thin lines then appear that indicate that the bar code for the item is on the way shortly. This pattern of lines is representative of the number 20786, which is the number assigned by the company to

The bar code on a can of green beans.
(Courtesy Bill Schulz)

a can of green beans. As the beam of the laser moves across this bar code, the light is reflected back to a detector when a white space appears, but is absorbed when a dark space appears. The detector senses these varying shades of light and darkness as the numbers described above and sends this information to a computer that looks up the price for the product and rings it up on the cash register.

---

the loudness of the sound at any interval is then converted to a binary number. For example, suppose the amplitude of the sound signal in Figure 12.16 at t_1, t_2, and t_3 is decimal 4, 5, and 4, respectively. The computer will translate decimal 4, 5, and 4 to binary 00000100, 00000101, and 00000100, respectively. This pattern of binary numbers is then inscribed on a laser disk: A series of pits and smooth places are burned onto the disk such that each pit is smaller than the dot over an "i" in this textbook. When light from the laser in Figure 12.15 falls on one of these smooth places on the disk, it is reflected back to the detector, which interprets the presence of the reflection as an "on"

signal or as a 1 in binary. When the light strikes a pit on the rotating disk, it is bounced off in some random direction and does not return to the detector. This is interpreted by the detector as an "off" or a zero in binary. Electrical equipment connected to the detector then drives a loudspeaker according to the size of the binary number it receives. Thus, the original sound signal (Fig. 12.16) that was used to etch the disk is reproduced.

A new entrant in the high-fidelity game is the digital audio tape (DAT). The method of recording information is essentially the same as that used on CDs in that the data are stored in the form of ones and zeros. The difference is that the medium is cassette tape. As noted in Chapter 9, information on cassette tapes is recorded by magnetizing the tape, and the same process is used in DAT technology. Magnetizing a small portion of the tape in one direction is interpreted as representing the number one, whereas magnetizing it in the opposite direction is interpreted as a zero. The fidelity of CDs and the convenience of cassettes would seem to make DATs the perfect audio medium.

SUMMARY

Dalton recognized that all atoms of the same element have the same mass and that chemical changes involve the transfer of whole atoms, not parts of atoms, from one substance to another. Atoms are not created or destroyed in ordinary chemical changes.

The negative charge in an atom is carried by electrons, which are much less massive than the atom itself. Alpha-scattering experiments showed that an atomic nucleus occupies only a small portion of the entire volume of an atom, even though it contains nearly all of its mass. **Atomic nuclei** consist of **protons** and **neutrons** (with the exception of hydrogen, which is a lone proton). A proton bears a positive charge equal but opposite to that of an electron. A neutron has approximately the same mass as a proton but is electrically neutral. The **mass number** of an atom is the total number of protons and neutrons. Atoms of the same element that have different mass numbers are called **isotopes.**

One of the first successful explanations of how an atom is constructed was the Bohr theory. This has largely been supplanted by a wave-mechanical model, in which the energy levels of electrons in atoms are described by a series of wave equations. An electron cloud diagram may be viewed as a description of the probability of finding an electron in any region of space around an atom.

We have found that light has a dual nature, but material objects also have a dual nature. The wavelength of a material object is given by the de Broglie equation.

Laser is an acronym for **l**ight **a**mplification by the **s**timulated **e**mission of **r**adiation. The light from a laser is characterized by the fact that it is **coherent;** this means that all the photons maintain a fixed phase relationship with each other.

EQUATIONS TO KNOW

Mass number = number of protons + number of neutrons

$$\lambda = \frac{h}{mv} \text{ (de Broglie wavelength)}$$

KEY WORDS

Atom	Cathode ray tube	Atomic number	Wave mechanics
Chemical change	Cathode	Mass number	Wave function
Element	Anode	Isotope	Laser
Molecule	Proton	Nucleus	Stimulated emission
Compound	Neutron	Bohr atom	Coherence
Electron	Alpha particle	De Broglie wavelength	Hologram

PROBLEMS AND CONCEPTUAL QUESTIONS

Problems requiring numerical work are identified with a blue number.

Atoms

1. The chemical formula for common table salt is NaCl, which means that sodium atoms are com-

bined with chlorine atoms. Is this an element or a compound?

2. Microscopic particles of matter suspended in a fluid are seen to undergo a random, haphazard motion, darting one way and then another. This type of motion is referred to as Brownian motion and can be explained by the atomic theory of matter. What is your explanation?

3. Based on your explanation of Brownian motion in problem 2, why would a cork floating on the surface of the water not be affected?

4. An individual is lost in a deep forest. A bloodhound is brought in to track him. Based on the atomic theory of matter, how does the bloodhound do its job?

5. Some ordinary salt is dissolved in water. Is this a chemical change?

Discovery of the electron

6. Explain the significance of the shadow cast in the cathode ray tube. In a cathode ray tube, the shadow is cast from the cathode to the anode. What conclusions would you draw if the shadow projected from the anode to the cathode? What conclusions would you draw if no shadow were cast at all?

7. What conclusion would you draw if the following observations were made during electrical experiments in evacuated tubes? (a) The rays come from the anode and are attracted to the negative plate of a perpendicular electric field. (b) The rays come from both electrodes and are split into two discrete beams by a perpendicular electric field. (c) The mass/charge ratio of the cathode rays depends on the particular metal used as the cathode; that is, there is one value when the cathode is copper, another when it is nickel, and so on. Defend your answers.

8. It is also observed in discharge tubes that some "rays" do move toward the cathode, not away from it. What is the sign of the electrical charge on these "rays"? The properties of the particles that make up these "rays" depend on the identity of the residual gas that is left in the partially evacuated tube. What conclusion can you draw from this fact?

Early models of the atom, inside the nucleus, and isotopes

9. Why did Rutherford use alpha particles, rather than electrons, for his scattering experiment?

10. Explain why the plum pudding model of the atom could never lead to large-angle scatterings.

11. If you could see an atom, to determine what kind it is, you could either count the number of protons or the number of electrons. Which would be the most reliable?

12. Explain the difference between the mass and the mass number of an atom.

13. Complete the accompanying table by substituting the correct numerical value where a question mark appears. (Use the table of elements on the inside back cover. Note that the table gives the symbols and atomic numbers of the elements, but not the mass numbers.)

ISOTOPE IN NUCLEUS	ATOMIC NUMBER	MASS NUMBER	NUMBER OF NEUTRONS
Oxygen-18	?	?	?
Strontium-90	?	?	?
Uranium-?	?	?	141
Iodine-?	?	131	?
?	17	35	?
?	?	226	138

14. What is the name of the element having atomic number 73?

15. Which one of the following has the most protons in its nucleus: iron (Fe), copper (Cu), or cadmium (Cd)?

16. What is the symbolic representation of (a) helium containing two protons and two neutrons and (b) helium containing two protons and one neutron?

17. What is wrong with the following headline story, "Scientists have discovered a new isotope of hydrogen that contains two protons in its nucleus."

18. About how many times more massive is an atom of neodymium (Nd) than one of neon (Ne)?

19. Experiments showed that nuclear charges are whole number multiples of the charge on an electron. What conclusions would you draw if the smallest nuclear charge were half the charge of an electron? Explain.

The Bohr atom

20. Explain why Bohr's model of the atom was an important step forward even though it was incomplete.

21. Describe what must happen to an electron in a Bohr atom in order for it to be ionized.

22. List some features of the Bohr atom that violate the realm of common sense in our everyday world.

23. Discuss some differences and similarities between photons and electrons.

24. The transitions in hydrogen that produce visible light are those in which an electron in an excited state drops from its level to the first excited state. In these transitions, photons of wavelength 656.3, 486.1, 434.1, and 410.2 nm are emitted. How much more energy do the excited states from which these photons originate have than the first excited state?

25. An electron in a hydrogen atom drops from the first excited state to the ground state and emits

a photon with a wavelength of 12.15 nm. (a) How much higher in energy is the first excited state than the ground state? (b) Is this an infrared, visible, or ultraviolet light photon? (c) What frequency photon could strike the electron in its ground state with just enough energy to kick the electron to its first excited state?

26. A hypothetical atom has its first excited state 3 units of energy higher than in the ground state, and in the second excited state, its energy is 3.5 units greater than in the ground state. What would happen to an electron in the ground state if struck by a photon having energy of (a) 2 units, (b) 3 units, (c) 3.2 units?

Particles as waves; wave mechanics; the hydrogen atom

27. Describe an experiment that would verify that light acts as a wave and one that would verify that an electron acts like a wave.

28. A rock is thrown straight up into the air. Discuss its de Broglie wavelength as it rises and as it falls.

29. Describe an experiment that would verify that light acts as a particle and one that would verify that an electron acts like a particle.

30. An electron and proton are both traveling with the same speed. Which has the longer wavelength?

31. Why are electron clouds used to describe the location of an electron in the wave mechanical view of the atom, while specified orbits are used in the Bohr theory?

32. (a) What is the wavelength of a proton moving with a speed of 10^5 m/s? (b) What is the wavelength of a 70 kg person jogging at 2 m/s?

Lasers

33. What are the differences between laser light and ordinary light from a tungsten filament light bulb with respect to (a) the wavelength of the light, (b) the means of production, (c) the phase relationships within the beam?

34. What is there about a laser beam that gives it the ability to cut through metal?

35. Before a laser can lase, there must be more atoms in excited states than in the ground state. Why is this the case?

36. Describe coherent light and explain why the light from an incandescent bulb is not coherent.

37. If cost were not a factor, do you think that laser light bulbs could be used efficiently to illuminate your kitchen or shop? Explain why or why not.

38. Why does a laser emit only one color, as opposed to a continuous band of colors?

39. Laser light is often compared to a battalion of soldiers marching, while ordinary light compares to these soldiers milling around at a break in the march. Discuss these analogies.

40. In order for a laser to work, both ends of the tube must be sealed with a mirror that is at least partially reflecting. Why would an ordinary piece of glass not be sufficient?

ANSWERS TO SELECTED NUMERICAL PROBLEMS

24. 3.03×10^{-19} J, 4.09×10^{-19} J, 4.58×10^{-19} J, 4.85×10^{-19} J

25. (a) 1.64×10^{-18} J, (c) 2.47×10^{16} Hz

32. (a) 3.97×10^{-12} m, (b) 4.74×10^{-36} m

CHAPTER 13

NUCLEAR PHYSICS

In Chapter 12, we discussed the experimental efforts of Rutherford and Bohr in establishing the planetary model of the atom. This model has some imperfections, but it is still useful in that it is successful in many applications, and its Solar-System–like structure provides us with a simple visual picture. In this chapter, we shall examine the nucleus of the atom. We shall investigate many of its features such as radioactivity, and we shall also discuss nuclear reactions. Among the types of nuclear reactions that we will investigate are fusion reactions of the type that power the Sun and therefore are responsible for our life here on Earth. If fusion reactions are ever successfully harnessed, our energy needs on this planet will be satisfied for thousands of years.

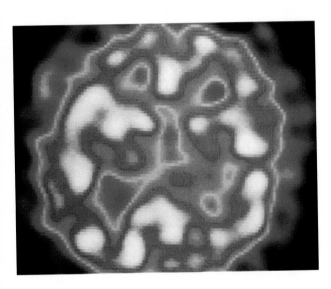

Noninvasive techniques such as magnetic resonance imaging have altered medical diagnostic procedures. This scan of a patient's brain is taken in an effort to detect signs of schizophrenia. (Courtesy Custom Medical Stock Photos)

13.1 NUCLEAR STABILITY

Why does a nucleus exist? At first thought, it seems quite improbable that a nucleus could possibly survive in nature. The reason for this assertion is that all nuclei except for hydrogen consist of positively charged protons packed closely together. As we saw in our study of electricity, charges of like sign repel one another via the Coulomb force, so why shouldn't the repulsion between protons be sufficiently great to cause the nucleus to fly apart? The reason that a nucleus can remain stable is because of a force that we have not yet discussed, called the **strong nuclear force.** Some of the characteristics of this force are illustrated in Figure 13.1. The force is an attractive force that acts between all nuclear particles. That is, as shown in Figure 13.1, the force acts between protons and protons, between protons and neutrons, and between neutrons and neutrons. This force of attraction is quite strong, much stronger than the Coulomb force of repulsion between charged particles. Thus, when protons and neutrons are assembled into a nuclear package, the

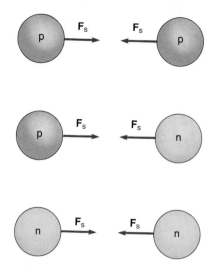

FIGURE 13.1   *The strong nuclear force* $\mathbf{F}_s$ *is short range and attractive. It attracts protons to protons, protons to neutrons, and neutrons to neutrons.*

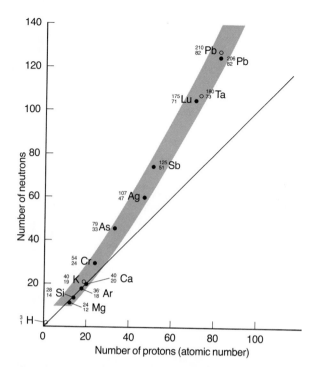

FIGURE 13.2   *The stability region, approximately represented by the shaded area. Several naturally stable nuclei are given to serve as reference points, shown by the solid circles. A few naturally occurring radioactive nuclei are indicated by open circles.*

force of attraction produced by the strong nuclear force binds the group together, overcoming the repulsive force between the protons. Other types of forces, such as the force of gravity and the Coulomb force, are noticeable in our everyday world because they are long-range forces that can act over considerable distances. The strong nuclear force is a *short-range force* that becomes negligible in magnitude for distances in excess of nuclear dimensions, that is, greater than 10^{-14} m.

If you took a field trip around our planet collecting all the various forms of stable nuclei that you could find, you would return with approximately 400 different species. In your journey, you would have found many nuclei that you would have had to reject—these are the radioactive nuclei. As we shall discuss in a later section, radioactive nuclei do not remain intact forever; instead, they spontaneously break up by emitting various types of radiation and becoming a nucleus of a different type. Figure 13.2 is a graph for the 400 or so particles.

The graph is a plot of the number of neutrons in the nucleus, along the vertical axis, versus the number of protons in the nucleus, along the horizontal axis. The plot shows that these nuclei cluster within the shaded region indicated in the figure. The graph also reveals some surprising features. First, note that most of the light, stable nuclei contain equal numbers of protons and neu-

trons. One of the most common of this type of nuclei is the $^{4}_{2}\text{He}$ nucleus, which contains two protons and two neutrons. As we see from the graph, however, the heavy stable nuclei have more neutrons than protons. We can understand why this should be so by examining the characteristics of the strong nuclear force once again. As the nuclei become heavier, there are more and more protons that must be contained. As a result, the repulsive force that all the other protons exert on one individual proton becomes great, and it gets harder and harder to keep the nucleus from flying apart. To compensate for this repulsion, more neutrons must be added to the nucleus so that the attractive strong nuclear force between neutrons and protons can maintain equilibrium. There is a limit, however, to what the strong nuclear force can do. Eventually the number of protons becomes so great that stability cannot be maintained by the addition of extra neutrons. This occurs when the number of protons reaches 83. All elements with atomic number greater than 83 do not exist as stable nuclei. There are elements with atomic numbers greater than 83, of course, but all of these substances are radioactive.

A final feature of nuclear stability needs to be pointed out that may not be obvious from the graph of Figure 13.2. It is found that isotopes seem to be most stable when there are an even number of both protons and neutrons. There are few stable nuclei with an odd number of both protons and neutrons. For example, there are 165 stable even-even nuclei and only 4 odd-odd nuclei. The reason for this occurrence is that protons and neutrons tend to group together in pairs inside the nucleus, with protons pairing up with protons and neutrons pairing up with neutrons. (Protons do not pair up with neutrons in this way.) Thus, a nucleus with even numbers of both tends to be more stable. The four odd-odd nuclei are low mass number isotopes, ^2_1H, ^6_3Li, $^{10}_5B$, and $^{14}_7N$. There are 105 stable nuclei that are even-odd nuclei. This means that there are an even number of either protons and neutrons and an odd number of the other.

CONCEPTUAL EXERCISE

(a) Use Figure 13.2 to explain why the isotope $^{142}_{71}Lu$ could not be stable if it existed.

Answer For nuclei with higher atomic numbers, the number of neutrons must greatly exceed the number of protons, for stability to occur. These nuclei would fall well outside the stability range indicated in Figure 13.2.

(b) Explain why uranium-238 cannot be stable.

Answer Uranium has an atomic number of 92, and there are no stable nuclei with atomic numbers greater than 83.

13.2 RADIOACTIVITY

In a sense, the study of nuclear physics actually began in 1896, when Henri Becquerel accidentally discovered radioactivity. He was working with a uranium-bearing ore when he discovered that a photographic plate in the laboratory had become exposed. In fact, he found that even if he wrapped the film in black paper so that no light could reach it, the film would still be exposed by the presence of the ore. After some careful experimentation, Becquerel found that it was the uranium that was producing the effect, and he proposed the idea that the uranium was spontaneously emitting some kind of radiation. *This emission of radiation from an element is now called* **radioactivity.**

Several significant facts emerged from Becquerel's studies. The pure uranium compounds that were extracted from the mineral were less radioactive than the crude mineral itself. This differ-

FOCUS ON
RADON POLLUTION

The Curies were the first to notice that the air in contact with radium compounds becomes radioactive. It was shown that this radioactivity came from the radium itself, and the product was therefore called "radium emanation." Rutherford and Soddy succeeded in condensing this "emanation" to a liquid and thus confirmed the fact that it is a real substance—the inert, gaseous element now called radon, Rn.

It is now known that the air in uranium mines is radioactive because of the presence of radon gas.

The mines must therefore be well ventilated to help protect the miners. However, the fear of radon pollution now extends to our homes. Many types of rocks, soils, as well as some brick and concrete, contain very small quantities of radium. Some of the resulting emissions of radon find their way into homes and other buildings. The most serious problems arise from leakage of radon from the ground into the house. A practical remedy is to exhaust the air through a pipe just above the underlying soil or gravel directly to the outdoors by means of a small fan or blower.

FOCUS ON
THE GEIGER COUNTER

Detection of radiation by allowing it to expose photographic film was one of the first methods for establishing the presence of radioactive decay. Most of the newer methods depend on the fact that as radiation passes through a gas (or any material), it ionizes some of the atoms of the gas. This means that it knocks electrons from their orbits about the nucleus and frees them from the atom. For example, suppose a beta particle, in passing through a gas, comes close to one of the electrons in an atom of the gas. Since both electrons have a negative charge, they repel one another, and the fast-moving beta may knock the electron out of its orbit. Thus, in its trail, a beta particle leaves a wake of free electrons and positively charged atoms, called ions. The detection of this trail is the objective of radiation detectors such as the **Geiger counter.**

A Geiger tube, shown in the diagram, consists of a metal tube with a wire down the center. One end of the tube is covered with a thin shield so that radioactive particles can pass into the tube easily. A high voltage is maintained such that the wire is held about 500 V higher than the grounded case of the tube. When a beta particle passes through argon gas in the tube, it leaves electrons and positive ions along its path. These charged particles respond to the pull of the high voltage, with the electrons moving toward the wire and positive ions toward the cylindrical housing. As each electron heads for the center, it gains more and more speed. During the trip, it encounters electrons in orbit about other gas atoms and knocks them loose. These electrons, in turn, free other electrons, and an avalanche is produced heading toward the center wire. When these electrons are collected by the wire, a burst of current is created in external circuits, and this current drives a counter to "count" the number of incoming beta particles. It is by this way that we are able to obtain half-life data such as those discussed in this chapter.

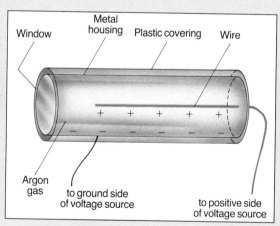

Schematic of a Geiger tube.

A hand-held Geiger counter. (Courtesy Bill Schulz)

ence implied that there were other more highly radioactive substances mixed with the uranium. A series of careful, tedious separations carried out by Marie and Pierre Curie (wife and husband) re-

sulted in the discovery of new radioactive elements, the most important of which was radium.

The Curies also learned that the radioactivity of substances is associated with the elements, not

FOCUS ON
SMOKE DETECTORS

Most simple smoke detectors used in the home use radioactive sources to perform their task. To see how they work, consider the figure. A battery is connected between two plates inside the chamber of the detector, and also in the circuit is a sensitive current detector and an alarm. A radioactive source near the two plates ionizes the air around it, and the charged particles created are drawn to the plates by the voltage of the battery. This sets up a small but detectable current in the external circuit. As long as this current flows, the alarm is deactivated. If there is a fire in the house, the smoke created drifts into the chamber of the smoke detector, and the ions become attached to the particles present in the smoke. These heavier particles

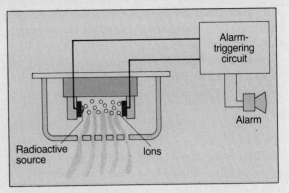

An ion smoke detector.

do not drift as readily between the plates as do the lighter ions, and thus the current in the external circuit drops. The external circuit detects this decrease in the current and sets off the alarm.

with their compounds. Thus, a gram of radium has the same radioactivity in the form of a pure metal, Ra, as in the form of any of its compounds, such as radium carbonate, $RaCO_3$. In any chemical bonding process such as this, it is the electrons of the atoms that are primarily responsible for holding the various atoms together. Thus, because the electrons did not seem to be playing a role in the radioactive decay of radium, scientists were led to the conclusion that

radioactivity is associated with the atomic nucleus.

When a naturally radioactive source such as uranium is placed at the bottom of a long, narrow hole in a block of lead, most of the radiation is absorbed by the lead, but a thin beam comes out of the hole. To investigate the nature of the radiation emerging from the hole, charged plates were placed near the beam, as shown in Figure 13.3. It was found that some of the radiation is bent toward the positive plate, some is bent toward the negative plate, and some travels in a straight-line path, unaffected by the plates. As shown in Figure 13.3,

FIGURE 13.3 *The behavior of alpha, beta, and gamma emissions in an electric field.*

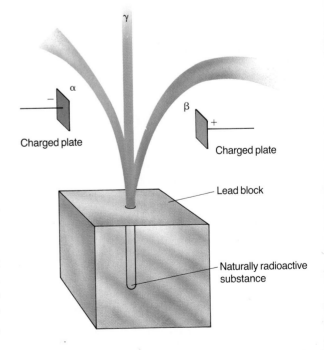

these three different types of emission were initially given the names alpha (α) rays, beta (β) rays, and gamma (γ) rays.

The alpha beam is drawn toward the negative plate. This observation shows that the alpha rays consist of positively charged particles. Eventually, it was found that these particles would interact with electrons to form neutral helium gas. Thus, the alpha ray, or **alpha particle,** is the nucleus of a helium atom. Symbolically, we represent this particle as $^4_2$He.

The beta rays are attracted toward the positive plate. Moreover, the beam of betas is more sharply bent than is the beam of alpha particles. These observations led to the conclusion that beta rays consist of negative particles that are light. Further evidence shows that these particles are electrons. Symbolically we represent electrons as $^0_{-1}$e. This notation is indicative of the fact that the electron, e, has a negative charge equal to that of the proton (as shown by the subscript -1) and that its mass is so small relative to that of nuclear particles that it is negligible (as shown by the 0 superscript).

The gamma rays are not deflected at all by the charged plates; therefore, they have no charge. This beam could be composed of neutral particles, such as neutrons. However, some experimentation with the penetrating power of these various particles led to other conclusions. It was found that the alpha particles would be stopped by a sheet of paper or by a few inches of air. Beta particles would penetrate a thin sheet of steel. But the gammas could pass through several feet of concrete before being absorbed. The conclusion was drawn that gamma rays are a form of electromagnetic radiation, even higher in frequency and energy than X-rays.

13.3 THE DECAY PROCESS

As we noted in the last section, radioactive emissions are of three types, alpha particles, beta particles, and gamma rays. Let us examine in more detail the three decay processes that lead to these emissions.

Alpha Decay

Consider what happens to the atomic nucleus of radioactive elements. If a particle is emitted by the nucleus, the nucleus must break apart and lose its identity in the process. Thus, the nucleus of one atom is converted into the nucleus of another. This fact was discovered by Ernest Rutherford and Frederick Soddy in 1902. In particular, they were working with radium in an attempt to find what had happened to it chemically after an alpha particle emission occurred. Their chemical analysis showed that a lighter element, radon, was appearing in their initially pure sample of radium. We can write the nuclear process occurring as follows:

$$\text{radium} \longrightarrow \text{radon} + \text{alpha particle}$$

It is of some historical interest to note that the results of their investigation produced a considerable amount of consternation in the two experimenters. This concern arose because **alchemy** had fallen into ill-repute by the 1900s. Alchemy is often called the chemistry of the Middle Ages. One of its fundamental assumptions was that ordinary metals could be *transmuted*, or changed, into gold. This was of obvious interest to kings and the well-to-do; therefore, much time was spent on scientific research devoted to determining how to make the desired transformation. In fact, Isaac Newton devoted much of his time and attention to alchemy before he gained fame via the law of universal gravitation and his laws of motion. The alchemists were never successful because all the processes used were those in which chemicals were mixed together in various combinations. In such chemical reactions, the nucleus remains intact; as a result, a new element is not formed. Rutherford and Soddy, however, had now found that one element, radium, was being transmuted into another, radon. According to a popular account, Soddy turned to his colleague and blurted, "Rutherford, this is transmutation!" Rutherford rejoined, "For Mike's sake, Soddy, don't call it transmutation. They'll have our heads off as alchemists." Rutherford and Soddy were careful to use the term "transformation" rather than "transmutation" in describing their results.

In any radioactive disintegration, the element that decomposes is called the **parent,** and the new element is called the **daughter.** Thus, in the decay of radium to radon, the parent is radium and the daughter is radon.

We can represent the radioactive decay of radium into radon in symbolic form by the notation

$$^{226}_{88}\text{Ra} \longrightarrow {}^{222}_{86}\text{Rn} + {}^4_2\text{He} \qquad \textbf{(13.1)}$$

This is a shorthand way of noting the changes that occur. The arrow indicates that the $^{226}_{88}$Ra nucleus decays into a $^{222}_{86}$Rn and an alpha particle, $^{4}_{2}$He. Note the following about the decay process: (1) The total amount of charge (the atomic numbers) on each side of the arrow (before and after) is the same ($88 = 86 + 2$). (2) The total mass number on each side of the arrow is the same ($226 = 222 + 4$). Regardless of the decay process, we shall find that atomic number and mass number are conserved.

EXAMPLE 13.1   TIME FOR A CHANGE
Uranium, $^{238}_{92}$U, decays by alpha emission. What is the daughter element formed?

Solution The decay can be written symbolically as

$$^{238}_{92}U \longrightarrow X + ^{4}_{2}He$$

We have used the symbol X to represent the unknown daughter element. To identify the daughter, we first note that mass numbers must add up on each side of the arrow. Thus, the mass number of X must be equal to 234 ($238 = 234 + 4$). The atomic number of X is found from the fact that the atomic numbers also must balance in the equation. The atomic number is thus found to be 90 ($92 = 90 + 2$). Therefore, our reaction is

$$^{238}_{92}U \longrightarrow ^{234}_{90}X + ^{4}_{2}He$$

The periodic table on the inside back cover shows that the nucleus with atomic number 90 is thorium, Th. Thus, the process may finally be represented as

$$^{238}_{92}U \longrightarrow ^{234}_{90}Th + ^{4}_{2}He$$

Beta Decay

The basic process in beta decay is one in which a parent nucleus emits an electron and a daughter nucleus is formed. The basic details discussed in alpha decay also apply here. That is, mass numbers and atomic numbers are conserved. A typical beta decay is that of an isotope of iodine, I, decaying into xenon, Xe, according to the following equation:

$$^{131}_{53}I \longrightarrow ^{131}_{54}Xe + ^{0}_{-1}e \tag{13.2}$$

This isotope is often used as a radioactive tracer to detect the uptake of iodine by the thyroid gland.

At first thought, the beta decay process seems to contradict some of the facts that we have discovered about the nucleus. We have stated many times in the text that the nucleus is composed of protons and neutrons, yet in beta decay, an electron is emitted from the nucleus. From where does it come? The answer is that the electron is created inside the nucleus in a process by which a neutron is changed to a proton. Symbolically, this is

$$^{1}_{0}n \longrightarrow ^{1}_{1}p + ^{0}_{-1}e \tag{13.3}$$

Based on this transformation of a neutron into a proton, let us re-examine the beta decay process represented by Eq. 13.2. We note that the number of nuclear particles is the same before and after the reaction, 131. One of these nuclear particles, however, has been changed from a neutron into a proton. Thus, the atomic number should be one greater after the process than before, and it is (53 before, 54 after).

EXAMPLE 13.2   THE CASE OF THE DISAPPEARING SULFUR
Sulfur-37 decays by beta emission. Find the daughter nucleus.

Solution The decay process may be written as

$$^{37}_{16}S \longrightarrow X + ^{0}_{-1}e$$

where X is the unknown decay product. Balancing mass numbers, we find that X must have a mass number of 37 ($37 = 37 + 0$). It also must have an atomic number of 17 ($16 = 17 - 1$). Thus, the daughter has the representation $^{37}_{17}X$. In the periodic table, we find that the element with an atomic number of 17 is chlorine, Cl. Thus, the complete decay process is

$$^{37}_{16}S \longrightarrow ^{37}_{17}Cl + ^{0}_{-1}e$$

Gamma Decay

The emission of gamma rays by a nucleus is similar to the emission of light by an atom. In an atom, photons are emitted when an electron in an excited state returns to the ground state. In its downward fall, the electron gets rid of its excess energy by emitting photons, which are usually in the infrared, visible, or ultraviolet portion of the electromagnetic spectrum. A nucleus also may have an excess of energy following a nuclear event, and it de-excites, or releases this pent-up energy, by emitting a photon. The amount of energy carried away by the photon, however, is generally considerably greater than that released in an atomic process. Thus, the frequencies associated with the emitted photons are much higher, and they fall into the region of the electromagnetic spectrum called gamma rays.

A typical nuclear event that may lead to the emission of a gamma ray is that of alpha or beta decay. Often, following these types of decay, the nucleus is left in an excited state, and the alpha or beta decay is then shortly followed by the emission of a gamma ray. As an example, consider the beta decay of boron to carbon represented as

$$^{12}_{5}B \longrightarrow {}^{12}_{6}C^* + {}^{0}_{-1}e$$

The asterisk following the symbol for carbon is used to indicate that the carbon nucleus is in an excited state after the decay. The carbon nucleus rids itself of the excess energy by a gamma ray emission as follows

$$^{12}_{6}C^* \longrightarrow {}^{12}_{6}C + \gamma$$

Note that gamma ray emission does not change the parent nucleus to a different daughter nucleus. The mass number and atomic number are the same before and after.

13.4 HALF-LIFE

Radioactive nuclei can decay by alpha, beta, or gamma emission, each of which produces drastically different effects on the original nucleus. For example, alpha decay reduces the mass number of the parent by four units and its atomic number by two units. Beta decay does not affect the mass number, but it increases the atomic number by one unit. Thus, the processes produce radically different results. But there is one feature of all types of decay that is common to the various forms of decay processes. *After a certain interval of time, half of the original number of radioactive nuclei will have decayed. This time is called the* **half-life**. Half-lives of radioac-

tive substances vary from long times, such as 4.5 billion years for uranium-238, to 10^{-21} s for lithium-5.

EXAMPLE 13.3 COUNTING CARBON NUCLEI
A continual rain of particles, called cosmic rays, falls on the Earth each day. These particles come to us from nearby sources such as the Sun and from more distant heavenly objects. Cosmic ray activity in the upper atmosphere produces carbon-14 nuclei, which decay by beta emission with a half-life of 5730 years. If you start with a sample of 1 million carbon-14 nuclei, how many will still be around in 22,920 years?

Solution In 5730 years, half of the original 1 million carbon-14 nuclei will have disappeared through beta emission. Thus, you will have 500,000 remaining. In another 5730 years (total elapsed time of 11,460 years), half of this number will have decayed, leaving you with 250,000 carbon-14 nuclei. In another half-life (total time = 17,190 years), the sample contains 125,000 carbons. Finally, in another half-life (total time = 22,920 years), there are 62,500 remaining. (In Section 13.8, we shall discuss an application of radioactive decay that enables us to find the age of organic relics by use of carbon-14.)

Please be aware that the circumstances outlined here are ideal and unrealistic. Radioactive decay is governed by the laws of probability and is an averaging process. One million initial atoms seems like a lot, but in a real situation, there would be many more than this in a given radioactive sample. Thus, for this case, if we were to be able to count the number remaining after one half-life, we would not get exactly 500,000. As the number in our original sample increases, however, the probability of getting extremely close to exactly one-half the original number remaining after one half-life increases dramatically.

13.5 NEUTRINOS

Let us briefly return to our discussion of beta decay because there are some unusual features of it that initially brought into question one of the most fundamental laws of nature, the conservation of en-

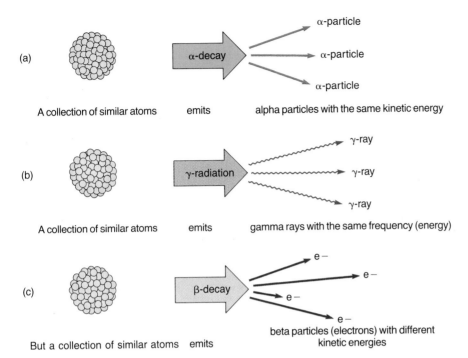

FIGURE 13.4 *All the alpha particles from the nuclei of a given radioactive isotope fly off with the same kinetic energy. Similarly, gamma rays from a particular type of decay all carry the same energy (frequency). However, beta particles from a specific decay have varying amounts of energy ranging from almost zero up to some maximum amount. This observation led to the discovery of neutrinos.*

ergy. The resolution of these difficulties led to the discovery of one of nature's most unusual creations, the **neutrino.**

In our study of the Bohr atom, electrons were shown to occupy various energy levels. These energy levels are easy to visualize because an electron in a higher energy level spends most of its time relatively far away from the nucleus, and an electron in a lower level is likely to be closer to the nucleus. Thus, the terms "falling from a higher energy level to a lower one" and "rising from a low one to a high one" bring obvious terrestrial analogies to mind. There are discrete energy levels in a nucleus as well. These are a little harder to visualize, but they are real, nevertheless. To demonstrate that this is true, consider the process of alpha emission. Physicists observed that all the alpha particles emitted by a specific type of radioactive isotope travel outward from the nuclei with the same kinetic energy. This is illustrated in Figure 13.4a, where we see that all the alphas emitted carry the same amount of energy away from the nucleus as the nucleus de-excites. This also occurs in gamma emissions from nuclei. That is, when an excited nucleus de-excites via gamma emission, the gamma emitted has a specific amount of energy, as shown in Figure 13.4b.

Radioactive decay via beta emission, however, is different, as shown in Figure 13.4c. A specific type of nucleus decaying via the emission of beta particles was found to release the betas with a wide range of kinetic energies. For example, in one particular beta decay event, the fastest betas might carry away five units of energy, but most of the betas would carry less than this maximum amount. There would be some with four, some with three, down the line, and all fractional energies in between would also be seen. Thus, there would be some with 3.13, some with 3.14, and so forth. Theoretically the fastest ones would carry the energy equal to the total energy released when the nucleus de-excited, but what about the slower ones? Was energy somehow being lost in the decay process? Some prominent physicists speculated that the law of conservation of energy might not be valid for beta decay. To compound the difficulties, it was also found that momentum was apparently not being conserved in beta decay processes.

In 1930, W. Pauli proposed that a third particle must be present to carry away the "missing energy" and to conserve momentum. This particle was later given the name **neutrino** (little neutral one) by Enrico Fermi. Thus, in all beta decay processes, the nucleus emits a neutrino (symbol v) in addition to the electron. As a result, the decay of platinum to gold via beta emission should be written to include the presence of the neutrino as

$$^{197}_{78}\text{Pt} \longrightarrow \ ^{197}_{79}\text{Au} + \ ^{0}_{-1}\text{e} + v$$

The neutrino is a curious and elusive particle. It was predicted to have the following properties.

1. It is electrically neutral.
2. It has an extremely low mass. In fact, its rest mass may be zero, although theories suggest that it does have some mass.
3. It interacts very weakly with other particles of matter.

When one considers the ramifications of these properties, it is not surprising that the neutrino was difficult to detect. Because it is neutral, it could not be deflected by electric or magnetic fields. Its small mass as well as the fact that it interacts weakly with matter means that it could travel long distances without encountering another particle and being captured.

To appreciate the order of scale of the neutrino and of the structure of the atom, let us return to our analogy of shooting an arrow into a cloud in our fantasy world described in Chapter 12. Imagine that there were a forest inside the cloud. If you shot an arrow blindly into this forest, the probability that the arrow would travel a given distance through the forest would depend on (1) how closely the trees were packed together, (2) the diameter of the trees, and (3) the diameter of the arrow. To take this analogy to the atomic scale, consider atoms to be the trees and a neutrino to be the arrow. The question is: If you shoot a neutrino into a block of lead, how thick a piece of lead would you need to have a 50 percent chance of stopping, or trapping, a neutrino? The answer isn't measured in millimeters, centimeters, meters, or even kilometers. It is measured in light-years (Fig. 13.5). If it were possible to construct a tube of solid lead extending all the way from the Earth to the nearest star, 4 light-years away, and a neutrino were shot into the tube, chances are that it would emerge at the other end without having hit *anything at all* along the way. As elusive as is the neutrino, it was finally detected experimentally in 1950. Theorists believe that neutrinos are all around us all the time. For example, nuclear processes occurring inside the Sun produce neutrinos in abundance that es-

FIGURE 13.5   *If a solid lead tube extended from the Earth to the closest star, 4 light-years away, and a neutrino flew off into the tube, there is a high probability that it would pass through to the other end without hitting anything along the way.*

cape from the Sun and go pouring out into space. There may be many more neutrinos in the Universe than there are protons and neutrons. These neutrinos are speeding through our Solar System, our planet, and even through you as you read this, but they seldom touch anything along the way.

13.6 NUCLEAR REACTIONS

In 1919, Ernest Rutherford allowed alpha particles from a radioactive source to pass through nitrogen gas. He found that some particles were emitted from the gas that were more penetrating than the alpha particles. Finally, by passing these new particles between charged plates, he was able to show by the way that they were deflected that they were protons. Based on his observations, Rutherford concluded that a nuclear reaction had occurred that could be represented as follows:

$$^{4}_{2}\text{He} + \ ^{14}_{7}\text{N} \longrightarrow \ ^{17}_{8}\text{O} + \ ^{1}_{1}\text{H}$$

This reaction equation is used to indicate that an alpha particle bullet, ^4_2He, strikes a nitrogen nucleus, $^{14}_7\text{N}$, and an interaction occurs such that the debris consists of a proton, ^1_1H, and an oxygen nucleus, $^{17}_8\text{O}$. This type of nuclear reaction is called a **bombardment reaction.** Note that mass numbers and atomic numbers add up on each side of the arrow, just as they did for radioactive decay processes. Since the time of Rutherford, thousands of nuclear reactions have been observed, particularly with the advent of particle accelerators in the 1930s. The study of nuclear reactions has been one of the most profitable techniques used by experimentalists in physics to probe the secrets of the nucleus. We shall examine many of these reactions in the remainder of this chapter.

EXAMPLE 13.4 THE ELUSIVE NEUTRON

(a) In 1932, a nuclear reaction was produced by James Chadwick that is of significant historical importance. In the experiment, Chadwick bombarded ^9_4Be with alpha particles. He found that there were two reaction products, one of which was $^{12}_6\text{C}$; what was the other?

Solution The reaction process may be represented as follows:

$$^4_2\text{He} + ^9_4\text{Be} \longrightarrow ^{12}_6\text{C} + X$$

Balancing mass numbers and atomic numbers, we see that X has the representation ^1_0X. The mass number of 1 indicates that the unknown particle has a mass approximately equal to that of a proton, but the atomic number 0 indicates that the particle has no charge. The unknown particle is a neutron, ^1_0n.

This was the first experiment to provide positive proof of the existence of neutrons. The experiment had the additional advantage that it added another type of "bullet" to those that could be used to bombard nuclei. Alpha particles used as bullets have the disadvantage of being positively charged. This means that when they are used as projectiles to cause reactions with heavy nuclei (with a lot of protons), the alphas must have a large energy to overcome the repulsive Coulomb force exerted on them by the nucleus. The neutral neutron, produced in reactions such as the one just described, does not suffer this disadvantage.

(b) A neutron-induced reaction that would have been of interest to the alchemists of the Middle Ages is shown. What is the reaction product X?

$$^1_0\text{n} + ^{198}_{80}\text{Hg} \longrightarrow X + ^2_1\text{H}$$

Solution Balancing mass numbers and atomic numbers, we see that X is represented as $^{197}_{79}X$. A search through the periodic table of elements on the inside back cover reveals that this particle is gold (symbol Au). Thus, the complete reaction is

$$^1_0\text{n} + ^{198}_{80}\text{Hg} \longrightarrow ^{197}_{79}\text{Au} + ^2_1\text{H}$$

The alchemists' dream has come true with this reaction, but be aware that the expense involved in producing gold in this way is prohibitive.

13.7 ARTIFICIALLY PRODUCED NUCLEI

In 1934, 15 years after Rutherford produced the first nuclear reaction, Irene and Frederic Joliot-Curie, Mme. Curie's daughter and son-in-law, converted boron to nitrogen-13, which is radioactive. This was the first *artificially produced* radioisotope. The reaction is

$$^4_2\text{He} + ^{10}_5\text{B} \longrightarrow ^{13}_7\text{N} + ^1_0\text{n}$$

Uranium is the element with the largest atomic number that exists in nature in any appreciable amount. In 1940, researchers became able to extend the list of known elements beyond uranium. The first man-made element was produced by bombarding uranium with neutrons. The process that ensued is described by the following:

$$^1_0\text{n} + ^{238}_{92}\text{U} \longrightarrow ^{239}_{92}\text{U} + \gamma$$

The uranium-239 produced is radioactive and decays via beta emission as

$$^{239}_{92}\text{U} \longrightarrow ^{239}_{93}\text{Np} + ^{\ \ 0}_{-1}\text{e}$$

Thus was produced neptunium, Np, the first man-made element. The process did not end there, however, because neptunium is also radioactive and decays by beta emission as

$$^{239}_{93}\text{Np} \longrightarrow ^{239}_{94}\text{Pu} + ^{\ \ 0}_{-1}\text{e}$$

Following this reaction, the number of man-made elements beyond uranium had grown to two. Plutonium, Pu, proved to be of great importance in the history of mankind because, as we shall see, it played a role for the good through its use in nuclear power plants and for the bad in the production of nuclear weapons.

The reactions indicated here were the first to be performed in a laboratory setting to produce elements with an atomic number above that of uranium. Neutrons arrive from outer space as a part of the cosmic ray barrage. Infrequently, one of these neutrons strikes a uranium-238 nucleus and initiates the chain of events leading to the production of neptunium and plutonium. Only trace amounts of these elements are found in nature, however, and the development of devices that use these elements as integral parts would have been impossible without the production by man-made avenues.

13.8 CARBON DATING

Much important information concerning the time line of life on this planet has been discovered because of a nuclear reaction that occurs high in the Earth's atmosphere. Let us first look at the sequence of events that occurs; then we shall examine how these events have been used to unravel many of the secrets about our past. The process begins with high-energy neutrons, which constitute a part of the cosmic ray bombardment falling on the Earth from outer space. The reaction that occurs when one of these neutrons strikes a nitrogen nucleus is

$$\,^{1}_{0}n + \,^{14}_{7}N \longrightarrow \,^{14}_{6}C + \,^{1}_{1}H$$

The end-products of the reaction are a proton and a radioactive isotope of carbon, carbon-14. Carbon-14 is radioactive and decays by beta emission as

$$\,^{14}_{6}C \longrightarrow \,^{14}_{7}N + \,^{0}_{-1}e$$

As far as scientists have been able to determine, carbon-14 has been produced at a fairly uniform rate in the upper atmosphere for at least the past 50,000 years, and just as regularly, this carbon has been decaying with a half-life of 5730 years to the stable isotope nitrogen-14. The rate of production of carbon-14 is such that there is about one nucleus of carbon-14 for every 10^{12} nuclei of the more common carbon-12 in our environment. Now, let us follow the processes that occur in nature that enable us to use carbon-14 as a clock to date the age of organic relics.

Carbon is a natural constituent of all organic materials; as a result, plants absorb carbon from the air and the earth. These plants then are eaten by animals, and the carbon becomes a part of their bones and flesh. The end result is that all organic materials, when alive, have the same ratio of carbon-14 to carbon-12 as that found in nature. When an animal dies, the stable carbon-12 stays around, but the radioactive carbon-14 begins to disappear by beta emission. Thus, the *ratio* of carbon-14 to carbon-12 in the remains begins to change. To understand how this fact enables us to date organic relics, consider the following scenario. At some time in the past, a hunter kills a deer and buries its bones near his campfire. An archaeologist then discovers the deer bones in the present day and analyzes them to determine the amount of carbon-14 still present. If he finds only half as much carbon-14 as that which would be found in a living deer, he knows that the deer died 5730 years ago.

There are limitations on how far back into the past carbon dating can take us. After about 30,000 years, the amount of carbon-14 in a sample has decayed to the point that reliable estimates of its age cannot be determined. One interesting application of carbon dating concerns the Dead Sea Scrolls. These were a group of manuscripts found in 1947 by a young boy in a cave near the ancient city of Qumran. These scrolls contained most of the books of the Old Testament, including a copy of the Book of Isaiah. Because the authenticity of some of the passages in this Book were disputed by scholars, it became important to determine the age of the manuscripts to see if certain material was added by later writers. The manuscripts had been wrapped in linen, and this substance was carbon dated. The age of the fragments was found to be about 1950 years. This age did not satisfactorily resolve the controversy, but it did involve a worthwhile application of the dating process.

SUMMARY

Nuclear stability is caused by a short-range force of attraction called the **strong nuclear force.** This force acts between protons and protons, neutrons and neutrons, and protons and neutrons.

Some isotopes are unstable, giving off particles and radiation. Such decomposition, which is called radioactive decay or **radioactivity**, may occur in a series of steps, ending when a stable isotope is produced. All radioactivity releases energy. The particles released in radioactive decay are the alpha particle, the nucleus of the helium atom, $_2^4He$; the beta particle, an electron, $_{-1}^0e$; or a

gamma ray, a highly energetic photon of electromagnetic radiation. In all beta-decay processes, a **neutrino** is also released. The **half-life** of a radioactive isotope is the time required for half of the nuclei in a sample to decompose.

Nuclear reactions are produced when high-energy bullets are directed against a target material. In the reaction, new products are formed. In both decay and bombardment reactions, both charge and mass number are conserved. **Carbon dating** is a process for finding the age of organic relics.

KEY WORDS

Strong nuclear force	Gamma ray	Half-life	Nuclear reactions
Radioactivity	Daughter	Neutrino	Artificially produced nuclei
Alpha particle	Parent	Isotope	Carbon dating
Beta particle			

PROBLEMS AND CONCEPTUAL QUESTIONS

Problems requiring numerical work are identified with a blue number.

Nuclear stability

1. Which of the following nuclei would you expect to be stable? (a) $_{14}^{28}Si$, (b) $_{90}^{232}Th$, (c) $_3^6Li$, (d) $_3^9Li$

2. Why was it necessary to postulate the existence of a strong nuclear force?

3. Explain why gravitational forces were not used in the explanation of how protons are held together in a nucleus. Are protons attracted to each other by gravitation?

4. Why do heavier elements require more neutrons to remain stable?

Radioactivity

5. Which of the three types of radiation, alpha, beta, or gamma, would not be deflected by the Earth's magnetic field? Defend your answer.

6. How could you prove that radioactivity cannot be affected by a chemical reaction?

7. Explain the significance of the fact that the radioactivity of an element does not depend on its chemical bonding.

8. (a) A Geiger counter registered 256 counts per second (cps) near a sample of polonium-210; 276 days later, the counter registers 64 cps. What is the half-life of polonium-210? What will the counter register after another 276 days?

(b) Polonium-210 decays in one step to lead-206, which is not radioactive. If you were asked to give a rough estimate of the length of storage time needed to reduce the radioactivity of polonium-210 to a safe level, would you say it is a matter of months, years, decades, or centuries? Would you be concerned about any radioactive progeny that might be produced?

9. Iodine-131 is a radioactive nuclear waste product with a half-life of 8 days. How long would it take for 2000 mg of iodine-131 to decay to 125 mg? Would it be correct to say that iodine-131 is no environmental hazard because its half-life is so short? Defend your answer.

10. 24 mg of tritium ($_1^3H$) decays to 1.5 mg in 49 years. What is the half-life of tritium?

11. Strontium-90, produced in nuclear explosions and present in radioactive fallout, has a half-life of 29 years. If the activity of a "bomb-test" Sr-90 sample collected in 1957 was about 80 disintegrations per second (dps), how many years would it take to reduce the activity to the natural background count of 2.5 dps? (*Note:* The dps count is registered as clicks on a Geiger counter and is proportional to the amount of Sr-90 present in the sample.) In what year would that count be reached?

12. All the uranium from a sample of uranium ore is extracted and purified. The uranium is less ra-

dioactive than the ore from which it came. Explain.

13. Radon-222 has a half-life of 3.8 days. Radon gas is emitted from some rocks and soils. Since the half-life of radon is short, why is there any left on Earth?

14. Complete the following radioactive decay equations.

(a) $^{238}_{92}U \longrightarrow ? + ^{4}_{2}He$

(b) $? \longrightarrow ^{14}_{7}N + ^{0}_{-1}e$

15. Complete the following radioactive decay equations.

(a) $^{12}_{5}Bi \longrightarrow ? + ^{0}_{-1}e$

(b) $^{144}_{60}Nd \longrightarrow ? + ^{4}_{2}He$

16. A moving particle with a large charge is deflected more by charged plates than if it had a lesser charge. Why, then, are alpha particles bent less than beta particles?

17. If photographic film is kept in a cardboard box, alpha particles produced outside the box cannot expose the film, but beta particles can. Why?

Neutrinos

18. The mass-to-charge ratio of an electron can be determined by measuring its deflection in a magnetic field. Can the mass-to-charge ratio of neutrons and neutrinos be measured in the same way? Why or why not?

19. Why was it necessary to postulate the existence of the neutrino?

20. Why is it unlikely that neutrinos would be responsible for genetic mutations?

21. Both electrons and neutrinos are thought to be point masses. Why is a neutrino so much harder to capture than an electron?

22. Do you think it would be possible for a neutrino to pass through the nucleus of an atom? Why or why not? Would a positively charged particle the size of a neutrino pass equally well through a nucleus? Explain.

Nuclear reactions and carbon dating

23. Complete the following nuclear reactions.

(a) $^{29}_{13}Al + ^{4}_{2}He \longrightarrow ? + ^{1}_{0}n$

(b) $^{95}_{42}Mo + ^{1}_{0}n \longrightarrow ? + ^{1}_{1}H$

24. An archaeologist unearths a beautiful copper statuette. Can carbon dating be used to find its age? Why or why not?

25. Complete the following nuclear reactions.

(a) $^{7}_{3}Li + ^{1}_{1}H \longrightarrow ? + ^{4}_{2}He$

(b) $^{27}_{13}Al + ^{4}_{2}He \longrightarrow ? + ^{30}_{15}P$

26. Why is a neutron a more effective bullet for penetrating a nucleus than an alpha particle? How would an alpha particle compare to a proton bullet?

27. $^{10}_{5}B$ is struck by an alpha particle. A proton and a product nucleus are released. What is the product nucleus?

28. Oxygen-18 is struck by a proton, and fluorine-18 and another particle are produced. What is the other particle?

29. Carbon dating relies on the assumption that the production of carbon-14 has been essentially constant over the last 20,000 years. If it should be determined that the production were considerably higher 20,000 years ago than it is now, would that increase the predicted life of relics or decrease it?

30. Why is the technique of carbon dating not accurate beyond about 30,000 years?

ANSWERS TO SELECTED NUMERICAL PROBLEMS

8. (a) 138 days; 16 cps

9. 32 days

10. 12.3 years

14. (a) $^{234}_{90}Th$, (b) $^{14}_{6}C$

15. (a) $^{12}_{6}C$, (b) $^{140}_{58}Ce$

23. (a) $^{32}_{15}P$, (b) $^{95}_{41}Nb$

25. (a) $^{4}_{2}He$, (b) $^{1}_{0}n$

27. $^{13}_{6}C$

28. $^{1}_{0}n$

CHAPTER 14

RELATIVITY AND ELEMENTARY PARTICLES

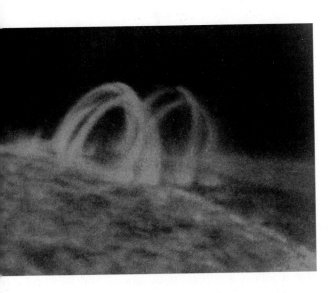

Advances in nuclear fusion research have increased the hope that someday the fusion reactions that power the Sun will provide us with almost limitless energy here on Earth. (Courtesy NASA)

Occasionally in our study of physics, we have encountered a concept that seems to violate common sense. The dual natures of both light and particles are two examples. Common sense says that light should be either a wave or a particle but not both; yet that is the way it is. Likewise, an electron should always behave as either a particle or a wave, but it does not. These are modern examples of violations of how the world looks to us, but there are also historical precedents. In the early history of astronomy, it seemed obvious to the average individual that the Sun, planets, and all other heavenly bodies revolved around the Earth. This idea was promoted by Aristotle and others to the point that it became a part of religious doctrine. In the sixteenth and seventeenth centuries, however, these ideas about the place of humans and the Earth began to change. Even though it looks like the Sun revolves around the Earth from our vantage point, Kepler, Copernicus, Galileo, and others demonstrated in quite convincing fashion that common sense had been violated. We were relegated to the position of an obscure planet, far from the center of our galaxy, revolving around an ordinary star.

Thus, it is not uncommon in the history of mankind to find that "what seems to be" is not always true. Perhaps no other concept in the field of science demonstrates this in a more profound way than does the theory of relativity, which we shall examine briefly in this chapter. It seems that there should be certain absolutes in the world around us. Hefting a rock indicates that it has mass, and common sense tells us that nothing is likely to change the mass of that rock, especially not something as mundane as throwing it at a high speed. Likewise, the length of a piece of wood or the periodic ticking of a clock should also be sacrosanct and not subject to change just because we cause these objects to move at a high rate of speed. Yet, according to the theory of relativity, all of these things *do* change as their speed changes.

The man responsible for this change in our viewpoint of the world around us is Albert Einstein. Einstein is an unlikely personage to have wrought such changes. He was born in Germany and entered grade school in Munich, where he met with little success. At first, it was feared that he was retarded; in fact, a teacher once told him quite forthrightly that he would never amount to anything. Following his graduation from a school in Switzerland, he had trouble finding a job, but through the intervention of some friends, he found employment in a patent office in Bern. The work was undemanding, and he was able to spend several hours each day "at play" with such scientific problems as that of the photoelectric effect. His theory, which resolved the difficulties associated with this scientific dilemma and later won him a Nobel Prize, has been discussed earlier in this book. This certainly would have been sufficient to have gained him immortality in physics, but the full flower of his genius became known with his special theory of relativity. It works, it is true, and it has had consequences of immense proportions in our world, but do not expect it to be within the realm of common sense.

14.1 RELATIVITY BEFORE EINSTEIN

When you hear the word "relativity" used, you probably think of Einstein and the theory of relativity, but relative motion was of interest to scientists even as early as the days of Newton. In fact, one of the questions that these early scientists had about relative motion was intimately connected to the laws of motion. The question they raised is: Are the laws of mechanics the same for all observers? Let us examine this question more precisely.

When we are performing an experiment, we always choose a **reference frame** from which we make our observations. For example, if you are doing an experiment in a laboratory, the reference frame you pick is one that is at rest with respect to the laboratory. Now suppose someone passing by in a car moving at a constant velocity were to observe your experiment. Would the observations made by the observer in the reference frame of the car differ dramatically from yours?

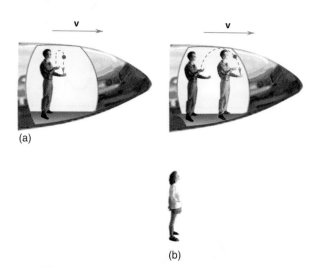

FIGURE 14.1 *(a) The observer on the plane sees the ball move in a vertical path. (b) The observer on the Earth sees the ball move as a projectile.*

To answer this question, let us consider a slight variation of this question, as shown in Figure 14.1a and b. Here we see two observers, one in an airplane moving at a constant velocity and another at rest on the Earth. The passenger on the airplane decides that he will perform a simple experiment of tossing a baseball into the air to see what will happen. He throws the ball straight up, and he finds that it follows a vertical path upward and falls vertically back to his hand (Fig. 14.1a). This is exactly the same thing that he would observe happening if he were on Earth. The law of gravity is obeyed, and the equations of motion with constant acceleration can be used to tell the plane rider about the details of the motion. The observer on Earth, however, sees the experiment played out somewhat differently. This observer sees the path of the ball to be like that of a projectile (Fig. 14.1b). While the ball was in the air, the passenger and the plane had moved to the right, so in order for the ball to be caught by the passenger, according to the Earth-bound observer, the ball would have to follow the dashed path shown. The two observers thus disagree on certain aspects of the motion of the ball, but both agree that the ball obeys the law of gravity and the laws of motion. As a result of experiments such as this, we draw the following conclusion:

The laws of mechanics are the same in all reference frames moving at constant velocity with respect to one another.

A second type of question connected with relativity that was of interest to early physicists concerned absolute motion. To understand what is meant by absolute motion, consider the following set of circumstances. Imagine that a jet airplane is traveling at 400 miles per hour, and following behind is a prop plane flying with a velocity of 100 miles per hour in the same direction as the jet. The first thing that we must note about the statement of all these velocities is that they are all measured relative to the Earth. In every measurement of a velocity, you have to specify the reference frame with respect to which it is measured. For example, if we took our reference frame to be that of the prop plane, that pilot would consider himself to be moving at a velocity of 0 miles per hour, and the jet would be moving relative to him with a velocity of 300 miles per hour. If the stationary reference frame were considered to be that of the jet plane, the pilot of the jet would consider herself to be at rest, and the prop plane would be moving with a velocity of 300 miles per hour, but *backward*. The point to be made with these observations is that

all motion is relative to some reference frame.

Thus the question arises, When is something at rest? We have often assumed that the Earth is at rest in our problems. It obviously is not, however, because the Earth rotates on its axis, and it revolves about the Sun. Is the Sun at rest? If so, we could state that an object in the Universe is also at rest if it is stationary with respect to the Sun. Again the answer is no. Our Sun moves with the Solar System in revolution about the center of our galaxy, and our galaxy moves with respect to other galaxies. This search for a reference frame that could be considered to be at absolute rest led physicists in a most unusual and interesting chase, as we shall see in the next section.

CONCEPTUAL EXERCISE

Suppose that you do an experiment in lab that proves that energy is conserved and that momentum is conserved. Would someone traveling by your lab at a speed near that of light agree with your results?

Answer One of the primary conclusions of all observations based on relativity is that the laws of mechanics are true for all observers. Thus, if you find that energy and momentum are conserved in the lab, someone peeking over your shoulder from a spaceship will agree with your results.

14.2 THE REST FRAME OF THE ETHER

In the early history of the study of light, several experiments, including Young's double-slit experiment, had led to the conclusion that light is a wave. The statement that light is a wave led to some serious fundamental problems for the investigators, however. The reason for their perplexity is that they knew quite a bit about several different types of wave motion, such as that of water waves, sound waves, waves on strings, and so forth, and all of these types of waves had one thing in common that waves of light did not seem to share: All of these different classifications of waves have a substance through which they move. This medium of propagation is obviously water for a water wave; a solid, liquid, or gas for a sound wave; and so on. Now scientists were in the position of calling light a wave, but there did not seem to be a medium for it to travel in. For example, if you place a glowing light bulb in a container and then pump all the air out of the container, you can still see the glowing bulb inside. Likewise, you can see the Sun, stars, and other distant objects in the Universe without there apparently being any material substance in the intervening space to carry this light to Earth. Scientists were uncomfortable with the concept of a wave without a medium for its propagation, so they conceived of a medium and gave it the name, **the luminiferous ether.**

This ether would have to have some quite unusual properties. It would have to permeate all of space because we can see distant stars as the light moves through the ether from them to us. Also, it would have to permeate matter because light can be transmitted through a transparent object. Yet, as all-pervading as is the ether, the Earth and other heavenly objects move through it apparently unaffected by its presence. As the Earth swings around the Sun, it must move through the ether of space without being slowed by it or having its motion altered by it in any way. Thus, the ether would have to be a most unusual material.

The important attributes of the ether as far as the study of relativity is concerned were that it was

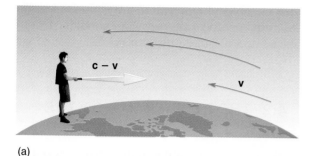

(a)

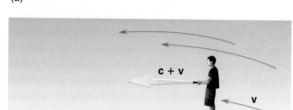

(b)

(c)

FIGURE 14.2   *(a) A beam of light traveling into the ether wind was expected to have a speed **c** − **v**. (b) A beam traveling with the ether wind was expected to have a speed **c** + **v**. (c) A beam traveling perpendicular to the ether wind was expected to take longer to reach its destination than it would in the absence of the wind.*

an all-pervasive medium that existed throughout the Universe and, just as important, that this medium was *stationary*. As a result of these unique characteristics, the ether took on the aspect of a privileged frame of reference that could be used to determine the absolute motion of an object. We could cease worrying about having to specify the

velocity of an object with respect to the Earth or anything else; instead, we would always specify its velocity with respect to the ether. Thus, if an object is at rest with respect to the ether, it is in a state of absolute rest. Also, we could avoid stating that the velocity of a baseball is 30 miles per hour with respect to the Earth; instead, we would specify its velocity with respect to the ether. Thus, the search for the ether became an important research activity for scientists of about 100 years ago. In the next section, we will look at an experiment conducted by Albert Michelson and E. W. Morley in an attempt to find the ether.

14.3 THE MICHELSON-MORLEY EXPERIMENT

In 1883, Michelson and Morley devised an experiment to measure the Earth's speed through the stationary ether. They reasoned that as the Earth moves through space, it must pass through the ether and that this would cause an ether "wind" to blow across the Earth. According to the accepted ideas of the time, a beam of light sent against the ether wind, as in Figure 14.2a, would travel with a speed of $c − v$, where c is the speed of light and v is the speed of the ether wind. The speed v is the same as the speed of the Earth in its orbit, since the ether is assumed to be in a state of absolute rest. By similar reasoning, a beam of light traveling with the ether wind, as in Figure 14.2b, would travel with a speed of $c + v$. But what about a beam of light sent on a path perpendicular to the ether wind, as in Figure 14.2c? An analysis of this trip would show that this beam would require more time to make the trip than it would if there were no ether wind.

Figure 14.3 shows the details of the Michelson-Morley experiment. Two beams of light are sent on a race, with beam A moving parallel to the direction of the wind and beam B moving perpendicular to the wind. The beams start at the light source, break into two parts at the partially silvered mirror, and then move their respective paths to reflect from mirrors M_1 and M_2. The rays then retrace their paths and come together at the position of the observer. When the two beams recombine, an interference pattern like that shown in Figure 14.4 is formed. Suppose the experiment is started as in Figure 14.3, with path A aligned with the wind. A pattern like that shown in Figure 14.4 is observed.

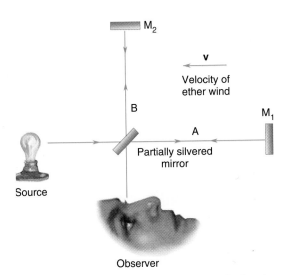

FIGURE 14.3 *The light from the source is broken into two parts by the partially silvered mirror. The light that follows path A hits mirror M_1 and is reflected back to the observer. The light along path B is reflected by mirror M_2 and also returns to the observer, where an interference pattern is formed.*

Michelson and Morley then rotated the device until path 2 was along the breeze, as shown in Figure 14.5. The two arms have, in effect, changed places,

and a slight change in the interference pattern should be observed. (The rings should move slightly.) The amount of this movement was predicted to be quite small, but Michelson and Morley had calculated how much movement they would be able to detect, and they were convinced that the change in the pattern would be noticeable. *The end result of this investigation was that they saw no movement of the pattern at all.*

The scientific community had expected Michelson and Morley to verify the existence of the ether wind and thereby to verify the existence of the ether. The results, however, were conclusive; the ether wind did not show itself. Much time and effort were exerted by many scientists in an attempt to determine why the ether did not make its presence known. The final conclusion that was reluctantly drawn by scientists was that the ether did not affect the light at all. This means that *light travels with a speed of* c *regardless of the motion of the ether.* Since the primary purpose for assuming that there was an ether was that it was the medium through which light traveled, and if light's own medium was not affecting it, there was no need to assume the existence of the medium. Thus, the final result of the Michelson-Morley experiment was a negative one: *There is no ether.*

FIGURE 14.4 *The beams reflected from mirrors M_1 and M_2 recombine at the eye of the observer to form an interference pattern like this one.*

FIGURE 14.5 *After rotation, path B is parallel to the ether wind and A is perpendicular to it. This should produce a change in the interference pattern.*

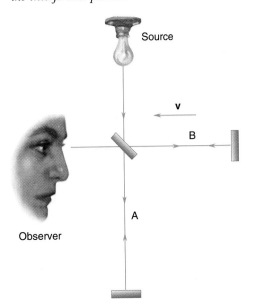

FIGURE 14.6 *Common sense says the speed of the beam of light from spaceship A should have a value of* **c** + **v** *as seen by the Earthbound observer, and the speed of the beam from spaceship B should be* **c** − **v**. *Special relativity, however, says that the speed will be* c *in both instances.*

14.4 EINSTEIN'S POSTULATES

Einstein's theory of relativity is more specifically referred to as his theory of **special relativity.** The word "special" is used to mean that it applies only in a special case—that of objects that move with constant velocity with respect to one another. The more general case, in which accelerations are allowed, is not included in the situations that we discuss here. Einstein based his theory on two postulates.

1. All the laws of physics are the same for all observers moving at constant velocity with respect to one another.
2. The speed of light in a vacuum is the same for all observers regardless of the motion of the source of light or the motion of the observer.

We have already pointed out that the accepted theories of relative motion at the time of Einstein had predicted that the laws of mechanics would be the same for all observers moving at constant velocity with respect to each other. Einstein's first postulate extended this idea to cover *all* the laws of physics.

The second postulate explains why the Michelson-Morley experiment failed, although it is reasonably certain that Einstein did not know of this experiment when he made the postulate. This second postulate also deserves some elaboration because it violates our laws of common sense. To see why this second statement is so out of step with the

way we believe nature should behave, consider Figure 14.6. There we see two space travelers sending out a beam of light from their spaceships and an observer at rest on the Earth watching the action. Spaceship A is moving toward the observer, and the question is: What does the Earthbound observer measure for the speed of the beam of light coming toward him? If the speed of the ship is *v*, common sense tells us that the speed of the ship should be added to that of the oncoming light, and the net speed measured should be *c* + *v*. Not so, says Einstein. According to him, the motion of the source or observer does not affect the speed of light. *The Earthbound observer will measure the speed to be c.* Likewise, common sense tells us that the velocity measured for the beam from B emitted by the ship moving away from the observer should be *c* − *v*. Again, not so. The speed in this instance is also *c*. In fact, if the observers on each spaceship look at the beam from the other, they will measure these speeds to be *c* also.

The details of where these postulates led Einstein and the complex calculations required to move through the theory will not be examined here. Instead, we will focus on some of the strange predictions that are based on these starting points. The end results of Einstein's work profoundly altered our view of space, time, and matter. As strange as we shall find these results to be, you should keep in mind that the predictions of special relativity have been verified time and time again. There is no escaping their validity.

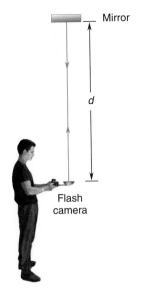

FIGURE 14.7 *An observer in a spaceship sees the light from the flash camera strike the mirror and retrace its path. The distance traveled in the round trip is 2d, and the time for the trip is* $\Delta t' = 2d/c$.

14.5 EINSTEIN AND TIME

The fact that the speed of light is a constant for all observers leads to some surprising conclusions concerning time intervals measured in different frames of reference. We will find that if a person at rest on Earth says his heart is beating at 70 beats per minute, an observer moving with respect to the Earth will not agree. His conclusion will be that the clock used by the person on Earth, who is in motion relative to him, is running slowly. To understand why different observers will measure different times for specific events to occur, consider the experiment illustrated in Figure 14.7. An observer in a spaceship moving at a high speed holds a camera with a flash attachment directly beneath a mirror. When the flash goes off, the light leaves the bulb, travels to the mirror a distance d above him, and returns. Let us call the time for the round trip of the light $\Delta t'$ (read as delta t-prime). We can easily find this time interval from the definition of velocity as

$$\Delta t' = \frac{\text{distance traveled}}{\text{speed}} = \frac{2d}{c}$$

where the total distance traveled by the beam of light is $2d$, and the speed of the light is $c = 3.00 \times 10^8$ m/s.

Now consider this same sequence of events as seen by an observer on the Earth. Because the spaceship is moving, he finds that the light will have to travel a different path to strike the mirror and return (Fig. 14.8). From his point of view, the mirror will have moved to the right, from position A to position B, by the time the light from the flash-

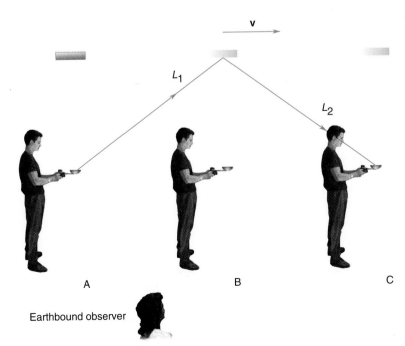

A

Earthbound observer

B

C

FIGURE 14.8 *The Earthbound observer sees the light from the flash camera travel a longer path because of the motion of the spaceship. The ship moves from A to B while the light moves toward the mirror and from position B to C while the light returns.*

FOCUS ON ATOMIC BOMBS

An explosion develops a sudden pressure on its surroundings by the rapid production of gas and by the further expansion of the gas as the explosive energy heats it. Chemical explosives produce gases rapidly by the decomposition of their molecules. This effect is called a blast. If a stick of dynamite explodes several hundred feet from you, the blast effect feels like a thump on your chest.

The main chemical high explosives of modern warfare have been TNT (trinitrotoluene), picric acid, and cyclonite (the explosive ingredient of "plastic explosive"). Nitroglycerin is the major explosive ingredient of dynamite, used mainly for blasting in construction and mining. The heaviest chemical bombs dropped by aircraft in World War II (the "blockbusters") contained about 1000 kg of high explosive.

The nuclear explosive in an atomic (fission) bomb is pure or highly concentrate fissile material: uranium-235 or plutonium-239. Such a material leaves only two significant fates for neutrons—fission capture or escape. The factor that determines which of these two fates will predominate is size, or mass; the minimum mass required to support a self-sustaining chain reaction is called the **critical mass.** To set off an atomic bomb, therefore, subcritical masses of uranium-235 or plutonium-239 are slammed together by precisely shaped chemical high explosives to make supercritical mass. The chain reaction instantly branches, and the mass explodes.

A hydrogen (fusion) bomb derives the major portion of its energy from a nuclear fusion reaction. The most powerful bombs ever exploded have been hydrogen bombs—in tests, not warfare.

The explosive effect of a fission or fusion bomb is rated in terms of its TNT equivalent. Thus, a "1 megaton" bomb is a nuclear bomb that is equivalent to 1 megaton (10^9 kg or about 2.2 billion lb) of TNT. This equivalence, however, refers only to the blast effect. Nuclear bombs have other consequences that are not produced by chemical high explosives. These other nuclear effects include extremely high temperatures that start fires at considerable distances, prompt radiation, electromagnetic pulses that can knock out electronic systems, and climatic changes.

The question is sometimes asked: "Can a nuclear reactor explode like an atomic bomb?" Opponents of nuclear energy have complained that no one ever made such an accusation; therefore, the question just diverts attention from more credible hazards. Nonetheless, the reader should consider that the fuel in a nuclear reactor contains no concentrations that even approach the levels of bomb-grade material and that an atomic explosion therefore cannot occur.

---

bulb reaches it. Thus, according to the Earthbound observer, the flashbulb will have to be held at an angle if the light from it is to hit the mirror. The light will then bounce off the mirror, obeying the law of reflection, and return to the level of the flash just as the motion of the spaceship brings the astronaut to this location (from position B to position C). Let us call the time for all this to happen Δt. Again, we can use our defining equation for speed to find the elapsed time.

$$\Delta t = \frac{\text{distance traveled}}{\text{speed}}$$

$$= \frac{\text{length L1} + \text{length L2}}{c}$$

According to Einstein's postulates, the speed of light remains at c, but in this case the distance traveled, length L1 plus length L2, will be greater than it was from the point of view of the person riding in the spaceship. Thus, Δ*t* *will be greater than* Δ*t′*. To summarize, Δ*t′* is the time interval measured by an observer at rest with respect to our experiment, while Δ*t* is the time interval measured by a person who is observing the experiment while it is in motion.

We conclude that moving clocks run slowly.

We shall not repeat the details of Einstein's analysis of the exact relationship between Δ*t* and Δ*t′*, but he found the two are related as

$$\Delta t = \frac{\Delta t'}{\sqrt{1 - \dfrac{v^2}{c^2}}} = \gamma \Delta t' \qquad (14.1)$$

where

$$\gamma = \frac{1}{\sqrt{1 - \dfrac{v^2}{c^2}}}$$

We have always assumed in our previous study of physics that a time interval seen by one individual is the same as that seen by another. According to special relativity, this really isn't so. Moving clocks slow down. But what is a clock? It could, of course, be an ordinary mechanical time-keeping device, but the definition of a clock can be extended to include biological processes also. Thus, the heart rate of the moving astronaut will be seen to have slowed from the point of view of the Earthbound observer. In fact, all biological processes will have slowed. Thus, from the stand-point of the Earthbound observer, the moving astronaut will age at a slower rate than does the observer. The astronaut will not have any sensation of time slowing for him. In fact, from his point of view, he will consider himself at rest, while it is the observer on the Earth who is in motion. As a result, he will consider the clock of the person on Earth to be running slowly.

CONCEPTUAL EXERCISE

A group of students in a rocket ship are taking an exam that is supposed to last for exactly 1 hour. The exam is being timed by an instructor at rest on the Earth. When the instructor says that time is up, will the students be pleased with the amount of time they have had or be unhappy?

Answer The students will consider themselves at rest and the instructor as the person who is moving. Thus, the students will consider the instructor's clock to run slowly. They will be pleased with the extra time for the exam.

EXAMPLE 14.1   HOW TIME FLIES

(a) An astronaut in a spaceship moving at 0.9 c watches the pendulum of a grandfather clock on the spaceship swinging back and forth. He finds that the pendulum makes one complete vibration every second. How long does an earthbound observer find that it takes for the pendulum to make one complete swing?

Solution The time interval Δ*t′* is the time interval as measured by the observer at rest with respect to the clock. Thus, Δ*t′* = 1 s. The time interval Δ*t* is the time interval as seen by an observer watching the moving clock. This time interval can be found from Eq. 14.1 with

$$\gamma = \frac{1}{\sqrt{1 - \dfrac{v^2}{c^2}}} = \frac{1}{\sqrt{1 - \dfrac{(0.9\ c)^2}{c^2}}} = 2.29$$

Thus,

$$\Delta t = \gamma \Delta t' = 2.29(1\ \text{s}) = \boxed{2.29}$$

This says that the Earthbound observer will find that it takes 2.29 s for the pendulum to make its complete vibration. Thus, since the clock is supposed to be "ticking" once each second, we must conclude that moving clocks run slowly.

(b) If the Earthbound observer has a pendulum that makes one vibration in 1 s, how long will the astronaut in the ship moving at 0.9 c say it takes to make one complete swing?

Solution In this case, Δ*t′* is the time as measured by the observer on Earth, who is now at rest with respect to the clock. The astronaut's point of view is that he is the one at rest, while the Earth is moving backward at 0.9 c. Thus, the same calculations as done previously indicate that the astronaut will say that it is the Earthbound clock that runs slowly. It will take, from his viewpoint, 2.29 s to make one vibration.

14.6 LENGTH CONTRACTION

The fact that time intervals are not absolutes but instead depend on the frame of reference used is the first of the surprises that stem from Einstein's theory of relativity. Just as surprising is the fact that lengths are not absolutes either. The results of special relativity indicate that *lengths of objects contract in the direction of motion when they are in motion with respect to an observer*, as shown in Figure 14.9. The length of an object moving with a speed **v** with respect to an observer is given by

$$L = L' \sqrt{1 - \frac{v^2}{c^2}} \text{ or } L = \frac{L'}{\gamma} \qquad (14.2)$$

where L' is the length as seen in a reference frame at rest with respect to the object and L is the length as seen in a reference frame in motion with respect to the object.

An additional consequence of special relativity is that length contraction takes place only along the direction of motion. *Lengths perpendicular to the direction of motion are unchanged by relative motion.* Thus, a spherical spaceship moving at a speed near that of light might change from the shape shown in Figure 14.10a to that of Figure 14.10b as its speed increases.

An interesting piece of experimental evidence that confirms both the slowing down of moving clocks and length contraction involves muons, particles that can be produced high in the atmosphere in collisions between cosmic rays and atoms in the

(a) At rest

(b) Shorter ruler in motion relative to an observer

FIGURE 14.9   (a) *The ruler has a certain length when it is at rest relative to an observer.* (b) *When the same ruler moves relative to an observer, its length is shorter.*

air. Muons are particles with a charge equal to that of an electron and a mass 207 times that of an electron. They are radioactive and decay with a half-life of 2.2 μs when the half-life is measured in a reference frame that is at rest with respect to them. A typical speed for these muons when they are produced in the atmosphere is about 0.97 c. At this speed, a muon would travel a distance of about 640 m during a time equal to its half-life. Thus, if 1000 muons were produced by cosmic rays at a height of 640 m above the Earth, we would expect to find approximately 500 of them reaching Earth. Many more than this predicted number, however, actually survive and make it to the Earth. Why?

The question posed in the last paragraph can be answered either from the point of view of Einstein's time expansion, Eq. 14.1, or from the standpoint of length contraction. Let us examine these in turn. From a frame of reference on the Earth, the half-life of a muon moving at 0.97 c will not be

(a)

(b)

FIGURE 14.10   (a) *If a spaceship in the shape of a sphere moves by a stationary observer at a speed near that of light, its shape will change to that of part (b). Its vertical dimension is unchanged, but its horizontal dimension is contradicted.*

2.2 μs. Instead, its half-life will be

$$\Delta t = \frac{\Delta t'}{\sqrt{1 - \dfrac{v^2}{c^2}}} = \frac{2.2 \ \mu s}{\sqrt{1 - \dfrac{(0.97 \ c)^2}{c^2}}} = 9.05 \ \mu s$$

At a speed of 0.97 c, a particle with a half-life of 9.05 μs would travel a distance of about 2630 m before decaying. Thus, based on the theory of special relativity, one would expect to find many more muons reaching the Earth because the time before decay is longer.

An alternative explanation based on special relativity, which leads to the same result, relies on length contraction. To see how this approach works, consider an observer on Earth holding a measuring rod of length 640 m that reaches into the sky to the point where our 1000 muons are produced. The length of this rod, however, is 640 m from the point of view of the observer on the Earth. A muon considers itself to be at rest and the Earth flying up to meet it at a speed of 0.97 c. Thus, in the reference frame of the muon, the rod is not really 640 m long; its length is given by Eq. 14.2 as

$$L = L' \sqrt{1 - \frac{v^2}{c^2}}$$

$$= (640 \ m) \sqrt{1 - \frac{(0.97 \ c)^2}{c^2}} = 156 \ m$$

Thus, from the point of view of length contraction, the distance that a muon has to travel before it decays is considerably shortened. As a result, many more muons than expected will reach the Earth. We have approached this problem from two different aspects of special relativity. In each case, however, the results are the same. More muons reach the Earth than would be expected, and the theory of relativity achieves a major victory.

CONCEPTUAL EXERCISE

A spaceship in the shape of a cube moving at a speed near that of light passes an observer at rest on Earth. Describe the shape of the cube as seen by the Earthbound observer.

Answer The ship will have the same height and width as it would if it were at rest, but the length along the direction of travel would be contracted.

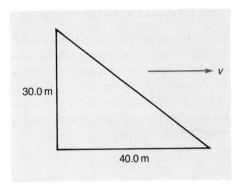

FIGURE 14.11

EXAMPLE 14.2   PLEASE MOVE TO THE REAR OF THE SPACESHIP

A spaceship is made in the shape of a triangle shown in Figure 14.11 with lengths of 40.0 m and 30.0 m when the ship is at rest. The spaceship flies by a stationary observer at 0.90 c in the direction shown. Find the shape of the spaceship as seen by the stationary observer.

Solution The 40.0 m length along the direction of travel is contracted, and this new length is given by

$$L = L' \sqrt{1 - \frac{v^2}{c^2}} = (40.0 \ m) \sqrt{1 - \frac{(0.90 \ c)^2}{c^2}}$$

$$= 17.4 \ m$$

The 30 m height of the ship is not changed because this length is perpendicular to the direction of travel. Thus, the shape of the ship from the observer's point of view is as shown in Figure 14.12.

FIGURE 14.12

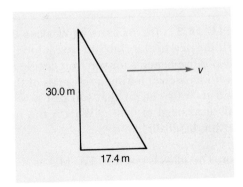

14.7 RELATIVITY AND MASS

Time intervals and lengths have been found to be subject to the frame of reference in which they are observed. Another sacrosanct entity in physics, the mass of an object, also undergoes a change with its speed. *The mass of an object is found to increase as its speed increases,* and the Einstein relationship that predicts this change is given by

$$m = \frac{m_0}{\sqrt{1 - \dfrac{v^2}{c^2}}} = \gamma m_0 \qquad (14.3)$$

where m_0 is the mass of an object as measured by an observer at rest with respect to it, and m is the mass of the object when in motion with respect to an observer.

This equation also points out that the greatest speed that an object can attain can never exceed the speed of light. To see that this is the case, consider what happens to the mass of an object as its speed, v, approaches c. The denominator of Eq. 14.3 approaches zero; thus, the mass m becomes infinitely large. An infinite amount of energy would be required to accelerate an infinite mass, and, as a result, the speed of the object cannot increase to that of light. *No material object can attain the speed of light.*

CONCEPTUAL EXERCISE

A baseball is dropped from the top of a tall building. Does it have the same mass just before it strikes the ground as it did when released?

Answer The speed of the ball has been increasing as it accelerates under gravity. Thus, its mass will also be increasing.

EXAMPLE 14.3   THE BENEFITS OF WORKING OUT

To see if the new health club you have joined is having any positive benefits for you, you decide to play catch with a baseball having a rest mass of 0.15 kg. You find to your delight that you can now throw the ball at a speed of 0.90 c. What is the mass of the moving baseball?

Solution The mass is given by Eq. 14.3 as

$$m = \frac{m_0}{\sqrt{1 - \dfrac{v^2}{c^2}}} = \frac{0.15 \text{ kg}}{\sqrt{1 - \dfrac{(0.90 \text{ c})^2}{c^2}}}$$

$$= 0.344 \text{ kg}$$

14.8 MASS AND ENERGY

One of the most basic principles of physics is that of the conservation of energy, which we studied in Chapter 3. There we found that energy can exist in a variety of forms, such as gravitational potential energy and kinetic energy. It may change from one of these forms to another, yet it never appears or disappears. Einstein's work with the theory of special relativity says that this idea of the conservation of energy must be modified. In particular, it says that

mass can be converted into energy and that energy can be converted into mass.

The relationship between energy E, mass m, and the speed of light c is given by the well-known equation

$$E = mc^2 \qquad (14.4)$$

We will see later in this chapter that this equation has had a tremendous impact on the course of human history in that its first application was a devastating one. The atomic bombs exploded over Hiroshima and Nagasaki provided the first dramatic evidence of the validity of this equation. More humane applications include the conversion of mass into energy to provide electrical power. In the next section, we shall look at another unusual verification that can be observed only within the confines of a laboratory.

EXAMPLE 14.4   MASS INTO ENERGY

Suppose that the entire mass of an electron (9.11×10^{-31} kg) could be converted into energy. How many joules would be produced?

Solution This is a direct application of Eq. 14.4. We have

$$E = mc^2$$
$$= (9.11 \times 10^{-31} \text{ kg})(3.00 \times 10^8 \text{ m/s})^2$$
$$= 8.20 \times 10^{-14} \text{ J}$$

The disappearance of a single electron does not produce a tremendous amount of energy, but a similar calculation would show that if 0.5 kg of matter were converted into energy, enough energy would be produced to keep a 100 W light bulb burning for approximately 14 million years. Thus, significant quantities of energy are produced with the disappearance of only small amounts of mass.

14.9 PAIR PRODUCTION AND ANNIHILATION

The atomic bomb and nuclear power plants provide vivid evidence of the fact that mass can be converted into energy, but does the process go in the reverse order? Can energy be converted into mass? Before we investigate the answer to this question, let us digress to examine a rather strange form of matter called **antimatter.**

In the 1920s, P. A. M. Dirac (1902–1984) developed a theory that incorporated the concepts of quantum mechanics and relativity. One of the outgrowths of the theory was the prediction that *for every type of particle, there is an antiparticle.* To investigate the properties of these particles, which are now known to exist, let us consider a special kind of decay process. Identify the missing element in the radioactive decay symbolized as

$$^{12}_{7}\text{N} \longrightarrow {}^{12}_{6}\text{C} + X$$

What is X? Balancing mass numbers and atomic numbers reveals that we must have a particle symbolized as $^{0}_{1}X$. The mass number of zero reminds us of the electron, but the particle has a positive charge. In fact, this particle is indeed like the electron in all respects except for its charge. Thus, we symbolize it as $^{0}_{1}\text{e}$. This is the **positron,** and it is said to be the **antiparticle** of the electron. The positron is seen often in decay processes, but it was only discovered in 1932 by Carl Anderson, in the same year that the neutron was discovered. Positrons are often produced in high-energy cosmic ray reactions

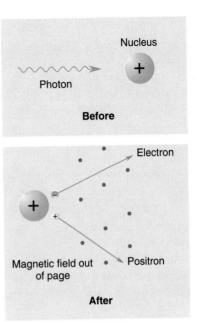

FIGURE 14.13 *An incoming photon is converted to an electron-positron pair when it interacts with a nucleus. The pair can be separated by a magnetic field.*

with nuclei of the atmosphere, and it was by studying such reactions that Anderson made his discovery.

One of the most common processes by which positrons are produced is through a mechanism called **pair production.** In this process, a gamma ray with sufficiently high energy collides with a nucleus; the gamma ray disappears, and in its place an electron-positron pair is created. This process is pictured in Figure 14.13, where the incoming gamma ray is shown to interact with a heavy nucleus. The electron and positron can be separated by a magnetic field that causes them to bend in opposite directions. There is a lower limit on the energy that the gamma ray can have before this process can occur. The minimum energy of the gamma is $E = 2m_0 c^2$, where m_0 is the mass of the positron or electron (they are the same). This equation says that the energy must be sufficient to produce a net mass equal to the sum of the masses of the positron and electron. This relationship provides a striking confirmation of the equivalence of mass and energy.

Another confirmation of Einstein's mass-energy relationship is provided by a process that is the

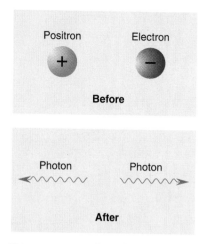

FIGURE 14.14 *A positron and an electron meet (before) and annihilate into gamma rays (after).*

reverse of pair production. Antiparticles cannot exist in our world for long because a particle such as the positron will soon encounter its counterpart in the world of ordinary particles, and the two will annihilate one another. The process is referred to as **pair annihilation.** This process is pictured in Figure 14.14. Note that two gamma rays are always produced so that momentum can be conserved, and the combined energy of the two gammas must be at least equal to $E = 2m_0c^2$.

Every ordinary particle has its antiparticle counterpart. Thus, the proton is matched with the **antiproton,** which is identical in all respects to the proton except it has a negative charge. An antiproton suffers the same fate as a positron when it encounters normal matter. The proton and the antiproton annihilate, and gamma rays having a combined energy equal to the mass-equivalent of two protons are created.

Because all particles have their antiparticles, a natural question arises: Is there, perhaps somewhere in the Universe, a world made up solely of antiparticles, as our world is made up of particles? If so, this world would consist of atoms having antiprotons and antineutrons in its negatively charged nucleus and circled by positively charged positrons. There would be no way to differentiate such a world via telescope because it would behave in isolation exactly as would any other heavenly body. If this object should come into contact with ordinary matter falling into it, however, we might be able to detect it because of the annihilation events going on. The most distant objects in our Universe are strange creations called quasars, which are roughly the same size as a star, yet which are pouring out into space tremendous amounts of energy. For example, some quasars are emitting as much radiation as that produced in a galaxy consisting of billions of stars. What is the source of energy that drives a quasar? Several alternatives exist that might be able to explain this fantastic energy emission, but one that some astronomers have envisioned is that of matter-antimatter collisions within the quasar. (It should be noted here that other theories seem to fit the observations better than this one.)

EXAMPLE 14.5   GONE IN A PUFF OF GAMMA RAYS
A proton and an antiproton, essentially at rest, come together and annihilate. Find the energy in joules of all the gamma rays produced.

Solution The energy produced is equal to the amount of mass that disappears. The mass of the proton and antiproton are both 1.67×10^{-27} kg. Thus, the net energy of the gammas is

$$E = 2m_0c^2$$
$$= 2(1.67 \times 10^{-27} \text{ kg})(3.00 \times 10^8 \text{ m/s})^2$$
$$= 3.01 \times 10^{-10} \text{ J}$$

14.10 NUCLEAR FISSION

Recall from Chapter 13 that many studies of the nucleus or of nuclear particles are carried out by bombarding one particle with another. The idea that neutrons might be used to bombard and alter atomic nuclei was exciting to all the scientists who were studying nuclear reactions. The reason is that a neutron, which does not bear any charge, is not repelled by positively charged atomic nuclei and can therefore travel in a straight line until it hits one. If the neutron is absorbed by the nucleus, the ratio of neutrons to protons is changed, and so the stability of the nucleus is also changed.

In 1939, three scientists, Otto Hahn, Fritz Strassman, and Lise Meitner, discovered that when a neutron hits and is captured by a uranium nucleus, the nucleus splits into two roughly equal fragments (Fig. 14.15). This splitting is called **nuclear fission.** Further studies showed that the isotope un-

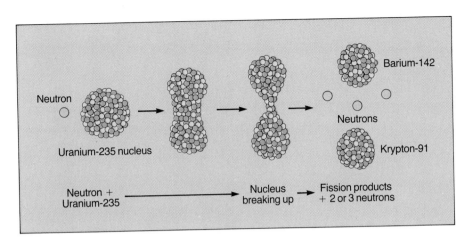

FIGURE 14.15 *A neutron strikes a uranium-235 nucleus, causing it to split into two nearly equal fragments (shown here as barium-142 and krypton-91) and extra neutrons.*

dergoing fission is uranium-235, which makes up less than 1 percent of natural uranium. The abundant form, uranium-238, does not undergo fission.

It was also learned that extra neutrons were released in the fission reaction of uranium-235. If the reaction is *started* by neutrons and then also *releases* neutrons, a new possibility arises that is different from anything discussed so far. It is the opportunity for a **chain reaction.** This discovery changed nuclear science from a study of purely theoretical interest to an issue of utmost importance to everyone.

A chain reaction is a series of steps that occur one after the other, in sequence, each step being added to the preceding step like the links in a chain. An example of a chemical chain reaction is a forest fire. The heat from one tree may initiate the reaction (burning) of a second tree, which, in turn, ignites a third, and so on. The fire will then go on at a steady rate. But if one burning tree ignites, say, two others, and each of these two ignite two more, for a total of four, and so on, the rate of burning will speed up. Such uncontrolled, runaway chain reactions are at the heart of the explosion created by a nuclear bomb (Fig. 14.16).

Fission is initiated when one neutron strikes a uranium-235 nucleus and can proceed in a number of different ways. For example, as shown in Figure 14.15, uranium-235 struck by a neutron can produce barium-142 and krypton-91, while releasing three neutrons. In equation form, the fission reaction here is

$$\,_0^1 n + \,_{92}^{235} U \longrightarrow \,_{56}^{142} Ba + \,_{36}^{91} Kr + 3\,_0^1 n$$

Note the following important points about this reaction:

1. The reaction is started by one neutron but produces three neutrons. These neutrons can, ideally, initiate three new reactions, which, in turn, produce more neutrons, and so forth. Thus, a chain reaction evolves.

2. The uranium-235 nucleus is split roughly in half by this reaction. The total mass of all the fission products (those on the right side of the arrow) is slightly less than the sum of the masses of the original uranium-235 atom and the incident neutron. This loss of mass is converted into energy in accordance with the equation $E = mc^2$. The energy appears mostly in the form of the kinetic energy of the fission products as they fly apart. These flying fragments then slow down as they hit other atoms, and in so doing they transfer their energy to these atoms in random patterns. This is the way in which nuclear fission releases heat. As has been pointed out, the energies involved in nuclear transfor-

FIGURE 14.16 *A chain reaction.*

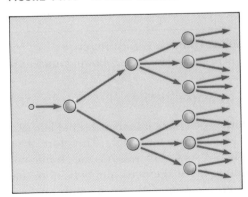

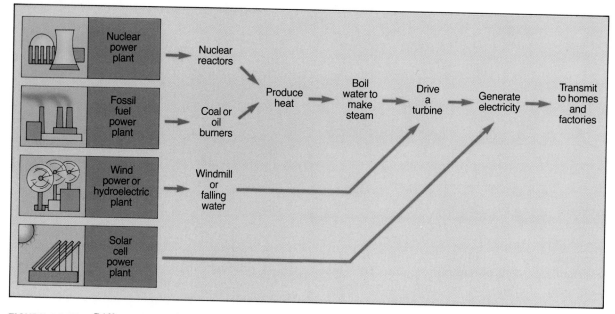

FIGURE 14.17   *Different types of power plants use different systems to generate electricity.*

mations are much greater than those in chemical reactions. If the chain reaction continues at a rapid rate, energy is released at an accelerating rate, and an explosion can result. (This is what happens in the detonation of a nuclear bomb.) If the chain reaction is controlled, energy can be released more slowly, and the heat produced can be used to make steam, which can then be used to drive a turbine and produce electricity.

3. Fission reactions produce radioactive wastes. Barium-142 and krypton-91, the products shown in the preceding equation, are both radioactive. Furthermore, the reaction represented by this equation is only one of many that occur in nuclear fission.

EXAMPLE 14.6   A FISSION REACTION

Complete and balance the following nuclear fission reaction:

$$\mathrm{{}^{1}_{0}n} + \mathrm{{}^{235}_{92}U} \longrightarrow \mathrm{{}^{97}_{39}Y} + \mathrm{?} + 2\mathrm{{}^{1}_{0}n}$$

Solution The atomic numbers on the left (92 + 0) must equal those on the right. Therefore, 92 + 0 = 39 + 0 + Z, so Z = 53, which is the atomic number of iodine, I. For the mass numbers, note that two neutrons are produced, which count for two mass

numbers. Then, 1 + 235 = 97 + 2 + N, from which N = 137, and the missing isotope is therefore $\mathrm{{}^{137}_{53}I}$.

14.11 NUCLEAR POWER PLANT CONSIDERATIONS

The purpose of a power plant is, of course, to generate electricity. Different kinds of power plants depend on different sources for their energy. A wind power or hydroelectric plant uses the mechanical energy of wind or falling water. A solar power plant generates electricity directly from solar cells. In a coal-fired plant, the energy is released from the chemical combustion of the coal. In a nuclear plant, energy from nuclear fission reactions is used to heat water and to produce steam. The steam then drives a turbine to generate electricity, just as in a coal-fired plant (Fig. 14.17).

Nuclear fission reactors require fuel, and the fuel must be a substance whose nuclei can undergo fission. There are two significant nuclear fuels, uranium-235 and plutonium-239. There are a few other fissionable materials, but these are the ones on which the current nuclear energy program in the United States is based.

Uranium-235 occurs in nature, but it constitutes only 0.7 percent of natural uranium. The remaining 99.3 percent is the heavier isotope uranium-238, which does not undergo fission in a reactor.

The second fuel, plutonium-239, does not occur in nature; it is produced by bombarding uranium-238 with neutrons. Thus, the two important naturally occurring sources of fission energy are uranium-235 (which fissions but is not abundant) and uranium-238 (which does not fission but is abundant and convertible into plutonium-239).

Nuclear reactors require another essential ingredient besides fuel: neutrons. In fact, the chain reaction is initiated by neutrons. The design and operation of reactors, as well as their safety, depend on how the neutrons are managed and controlled. There are four possible events that can happen to a neutron in a reactor:

1. A neutron can be captured by a uranium-235 nucleus, which then undergoes fission. The reaction releases fast neutrons. However, slow neutrons are more readily captured by uranium-235 nuclei. The fast neutrons can be slowed down by colliding with some other particles with which they can exchange momentum and energy. The most effective particles are those with about the same mass as a neutron. As an analogy, a moving billiard ball is best slowed down by colliding with another billiard ball, not by hitting a dust particle, which will hardly affect it, or by hitting a boulder, from which it will bounce back with little loss of energy. The particles closest in mass to neutrons are hydrogen nuclei, and that is why water, H_2O, which is a convenient source of hydrogen, is a good choice. A medium that slows down neutrons is called a **neutron moderator.**

2. A neutron can be captured by a uranium-238 nucleus, producing plutonium-239. As you recall from Chapter 13, this reaction takes place in two steps:

$$\mathstrut_0^1n + \mathstrut_{92}^{238}U \longrightarrow \mathstrut_{93}^{239}Np + \mathstrut_{-1}^{0}e$$

followed by the beta decay of Np as

$$\mathstrut_{93}^{239}Np \longrightarrow \mathstrut_{94}^{239}Pu + \mathstrut_{-1}^{0}e$$

Plutonium-239 can undergo fission. Thus, the production of plutonium is, in effect, a "breeding" of new fuel and is therefore attractive as a means of using uranium resources more completely. The choice of whether or not to favor the breeding of plutonium determines, in large part, the design of the reactor.

3. A neutron can be captured by impurities. This causes loss of neutrons and slowing down, or "damping," of the chain reaction. It is thus necessary to have a controlled means of absorbing neutrons to regulate the reaction. The most direct method is to insert a stick of neutron-absorbing impurity. Devices used in this fashion are called **control rods;** they usually contain cadmium or boron and other elements, and they can be inserted into or withdrawn from the reactor core to regulate the neutron flow with great precision.

4. A neutron, traveling as it does in a straight line, may simply miss the other nuclei in the reactor and escape. If the reactor were very small, too many neutrons would escape and the chain reaction would not be sustained. This circumstance imposes lower limits on reactor size; there will never be a pocket-sized fission generator nor even fission engines for motorcycles. This tendency to escape also demands adequate shielding to prevent neutron leakage into the environment.

The following section discusses how all of these requirements are taken into consideration in a typical nuclear power plant in the United States.

14.12 THE DESIGN OF A NUCLEAR POWER PLANT

The heart of a nuclear power plant is the reactor core, in which the essential components are (1) the nuclear fuel, (2) the moderator, (3) the coolant, and (4) the control rods (Fig. 14.18).

1. The nuclear energy source is the uranium-235 isotope, but in pure form it could serve as a nuclear explosive, not as a practical fuel. The uranium actually used is natural uranium, enriched up to 3 percent with fissionable uranium-235. Furthermore, the material used is not metallic uranium, but rather uranium dioxide. This compound is fabricated in a ceramic form that is much better than the pure metal in its ability to re-

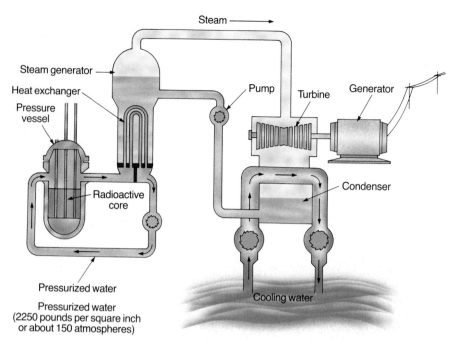

FIGURE 14.18 *Schematic illustration of a nuclear plant powered by a pressurized water reactor.*

tain most fission products, even when overheated. The fuel is inserted in the form of pellets into long, thin tubes, called the "fuel cladding," made of stainless steel or other alloys. These "fuel rods" are then bundled into assemblies that are inserted into the reactor core (Fig. 14.19).

2. The moderator serves to slow down the neutrons. As mentioned earlier, fission is more likely to occur with slow neutrons, but the fission reaction releases fast neutrons, which must be slowed to maintain a chain reaction. Water is convenient for this purpose because

FIGURE 14.19   *(a) A fuel pellet. (b) A fuel rod. (c) An assembly of fuel rods. (d) A fuel assembly being lowered into place in a reactor.*

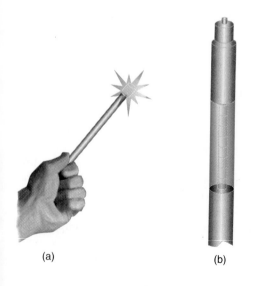

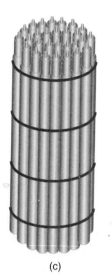

(a)   (b)   (c)   (d)

it can also serve as the coolant. In the United States, almost all commercial reactors use ordinary, or "light," water, H_2O, in which the hydrogen atoms are the ^1_1H isotope. This is cheap but not ideal because H_2O molecules do capture neutrons to some extent. The result of such capture is that natural uranium, which contains only 0.7 percent uranium-235, cannot be used in light-water reactors. Instead, the concentration of uranium-235 must be enriched to about 3 percent to make up for the neutron loss to the coolant water.

3. When water is the moderator, it is also the coolant. In some designs, the coolant water actually boils, and the steam it produces drives the turbines. (You may think it strange for boiling water to be a coolant, but remember that heat always flows spontaneously from a higher to a lower temperature, and even boiling water is much cooler than the core of a nuclear reactor. The heat therefore flows from the reactor core to the boiling water.) In most of the commercial reactors in the United States, the water is kept under high pressure, as in a pressure cooker, and little boiling actually occurs. This is the design shown in Figure 14.18. Note that the pressurized water flows through a heat exchanger, where it transfers its heat to a secondary loop of water that actually boils and delivers steam to the turbine.

4. Interspersed into the matrix of fuel, moderator, and coolant are the control rods (Fig. 14.20), which serve to regulate the flow of neutrons. A nuclear reactor that generates power at a constant rate must operate at a critical condition, with as many neutrons being produced as are lost by capture or escape. Because no design is that perfect, the control rods provide a means of fine-tuning the operation. The control rods serve other important functions, however. Recall that fission reactions produce impurities that absorb neutrons. Therefore, in the absence of any neutron regulation, the neutron flow would gradually slow down, and the fission reactions would die out before much fuel is exhausted. For this reason, there must be a provision to increase the neutron flow gradually during the life of the fuel to make up for the loss caused by impurities. The way to do this is to design for an extra large neutron flow at the outset but to limit the actual flow by control rods. As fuel is consumed, the control rods are gradually withdrawn to compensate for the accumulation of neutron-absorbing impurities. This amounts to a neat balancing of impurities to maintain steady power production. The other purpose of the control rods is to serve as an emergency shut-off system. If an emergency occurs and the fission reaction must be quenched, the rods are pushed rapidly all the way into the core.

14.13 DISPOSAL OF RADIOACTIVE WASTES

Nuclear fuels and their radioactive waste products will become more intimately involved with human activities as the nuclear industry grows. The handling and ultimate disposal of these materials are complex and serious problems.

The half-lives of radioactive wastes range from a fraction of a second to thousands of years. Furthermore, these wastes produce generations of new radioactive substances as they decay, one following the other, before a final stable isotope is produced. The handling of these radioactive wastes involves a sequence of three steps: First, the spent rods are held underwater in cooling ponds for several months to allow the initial, intense radioactivity to die down. Second, the wastes must be "reprocessed" to convert them into a form suitable for the final

FIGURE 14.20 *Control rod mechanism of a nuclear reactor.* (Courtesy U.S. Council for Energy Awareness)

step. Third, and (it is hoped) last, the reprocessed wastes must be "permanently" stored somewhere. Reprocessing and final storage are discussed here.

Reprocessing

The spent fuel rods are first dissolved in strong acid. The resulting solution is radioactive, hot (the radioactivity keeps the temperature up), and corrosive. The dissolved wastes must next be chemically separated. Remember that some useful uranium or plutonium remains in the spent fuel, and this must be recovered and reused. The recovery of plutonium presents a particular danger because it is intensely carcinogenic and because if it is stolen, it can be used to make illicit fission bombs. In fact, the dangers associated with handling plutonium make it a difficult material to steal.

The technical aspects of reprocessing are not theoretically difficult, but the U.S. commercial experience has not been good. The first such reprocessing plant, near Buffalo, New York, opened in 1963 but was closed in 1972 when reprocessing was determined to be economically unfeasible. Reprocessing plants, however, are operating in France, England, Japan, and India; pilot plants are at work in Germany, Italy, and Belgium.

Final Disposal

The third and final step concerns safe, "permanent" storage. How long is permanent? It has been suggested, as a reasonable target, that the wastes should be reduced to the same radioactivity level as that of a natural uranium mine. (Not everyone would consider that sufficient; uranium mining is a contributing cause of lung cancer). If that target is accepted, the time needed for "permanent" storage depends on whether reprocessing is done. If the uranium and plutonium are removed, only the fission products need be permanently stored. These wastes decay more rapidly and would reach the radioactivity level of a uranium mine in about 500 years. If unprocessed spent fuel is to be stored, it is estimated that about 7000 years would be necessary. There have been various suggestions for storage places: deep in the sediments under the ocean floor, under the Antarctic ice cap, or even in outer space. All major current effort, however, is devoted to developing underground, geologically stable

waste sites. At present, several different types of rock formation have been proposed, including salt mines, desert sandstones, and granite or basalt formations, with salt mines the most frequently used location.

Various types of impervious containers have been designed. After the container is sealed in the rock, it continues to give off heat from its radioactivity. Its temperature and that of the surrounding rock remain elevated for about 1000 years. There are two ways in which the radioactive waste could get out: It could be physically exposed (by geological or human activities), or it could be carried out by groundwater. The risks of damage to the host rock by the evolved heat, of erosion by streams or glaciers, of transport by groundwater, and of human intervention have all been studied, and scientists who work on these problems have concluded that the risks are small and acceptable. Critics of the nuclear program have disagreed; they point out that political systems, on which we depend for reliable continuity of any public policy, do not last as long as radioactive isotopes. Furthermore, the assumptions about the permanent segregation of the radioactive containers in salt caverns or rock formations are not proven; they are only predictions.

One other matter regarding the ultimate disposal of reprocessed high-level wastes is the total quantity involved. By comparison with various other wastes, the volume is not large. For example, accumulations of wastes from mining ("mine spoil") and from domestic garbage look like mountain ranges. By contrast, it has been estimated that the volume produced by all commercial nuclear power plants in the year 2000—if concentrated and preprocessed into glassy solids—would make a pile only about 13 feet high on one football field. Translated to a personal basis, that would mean a cube about 4.5 cm on each side, a convenient size for a paperweight, for every individual in the United States. These figures do not mean that the problems described here are trivial. Rather, they are presented to show that it is not the ultimate volume of the wastes, but rather their radioactivity, that is the problem.

14.14 NUCLEAR FUSION

Fission reactions occur when heavy nuclei split apart. *Nuclear fusion reactions occur when nuclei of light elements are joined together.* The energy that is derived from stars in the prime of their life, such as the Sun, comes from hydrogen fusion reactions. Controlled nuclear fusion would provide abundant energy with much less environmental danger than is faced from fission reactors. No useful fusion reactor, however, has yet been developed.

In contrast to the fission reaction, fusion cannot be triggered by neutrons. Instead the nuclei to be fused must be brought into contact with each other. Positive nuclei repel each other at normal interatomic distance, but if they are close together, the strong force predominates and binds them together. To overcome the electrical repulsion and bring the nuclei close enough together for the strong force to take over, the nuclei must be moving rapidly, or, in other words, they must be elevated to high temperatures. The resulting fusion is therefore called a **thermonuclear reaction.** If a large mass of hydrogen isotopes fuses in a short time, the reaction cannot be contained, and it goes out of control; this is the explosion of the "hydrogen bomb." Useful energy could be extracted from fusion if it were possible to devise a controlled thermonuclear reaction.

Any fusion reactor would use hydrogen nuclei. There are three isotopes of hydrogen: "ordinary" hydrogen, ^1_1H; deuterium, ^2_1H; and tritium, ^3_1H. Some information is given about them in Table 14.1. Fusion reactions can occur between any two hydrogen isotopes. The lighter isotopes are more abundant but require much higher temperatures to initiate fusion.

For example, to start the reaction between two deuterium nuclei, the temperature would have to be raised to about 400 million degrees Celsius. The problems imposed by this requirement are so severe that it is not even being attempted. Instead, all

efforts to control fusion are being directed to a cooler reaction (only about 40 million degrees Celsius): the fusion of deuterium with tritium to give helium-4 plus a neutron:

$$^2_1H + {}^3_1H \longrightarrow {}^4_2He + {}^1_0n$$

This process requires a source of tritium, which is not naturally available on Earth. Tritium is produced artificially by neutron bombardment of lithium. Even though this process is expensive, it is the only feasible fusion reaction and is the one being studied at the present time.

Now think for a moment about high temperatures. An iron bar turns red hot at about 600°C, white hot around 1100°C, melts around 1500°C, and the molten iron boils at 2885°C. Other solids survive to higher temperatures—carbon and tungsten, for example, to about 3500°C—but no solids, liquids, or even any chemical bonds survive about 5000°C. At these temperatures, everything is a gas consisting of lone atoms that are much too energetic to combine chemically with each other.

At the temperatures involved in nuclear fusion reactions, measured in millions of degrees, not even atoms survive because electrons are stripped away from the nuclei. Such an extremely hot mixture of independently moving electrons and nuclei is called a **plasma.** Obviously, no container exists that can survive long enough to confine a plasma for the useful production of thermonuclear energy. Instead, what is envisaged is a sort of "magnetic bottle," which does not consist of a physical substance at all but rather is a magnetic field so designed that it will confine the charged particles of the plasma in which the thermonuclear reaction is going on. Research efforts have shown promise, in a modest way. On November 3, 1984, scientists at the Massa-

TABLE 14.1	**Isotopes of hydrogen**		
ISOTOPE	NAMES	RADIOACTIVE?	NATURAL ABUNDANCE (%)
^1_1H	"Ordinary" hydrogen "Light" hydrogen Hydrogen Protium	No	99.985
^2_1H or ^2_1D	"Heavy" hydrogen Deuterium	No	0.015
^3_1H or ^3_1T	Tritium	Yes (12 year half-life)	Almost none

chusetts Institute of Technology maintained a controlled nuclear fusion for about 50 ms (about the blink of an eye), and the brief process actually yielded as much energy as was put into it.

The useful energy, once liberated, will have to be extracted in the form of the kinetic energy of the evolved neutrons. Because the neutrons carry no charge, they will pass through the magnetic field and escape from the plasma. The energy of the speeding neutrons can then be extracted by a moderator. If the moderator were water, the energy would create steam that could drive a turbine. The entire fusion reactor would be encased in a sheath or blanket in which molten lithium would be continuously circulated. The lithium would absorb the neutrons, supply the tritium, and then release its heat to water in a heat exchanger (Fig. 14.21).

Could a fusion reactor get out of control and go off like a hydrogen bomb? Nuclear scientists are entirely confident that the answer is no, an explosion could not occur. The reason is that the hydrogen isotopes are continuously fed into the reactor and continuously consumed; they do not accumulate. The total quantity of fuel in the plasma of a "hot hydrogen" reactor at any one time would be small—about 2 g—far below the critical mass required for a runaway reaction. If the temperature dropped or the plasma somehow dispersed, the reaction would stop; in effect, the fusion would turn itself off.

Would there be a problem with environmental radioactivity? The answer here is yes because both tritium and neutrons could be released. Tritium is radioactive (half-life, 12 years) and can combine with oxygen to form radioactive water. The beta particles emitted by tritium, however, have so little penetrating power that it is virtually harmless to living organisms as long as its source is outside the body. The neutrons released from the reactor can be absorbed by atomic nuclei in the reactor's shielding material, and the new isotopes thus produced may be radioactive. As a result, there could be substantial quantities of radioactive matter to be disposed of. We are so far from a practical fusion reactor that we have hardly begun to study the problems of handling the wastes.

EXAMPLE 14.7   FUSION IN THE SUN

The fusion reactions that power the Sun are be-

lieved to occur in a three-stage process, with energy being released in each of the stages. You fill in the blanks in each of the stages outlined.

(a) In the first stage of the cycle, two protons collide and fuse to form a positron, a neutrino, and another particle. What is the unknown particle?

Solution The reaction can be written symbolically as

$$^1_1H + ^1_1H \longrightarrow X + ^0_1e + \nu$$

Balancing mass numbers and atomic numbers, show that the unknown substance is ^2_1H, deuterium.

(b) In the second stage of the series, a deuterium nucleus produced by the reaction in part (a) fuses with a proton to produce a gamma ray and another nucleus. What is this nucleus?

Solution The reaction is

$$^2_1H + ^1_1H \longrightarrow X + \gamma$$

Balance the mass numbers and atomic numbers to show that X is ^3_2He.

(c) Finally, two of the helium nuclei produced by reactions like those of part (b) fuse to form helium-4 and two other identical particles. What are the identical particles?

Solution The reaction is

$$^3_2He + ^3_2He \longrightarrow ^4_2He + 2X$$

Again, we leave it to you to balance the reaction and to show that the two particles are symbolized as ^1_1H.

(d) Often, these reactions are described figuratively by saying that hydrogen is the fuel and helium-4 is the ash. Start with part (a) of the series and count the total number of hydrogens that have been burned in one series.

Solution The reaction in part (a) had to occur twice so that we would eventually end up with two ^3_2He nuclei to fuse in part (c). Thus, part (a) of the series will consume four hydrogens. The reaction in (b) also must occur twice, so we use a total of two more hydrogens for this reaction. The grand total

FIGURE 14.21 *(a) Cut-away view of a nuclear fusion reactor. (b) Schematic drawing of a thermonuclear power plant.* (Courtesy Department of Energy)

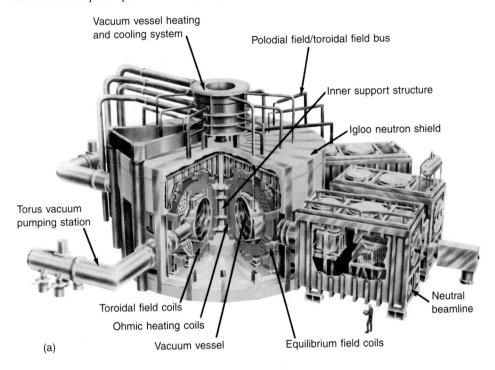

Vacuum vessel heating and cooling system

Polodial field/toroidal field bus

Inner support structure

Igloo neutron shield

Torus vacuum pumping station

Neutral beamline

Toroidal field coils

Ohmic heating coils

Vacuum vessel

Equilibrium field coils

(a)

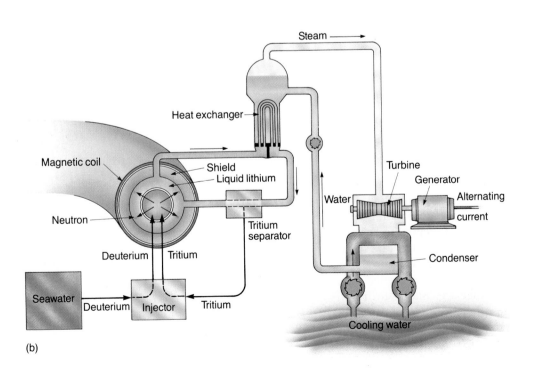

Steam

Heat exchanger

Magnetic coil

Shield
Liquid lithium

Turbine

Generator

Water

Alternating current

Neutron

Tritium separator

Deuterium   Tritium

Condenser

Seawater   Deuterium   Injector   Tritium

Cooling water

(b)

TABLE 14.2   A few of the many elementary particles

CATEGORY	PARTICLE NAME	SYMBOL	ANTIPARTICLE
Photon	Photon	ν	self
Leptons	Electron	e^-	e^+
	Neutrino	ν	$\overline{\nu}$
Muon	μ^-	μ^+	
	Tau	τ^-	τ^+
Hadrons			
Mesons	Pion	π^+	π^-
	Kaon	K^+	K^-
Baryons	Proton	p	$\overline{p}$
	Neutron	n	$\overline{n}$
	Lambda	λ^0	$\overline{\lambda}^0$
	Omega	Ω^-	Ω^+

used is now six. The end result of the reaction in part (c), however, is that two hydrogen nuclei are left over. Thus, the total number of hydrogens used is four, so it may be said that the overall result of this series of reactions is that four hydrogen nuclei have been fused into one helium-4 nucleus. The mass of helium-4 is less than the combined mass of four hydrogen nuclei. By Einstein's mass-energy relationship, this mass is converted into energy.

(lepton means light), of which the electron and the neutrinos are the most familiar. Note that all the particles have a corresponding antiparticle. The third grouping, the hadrons, are made up of two groups called mesons and baryons. Our old friends the proton and neutron are baryons.

The leptons seem to be fundamental particles on their own, but the mesons and baryons are now thought to be composed of more fundamental particles called **quarks.** Thus, perhaps one should not

14.15 QUARKS

Recall that scientists once believed that the atom was a structureless, indivisible entity—the smallest indivisible particle imaginable. As we have noted, this picture has now been considerably modified. This development of the understanding of the atom suggests another line of questions. Is the proton itself really a structureless, indivisible entity, or is it composed of particles that are smaller yet? Particle accelerators were built shortly after Word War II that were capable of producing high-energy nuclear reactions, and it was found that some esoteric particles were being produced that seemed to have no purpose in the world of the atom. Table 14.2 shows a few of the inhabitants of this nuclear zoo and a classification scheme for them. One particle, the photon, has properties that are unique to it and stands apart from the others as a classification of one. A second grouping of particles is the leptons

TABLE 14.3   Some properties of quarks and antiquarks

NAME	SYMBOL	CHARGE
QUARKS		
Up	u	$+\frac{2}{3}e$
Down	d	$-\frac{1}{3}e$
Strange	s	$-\frac{1}{3}e$
Charmed	c	$+\frac{2}{3}e$
Bottom	b	$-\frac{1}{3}e$
Top	t	$+\frac{2}{3}e$
ANTIQUARKS		
Up	$\overline{u}$	$-\frac{2}{3}e$
Down	$\overline{d}$	$+\frac{1}{3}e$
Strange	$\overline{s}$	$+\frac{1}{3}e$
Charmed	$\overline{c}$	$-\frac{2}{3}e$
Bottom	$\overline{b}$	$+\frac{1}{3}e$
Top	$\overline{t}$	$-\frac{2}{3}e$

ask the question of why a baryon called an omega particle exists. The answer well may be that the only reason it exists is that it is a particle that *can* be produced by a combination of quarks. Table 14.3 shows that there are now thought to be six quarks, the up (u), down (d), strange (s), charmed (c), bottom (b), and top (t). Associated with each of these are their antiparticles. For example, the anti-up has the symbol $\bar{u}$. Also, note from the table that the quarks are predicted to have the unique feature of having fractional electronic charges. That is, the u, c, and t quarks have charges of $\frac{2}{3}e$, whereas the d, s, and b quarks have charges of $-\frac{1}{3}e$.

All elementary particles falling under the heading of mesons are believed to be composed of a collection of two quarks bound together as a unit. For example, one of the most commonly produced mesons in nuclear events is the pi meson, or pion, represented by the symbol π^+. The plus sign as a superscript means that it has the same charge as a proton. The quark combination that produces the π^+ is the combination u $\bar{d}$. Let us check the charge

on the resulting particle composed of this combination to see if it has a chance of producing the pion. We find the net charge is the sum of that due to the u ($\frac{2}{3}e$) and the $\bar{d}$ ($\frac{1}{3}e$) to produce a grand total of e, the desired charge.

Baryons are believed to be combinations of three quarks joined together. For example, the proton is thought to be a uud grouping. Note that the charge adds up properly, $\frac{2}{3}e + \frac{2}{3}e - \frac{1}{3}e = e$.

The neutron is composed of two downs and one up. You check this out to see if the charge adds up correctly.

Elementary particle physics is an active field of investigation with many questions to be answered. Whey are some particles charged and others neutral? Whey do quarks carry a fractional charge? What determines the masses of the quarks? Can isolated quarks exist? The questions go on and on, and because of the rapid advances and new discoveries in this field, by the time you read this book, some of these questions will likely be resolved and others may have emerged.

SUMMARY

Ether was a material proposed as the medium through which electromagnetic waves would travel. Since it was considered to be a privileged frame of reference for the measurement of velocities, extensive research was done in an effort to detect it. This culminated with the Michelson-Morley experiment, which produced the result that the ether does not exist.

Einstein's special theory of relativity is true only for reference frames that are moving with constant velocity with respect to one another. The fantastic results of this theory indicate that moving clocks run slowly, that lengths contract for objects in motion, and that mass increases for objects in motion. Finally, it predicted that energy could be converted into mass, and vice versa. The processes of **pair production and annihilation** verify this as well as does the use of nuclear reactors and bombs.

In **nuclear fission,** an isotope of a heavy element is split into lighter elements, releasing a large amount of energy. The important naturally occurring fissionable isotope is uranium-235. The fission is triggered by a neutron, and each atom releases two or three neutrons. The result can be a **chain reaction,** in which a series of steps occurs. Nuclear fission can be used in an atomic bomb

or in a nuclear power plant. The difference between the two applications depends on the neutrons. There are four things that can happen to a neutron:

1. Fission capture by uranium-235 to yield energy and fast neutrons. But the reaction is favored by slow neutrons. Therefore, a **moderator,** which is a substance that slows down neutrons, can be used.
2. Nonfission capture by uranium-238, thereby producing plutonium-239, which can undergo fission.
3. Nonfission capture by impurities, which causes loss of neutrons. This action can be used to control the fission process.
4. Escape. A neutron might miss everything and get lost.

In **nuclear fusion,** the nuclei of the light isotopes deuterium and tritium combine to produce helium and release energy. Most subatomic particles are now thought to be made up of particles, called **quarks,** that have fractional electronic charges. There are six of these, called the up, down, strange, charmed, bottom, and top quark.

EQUATIONS TO KNOW

$$\Delta t = \frac{\Delta t'}{\sqrt{1 - \dfrac{v^2}{c^2}}} \quad \text{(time dilation)}$$

$$m = \frac{m_0}{\sqrt{1 - \dfrac{v^2}{c^2}}} \quad \text{(mass increase)}$$

$$L = L' \sqrt{1 - \frac{v^2}{c^2}} \quad \text{(length contraction)}$$

$$E = mc^2 \qquad \text{(mass-energy equivalence)}$$

KEY WORDS

Reference frame
Luminiferous ether
Michelson-Morley experiment
Special relativity
Length contraction

Mass increase
Energy-mass equivalence
Antimatter
Pair production
Pair annihilation

Positron
Antiproton
Nuclear fission
Chain reaction
Moderators

Control rods
Nuclear fusion
Plasma
Quark

PROBLEMS AND CONCEPTUAL QUESTIONS

Problems requiring numerical work are identified with a blue number.

Relativity before Einstein

1. A person driving from Los Angeles to San Francisco is said to be moving northwest. But such a simple description ignores the fact that the person is rotating with the Earth, orbiting the Sun, and simultaneously flying through intergalactic space. Would it be more accurate to define the net motion of the car as the sum of all these independent earthly celestial movements? Explain.

2. A hiker is walking northward along the shore at a rate of 2 km per hour, while a canoeist is paddling south at 2 km per hour. (a) What are their relative velocities with respect to each other? (b) What are their velocities relative to a stationary observer on the riverbank?

3. Rocket ship X, traveling from the Moon to Earth, is flying at 1000 km per hour relative to the Earth. An astronaut in that rocket ship is floating forward at 1 km per hour relative to the vessel. A second rocket, Y, is flying from the Earth to the Moon directly toward rocket X at a rate of 1500 km per hour. The pilot of the second rocket is stationary relative to his rocket. Calculate the velocity of (a) rocket X with respect to the Earth, (b) rocket X with respect to

rocket Y, (c) the astronaut in rocket X with respect to the Earth, (d) the astronaut in rocket X with respect to the astronaut in rocket Y.

4. Would it be possible to play a game of pool on an airplane moving at 1000 km per hour? 500 km per hour? Would it be possible to play a game of pool on an airplane that is accelerating? Turning a sharp corner? Explain and discuss.

5. If velocity is a relative quantity, is kinetic energy also relative to the reference frame used? Explain.

The Michelson-Morley experiment

6. What was the ether thought to be? How did scientists search for it? What is the significance of their findings?

7. In the situation of problem 3, a beam of light is directed at ship X from Earth. What is the speed of this light as measured by the astronaut in rocket X?

8. An extremely fast rocket is traveling toward Earth at a rate of 2×10^8 m/s. The pilot of that ship is moving forward at a rate of 1 m/s with respect to the vessel. A radio operator on Earth sends a message to this rocket. What is the speed of the radio signal with respect to (a) the Earth, (b) the moving rocket, and (c) the person in the rocket? Explain, and discuss the implications of your answer.

9. For the passage of light, the ratio of distance to time is a constant throughout the Universe. What is the value of this ratio?

10. What is the speed of light measured here on Earth from the following sources: (a) radiated directly from the Sun, (b) reflected off the Moon, and (c) originating from a distant galaxy that is flying away from us at a rate of 2×10^8 m/s?

11. A question that supposedly perplexed Einstein as a child is: What would happen if someone runs at the speed of light while carrying a mirror in his hand? What do you believe the runner would see in the mirror?

12. A passenger standing at the rear of a north-bound train moving at 50 m/s throws a ball at a speed of 50 m/s southward. Where does the ball hit the ground?

13. In our discussion of refraction, we stated that the speed of light in glass is about 2×10^8 m/s. Is this a contradiction of the results of the Michelson-Morley experiment?

Einstein and time, length contraction, relativity and mass, and mass and energy

14. If you were traveling in a rocket ship at close to the speed of light, could you detect any contraction in length or a slow-down in time within your environment? Explain and discuss.

15. Two observers watching a baseball game see a batter hit a fly ball to the outfield. One of the observers is in the stands, and the other is in a rocket ship passing by the stadium. Which of the following do the observers agree on, the height to which the ball rises or the distance it travels?

16. Would one be able to observe length contraction or time dilation if the speed of light were infinite? Use the equations for these effects to support your answer.

17. What is the reasoning behind the conclusion that mass and energy are related?

18. While working on a project in a rocketship traveling at 0.90 c, the heartbeat of an astronaut increases to 100 beats/min as measured on the ship. (a) What would an observer monitoring the heart rate from Earth measure? (b) If the observer on Earth also has a heart rate of 100 beats/min, what heart rate will the astronaut monitor for him?

19. At what speed would a rocketship have to be traveling so that its rest length of 20 m is reduced to 10 m?

20. A 0.5 kg football is kicked with a speed of 0.95 c. What is its mass at this speed?

21. (a) How fast would a proton have to travel before its mass is doubled? (b) Repeat for an electron.

22. If all the mass of a proton could be converted into energy, how many joules would be produced?

23. (a) If 1 g of matter should be converted into energy, how many joules would be produced? (b) A 100 W light bulb uses energy at the rate of 100 joules/s. How many seconds would the 1 g of matter keep this light bulb burning?

Nuclear fission and nuclear power plants

24. Use the periodic table of the elements to find some possible products for the fission of U-235. Write your results in the form of a nuclear reaction.

25. You have been assigned the task of explaining the difference between fission and fusion to someone who has never had a physical science class. Write down your answer outlining what you would tell this person. Remember you cannot use any words outside of the person's common experience.

26. A neutron strikes U-235 and the fission products are Sr-88 and Xe-136. How many neutrons are produced?

27. What are the essential features of a nuclear fission reactor? Explain the function of each feature.

28. List the possible fates of neutrons in a fission reactor. Which of these events should be favored and which should be inhibited to (a) shut down a reactor, (b) breed new fissionable fuel, and (c) produce more energy?

29. Could a nuclear reactor ever be miniaturized to provide long-term power for your wristwatch? Your camera? Your pocket radio? (Assume that the proper shielding against radioactivity could be provided.) Defend your answer.

30. If no emergency ever occurred, and if a nuclear reactor always operated at steady power production, would control rods still be needed? Defend your answer.

Fusion

31. Suppose that someone claims to have found a material that can serve as a rigid container for a thermonuclear reactor. Would such a claim merit examination, or should it be ignored as a

"crackpot" idea not worth the time to investigate? Defend your answer.

32. Outline the reasons why fusion reactors are expected to be far less serious sources of radioactive pollutants than fission reactors.

33. Write the following reaction in equation form. Two deuterium nuclei collide to produce tritium and an additional particle. What is the particle?

34. Write the following reaction in equation form. Deuterium and tritium collide to produce He-4 and another particle. What is the particle?

Quarks

35. Which of the following quark combinations could produce the π^- meson: $u\bar{s}$, $\bar{u}d$, or uds?

36. Which of the following quark combinations could produce the Ω^- particle, $u\bar{s}$, $\bar{u}d$, sss, or uus? *Hint:* This particle is not classified as a meson.

37. The study of elementary particles is a realm of scientific investigation in which there is unlikely to be a payoff in new products for consumers. In light of this, how can such research be justified? Take one side or the other and defend your answer.

ANSWERS TO SELECTED NUMERICAL PROBLEMS

2. (a) 4 km per hour, (b) 2 km per hour for each
3. (a) 1000 km per hour, (b) 2500 km per hour, (c) 1001 km per hour, (d) 2501 km per hour
7. 3×10^8 m/s
18. 229 beats/min, (b) 229 beats/min
19. 0.87 c
20. 1.60 kg
21. (a) 0.87 c, (b) 0.87 c
22. 1.50×10^{-10} J
23. (a) 9×10^{13} J, (b) 9×10^{11} s (about 30,000 years)
26. 11
33. ^1_1H
34. ^1_0n

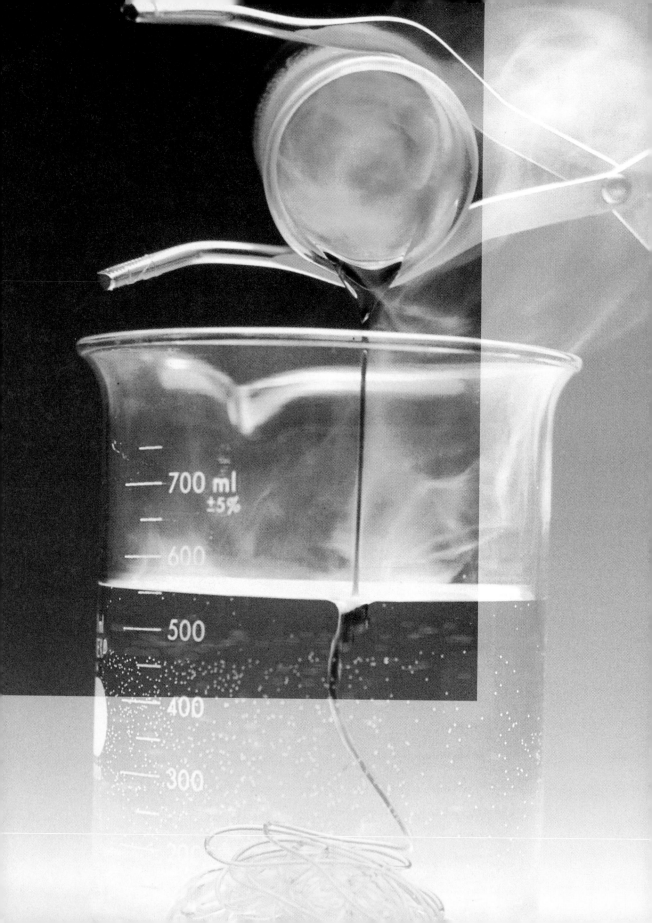

PART TWO
CHEMISTRY

The physical sciences break up neatly into the divisions of physics, chemistry, earth science, and astronomy only in the pages of a textbook. We will see when we study chemistry that we will often rely on our knowledge of physics. For example, to understand chemical reactions we must understand the forces that arise between atoms when they are brought into close proximity. This requires a general knowledge of such physics topics as forces and electricity because the forces between atoms are frequently electrical interactions. To understand why gases behave as they do requires a knowledge of how atoms behave and how this behavior explains such large-scale properties as pressure, volume, and temperature.

As we proceed beyond our study of chemistry to address earth science and astronomy we will see that an understanding of complex chemical reactions is essential to the study of the atmosphere, waters, and layers of the earth. And as we move on to discuss astronomy we will learn something of the origins of the atoms and molecules that we have studied.

When sulfur is heated, a transformation takes place that results in the formation of long chains of sulfur atoms called "plastic sulfur." Following your study of chemistry you will be able to visualize why and how such changes occur. (Courtesy Charles D. Winters)

JACQUELINE K. BARTON

INTERVIEW

Jacqueline K. Barton is a native New Yorker and was educated in that city. She received her B.A. degree from Barnard College in 1974 and her Ph.D. from Columbia University in 1979. Following her Ph.D. work, Dr. Barton did further research at Yale University and Bell Laboratories and then joined the faculty of Hunter College. In 1983 she returned to Columbia University where she rose rapidly to the rank of full professor. In the fall of 1989 she assumed her present position as Professor of Chemistry at the California Institute of Technology.

In spite of having been a research scientist for a relatively short time, she has done important new work and has received many honors. In 1985 she received the Alan T. Waterman Award of the National Science Foundation as the outstanding young scientist in the United States. In 1987 she was the recipient of the American Chemical Society's Eli Lilly Award in Biological Chemistry, and the following year she received the Society's Award in Pure Chemistry. That same year, 1988, she also received the Mayor of New York's Award of Honor in Science and Technology.

A BACKGROUND IN MATH—BUT NO CHEMISTRY

I never took chemistry in high school. Maybe one shouldn't publicize that, but it's the truth. However, I was always very interested in mathematics, so I took a lot of calculus when I was in high school. I also took a course in geometry, and that interest in geometry has carried over into my research, since the sort of science I do now is very much governed by structures and shapes.

When I went to college I thought that, in addition to taking math, I should take some science courses. I walked into the freshman chemistry class, and there were about 150 people there. However, there was also a small honors class with about 10 students. Even though I hadn't had chemistry before I thought I would try it—and loved it. What chemistry allowed me to do was to combine the abstract and the real. I was very excited by it.

The interest of my group is to exploit inorganic chemistry as a tool to ask questions of biological interest and to explore biological molecules. A lot of the work in bioinorganic chemistry thus far has been the exploration of metal centers in biology. Why is blood red? Why does the iron [in heme] do what it does? That's just one example, but there are hundreds of others. Many enzymes and proteins within the body in fact contain metals, and the reason we've looked at blood and then the heme center within it has been because it's colored. An obvious tool that transition metal chemistry provides is color, and so things change color when reactions occur. That is one of the things that fascinated me in the first place.

Another wonderful thing about transition metal chemistry is that it allows us to build molecules that have interesting shapes and structures depending upon the coordination geometry. In fact, you can create a wealth of different shapes, several of which are chiral, and that's something we take advantage of in particular. What we want to do is make a variety of molecules of different shapes, target these molecules to sites on a DNA strand, and then ask questions such as 'Does DNA vary in its shape as a function of sequence?' If we think about how proteins bind to DNA, do they also take advantage of shape recognition in binding to one site to activate one gene or turn off another gene? When scientists first wondered about these and other such problems, they would write down a one-dimensional sequence of DNA and would think

IT'S AN EXCITING TIME TO BE DOING CHEMISTRY, AND THAT IS WHY I SEE IT AS A NEW FRONTIER AREA.

about it in one-dimensional terms. How does the protein recognize a particular DNA sequence? DNA is clearly not one dimensional. It has a three-dimensional structure, and different sequences of bases will generate different shapes and different forms. Therefore, we think we can build transition metal complexes of particular shapes, target them to particular sequences of bases in DNA, and then use these complexes to plot out the topology of DNA. We can then ask how nature takes advantage of this topology. We want to develop a true molecular understanding, a three-dimensional understanding, of the structure and the shapes of biologically important molecules such as DNA and RNA.

A REVOLUTION IN CHEMISTRY IN THE LAST TEN YEARS

I think our work may be an example of where chemistry is going in general. I think there has been a revolution in chemistry in the past 10 years. The revolution is the interface between chemistry and biology where we can now ask chemical questions about biological molecules. First of all, we can make biological molecules that are pure. I can now go to a machine called a DNA synthesizer, and I can type in a sequence of DNA; from that sequence I can synthesize a pure material, with full knowledge of

where all of the bonds are. Then I can run it through a HPLC and get it 100% pure.[1] Therefore, I can now talk about these biopolymers in chemical terms as molecules rather than as impure cellular extracts. I couldn't do that before.

The development of new techniques allows us to make a bridge between chemistry and biology and ask chemical questions with molecular detail. It's an exciting time to be doing chemistry, and that is why I see it as a new frontier area.

NEW CHEMISTRY CURRICULA

Chemists are making new materials and making and exploring biological systems. It's the chemist who looks at questions of molecular detail and asks about structure and its relationship to function. We are going to have to stop making divisions betwen inorganic, physical, analytical, and organic chemistry. We must all do a little bit of each.

What is important in the education of scientists is to get across the excitement that now we can know what biologically important molecules look like. And, from knowing what they look like, we can manipulate them and change them a little. Then we ask how these changes affect the function, so we can relate the structure of the molecule and its macroscopic function.

A protein molecule of average size is so small you could put more than a billion billion of them on the head of a pin. We now know we can manipulate

---

[1]An HPLC is a "high pressure liquid chromatograph," an instrument capable of separating one type of molecule from another.

molecules that are of those dimensions and can know exactly what they look like. I can't imagine that we can't get people interested in chemistry if we can get across the excitement that comes from the realization that we are looking at things so small and yet can do surgery on them.

CHEMISTRY IS FUN

The bottom line is that chemistry is fun, it's addictive, and, if one has a sense of curiosity, it can be tremendously entertaining and appealing. And it is not so difficult. It's difficult when one thinks about it as rote memorization, which is difficult and boring. But that isn't what chemistry is. Chemistry is trying to understand the world around us in some detail. For example, we are interested in knowing such things as what makes skin soft, what makes things different in color, why sugar is sweet, or why a particular pharmaceutical agent makes us feel better.

WOMEN IN SCIENCE. ARE THERE SPECIAL OPPORTUNITIES FOR WOMEN? ARE

CHEMISTRY IS TRYING TO UNDERSTAND THE WORLD AROUND US IN SOME DETAIL.

THERE PROBLEMS THAT WOMEN NEED TO OVERCOME OR BE AWARE OF?

Because I am a woman, and there are so few women currently in professional positions in chemistry, I'm asked those questions often. First of all, I am not an expert on the subject. What I like to think my best contribution to women in chemistry can be is to do the best science I can, and to be recognized for my science, not for being a woman in science. I think that it is generally important when women go into science that they should appreciate that there are no special opportunities; that is, you will be treated like any other person doing science. But just as there should be no special opportunities in that respect, happily—maybe this is naive of me—I think there are also no special detriments or obstacles that one need to consider

in this day and age. One shouldn't think that 'because I am a woman I can't do it.' That's patently false. In fact, everyone is extremely supportive of women who do science. However, I remember talking to Bernice Segal, my former teacher at Barnard College, and having her explain to me that when she was a graduate student she had to do things behind a curtain, because the women weren't supposed to be doing chemistry. Mildred Cohn, another one of my role models, took over 20 years to have her own independent position as a professor, as opposed to being a laboratory assistant working for someone else. The bottom line is that I don't have a story like that to tell. That's the good news. In my generation there are few such stories of blatant discrimination. Now the world is a much better place for a woman to do science.

This interview was conducted by John C. Kotz and Keith F. Purcell and originally appeared in their text *Chemistry and Chemical Reactivity, Second Edition,* Saunders College Publishing, 1991. The interview appears here in an abbreviated form.

CHAPTER 15

ELEMENTS, COMPOUNDS, AND THREE STATES OF MATTER

Some common chemical elements and chemical compounds. (Charles D. Winters)

15.1 STUDY OF CHEMISTRY

Chemistry is the study of matter and the changes it undergoes. *Matter*—defined as anything that occupies space and has mass—is the material of the universe. Matter includes the air we breathe, the water we drink, the food we eat, the clothes we wear, and the countless things we use in our modern society. Thus, everything we see or use has a "chemical" connection. A transformation that produces a new substance is called a **chemical change** or **chemical reaction.** For example, the burning of wood is a chemical change that produces ash, smoke, and other substances that we cannot see, such as carbon dioxide. Other examples of chemical reaction are photosynthesis, digestion, fermentation, and rusting. A transformation that does not produce a new substance is called a **physical change.** For instance, when we make ice cubes and boil water we are creating physical changes because water, ice, and steam are different states of the *same* substance.

Antoine Laurent Lavoisier (1743–1794) is often called the father of modern chemistry. His most important contribution was his law of conservation of matter, which established chemistry as a quantitative science. The **law of conservation of matter** states that matter is neither lost nor gained in a chemical reaction. As an example of Lavoisier's research, consider the chemical decomposition of mercury oxide (HgO) into mercury and oxygen. The chemical reaction is

$$2HgO \longrightarrow 2Hg + O_2$$

Lavoisier's careful weight measurements showed that the total weight of all the chemicals remains constant during the process of a chemical change. This means that if one finds the weight of all the

Chemical changes. (a) Iron rusts, (b) plants grow, (c) rocket fuel burns, (d) an egg hardens, (e) muscle action consumes sugar. [(b) and (e) courtesy Greg Perry; (c) courtesy NASA]

matter before a chemical reaction, one will find the same weight of matter after the reaction takes place. This was definite proof that *substances* can be produced or destroyed, for example, mercury oxide in the equation, but *matter* cannot. You have the same weight of mercury and oxygen atoms after the reaction as you had before. Thus, Lavoisier showed that a chemical reaction is just a recombination of atoms; it is not a process by which matter is created or destroyed.

Properties that depend on the size or shape of a sample of matter are called **extensive properties.** Examples are the length of a ruler, the mass of a rock, and the volume of oil in a barrel. Properties that are independent of the size or shape of a sample are **intensive properties.** Thus the temperature of water, the density of lead, and the color of emerald are all examples of intensive properties.

15.2 ELEMENTS

An **element** is a substance that cannot be separated into simpler substances by chemical means. In Chapter 12, we saw that an element is made up of only one kind of atom. Obviously, then, carbon and oxygen are different elements because they consist of different kinds of atoms. But what about isotopes,

such as $^{12}_{6}C$ and $^{14}_{6}C$? These atoms have different mass numbers, and, furthermore, carbon-12 is stable, whereas carbon-14 is radioactive. Nonetheless, they are classified as the same element—carbon—because they exhibit the *same* chemical properties, which are dependent only on the atomic number.

When you think about a particular element, you may think either about its atoms or about the substance made up of those atoms. For example, if you are asked to say something about carbon, you might answer that it has an atomic number of 6, that the six nuclear protons are electrically balanced by six electrons outside the nucleus, and that the most abundant isotope has six neutrons in the nucleus, adding up to a mass number of 12. If you said those things, you would have been thinking about the carbon atom. But you might also answer, equally correctly, that carbon is a soft, black solid commonly known as graphite. Then you would have been thinking about the *substance* called carbon, which is made up of carbon atoms. Someone else, however, might offer a different answer, saying that carbon is a hard, brilliant substance, not a soft, black solid. That description, too, is correct, because it refers to diamond, which is also made up of carbon atoms.

FOCUS ON
BUCKYBALL—A NEW ALLOTROPE FOR CARBON

For many years, the only two familiar allotropes of carbon were graphite and diamond. The situation changed dramatically in 1985. That year a team of chemists at Rice University in Texas used a high-powered laser to vaporize graphite, in an attempt to create unusual molecules believed to exist in interstellar space. Among the products formed was a stable molecule whose mass corresponds to the formula C_{60}. Working with paper, scissors, and Scotch tape, the chemists assembled a closed sphere from 20 paper hexagons and 12 paper pentagons. This sphere contains 60 vertices, which corresponds to 60 carbon atoms.

The chemists named the new molecule "buckminsterfullerene," in honor of the late R. Buckminster Fuller, whose engineering and architectural designs often incorporated geodesic domes. But buckminsterfullerene is something of a mouthful, and the new molecule quickly became known as "buckyball," because of its resemblance to a soccer ball. Subsequent spectroscopic and X-ray measurements have shown that the predicted geometry is indeed correct.

The buckyball discovery has generated tremendous interest in the scientific community. After all these years, we now have another common allotrope of carbon, in addition to diamond and graphite. It turns out that buckyball can be

A representative of a third form of carbon, a new class of molecules called the fullerenes. One of the most important is C_{60}, buckminsterfullerine or "buckyball."

prepared under less extreme conditions, and it is a component of soot. Furthermore, chemists have created similar caged molecules with 70, 76, and even larger numbers of carbon atoms. In 1992, the fullerenes C_{60} and C_{70} were discovered in a mineral called *shungite*.

Buckyball and its relatives are more than just an interesting class of molecules. They represent a whole new concept of molecular architecture. Also, there are indications that these molecules and compounds derived from them can act as lubricants, high-temperature superconductors, and even antiviral drugs.

These two correct descriptions raise the question: Can one element exist in the form of different substances? The answer is yes. The definition of an element does not specify how its atoms are arranged or bonded to each other. Different arrangements or different bondings make different substances. Such different forms of the same element are called **allotropes.** Graphite and diamond are allotropic forms of the element carbon. Different types of chemical bonding are described in the

next chapter, but a brief explanation of the allotropy of carbon is helpful here. The carbon atoms in graphite are arranged in layers that can easily slip past each other, making graphite soft and slippery, whereas the carbon atoms in diamond are strongly bonded in three dimensions and are difficult to separate, making diamond a hard substance. Do you know that diamond, being a form of carbon, can burn? It's not easy, but if you heat a diamond to an extremely high temperature in an atmosphere of pure oxygen, it will burn with a bright glow until it is all gone.

The physical states of matter, such as solid, liquid, and gas, will be discussed later in this chapter. For now, note that the different physical states of an element, such as liquid mercury and solid mercury (below $-39°C$), are not considered to be allotropes. Therefore, a proper definition of allotropes is that they are two or more forms of the same element that differ significantly in chemical and physical properties. Do not confuse allotropes with isotopes. Carbon-12 and carbon-14, for example, are isotopes, not allotropes. Carbon-12 can exist as graphite or diamond; so can carbon-14. A number of other elements also exist in allotropic forms. An interesting and important example is oxygen, whose allotropes will be described in the next section.

How Many Elements Are There?

A total of 109 elements occur naturally on Earth or have been synthesized in nuclear reactors and in high-energy accelerators. Because nuclear chemistry is an active field, it seems most likely that the total number will increase in the future. The naturally occurring elements range from atomic number 1 (hydrogen) to 92 (uranium), but three elements within this set are synthetic elements. They are technetium (43), promethium (61), and francium (87). All the elements beyond uranium, called transuranium elements, are synthetic.

Abundance of the Elements

The most abundant element in the universe is hydrogen, which makes up about 75 percent by mass of all matter known to exist. Helium is next, with about 25 percent. The remaining small proportion is made up of the rest of the elements. This universal abundance reflects, for the most part, the compositions of stars and intergalactic dust and gas, not the planets, since planetary material composes only a minute portion of the total universe.

The planet Earth consists mostly of iron, oxygen, silicon, and magnesium, but these elements

FIGURE 15.1 *Abundance of the elements, in percent by mass; (a) in the Universe; (b) in the Earth; (c) in Earth's crust; (d) in the human body.*

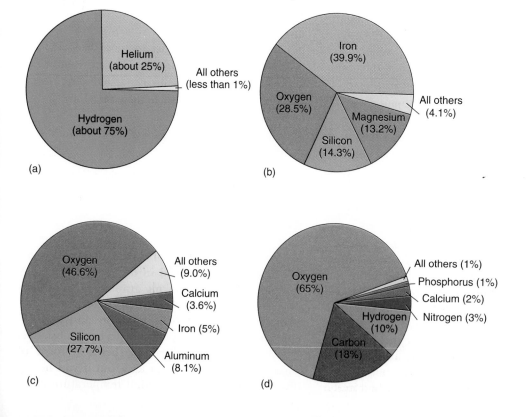

are by no means uniformly distributed. Most of the iron is in Earth's core and therefore not available for use. The most abundant elements in Earth's crust, which is the relatively paper-thin layer that serves as the environment for living organisms, are oxygen, silicon, aluminum, iron, and calcium. Ninety-nine percent of the mass of the human body consists of oxygen, carbon, hydrogen, nitrogen, calcium, and phosphorus.

The elemental abundances are shown in Figure 15.1.

Physical and Chemical Properties of the Elements

About 75 percent of the elements are metals. Metals are good conductors of heat and electricity, and they have a shiny appearance. There are 17 nonmetallic elements and 8 elements that are somewhere in between. These elements, called *metalloids*, have properties that lie between metals and nonmetals. Metalloids such as silicon and germanium are important in the semiconductor industry for the manufacture of transistors and computer chips.

Of the 109 elements, 11 are gases at ordinary temperatures and pressures, and only 2 are liquids (bromine and mercury). On a hot summer's day, two other metallic elements, cesium and gallium, would melt and join the company of liquids.

Among the metals, only a few are found in the elemental (that is, chemically uncombined) state in nature. These elements include the precious metals gold, silver, and platinum. Think of what this means. The fact that elemental gold can be found among the grains of sand in a stream bed means that it does not react chemically with water or air. Gold is chemically unreactive but not totally inert like helium, since it would combine with certain elements under drastic conditions. Most of the other metals are found only in a chemically combined state, as in minerals. As an example of this second category, consider calcium. Calcium is the fifth most abundant element in Earth's crust, but you could search all the natural places on the surface of the planet and never find a speck of metallic calcium. All the calcium is chemically locked up as compounds in formations such as limestone mountains, coral reefs, animal bones, sea shells, and pearls. Metallic calcium can be prepared in the laboratory, but if it is exposed to air or water, it starts to react at a steady but not violent rate. Some other metals

TABLE 15.1	Some important elements	
ATOMIC NUMBER	NAME	SYMBOL
1	Hydrogen	H
2	Helium	He
6	Carbon	C
7	Nitrogen	N
8	Oxygen	O
9	Fluorine	F
10	Neon	Ne
11	Sodium	Na
12	Magnesium	Mg
13	Aluminum	Al
14	Silicon	Si
15	Phosphorus	P
16	Sulfur	S
17	Chlorine	Cl
18	Argon	Ar
19	Potassium	K
20	Calcium	Ca
26	Iron	Fe
29	Copper	Cu
47	Silver	Ag
78	Platinum	Pt
79	Gold	Au
80	Mercury	Hg
82	Lead	Pb
86	Radon	Rn
88	Radium	Ra
92	Uranium	U
93	Neptunium	Np
94	Plutonium	Pu

react more rapidly, others more slowly. A piece of metallic sodium or potassium, if it is dropped into water, reacts violently. The reacting substances get so hot that the evolved hydrogen gas would catch fire. Iron reacts much more slowly; for example, it takes days for rust to form. Nevertheless, iron is also never found in the elemental state in nature.

You will not be expected to memorize the names and symbols of all the elements, but it would be helpful to know at least the common ones shown in Table 15.1. (You need not memorize the atomic numbers, however.) The complete list appears inside the back cover of this book.

EXAMPLE 15.1 ISOTOPES AND ALLOTROPES

Which of the following pairs are isotopes, which are allotropes, and which are different elements? **(a)** "White tin," a soft metal, and "gray tin," a crumbly nonmetallic powder; **(b)** uranium-238 and uranium-235; **(c)** potassium $^{40}_{19}K$ and argon $^{40}_{18}Ar$.

Solution (a) The two forms of tin have different chemical properties (one is metallic, the other is not); they are allotropes. **(b)** The two uraniums have different mass numbers; they are isotopes. **(c)** Potassium and argon are different elements because they have different atomic numbers; the fact that they have the same mass number does not make them the same element; it just means that the sum of the nuclear protons and neutrons happens to be the same.

15.3 MOLECULES

The atoms of most elements tend to combine with other atoms. These combinations of atoms, called molecules, are held together by electrical forces. In this context, the electrical forces are called chemical bonds. A **molecule** may therefore be defined as an aggregate of at least two atoms in a definite arrangement held together by chemical bonds.

A **chemical formula** represents the composition of a molecule in terms of the symbols for the atoms of the elements involved. A subscript after each symbol indicates the number of the atoms of that element present. When there is only one atom present, the subscript 1 is not shown. Groups of symbols sometimes appear in parentheses; a subscript after the end parenthesis then refers to all the symbols inside. Here are some examples:

Formula	Molecule
O_2	Oxygen—there are two oxygen atoms per molecule
O_3	Ozone—there are three oxygen atoms per molecule. Oxygen and ozone are allotropes
H_2O	Water—there are two H atoms and one O atom per molecule
NH_3	Ammonia—there are one N atom and three H atoms per molecule
C_3H_8	Propane (used as a fuel)—there are three C atoms and eight H atoms per molecule
$(NH_2)_2CO$	Urea (used as a fertilizer)—there are two N atoms, four H atoms, one C atom, and one O atom per molecule

Molecules change as the bonds between their atoms are made or broken; that's chemistry. Eat a piece of bread, and the starch molecules in it break down to sugars. The sugars react with the oxygen you breathe and eventually are converted to the carbon dioxide (CO_2) and water you exhale. These molecules, in turn, may be converted to parts of the molecule of a blade of grass, which are then converted to something else by the cow or horse that eats the grass. Thus, molecules are broken up, rearranged, and recombined during chemical reactions, but atoms (all but those of the radioactive isotopes) do not change. Any particular oxygen atom in a water molecule in your body may once have been the oxygen gas inhaled by William Shakespeare or part of a dinosaur bone 100 million years ago.

15.4 COMPOUNDS AND MIXTURES

If you buy several brands of granulated white table sugar, perhaps from different stores or at different times of the year, you will find them all to be basically alike. On a large-scale level, for example, you might judge them all to be equally sweet. Additionally, they will all have the same composition corresponding to the molecular formula $C_{12}H_{22}O_{11}$. Table sugar or cane sugar, the chemical name for which is sucrose, is a **substance,** which is a form of matter that has a definite composition (the number and type of atoms present) and distinct properties. A substance whose molecules consist of more than one kind of element is called a **compound.**

Quartz, which in the form of small particles is called sand, is the compound known as silicon dioxide, SiO_2. If you stirred some sand into the cane sugar, the result would be a **mixture,** which is a combination of two or more substances in which the substances retain their identity.

Now, if you spread out the mixture of sand and sugar on a smooth surface and used a magnifying glass and a fine tweezers, and were patient, you could separate them because the sugar grains have different shapes than the grains of sand. Such a mixture, whose properties are not the same throughout, is said to be **heterogeneous.** A mixture that is uniform, having the same properties throughout, is said to be **homogeneous.**

Now consider this question: How can we create a homogeneous mixture? Simply by dissolving sugar in a beaker of water and stirring the liquid until the

A heterogeneous mixture of fruit.

dissolved sugar is evenly distributed throughout. Such a homogeneous mixture of water and sugar is called a **solution.** For a liquid solution, the **solute** is the substance present in smaller amount in a solution, whereas the **solvent** is the substance present in larger amount in a solution. Thus, sugar (the solute) dissolves in water (the solvent).

Gases, too, dissolve in liquids. An example is ammonia water, a household cleaning fluid. It is prepared by dissolving ammonia gas in water. When liquids dissolve in each other, the distinction between solvent and solute becomes less clear. Gasoline, for example, is a solution made of many different liquids. It is a homogeneous mixture.

Finally, solids can be dissolved in each other. The most important examples are solid solutions of metals, called **alloys.** Some familiar ones are bronze (an alloy of copper and tin), brass (copper and zinc), and plumber's solder (lead and tin). Outlined in Figure 15.2 is a classification that shows the relationships among various terms that have been discussed so far in this chapter.

EXAMPLE 15.2   ELEMENTS AND COMPOUNDS

Which of the six substances whose formulas were listed earlier, O_2, O_3, H_2O, NH_3, C_3H_8, and $(NH_2)_2CO$, are compounds?

Solution H_2O, NH_3, C_3H_8, and $(NH_2)_2CO$ all have more than one kind of element per chemical formula; they are compounds. O_2 and O_3 are elements; they exist as molecules, but not as compounds.

15.5 CHEMICAL FORCES

Having introduced chemical change, elements, and molecules, and before going on to the physical states of matter, it is appropriate to consider the nature of chemical forces, or chemical bonds.

Electromagnetism is the force that is involved in all chemical behavior. This includes the chemical bonds that hold the atoms together within a molecule as well as the forces that hold molecules to each other to make liquids and solids (called intermolecular forces). The electrostatic attraction or repulsion that operates between a pair of individual points of electrical charge can be calculated directly from Coulomb's law (see Chapter 8), but atoms and molecules are not point charges. Atoms consist of positive nuclei and negative electrons, and a molecule may contain many atoms, so that the *net* forces involved are the resultants of many individual electrostatic attractions and repulsions. To make matters even more complicated, an individual entity such as an atom or a molecule is not a static object. The electronic structure of an atom

FIGURE 15.2   *The chemical classification of matter.*

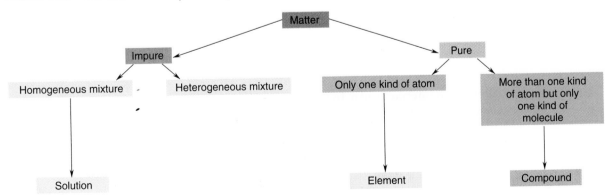

can take on various energy states, and even its symmetry can be distorted by nearby charged bodies. Molecules change their shape, and therefore their distribution of electric charges, when they vibrate by stretching, bending, or twisting. These distortions are caused by collisions with neighboring molecules or by absorption of energy from photons. All of these changes can affect the electrostatic forces between atoms and molecules.

How can two electrically neutral molecules, such as hydrogen chloride (HCl), attract each other? The answer is that although the HCl molecule as a whole is neutral, the electrical charges within the molecule are not uniformly arranged. The negative charges (the electrons) are displaced toward the chlorine atom. The result is that each HCl molecule has a positive side (the H atom) and a negative side (the Cl atom), and the oppositely charged sides attract each other: (H—Cl)(H—Cl), and so forth. (The line between H and Cl represents a chemical bond; more on this in Chapter 16.)

Some molecules, however, do not have separated positive (+) and negative (−) sides, yet the substance can be liquefied and solidified, which must mean that there is attraction to hold the molecules together. An example is the H_2 molecule. Here the explanation is more subtle. The electrical symmetry of the H—H molecule is its *average* condition. During a collision, the electrons and nuclei move away from their average pattern from instant to instant. Thus, a separation of positive and negative charges does occur momentarily, and during such moments, one distorted molecule will attract another, much like a magnet attracting an iron nail. Of course, such attractions are quite weak, which accounts for the fact that hydrogen molecules must be cooled to −253°C to liquefy it and to −259°C to freeze it.

The result of all these factors is that chemical forces are generally too complex to be predicted by mathematical calculations, and for that reason chemistry is, to a large extent, an empirical science; that is, our knowledge is largely derived from experiments and observations. The important thing to remember is that chemical bonds of all sorts as well as forces between molecules that account for the existence of liquids and solids are all the *net* resultants of many electrostatic attractions and repulsions.

15.6 STATES OF MATTER: CLASSIFICATION OF PHYSICAL STATES

Atoms bond to each other to form molecules. Molecules, too, attract each other, but the forces between molecules (intermolecular forces) are much weaker than the forces between atoms in a molecule. These relationships are summarized in Table 15.2.

The physical states of matter are generally grouped into three main categories: gases, liquids, and solids. One state can often be changed to another by the transfer of energy. For example, water can be frozen to ice by removing heat from the water; ice can be melted or water can be evaporated by adding heat to the ice or water.

Gases are substances that have no definite shape. Instead, they disperse rapidly in space and occupy any volume available to them. Most gases, such as hydrogen and nitrogen, are colorless. Some are colored; examples are chlorine (pale green) and nitrogen dioxide (NO_2), whose brown color can sometimes be seen on a smoggy day in cities like New York or Los Angeles. The attractions between the molecules of a gas are not strong enough to bind them to each other at ordinary temperatures.

Liquids do not disperse in space, but they do change shape easily. A liquid, therefore, occupies a definite volume. The molecules of a liquid attract each other strongly enough to prevent their dispersal in space but not enough to prevent the liq-

TABLE 15.2	Interatomic and intermolecular forces	
	FORCES BETWEEN ATOMS WITHIN A MOLECULE	FORCES BETWEEN MOLECULES
Name	Chemical bonds	Intermolecular forces
Relative strength	Stronger	Weaker
Effect	Formation of molecules	Formation of liquids and solids

uid from flowing. For this reason, a liquid, like a gas, is described as a fluid.

Solids possess both definite shape and volume. Their intermolecular forces are strong enough to maintain the rigidity of the substance.

Two other terms refer to specific states of matter.

Liquid crystals are substances that have some properties common to solids and some common to ordinary liquids. Liquid crystals can maintain a definite shape but can also be made to flow with only small inputs of mechanical or electrical energy.

Recall from our discussion of nuclear physics that a **plasma** is matter at such a high temperature that all chemical bonds are broken, and even the electrons are stripped away from the atomic nuclei. A plasma is, therefore, a state of matter composed of a mixture of electrons and positive ions or nuclei. The Sun and other active stars are largely plasmas of hydrogen and helium, and the plasma state is the most common of all when the entire Universe is considered.

15.7 PROPERTIES OF GASES

Gases are common to our daily experience. Air is a gas and so is the "natural gas" (mostly methane, CH_4, plus some ethane, C_2H_6, and propane, C_3H_8) that is used for heating and cooking. The aroma of a flower and the stench of a rotten egg are gases.

If you grind a solid into a very fine powder, the individual particles may be small enough to be car-ried aloft on air currents. The same may be true of tiny liquid droplets. Such particles or droplets, however, are still much larger than gas molecules; therefore, they are not gases. Smoke is such a mixture of airborne particles; it is not a gas. Fog and mists consist of liquid droplets, and they, too, are not gases.

Steam is the gaseous state of water, but the word is often misused. What you see coming out of a teapot of boiling water is not the steam but the mist of water droplets that forms after the steam cools a little; the invisible matter between the spout and the mist is the steam (Fig. 15.3). The white smoke that is discharged from many industrial chimneys is mostly a water mist (with or without other waste products). A mist droplet, like a grain of sugar, contains many molecules.

Gases show little resistance to flow compared with liquids or solids; they spread out rapidly in space and flow through small openings, such as the tiniest pinprick in a balloon or an automobile tire. All gases can be mixed with each other in any proportions, assuming that they do not react chemically with each other. Once mixed, gases never separate from each other spontaneously, nor can they be separated by filters. For example, carbon monoxide is a gas found in all cigarette smoke; no filter can remove it.

If you squeeze an inflated balloon, its volume decreases, which means that the gas inside has been compressed. It is not easy to do the same with a rock or a piece of wood, nor can it be done with liquids. If you had a strong, sealed leather pouch

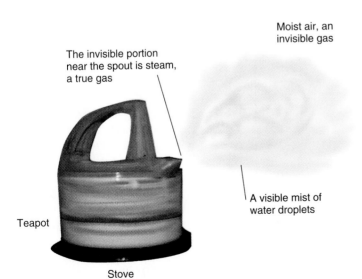

The invisible portion near the spout is steam, a true gas

Moist air, an invisible gas

A visible mist of water droplets

Teapot

Stove

FIGURE 15.3 *Steam is an invisible gas.*

completely filled with water, you would not be able to squeeze it down to a noticeably smaller volume.

A gas occupies the entire space of its container. For example, if the room you are in right now has a volume of 30 m³, it contains 30 m³ of air. If all that air were cooled down to about − 200°C, it would condense to a pale blue liquid that would occupy about 0.03 m³ (30 L), or about a tenth of 1 percent of its original volume. A reasonable conclusion from these facts is that much of a gas is empty space. Now think of any particular sample of gas, such as the hydrogen in a balloon, the air in a bicycle tire, or the helium-oxygen mixture in a diver's breathing tank. Each of these samples may be characterized by four physical properties that are related to each other. If any one of these properties is changed, there *must* be a change in at least one of the other three. These four properties are:

1. Volume of the gas.
2. Pressure of the gas.
3. Temperature of the gas.
4. Number of molecules of the gas.

These are all physical properties because a change in any of them does not produce a new chemical substance.

Now consider what may happen if the air in the bicycle tire is heated. The pressure may go up, the tire may swell so that the volume of the gas expands, some air may leak out so that the number of molecules of air in the tire decreases, or there may be some combination of these effects. No matter what, something must happen to at least one of the other three properties. The relationships among these variable properties of gases were not discovered all at once. Instead, different scientists studied these variables two at a time and came up with a set of simple laws that led to great insights into the nature of gases and the molecules that compose them.

15.8 BOYLE'S LAW

Robert Boyle (1627–1691) discovered the first of the laws relating two of the four variable properties of gases—pressure and volume. His method was to trap a volume of air in a glass tube and then to compress the air by forcing mercury into the tube, as shown in Figure 15.4. During the experiments, he measured the changes in the pressure and volume of the air. After he got all his results, he did something with them that set his name down forever in the history of science: He multiplied the pressure by the volume obtained in each experiment and noted that the product was always the same (within experimental error). This observation, that *the pressure times the volume of a fixed mass of gas at a given temperature is a constant,* became known as Boyle's law.

Mercury

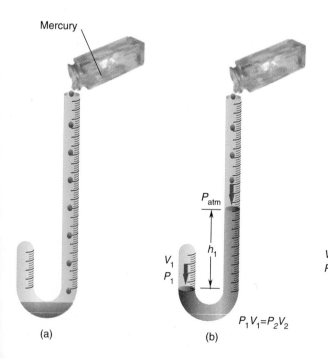

$P_1V_1 = P_2V_2$

(a)   (b)

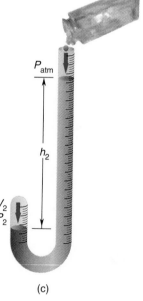

(c)

FIGURE 15.4   *Boyle's experiment. The liquid is mercury. The pressure is measured in terms of the heights of the columns, plus the atmospheric pressure exerted on the open end of the tube. (a) Mercury is poured in until (b) the air in the left column is trapped at volume V_1 and pressure P_1. (c) More mercury is poured in. Now $P_2V_2 = P_1 V_1$.*

FIGURE 15.5 *A simple engine first developed by Hero of Alexandria.*

Another way of expressing the same law is:

The volume of a sample of gas (at a fixed temperature) is inversely proportional to the pressure.

(See Appendix C for a discussion of proportionality.) This means, for example, that if you double the pressure on a gas, its volume is reduced by half. Conversely, if you squeeze a gas enough to cut its volume in half, its pressure doubles. This relationship can be expressed in equation form as

$$P_1 V_1 = P_2 V_2 \quad \text{(fixed mass and temperature)} \quad \textbf{(15.1)}$$

where P_1 = initial pressure of the gas, P_2 = final pressure of the gas, V_1 = initial volume of the gas, and V_2 = final volume of the gas.

The two values of pressure must be expressed in the same units, and the same is true for the two values of volume. For example, the volume units may both be expressed in liters or both in milliliters, but not one in liters and the other in milliliters. (Why not?)

EXAMPLE 15.3   BOYLE'S LAW

A sample of neon occupies 30 L at 1.5 atm pressure. What volume will this sample occupy at 4.5 atm? Assume no change in temperature.

Solution Initial volume, V_1 = 30 L; initial pressure, P_1 = 1.5 atm; final pressure, P_2 = 4.5 atm; final volume, V_2 = ?

Solving the Boyle's law equation (Eq. 15.1) for V_2 gives

$$V_2 = V_1 \frac{P_1}{P_2} = 30 \text{ L} \frac{1.5 \text{ atm}}{4.5 \text{ atm}} = \boxed{10 \text{ L}}$$

CONCEPTUAL EXERCISE

A weather balloon expands as it rises in the air. Explain. (Assume temperature to remain constant.)

15.9 CHARLES' LAW

The observation that a gas expands when it is heated is not new. It may surprise you that even rocket engines, which depend for their action on the expansion of heated gases, were known in the later ages of ancient Greece. These engines did not launch satellites, of course; they were merely pinwheel toys containing water. When the water was heated, steam escaped from opposing nozzles, making the toys spin in accordance with Newton's third law of motion, as shown in Figure 15.5.

A more precise study of the expansion of a heated gas was not done until about 1787, when Jacques Charles carried out his experiments. An easy way to show this relationship is sketched in Figure 15.6. A droplet of mercury is suspended in a thin glass tube that is closed at the bottom and open to the atmosphere at the top. Because the droplet is so small, it does not break apart and fall to the bottom of the tube. (A droplet of water in a drinking straw would behave in the same way.) The air below the mercury is therefore trapped; it cannot get out through the glass or go through the mercury. The tube is placed in a beaker of boiling water at 100°C, as shown in Figure 15.6a, and the volume of the trapped gas below the mercury is noted. When the temperature is allowed to drop, the volume of the trapped gas decreases, and the mercury plug goes down. It is observed that the decrease in the volume of the gas as the temperature drops from 100°C to 50°C is the same as the decrease when the temperature drops from 50°C to 0°C, as shown in Figure 15.6b and c. In fact, for every equal drop in temperature, a gas shrinks by the *same constant loss of volume*. This observation implies that if a gas were cooled sufficiently, it would shrink to nothing. Such a miracle does not occur because the molecules themselves occupy some volume, and besides, cold gases liquefy first. But if the constant shrinking were to continue without liquefaction, and if the molecules were merely points in space,

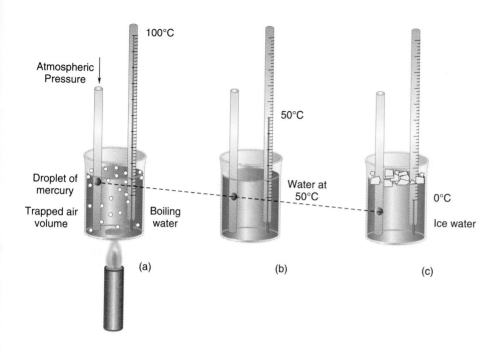

FIGURE 15.6   *Charles' law. (a) A droplet of mercury traps a sample of air below it at 100°C and atmospheric pressure. (b) The same air sample at the same pressure, at 50°C. (c) Same air sample and pressure at 0°C.*

zero volume would be reached at −273°C (Fig. 15.7). This temperature is therefore the lowest limit that could possibly be reached by cooling anything because no gas, even theoretically, could have less than zero volume. This temperature is −273.15°C, but we will use the conveniently approximate value of −273°C. As we saw in Chapter 5, this lowest temperature is the basis of the *Kelvin scale of absolute temperature.* Kelvin temperature starts at 0 K, which is called *absolute zero.*

$$-273°C = 0 \text{ K}$$

and

$$0°C = 273 \text{ K}$$

Therefore, Kelvin temperature is always 273° higher than Celsius temperature, and it is always positive.

$$T_K = T_C + 273 \qquad \textbf{(15.2)}$$

The relationship between the volume of a gas and its Kelvin (absolute) temperature is known as **Charles' law,** which states that

the volume of a gas (at constant pressure) is directly proportional to its absolute temperature.

This law can be expressed in equation form as

$$\frac{V_1}{T_1} = \frac{V_2}{T_2} \quad \text{(fixed mass and pressure)} \qquad \textbf{(15.3)}$$

where T_1 = initial temperature of the gas in kelvins; T_2 = final temperature of the gas in kelvins, V_1 = initial volume of the gas, and V_2 = final volume of the gas. If the temperatures are given in °C, they must be converted to kelvins, using Eq. 15.2.

FIGURE 15.7   *A Charles' law plot (graph) of volume versus temperature of a gas shows that gases expand when heated and contract when cooled, at constant pressure. Note the dashed line indicates that the temperature approaches −273°C as the volume approaches zero.*

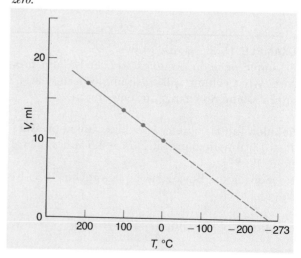

EXAMPLE 15.4 CHARLES' LAW

A sample of hydrogen gas occupies 20 mL at 300 K. What volume will the hydrogen occupy at 750 K, assuming that the pressure remains constant?

Solution Solving the Charles' law equation (Eq. 15.3) for V_2, the final volume, gives

$$V_2 = V_1 \frac{T_2}{T_1} = 20 \text{ mL} \frac{750 \text{ K}}{300 \text{ K}} = \boxed{50 \text{ mL}}$$

CONCEPTUAL EXERCISE

A dented ping-pong ball can often be restored to its original shape by immersing it in very hot water. Why?

15.10 AVOGADRO'S LAW

If more gas is introduced into a given sample of gas and if the temperature and pressure are kept constant, the volume of the gas increases. More gas means more molecules; therefore, the volume of a gas depends on the number of molecules it contains. The quantitative expression of this relationship is known as **Avogadro's law** (after Amedeo Avogadro, 1776–1856):

Equal volumes of all gases (at the same temperature and pressure) contain the same number of molecules.

CONCEPTUAL EXAMPLE 15.5 AVOGADRO'S LAW

Imagine that you are selling balloons to children at a county fair. You offer balloons of two sizes: 1 L or 2 L in volume. Also, you may fill the balloons with

FOCUS ON THE PRESSURE OF GASES

One of the properties of gases and liquids that chemists must use frequently is that of pressure. We have defined pressure in Chapter 2 as the force per unit area, $P = F/A$, and the units which we used were N/m^2. For historical reasons and for reasons of convenience, however, chemists often resort to other systems of units. Let us examine some of the background for these various measurement techniques.

Almost everyone is aware that the pressure increases with depth as one goes beneath the surface of a pool of water. In a lake, you can recognize this by a popping in your ears when you descend, but if you dive far beneath the surface of water, the increase of pressure can do far more than create a mildly unpleasant sensation on your eardrums. For example, to reach great depths under the sea, carefully designed diving bells capable of surviving tremendous pressures must be constructed. Fish living deep under water are unaware of these great pressures because they were conceived and have lived under such conditions all their lives. Like the fish, land dwellers also live on the bottom of a great ocean—an ocean of air—and as a result, we also unknowingly exist under high pressures. As we move about here on Earth, a force is exerted on us caused by the weight of a column of air that extends to the "top" of the atmosphere. In SI units, standard atmospheric pressure at sea level is 1.013×10^5 N/m^2 (14.7 lb/in^2). Thus, like Atlas of mythology, who supported the world on his shoulders, each square meter of surface area on the Earth has a force of 1.013×10^5 N exerted on it by the atmosphere. An alternative way of discussing pressures of gases or liquids is in terms of a unit called an atmosphere (atm), where 1 atm = 1.013×10^5 N/m^2. Thus, a gas under 2 atm of pressure has a pressure exerted on it that is twice as great as that which air would exert on it at sea level.

air or with the much lighter gas, helium. All the balloons are filled to a pressure of 1.1 atm. If the 1 L balloon of helium contains X molecules, how many molecules are in the 1 L balloon of air; the 2 L balloon of helium; the 2 L balloon of air?

Solution Avogadro's law applies to all gases, so the answers must be the same for the air and for the helium. Both 1 L balloons, therefore, contain X molecules and the 2 L balloons contain $2X$ molecules.

CONCEPTUAL EXERCISE

Which of the gas laws most closely explains the reason why a hot air balloon can rise into the air?

15.11 KINETIC THEORY OF GASES

Recall the properties of gases described in the previous sections: Gases are transparent, they spread out readily in space, they can flow through small openings, and they are compressed much more easily than liquids and solids. In 1738, Daniel Bernoulli proposed a concept to account for these properties. What could Bernoulli have known in 1738? Boyle had done his experiments on pressure and volume in 1662, but modern ideas about chemical change were not to come until the work of Lavoisier in 1780 and Dalton's atomic theory in 1803. Charles' law and Avogadro's law were also not to be formulated until early in the nineteenth century. The idea that matter consists of small fundamental particles was fairly well appreciated, but the distinction between molecules and atoms was not so clear. Bernoulli's theory of gas behavior was based

on his concept of the *motion* of molecules. This idea was quite novel and was not accepted until it was revived about a century later and gradually won out over rival theories.

Today, Bernoulli's concept is called the **kinetic theory of gases.** This theory accounts for the relationships among volume, pressure, temperature, and number of molecules of a gas. The theory makes the following assumptions:

1. Gases consist of small particles called molecules.
2. The molecules have mass and are in constant motion.
3. The volume of the molecules themselves is insignificant compared with the total space they occupy. Therefore, a gas is mostly empty space, as shown in Figure 15.8.
4. The molecules do not attract nor repel each other. Each molecule moves independently, in a straight line, until it collides with another molecule or with the walls of the container. Furthermore, the collisions are perfectly elastic, which means that the average kinetic energy of the molecules remains unchanged at a constant temperature.

These assumptions refer to an ideal situation. Actually, there are intermolecular forces even in

FIGURE 15.8 *The kinetic theory model of a gas. A schematic representation of molecules in the space of 1 L. The volume of the gas is 1 L, but the volume of the molecules is much smaller. Therefore, most of the space in the container is empty.*

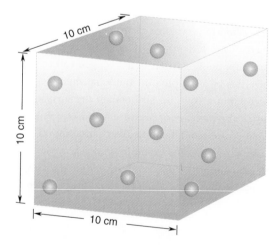

gases, but their effect is small compared with the effects in liquids and solids.

Now let us see how the theory accounts for the properties of gases.

Boyle's Law

The pressure of a gas results from the collisions of the molecules with the walls of the container. If the volume is reduced by half, the molecules are twice as crowded and hit the walls twice as often. Twice as many hits on the wall means that the pressure doubles.

Charles' Law

The temperature of a gas is a manifestation of the average energy of motion (average kinetic energy) of its molecules. As the temperature increases, the molecules speed up. As a result, the molecules hit the walls more frequently, and each collision on average is more energetic. If the opposing pressure that the walls can exert remains constant (for example, by a droplet of mercury in a capillary tube or by a movable piston in a cylinder at atmospheric pressure), the walls of the "container" are pushed out. This means that the volume of the gas increases, or, in other words, the gas expands. Conversely, as the temperature decreases, the gas molecules slow down, and their impacts on the walls become less frequent and less energetic. If the walls exert a constant pressure, they close in on the gas, and the volume decreases.

Avogadro's Law

Note that none of the statements of the kinetic theory says anything about the *kinds* of molecules in the gas. Think of two balloons of equal volume, one containing hydrogen, the other nitrogen, and both at the same temperature and pressure. Avogadro's law tells us that the number of hydrogen molecules and the number of nitrogen molecules must be equal. If their temperatures are the same, their average kinetic energies must be equal. How can that be? A molecule of hydrogen or of nitrogen has kinetic energy because it has mass and speed. Recall that kinetic energy = $\frac{1}{2}mv^2$. Therefore, $\frac{1}{2}mv^2$ for hydrogen must be the same as $\frac{1}{2}mv^2$ for nitrogen at the same temperature. But a molecule of hydrogen has less mass than a molecule of nitrogen. The only way for its mass to be less while its kinetic energy is equal to that of the nitrogen molecules is for its speed to be greater. Thus, lighter gases move more rapidly than heavier gases at the same temperature.

CONCEPTUAL EXAMPLE 15.6   FILL 'ER UP

Imagine that you drive into a service station that offers a choice of gases to fill your tires. Being in an adventurous state of mind, you say, "Put helium in one, air in another, neon in a third, and hydrogen in the fourth, all at the same pressure." The attendant does as you request. Assume that all four tires have the same volume and are at the same temperature. Hydrogen is the lightest (least dense) gas. Answer true or false for each of the following statements and defend your answers: **(a)** All four tires are at the same pressure. **(b)** All four tires contain the same number of molecules. **(c)** The molecules in all four tires have the same average speeds. **(d)** If each of the tires had the same tiny leak, they would lose pressure at the same rate.

Solution (a) True. They were filled to the same pressure; the gauge read the same for each tire, so all the pressures are the same.

(b) True. The volume, temperature, and pressure in each tire are the same; therefore, the number of molecules in each is the same. This is the relationship expressed by Avogadro's law.

(c) False. Since all the gases are at the same temperature, the average kinetic energy of their molecules must be the same. To make up for their lower mass, the lighter molecules must have greater speeds.

(d) False. The speedier molecules reach the walls more frequently and find the hole more often and so escape faster. The tire with hydrogen therefore loses pressure most rapidly.

15.12 LIQUIDS

Electrical force, which is responsible for the attractions between molecules, does not change with temperature. What does change with the temperature of a substance is the kinetic energy of its molecules. As the temperature drops, the molecules slow down

Zero gravity – spherical blobs
of orange juice
(a)

Falling – rain drops
(b)

Fallen – in the shape
of the container
(c)

FIGURE 15.9   *Properties of liquids.*

and thus lose kinetic energy. When the temperature is low enough, the intermolecular forces can begin to hold them together, either as a liquid or as a solid. We will consider liquids first.

Liquids like water, gasoline, and olive oil are familiar to us in our daily lives. Nevertheless, the liquid state is, in some ways, a rather strange condition of matter. Think of it this way: If the atoms or molecules have enough kinetic energy to move about independently in the space available to them, the substance is a gas. If the atoms or molecules are strongly attracted or bonded to each other and do not have enough kinetic energy to break loose, they maintain definite positions relative to each other in space, and the substance is a solid. But liquids? The atoms or molecules must be attracted to each other strongly enough not to break loose, but they are not held together strongly enough for the substance to maintain a rigid shape. Therefore, if an astronaut in a space capsule, under zero gravity, squeezes some orange juice out of a bottle, the juice does not fly apart but remains suspended as a blob. Does the blob have a characteristic shape? Yes, it is spherical; but when the capsule returns to Earth, and the round blob of juice is put into a glass or bottle, it takes the shape of its container (Fig. 15.9). Thus, the intermolecular forces in a liquid are strong enough to hold the substance together but not strong enough to resist deformation of the substance by gravity.

Some liquids, such as gasoline, flow readily. Others, such as honey, flow slowly. The resistance to flow of a liquid is called its **viscosity.** The viscosity of a liquid is related to its intermolecular forces, to the actual shapes of the molecules, and to the mechanical interference they exert when they flow past each other.

Thus, we may conclude that the properties of liquids result mainly from two factors: (1) the kinetic energy of the molecules, which enables them to slide past each other, and (2) the forces of attraction between them, which hold them together. Because the cohesive forces are too weak to hold the molecules in rigid positions, the molecular arrangements in liquids shift from moment to moment. As a result, the distribution of molecules is random, or disorderly.

FIGURE 15.10   *Shapes on which crystal systems are based.*

Cubic   Tetragonal   Rhombic   Monoclinic

Triclinic   Rhombohedral   Hexagonal

15.13 SOLIDS

A **crystalline solid** is a substance whose atoms or molecules are arranged in orderly or repeating patterns. These orderly arrangements are often suggested by the shapes of large particles of the substances, which are called **crystals.** The atomic or molecular patterns in such solids are based on shapes such as those shown in Figure 15.10. Imagine atoms located at the corners, sides, edges, or interiors of these imaginary shapes, and then think of a solid structure in which one such unit repeats itself many times in all directions. Such a structure is called a **crystal lattice;** crystalline solids are made up of such lattice structures. The different basic re-peating shapes and the different locations of atoms within these shapes give a variety of crystal lattices. Three that are based on the cubic shape are shown in Figure 15.11. Furthermore, a given lattice can grow in various directions and reach different sizes, so there are limitless possible shapes of crystals of even the same substance. This circumstance is the basis for the oft-repeated statement that of all the snowflakes that have fallen in the Earth's history, no two were ever exactly alike. One such snowflake crystal is shown in Figure 15.12; note its character-istic hexagonal pattern.

A substance made of large crystals is visually more interesting than a powder consisting of ag-gregates of tiny crystals whose shapes are not visi-ble to the naked eye.

The density of a crystal depends on the masses of its individual atoms and on how closely the atoms are packed together (not on how strong the bonds are nor on the sizes of the atoms). The closest pack-ing of atoms occurs in metals, which is why the dens-est solids we know are metals. The densest one of all is osmium (22.5 g/cm^3); a sphere of osmium the size of a grapefruit, about 15 cm in diameter, would surprise anyone who tried to pick it up, for it would have a mass of about 40 kg and so would weigh close to 90 lb. Osmium atoms are tightly packed in a lat-tice of the type shown in Figure 15.11c. Diamond is an example of a solid whose atoms are strongly bonded to each other but are not so tightly packed. The carbon atoms in a diamond are arranged like

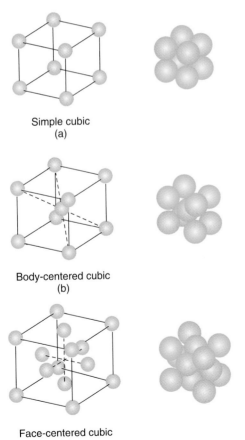

FIGURE 15.11 *Three types of crystal lat-tices based on the cubic system. (a) Simple cubic. (b) Body-centered cubic. (c) Face-centered cubic.*

Simple cubic
(a)

Body-centered cubic
(b)

Face-centered cubic
(c)

FIGURE 15.12 *Snowflake crystal.*

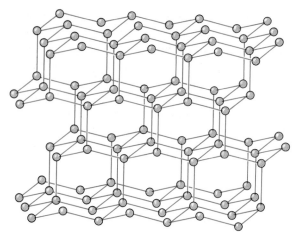

FIGURE 15.13   *Crystal structure of diamond.*

Recall that at low temperatures the molecules of a substance may slow down so much that the attractive forces among them hold them together rigidly. There is no guarantee, however, that crystallization will occur. It is, to some extent, a matter of chance. Instead, the molecules may become rigid in their randomness, like frigid tar or some sugar candies. The general name for such a material is a **glass,** or **glassy solid,** which is defined as a solid in which the component atoms or molecules are randomly arranged. This category includes the common material whose specific name is glass (as in windows), which consists of uncrystallized silicate compounds. Glasses do flow slowly; as a result, the bottoms of windowpanes found in old cathedrals are somewhat thicker than the tops. Crystals do not flow like this.

stacks of cubes with every other corner empty, as shown in Figure 15.13. The strong bonds are reflected in the fact that diamond remains solid even above 3500°C, which is about 800°C higher than the melting point of osmium. A research team in the Department of Geology at Cornell University reported in 1984 that a diamond surface was melted at a high pressure with the use of a laser beam. The temperature was around 4000°C. The open packing of the carbon atoms, however, gives diamond a low enough density (3.5 g/cm^3) for it to float easily on mercury (density 13.4 g/cm^3). Combinations of such factors engender an extensive variety of properties among solids. Lead bends, quartz is brittle, talc is soft, and steel is strong. All of these attributes are related in some way to the forces that hold the atoms or molecules of these solids together.

15.14 LIQUID CRYSTALS

Recall from our discussion of the physical states of matter that liquid crystals can maintain a definite shape but can also be made to flow. How can a substance be, as it were, both a solid and a liquid? Must it not be one or the other? The answer is this: The atoms or molecules in a crystalline solid are arranged in orderly patterns; the regularity extends throughout the crystal in three dimensions. The molecules in a gas are random; there is no regularity at all. Between these extremes, there are various degrees of partial order—less than in a true crystal but more than in a gas. The liquid state represents one such condition but not the only one. It is possible for the molecules of a substance to be more orderly than in a liquid but less so than in a

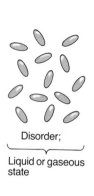

Disorder;

Liquid or gaseous state

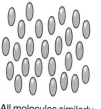

All molecules similarly aligned, but no uniform rows or layers

All molecules similarly aligned, and layers of molecules parallel to each other, but not uniform within each layer

Molecules, rows, and layers uniform

Crystalline solid

Liquid crystals

INCREASING ORDER ⟶

FIGURE 15.14   *Order and disorder in molecular arrangements.*

true crystalline solid (Fig. 15.14). The forces that hold the molecules together in such a state are just barely strong enough to enable the substance to maintain a definite shape, so it is reasonable to call it a crystal. But only slight inputs of mechanical or electrical energy can disrupt these weak intermolecular forces and make the substance flow, so it is also reasonable to call it a liquid. Hence the name, liquid crystal.

One practical application of liquid crystals takes advantage of the fact that light reflected from a thin layer of liquid, such as a slick of oil of only a few molecular layers floating on water, is colored. (Did you ever notice the rainbow effect on the surface of an oily puddle near a garage?) The reason for this effect is that some rays of light are reflected from the upper surface of the oil and some from the surface of the water on which the oil floats. When these rays recombine, they produce an interference pattern that can be either constructive or destructive, depending on the thickness of the oil film. The particular wavelengths that undergo constructive interference determine the color of the reflected light. Because the thickness of the oil layer determines which wavelengths reinforce each other, the color depends on the thickness. In some liquid crystals, the distance between the layers is sensitive to the temperature. This property can be used in "color-mapping" a portion of the human body. When a section of skin is coated with an ap-

FIGURE 15.15 *A liquid crystal display on a laboratory instrument.*

propriate liquid crystal, the warm areas over blood vessels and certain organs or diseased regions are different in color from the neighboring cooler areas, and their precise locations are thus identified.

Another application is the familiar displays on pocket calculators, wristwatches, and other instruments (Fig. 15.15). Here the liquid crystals exist as a very thin layer between thin sheets of transparent glass or plastic that have electrical conductors embedded in them. These electrodes are arranged in patterns that can represent letters or numerals. When a voltage is applied in a particular pattern, the adjacent liquid crystals are energized and converted to a different molecular orientation—one that combines with the thin sheets around them to become opaque. These opaque areas are seen as the desired letters or numerals.

SUMMARY

A transformation that produces a new substance with different properties is a **chemical change;** one that does not produce a new substance is a **physical change.** Different forms of the same elements are called **allotropes.** There are 109 known elements, 88 of which occur naturally on Earth; others have been synthesized. Most elements are metals. A **molecule** is an aggregate of at least two atoms in a definite arrangement held together by chemical bonds. A substance whose molecules consist of more than one kind of element is called a compound. A chemical formula is used to express the composition (number and type of atoms) of a molecule or compound. A solution is a homogeneous mixture of two or more substances, in which the dissolved substance is the **solute** and the medium is the **solvent.**

Gases tend to disperse in space.

Liquids do not disperse but assume the shape of their containers.

Solids maintain a rigid shape.

Liquid crystals can maintain a definite shape and can also be made to flow.

Plasmas are gases that contain electrons and bare nuclei of atoms.

The four properties of a sample of gas that are related to one another are pressure, volume, temperature, and number of molecules. The laws that relate these properties are as follows:

Boyle's law: *The volume of a given amount of gas at a constant temperature is inversely proportional to pressure.*

Charles' law: *The volume of a given amount of gas at a constant pressure is directly proportional to absolute (Kelvin) temperature.*

Avogadro's law: *Equal volumes of gases at the same tem-* *perature and pressure contain the same number of molecules.*

*The **kinetic theory of gases** interprets the gas laws in terms of the masses and kinetic energy of the molecules and their collisions with each other and with the walls of the container.*

EQUATIONS TO KNOW

$$P_1 V_1 = P_2 V_2 \qquad \text{(Boyle's law)}$$

$$\frac{V_1}{T_1} = \frac{V_2}{T_2} \qquad \text{(Charles' law)}$$

KEY WORDS

Substance	Element	Solvent	Liquid crystal
Heterogeneous	Allotropes	Solute	Plasma
Homogeneous	Molecule	Gas	Boyle's law
Chemical change	Chemical formula	Liquid	Charles' law
Physical change	Compound	Solid	Avogadro's law
Extensive property	Mixture	Crystal	Kinetic theory of gases
Intensive property	Solution	Crystal lattice	Viscosity

PROBLEMS AND CONCEPTUAL QUESTIONS

Problems requiring numerical work are identified with a blue number.

Chemical and physical changes

1. In each of the following sequences, which transformations are chemical changes and which are physical changes? (a) Oxygen gas is liquefied; some of this liquid oxygen is used as part of the fuel to launch a spaceship; some of it is allowed to vaporize into the passenger space; the gaseous oxygen is then used in respiration by the passengers and becomes part of CO_2 and H_2O molecules. (b) Beef fat is warmed in a pan until it melts; as it is heated further, it begins to darken and give off pungent gases; upon still further heating, it bursts into flame and burns.

Elements

2. White phosphorus is an extremely toxic substance that ignites spontaneously in air. Red phosphorus does not ignite spontaneously and is much less toxic than the white variety. Both substances consist entirely of $^{31}_{15}P$ atoms. What can account for the difference in their properties?

3. There is something wrong with each of the following definitions of an element. Explain what is wrong in each case. Supply a better definition. (a) An element is a substance all of whose atoms are exactly alike. (b) An element is a substance all of whose atoms have the same mass number. (c) An element is a substance that cannot be decomposed into simpler substances under any circumstances.

4. (a) Name two metals that are unreactive enough to survive as elements for long periods of time in contact with air and moisture. (b) Name a metal that is liquid at ordinary temperatures. (c) Name five elements that are gases at ordinary temperatures.

5. For the elements listed, identify those that (a) occur naturally on Earth as stable isotopes, (b) occur naturally on Earth but only as unstable isotopes, (c) do not occur naturally on Earth (the atomic numbers precede the names): 92-uranium, 6-carbon, 8-oxygen, 94-plutonium, 99-einsteinium.

6. Supply examples of the following types of elements: (a) five metals; (b) three nonmetals; (c) one semimetal; (d) one noble gas; (e) one

poisonous gas; (f) two elements that occur naturally in their uncombined states.

7. Which of the following pairs are isotopes, which are allotropes, and which are different elements? (a) hydrogen-2 (deuterium) and hydrogen-3 (tritium); (b) rhombic sulfur (prism-like crystals) and monoclinic sulfur (needle-like crystals); (c) radon-215 and radium-215.

Molecules

8. Calculate the number of atoms of each element and the total number of atoms per molecule in (a) $C_{12}H_{22}O_{11}$ (sucrose); (b) $C_7H_5(NO_2)_3$ (trinitrotoluene, TNT).

9. State whether each of the following particles can or cannot be classified as a molecule. Defend your answers. (a) An atom of neon. (Neon atoms do not enter into chemical combinations). (b) An atom of carbon in a one-carat diamond. (The carbon atoms in a diamond are all chemically bonded to each other in a continuous three-dimensional network.) (c) An entity represented by the formula CO_2. (d) An entity consisting of 20 atoms, represented by the formula $C_6H_8O_6$, in a crystal of vitamin C. (Each such entity is held to others by intermolecular forces.)

10. The element neon is often said to be "inert." What is meant by this term?

Compounds and mixtures

11. Classify each of the following substances as an element, a compound, or a mixture: (a) Dry Ice, which can be produced from combustion gases or fermentation gases, is 27.3 percent carbon and 72.7 percent oxygen by mass, and different commercial samples all have the same properties. (b) Honey has a flavor that depends in part on the bees that produce it. If honey is frozen and placed in a vacuum, water vapor escapes and sugary crystals remain behind. (c) Brass is a metallic substance composed of copper and zinc. Red brass contains 85 to 90 percent copper by mass, and yellow brass is about 67 percent copper. (d) Tantalum is another metallic substance. It has high melting and boiling points but does not separate or decompose into any other substances even at extremely high temperatures.

12. What is a solution? Can a solution be a liquid? A solid? A gas? Which of the following substances are solutions, and which are heterogeneous mixtures? (a) Cow's milk, in which the cream rises to the top; (b) antifreeze used in automobile radiators, which is a clear mixture containing water and ethylene glycol; (c) cooking oil, which is a clear liquid containing various different liquid fats; (d) silver amalgam, used for tooth fillings and consisting of silver in which mercury atoms are uniformly dispersed; (e) topsoil; (f) mud.

States of matter

13. (a) Another way of describing the three most common states of matter is to say that state X maintains a definite volume and shape, state Y does not maintain a definite volume or shape, and state Z maintains a definite volume but not a definite shape. Which common state of matter corresponds to X? To Y? To Z? (b) Fluids are substances that flow readily. Which of the various states of matter described in the text are fluids?

14. As you know, the three states of a substance (solid, liquid, and gas) can be interconverted. Give an example for each of the following physical transformations: (a) a liquid converting to a solid, (b) a gas (or vapor) condensing to a liquid, (c) a solid directly converting to a vapor.

15. (a) Name the four properties of a sample of gas that are related to each other. (b) A vendor of compressed gases receives an order for a specific volume of oxygen at a specific pressure, containing a specific number of molecules, to be delivered at a specified temperature. Explain why it is unlikely that such an order could be filled.

16. The warning on a gas tank reads: "Caution. Contents of this tank are under pressure. Do not store in direct sunlight. Do not use or store near heat or open flame." What is the reason for this warning?

17. A novice diver asks the old-timer, "How full is this tank of compressed air?" The old-timer taps thoughtfully up and down the tank, stops about halfway, and answers jokingly, "It's down to here!" Explain what is foolish about the novice's question and how the old-timer is joking. How should the novice have asked the question to get a straight answer?

18. A sample of helium occupies 8.0 L at 10 atm pressure. (a) What volume will this sample occupy at 20 atm? (b) At 5.0 atm? (c) At what pressure will the gas occupy 2.0 L? (d) 32 L? Assume no change in temperature.

19. Another statement of Boyle's law is that the pressure multiplied by the volume of a gas is constant at constant temperature, or $PV = k$. Assume that the value of k is 12, and complete the table shown by solving for P. Now plot your results on a graph of P (12 units on the x-axis) versus V (12 units on the y-axis). What is the shape of the

curve? If you have ever pumped air into a bicycle tire by hand, can you say whether any single stroke is harder to push near the beginning or near the end of the stroke? Explain how the shape of your curve expresses this experience. Remember, you are compressing the air from a larger volume (tire + pump) into a smaller volume (tire alone).

$$PV = 12$$

V	P
1	
2	
3	
4	
6	
12	

20. (a) A sample of argon gas occupies 16 L at 100 K. What volume will the argon occupy at 200 K? At 400 K? (Assume that the pressure remains constant.) (b) A sample of air occupies 200 mL at 0°C. What volume will the air occupy at 273° at the same pressure? (*Hint:* First convert Celsius to Kelvin temperatures.)

21. Imagine that an automobile tire is filled with cold compressed air to a certain pressure. You now drive the car rapidly on a warm road; the air in the tire warms up, and the pressure increases. To reduce the tire pressure back to its original value, you bleed out some air and catch the leaked air in a balloon (see fig.) Would the volume of air in the balloon tell you anything about the warming of the air in the tire? Explain.

22. Look at Figure (a) on p. 359, which shows a gas thermometer. The leveling bulb can be raised or lowered so as to keep the two liquid levels equal and thus maintain atmospheric pressure inside the gas globe. Describe how you could use this

(Question 21)

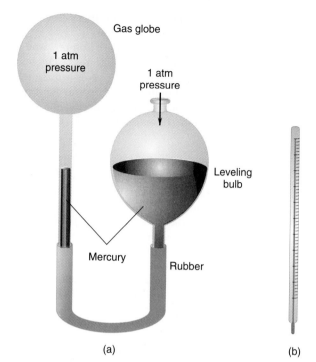

(Question 22)

device as a thermometer. Where would you place the temperature markings? Do you think that such a thermometer would be more sensitive than the ordinary liquid-in-glass thermometer (Fig. b) or less sensitive? Explain.

23. Give the name of the gas law that is illustrated by each of the following observations: (a) As a bubble of air rises from a diver's helmet to the surface of the water, it continuously expands. (b) As a rubber balloon filled with air cools during the night, it shrinks in size. (c) A 1 L volume of neon or of helium at 0°C and 1 atm pressure contains 2.7×10^{22} molecules.

24. Explain the following phenomena in terms of the theory of gases: (a) A bottle of perfume is overturned and spills onto the floor. A person standing nearby smells it right away, yet someone at the other end of the room does not smell it at once but does smell it a little while later. (b) Two equivalent gas-tight balloons are inflated with helium to the same pressure. One is kept at room temperature, while the other is stored in the freezer. The cold balloon shrinks, while the volume of the one at room temperature remains constant. (c) Two equivalent balloons made of slightly porous rubber are inflated to the same pressure, one with hydrogen, H_2, and the other with helium, He. The hydrogen leaks out more rapidly than the helium.

25. Often Boyle's law and Charles' law are combined into a single statement as PV/T = constant. Explain how this transition can be made.

26. According to the equation of question 25, does the temperature of a gas increase or decrease if the pressure is raised without changing the volume? Why? What happens to the volume if the temperature is raised without changing the pressure? Why?

27. When a closed bottle is thrown into a flame, it often shatters. Which of the gas laws best describes this situation?

Liquids and solids

28. What two factors account for the general properties of liquids? Which factor tends to make a liquid more like a gas? Which factor tends to make it more like a solid?

29. (a) The density of a solid depends on the densities of its atoms and on the geometry of their packing arrangements. Does it also depend on the sizes of the atoms? Explain. (b) A box is filled with pea-sized spheres of lead packed as closely as possible. Another box of equal size is filled with spherical grains of sand, also packed as closely as possible. Sand consists of quartz, which is less dense than lead. From this information alone, can you tell which box holds the greater mass?

30. A zinc metal rod and a glass rod are each heated in the absence of air. At a sufficiently high temperature, liquid zinc begins to drip from the zinc rod, although the rod itself remains rigid. The glass rod sags, but does not drip. Account for these phenomena.

ANSWERS TO SELECTED NUMERICAL PROBLEMS

8. (a) 12 atoms carbon, 22 atoms hydrogen, 11 atoms oxygen—45 total atoms; (b) 7 atoms carbon, 5 atoms hydrogen, 3 atoms nitrogen, 6 atoms oxygen—21 total atoms

18. (a) 4.0 L, (b) 16 L, (c) 40 atm, (d) 2.5 atm

19.

V	P
1	12
2	6
3	4
4	3
6	2
12	1

20. (a) 32 L at 200 K, 64 L at 400 K; (b) 400 mL

CHAPTER 16

THE PERIODIC TABLE AND CHEMICAL BONDS

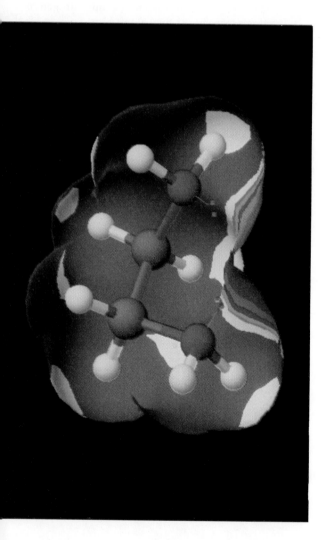

A computer model of a molecule showing chemical bonding. (Courtesy Charles D. Winters and CAChe Scientific, Inc.)

16.1 ELECTRONIC STRUCTURES OF ATOMS

The atom was the subject of Chapter 12, and we recall from that discussion that only certain fixed energies are possible for the electrons of an atom. These fixed energies are designated by various quantum numbers. Here we will consider only the **principal quantum number** n, where $n = 1, 2, 3, \ldots$, etc. According to this quantum number designation, an electron in the lowest available energy level is said to be in the $n = 1$ state (the ground state). The next lowest energy level is the $n = 2$ state, and so forth. All the states labeled $n = 2$ or higher are called excited states. We should remember two things about electron energy levels: (1) each level can accommodate only a limited number of electrons; (2) the electrons at a given principal energy level make up an **electron shell.**

The higher the shell number (n), the more distant are its electrons from the nucleus and the greater is the energy of the electrons. The electron that can be separated most easily from its atom is therefore an electron in the highest shell. Electrons can be promoted to higher energy levels or even knocked free from their atoms in a variety of ways. The electrons in the highest shells are the ones usually involved in chemical changes; these shells are called the **valence shells,** and the electrons in them are called **valence electrons.** All chemical reactions involve the sharing of valence electrons between atoms or transfer of valence electrons from one atom to another.

The larger atoms have larger clouds of electrons around them. The higher the principal quantum number, the more electrons can be accommodated within the shell. The relationship is

maximum number of
electrons per shell $= 2n^2$ (16.1)

where n is the principal number.

Table 16.1 shows the distribution of electrons in the shells of the first 18 elements. Note that the first shell is full before electrons enter the second shell, and the second shell is full before electrons enter the third. It should be noted here, however, that this pattern is not followed throughout all the shells. For example, the third and higher shells do not completely fill before some electrons go into still higher shells.

EXAMPLE 16.1   THE SHELL GAME
What is the maximum number of electrons in the shell for which $n = 1$? For the $n = 2$ shell? For the $n = 3$ shell?

Solution The maximum number is $2n^2$. Therefore, when $n = 1$, the number of electrons is $2(1)^2 = 2$. For $n = 2$, it is $2(2)^2 = 8$. For $n = 3$, it is $2(3)^2 = 18$.

16.2 VALENCE—CHEMICAL COMBINING CAPACITY

Consider the following formulas of some well-known chemicals:

HCl	hydrogen chloride
NaH	sodium hydride
NaCl	sodium chloride
CaO	calcium oxide
H₂O	water
NH₃	ammonia
CH₄	methane
CaH₂	calcium hydride

In the first formula, the H and the Cl are combined in equal atomic ratios, one atom of H to one atom of Cl. They are said to have equal chemical *combining capacities*. The same is true for the atoms in the next three formulas, NaH, NaCl, and CaO. The formulas H_2O, NH_3, CH_4, and CaH_2, however, clearly show that not all atoms are equal in combining capacity. In fact, these formulas show that more than one atom of hydrogen can combine with

---

TABLE 16.1   Electron configurations of elements 1 to 18[*]

ATOMIC NUMBER	ELEMENT AND SYMBOL	NUMBER OF ELECTRONS IN EACH SHELL			TOTAL NUMBER OF ELECTRONS
		FIRST	SECOND	THIRD	
1	Hydrogen, H	1			1
2	Helium, He	2 (full)			2
3	Lithium, Li	2	1		3
4	Beryllium, Be	2	2		4
5	Boron, B	2	3		5
6	Carbon, C	2	4		6
7	Nitrogen, N	2	5		7
8	Oxygen, O	2	6		8
9	Fluorine, F	2	7		9
10	Neon, Ne	2	8 (full)		10
11	Sodium, Na	2	8	1	11
12	Magnesium, Mg	2	8	2	12
13	Aluminum, Al	2	8	3	13
14	Silicon, Si	2	8	4	14
15	Phosphorus, P	2	8	5	15
16	Sulfur, S	2	8	6	16
17	Chlorine, Cl	2	8	7	17
18	Argon, Ar	2	8	8 (full)	18

[*]Among the heavier elements, the distribution of electrons becomes more complicated because of the division of shells into sub-shells.

one atom of the other element. Other elements can thus have higher combining capacities than hydrogen.

The combining capacity of an atom is known as its **valence** (from Latin, *valentia*, meaning "capacity"). Valence is, of course, a chemical property. By convention, hydrogen has been assigned a valence of 1. Then *the valence of an element is defined as the number of hydrogen atoms that combine with one atom of that element.* It follows that the valences of the other elements in the formulas listed must be 1 for Na (sodium) and Cl (chlorine); 2 for O (oxygen) and Ca (calcium); 3 for N (nitrogen); and 4 for C (carbon).

These assigned valences make it possible to predict other formulas. The rule is that when two elements combine, they do so in such a way that the total valences of each element are equal. For example, calcium hydride is a compound of calcium and hydrogen. What is its formula? The valence of H is 1, and the valence of Ca is 2. In the formula, the total valences of H must equal the total valences of Ca. The formula must be

$$CaH_2$$

where the valence of Ca is 2 and the total valence of H_2 is $1 + 1 = 2$. The subscripts are generally reduced to their least common denominators. Thus, for calcium oxide we write CaO, not Ca_2O_2. This reduction to the lowest common denominator is not always done. When the molecule has a definite and known structure, the formula includes all the atoms in one molecule. An example is hydrogen peroxide, H_2O_2.

The prediction of a formula from valences does not guarantee that the compound exists. For example, the valences of 4 for C and 2 for Ca predict the formula Ca_4C_2 or Ca_2C. However, there is no such compound. Conversely, many compounds whose formulas cannot be predicted from valences do exist. Therefore, valences are a useful, but not infallible, guide to the prediction of formulas of simple substances composed of two elements.

EXAMPLE 16.2   NAME THAT COMPOUND

Write the formulas for the compounds formed by the combination of **(a)** Ca and Cl; **(b)** N and Cl; **(c)** C and Cl; **(d)** Ca and N; **(e)** C and O.

Solution Since Cl has a valence of 1, its subscript in a formula will be the valence of the other element. Thus, **(a)** $CaCl_2$; **(b)** NCl_3; **(c)** CCl_4. **(d)** Ca has a valence of 2 and N has a valence of 3. For the total valence of each element to be equal, 3 Ca atoms must combine with 2 N atoms giving both Ca and N a total valence of 6. This gives the formula Ca_3N_2. **(e)** The same procedure yields a formula of C_2O_4, which reduces to CO_2. This formula actually corresponds to the composition of the molecule.

16.3 PERIODIC TABLE

The fact that elements can be classified by their properties into various types, such as metals and nonmetals, or into groups that are characterized by their chemical properties interested chemists for many years. In 1864, John Newlands in England noted that when the elements were arranged in the increasing order of their atomic weights, various sets of chemical properties tended to repeat themselves. This trend was best noted in the elements from lithium to chlorine. (Helium and neon had not yet been discovered, so these elements are omitted. Modern periodic arrangements are based on atomic numbers, not weights, but the two sequences usually match each other.) We start with lithium (Table 16.2) because the first element, hydrogen, is really in a class by itself; there is nothing very much like it.

Going across the first row, from lithium to fluorine, the properties of the elements change considerably from one to the next. Lithium is an active metal, beryllium is a less active one, and boron is hardly metallic at all. These large changes continue from one element to the next going to the right: carbon, nitrogen, oxygen, and fluorine are all nonmetals but are different from one another in chemical reactivity. The reactivity increases sharply all the way to fluorine, which is a pale yellow, violently reactive gas. But next comes sodium, which is a metal similar to lithium, and, progressing on through the following several elements all the way to chlorine, many of the properties of the first sequence are repeated. For example, chlorine is a pale greenish yellow gas closely resembling fluorine. When Newlands suggested an analogy to the musical scale, where the frequencies establish repeating patterns,

TABLE 16.2	**Repeating patterns of chemical properties**						
lithium $\longrightarrow$	beryllium $\longrightarrow$	boron $\longrightarrow$	carbon $\longrightarrow$	nitrogen $\longrightarrow$	oxygen $\longrightarrow$	fluorine
sodium $\longrightarrow$	magnesium $\longrightarrow$	aluminum $\longrightarrow$	silicon $\longrightarrow$	phosphorus $\longrightarrow$	sulfur $\longrightarrow$	chlorine

by calling these chemical relationships the "law of octaves," he was ridiculed by his contemporaries.

The classification of the elements reached the status of a serious and valuable concept in 1869, when Dmitri Ivanovich Mendeleev in Russia and Julius Lothar Meyer in Germany published independent versions of a **periodic table of the elements.** (Mendeleev left "holes" in his chart, predicting the properties of the unknown elements that would someday fill these blank spots. Because his predictions were on target, he is given the lion's share of the credit for the development of the periodic table.) Chemists' views of the idea of such a table were transformed. Instead of looking on it as an idle curiosity, they recognized it as a broad and useful concept that could correlate a wide range of physical and chemical properties of the elements. Modern forms of the periodic table, such as that of Figure 16.1, are based on the electron configurations of the atoms. (A complete list of the elements and their atomic weights and numbers is found in Appendix E.)

1. Each box contains the symbol of the element and its atomic number.
2. The elements appear in the increasing order of their atomic numbers (except for two long sequences that are set below as separate rows).
3. The vertical columns, called **groups,** fall into two categories: the so-called main-group elements, where the groups are numbered from 1 to 8, and the transition elements, which lie between groups 2 and 3 of the main-group elements starting with the fourth period. *The numbers of the main group correspond to the number of electrons in the highest shell of the atom, the number of valence electrons.*

Thus, all the elements in Group 1, from hydrogen on down, have one electron in the highest shell of the atom; the elements of Group 2 have two electrons in the highest shell, the elements in Group 3 have three electrons in the highest shell, and so on. The transition elements are different; the important thing to remember is that there are two electrons in the highest shell for most of the transition elements.

4. *The horizontal rows, called **periods,** correspond to the number of occupied electron shells in the atom.*
5. The elements set off below the main table (the lanthanoids and actinoids) also usually have two electrons in their highest shells. They, similar to the transition elements, differ from one another in their inner electron shells.

Of course, Mendeleev did not base his periodic table on electron shells; no one knew anything about atomic structure then. The early tables were based on the chemical properties of the elements, especially valence. Some of these relationships were clear enough, such as the resemblances between fluorine and chlorine or between sodium and potassium, but others were not so clear, so there was a certain amount of confusion. Now, with the benefit of our knowledge about electron shells, we can use the modern periodic table to help us understand the chemical relationships among the elements.

The first thing to do is to consider the sets of main-group elements (also called representative elements) that belong to given groups. Remember that the elements in a given group all have the same number of electrons in their highest shells, but they differ in the number of shells they have. We will start on the right side of the table with Group 8, which consists of helium, neon, argon, krypton, xenon, and radon. These elements are all gases, and they are chemically completely inert or nearly so. They are known as the **noble gases.** The molecules of these gases consist of a single atom, so their chemical formulas are the same as their atomic sym-

FIGURE 16.1 *The periodic table.*

FIGURE 16.1 The periodic table.

Key:

Atomic number
Symbol
Mass number

92
U
Uranium
238.03

State:
- S Solid
- L Liquid
- G Gas
- X Not found in nature

- Metals
- Transition Metals
- Nonmetals
- Noble gases
- Lanthanide series
- Actinide series

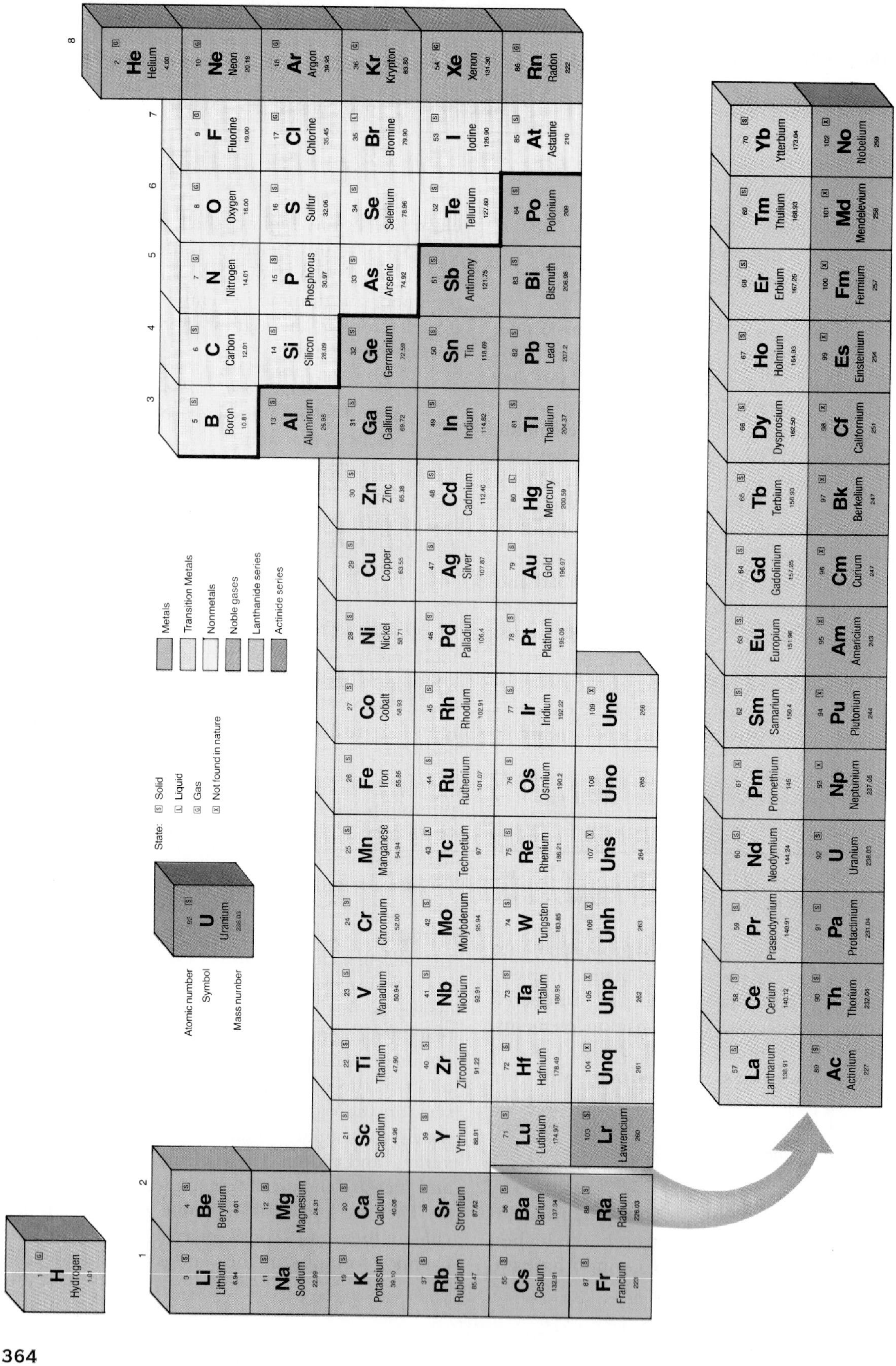

bols: He, Ne, Ar, Kr, Xe, and Rn. Except for helium, which has only two electrons, the number of electrons in the highest shells of these Group 8 elements is eight. This fact leads to the conclusion that there is some relationship between having eight electrons in the highest shell and being chemically inert. This conclusion is an important one that will be dealt with in more detail later in this chapter.

Next, we move on to Group 7, called the **halogens,** which consists of fluorine, chlorine, bromine, and iodine. (Astatine, the last element in Group 7, is omitted because there is so little of it on Earth, and chemists generally never deal with it.) All these elements are colored, toxic substances. Fluorine and chlorine are yellow and greenish yellow poisonous gases. Chlorine was the first poison gas used in World War I, in 1915, on the Western Front. Bromine is a liquid that evaporates readily to a reddish brown vapor, and iodine is a dark solid that evaporates at room temperature to give a purple vapor. These vapors, too, are toxic. The molecules of all these elements consist of two atoms; their formulas are F_2, Cl_2, Br_2, and I_2. They all show a valence of 1 in their combinations with hydrogen; the formulas are HF, HCl, HBr, and HI. So, for now, we can say that elements in the same group have similar chemical properties, including valence. Because the elements in a given group all have the same number of electrons in the outer shell, this number of outer electrons seems to be related to the properties of the elements. That statement sounds good, but a look at Group 6 shows that the situation is not so simple.

Group 6 includes oxygen, sulfur, selenium, tellurium, and polonium (look for the symbols in Figure 16.1). Here the differences are more striking than the similarities. The first element, oxygen, is a nonmetal and a gas. All the other elements are solids, and the last one, polonium, is a metal. Selenium and tellurium, which are in between, have some semimetallic properties. What is evident here is a *progression* from nonmetallic to metallic character. Yet all these elements exhibit a valence of 2. For example, oxygen combines with hydrogen to form water, H_2O, and compounds with similar formulas are formed by three of the other elements of Group 6: H_2S (hydrogen sulfide, which has the odor of rotten eggs), H_2Se, and H_2Te (both of which stink much worse than H_2S). Polonium, the last element of this group, does not form a stable compound with hydrogen, but it does form $PoCl_2$.

You are not expected to memorize the periodic table, but the general trends within groups and periods illustrate important chemical principles (to be summarized later). To continue this overview, examine Table 16.3 and compare the trends in Groups 4 and 5 with those in Group 6.

Again, progression from nonmetallic to metallic character is evident as we go down the groups. The elements in each group, however, show a common typical valence. In Group 5, the typical valence is 3, as shown in the compounds NH_3, PH_3, AsH_3, SbH_3, and $BiCl_3$. In Group 4, it is 4, as in CH_4, SiH_4, GeH_4, $SnCl_4$, and $PbCl_4$.

In Group 3, all the elements but the first one, boron, are metals, and in Group 2, they are all metals, with no exceptions. Group 1 starts with hydrogen, which is really in a class by itself. Other than that, all the Group 1 elements are metals, and are called the **alkali metals.** The typical valences are 1 for Group 1, 2 for Group 2, and 3 for Group 3.

The most important property of the transition elements can be stated simply: They are all metals. If someone asked you to name some metals, most of those you would think of would be transition elements. Try it. The six metals that have been known since ancient times are: gold, silver, copper, iron, lead, and tin. The first four of these are transition

TABLE 16.3	Comparison of groups in the periodic table	
GROUP 4	GROUP 5	GROUP 6
carbon, C	nitrogen, N phosphorus, P	oxygen, O sulfur, S } nonmetals
silicon, Si germanium, Ge	arsenic, As	selenium, Se tellurium, Te } semimetals
tin, Sn lead, Pb	antimony, Sb bismuth, Bi	polonium, Po } metals

elements. (Bronze, after which an entire age of ancient history is named, is an alloy of copper and tin.) The 14 elements appearing in the two lower sections of Figure 16.1 are all metals; they include all the possible nuclear fuels: uranium, plutonium, and thorium.

We may now make four simple summarizing statements about the properties of the main-group elements and their positions in the periodic table:

1. Going from left to right within a given period (the horizontal rows), the elements tend to become *less* metallic.
2. Going down within a given group (the vertical columns), the elements tend to become *more* metallic except for Groups 7 and 8, which are all nonmetals.
3. Within a given period, the typical valences of the elements change by one unit from one group to the next.
4. The elements within the same group have the same typical valence.

Note the heavy zig-zag line in Figure 16.1, which is an approximate separation between the metallic and nonmetallic elements. The nonmetals are to the right and up. The metals are to the left and down and include all the transition elements and the lanthanoids and actinoids.

EXAMPLE 16.3 SHELLS AND ELECTRONS

State the number of electron shells and the number of electrons in the highest shell of $_7$N, $_{55}$Cs, $_{22}$Ti, $_{77}$Ir, and $_{35}$Br.

Solution $_7$N is in the second period, so it has two electron shells. It is in Group 5, so there are five electrons in its second, or highest, shell. $_{55}$Cs is in the sixth period, so it has six shells. It is in Group 1, so its highest (sixth) shell has one electron. $_{22}$Ti, in the fourth period, has four electron shells. But it is a transition element, for which the general rule is two electrons in the highest shell. $_{77}$Ir, in the sixth period, has six electron shells. It is a transition element, so it has two electrons in its sixth or highest shell. $_{35}$Br, in the fourth period, has four electron shells. It is in Group 7, so its highest, or fourth, shell has seven electrons.

16.4 CHEMICAL EQUATIONS AND TYPES OF CHEMICAL BONDING

Chemical changes occur all around us—in biological systems, in geological processes, in the cooking of food, in chemical manufacturing, and in fires and explosions. It would be convenient to be able to represent these changes by some kind of chemical shorthand, just as symbols represent atoms and formulas represent molecules and compounds. The key to the problem is the fact that

chemical transformations involve the making and breaking of chemical bonds, but not the creation or destruction of atoms.

Therefore, the chemical formulas change, but the kinds and numbers of atoms stay the same. Chemical reactions are represented by **chemical equations,** which show the formulas of the starting substances (the reactants) and final substances (the products) as well as the relative numbers of molecules involved:

$$\text{reactants} \longrightarrow \text{products}$$

where the arrow means "to yield." The law of conservation of mass states that matter can neither be created nor destroyed. Therefore, the products must contain the same number and kinds of atoms as the reactants. An equation that obeys this law is called a **balanced equation.**

To learn how to balance equations, let us start with a reaction expressed in words: Hydrogen molecules react with chlorine molecules to produce hydrogen chloride molecules, or

$$H_2 + Cl_2 \longrightarrow HCl$$

where the "+" sign means "reacts with." This equation is incomplete because there are two H atoms and two Cl atoms on the left and only one of each on the right. The equation must be balanced by changing the numbers of molecules, never by changing the kinds of molecules:

$$H_2 + Cl_2 \longrightarrow 2HCl$$

This reaction is summarized in Table 16-4.

It will not be necessary to use diagrams to balance equations. You need only adjust the number of molecules. To balance equations, it is often necessary to add a coefficient in front of one or more

TABLE 16.4   Three ways of representing the reaction between hydrogen and chlorine molecules

A hydrogen molecule + A chlorine molecule
$\longrightarrow$ Two hydrogen chloride molecules

$H_2 + Cl_2 \longrightarrow 2HCl$

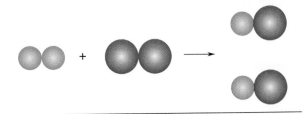

---

of the chemical formulas. For example, the number 2 in front of HCl in the equation is the coefficient.

The following are some useful rules to apply in balancing chemical equations:

1. We must first identify all reactants and products and write their correct formulas.

2. Look for elements that appear only once on each side of the equation and with equal numbers of atoms on each side—the formulas containing these elements must have the same coefficient. There is no need to balance these elements.

3. Next, look for elements that appear only once on each side of the equation but in *unequal* numbers of atoms. Balance these elements. Finally, balance elements that appear in two or more formulas on the same side of the equation.

EXAMPLE 16.4   BALANCING EQUATIONS
Balance the following equations:

(a) hydrogen molecules + oxygen molecules $\longrightarrow$ water molecules

(b) methane molecules + oxygen molecules $\longrightarrow$ carbon dioxide molecules and water molecules

(c) iron oxide + carbon monoxide $\longrightarrow$ iron + carbon dioxide

Solution   **(a)** The unbalanced equation is

$H_2 + O_2 \longrightarrow H_2O$

We note that H appears once on both sides of the

equation and in equal numbers of atoms, so we do not need to balance them. To balance the O atoms, we place a 2 in front of H_2O.

$H_2 + O_2 \longrightarrow 2H_2O$

Now the H atoms are unbalanced. Finally, we place a 2 in front of H_2 and obtain the balanced equation

$2H_2 + O_2 \longrightarrow 2H_2O$

(b) The unbalanced equation is

$CH_4 + O_2 \longrightarrow CO_2 + H_2O$

The C atom appears once on each side and in equal numbers. The H atom also appears once on each side but in unequal numbers, so we place a 2 in front of H_2O:

$CH_4 + O_2 \longrightarrow CO_2 + 2H_2O$

Finally, we place a 2 in front of O_2 to obtain the balanced equation:

$CH_4 + 2O_2 \longrightarrow CO_2 + 2H_2O$

(c) The unbalanced equation is

$Fe_2O_3 + CO \longrightarrow Fe + CO_2$

To balance Fe we place a 2 in front of Fe:

$Fe_2O_3 + CO \longrightarrow 2Fe + CO_2$

To balance the O atoms, note that the coefficients for CO and CO_2 must be the same (since both contain one C atom). Therefore, we must increase both molecules equally to add more O atoms to the right side. A coefficient of 2 does not work, but a "3" works, giving us the balanced equation:

$Fe_2O_3 + 3CO \longrightarrow 2Fe + 3CO_2$

EXERCISE

Balance the equation

$$H_2O_2 \longrightarrow H_2O + O_2$$

Naming Compounds

In our study of chemistry, we encounter a variety of chemical compounds with strange sounding names. As unusual as some of these names may appear to be, there is a method to naming compounds. There are a variety of examples that one may find that violate the simple rules given here, but the technique works most of the time.

1. Compounds that contain only two elements are referred to as binary compounds. These are named by first naming the more metallic of the two elements, then giving the name of the second element followed by the suffix "ide." For example, common table salt is NaCl. Na is the more metallic of the two, so the compound is called sodium chloride.

EXERCISE

Use the information given in (1) to name the compounds **(a)** HCl (found in stomach acid) and **(b)** H_2S (the gas that gives rotten eggs their distinctive odor).

Answer (a) HCl is called hydrogen chloride. **(b)** H_2S is called hydrogen sulfide.

2. When compounds are formed that contain polyatomic ions (see Table 16.5 on page 371), the same basic procedures are used as in (1) except that the polyatomic ion usually retains its name. For example, NH_4Cl is called ammonium chloride.

EXERCISE

Name these compounds containing polyatomic ions: **(a)** $CaCO_3$, **(b)** $MgSO_4$.

Answer (a) This is calcium carbonate (limestone or marble). **(b)** This is magnesium sulfate (Epsom salt).

3. In many binary compounds, especially of nonmetals, the names are derived according to the number of atoms that occur within the molecule. The prefixes used to distinguish the number of atoms are "mono" = one, "di" = two, "tri" = three, "tetra" = four, "penta" = five, "hexa" = six, "hepta" = seven, and "octa" = eight. For example, the compound CO is called carbon monoxide, and the compound CO_2 is carbon dioxide.

EXERCISE

(a) The compound N_2O is laughing gas. Use the techniques of (3) to give its chemical name.

(b) P_4S_3 is the white tip on wooden matches. Name it chemically.

Answer (a) N_2O is dinitrogen monoxide. **(b)** P_4S_3 is tetraphosphorus trisulfide.

Now that we have learned to write chemical formulas, balance chemical equations, and name compounds, our next step is to study chemical bonding. The following sections investigate the different types of chemical bonds by considering four major categories of substances: ionic substances, covalent substances, network covalent substances, and metallic substances.

16.5 IONIC SUBSTANCES

Chapter 15 pointed out that chemical bonds are the result of electrical forces that hold atoms together in the form of molecules. An important characteristic of each bonding type will be illustrated by the description of a simple electrical experiment carried out with familiar substances. The first substance is ordinary table salt (sodium chloride, NaCl), which will be used to illustrate the ionic bond.

Electrical Experiment with Salt

Safety Note. The electrical experiments outlined in this chapter are described only to help explain chemical bonding. They are *not* for the reader to carry out. **Do not attempt to do so.** The reason

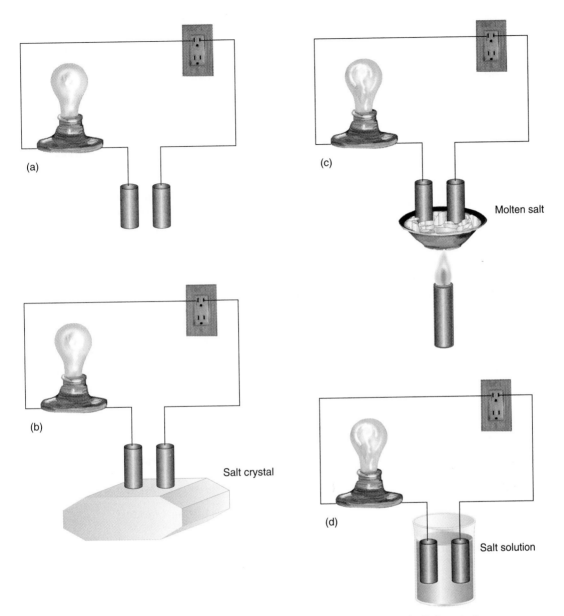

FIGURE 16.2 *Experiments with electrical conductivity. (a) Open circuit. (b) Salt crystal does not conduct. (c) Molten salt does conduct. (d) Salt water does conduct.*

is that exposed electrodes with a potential difference of 110 volts could, if handled improperly, produce a serious electric hazard. Depending on the location of contact to the human body, the resulting electric shock could be fatal.

The experiment uses the simple electric circuit shown in Figure 16.2a, which consists of an ordinary light bulb plugged into a household socket (110 volts). There is a gap in the circuit, however, so that no current flows and the bulb does not light.

The two ends of wire at the gap are attached to sticks of graphite (carbon). These pieces of graphite are called the **electrodes.** If the gap between the electrodes is bridged by anything that conducts electricity, the bulb will light. Figure 16.2b shows the two electrodes touching a large crystal of salt. The bulb does not light, which means that the salt crystal does not conduct electricity. Next the salt is placed in a porcelain dish and melted, which requires a temperature of about 800°C. The electrodes are now inserted into the hot liquid salt, as shown in Figure 16.2c. The bulb lights up, which means that an electric current is flowing through the molten salt. Something else also happens if direct current is used: Chlorine (Cl_2), a greenish yellow gas, is bubbling out at the positive electrode, and sodium, a shiny metal, is being produced at the negative electrode.

Finally, some salt is dissolved in water and the electrodes are dipped into the salt water, as shown in Figure 16.2d. Again the bulb lights up, and the action at the electrodes shows that a chemical change is taking place.

The results of this experiment are interpreted as follows:

1. An electric current is a movement of electric charge. Because sodium chloride conducts electricity when it is molten or dissolved in water, it must contain electrically charged particles that are free to move between the electrodes in response to an applied voltage. Solid sodium chloride, however, does not conduct electricity. Therefore the electrically charged particles must be locked in place in the solid crystal.
2. Because the conduction of electricity by sodium chloride is accompanied by chemical change, and chemical changes involve atoms, the charged particles in NaCl must be charged atoms. The sodium is produced at the negative electrode, so it must have a positive charge, and the chlorine is produced at the positive electrode, so it must have a negative charge.

When electrons are removed from or added to a neutral atom, a charged particle called an **ion** is formed. An ion that bears a net positive charge is called a **cation;** an ion whose net charge is negative is called an **anion.** Sodium chloride contains sodium cations (Na^+) and chloride anions (Cl^-). The atomic number of sodium is 11, which means that the nucleus has 11 protons and the neutral Na atom has 11 electrons. The sodium cation thus has 11 protons and only 10 electrons, for a net charge of $+1$:

NA ATOM	NA$^+$ ION
11 protons	11 protons
11 electrons	10 electrons

Similarly, chlorine, with an atomic number of 17, has 17 nuclear protons and 17 electrons. The chloride anion, with one extra electron, has 17 protons and 18 electrons, for a net charge of -1:

CL ATOM	CL$^-$ ION
17 protons	17 protons
17 electrons	18 electrons

NaCl is a stable substance because the oppositely charged ions attract each other strongly. The solid does not conduct electricity because the ions, although charged, are held rigidly in the crystal structure and do not migrate. Figure 16.3 shows a model of the NaCl crystal lattice, in which the ions are in a cubic pattern. Note that the Na^+ and Cl^- ions alternate with each other, so that oppositely charged ions, which attract each other, are always close together. Such an array is stable. When the salt melts, however, or dissolves in water, the ions are free to move. If a voltage is applied, the positive ions go one way, and the negative ions go another. The liquid then conducts electricity.

The chemical changes that occur when molten salt is electrolyzed are as follows.

At the negative electrode (cathode):

$$Na^+ + 1 \text{ electron} \longrightarrow Na \text{ (metallic sodium)}$$

At the positive electrode (anode):

$$Cl^- \longrightarrow Cl \text{ (a chlorine atom)} + 1 \text{ electron}$$

Chlorine atoms combine with each other to form molecules of Cl_2:

$$Cl + Cl \longrightarrow Cl_2$$

FIGURE 16.3 *Portion of the NaCl lattice.*

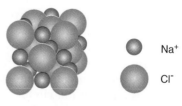

NaCl

TABLE 16.5	**Some common ions**
POSITIVE IONS	**NEGATIVE IONS**
Ammonium, NH_4^+	Bromide, Br^-
Potassium, K^+	Chlorate, ClO_3^-
Silver, Ag^+	Chloride, Cl^-
Sodium, Na^+	Fluoride, F^-
Barium, Ba^{2+}	Hydroxide, OH^-
Calcium, Ca^{2+}	Iodide, I^-
Copper, Cu^{2+}	Nitrate, NO_3^-
Lead, Pb^{2+}	Carbonate, CO_3^{2-}
Magnesium, Mg^{2+}	Oxide, O^{2-}
Zinc, Zn^{2+}	Sulfate, SO_4^{2-}
Aluminum, Al^{3+}	Sulfide, S^{2-}
	Sulfite, SO_3^{2-}
	Phosphate, PO_4^{3-}

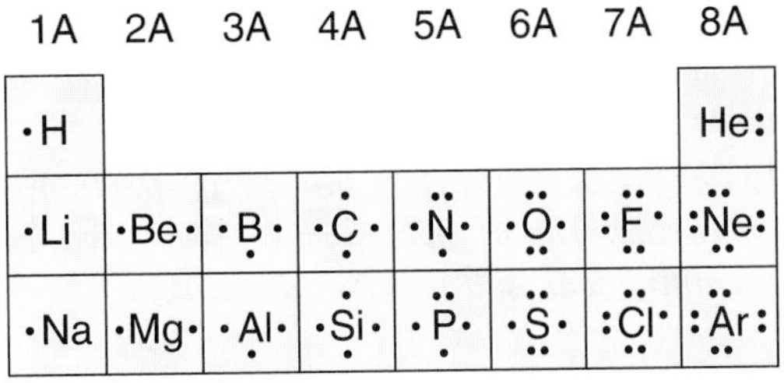

FIGURE 16.4   *Lewis dot symbols of the first three periods.*

Note that the electrolysis must be carried out with a direct current, not an alternating current as shown in Figure 16.2.

Not all ions have charges of +1 or −1. Some ions have higher charges, although rarely greater than +3 or −3. Furthermore, an ion need not be a single charged atom. Some groups of atoms occur together in many compounds, and the entire group may carry an electric charge. Examples of such polyatomic ions are ammonium ion, NH_4^+, and hydroxide ion, OH^-. Table 16.5 lists the names and formulas of some common ions, both positive and negative, simple and polyatomic.

The names and formulas of common ionic compounds can be obtained directly from Table 16.5 simply by combining the names and formulas of the separate ions. When you write a formula, remember that the total plus and minus charges must balance each other.

Now consider the questions: How can the charge on ions be accounted for? Why are some positive and some negative? Why are some charges greater than others? To explain the formation of ions and ionic compounds such as NaCl, chemists use a system of dots devised by the American chemist Gilbert Lewis, called Lewis dot symbols. A Lewis dot symbol consists of the symbol of an element and one dot for each valence electron in an atom of that element. Figure 16.4 shows the Lewis dot symbols of the representative elements in the first three periods of the periodic table. Note that except for helium the number of dots (or valence

electrons) each atom has is the same as the group number of the element. For example, Li is a Group 1A element and has one dot for one valence electron; Be, a Group 2A element, has two valence electrons (two dots); and so on. Except for Ne and Ar, each of these elements forms an ion. The transition metals, lanthanides, and actinides all have incompletely filled inner shells, and in general, we cannot write simple Lewis dot symbols for them.

If we examine the electronic structure of the anions and cations derived from the representative elements, an interesting pattern emerges. Take the Li^+ ion as an example. By losing an electron, the Li^+ ion has the same electronic structure as helium (see Table 16.1). Therefore, we say that Li^+ is **isoelectronic** with helium. Similarly, when a F atom accepts an electron to become the F^- ion, it has the same electronic structure as, or is isoelectronic with, the neon atom. The formation of lithium fluoride, an ionic compound like NaCl, can be represented as an electron-transfer reaction as follows:

$$\cdot L + : \ddot{\underset{\cdot\cdot}{F}} \cdot \longrightarrow Li^+ \; : \ddot{\underset{\cdot\cdot}{F}} \colon^-$$

On the same basis, we can also show that the Mg^{2+} and Al^{3+} cations are isoelectronic with Ar, and the O^{2-} and N^{3-} anions are isoelectronic with Ne.

Similar relationships exist among the ions of elements whose atomic numbers are close to those of the other noble gases in Group 8 of the periodic table. These observations lead to the following line of reasoning:

Electrons in atoms are involved in chemical bonds.

The noble gases are chemically inert. They do not readily gain or lose electrons, nor do they generally enter into chemical reactions. (Kr and Xe do form some compounds.) The filled shells of the noble gases must therefore represent stable arrangements of electrons.

Other atoms form ions by gaining or losing electrons so as to acquire the stable electronic arrangement of a noble gas.

When atoms react by electron transfer, the total number of electrons gained and lost must be equal because the resulting ionic salt is electrically neutral.

Notice that the number of electrons lost or gained by an atom in forming an ionic bond is equal to its valence. The magnitude of the charge on the ion formed from the atom is the valence of the element. Atoms that lose electrons to form positive ions are generally the metals. Atoms that gain electrons to form negative ions are generally the nonmetals.

The following paragraphs summarize these concepts about the formation of ionic compounds:

1. There are six noble gases: helium, neon, argon, krypton, xenon, and radon. They have stable electronic configurations, characterized by eight electrons in their highest shells (except for helium, which has, and can have, only two).
2. Elements whose atomic numbers are one or two (sometimes three) higher than that of a noble gas can enter chemical reactions in which they lose just enough electrons to become isoelectronic with their noble-gas neighbor. Thus, they become positive ions.
3. Elements whose atomic numbers are one or two (and sometimes three) lower than that of a noble gas can enter chemical reactions in which they gain just enough electrons to become isoelectronic with their noble-gas neighbor. Thus, they become negative ions.
4. In many cases, when elements in categories (2) and (3) are brought together, electron gains and losses occur in the same reaction, resulting in an *electron transfer* to form an ionic compound.
5. Electrons are not created or destroyed in chemical reactions, so electrons gained = electrons lost. For example,

$$Mg \longrightarrow Mg^{2+} + 2e^- \text{ (two electrons lost)}$$

$$F_2 + 2e^- \longrightarrow 2F^- \text{ (two electrons gained)}$$

The formula of the resulting ionic compound is therefore MgF_2.

EXAMPLE 16.5 IONIC COMPOUNDS

Write the names and formulas of the ionic compounds formed by a combination of (a) ammonium ion with bromide ion, with chloride ion, and with sulfate ion; (b) calcium ion with hydroxide ion, with carbonate ion, and with phosphate ion.

Solution (a) Combining the names of the + and − ions gives ammonium bromide, NH_4Br; ammonium chloride, NH_4Cl; ammonium sulfate—here the formula cannot be NH_4SO_4 because the charges do not balance. It must be $(NH_4)_2SO_4$. Note that the procedure for writing formulas from ionic charges is the same as that which was used for writing formulas from valences.

(b) The combinations give calcium hydroxide, $Ca(OH)_2$; calcium carbonate, $CaCO_3$; and calcium phosphate, $Ca_3(PO_4)_2$, in which six + charges neutralize six − charges.

EXAMPLE 16.6 ION SYMBOLS

Write symbols for ions of elements number 35, 37, and 38 that are isoelectronic with krypton.

Solution Krypton is number 36, so an element just before krypton in the periodic table will need one more negative charge to become isoelectronic with it. Elements after krypton will have to lose electrons and hence become positive. The ions are Br^-, Rb^+, and Sr^{2+}.

EXAMPLE 16.7 IONIC FORMULAS

Write formulas for the ionic compounds rubidium bromide and strontium bromide.

Solution To conserve charge, the formulas must be $RbBr$ and $SrBr_2$. In all ionic formulas, the total positive charge must equal the total negative charge.

16.6 COVALENT BOND

Electrical Experiment with Ice and Water

This experiment is carried out with the same apparatus as that previously used with salt. This time, however, the electrodes are first placed on a block of ice, as shown in Figure 16.5a, and then are immersed in pure water, as shown in Figure 16.5b. The bulb does not light in either case.

The results of this experiment are interpreted to mean, simply, that water is not ionic. If it were, the ions could have been made to move in the liquid water, current would have flowed, and the bulb would have lit up. No light means no ions and no ionic bond.

What then holds the atoms together in a water molecule? The modern theory of chemical bonding is based on the work of Gilbert Lewis in the 1930s. Lewis advanced the idea that a **covalent bond**

FIGURE 16.5   *Neither (a) ice nor (b) pure water conducts electricity, as is demonstrated by the unlit bulb.*

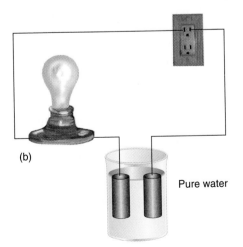

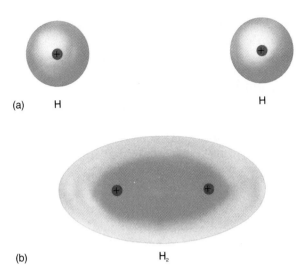

FIGURE 16.6   *(a) The nuclei of two H atoms repel each other; the electrons repel each other; and the nuclei and electrons attract each other. When the two H atoms are far apart, the attractions balance the repulsions, and there is no bonding. (b) At an optimal distance, the electronic charge is more concentrated between the nuclei, and the attractions exceed the repulsions. Then, a chemical bond exists, and the molecule H_2 is formed.*

between two atoms is formed by the sharing of electrons between the atoms. Consider the formation of an H_2 molecule from two H atoms, as shown in Figure 16.6. In Lewis' representation, the separate H atoms and the H_2 molecule are written as

ATOMS	MOLECULE
$\cdot$ H $\cdot$ H	H : H or H—H

where H : H or H—H is the Lewis structure for the H_2 molecule, and the single line represents the covalent bond.

The stability of the molecule depends on the attraction between each electron and each nucleus as well as the repulsion between the two electrons and between the two nuclei. If the sum of all the attractions exceeds the sum of all the repulsions, the molecule is stable. The force in a covalent bond, similar to that in an ionic bond, is electrical. The difference is that no ions are present in a molecule.

For molecules more complex than H_2, Lewis proposed the **octet rule** (the rule of eight), which states that an atom other than hydrogen tends to form bonds until it is surrounded by eight valence electrons, which resembles an inert noble gas atom. For compounds containing hydrogen, the hydrogen atom must be surrounded by 2 valence electrons, which resembles helium.

Consider the formation of a chlorine molecule from two chlorine atoms. Each chlorine atom has 7 valence electrons and does not satisfy the octet rule. By sharing a pair of electrons, however, the chlorine atoms now attain the stability of a noble gas (that is, argon):

ATOMS	MOLECULE
$\ddot{\underset{..}{Cl}}\cdot$ $\cdot\ddot{\underset{..}{Cl}}\cdot$	$\ddot{\underset{..}{Cl}}:\ddot{\underset{..}{Cl}}:$ or $:\ddot{\underset{..}{Cl}}-\ddot{\underset{..}{Cl}}:$

Note that the electrons between two atoms are counted twice, once for one atom and again for the other. By sharing a pair of electrons, the two chlorine atoms form a covalent bond. The other pairs of valence electrons are called lone pairs, and they are not involved in bond formation. A shorthand notation for the Lewis structure is Cl—Cl, where the single line represents the covalent bond, and the lone pairs are often omitted.

Similarly, we can write the Lewis structure for the hydrogen chloride molecule as:

ATOMS	MOLECULE
$\cdot H$ $:\ddot{\underset{..}{Cl}}\cdot$	$H:\ddot{\underset{..}{Cl}}:$ or $H-\ddot{\underset{..}{Cl}}:$

For the water molecule (H_2O), we have

ATOMS	MOLECULE
$\cdot H$ $\cdot H$ $\cdot\ddot{\underset{..}{O}}\cdot$	$H:\ddot{\underset{..}{O}}:H$ or $H-\ddot{\underset{..}{O}}-H$

Finally we should keep in mind that the octet rule, although useful, has a number of exceptions, especially for elements having atomic number greater than 18.

Double Bonds and Triple Bonds

All the covalent bonds discussed so far are single bonds; that is, there is only one pair of electrons

between two atoms. Many molecules contain *multiple bonds;* that is, there are two or three electron pairs between atoms, giving rise to double and triple bonds. Consider the carbon dioxide molecule (CO_2). The Lewis dot symbols for carbon and oxygen atoms are:

$$\cdot\ddot{\underset{..}{O}}\cdot \quad \cdot\underset{.}{C}\cdot \quad \cdot\ddot{\underset{..}{O}}\cdot$$

We know from experimental study that the C atom is bonded to the two O atoms. Thus we might pair the electrons between the atoms as follows:

$$\cdot\ddot{O}:\underset{.}{C}:\ddot{O}\cdot$$

This structure, however, does not satisfy the octet rule for either O or C. By pairing the remaining electrons, we obtain the correct Lewis structure, which contains two double bonds:

$$\ddot{O}::C::\ddot{O} \quad \text{or} \quad \ddot{O}=C=\ddot{O}$$

Ethylene (C_2H_4) is another example of a molecule containing a double bond. In ethylene, each C atom is bonded to two H atoms and to the other C atom, so we might write the Lewis structure as

$$\begin{array}{cc} H & H \\ & \overset{..}{C}:\overset{..}{C} \\ H & H \end{array}$$

To satisfy the octet rule, we form another bond between the C atoms:

$$\begin{array}{cc} H & H \\ \underset{.}{C}::\underset{.}{C} \\ H & H \end{array} \quad \text{or} \quad \begin{array}{cc} H & H \\ \diagdown C=C\diagup \\ H & H \end{array}$$

A triple bond is formed by the sharing of three electron pairs between two atoms. Nitrogen molecule contains a triple bond. The Lewis dot symbols for the N atom is

$$\cdot\ddot{N}\cdot \quad \cdot\ddot{N}\cdot$$

and the Lewis structure that satisfies the octet rule is

$$:N:::N: \quad \text{or} \quad :N\equiv N:$$

In acetylene (C_2H_2), each C atom is bonded to one H atom and the other C atom, so the correct Lewis structure is

$$H:C:::C:H \quad \text{or} \quad H-C\equiv C-H$$

EXAMPLE 16.8   LEWIS STRUCTURE

Write the Lewis structure for **(a)** the ammonia (NH_3) molecule and **(b)** the hydrogen cyanide molecule in which the C atom is bonded to the H atom and the N atom.

Solution (a) The Lewis dot symbols of H and N are

$$\cdot H \quad \cdot H \quad \cdot H \quad \cdot \ddot{N} \cdot$$

For the nitrogen atom to satisfy the octet rule, it must share three pairs of electrons with hydrogen as follows:

$$H \!:\! \ddot{N} \!:\! H \quad \text{or} \quad H \!-\! \underset{\underset{H}{|}}{N} \!-\! H$$
$$\;\; H$$

(b) The Lewis dot symbols of H, C, and N are

$$\cdot H \quad \cdot \dot{C} \cdot \quad \cdot \ddot{N} \cdot$$

To satisfy the octet rule for C and N, these two atoms form a triple bond as follows:

$$H \!:\! C \!:::\! \ddot{N} \!:\! \quad \text{or} \quad H \!-\! C \!\equiv\! N \!:$$

we expect the boiling points of a series of similar compounds to increase with increasing mass. Consider the Group 7A compounds HF (19.5°C), HCl (−84.9°C), HBr (−67.0°C), and HI (−35.4°C), where the temperatures in parentheses denote the boiling points. The lightest molecule, HF, has the *highest* boiling point because it alone can form hydrogen bonds and therefore exert greater intermolecular attraction among its molecules:

$$\text{---}H \!-\! F \text{---} H \!-\! F \text{---} H \!-\! F \text{---} H \!-\! F \text{---}$$

Why does an ice cube float on water? The answer is hydrogen bonding. Each water molecule can take part in four hydrogen bond formations—the two H atoms covalently bonded to the O atom can form two hydrogen bonds, and the two lone pairs on the same O atom can hydrogen bond with H atoms on two other water molecules. Consequently, ice has an extensive three-dimensional structure with much empty space in the lattice (Figure 16.7).

Recalling that density is mass divided by volume, we see that for the same amount of water molecules, ice has a smaller density than liquid water because it occupies a larger volume. This is a virtually unique property of water. For most other sub-

16.7 HYDROGEN BOND

The **hydrogen bond** is a special type of bonding between molecules. The prerequisite for hydrogen bond formation is that an H atom must be covalently bonded to one of the three atoms: N, O, or F. The same H atom can then be hydrogen-bonded to one of the same three atoms on *another* molecule. Examples are:

$$A \!-\! H \text{---} B \quad \text{or} \quad A \!-\! H \text{---} A$$

where *A* and *B* represent N, O, or F; *A*—H is one molecule or part of a molecule, and *B* is part of another molecule; and the dotted line represents the hydrogen bond.

The N, O, and F atoms are said to be *electronegative* because they have a strong tendency to attract electrons (in a covalent bond) toward themselves. This action causes the H atom bonded to these atoms to be slightly deficient of electrons and hence bear a positive charge. In the vicinity of another electronegative atom, a hydrogen bond is formed as shown.

Hydrogen bonding has an important effect on various physical properties. For example, normally,

FIGURE 16.7 *Three-dimensional structure of ice. The hydrogen bonds are shown in color dashed lines.*

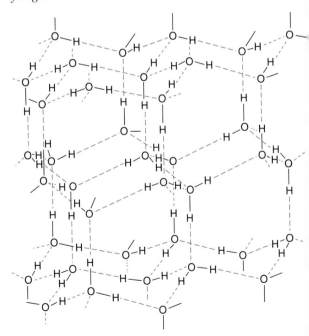

stances, the density of the solid is greater than that of the liquid. One important consequence of ice being lighter than water is that lakes freeze from top to bottom in winter. The layer of ice on top can then thermally protect the biological environment beneath it. If ice were more dense than liquid water, lakes would freeze from bottom to top and most aquatic life would not survive. Hydrogen bonds also play a crucial role in determining the three-dimensional structure of biological molecules, such as proteins and DNAs (deoxyribonucleic acids).

EXAMPLE 16.9   HYDROGEN BONDS
Draw hydrogen bonds between ammonia molecules.

Solution We see that an ammonia molecule (NH_3) qualifies for hydrogen bond formation because the H atoms are covalently bonded to an N atom. We can represent hydrogen bonds between three NH_3 molecules as follows:

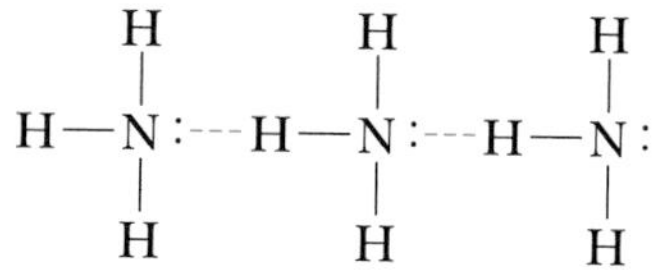

CONCEPTUAL EXERCISE

Can a methane molecule (CH_4) form a hydrogen bond with a water molecule?

16.8 ORGANIC CHEMISTRY

Carbon forms a unique series of compounds of extraordinary diversity. Moreover, living organisms on Earth consist largely of carbon compounds, and the processes of life involve, in large measure, the making and breaking of bonds to carbon atoms. For these reasons, *the chemistry of carbon compounds is called* **organic chemistry.**

The uniqueness of carbon lies, in part, in the ability of carbon atoms to bond to each other in sequences of extensive length and in continuous or branched patterns. These features can be illus-

trated with formulas of compounds that contain only carbon and hydrogen. Such compounds, which are called **hydrocarbons,** are the major components of natural gas, bottled gas, lighter fluid, gasoline, kerosene, diesel fuel, petroleum jelly (Vaseline), and paraffin candle wax. Illustrated here are structural formulas of propane, C_3H_8, the major component of bottled gas, and octane, C_8H_{18}, which is a component of gasoline. Recall that carbon has a valence of 4, as shown by the fact that each carbon atom in these compounds makes four bonds.

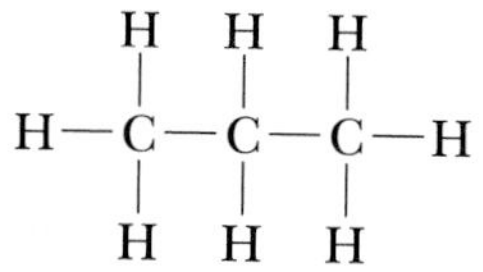

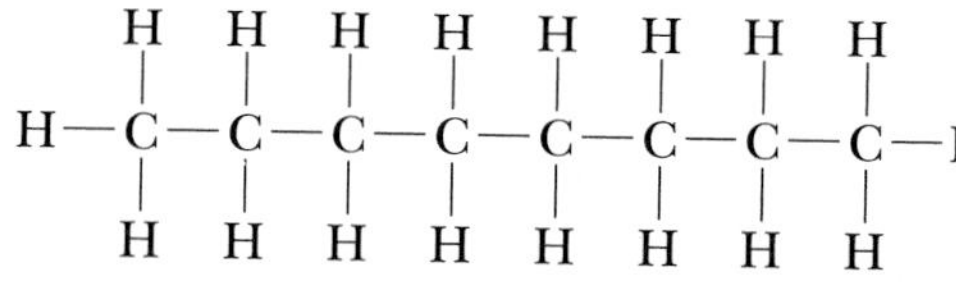

The octane pictured is, fortunately, only a minor component of gasoline. Too much of it would cause the gasoline to burn in an uncontrolled, almost explosive, manner. The result can be heard as "engine knock," which is the sound of an inefficient, possibly damaging mode of combustion within the cylinders.

Note the preceding statement that the sequence of carbon linkages can be branched. Branched-chain hydrocarbons have much less tendency to knock in gasoline engines. A particularly good hydrocarbon (the standard for 100-octane gasoline), known as "iso-octane," is

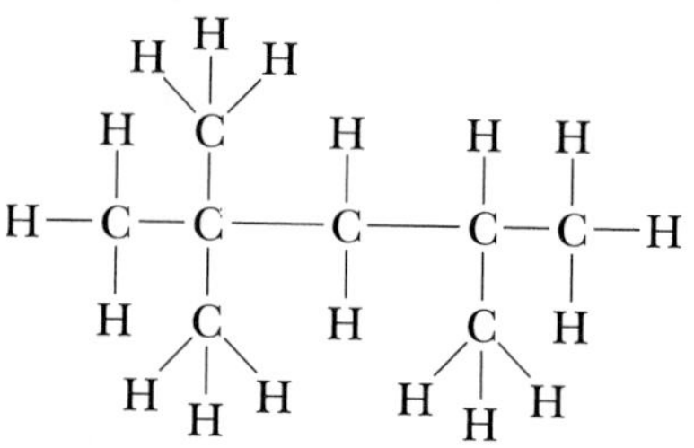

Count the carbons and hydrogens in this formula and note that they add up to C_8H_{18}, the same formula as that for the straight-chain octane. Molecules with different structural formulas make up different compounds, even if their molecular formulas are the same. The straight-chain C_8H_{18} knocks; the branched-chain one does not. They have different properties, so they are different substances. *Such substances with the same molecular formulas but different structural formulas are called* **isomers.**

EXAMPLE 16.10   STRUCTURAL FORMULAS

Write structural formulas for **(a)** C_4H_{10} and **(b)** C_2H_6O (two isomers of each).

Solution (a) The carbons must all be bonded to each other, whereas the hydrogens, with a covalence of 1, can be bonded only to carbon. The four carbon atoms can be linked in either a continuous or a branched chain.

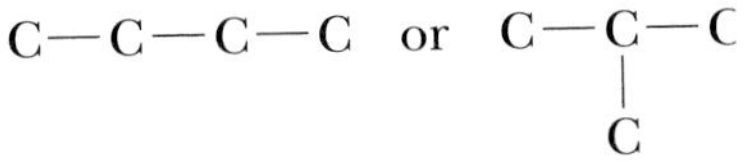

$$C—C—C—C \quad \text{or} \quad C—C—C \atop \quad\quad\quad\quad\quad\quad | \atop \quad\quad\quad\quad\quad\quad C$$

If enough bonds are now added to each carbon atom to reach a total covalence of 4, and an H atom is then attached to each bond, 10 H atoms are needed for each molecule:

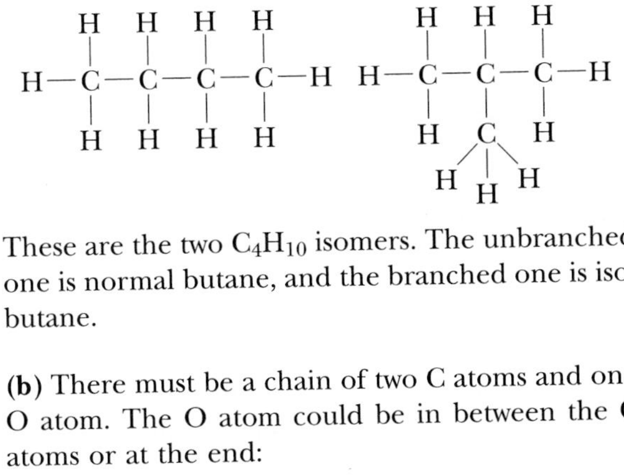

These are the two C_4H_{10} isomers. The unbranched one is normal butane, and the branched one is isobutane.

(b) There must be a chain of two C atoms and one O atom. The O atom could be in between the C atoms or at the end:

$$C—O—C \quad \text{or} \quad C—C—O$$

Bonds must now be added to the formulas to give each C atom a covalence of 4 and the O atoms a covalence of 2, and an H atom must be attached to each bond. Note that the O atom in the first chain already has two bonds, so that no more are needed:

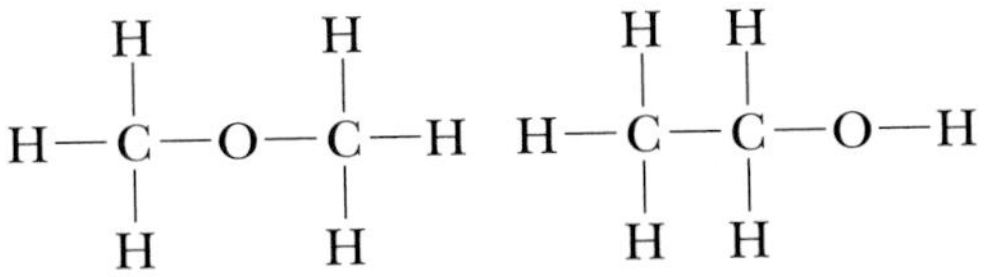

These are the two C_2H_6O isomers. The one on the

left is called dimethyl ether. The one on the right is ethyl alcohol.

Cyclic Structures

There is no reason why the ends of a chain cannot find each other and form a bond, provided that there is room for another bond on each end. The result is a cyclic structure. Note that a formula like C_3H_6 can represent either a doubly bonded or a cyclic structure:

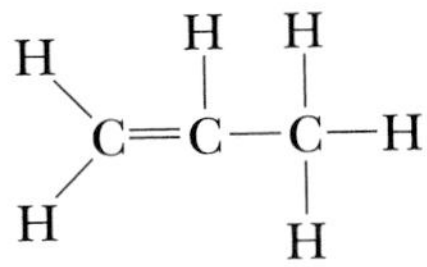

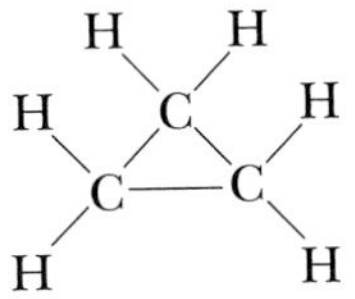

Ethylene, Acetylene, and Benzene

Three important hydrocarbons that contain multiple bonds are ethylene (C_2H_4), acetylene (C_2H_2), and benzene (C_6H_6). The Lewis structure of ethylene was discussed in Section 16.6. Ethylene is an extremely important substance because it is used in large quantities for the manufacture of the polymer (giant molecule) called polyethylene and in the preparation of many other organic compounds. As we saw earlier, acetylene is a molecule that contains a triple bond. When acetylene is burned in an "oxyacetylene torch," the temperature of the flame can reach about 3000°C, hot enough to weld metals.

The structure of benzene is interesting because it has a cyclic structure with three alternating single and double bonds. There are two ways to draw the structure of benzene, and together they represent the properties of benzene:

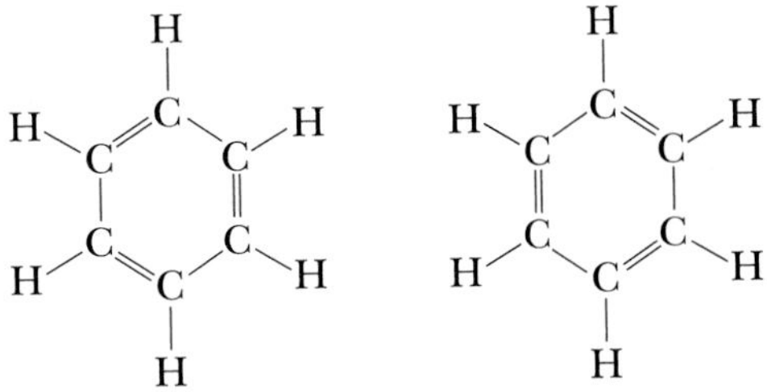

Benzene is used as a solvent and a starting material in organic synthesis. It is a carcinogen (cancer-causing substance).

Organic Compounds of Oxygen and Nitrogen

Carbon forms covalent bonds with many elements other than hydrogen. The most important ones in living systems are the organic compounds that contain oxygen and nitrogen. Some of the more common types of these structures are illustrated in the following:

An **alcohol** is a compound that contains the

$$-\overset{\displaystyle |}{\underset{\displaystyle |}{C}}-O-H$$

linkage, where the other three bonds from the carbon atoms must be linked to hydrogen atoms or to other carbon atoms. The simplest such compound is methyl alcohol,

$$H-\overset{\displaystyle \overset{H}{|}}{\underset{\displaystyle \underset{H}{|}}{C}}-O-H$$

sometimes also called "wood alcohol" because it can be produced by heating and decomposing wood in the absence of air. Methyl alcohol is sometimes found in adulterated alcoholic beverages; drinking it can cause blindness or death. The next compound in the series is ethyl alcohol, which is produced by fermentation:

$$H-\overset{\displaystyle \overset{H}{|}}{\underset{\displaystyle \underset{H}{|}}{C}}-\overset{\displaystyle \overset{H}{|}}{\underset{\displaystyle \underset{H}{|}}{C}}-O-H$$

The series can continue indefinitely by extending and branching the carbon chain.

Carboxylic acids are another important class of organic compounds that contain oxygen. They are characterized by the presence of a carboxyl group

$$-\overset{\displaystyle \overset{O}{\|}}{C}-O-H$$

where the fourth bond from the carbon atom must be linked to hydrogen or to another carbon. The first two in this series are:

$$H-\overset{\displaystyle \overset{O}{\|}}{C}-O-H \quad \text{and} \quad H-\overset{\displaystyle \overset{H}{|}}{\underset{\displaystyle \underset{H}{|}}{C}}-\overset{\displaystyle \overset{O}{\|}}{C}-O-H$$

formic acid   acetic acid

The most important class of organic nitrogen compounds is the **amines,** characterized by the linkage

$$-\overset{\displaystyle |}{\underset{\displaystyle |}{C}}-\overset{\displaystyle |}{N}-$$

where, again, the other bonds are to hydrogen or carbon atoms. The simplest such compound is therefore

$$H-\overset{\displaystyle \overset{H}{|}}{\underset{\displaystyle \underset{H}{|}}{C}}-\overset{\displaystyle \overset{H}{|}}{N}-H \quad \text{methylamine}$$

16.9 THE MILLER-UREY EXPERIMENT AND THE ORIGIN OF LIFE

Having seen how structural formulas are written, you may wonder whether these exercises are merely games chemists play or whether they reflect the properties of atoms. If you cut out all the squares of a crossword puzzle and tossed them into the air, you would hardly expect the letters to fall down again in the same word pattern or indeed in any arrangement that made sense. What happens then when some simple covalent compounds are energized with particles or photons of high energy to break them into various molecular fragments? Do the pieces bind themselves together again? Can new products be formed? Are the normal covalences reestablished?

An interesting experiment was carried out in 1953 by Stanley Miller and Harold Urey to explore what the chemistry of the Earth's atmosphere might have been before life existed. They had speculated that the primitive Earth's atmosphere consisted largely of hydrogen (H_2), water (H_2O), methane (CH_4), and ammonia (NH_3). Miller and Urey

FOCUS ON
BONDING IN DNA

DNA (deoxyribonucleic acid) is the primary molecule in chromosomes, which are the carriers of genetic information. This molecule has a double helix structure as shown in the figure, where the shape is maintained by bonds across the strands of the helix. In the figure, the bonds connecting the double helix are labeled A-T and C-G. These subunits are called nucleotide base pairs and are composed of adenine (A), thymine (T), cytosine (C), and guanine (G). The base pairs (A-T and C-G) are held together by hydrogen bonds. When a cell reproduces, the double helix comes apart at one end, as shown in the figure, and separates.

Each portion of the split double helix can now form a new double helix. The new helix must match the original exactly because all the A-T and C-G bonds must match perfectly throughout the strand. Genetic information is passed from one generation of cells to the next through this matching of hydrogen bonds.

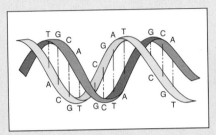

The DNA molecule.

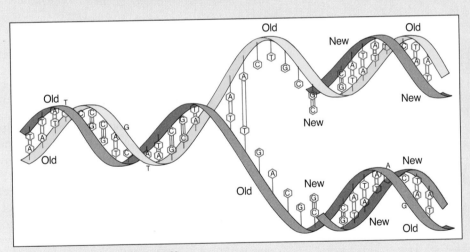

The DNA molecule replicating itself.

passed an electric spark through a mixture of these gases to imitate the action of lightning, as shown in Figure 16.8. The molecules broke into a variety of fragments. These fragments then recombined spontaneously to form various more complex molecules. The importance of this experiment lies in the fact that the spontaneously formed molecules were amino acids, which are the building blocks of proteins. The structural formulas of some of these amino acids are:

$$
\begin{array}{cc}
\text{glycine} & \text{alanine}
\end{array}
$$

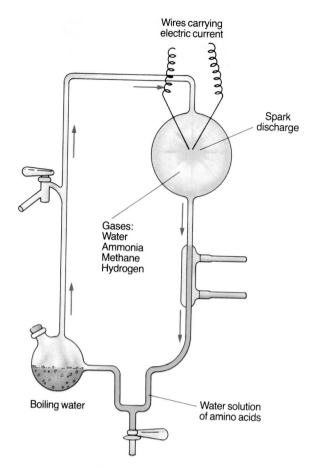

FIGURE 16.8   *The Miller-Urey experiment imitating the action of lightning.*

aspartic acid

Thus, the interplay of energy and probability (which molecular fragments hit each other and in what ways they collide) favors the formation of molecules of stable substances with normal valences from a chaotic mixture of molecular fragments. The atoms that make up the essential molecules of living matter, such as protein, cellulose, and DNA, are linked by the same types of bonds that exist in the simpler molecules discussed in this chapter. But this fact is not an answer to the question of how the molecules of life were formed. The Miller-Urey ex-

periment, interesting as it was, did not provide the answer either. For one thing, the Miller-Urey hypothesis on the composition of the primitive atmosphere has been greatly modified by more recent studies and speculations. It is generally recognized that we do not have the last word on what the early Earth was like. The synthesis of amino acids, even if it did occur spontaneously, is not the same as, or even close to, the synthesis of a living organism. Living things are characterized by the ability to use an outside source of energy to form complex structures and to reproduce themselves at the expense of materials from the environment. A virus, for example, is a complex organism, up to several hundredths of a micrometer in diameter. It is considered to be on the borderline of life because it can reproduce itself but cannot ingest food or grow. No such structures have ever been made in the laboratory, and there is no reliable way to estimate just how difficult such a synthesis would be or indeed whether it would ever be possible. We still do not know how life originated.

16.10 NETWORK COVALENT SUBSTANCES

Electrical Experiment with Sand

This experiment is carried out with the same apparatus as that previously used with salt, ice, and water. When the electrodes touch the sand, as in Figure 16.9, the bulb does not light. Sand, which is particles of quartz, does not conduct electricity. Any attempt to melt the sand by heating it with an ordinary gas flame would fail because the flame is not hot enough. The melting point of sand or quartz is about 1700°C.

Quartz, like water, is a covalent compound. The molecular formula of quartz, SiO_2, is analogous to that of carbon dioxide, CO_2. (Both Si and C have a covalence of 4.) But quartz is a hard mineral, and carbon dioxide is a gas at ordinary temperatures, as shown in Figure 16.10. With the same valences and analogous molecular formulas, why should they be so different? The answer lies in the fact that CO_2 has double bonds, whereas SiO_2 has only single bonds. Why should this fact be so important? In general, double bonds are shorter than single bonds. This means that doubly bonded atoms are closer to each other than are singly bonded atoms.

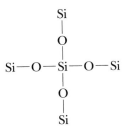

covalence of 2 for O. The remedy is to add an Si to each O atom,

$$
\begin{array}{c}
\text{Si} \\
| \\
\text{O} \\
| \\
\text{Si}-\text{O}-\text{Si}-\text{O}-\text{Si} \\
| \\
\text{O} \\
| \\
\text{Si}
\end{array}
$$

and then another O to each Si atom. As we continue, the formula grows larger, but the covalences of the outermost Si and O atoms are still not satisfied. To satisfy the covalences with single bonds would require endless writing. What this means is that the covalent bonding in SiO_2 continues indefinitely; it is not limited to a small molecule. Therefore, a crystal of silicon dioxide is not composed of molecules in the ordinary sense but rather is a continuous three-dimensional network of single Si—O covalent bonds. It may be considered to be a **giant molecule.**

There have been science fiction accounts of intelligent life elsewhere in the Universe that is based on the chemistry of silicon, in contrast to the carbon-based life on Earth. Is silicon-based life plausible? Most chemists think it is not because the Si—Si bond is much weaker than the C—C bond. Consequently, structures with Si—Si bonds would react rapidly with oxygen to form giant-molecule compounds with Si—O bonds as in SiO_2 or silicate rocks. Furthermore, Si does not tend to form double or triple bonds.

Life may exist elsewhere, and its forms may be different, but we can be sure of one thing. Only carbon has the variety of bonding to furnish the multitude of compounds needed for life as we know it.

FIGURE 16.9 *Sand does not conduct electricity.*

But large atoms cannot get as close to each other as smaller atoms can because at closer range, the repulsions of the more numerous electrons in the lower shells become too great. As a result, it is more difficult for the larger atoms to form double bonds. To see how important this difference is, try to write structural formulas for the oxides CO_2 and SiO_2, using double bonds for carbon but only single bonds for silicon. Carbon dioxide presents no problem. Using double bonds, one can write O=C=O, and the structural formula is complete. But using only single bonds for Si, we start with

$$
\begin{array}{c}
\text{O} \\
| \\
\text{O}-\text{Si}-\text{O} \\
| \\
\text{O}
\end{array}
$$

This satisfies the covalence of 4 for Si but not the

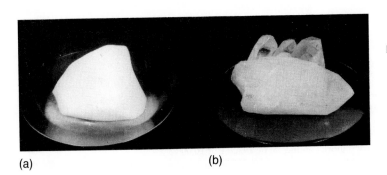

(a)   (b)

FIGURE 16.10 *(a) "Dry Ice" (solid CO_2); the temperature of the solid is about $-79°C$. It is rapidly turning to gaseous CO_2, and the cold fumes are generating a visible mist. Minutes after the photograph was taken, the Dry Ice was gone. (b) A quartz crystal will keep its shape indefinitely.*

16.11 METALLIC SUBSTANCES

Electrical Experiment with Silver

This experiment is again carried out with the previous apparatus. The electrodes now touch a piece of silver. The bulb lights up and stays lit, as shown in Figure 16.11. No chemical action is noted on the surface of the silver. In fact, if the silver is weighed before the experiment is started, and the current is then allowed to flow through it for a few hours, or days, or months, and the silver is then removed and reweighed, no change can be detected. Nothing is lost; all is the same as before. If the silver is melted, it still conducts electricity without chemical change. In these experiments, therefore, silver does not behave like salt, water, ice, or sand.

Silver is a metal. Similar to other metals, silver is an excellent conductor of electricity (one of the best, in fact) in either its solid or its liquid state. Moreover, metals are distinctive enough in other properties that we can recognize them as metals after only casual observation—by looking at them, feeling them, and hefting them. (You could not tell just by looking at a white powder, or some blue crystals, whether the substance is ionic or covalent.) What is it then that is unique about metals, and what kind of bond holds their atoms together? Let us first look more carefully at their physical properties.

The most conspicuous attribute of metal is **luster**—the bright, highlighted appearance of a substance that reflects light well. A smooth silver surface, for example, throws back over 90 percent of the light that shines on it. The adjective "silvery" is often used in the general sense of having luster. Two metals, gold and copper, are colored but

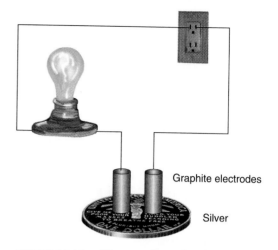

Graphite electrodes

Silver

FIGURE 16.11   *Silver conducts electricity.*

nonetheless lustrous. Of course, many metallic objects (such as cast-iron frying pans) look dark or dull, but if they are polished, the natural luster of the metal reappears, showing that the dullness was a coating of foreign, nonmetallic matter.

In the arctic winter, you don't dare place your bare hand on a piece of metal—it would freeze right onto the surface. And if one end of a metal bar is in a flame, don't touch the other end, for it is too hot. Metals are good conductors of heat.

The electrical conductivity of metals is far greater than that of molten or dissolved salts. Copper, for example, is about 60 million times better as a conductor than salt water (Table 16.6). Furthermore, when metals conduct electricity, no chemical change occurs. Thus, a silver bar or a copper wire can carry an electric current indefinitely without decomposition. Chemical changes occur

TABLE 16.6 Approximate electrical conductivities* at room temperature	
Silver or copper	600,000
Iron	100,000
Mercury	10,000
Salt water	0.01
Pure water	0.00000006
Rubber	0.000000000000001

*The units of electrical conductivity shown here are "reciprocal ohms/cm," but it is not the purpose of this table to emphasize such absolute values. Rather, attention should be given to the relationships among the different substances. Thus, if we compare samples of mercury and salt water *of equal physical dimensions*, we find that the mercury is more conductive by a factor of 10,000 to 0.01 or 1,000,000 to 1.

when electric currects pass through a solution containing dissolved ionic substances.

It was pointed out in Chapter 15 that the densest known substances are metals and that the densest one of all is osmium. Other dense metals include lead, 11 times as dense as water; mercury, almost 14 times as dense as water; and gold, 19 times as dense as water. By contrast, granite rock and concrete, which are commonly thought of as "heavy" materials, are only about two or three times as dense as water. Of course, there are "light" metals, like aluminum. Aluminum, however, would sink even in the dense waters of the Great Salt Lake or the Dead Sea, and a comparison of aluminum with solid nonmetallic elements of comparable or even greater atomic weight, such as sulfur or phosphorus, shows that aluminum is denser.

One of the most remarkable and unique properties of metals is their ability to maintain their crystal structure even when their shape is deformed. We say that metals are **malleable** (they can be hammered flat, as with a mallet) and **ductile** (they can be drawn into a wire). No other crystalline solids have these properties. Thus, if you strike a crystal of salt or sugar with a hammer, it will shatter, not flatten.

These sets of metallic properties are not at all consistent with our concepts of ionic or covalent bonding. The crucial difference is that the electrons involved in ionic or covalent bonds are more closely associated with specific bonded atoms. In the **metallic bond,** the electrons are thought to be somehow "loose"; that is, they are not associated with particular atoms.

The concept of "loose" or "free" electrons in metals has led to a theory of the metallic bond. One important clue is that metals crystallize in lattice structures in which the atoms are closely packed. In such close packing, one metal atom may be equally close to as many as 12 other atoms. In these arrangements, there are not enough valence electrons to provide ordinary two-electron covalent bonds between adjacent atoms. Therefore, a given valence electron is not associated with only two atoms but rather with all the atoms in a given sample of metal. A metallic crystal then has a large number of such "unattached" electrons scattered about the crystal. The energy levels of these loose valence electrons are closely spaced; that is, the energy differences among them are small. Taken together, the energy levels of these loose electrons make up

what amounts to a continuous band, called a **conduction band.**

Now, how does this idea of the metallic state account for the known properties of metals? The key must lie in the behavior of electrons in the conduction band. This behavior is related to the properties of metals in the following ways.

Density.   Electrons can establish more bonds when they are free to associate with many atoms than when they are localized between only two atoms. As a result, each atom in a metal can be closely linked to more atoms than would be possible with ionic or covalent bonding. In some metals, a given atom is closely associated with as many as 12 other atoms. (Twelve is the maximum number of spheres that can touch a given sphere of the same size.) This close packing is the reason metals are so dense.

Electrical Conductivity.   The loose electrons can be promoted to slightly higher energy levels within the conduction band when an electrical potential (voltage) is applied; they are then free to move in the direction of the applied potential. Hence, metals conduct electricity. When the voltage is removed, the electrons can fall back to the lower energy levels. No nuclei migrate anywhere. When the current stops flowing, the metal is in the same condition as at the start. Hence, no chemical change occurs.

Metallic Luster.   For a surface to reflect light, a photon must be absorbed and another one emitted. A photon is not like a rubber ball that literally bounces off a wall. A photon disappears when it is absorbed, and its absorption promotes an electron to a higher energy level. The loose electrons in the conduction band of metals are easily promoted to higher energies, and for most metals they fall back almost instantaneously to their original levels, emitting photons of the same frequency. That is why metals are shiny. The electrons in copper and gold, however, fall back to different levels, emitting photons of different frequencies, and therefore these metals are colored.

Conduction of Heat.   The electrons in the conduction band can also absorb thermal energy readily. Recall that thermal energy can be transmitted by photons of infrared frequency. Photons in this range are less energetic than photons of visible light, but because the energy levels in the conduction band are so closely spaced, it does not take much to promote electrons within the band. Infrared energy will do this. That is why metals conduct heat so well.

Malleability and Ductility.   Metals are ductile and malleable because under mechanical stress, the atoms of the crystal can move past each other with relatively little resistance and without breaking the metallic bonds. The bonds do not break because the valence electrons in the conduction band "belong" to all the atoms and are not localized to individuals or pairs of atoms.

EPILOGUE

Let us not forget that this chapter has dealt with chemical bonds that are common under conditions that are familiar to us on Earth. In the vast regions of outer space, matter is so sparse that a molecule such as OH or CN, with unsatisfied valences, which would react rapidly with something on Earth, might survive for a long time without meeting any other molecule. Toward the other end of the pressure scale, say at about a million Earth atmospheres in some cold stars and in a large planet like Jupiter, atoms of nonmetallic substances are squeezed together as tightly as are atoms in metals at ordinary Earth pressures, and a substance such as hydrogen becomes a metallic conductor. At still higher pressures, atoms collapse entirely and chemistry does not exist as we know it. High temperatures, too, change chemical bonds. The temperature of a gas flame is about 2000°C. The chemistry of flames is therefore different from that of substances at ordinary temperatures. Recall from the discussion of nuclear fusion in Chapter 14 that above about 5000°C, there are no chemical bonds at all and therefore no molecules and no chemistry.

SUMMARY

Electrons in the principal energy levels of an atom are said to be in **electron shells.** The shells corresponding to the highest energy levels include the electrons usually involved in chemical bonding and are therefore called the **valence shells.** Chemical formulas of simple compounds can often be predicted from valence.

The **periodic table** classifies elements into **periods,** which include elements with the same number of electron shells, and **groups,** which include elements with the same number of electrons in their valence shells. Elements within the same period differ markedly from one to the next, generally progressing from metallic to nonmetallic character as the atomic number increases. Elements in the same group have the same typical valences but generally progress to more metallic character as the number of shells increases.

Chemical equations express the formulas and relative numbers of molecules of reactants and products in a chemical reaction. The equation is balanced when the numbers and kinds of atoms are equal on both sides of the arrow.

Ionic, covalent, hydrogen, and metallic bonds are responsible for the different kinds of substances and their special properties we observe:

Ionic: Electrons are transferred from the metallic to the nonmetallic atom, producing ions that are isoelectronic with a noble gas atom. The oppositely charged ions attract each other. Ionic compounds conduct electricity only in the liquid state, either molten or dissolved in a solvent like water. The current is carried by the ions.

Covalent: Electrons are shared so that the number of valence electrons surrounding each atom other than hydrogen is eight (the octet rule). A single bond represents one pair of shared electrons, a double bond two pairs, and a triple bond three pairs. Covalent compounds do not conduct electricity. Carbon atoms can form covalent bonds with each other that can extend into long chains, branches, or cyclic structures.

Hydrogen: A hydrogen atom covalently bonded to one of the three electronegative atoms—N, O, or F—can form a hydrogen bond with one of the three same electronegative atoms on another molecule. Hydrogen bonding explains why some small molecules have unusually high boiling points and why ice is less dense than water.

Network covalent: The bonds are ordinary covalent bonds, but they extend indefinitely in various directions so that the molecule does not have an ordinary limited boundary. Such substances often have high melting points and are hard.

Metallic: The bonding electrons are not located between specific pairs of atoms but are "loose" and are grouped in tightly spaced energy levels called conduction bands. The characteristic properties of metals result from the ease with which the loose electrons can be promoted between energy levels in the conduction bands.

KEY WORDS

Electron shell

Valence shell

Valence

Group (of the Periodic Table)

Period (of the Periodic Table)

Main-group elements

Transition elements

Chemical equation

Ion

Chemical bond

Ionic bond

Isoelectronic

Covalent bond

Octet rule

Hydrogen bond

Isomers

Metallic bond

PROBLEMS AND CONCEPTUAL QUESTIONS

Problems requiring numerical work are identified by a blue number.

Electronic structure and the periodic table

1. (a) If the elements were arranged in the increasing order of their atomic weights rather than atomic numbers, would the sequence of elements generally be the same as in the present periodic table? (To answer this question, pick several sets of four or five consecutive elements from the periodic table, such as N, O, F, Ne, and check whether they are in the order of increasing atomic weights.) (b) Would the sequence always be the same? (Check the set Cl, Ar, K, Ca; the set Fe, Co, Ni, Cu; and the set Te, I, Xe.) (c) Explain your finding in part (b). (*Hint:* Remember isotopes?)

2. How many periods are there in the periodic table? How many groups? From this information, can you tell which is greater—the maximum number of shells that atoms can have in their stable (ground) states or the maximum number of electrons that some shells can have?

3. What is the maximum number of electrons in the $n = 4$ shell?

4. Predict the number of electrons in each shell for (a) $_{13}Al$; (b) $_{18}Ar$; (c) $_4Be$.

5. Predict the number of electrons in each shell for the main-group element that appears in the (a) second period, Group 7; (b) third period, Group 1; (c) first period, Group 8.

6. If you closed your eyes and poked your finger somewhere at the periodic table, then looked and saw you had "landed" on a nonmetal, in which direction would you move your finger to reach a metal: (a) Up or down within a group? (b) Right or left across a period? (c) From main-group to transition elements or from transition to main-group elements?

7. Which of the following properties is most generally typical of all the main-group elements in a given group: (a) metallic character; (b) nonmetallic character; (c) physical properties; (d) valence?

Valence and chemical equations

8. With the aid of the periodic table, write the formulas for the compounds formed by the combination of (a) Be and F; (b) Al and Br; (c) Na and S; (d) Ca and P.

9. Given the valences of 1 for H, F, and Na; 2 for O and Ca; and 3 for N and Al, calculate the valences of the other elements in the following compounds: (a) AsH_3; (b) BaO; (c) GeO_2; (d) PF_3; (e) $GaAs$; (f) BP.

10. Balance the following equations:
    (a) $Ag + O_2 \longrightarrow Ag_2O$ (oxidation of silver)
    (b) $KClO_3 \longrightarrow KCl + O_2$ (decomposition of potassium chlorate)
    (c) $Fe_3O_4 + C \longrightarrow Fe + CO_2$ (reduction of iron ore)
    (d) $H_2S + SO_2 \longrightarrow S + H_2O$ (production of sulfur in volcanos)

11. Complete the following chemical equations by filling in the blanks.
    (a) $2O_2 + 3$ _____ $\longrightarrow Fe_3O_4$
    (b) 4 _____ $ + 3O_2 \longrightarrow 2Al_2O_3$

12. Complete the following chemical equation by filling in the blank.
    _____ $ + 2LiOH \longrightarrow Li_2CO_3 + H_2O$

13. Balance each of the following chemical equations.
    (a) $Sc + HBr \longrightarrow ScBr_3 + H_2$
    (b) $NH_3 + O_2 \longrightarrow NO + H_2O$
    (c) $HClO \longrightarrow HCl + HClO_3$

14. Balance each of the following chemical equations.
    (a) $KMnO_4 + HCl \longrightarrow MnCl_2 + Cl_2 + KCl + H_2O$
    (b) $Fe_2O_3 + S \longrightarrow Fe + SO_2$

Chemical bonds

15. What are the four major types of chemical bonds?

16. Predict whether the structure of each of the following substances is ionic, covalent with small molecules, or covalent with a "giant" molecular network: (a) HCl, a gas at room temperature, liquefies at $-84°C$. The liquid is a poor conductor of electricity. (b) SiC. The covalency of

each atom is 4, but there are no multiple bonds. (c) $BaCl_2$, a solid that does not melt on a hot plate. A solution of $BaCl_2$ in water conducts electricity.

17. Predict the type of bonding in each of the following substances: (a) BN, borazon, a substance almost as hard as diamond; (b) $SbCl_3$, antimony chloride ("butter of antimony"), which melts in hot water to form a liquid that is a poor conductor of electricity; (c) $RaBr_2$, radium bromide, a solid that will not melt in a baker's oven but that dissolves readily in cold water. Radium bromide solution conducts electricity.

Ionic bonds

18. Write the formulas of (a) ions of elements number 15, 16, 17, 19, and 20 that are isoelectronic with argon; (b) ions of elements number 1 and 3 that are isoelectronic with helium.

19. Using the information obtained by answering question 18, write the formulas for the following ionic compounds: (a) lithium hydride, which is a compound of lithium and hydrogen; (b) calcium chloride; (c) potassium sulfide; (d) calcium hydride; (e) lithium chloride.

20. Use Lewis dot symbols to describe the formation of the ionic compounds in question 19.

21. Using the information in Table 16.5, write the names and formulas of the ionic compounds formed by a combination of (a) magnesium ion with oxide ion and with fluoride ion; (b) sulfate ion with ammonium ion, with potassium ion, and with aluminum ion.

22. Calcium ion, Ca^{2+}, has 20 protons in its nucleus. Exactly what gives it a 2+ charge?

23. What factor determines the number of electrons gained or lost by an atom in the formation of an ionic bond?

Covalent bonds

24. What is the octet rule?

25. Draw the Lewis structure of the following molecules that contain only single bonds: (a) H_2S, (b) SiF_4, (c) $CClF_3$, (d) PH_3, (e) C_3H_7Cl (two possibilities), (f) NH_2Cl, (g) $C_2H_3Cl_3$ (two possibilities), (h) C_2H_6S (two possibilities), (i) $C_2H_6O_2$ (two possibilities), (j) N_2H_4, (k) C_2H_7N (two possibilities).

26. Draw the Lewis structure of the following molecules that contain one or more double bonds: (a) C_3H_6, (b) $COCl_2$, (c) NOBr, (d) H_2CO, (e) N_2F_2, (f) H_2CO_3, (g) H_2C_2O, (h) C_3O_2, (i) C_2H_3Cl.

27. Draw the Lewis structure of the following molecules that contain one triple bond: (a) C_3H_4, (b) C_2H_3N (two possibilities), (c) C_4H_6 (two possibilities).

28. Draw the Lewis structure of the following molecules that contain single bonds and a cyclic structure: (a) C_4H_8, (b) C_4H_8O, (c) $C_5H_{10}S$, (d) C_4H_9N, (e) C_3H_6O. (*Note:* At least two possibilities exist for each, but you should show just one.)

Hydrogen bonds

29. Which of the elements (besides hydrogen) are needed for hydrogen bond formation?

30. Which of the following molecules can form hydrogen bonds among themselves: (a) C_2H_6 (ethane), (b) CH_3OH (methanol), (c) HCl?

31. Draw a structure showing a HF molecule being hydrogen bonded to two NH_3 molecules.

Metallic bonds

32. (a) In your own words, summarize the physical properties of metals. (b) Iron pyrite, FeS_2, is called "fool's gold" because many inexperienced prospectors have mistaken it for the real thing. What are the probable reasons for this confusion? What step would you, as an amateur prospector, take to ensure that your find was metallic?

33. Wax can be hammered into a sheet; organic matter can occur as filaments, such as hair or fibers, resembling thin wires; molten salt conducts electricity well; cinnabar (HgS) is eight times as dense as water. None of these substances is metallic. Explain in each case why the property described does not prove that the substance is a metal.

34. In your own words, summarize the band theory of metals and state how the theory accounts for the properties of metals.

35. Lithium crystallizes in a cubic arrangement in which each Li atom may be considered to be in the center of a cube, "touching" eight other Li atoms at the corners of the cube. Explain why the theories of neither ionic nor covalent bonding could account for such a structure.

ANSWERS TO SELECTED NUMERICAL PROBLEMS

3. 32

4. (a) 2, 8, 3; (b) 2, 8, 8; (c) 2, 2

8. (a) BeF_2, (b) $AlBr_3$, (c) Na_2S, (d) Ca_3P_2

9. (a) As = 3, (b) Ba = 2, (c) Ge = 4, (d) P = 3, (e) Ga = 3, (f) B = 3

10. (a) $4Ag + O_2 \longrightarrow 2Ag_2O$, (b) $2KClO_3 \longrightarrow 2KCl + 3O_2$, (c) $Fe_3O_4 + 2C \longrightarrow 3Fe + 2CO_2$, (d) $2H_2S + SO_2 \longrightarrow 3S + 2H_2O$

11. (a) Fe, (b) Al

12. CO_2

13. (a) $2Sc + 6HBr \longrightarrow 2ScBr_3 + 3H_2$, (b) $4NH_3 + 5O_2 \longrightarrow 4NO + 6H_2O$, (c) $3HClO \longrightarrow 2HCl + HClO_3$

14. (a) $2KMnO_4 + 16HCl \longrightarrow 2MnCl_2 + 2KCl + 5Cl_2 + 8H_2O$, (b) $2Fe_2O_3 + 3S \longrightarrow 4Fe + 3SO_2$

18. (a) P^{3-}, S^{2-}, Cl^-, K^+, Ca^{2+}; (b) H^-, Li^+

19. (a) LiH, (b) $CaCl_2$, (c) K_2S, (d) CaH_2, (e) LiCl

21. (a) MgO—magnesium oxide, MgF_2—magnesium fluoride, (b) $(NH_4)_2SO_4$—ammonium sulfate, K_2SO_4—potassium sulfate, $Al_2(SO_4)_3$—aluminum sulfate

25. (a) H—S—H, (b) F—Si—F (with F above and F below Si)

(c) F—C—Cl (with F above and F below C), (d) H—P—H (with H below P)

(e) H—C—C—C—Cl (with H's above and below each C), H—C—C—C—H (with H, Cl, H above and H, H, H below), (f) H—N—Cl (with H below N)

(g) H—C—C—Cl (with H, Cl above and H, Cl below), H—C—C—Cl (with Cl, Cl above and H, H below)

26. (a) H—C—C=C—H (with H, H, H above and H below), (c) O=N—Br

CHAPTER 17

PRINCIPLES AND APPLICATIONS

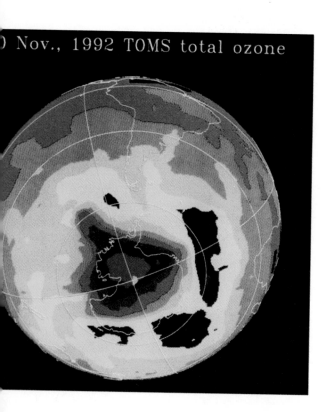

Nov., 1992 TOMS total ozone

Scientists have noticed that the ozone layer in the stratosphere over the South Pole has become thinner. This map, based on data collected in 1992, shows the depletion of ozone (in purple) covering an area about the size of the North American continent.
(*Source:* NASA)

17.1 ATOMIC MASS, MOLECULAR MASS, AND THE MOLE

We can define chemical substances as collections of small particles such as atoms, molecules, and ions. A single atom or molecule can be detected, but it cannot be handled in the way that ordinary samples of matter are dealt with in the laboratory. The smallest samples that chemists work with may be milligrams (10^{-3} g), micrograms (10^{-6} g), or even nanograms (10^{-9} g), but these small masses contain many constituent particles. A nanogram of copper, for example, which is a billionth of a gram, contains about 10 trillion (10^{13}) copper atoms. Chemists therefore find it convenient to express the amount of a substance in terms of a standard number of its atoms or other constituent particles. Some examples in other walks of life show us that this practice is not unusual. A cook who prepares food for large numbers of people deals with eggs by the dozen, not by the kilogram or the liter. A printer counts sheets of paper by the ream (1 ream = 500 sheets). Chemists use the **mole,** which is the amount of a substance that contains a certain number of fundamental particles. What is that certain number? The chemist answers the question in a way that may not satisfy you—the answer is this: Take *exactly* 12 g of carbon—not just any carbon, but specifically 12 g of the carbon-12 isotope. Those 12 g contain a certain number of atoms; that number is the number you are looking for, the number of fundamental particles in a mole. The problem is that the number is not known exactly. The number is large; it starts with 6 and is followed by 23 other digits. The best experimental value is 6.022045×10^{23}, but the approximation 6.02×10^{23} is often used. Therefore, a simpler expression is:

the mole is the amount of substance that contains 6.02×10^{23} basic particles.

(a)

(b)

(a) One mole of some common elements. Back row (left to right): bromine, aluminum, mercury, and copper. Front row (left to right): sulfur, zinc, and iron. (b) One-mole quantities of a range of compounds. The white compound is NaCl; the blue compound is $CuSO_4 \cdot 5H_2O$; the deep red compound is $CoCl_2 \cdot 6H_2O$; the green compound is $NiCl_2 \cdot 6H_2O$; and the orange compound is $K_2Cr_2O_7$. (Courtesy Charles D. Winters)

The basic particles may be atoms, molecules, or ions. Thus, a mole of copper, which is an element, is the amount that contains 6.02×10^{23} copper atoms, whereas a mole of water, which is a compound, is the amount that contains 6.02×10^{23} water molecules.

Neither of the two definitions is helpful to anyone who needs, say, a mole of nickel or a mole of water for an experiment. A chemist cannot count out the required number of atoms or molecules as a cook would count out eggs for a recipe. Fortunately, there is a third way of expressing what a mole is. Refer to the table of atomic masses in Appendix E. The atomic masses are not the same as mass numbers because they take into account the distribution of isotopes as they occur on Earth. Thus, the atomic mass of carbon is not 12 but 12.011 because on Earth there is a little carbon-13 (and other isotopes) mixed in with the preponderant carbon-12 atoms. The atomic masses, then, are average values, but they are still based relative to the assignment of 12 units of mass of the carbon-12 isotope. From the first definition of the mole given, a mole of carbon-12 weighs exactly 12 g. Because all other atomic masses are based on carbon-12, it follows that *the mass of a mole of atoms of an element is the atomic mass of the element expressed in grams.*

Thus, the atomic mass of nickel is 58.70, so a mole of nickel is 58.70 g.

What about molecules? The sum of the atomic masses in a molecular formula is the **molecular mass.** Therefore, the mass of a mole of molecules of a substance is the molecular mass of the substance expressed in grams.

It is difficult to visualize the meaning of such a large number as 6.02×10^{23}, the number of atoms or molecules in a mole. One way to think about it is to recognize that atoms in any ordinary sample of matter are so numerous, and chemical changes go on so constantly all around us, that it is likely that many of the atoms that were in the body of any ancient person you choose to name are now in your body. Or consider this: If you could identify every molecule in a cup of water and then poured that water in the ocean and stirred the water so thoroughly that your molecules were uniformly dispersed in *all* the world's oceans and then dipped your cup back in to refill it, you would recover over 1000 of your original molecules.

EXAMPLE 17.1 MOLECULAR MASSES

(a) Calculate the molecular mass of water, H_2O, and of carbon dioxide, CO_2. Use the following approximate atomic masses: O, 16; H, 1; C, 12.

(b) Calculate the mass of 1 mole of water and of 3 moles of carbon dioxide.

Solution (a) From the molecular formulas, the molecular mass of water is $1 + 1 + 16 = 18$ and of carbon dioxide is $12 + 16 + 16 = 44$.

(b) The mass of a mole of water is its molecular mass in grams, or 18 g. For carbon dioxide, 3 moles are $(3 \text{ moles})(44 \text{ g/mole}) = 132 \text{ g}$.

17.2 ACIDS AND BASES

The original meaning of **acid** is "sour," referring to the taste of substances such as vinegar, lemon juice, unripe apples, and old milk. It has long been observed that all acidic substances have some typical properties in common. For naturally occurring **acidic solutions,** in which the solvent is always water, these properties are:

1. Acids taste sour. (But *never try to taste acids found in the laboratory; the results could be fatal.*)
2. Acids speed up the corrosion, or rusting, of metals. When the attack on a metal by an acidic solution is vigorous, hydrogen gas, H_2, is evolved in the form of visible bubbles.
3. Acidic solutions conduct electricity. As the current passes through the solution, hydrogen gas is evolved at the negative electrode (cathode).
4. Acids affect the colors of certain botanical substances, which are known as **indicators.** For example, if you add lemon juice to a cup of tea, the color of the tea becomes lighter. Lemon juice is an acid; therefore tea is an indicator that changes color when an acid is added.
5. A characteristic reaction of acids is that they react with certain other substances, known as **bases.** This chemical reaction is called neutralization.

Bases do not occur in as many common materials as do acids, but a few common basic materials are ashes, soap (especially strong dishwasher soaps), borax, lime, and lye. The characteristic properties of bases and **basic solutions** are:

1. Bases taste bitter. *But don't ever try to taste the materials referred to here or basic solutions found in the laboratory.* Two common household products, drain cleaners and oven cleaners, commonly contain lye, which is a strong base, also called a caustic or an alkali. The chemical name is sodium hydroxide, $NaOH$. Products that contain this material should not be allowed near children.
2. Basic solutions in which the solvent is water conduct electricity, just as acidic solutions do. Oxygen gas, O_2, is given off at the anode (positive electrode).
3. Bases affect the colors of indicators, but the color changes are different from those produced by acids. For example, the natural indicator in purple grape juice, which is an acidic solution, turns green when enough dishwasher soap is added to make the juice basic. A common laboratory indicator is **litmus,** which is red in acid solutions and blue

in basic ones. In literary parlance, a "litmus test" is used in the general sense of proving that something is definitely either this way or that—no ifs, ands, or buts.

4. Bases neutralize acids. The reaction is so rapid that it is practically instantaneous, and it gives off considerable heat. If, for example, solutions of hydrochloric acid (a strong acid) and sodium hydroxide (a strong base) are mixed in just the right proportions, the product is neither acidic nor basic; it is a neutral solution of ordinary salt, NaCl. Other acids and bases produce other salts. In general,

neutralization is defined as the reaction between an acid and a base to produce a salt and water.

The properties of acids and bases are interpreted as follows:

The fact that hydrogen gas, H_2, is produced when acidic solutions react with metals leads to the conclusion that acidity is related to hydrogen.

Because acidic solutions conduct electricity, they must contain charged particles. The charged particles that migrate to the cathode (negative electrode) must be positive. Since hydrogen gas is given off at the cathode, there must be some molecules or ions that contain hydrogen. The simplest positive ion is H^+, which is a proton. A proton cannot exist as an independent particle in water, however, because it becomes chemically bonded to the oxygen atom of the water molecule. The resulting ion is formulated as $H(H_2O)^+$, or simply H_3O^+, and is called the hydronium ion. It is simpler to write H^+, as many books, including this one, do; regard this as an abbreviation. The reaction at the cathode may then be written as

$$2H^+ + 2e^- \longrightarrow H_2$$

Basic solutions contain hydroxide ions, OH^-. When an electric current is passed through such a solution, the OH^- ions migrate to the anode, where oxygen gas is liberated:

$$4OH^- \longrightarrow O_2 + 2H_2O + 4e^-$$

Hydroxide ions, OH^-, can neutralize H^+ ions by reacting with them to produce water, as indicated by the equation

$$H^+ + OH^- \longrightarrow H_2O$$

Even pure water ionizes to a slight extent. The concentration of H^+ and of OH^- in pure water at 25°C is 1.0×10^{-7} mole/L. This solution is said to be **neutral,** because the concentrations of the two ions are equal. When the hydrogen ion concentration is greater than 1.0×10^{-7} mole/L at 25°C, the solution is **acidic.** When it is less than this value, the solution is **basic.**

Hydrogen ion concentrations are usually expressed by a set of values called a **pH** scale (see next section). A neutral solution has a pH of 7. Acidic solutions have pH values below 7, and every decrease of one pH unit represents a *tenfold* increase in the concentration of H^+ ions. Basic solutions have pH values above 7, and every additional pH unit represents a tenfold *decrease* in the H^+ ion concentration (and a tenfold increase in hydroxide ion concentration).

Strong acids ionize completely in water, or nearly so. As a result, they furnish a large concentration of H^+ ions. An example in this category is nitric acid, HNO_3:

$$HNO_3 \longrightarrow H^+ + NO_3^-$$

In the case of a weak acid, such as acetic acid, only a small percentage of the molecules ionize, so that the concentration of H^+ ions remains low.

Figure 17.1 gives the pH values of various common materials.

17.3 THE PH SCALE

The pH values are based on logarithms to the base 10. The relationship is

$$pH = -\log_{10} \text{(hydrogen ion concentration)}$$

(17.1)

Recall that the logarithm of a number to the base 10 is the exponent to which 10 must be raised to give that number. For example, $10^2 = 100$, so log $100 = 2$, and $10^{-3} = 0.001$, so log $0.001 = -3$. The negative sign in the definition of pH changes the sign of the exponent. Thus, if the H^+ concentration (in mole/L) is 10^{-7}, the pH is 7. If it is 10 times this value, or 10^{-6}, the pH is 6, which is an acidic solution. If it is one-tenth as concentrated, or 10^{-8}, the pH is 8, which is basic.

In summary, a solution whose pH is less than 7 is acidic. If the pH is greater than 7, it is basic. A pH of 7 is neutral.

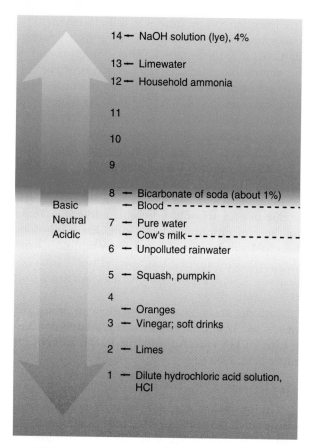

14 — NaOH solution (lye), 4%

13 — Limewater
12 — Household ammonia

11

10

9

8 — Bicarbonate of soda (about 1%)
— Blood - - - - - - - - - - - - - - - -

Basic
Neutral   7 — Pure water
Acidic   — Cow's milk - - - - - - - - - - - - - -
6 — Unpolluted rainwater

5 — Squash, pumpkin

4
— Oranges
3 — Vinegar; soft drinks

2 — Limes

1 — Dilute hydrochloric acid solution, HCl

FIGURE 17.1 *Approximate pH values of various substances.*

The term pH was derived from "powers of hydrogen," where "power" refers to the exponent.

EXAMPLE 17.2 CALCULATING PH

What is the pH of a commercial vinegar whose hydrogen ion concentration is 2×10^{-3} mole per liter?

Solution From Eq. 17.1, we can calculate the pH of the vinegar by writing

$$pH = -\log_{10} (2 \times 10^{-3})$$
$$= \boxed{2.7}$$

A pH of 2.7 is fairly acidic, which is consistent with the sour taste of vinegar.

17.4 OXIDATION AND REDUCTION

Oxygen is not only the most abundant element in the Earth's crust, but also one of the most versatile. In its common molecular form, O_2, it is the second most abundant constituent of the atmosphere. In its chemically combined forms, it is a major component of water, of all living organisms, and of most rocks and minerals. Oxygen forms bonds with every element except some of the noble gases. The formation of these bonds releases energy. A reaction with oxygen that releases energy rapidly enough to give off light is called **combustion.** The reacting chemical is said to *burn* in oxygen, or in a gas, such as air, that is rich enough in oxygen. You are familiar with the fact that materials such as wood and paper burn in oxygen or in air. It is not so widely known, however, that some metals, too, can burn.

FOCUS ON PURE WATER

Chemically pure water, which has a pH of 7, does not occur in nature. The reason is that water is a good solvent, and even in an unpolluted environment, it captures impurities by dissolving gases from the atmosphere and minerals from soil and rock. Sometimes these impurities are acidic, which reduces the pH below the neutral value of 7, and sometimes they are basic, yielding a pH higher than 7. When we speak of "pure drinking water" in ordinary conversation, we mean water that is wholesome, free from noxious chemicals or disease organisms, and usually containing small amounts of tasty and probably beneficial mineral matter. The preparation of chemically pure water requires great care. It must be distilled in the absence of air, to avoid carbon dioxide, and without using glass, from which the water would dissolve silicates.

A piece of magnesium ribbon burns in air to give the white solid magnesium oxide, MgO.
(Courtesy Charles D. Winters)

Magnesium, for example, is a shiny metal that burns rapidly in oxygen. If the magnesium is dispersed in the form of a powder, all of it burns practically at the same time, and the result is an explosion. The balanced equation for the reaction is:

$$2Mg + O_2 \longrightarrow 2MgO$$

*The formation of a chemical bond between oxygen and another element is called **oxidation**.*

Compounds, too, can be oxidized. The familiar reactions of respiration and of the burning of natural gas (methane) are oxidations. The equation for respiration is shown as the oxidation of glucose (blood sugar):

$$C_6H_{12}O_6 + 6O_2 \longrightarrow 6CO_2 + 6H_2O$$
Glucose

The equation for the oxidation of methane is:

$$CH_4 + 2O_2 \longrightarrow CO_2 + 2H_2O$$

In general, when substances containing carbon, hydrogen, or both oxidize completely, all the carbon is oxidized to carbon dioxide, CO_2, and all the hydrogen to H_2O. Thus, the same oxidation products are formed by the complete oxidation of glucose, methane, or octane (C_8H_{18}).

We now return to magnesium: Its oxidation product, MgO, is not metallic at all; it is a white,

powdery solid. Furthermore, it is an ionic compound, which can be represented as $Mg^{2+} + O^{2-}$. Therefore, the equation for the oxidation of magnesium can also be written as

$$2Mg + O_2 \longrightarrow 2Mg^{2+} + 2O^{2-}$$

Magnesium also reacts with chlorine, Cl_2, to produce magnesium chloride, $MgCl_2$, as shown by the equation

$$Mg + Cl_2 \longrightarrow MgCl_2$$

But magnesium chloride, too, is ionic, so the equation could be written as

$$Mg + Cl_2 \longrightarrow Mg^{2+} + 2Cl^-$$

Now note that the Mg becomes Mg^{2+} whether it reacts with oxygen or with chlorine, so if the reaction with oxygen is oxidation, wouldn't it be reasonable to call the reaction with chlorine oxidation, even if no oxygen is involved? Consider another question: If all that happens to the magnesium is its conversion to Mg^{2+}, could such ions be produced in any other way? The answer is yes, the magnesium could simply be the anode in an electrochemical cell, as shown in Figure 17.2. Then the magnesium loses electrons, producing positive ions, which dissolve in the solution:

$$Mg \longrightarrow Mg^{2+} + 2e^-$$

These observations lead to a second definition of oxidation:

Oxidation is the loss of electrons.

If oxidation is either the formation of a bond to oxygen or the loss of electrons, there must be an opposite process:

***Reduction** is defined as the breaking of a bond to oxygen or the gain of electrons.*

The word "reduce" comes from the Latin *reducere*, to bring back or restore, but this meaning has almost entirely disappeared from ordinary usage. In chemistry, however, the original sense is preserved in that reduction of a metallic oxide brings back or restores the metal:

$$2MgO \longrightarrow 2Mg + O_2$$

In ionic form, the equation showing the reduction of the Mg^{2+} ion by the gain of electrons is

$$Mg^{2+} + 2e^- \longrightarrow Mg$$

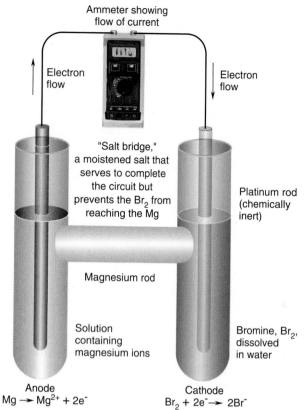

Ammeter showing flow of current

Electron flow

Electron flow

"Salt bridge," a moistened salt that serves to complete the circuit but prevents the Br_2 from reaching the Mg

Platinum rod (chemically inert)

Magnesium rod

Solution containing magnesium ions

Bromine, Br_2, dissolved in water

Anode
$Mg \rightarrow Mg^{2+} + 2e^-$

Cathode
$Br_2 + 2e^- \rightarrow 2Br^-$

FIGURE 17.2   *An electrochemical cell in which magnesium is the anode and loses electrons to become Mg^{2+}. At the cathode, bromine (Br_2) is being reduced by gaining electrons to become Br^- ions.*

One way to think about oxidation and reduction is to realize that oxidation generally turns useful materials into wastes, whereas reduction does just the opposite—it converts wastes or useless products into valuable materials. In the following examples, the substances referred to are all being oxidized:

> *Burning of wood or coal* $\longrightarrow$ CO_2, H_2O, *smoke, and ashes*
>
> *Metabolism of food (respiration)* $\longrightarrow$ *body wastes*
>
> *Corrosion of iron* $\longrightarrow$ Fe_2O_3 *(red iron rust)*

The benefit realized from these oxidations is the release of energy.

In the following chemical changes, substances are being reduced:

> CO_2 + *water (photosynthesis)* $\longrightarrow$ *sugars*
>
> *Reaction of iron ore, Fe_2O_3, with carbon* $\longrightarrow$ *iron*

These reductions are not without cost; they consume energy.

One more point about oxidation and reduction: They always occur together in chemical systems, never separately. If one thing is oxidized, something else must be reduced. The reason is that the electrons that are released must go somewhere. Returning to the first reaction given in this section,

$$2Mg + O_2 \longrightarrow 2MgO$$

If the magnesium is oxidized, what is reduced? Rewriting the equation in ionic form provides the answer:

$$2Mg + O_2 \longrightarrow 2Mg^{2+} + 2O^{2-}$$

The oxygen gains electrons and therefore is reduced:

$$O_2 + 4e^- \longrightarrow 2O^{2-}$$

CONCEPTUAL EXERCISE

Barium reacts with fluorine to form barium fluoride. Break down the reaction into an oxidation and a reduction step and identify the substances that are oxidized and reduced.

17.5 ACID RAIN

Acid rain has been recognized as a significant environmental issue. (The more general terms are acid precipitation or acid deposition, which include acidic snow and dust as well as rain.) It is important to realize, however, that even rain that falls through an unpolluted atmosphere is slightly acidic. The reason is related to the composition of the atmosphere, which is shown in Figure 17.3. The 1 percent of "other gases" includes carbon dioxide, which reacts to a slight extent with water to produce carbonic acid, a weak acid:

$$CO_2 + H_2O \longrightarrow \quad H_2CO_3$$
carbonic acid

In addition, some nitric acid is formed during lightning storms by the oxidation of nitrogen, N_2, in the presence of water:

$$2N_2 + 5O_2 + 2H_2O \longrightarrow \quad 4HNO_3$$
nitric acid

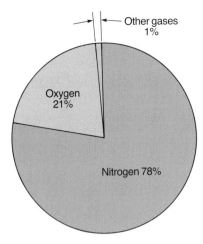

FIGURE 17.3 *Approximate gaseous composition of natural dry air.*

(a)

(b)

The effects of acid rain on the evergreen population atop Camel's Hump in Vermont's Green Mountain range. Part (a) was photographed 15 years earlier than part (b). (Courtesy U.S. Environmental Protection Agency)

Nitric acid is a strong acid, but not enough of it is formed by natural processes to introduce much acidity into rainwater. The combined effects of carbonic acid and nitric acid in unpolluted air make rainwater (or snow) slightly acidic, with a pH a bit below 6.

In recent years, however, rain and snow in many parts of the world have become considerably more acidic. Much of this acid precipitation has been between pH 4 and 5, but more severely acidic episodes occur from time to time. For example, a rainstorm in Baltimore in 1981 had a pH of 2.7, which is about as acidic as vinegar. Acid rain is a worldwide problem that has been estimated to cause billions of dollars worth of damage. Acids corrode exposed metallic structures, such as bridges and automobiles, and attack concrete and rock, particularly limestone and marble. Acids also cause the death of trees and fish and reduce the growth of certain agricultural crops.

Most of this excess acidity can be traced to a series of chemical reactions involving sulfur and nitrogen. The small concentrations of nitric acid that are formed by the action of lightning are augmented by the burning of fossil fuels. In the gasoline engine, for example, the electric spark generated by the spark plug serves as a substitute for lightning, and some nitrogen in the air in the cylinder is oxidized to NO,

$$N_2 + O_2 \longrightarrow 2NO$$

When the NO leaves the exhaust pipe and comes in contact with the air and moisture of the outside atmosphere, it is further oxidized to nitric acid:

$$4NO + 3O_2 + 2H_2O \longrightarrow 4HNO_3$$

All in all, however, the total atmospheric acidity that starts with sulfur is greater than that coming from nitrogen. The reason is that sulfur, being essential to life, existed in the organisms from which fossil fuels originated. As the remains of prehistoric plants and animals gradually became transformed to coal and oil, some of the hydrogen and most of the oxygen and nitrogen escaped, but much of the sulfur stayed put and is still present in these fuels. (So is the mineral content, which also contains sulfur compounds.) Coal is especially rich in sulfur. As coal is burned and the carbon is oxidized to CO_2, the sulfur is oxidized to sulfur dioxide, SO_2, which is also a gas:

$$C + O_2 \longrightarrow CO_2$$

$$S + O_2 \longrightarrow SO_2$$

The oxidation of sulfur to SO_2 occurs directly in the flame, and therefore SO_2 is discharged to the atmosphere from the smokestack. As the SO_2 is swept along by the prevailing winds, it is slowly oxidized at ordinary temperatures to SO_3:

$$2SO_2 + O_2 \longrightarrow 2SO_3$$

SO_3 then reacts rapidly with atmospheric moisture to form sulfuric acid, which is a strong acid:

$$SO_3 + H_2O \longrightarrow \quad H_2SO_4$$
$$\text{sulfuric acid}$$

Sulfuric acid is soluble in water and is therefore washed out by rain. It is for these reasons that acid rain often occurs at considerable distances from the sources where the sulfur dioxide is introduced into the atmosphere. The conversion of nitrogen oxides to nitric acid is also slow enough to carry these pollutants for some distances before they come down as acid rain.

Even before the acids are formed in the atmosphere, some of the oxides of nitrogen and sulfur become attached to dust particles and adhere to them as the particles fall to Earth. When they come in contact with rivers, lakes, or the moisture in soil, they react to form acidic solutions. These acidic dusts often precipitate closer to the pollution sources than the acid rain and snow (Fig. 17.4).

Various surveys in the Adirondack Mountains of upstate New York as well as in Canada and elsewhere showed that the pH of large numbers of ponds and lakes was less than 5 and that many of these had completely lost their fish populations. Plants, too, are affected. Mysterious blights that have killed increasing numbers of trees have been traced to acid rain. There is no simple relationship, however, between the amount of acid and the resulting environmental damage, for some interesting chemical reasons. Here are some of the complications:

1. Most of the acid precipitation that reaches a lake does not fall directly into it but, instead, falls on the land and then runs off into the water. If the soil is rich in limestone, it can neutralize most of the acid before the rain runs off the land:

$$CaCO_3 + H_2SO_4 \longrightarrow CaSO_4 + H_2O + CO_2$$
limestone sulfuric   calcium
                acid   sulfate

In some instances, the added sulfate even acts as an agricultural fertilizer. When the soil is poor in limestone, however, this mechanism does not operate.

2. Observations in West Germany's Black Forest and elsewhere indicated that trees on mountaintops and hilltops suffered a much greater percentage of loss than those at lower levels. This curious phenomenon is related to the prevalence of fogs at these summits. Recall from Chapter 15 that a fog or

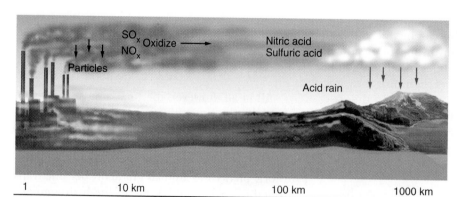

FIGURE 17.4   *Acid rain.*
SO_x means SO_2 or SO_3.
NO_x means NO, NO_2, or
other oxides of nitrogen.

mist is not a gas but a collection of tiny liquid droplets. Sulfuric acid and nitric acid are highly soluble in water, and therefore they dissolve in fog droplets, where their concentration becomes about ten times higher than their average atmospheric concentration. For this reason, fogs have been called the "vacuum cleaners of the atmosphere." These high acid concentrations cause the greatest damage to trees.

3. Another unexpected effect has been the greater damage done to pine trees by nitric acid as compared with sulfuric acid, even though the atmospheric concentration of nitric acid is much less than that of sulfuric acid. Here the damage may be caused by the nutrient effect of nitrogen, which fertilizes the pine needles and causes excess growth in the late fall, making them more susceptible to winter injury.

There are several approaches to the prevention of acid precipitation or the damage it creates. One is desulfurization, the removal of sulfur from fuel before it is burned. This method uses a combination of chemical and mechanical processes to remove sulfur, mainly in the form of calcium sulfate, $CaSO_4$, which can then be disposed of as a nonhazardous solid waste. Another approach is the removal of the sulfur and nitrogen oxides from the combustion gases before they enter the atmosphere. This method uses **scrubbers,** which are devices for bringing gases and liquids into close contact with each other (Fig. 17.5). Because the gases to be removed are acidic, the scrubbing liquid must contain a base to neutralize the acid. The cheapest basic material is limestone, $CaCO_3$. The base derived from limestone is calcium hydroxide, $Ca(OH)_2$, and the reactions with the sulfur oxides can be formulated as follows:

$$SO_2 + Ca(OH)_2 \longrightarrow CaSO_3 + H_2O$$
$$\text{calcium sulfite}$$

$$SO_3 + Ca(OH)_2 \longrightarrow CaSO_4 + H_2O$$
$$\text{calcium sulfate}$$

Finally, if the source cannot be controlled, the acidified area can be treated. An effective stop-gap method involves neutralizing the acid in a lake by adding the necessary quantity of lime, after which the lake can be restocked with fish.

17.6 CATALYSIS

A mixture of hydrogen and oxygen gases, ideally with twice as many hydrogen molecules as oxygen molecules, is potentially explosive. But if such a mixture is left alone at room temperature, nothing is seen to happen—no explosion, not even any slow reaction. If a lighted match is introduced into the mixture, however, or even the tiniest spark, the reaction starts and spreads so rapidly that it seems to happen all at once: It explodes. The equation is

$$2H_2 + O_2 \longrightarrow 2H_2O$$

If the mixture were in a strong steel cylinder, the kind that is designed to hold gases at hundreds of atmospheres of pressure, the cylinder would burst, and fragments of steel would be blown away for distances of hundreds of meters, some perhaps for kilometers. There is another way to set off this explosion without a match or a spark, in fact without raising the temperature of the mixture at all. All you would have to do would be to introduce a little finely divided platinum powder into the gas, and off it would go. If it were possible to gather all the products of the explosion, the platinum could be recovered and used again. The platinum has served as a **catalyst.**

FIGURE 17.5   *Schematic diagram of a scrubber.*

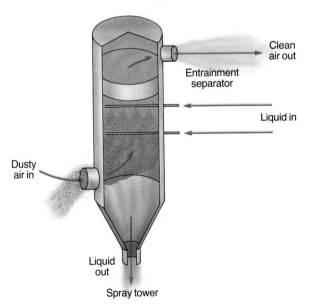

Clean air out

Entrainment separator

Liquid in

Dusty air in

Liquid out

Spray tower

A catalyst is a substance that increases the rate of a chemical reaction but is not consumed in the reaction.

The action of a catalyst may seem mysterious. How can a substance that is not used up in a reaction influence the rate of that reaction? Before addressing that question, we may ask why even a spark or a lighted match is needed to set off the hydrogen-oxygen explosion. After all, the reaction itself releases a lot of energy, so why is energy needed to start it? The process can be understood by considering the structural formulas of the substances involved: H—H for hydrogen, O=O for oxygen, and H—O—H for water, the product of the explosion. The important point is that there are no H—H or O=O bonds in a water molecule. Therefore, for the reaction to occur, the H—H bonds in hydrogen and the O=O bonds in oxygen must be broken. Some outside source of energy is needed to break those bonds. If those bonds are not broken first, nothing happens. Once the reaction does start, however, it provides more than enough energy to keep itself going. The energy released when the H—O—H bonds of water are formed serves to decompose more O_2 and H_2 molecules. There is enough energy released, in fact, to initiate a branching chain reaction, which becomes an explosion.

Now we can understand the action of the catalyst. Because the reaction cannot start before some bonds are broken, the catalyst must somehow substitute for the spark or the lighted match. What does actually happen in this case is that the hydrogen molecules form bonds to the platinum atoms, in an arrangement that may be represented as

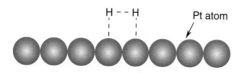

This action weakens the bonds between the hydrogen atoms, and this weakening is just enough to get the reaction started. The start is all that is needed; the chain reaction is then on its way.

Does a catalyst participate in the reaction that it speeds up? The answer is yes. Return to the definition of a catalyst, and note that it does not say that the catalyst is uninvolved, only that it is not consumed. The platinum has to react (that is, form bonds) with the hydrogen for the catalysis to occur.

When the H—H bonds break and the H atoms start to react, the platinum remains behind. It has not been consumed, which is a necessary condition for a catalyst.

Catalysis occurs in many naturally occurring chemical reactions, both in living and in nonliving systems, and is also used extensively in the chemical industry. Catalysts can operate in the gaseous, liquid, or solid states of matter as well as in reactions that involve more than one state. The use of platinum to catalyze the reaction between hydrogen and oxygen is an example that involves more than one state: The platinum is a solid, and the hydrogen and oxygen are gases.

Humans and other animals obtain their energy by the oxidation of carbohydrates, fats, and proteins. If you were given some sugar or beef fat in the laboratory and told to oxidize it, you would have to burn it; the oxidation would then take place at the temperature of the flame, which can be somewhere between 500 and 1000°C. In air at 37°C, the reaction would be too slow to measure. But in our bodies, the oxidation does take place at 37°C, for that is normal human body temperature. Speeding up the reaction from the immeasurably slow rate in outside air to our normal metabolic rates requires powerful catalysts. These body catalysts are called **enzymes,** which are a particular group of proteins.

EXAMPLE 17.3   WHAT IS THE CATALYST?
The oxidation of SO_2 to SO_3 in air is slow. The equation is

$$2SO_2 + O_2 \longrightarrow 2SO_3$$

The following two reactions, however, are fast:

$$2NO + O_2 \longrightarrow 2NO_2$$

$$2NO_2 + 2SO_2 \longrightarrow 2NO + 2SO_3$$

What is the net effect of the last two reactions? What substance is the catalyst?

Solution Add the two reactions by combining all the formulas to the left of both arrows and showing that they yield all the formulas to the right of both arrows:

$$2NO + O_2 + 2NO_2 + 2SO_2 \longrightarrow$$
$$2NO_2 + 2NO + 2SO_3$$

Note that the 2NO and 2NO$_2$ formulas appear on both sides of the equation and can therefore be crossed out. The result will be the net reaction

$$O_2 + 2SO_2 \longrightarrow 2SO_3$$

This is the same reaction as the slow oxidation shown at the beginning of this example, but now it has taken place in two fast steps, which are faster than one slow one. The new chemical that was added was NO, but that was also recovered, not used up, so NO was the catalyst.

17.7 OZONE IN THE ATMOSPHERE

In the Stratosphere

Ozone, O$_3$, is different from oxygen, O$_2$. Pure ozone is a blue, explosive, poisonous gas. Few people ever see it, but most have smelled it. The pungent odor from electric sparks, such as from a worn, sparking electric motor, or the slightly pungent odor of the atmosphere after a lightning storm is the odor of ozone. Small amounts of ozone are produced naturally in the stratosphere by the action of sunlight on oxygen. (The stratosphere is the section of the atmosphere about 20 km above sea level. The structure of the atmosphere is described in Chapter 18.) Oxygen molecules, O$_2$, are first broken down by ultraviolet (UV) radiation to oxygen atoms. In a second step, the oxygen atoms combine with oxygen molecules to produce ozone. The equations are:

$$O_2 + UV \longrightarrow 2O$$

$$O_2 + O \longrightarrow O_3 + \text{infrared (IR) radiation}$$

The prevailing ozone concentration in the stratosphere from these processes is about 0.1 ppm.

Stratospheric ozone is involved in an interesting transition of solar radiation from a higher to a lower energy level. Specifically a large portion of the solar UV radiation that reaches the stratosphere is converted to IR radiation before it reaches the surface of the Earth. The conversion involves the following steps:

Step 1: Ozone absorbs UV radiation and is decomposed:

$$O_3 + UV \longrightarrow O_2 + O$$

Step 2: The O$_2$ and O recombine, releasing IR radiation (thermal energy):

$$O_2 + O \longrightarrow O_3 + IR$$

Adding steps 1 and 2 gives:

$$O_3 + UV + O_2 + O \longrightarrow$$
$$O_2 + O + O_3 + IR$$

All the chemical formulas cancel out, and the net result is

$$UV \longrightarrow IR$$

Thus, stratospheric ozone provides a chemical pathway for converting some (not all) of the solar UV radiation into IR radiation.

The chemistry of ozone in the stratosphere has attracted attention because the ozone layer is considered to be an important barrier that protects life on Earth, and some gases newly introduced into the atmosphere by human activities threaten that protective barrier. The protection offered by stratospheric ozone lies in its action in converting UV to IR radiation. Photons in the UV range are energetic enough to promote electrons in various organic molecules to excited states and thus trigger chemical reactions. The UV content of sunlight can thus affect human skin in various ways, some good and some bad. The good ones include the attractive tanning of a pale skin and, more important for health, the conversion of ergosterol (a chemical naturally present in skin) to vitamin D. The bad effects may include a painful sunburn and, more seriously, skin cancer. Also, other organisms, plants as well as animals, are affected in various ways by UV radiation. The life forms that now exist on Earth have adapted to the present ranges of UV intensities. These intensities, however, differ from place to place and are about seven times greater in the tropics than in the arctic. People who migrate from high to low latitudes may be damaged by the more intense UV exposure, and it is possible that migrants in the other direction may suffer from a deficiency of UV. These complexities make it difficult to predict accurately what would happen if a depletion of the ozone layer resulted in a general increase of UV intensity on the surface of the Earth. Earlier reports predicted large increases in the incidence of skin cancer in humans as well as retardation of the growth of some food crops. More recent estimates are less certain

FOCUS ON
EXPRESSIONS OF FRACTIONAL CONCENTRATIONS

The expressions ppm and ppb as well as percent are commonly used to refer to small concentrations. The meanings are:

$1\% = 1$ part per hundred $= 1/100 = 10^{-2}$

1 ppm $= 1$ part per million $=$
    $1/1{,}000{,}000 = 10^{-6}$

1 ppb $= 1$ part per billion $= 1/1{,}000{,}000{,}000$
    $= 10^{-9}$

These expressions do not, by themselves, state what "parts" are referred to. For gases, the reference is always to parts by volume. Because the volume of a gas is proportional to the number of molecules it contains (at a definite temperature and pressure), parts by volume is the same as parts by number of molecules. Thus, the statement that the atmospheric concentration of CO_2 is 350 ppm means that there are 350 molecules of CO_2 in every million molecules of air. A concentration of ozone of 0.1 ppm means that there is 1 molecule of ozone for every 10 million molecules of air.

For solids, the reference is usually to parts by mass, but for liquids it could be either, so it should be specified.

Concentrations expressed in parts per million or parts per billion seem quite small. In terms of atoms or molecules per unit mass or volume, however, the numbers are large. For example, if there is 1 ppb (10^{-9}) by mass of, say, lead in water, each gram of water contains about 3×10^{12}, or 3 trillion, lead atoms. It is not just the magnitude of a number—small for a concentration but large when you count atoms—that is significant. Rather, it is the effect of the components on the properties of the mixture that is important.

---

of what the actual damage would be. There is no question, however, that UV radiation can cause cancer, that there is no "safe" level of UV, and that any increase over the present levels is therefore a potential hazard.

The possibility that human activities may reduce the natural stratospheric concentration of ozone arises from the introduction of certain atmospheric pollutants. Estimates of the seriousness of this threat are uncertain, but findings of greatly lowered stratospheric ozone in the antarctic are alarming. Two possible chemical processes are described here. One of these processes involves nitric oxide (NO).

$$NO + O_3 \longrightarrow NO_2 + O_2$$
$$NO_2 + O \longrightarrow NO + O_2$$
Net equation $O_3 + O \longrightarrow 2O_2$

Note that the sum of the two equations is the de-struction of ozone. Furthermore, the NO is not consumed in the process. It acts as a catalyst, and therefore small quantities of NO can destroy large quantities of ozone. These reactions have implied that a fleet of supersonic transport (SST) aircraft, flying in the stratosphere, might upset the ozone balance because the NO in the jet exhaust could initiate the ozone depletion sequences shown here.

Another mechanism for depletion of the ozone layer involves chlorine atoms. As shown in the following equations, chlorine atoms catalyze the ozone depletion reaction:

$$Cl + O_3 \longrightarrow ClO + O_2$$
$$ClO + O \longrightarrow Cl + O_2$$
Net equation $O_3 + O \longrightarrow 2O_2$
(same reaction catalyzed by NO)

The important stratospheric sources of atomic chlorine are the chlorofluorocarbons, or CFCs. These

TABLE 17.1   Two Types of Smog

LOS ANGELES SMOG (PHOTOCHEMICAL)*	LONDON SMOG (SOOT AND SULFUR)
Begins only during daylight	Begins mostly at night
Smells something like ozone; can also irritate the nose	Smells smoky
Looks yellow to brown	Looks gray to black
Damages certain crops such as lettuce and spinach	Damages stone buildings, especially limestone and marble
Irritates the eyes, causes blinking	Can be responsible for acute respiratory illnesses
Makes rubber crack	Causes acid rain

*The association of the type of smog with the city is historical and does not necessarily refer to typical conditions now or in the future. Major efforts at improving air quality have been made in various areas that were once polluted, and, conversely, air quality has deteriorated in other areas that were once pristine.

compounds contain covalently bonded C, Cl, and F atoms, and they are chemically stable in the lower atmosphere. One of them, $CFCl_3$, has been used as a propellant in aerosol cans. Under compression, it provides the pressure that propels the liquid out as a fine mist. The $CFCl_3$ itself simply escapes as a gas into the atmosphere. The other important chlorofluorocarbon, CF_2Cl_2 (Freon), is the working substance that transfers energy in refrigerators and air conditioners. When one of these appliances breaks or wears out, the CF_2Cl_2 leaks out, and it, too, escapes as a gas into the atmosphere. Because these compounds are stable, they persist long enough to diffuse into the stratosphere. There they become exposed to solar UV radiation that is energetic enough to break the C—Cl bonds and release Cl atoms. The chlorine pathway is the more effective one for the removal of atmospheric ozone. C—F bonds are too strong to be broken by UV photons, so F atoms are not involved in this process.

In the Lower Atmosphere

The natural concentration of O_3 in the lower atmosphere is about 0.02 ppm. At one time, you could buy home "air purifiers" that were supposed to make your air fresher by producing ozone. Ozone is a toxic gas, however, even though small concentrations of it in air do give a sensation of freshness. Therefore these devices did not purify the air; they polluted it.

17.8 POLLUTION BY AUTOMOBILE EXHAUST

By the early 1900s, many industrial cities were heavily polluted. The major sources of pollution were no mystery. The burning of coal was number one. Other specific sources, such as a steel mill or a copper smelter, were readily identifiable. The major air pollutants were mixtures of soot and oxides of sulfur, together with various kinds of mineral matter that make up fly ash. When the pollution was heavy, the air was dark. Black dust collected on window sills and shirt collars, and newly fallen snow did not stay white long.

Air pollution in Los Angeles, however, seemed to have different qualities. Especially in the years after World War II, when population boomed and automobiles became almost as numerous as people, the quality of the atmosphere began to deteriorate in a strange way. It was certainly air pollution, but it did not resemble the smog in London or Pittsburgh. The differences are summarized in Table 17.1.

Scientists looked for possible sources of the mysterious air pollution. There was relatively little heavy industry in Los Angeles in those years, the warm climate minimized the need for home heating, and people did not burn coal. Domestic garbage was commonly burned in open backyard incinerators, however, which were inevitably smoky, so these were banned. Nevertheless, the pollution persisted. Finally, A. J. Haagen-Smit, a chemist,

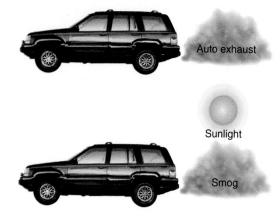

FIGURE 17.6   *Smog is produced when automobile exhaust is exposed to sunlight.*

turned his attention to automobiles, and in 1951 he reported on his crucial experiments with them. He piped automobile exhaust into a sealed chamber equipped with UV lamps, as shown in Figure 17.6. The chamber contained various green plants and pieces of rubber. The chamber was also provided with little mask-like windows that permitted people to stick their faces in and smell the inside air. His results are summarized in Table 17.2. These were the crucial experiments that identified the source of Los Angeles smog and eventually caused its name to be changed to **photochemical smog.** (It isn't just in Los Angeles.) Haagen-Smit's results showed that the combination of auto exhaust and sunlight is responsible. Solar UV radiation promotes the formation of ozone, which must be involved in the production of the smog, as shown by the results of Experiment 3. Haagen-Smit and other chemists then searched for reactions that caused this pollution.

Gasoline is a complex mixture of chemicals, mostly hydrocarbons containing seven or eight carbon atoms per molecule. A typical formula is C_8H_{18}, which represents a set of isomers known as "octanes." In an ideal combustion, the octanes would all burn completely to CO_2 and water, as shown by the equation

$$2C_8H_{18} + 25O_2 \longrightarrow 16CO_2 + 18H_2O$$

The molecular structure of the hydrocarbon includes various C—C and C—H bonds, as illustrated by the structural formula for "iso-octane" given in Chapter 16.

There are no C—C bonds in CO_2, however. Therefore, all the bonds in the octane must be broken for the reaction to go to completion. This decomposition does not occur instantaneously; it takes a little time. When an automobile engine is running, the piston moves so rapidly (typically 1500 to 3000 strokes per minute) that the gasoline vapor spends little time in the cylinder. There is not enough time, in fact, for the octanes to break down completely. The incompletely decomposed fragments, however, do oxidize to some extent, and these partly oxidized fragments are released through the exhaust pipe into the atmosphere. Here they react with ozone or other chemically active forms of oxygen to form, among other products, peroxyacetyl nitrate (PAN):

H—C—C—O—O—NO₂ (structural formula with H, H, O shown)

PAN is a powerful lachrymator, or tear producer, and causes breathing difficulties.

Ozone attacks the C=C linkage in rubber:

TABLE 17.2	**Results of Automobile Exhaust Experiments**	
EXPERIMENT	CONDITION	RESULTS
1	Auto exhaust piped into chamber; UV lamps turned off	Smelled like auto exhaust but not like smog; smog effects not evident.
2	Auto exhaust piped into chamber; UV lamps turned on	Smog. Plants were damaged and rubber was cracked; if you stuck your head in the window, your eyes became irritated
3	Auto exhaust piped into chamber; UV lamps turned off; ozone added to chamber	Smog again, as in Experiment 2

$$\underset{R}{\overset{R}{>}}C=C\underset{R}{\overset{R}{<}} + O_3 \longrightarrow \underset{R}{\overset{R}{>}}C\underset{O-O}{\overset{O}{<}}C\underset{R}{\overset{R}{<}} \xrightarrow{H_2O}$$

$$\underset{R}{\overset{R}{>}}C=O + O=C\underset{R}{\overset{R}{<}} + H_2O_2$$

where the R groups represent other parts of the molecule. This reaction causes cracks in automobile tires. Similar reactions are also damaging to lung tissues and other biological substances.

The detailed chemistry of photochemical smog is complex. Many separate chemical steps are involved. The rates of these various reactions change during the course of the day as the intensity of the sunlight changes. In fact, the whole story—all the reactions and their rates—is not yet known. One thing, however, is clear. If hydrocarbons and other organic compounds were not introduced into the atmosphere, the polluting process would not take place. The best remedy would be to burn the gasoline completely within the cylinder to CO_2 and H_2O. This objective can be approached, but not fully realized, by improving the design of the engine. Any such modification also enhances engine efficiency and results in substantial fuel savings.

Another approach is to oxidize unburned fuel after it leaves the cylinder but before it is released into the air. This objective is achieved by using the catalytic converter (Fig. 17.7). The best catalysts for speeding up the oxidation of organic molecules and CO to CO_2 and H_2O are certain heavy precious metals, especially platinum and palladium. (Recall how platinum catalyzes the oxidation of hydrogen.) Such catalytic converters are now required by law in the United States for gasoline engines.

Two major problems arise, however. For many years, a lead compound, tetraethyl lead, $Pb(C_2H_5)_4$, was added to gasoline to improve engine performance. But the lead poisons the catalyst, destroying its effectiveness. For this reason, automobiles equipped with catalytic converters must use unleaded gasoline. The second problem is that the catalytic oxidation of the gasoline hydrocarbons generates heat within the catalytic converter that promotes other environmentally unfavorable oxidations. Probably the most harmful of these is the increased conversion of N_2 to NO and NO_2. The production of oxides of nitrogen is just what we want to avoid because these compounds are also involved in the photochemical smog sequence. For this reason, a second chamber containing a different catalyst (again a transition metal or a transition metal oxide) operating at a lower temperature is required to dissociate NO into N_2 and O_2 before the exhaust is discharged through the tailpipe.

17.9 SOAPS AND DETERGENTS

We have looked at three case histories: acid rain, ozone depletion, and photochemical smog. All of these are examples in which chemicals have intruded into our daily lives in a less than desirable manner. We should not forget, however, that chemicals have a good side also. In fact, our present standard of living, our health, and many of our creature comforts can be traced to the use of chemicals in desirable ways. We will end this chapter by looking at one example of how some ordinary chemicals like soaps and detergents do what they do. It

FIGURE 17.7 *Cutaway view of catalytic converter, showing catalyst pellets.*

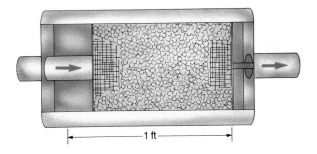

is only one example from millions of how chemicals are used in a worthwhile, constructive manner.

Soaps are made by heating sodium hydroxide (NaOH) with animal fat or vegetable oil. They are sodium salts of organic molecules that have a long hydrocarbon chain at one end and a —COO⁻ group at the other (Fig. 17.8). The nonpolar hydrocarbon chain tries to avoid the polar water molecules and is said to be **hydrophobic** (meaning water-fearing). The —COO⁻ group interacts favorably with the polar water molecules and is described as **hydrophilic** (water-liking). The cleansing action of soap is the result of the dual nature of the hydrophobic hydrocarbon chain and the hydrophilic end group. The hydrocarbon tail is readily soluble in oily substances, which are also nonpolar, while the ionic —COO⁻ group remains outside the oily surface. When enough soap molecules have surrounded an oil droplet, as shown in Figure 17.9, the entire system becomes soluble in water because the exterior portion is now largely hydrophilic. This is how greasy substances are removed by the action of soap.

The action of soap is adversely affected by water containing certain metal ions such as Ca^{2+} or Mg^{2+}. Water containing these ions is called **hard**

(a)

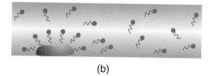

(b)

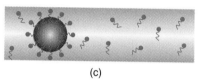

(c)

FIGURE 17.9 *The cleansing action of soap. A greasy spot (a) can be removed by soap (b) because the hydrophobic tails of soap molecules dissolve in the grease, and the whole system then becomes soluble in water (c).*

water, whereas water that is mostly free of these ions is called **soft water.** The Ca^{2+} and Mg^{2+} ions in hard

FIGURE 17.8 *(a) Structure of sodium stearate, a soap molecule. (b) A simplified representation of a soap molecule. The black sphere denotes the hydrophilic head, and the zigzag line represents the hydrophobic tail.*

Sodium stearate $(C_{17}H_{35}COO^-Na^+)$

(a)

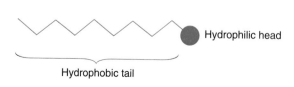

Hydrophilic head

Hydrophobic tail

(b)

$$CH_3CH_2CH_2CH_2CH_2CH_2CH_2CH_2CH_2CH_2CH \underset{\underset{CH_3}{|}}{} \hspace{-0.5em} -\!\!\!\bigcirc\!\!\!- SO_3^- Na^+$$

water react with soap to form insoluble salts or curds, which often appear in the form of "bathtub rings."

$$2C_{17}H_{35}COO^- Na^+ + Ca^{2+} \longrightarrow$$
$$Ca^{2+}(C_{17}H_{35}COO^-)_2 + 2Na^+$$

This process continues until most of the Ca^{2+} and Mg^{2+} ions are removed. Thus, more soap is needed in hard water than in soft water, and, in areas of extremely hard water, the cost of extra soap can become significant. The removal of curds from clothes is also a considerable problem.

Detergents are an alternative to soap. Because they are synthetic cleaning compounds, detergents are often referred to as "syndets." A well-known detergent has the following structure:

$$\underset{\hspace{2em}CH_3 \hspace{5em} CH_3}{} $$
$$H_3C-\overset{|}{CH}-(CH_2-\overset{|}{CH})_3 \hspace{-0.5em} -\!\!\!\bigcirc\!\!\!- SO_3^- Na^+$$

sodium alkylbenzenesulfonate (ABS)

The cleansing action of detergents is quite similar to that of soaps. (Note that they also contain a polar head and a nonpolar tail.) In contrast to soap, it does not form precipitates with Ca^{2+} or Mg^{2+} ions, so it can be used in both soft water and hard water.

A disadvantage of ABS is that it is not biodegradable, that is, the branched-chain structure of ABS is not readily broken down by microorganisms in the sewage treatment plants. As a result, foams began to appear in the rivers and streams. To solve this problem, the detergent industry came up with a soap molecule having a linear chain of carbon atoms which is biodegradable, shown in the formula at the top of this page. This is one of the major types of detergent used today.

SUMMARY

A **mole** is the amount of substance that contains 6.02×10^{23} basic particles. The mass of 1 mole of atoms of an element is the **atomic mass** of the elements (in grams). Similarly the **molecular mass** refers to the mass in grams of 1 mole of a molecule.

Acids taste sour and produce H^+ ions in an aqueous solution. **Bases** taste bitter and produce OH^- ions in an aqueous solution. Acids and bases change the colors of indicators and neutralize each other to form a salt and water. The acidity or basicity of a solution is expressed by pH values. A neutral solution has equal concentrations of H^+ and OH^- ions and a pH of 7. The more basic the solution, the higher is the pH above 7; the more acidic the solution, the lower is the pH below 7.

Oxidation refers to the combination with oxygen or, more generally, to the loss of electrons. **Reduction** refers to the loss of oxygen or, more generally, to the gain of electrons. A **catalyst** speeds up a chemical reaction but is not consumed in the reaction.

Acid rain refers to the excess acidification resulting from the oxidation of nitrogen, ultimately to nitric acid (HNO_3), and of sulfur, ultimately to sulfuric acid (H_2SO_4). Acid rain can be controlled by neutralizing it

with a base such as calcium hydroxide [$Ca(OH)_2$]. **Ozone** (O_3) in the stratosphere acts as a converter of solar ultraviolet (UV) radiation to the less energetic infrared (IR) radiation. Nitrogen oxides and chlorofluorocarbons can act as catalysts to convert ozone back to oxygen (O_2), thereby increasing the amount of UV radiation that reaches Earth's surface. Automobile exhaust contains oxides of nitrogen and unburned as well as partly burned hydrocarbons. This mixture reacts in the presence of sunlight to produce **photochemical smog.** The effect can be prevented by the use of a catalytic converter that helps to complete the oxidation of the partly burned hydrocarbons and carbon monoxide.

Soaps contain a polar head and a long nonpolar tail. The cleansing action of soap is due to the favorable interaction between the nonpolar tails dissolving in nonpolar substances such as grease and the polar heads on the exterior, facing water. Soaps perform better in **soft water** than in **hard water** because the latter contains Ca^{2+} or Mg^{2+} ions that form precipitates with soap. **Detergents** are synthetic compounds that resemble soaps in structure but do not form precipitates with Ca^{2+} or Mg^{2+} ions.

EQUATIONS TO KNOW

pH = $- \log_{10}$(hydrogen ion concentration)

KEY WORDS

Mole	pH	Catalyst	Hydrophilic
Acid	Oxidation	Enzyme	Hydrophobic
Base	Reduction	Ozone	Hard water
Indicator	Acid rain (also, acid precipitation	Photochemical smog	Soft water
Neutralization	or acid deposition)	Soap	Detergent

PROBLEMS AND CONCEPTUAL QUESTIONS

Problems requiring numerical work are identified with a blue number.

Molecular weights and the mole

1. (a) Calculate the molecular mass of (i) ozone, O_3; and (ii) sulfur dioxide, SO_2. Use the atomic mass from the table in Appendix E, and carry the answers to one decimal place. (b) Which has more mass, 1 mole of O_3 or 1 mole of SO_2? (c) Which contains more moles, 1 kg of O_3 or 1 kg of SO_2? (d) Which contains more molecules, 1 kg of O_3 or 1 kg of SO_2? (e) Which contains more molecules, 1 mole of O_3 or 1 mole of SO_2?

2. What is the mass of a mole of (a) argon, (b) carbon dioxide, CO_2?

3. Sulfur boils at 445°C to form a vapor consisting of S_8 molecules. At still higher temperatures, this vapor decomposes to form S_4 molecules:

$$S_8 \longrightarrow 2S_4$$

As this change occurs, which of the following quantities remain constant, which increase, and which decrease? (a) mass; (b) number of molecules; (c) number of atoms; (d) number of moles; (e) atomic mass of sulfur; (f) molecular mass of the sulfur vapor.

4. Container X holds 1 mole of carbon monoxide gas, CO, at 100°C and at a pressure of 2 atmospheres. Container Y holds 1 mole of carbon dioxide gas, CO_2, at the same temperature and pressure. Which container, if either, has (a) the greater number of molecules; (b) the greater mass of gas; (c) the greater volume of gas?

Acids and bases

5. (a) Summarize the properties of acidic and basic solutions. (b) Imagine that you are hiking in the wilderness in summer and come upon a small pool of water and want to determine whether the water is strongly acidic, strongly ba-

sic, or fairly close to neutral. You have no chemical indicator papers, but your pack does include lemonade and soap. There are various berry bushes nearby, but the varieties are not familiar to you. You may handle the unknown water, but assume that you are not willing to taste it. Describe how you could test the water.

6. From their names or other evidence, tell whether each of the following substances is acidic, basic, or neutral: (a) caustic potash; (b) sour salt; (c) pickling liquor (used to etch metals); (d) Alka-Seltzer; (e) a solution of hydrogen iodide, HI, which ionizes in water; (f) a solution of sugar, which does not ionize.

7. Characterize each of the following solutions, from their given pH values, as weakly acidic, strongly acidic, weakly basic, strongly basic, or neutral; (a) 7.1; (b) 1.1; (c) 7.0; (d) 6.85; (e) 13.7.

8. What is the pH of a solution in which the H^+ concentration in moles per liter is (a) 10^{-5}; (b) 10^{-2}; (c) 10^{-9}; (d) 10^{-12}?

9. (a) Could an acidic solution be so concentrated that its pH is 0? (b) Could the pH be a negative number? (c) What is the pH of a solution whose H^+ concentration is 1 mol/L? (d) 10 mol/L?

Oxidation and reduction

10. Give two definitions of oxidation and two of reduction.

11. In each of the following oxidation-reduction reactions, state which substance is being oxidized and which is being reduced:

(a) $H_2O_2 + Fe \longrightarrow FeO + H_2O$ (reaction of hydrogen peroxide with iron)

(b) $Zn + F_2 \longrightarrow Zn^{2+} + 2F^-$ (zinc + fluorine $\longrightarrow$ zinc fluoride)

(c) $2Na^+ + 2Cl^- \longrightarrow 2Na + Cl_2$ (electrolysis of molten salt)

(d) $2H_2 + O_2 \longrightarrow 2H_2O$ (burning or explosion of hydrogen in oxygen)

12. In each of the following cases, state whether the material referred to is being oxidized or reduced: (a) A fallen tree slowly rots. (b) Copper ore is processed to produce the pure metal. (c) Bamboo is eaten and metabolized by a panda. (d) Propane gas leaks out to the atmosphere through a faulty valve, comes in contact with a spark, and explodes.

13. Can an oxidation occur without a reduction? If your answer is yes, show how. If it is no, explain why not.

Acid rain

14. What are the sources of the slight acidity of rainwater in unpolluted atmospheres?

15. How can automobile exhaust contribute to the acidity of rainwater? (Assume that the automobile uses sulfur-free gasoline.) What happens in the cylinders? What happens in the outside atmosphere?

16. How can sulfur in coal contribute to the acidity of rainwater? What happens in the flame when coal is burned? What happens in the outside atmosphere?

17. Sulfur dioxide emitted from a stack is responsible for acid dusts and acid rain. Which is more likely to fall to Earth closer to the stack? Explain.

18. What are the two basic approaches to the prevention of acid precipitation from sulfur compounds?

Catalysis

19. What is wrong with defining a catalyst as a substance that speeds up a reaction without entering into the reaction? Suggest a better definition.

20. What is the catalyst in the following transformation? Unsaturated oil + nickel + hydrogen ⟷ nickel + saturated fat.

21. The catalytic oxidation of vanadium from V^{3+} to V^{4+} occurs as follows:

$$V^{3+} + Cu^{2+} \longrightarrow V^{4+} + Cu^+$$

$$Cu^+ + Fe^{3+} \longrightarrow Cu^{2+} + Fe^{2+}$$

What is the net effect of these two reactions? What is the catalyst?

Ozone

22. Stratospheric ozone protects us from excessive UV irradiation. If you were trying to get a suntan by using UV sunlamps, do you think it would be a good idea to use an ozone-producing device in your room to protect you against a burn from excessive exposure? Defend your answer.

23. The law of conservation of energy tells us that energy cannot be created or destroyed. UV radiation is a form of energy. How, then, can stratospheric ozone reduce the solar UV radiation that reaches the surface of the Earth? What happens to it?

24. Explain how a supersonic airplane or a can of aerosol spray might reduce the level of stratospheric ozone.

25. Which of the following compounds could threaten the ozone layer if they reached the stratosphere? (a) Carbon tetrachloride, CCl_4, formerly used as a cleaning solvent but now banned because of its high toxicity; (b) benzene, C_6H_6, another toxic solvent; (c) carbon tetrafluoride, CF_4, a refrigerant; (d) methyl bromide, CH_3Br, an agricultural fumigant. (C—Br bonds are weaker than C—Cl bonds. C—F bonds are stronger than C—Cl bonds.)

Smog

26. Write the balanced equation for each of the following "smog" reactions: (a) the formation of nitrogen dioxide and atomic oxygen from NO and O_2; (b) the formation of nitrogen dioxide from NO and ozone; (c) the decomposition of one molecule of NO_2 to form atomic oxygen and another product; (d) the formation of ozone from two other forms of oxygen.

27. Gasoline vapor and UV lamps do not produce the same smog symptoms that auto exhaust and UV lamps do. What do you think is missing from gasoline vapor that helps to produce smog?

28. Why is it illegal as well as harmful to use leaded gasoline in a modern automobile?

Soaps and detergents

29. What special features of soap molecules make them effective cleansing agents?

30. What is hard water? How does it affect the cleansing action of soap?

ANSWERS TO SELECTED NUMERICAL PROBLEMS

1. (a) 48.0, 64.1, (b) SO_2, (c) O_3, (d) O_3, (e) same
2. (a) 39.9 g, (b) 44.0 g
4. (a) both have same number, (b) container Y, (c) both have same volume
7. (a) weakly basic, (b) strongly acidic, (c) neutral, (d) weakly acidic, (e) strongly basic
8. (a) 5, (b) 2, (c) 9, (d) 12
9. (a) yes, (b) yes, (c) 0, (d) −1
11. (a) Fe oxidized, H_2O_2 reduced, (b) Zn oxidized, F_2 reduced, (c) Cl^- oxidized, Na^+ reduced, (d) H_2 oxidized, O_2 reduced
26. (a) $NO + O_2 \longrightarrow NO_2 + O$, (b) $3NO + O_3 \longrightarrow 3NO_2$, (c) $NO_2 \longrightarrow O + NO$, (d) $O_2 + O \longrightarrow O_3$

PART THREE
EARTH SCIENCE

If you throw a ball into the air, at a known angle and speed, a physicist can calculate exactly when and where the ball will land. But now what happens if you throw a billion balls into the air? In theory, all of their trajectories are predictable, but so many collisions occur that the problem becomes too complicated to solve. The atmosphere is composed of a nearly uncountable number of molecules. We will see that meteorologists begin with the fundamental laws of physics and chemistry and apply these principles to the immensely complex system of molecules that make up the atmosphere.

While chemists study reactions that occur in seconds, minutes, or hours, a geologist might study chemical reactions in rock that occur over tens or even hundreds of millions of years. Geologists also study changes in the structure of the Earth and its landforms. For example, hot rock deep beneath the Earth's surface establishes currents in solid rock. As the rock flows, it transports continents, raises mountain ranges, and causes earthquakes and volcanoes. We will see that geophysicists attempt to understand these processes through studies of fluid dynamics, wave behavior, heat transfer, and related fields.

Lightning storm near Lava Beds National Monument, California. (© Barbara Filet, Tony Stone Images)

STEPHEN SCHNEIDER
INTERVIEW

Stephen Henry Schneider was born in 1945 in New York City. He received his B.S. and M.S. in mechanical engineering and his Ph.D. in mechanical engineering and plasma physics at Columbia University. Following a brief stint with NASA's Goddard Space Center, Dr. Schneider in 1972 signed on with the National Center for Atmospheric Research in Boulder, Colorado, and from 1987 to 1992 served as the head of the Interdisciplinary Climate Systems Section there. In 1992 he became a professor of biological science and international studies at Stanford University.

The author of several books on climate and climate change, and coauthor of many others, Dr. Schneider has focused his research on climate modeling and the forecasting of the implications of climate change on our environment. He applies his comprehensive knowledge of the field as an editor of the international journal **Climatic Change** and of **The Encyclopedia of Climate and Weather.** As a result of his nearly 100 articles in publications as diverse as **Scientific American** and **Good Housekeeping,** and his appearances on several television programs, in 1991 he was given the American Association for Advancement of Science/Westinghouse Award for Public Understanding of Science, and in 1992, a MacArthur Foundation Fellow for creativity.

Included as one of **Science Digest's** "One Hundred Outstanding Young Scientists in America" in 1984, Dr. Schneider has proven worthy of the honor through his many contributions to both the scientific and global communities.

WHERE WERE YOU BORN AND RAISED?

I grew up on the south shore of Long Island in a town called Woodmere. What I remember enjoying a lot about Long Island was going to a square-mile acre of woods, where I would run around and just enjoy streams and nature. One day a hurricane came by, and I went back to the forest, and half the trees were knocked down and it all had been disturbed. Even at the age of nine, I realized that ecology and climate and soils were all connected systems. Later on, I suppose, this was an emotional driver for getting involved in Earth sciences.

AT WHAT POINT DID YOU BECOME INTERESTED IN, AND THEN COMMIT YOURSELF TO, ATMOSPHERIC RESEARCH?

I liked racing cars when I was in high school. So I went to Columbia University's engineering school to learn how to build the fastest race car. I ended up a mechanical engineering student studying fluid mechanics, then called engineering physics. So my initial practical notion to build race cars was completely dashed by my own choice to go into more theoretical parts of engineering.

My interest in atmospheric research began around 1970, when I attended the first Earth Day celebrations at Columbia. I was a little bored working on a plasma physics problem. I wanted to do something enviromentally useful. At one of the Earth Day presentations, somebody (I think it was Barry Commoner) said, "What if pollution could change the climate—it could either heat it up if it's greenhouse gases or cool it down if it's sulfur injections," and I didn't believe it.

There was an atmospheric sciences course taught at Columbia by Ichtiaque Rasool. He went over the difference between Mars, Earth, and Venus. He said Venus is very hot with its very thick atmosphere, a super greenhouse effect. Mars is very cold with a very thin atmosphere, a weak greenhouse effect. Earth is right in the middle; water is what makes us different, and pollution could, in fact, dirty the greenhouse window. This was fascinating to me. Rasool said, "I will give you your postdoctoral fellowship if you'll leave plasma physics to convert to atmospheric science—to mathematically modeling the climate." So I took up Rasool's offer and became a post-doc at the Goddard Institute for Space Studies, a NASA laboratory at Columbia.

WHAT DID YOU DO NEXT?

In 1972 I went to the National Center for Atmospheric Research (NCAR) to help them start what we dubbed the Climate Project, and began doing climate research in Boulder. Climate involves the integration of materials from many disciplines—from oceanography, ecology, geography, meteorology, chemistry, and physics. Critics argued that such an integration was premature because each of these subtopics is not yet understood to the satis-

---

CLIMATE THEORY IS NOT A PERFECT REPLICA OF NATURE; IT'S A MODEL.

---

faction of the practitioners. And my answer is that the world has to have the answer whether the disciplines are ready or not, so why don't we take halting steps to try to see how the system is integrated and connected—what we now call Earth systems science?

MUCH OF YOUR WORK IN ATMOSPHERIC RESEARCH IS BASED ON COMPUTER MODELING OF THE ATMOSPHERE. CAN YOU TELL US HOW THAT WORKS?

Climate theory is not a perfect replica of nature; it's a model. And it's not a physical model in the laboratory, because you cannot make a physical model in a laboratory that includes enough of the important complexity. For example, the single most important component of the Earth's climate—the evaporation of water at the surface, and the recondensation in clouds—can't be done meaningfully in the lab. So the lab is very limited. Thus, the lab that you have to have for your model, literally, is a computer Earth. It's an Earth that we can pollute—it's an Earth we can modify by just changing something in the computer, and then resimulate the climate under these new conditions.

TELL US MORE ABOUT YOUR MODELING TECHNIQUE.

You break the atmosphere up into a bunch of boxes, or what we call grid squares. You break it up into a latitude-longitude grid of

---

$4\frac{1}{2}$ degrees latitude and $7\frac{1}{2}$ degrees longitude. There are 20,000 of these grid squares around the world if this grid is piled into ten vertical layers.

Now, if you have a box 500 by 500 kilometers—say, the size of the state of Colorado typically—you can't, obviously, resolve an individual cloud. Nobody's ever seen a cloud the size of Colorado. So we've got a problem, because clouds are the venetian blinds of the Earth. They control, more than any other elements, the amount of solar energy absorbed, which is how much the planet is heated, and the amount and distribution of infrared radiative energy that escapes back to space—the so-called greenhouse effect. They are more important radiatively than water vapor, carbon dioxide, chlorofluorocarbons, methane, and all those things we argue about for the human-induced greenhouse effect.

Yet our models cannot explicitly resolve clouds, because they are too small. They are not 500 by 500 kilometers square. They are 5 by 5, maybe, or less. So what do we do? Some people say, "Well, your models are no darn good. Throw them away." But you don't need to know every detail in order to make a prediction about how something works. I mean, you don't have to predict what happens in every play to know that the 49ers would beat Stanford in a football game.

So the question, then, is: How can we get the *average effects* of clouds at the grid scale, even though we're not explicitly calculating individual clouds? We use a technique called parameterization, which is a short contraction for parametric representation.

Say the model produces humidity, temperature, and wind at the grid box. We know the relative humidity of, say, Colorado even if we don't know how many clouds there are. So we can make a rule, a parametric representation, which says if the air is humid it's more likely to be cloudy than if the air is dry. Every farmer knows that. You don't need a Ph.D. in atmospheric science to know that. So then you have a parametric form which says cloudiness is equal to a number, a parameter, times the relative humidity.

But it's more sophisticated than that. Is it more likely to be cloudy if the air is rising or sinking? Think about it for a second. If air is rising, it would be drawing in the humid air from below. If air is sinking it's starting from a high place, which is dry, and coming down. Plus it's heating when it's going down, which tends to cause evaporation. And when it's rising, the air is expanding, which means it's cooling, which means it would tend to make condensation. So then you can say, "All right, my cloud parameterization will get more sophisticated. The cloudiness in the grid box is equal to a parameter times the relative humidity plus another parameter times the vertical velocity, both of which are calculated at the grid box average." The point is, you can get more and more sophisticated without resolving the explicit nature of the details. You are able to get the averages.

The problem is, how well are we doing this? And that's why there's an endless debate among the practitioners as to whether it's been done well, or medium well, or terribly. And how do we validate it? The Earth warms up somewhere between $1\frac{1}{2}$ and $4\frac{1}{2}°C$ for almost all of the models that have been run for carbon dioxide doubling. Now, there are some that aren't quite as hot. But the bulk of them fall in that range. So there's an uncertainty factor of 3 (between $1\frac{1}{2}$ and $4\frac{1}{2}$). So our models could be off by a factor of 3 global average temperatures. And this factor of 3 is largely because of differences in models' cloudiness parameterizations.

The reason we worry about CO_2 is, we know beyond a doubt that if you double CO_2, you're going to trap something like 4 watts of energy over every square meter of Earth. Now, since the industrial revolution, we've added 25 percent more CO_2 and doubled the methane; that's known beyond doubt. The point is that because we've added all these chemicals, we've trapped about 3 extra watts of energy in the last 100 years in the Earth's surface layers. That's not the debate. The debate is should 3 watts of energy warm up the Earth a quarter of a degree, 2 degrees, 4 degrees? And in order to answer that question, you've got to then say, "Now, what happens to the natural system? Do the trees get darker and greener, which would accelerate the warming? Or do they get slightly lighter and brighter, which would retard the warming? What do clouds do?" All these so-called "feedback processes" are endlessly debated and cause that uncertainty factor of 3 or so that we argue about.

The point is, there's positive and negative feedback in nature and in a physical and chemical world. And the sum of the positive and negatives, as to who wins, is not known. That's the debate. So the fact that clouds are a very large factor doesn't mean that factor will necessarily be a positive or a negative feedback. We simply don't know. If you double CO_2 as is forecast sometime in the next century, from growing populations using fossil fuel at a higher and higher rate and continuation of deforestation, then what's going to happen is, we're going to trap another couple of watts of energy.

The debate, then, is translating that into temperature change. Is it going to end up just a half a degree to a degree more warming? Or are we going to get 3 or 4 degrees? In other words, are we going to have a mild change or a catastrophic change—because that's the range. And the debate goes on between mild and catastrophic without any clear and obvious answer right now, because we do not have either the theory or the measurements to validate the overall effect on the globe to much better than a factor of 3.

The rate at which nature changes is on the order of 5°C. That's how much colder an ice age is than a so-called interglacial period, on a global average basis. Say it takes nature 5000 years to end the ice age, then we end up with a rate that's about 1°C per millennium as natural average rate of change.

The conservative forecast for global warming is a degree a century, or ten times faster than the natural average rate of change. The radical forecast says we could warm up 10 degrees in a century. That's a hundred times faster than natural rates.

Pick a middle number, and it's something like 2 to 5 degrees' warming projected to occur over the next century. That would be something like 20 to 50 times faster than natural rates of change.

WHY DO YOU THINK THERE'S SUCH DIFFICULTY IN PRODUCING A SOCIAL, POLITICAL, AND HUMAN RESPONSE TO THIS PROBLEM EVEN THOUGH THE LOGICAL CONCLUSION IS THAT IF THE GLOBAL CLIMATE WARMS SIGNIFICANTLY IN A SHORT TIME, THE EFFECTS WILL BE MORE NEGATIVE THAN POSITIVE?

It's not like chlorofluorocarbons and the ozone hole, because we all admit we can do without the spray cans and fluorocarbon refrigerators and change what's in the air conditioners in our cars. It might cost a few percent more, but very few people are willing to risk skin cancer and disruption of nature for a few chemicals that are substitutable.

When we're talking about global warming, we're talking about methane produced by agriculture, coal mining, natural gas, and landfills. We're talking about CO_2 produced by coal, oil, and gas, which the Third World is expecting to use to power their industrial revolution just the way the western countries did in the Victorian era. And people do not want to hear that these mainstays of economic growth that permit growing populations to increase standards of living have side effects that might be dangerous.

So the problem is, it's not easy to get political agreement to slow down emissions, because it could be painful. Thus, people use the honest and legitimate scientific debate as an excuse to wait and see. They'll say, "Well, we're really not sure."

No honest scientist says that we know the answer. The degree of uncertainty ranges from mild to catastrophic, and we don't know where in this range the actual outcome is going to happen. And we're not going to know in the next 5, 10, or probably even 20 years. Whether to take the chance is not a scientific question per se, but a personal value judgment about which you fear more—investing *present* resources as the hedge against potentially catastrophic change, even though that change may not be so bad if you're lucky, or *not* investing present resources to find that you might be unlucky and, in not trying to slow it down, you have gotten a really whopping big dose that will be impossible to stop without irreversible damage.

SO WHAT SHOULD WE DO?

Study after study has shown that, depending upon how much we're willing to invest to buy these new materials, we can cut somewhere between 10 and 40 percent of our greenhouse emissions by replacing existing inefficient technology, and do it at below zero net cost. In other words, the amount of money you'll save in reduced energy costs actually will pay for itself without even counting the free extras, like reducing acid rain and the threat of disrupted ecosystems.

So my personal opinion is, why don't we do those things first that are free-standing and make sense anyway? Get rid of the fluorocarbons, because they not only trap 25 percent of the heat, but they also help cause ozone depletion. That's the easiest and the first thing to do. Let's use energy efficiently, because not only does it reduce global warming, but it also reduces acid rain and reduces dependency on foreign supplies, which are expensive and sometimes militarily dangerous to protect. It reduces local air pollution. It also makes our products more competitive in the long run. All of these things have to take place, and can, and are, but we need to push them harder. We can also pursue vigorously research and development on non-fossil-fueled energy systems. And then in the time frame of 5 to 15 years, while we're slowing down the rate at which we're making the system change, and therefore slowing down the rate at which nature will have to adapt, we can buy time to have the scientists determine what is likely to happen. To use the excuse that the thing is now very uncertain and therefore we shouldn't act is to say we should never have insurance, the police, or the military.

413

What students need to recognize is, we're not talking about the environment *or* the economy. What we're talking about is, there are trade-offs between the two, but that if we use our brains and hands cleverly we can find ways to have the economy grow in a much less environmentally destructive way.

Problems are increasingly cross-cutting. Environmental problems are only one example. Health problems are another example. And the way we're organized is not set up to deal with problems that crosscut. In universities we learn disciplines; in governments we deal with departments. Real issues, like global warming, involve solutions that have a little bit in population, forestry, agriculture, energy, and foreign policy. In or-der to manage that problem, each group has to do a little, and *it has to be coordinated*. And therefore an Earth systems science approach, such as this book is trying to take by integrating the disciplines, is the only way to be in tune with the way the world's real problems are.

This interview originally appeared in *Earth Science and the Environment* by Graham R. Thompson and Jonathan Turk, Saunders College Publishing, 1993.

CHAPTER 18

THE ATMOSPHERE AND METEOROLOGY

The Hawaiian Islands chain perturbs the prevailing northeasterly winds, producing extensive cloud wakes in the lee of the islands. The haze in the wake of the islands is probably a result of eruptions from the Kilauea volcano on the southeast coast. (Courtesy NASA)

In this chapter we will be concerned with our atmosphere. Why is it like it is, and what factors have caused it to evolve into its present form? We will also study both climate and weather. **Climate** is a description of long-term meteorological conditions. For example, New York City has a temperate climate with moderate rainfall; one can always expect warm summers and cold winters there. **Weather** is a description of relatively unpredictable short-term conditions. Storms, a heat wave, or a cold spell are all part of the weather in a given area.

Why does a particular region have the climate that it does? What causes rain, clouds, heat, and cold, and how can one predict the weather? **Meteorology** is the study of the atmosphere and its associated activities, including weather and climate.

18.1 EVOLUTION OF THE ATMOSPHERE

Since the start of the space age, astronomers have studied the Solar System intensely. By 1994, spacecraft had landed on or flown past all of the planets except Pluto and many of the planetary moons. Many of these missions will be discussed in more detail in Chapter 22. At this point, two important conclusions are relevant. First, many of the planets and moons have an atmosphere. Second, no other object in our Solar System has an atmosphere that is similar to our own.

Most scientists agree that the Earth's atmosphere today is quite different from its original, primordial atmosphere. At first, the Earth was a loosely collected mass of dust and gases, mostly hydrogen. Over the millennia, the dust coalesced into a solid sphere, most of the hydrogen escaped into space, and a secondary atmosphere was formed. This secondary atmosphere was produced mainly by volcanic ejection of gases that had been trapped within

(a)

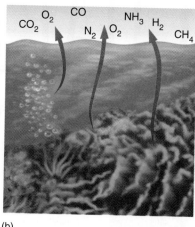
(b)

(c)

FIGURE 18.1   *Evolution of the atmosphere. (a) The primitive atmosphere contained carbon dioxide, nitrogen, and other gases. (b) As plants evolved, the composition of the atmosphere began to change. Oxygen, released during photosynthesis, began to accumulate. (c) The modern atmosphere is composed mainly of nitrogen and oxygen, with smaller concentrations of water, carbon dioxide, and other gases. The ratio of oxygen to carbon dioxide is maintained by dynamic exchange among plants and animals.*

the planet as it formed and by accretion of gases from comets and meteorites that bombarded the Earth. By comparing the Earth with its two nearest neighbors, Venus and Mars, and by studying the composition of old rocks on the Earth's surface, scientists now believe that the original gases consisted mainly of carbon dioxide (CO_2), nitrogen (N_2), and water vapor (H_2O), with smaller concentrations of methane (CH_4), ammonia (NH_3), hydrogen (H_2), and carbon monoxide (CO). The best evidence indicates that oxygen was present in trace quantities only.

Living creatures that exist today would not survive in a carbon dioxide, nitrogen, and water atmosphere. How was the modern atmosphere formed? Although some theorists believe that geological processes altered atmospheric composition, most scientists believe that living organisms and a favorable atmospheric environment evolved together. In the beginning, when there was little free gaseous molecular oxygen (O_2), there would be no ozone (O_3). Without a protective ozone layer, intense ultraviolet radiation from the Sun could reach the Earth. Ironically, these rays, which could harm or destroy life today, may have been responsible for the formation of the first organic compounds and living cells. One theory suggests that the synthesis of simple organic molecules was initiated by the action of energetic ultraviolet light on the molecules

of the primitive atmosphere. Once the first organic molecules were formed, they presumably, in time, combined to form proteins and complex molecules that carry hereditary information. These large molecules then joined together to form simple living organisms.

Not all scientists agree that life evolved from reactions involving atmospheric gases. One alternative theory suggests that perhaps the first living organisms were formed in the vicinity of underwater volcanic vents, where carbon, nitrogen, and sulfur compounds were more concentrated. Yet another theory postulates that the earliest forms of life evolved on the surface of certain clay minerals.

In any case, bacteria, the first living creatures preserved in the fossil records, must have lived in water. Since there was little oxygen, they could not have metabolized their food as most organisms do today but must have lived by some **anaerobic** (without oxygen) process. Up until this point, there was little free oxygen in the atmosphere, and the organisms required none. The next evolutionary step was crucial. Blue-green algae evolved, able to synthesize their own complex organic molecules by combining simple organic molecules in the presence of sunlight, as discussed in Chapter 17. During this process, carbon dioxide and water are combined in the presence of sunlight to form glucose (a sugar) and oxygen.

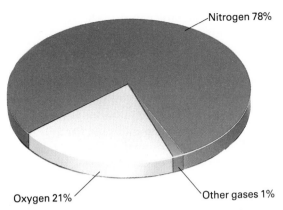

Nitrogen 78%

Oxygen 21%

Other gases 1%

FIGURE 18.2 *The gaseous composition of the atmosphere.*

Most scientists believe that the excess oxygen released by these first photosynthetic organisms accumulated slowly over the millennia until its concentration reached about 0.6 percent of the atmosphere. Most multicellular organisms require oxygen to survive and could have evolved only at this point. The emergence of various multicelled organisms about 1 billion years ago triggered an accelerated biological production of oxygen (Fig. 18.1). At present, the atmosphere is composed of 21 percent oxygen (Fig. 18.2).

If the oxygen concentration in the atmosphere were to increase by even a few percent, fires would burn uncontrollably across the planet; if the carbon dioxide concentration were to rise by a small amount, plant production would increase appreciably and the Earth's temperature would probably rise. Since these apocalyptic events have not occurred, the atmospheric oxygen may have been balanced to the needs of the biosphere during the long span of life on Earth. By what mechanism has this gaseous atmospheric balance been maintained? Some scientists believe that it is maintained by the living systems themselves, as shown in Figure 18.3. According to this theory, not only is the delicate oxygen–carbon dioxide balance biologically maintained, but also the presence of oxygen in our atmosphere can be explained only by biological activity. If all life on Earth were to cease and the chemistry of our planet were to depend solely on inorganic processes, oxygen would become a trace gas, and the atmosphere would revert to its primitive condition and be poisonous to any complex plants and animals that were reintroduced.

If it is true that the required atmospheric oxygen concentration of about 20 or 21 percent is maintained by biological processes, the Earth's at-

Respiration

Plants and animals consume oxygen (O_2) and sugars during respiration

Plants produce oxygen and sugars during photosynthesis

Plants and animals produce carbon dioxide and water during respiration

Plants consume carbon dioxide and water during photosynthesis

Photosynthesis

FIGURE 18.3 *Exchange of gases and nutrients among plants and animals.*

mosphere is not in danger of disastrous changes so long as living species survive. An alternative theory claims that our physical environment has evolved through a series of inorganic reactions and that biological evolution and physical evolution were independent. If the biological theory is correct, a large ecological catastrophe, such as the death of the oceans or the destruction of the rain forests in the Amazon Basin, could cause reverberations throughout our physical world that might create an inhospitable environment for life on Earth. Alternatively, if the physical world did evolve independently of the biological and is now controlled by inorganic processes, such a prediction concerning oxygen balance might be considered unnecessarily alarming.

18.2 THE ATMOSPHERE TODAY

Did you ever wonder what keeps the atmosphere up there? Consider, for example, a molecule of oxygen at the level of your nose. It is being pulled downward by the force of gravity, so why doesn't it fall—like Newton's apple—and land on the floor?

The reason is that molecules are always moving and colliding with each other, and the hotter the gas is, the faster they go. At 0°C, oxygen molecules travel at an average speed of about 425 m/s, although at any one instant some go much faster while others move more slowly. They travel randomly in all directions, but the average distance between collisions is short. Thus, each molecule has at least two kinds of motion: the thermal motion influenced by collisions with the other gas molecules, and a downward acceleration caused by gravity at 9.8 m/s². If a box of gas were floating about in free space, where there are no gravitational forces, the gas would be distributed uniformly throughout the box. But if this container were placed on Earth, the gravitational acceleration would draw molecules downward, and there would be more gas molecules in a given volume at the floor level than at the ceiling (Fig. 18.4).

Thus, the Earth's atmosphere grows less dense with increasing altitude. Anyone who has ever climbed a high mountain has experienced the effects of this atmospheric thinning with height. At about 3000 m (about 10,000 feet), even a person in good physical condition readily notices that

FIGURE 18.4 *Gas molecules under the influence of gravity in the absence of wind and air currents.*

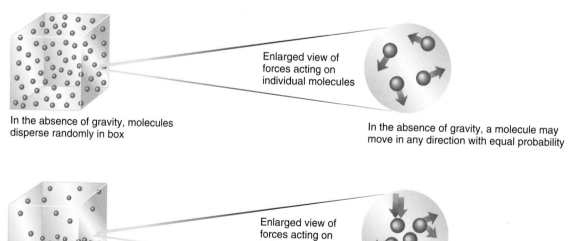

In the absence of gravity, molecules disperse randomly in box

Enlarged view of forces acting on individual molecules

In the absence of gravity, a molecule may move in any direction with equal probability

In a gravitational field, molecules concentrate near the bottom

Enlarged view of forces acting on individual molecules

In a gravitational field, all molecules accelerate downward (red line shows movement in the absence of gravity; blue line shows velocity in a gravitational field)

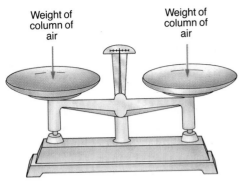

FIGURE 18.5 *You cannot measure the weight of the atmosphere with a double balance because the weight of air is equal on both pans.*

exertion is more difficult than it is at sea level. At 4500 m, a person's actions are slowed considerably, and above 6000 m, climbers find that they move surprisingly slowly and lose their breath quite quickly.

FIGURE 18.6 *A barometer. In a barometer, the weight of the column of air on the mercury in the dish is not balanced by any air pressure from within the tube because the upper region is evacuated. Instead the air pressure is balanced by the weight of the mercury in the tube. As a result, the height of the mercury in the column is a measure of the outside air pressure.*

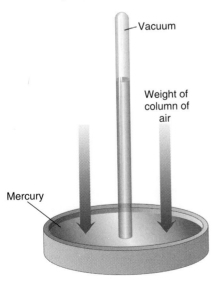

The blanket of air resting above the Earth has a great deal of mass. How can this column of air be weighed? If a two-pan balance is resting on a table, it reads zero, not because the air is weightless but because the air pushes down equally on both pans, as shown in Figure 18.5. But if one end of an open glass tube is placed in a dish of a liquid, such as mercury, and the air is evacuated from the other end, the liquid rises in the tube. Why? The reason is that the air column pushes down on the liquid in the dish, while there is no equivalent force from within the evacuated tube, as shown in Figure 18.6. This device is called a **barometer.** The height to which the liquid rises is a measure of the downward force exerted by the air. At sea level, the mercury will rise approximately 76 cm, or 760 mm, into the tube. The effect of the column of air is generally expressed in pressure units (pressure = force/area) and is referred to as **barometric pressure,** as explained in Chapter 2. The pressure exerted by the Earth's atmosphere at sea level fluctuates with atmospheric conditions. On the average, this pressure supports a column of mercury 76 cm high, as noted previously. One standard atmosphere is defined as the atmospheric pressure that would cause a column of mercury to stand at a height of 76 cm.

Meteorologists often use the units **bars** and **millibars** to describe pressure. One bar is nearly equal to 1 atmosphere, and 1 millibar equals 1/1000 bar.

EXAMPLE 18.1   BARS AND MILLIBARS

Using Figure 18.7, calculate the pressure, in atmospheres, at sea level and at the summit of Mount Everest.

Solution According to the graph, the pressure at sea level (0 km) is 1000 millibars, or 1 bar. Thus the pressure at sea level is 1 atmosphere, which agrees with our definition. At the summit of Mount Everest, the pressure is about 300 millibars, or about 0.3 atmosphere. Climbers have difficulty obtaining enough oxygen to survive at this altitude, and many have died.

Figure 18.7 is a graph of the atmospheric pressure as a function of altitude. It shows that pressure decreases throughout the whole atmosphere. If the change of temperature with altitude is studied, no such smooth curve is observed. Rather, as shown in

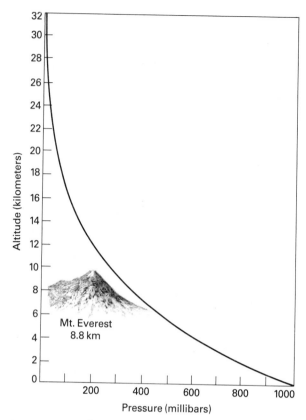

FIGURE 18.7 *Decrease of atmospheric pressure with altitude.*

our world, nighttime temperatures are maintained at a fairly high level because the atmosphere absorbs and retains a great deal of thermal energy before it can escape to outer space. At higher elevations in the troposphere, the atmosphere becomes thinner, its insulation properties decrease, and the average temperature decreases, as shown in Figure 18.8. Thus mountaintops are generally colder than valley floors, and pilots flying at high altitudes must keep their cabins well heated.

The steady decline of temperature with increasing altitude ceases abruptly about 12.5 km above the Earth. If we ascend farther, a gradual warming trend is observed, and above that level the temperature increases rapidly with altitude. The layer of air of fairly constant and then rising temperature is called the **stratosphere.** Throughout the stratosphere and the lower portion of the mesosphere, where the air is too thin to support life, high-energy ultraviolet rays are absorbed by ozone. The primary ozone layer lies at a height of 20 to 40 km. As noted in Chapter 17, this layer is responsible for filtering out the dangerous ultraviolet radiation from the Sun. This absorption of ultraviolet radiation is largely responsible for the temperature increase at these altitudes. Temperature falls again in the **mesosphere,** which lies above the stratosphere. In the upper regions of the mesosphere, little radiation is absorbed, and the thin air is extremely cold. Starting at about 80 km above the Earth, the temperature again starts to rise rapidly in a region known as the **thermosphere.** Here high-energy X-rays and ultraviolet radiation from the Sun are absorbed by atoms and molecules in the atmosphere. High-energy reactions result, which strip electrons from atoms and molecules to produce ions. Subsequent reradiation of infrared produces a warming effect.

Figure 18.8, the temperature profile alternates. If a rocket with a recording thermometer mounted on it rose straight upward from the surface of the Earth, it would register first a cooling trend, then a gradual warming, a rapid warming, another cooling trend, and finally a pronounced warming.

The layer of air closest to the Earth is known as the **troposphere.** This is where we live, where our weather occurs, and here the air is close to the land and oceans. When sunlight strikes the Earth, the energy is readily absorbed by soil, rock, water, and living organisms, and thus the surface of the planet becomes warm. But just as a piece of rock or soil can absorb thermal energy, it also emits radiation (at infrared wavelengths) and thus loses energy, which goes back out to space. If there were no atmosphere above us, the loss of energy by radiation would be extremely rapid; at night, when this loss is not compensated by thermal energy received from the Sun, the surface would cool drastically. In

18.3 ENERGY BALANCE OF THE EARTH

In its daily and seasonal variations, the climate of any given region of the Earth follows a recurring pattern from year to year. This pattern results from a balance of opposing processes. The Earth receives a continuous influx of energy from the Sun. If there were no energy losses, the Earth would get hotter

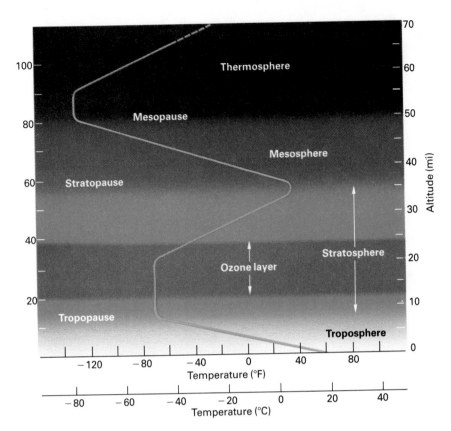

FIGURE 18.8 *Temperature change with altitude and the major layers of the atmosphere.*

and hotter until rocks melted and vaporized. Since these catastrophes have not occurred for billions of years, we know that energy must also be radiated away from the Earth at a constant rate. This opposition of inflow and outflow is called the **energy balance of the Earth.**

Figure 18.9 shows how the balance is distributed. Only 21 percent of the incident solar radiation strikes the Earth directly. The other 79 percent is intercepted by the atmosphere—the clouds, gases, and small particles. For example, much of the ultraviolet light is absorbed by ozone in the atmosphere. Some of this intercepted radiation is reflected back to space (31 percent), some is absorbed as heat (19 percent), and some is rescattered down to Earth (29 percent). On a global average, just about half of the energy received from the Sun reaches the surface of the Earth (21 percent direct and 29 percent scattered). Under equilibrium conditions, all of the energy that reaches the Earth must be returned to space. If not, the temperature of the planet would change. That is, if less energy were returned to space than is received, the temperature of Earth would rise. Thus the Earth must

be returning all radiation it receives to space, as shown on the right side of Figure 18.9.

The temperature of the Earth's surface and its surrounding atmosphere depends on the total rate of absorption and reflection. Thus, the reflectivity of the Earth and its atmospheric components is an important factor controlling temperature. Some surfaces are better reflectors than others. The **albedo** is a measure of the reflectivity of a surface. Clouds, snowfields, and sparkling glaciers reflect sunlight efficiently and are said to have a high albedo. City smog and dark, rough surfaces do not reflect light well and have a low albedo. If the albedo of the Earth were to increase, as it would, for example, if the ice cover grew, the surface of the planet would cool; alternatively, a decrease in albedo would cause a gradual warming trend.

CONCEPTUAL EXERCISE

As the winter ends, snow generally starts to melt around trees, twigs, and rocks. The line of melting radiates outward from these objects. The snow in open areas melts last. Explain.

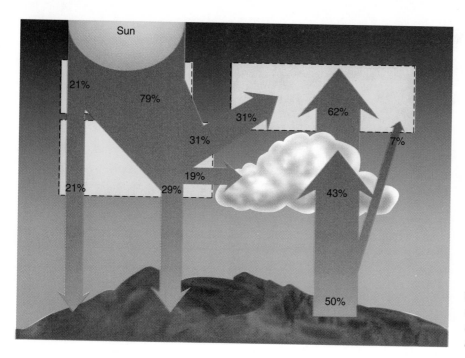

FIGURE 18.9 *Energy balance of the Earth. The sets of numbers in the dashed areas total 100 percent.*

Answer Snow has a high albedo and reflects much of the solar heat back into the atmosphere. Trees, twigs, rock, and soil absorb sunlight and become warm, thereby melting adjacent snow.

This discussion of energy balance does not explain why some regions of the Earth are warm and others are generally much colder. Temperate zones experience distinct summer and winter seasons, and the polar regions are always cooler than the equatorial ones. To understand the reasons behind these temperature differences, consider what happens if a flashlight is shined onto a flat board. If the light is held directly overhead and the beam is shined vertically downward, the light illuminates a smaller area than it would if the beam were shined onto the board from an angle (Fig. 18.10). Of the

FIGURE 18.10 *If a light is shined from directly overhead, the radiation is concentrated on the surface. However, if the source of the light (or the surface) is tilted, the light spreads over a larger area, and the radiation is dispersed.*

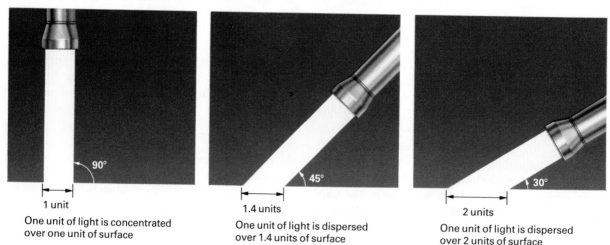

1 unit
One unit of light is concentrated over one unit of surface

1.4 units
One unit of light is dispersed over 1.4 units of surface

2 units
One unit of light is dispersed over 2 units of surface

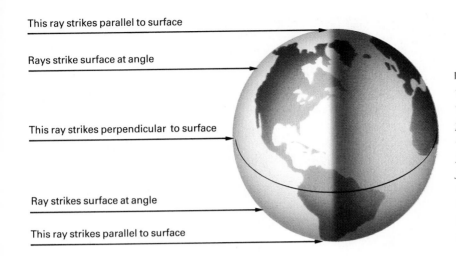

This ray strikes parallel to surface

Rays strike surface at angle

This ray strikes perpendicular to surface

Ray strikes surface at angle

This ray strikes parallel to surface

FIGURE 18.11 *If the Earth were not tilted, the Sun's rays would always strike perpendicular to the Earth's surface at the Equator and parallel at the poles. Therefore, the equatorial regions would receive the most intense solar radiation, and the polar regions would receive hardly any.*

three positions shown in the figure, the one aimed at the shallowest angle illuminates the greatest area. Thus, the more the flashlight is tilted, the less concentrated is the light on the board, and the lower is the temperature of any point in the illuminated area. (The same results would be observed if the board were tilted instead of the flashlight.)

With this in mind, let us consider what happens when light is beamed onto a spherical surface such as the Earth. First, imagine that the globe were held in a fixed vertical position. The top-most and bottom-most points of the globe are the poles, the imaginary line that encircles the middle is the Equator, and the imaginary straight line through the Earth from pole to pole is the Earth's axis. If a light is shined perpendicular to the Equator, all the other surfaces receive light at an angle. Specifically the polar surfaces are angled farther and farther from the perpendicular until at the pole itself the surface of the sphere is parallel to the light, as shown in Figure 18.11. Thus, the light intensity per unit area *decreases* from the Equator to the poles. At the poles, the light is parallel to the surface, and no direct radiation strikes the area at all. A uniform light source shining perpendicular to the Equator delivers the most radiant energy to the Equator and the least to the poles.

It is now obvious why the hottest climates on Earth are found near the Equator and why climates become progressively cooler north or south of that line. This explanation accounts for general climatic regions but does not tell us why summer and winter seasons occur in the higher latitudes. To understand these effects better, it is necessary to consider the Earth's rotation around an imaginary axis

running through the North and South Poles. If this axis were perpendicular to the plane of the Earth's orbit, the Sun would always be directly above the Equator, and there would not be any seasonal temperature variations. In reality, the Earth's axis is tilted at an angle of 23.5° with respect to a line drawn perpendicular to the plane of its orbit. As a result of this tilt, the angle of incidence of the Sun's rays on the Earth changes as the planet moves around in its orbit.

On June 21, the Earth is located so that the North Pole leans the full 23.5° toward the Sun, as shown in Figure 18.12. This condition is called the **summer solstice.**[*] The Sun strikes the Earth directly overhead at a location 23.5° north of the Equator and not at the Equator itself. The northern latitudes receive more direct sunlight than the southern ones, and the North Pole, tilted toward the Sun, receives a continuous 24 hours of daylight. Polar regions are often called "lands of the midnight Sun" because the Sun never sets in the summertime (Fig. 18.13). While it is summer in the Northern Hemisphere, the South Pole is tilted away from the Sun and lies in continuous darkness. June 21 marks the first day of winter in the Southern Hemisphere. Six months later, on December 22, the situation is reversed: The North Pole is tilted away from the Sun and lies in continuous darkness, whereas the South Pole is bathed in constant light. It is summer in the Southern Hemisphere and winter in the Northern.

[*]In some years, the solstices and equinoxes occur 1 day earlier or later than mentioned here. Thus, the summer solstice sometimes occurs on June 22.

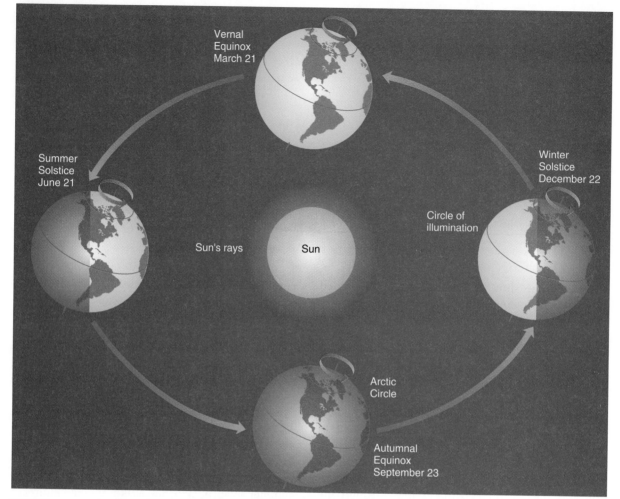

FIGURE 18.12   *A schematic view of the Earth's orbit showing the progression of the seasons.*

FIGURE 18.13   *The midnight Sun. This photograph was taken at midnight in July, at 70° north latitude in the Canadian Arctic.*

What happens midway between these two extremes, on March 21 and September 23? Although the Earth is still tilted at 23.5°, the tilt aims the North Pole neither toward nor away from the Sun but rather at right angles to it. Because no part of the Earth is angled toward or away from the Sun, the most direct light shines at the middle, that is, at the Equator, and the North and South Poles each receive equal periods of night and day. In fact, when the Earth is in either of these two positions, which are called spring and autumn **equinoxes,** every portion of the globe receives 12 hours of daylight and 12 hours of darkness. This does not mean that all areas of the globe receive equal quantities of solar energy. Locations on the Equator receive the most, whereas the polar regions receive hardly any direct radiant energy at all; it is just that all areas are in daylight for the same length of time, 12 hours.

It is interesting that all areas of the globe receive the same total number of hours of sunlight every year. The North and South Poles receive their sunlight in dramatic opposition—6 months of continuous light and 6 months of continuous darkness—whereas at the Equator each day and night is close to 12 hours long throughout the year. Once again, however, although the poles receive the same number of hours of sunlight as do the equatorial regions, the sunlight reaches the poles from a shallow angle and therefore delivers much less total radiant energy.

18.4 WIND SYSTEMS

A steamship chugging across the ocean derives its power from the thermal energy released when coal or oil is burned to heat the boilers. Where does the energy come from that powers a sailboat across the ocean? From the wind, of course. But wind is not an energy source like coal or oil; it is more closely analogous to a working substance like the steam that drives a turbine. So the question remains: What energy source drives the winds?

The wind systems of our globe represent the functioning of a great, natural heat engine that is analogous in many respects to a mechanical heat engine. In the engine room of a steamship, fuel is used to boil water and to heat the resultant steam. The moving steam then forces the blades of the turbine to rotate.

The power source of our natural wind systems is the Sun. When the Sun heats one part of the atmosphere more than another, this warm air expands. The expansion makes the heated air less dense than the surrounding air, and the warm air rises. Thus, heat is converted to motion. As the less dense air rises, colder, denser air moves along the surface to replace it. This surface movement is **wind.** As you can see, wind systems operate much like convection currents in a room. Recall from our discussion of convection currents in Chapter 5 that if a heater is placed in one corner of a room, it heats the air adjacent to it. The heated air expands, becoming less dense. This light air rises and is replaced by denser cool air moving along the floor, and an air current is thus established. Wind systems are large convection currents that operate on local, continental, or global scales.

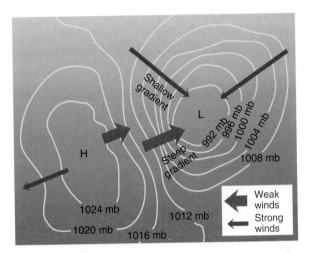

FIGURE 18.14   *Along a steep pressure gradient, winds are strong. Likewise, they are weak along a shallow gradient.*

Pressure Gradients and Winds

The pressure at the Earth's surface is not constant from place to place or from time to time. Daily weather maps show the positions of areas of high and low pressure. These variations in pressure within the atmosphere act as the primary mechanism to drive winds. The actual variations at any one time in a particular region are best discussed in terms of a pressure map like that of Figure 18.14. To produce one of these maps, all pressures are reduced to the value they would have if the altitude were sea level. This is necessary because pressure also declines with increasing altitude, and this decline is far greater than that which would be produced by regional differences in pressure.

Once these atmospheric pressures are determined, a map is plotted (see Figure 18.14). These maps have contour lines, called **isobars.** Isobars are lines of equal pressures. At regions where the isobars are close together, large changes in pressure occur as one moves between these locations. These regions are said to exhibit large **pressure gradients.** If the isobars are spaced far apart, the changes in pressure from one location to another within this region are not great. Such regions have low pressure gradients. Notice on these maps that the isobars form a closed pattern. The central part of this region is referred to as either a **high pressure** or a **low pressure region.** The importance of these maps is that wind tends to move from high pressure regions to low pressure regions, and the winds are

FOCUS ON
THE EFFECTS OF THE SPEED OF THE WIND—THE WIND CHILL FACTOR

Wind can affect our comfort and safety in a variety of ways, but the two that we shall consider in this section are those produced by the speed of wind. One obvious effect that wind speed can have is that high wind velocities can cause damage to structures. How fast does a wind have to be to produce certain kinds of damage? The question is best answered by referring to the table, which shows the effect that winds of different speeds can have on our environment.

Wind speed also has an effect on our comfort. For example, on a hot day, a slight breeze may bring a welcome relief. The reason for this is that a wind blowing across the body enhances the evaporation of perspiration, thus cooling the body. (See the box on relative humidity and comfort later in this chapter.) However, a strong wind during a cold day may quickly cause injury to exposed body tissue. The wind speed and the temperature are related through the **wind chill index.** The wind chill index is best illustrated by way of a table, which shows air temperatures along the horizontal axis and wind speeds along

SPEED (mph)	EFFECT
0	Smoke rises vertically from a chimney
1–3	Smoke is deflected in the direction of the wind
4–7	Slight rustling of leaves on trees
8–12	Leaves move perceptibly
13–18	Dust is kicked up and small branches are set into motion
19–24	Small trees sway
25–31	Large branches on trees move
32–38	Large trees sway
39–46	Walking is difficult
47–54	Roofs may be damaged
55–63	Trees are uprooted
64–74	Structures are damaged considerably
74+	Hurricane force winds

the vertical. Let us examine what this table tells us about a day when the air temperature is 5°F and the wind speed is 10 mph. Find these values in the table; the point where the row and column intersect is the wind chill index, or the effective temperature of this day. We find this value to be −15°F, a bitterly cold day. Also, notice that if the wind picks up only slightly to 15 mph at the same temperature, the effective temperature is −25°F, an extremely cold temperature.

MPH	Equivalent Temperature* of Wind Chill Index (°F)																
Calm	35	30	25	20	15	10	5	0	-5	-10	-15	-20	-25	-30	-35	-40	-45
	Cold																
5	33	27	21	16	12	7	1	-6	-11	-15	-20	-26	-31	-35	-41	-47	-54
10	21	16	9	2	-2	-9	-15	-22	-27	-31	-38	-45	-52	-58	-64	-70	-77
15	16	11	1	-6	-11	-18	-25	-33	-40	-45	-51	-60	-65	-70	-78	-85	-90
	Very Cold			Bitterly Cold									Extremely Cold				
20	12	3	-4	-9	-17	-24	-32	-40	-46	-52	-60	-68	-76	-81	-88	-96	-103
25	7	0	-7	-15	-22	-29	-37	-45	-52	-58	-67	-75	-83	-89	-96	-104	-112
30	5	-2	-11	-18	-26	-33	-41	-49	-56	-63	-70	-78	-87	-94	-101	-109	-117
35	3	-4	-13	-20	-27	-35	-43	-52	-60	-67	-72	-83	-90	-98	-105	-113	-123
40	1	-4	-15	-22	-29	-36	-45	-54	-62	-69	-76	-87	-94	-101	-107	-116	-128
45	1	-6	-17	-24	-31	-38	-46	-54	-63	-70	-78	-87	-94	-101	-108	-118	-128
50	0	-7	-17	-24	-31	-38	-47	-56	-63	-70	-79	-88	-96	-103	-110	-120	-128

*Wind speeds greater than 40 mph have little additional chilling effect. **The Wind Chill Index** (ESSA, Washington, D.C.)

Equivalent temperature of wind chill index.

fastest and strongest along steep gradients in pressure, as shown in the figure.

Wind does not flow in a straight line from high pressure to low pressure. The direction of wind flow is affected by the Coriolis effect.

Global Wind Systems and the Coriolis Effect

Sunlight reaches the polar regions at such an acute angle, even in summertime, that the quantity of radiation received is quite low. If the Sun were the only source of heat, these regions would be cold in summer and near absolute zero ($-273°C$) in winter, when no sunlight at all is received. Throughout the year, polar regions are considerably warmer than can be accounted for by the amount of sunlight received. Measurements and calculations show that the equatorial regions of the planet receive so much radiant energy that they should be much hotter than they actually are. Winds and ocean currents carry heat from the tropics to the poles.

To understand global wind systems, consider what would happen if the Earth did not rotate about its axis. The intense sunlight at the Equator would heat the air there, causing it to rise. As the air rose, cooler air would move in from the polar regions, and a set of convection currents would be established, as shown in Figure 18.15. According to

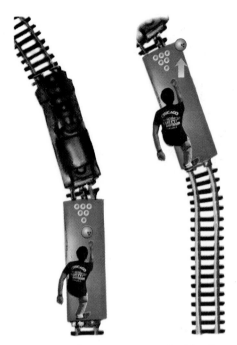

FIGURE 18.16 *Observed path of bowling ball thrown in a straight line while train turns to the left.*

this model, in the Northern Hemisphere, the predominant surface winds would blow southward, while in the upper atmosphere, warm air would move poleward. In the Southern Hemisphere, the situation would be reversed.

Our static model is too simple because the Earth rotates. To visualize the effect of this rotation, imagine what it would be like if there were a bowling alley on a moving train. Suppose that you were riding due north on the train and that you rolled your bowling ball directly down the center of the alley on a perfect path toward the strike zone. But just as the ball left your hand, the train reached a curve and started to turn toward the west (left). The ball continued to travel due north, but the target was moving away toward the left, so the ball appeared to veer to the right, thereby missing the pins. The observed path, shown in Figure 18.16, results from the motion of the ball moving straight along an alley that is moving to the left. The bowling ball situation is an example of relative motion. To an observer on the train, the ball has moved off to the right, regardless of its motion relative to some other frame of reference such as a compass.

What does this have to do with the wind? The distance around the Earth, measured parallel to the

FIGURE 18.15 *A hypothetical model of global circulation patterns presuming that the Earth does not rotate about its axis.*

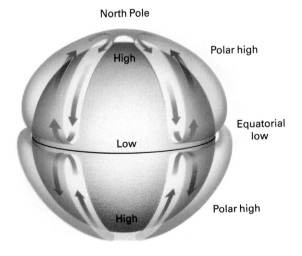

North Pole

Polar high

High

Equatorial low

Low

Polar high

High

South Pole

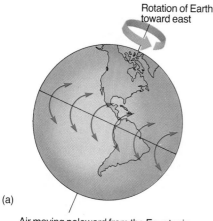

(a)

Air moving poleward from the Equator is traveling east faster than the land beneath it and veers to the east (turns right in the Northern Hemisphere and left in the Southern Hemisphere)

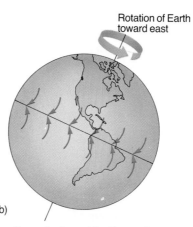

(b)

Air moving toward the Equator is traveling east slower than the land beneath it and veers to the west (turns right in the Northern Hemisphere and left in the Southern Hemisphere)

FIGURE 18.17 *The Coriolis effect. Air moving north or south will be deflected by the rotation of the Earth.*

Equator, is greatest at the Equator and decreases toward the poles. But all parts of the planet make one complete rotation every day. Thus, a point on the Equator must travel farther than any other point on the Earth in 24 hours. The equatorial region must move faster. At the Equator, all objects move eastward with a speed of about 1600 km per hour; at the poles, there is no eastward movement at all, and the speed is 0 km per hour. Now imagine a parcel of air located at the Equator. It is traveling eastward at 1600 km per hour and then gets heated and moves upward and starts to travel poleward. Let us say that this parcel moves north. As it starts, there are two components to the velocity, an eastward component and a northward component. At any distance north of the Equator, it is traveling eastward *faster* than the Earth beneath it. Thus, the motion of the air relative to Earth is curved toward the east, or to the right, of the direction of motion, as shown in Figure 18.17a.

CONCEPTUAL EXAMPLE 18.2 CORIOLIS EFFECT
Predict the motion of a parcel of air moving southward from the North Pole to the Equator.

Solution The ground at the North Pole has no east-west velocity. If air from the North Pole moves southward, it is moving more slowly than the ground beneath it, and therefore it lags behind. Because the Earth moves in an easterly direction,

the parcel of air veers toward the west, or to the right of the direction of motion as shown in Figure 18.17b.

As a general rule, when one looks along the direction of motion of the wind, all winds veer to the right in the Northern Hemisphere. In the Southern Hemisphere, winds veer toward the left. The deflection of airflow caused by the rotation of the Earth is called the **Coriolis effect.**

Figure 18.18 shows how the Coriolis effect affects the wind moving away from a high pressure region. For simplicity, we have shown the isobars as concentric circles about a high pressure region in Figure 18.18a. As the wind moves down the pressure gradient, it is deflected toward the right in the Northern Hemisphere. Thus the wind moves in a clockwise spiral about a high pressure region. As the wind moves toward the center of the low shown in Figure 18.18b, the deflection caused by the Coriolis effect produces a counterclockwise spiral about the center of the low. These flow patterns are reversed in the Southern Hemisphere. That is, the winds move in a counterclockwise spiral about a high and in a clockwise spiral about a low.

Atmospheric Circulation

The simplified global wind model shown in Figure 18.15 is called the Hadley model after British meteorologist George Hadley, who proposed it in

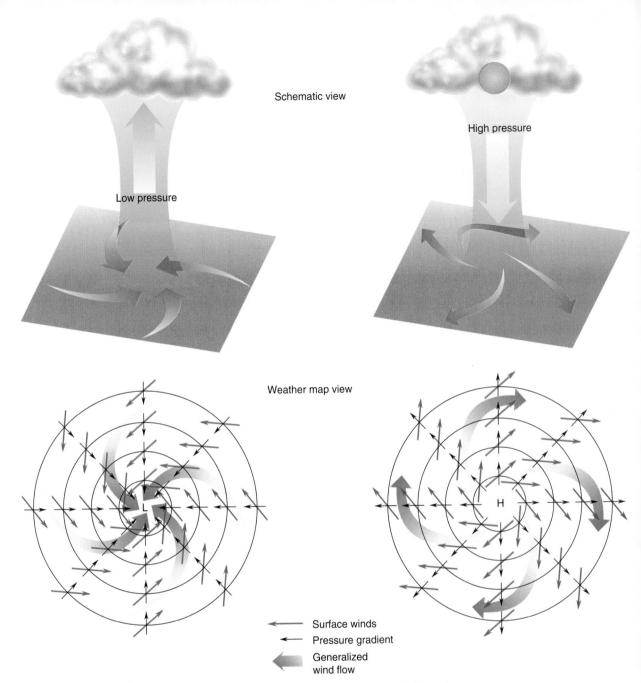

Schematic view

Low pressure

High pressure

Weather map view

Surface winds

Pressure gradient

Generalized wind flow

FIGURE 18.18 *Schematic views and weather map views of (a) cyclones and (b) anticyclones in the Northern Hemisphere. In the absence of the Coriolis effect, winds would follow the pressure gradient indicated by the black arrows. The Coriolis deflection, however, causes the winds to spiral into a low and away from a high, as shown by the arrows labeled "generalized wind flow."*

1735. However, when meteorologists began to map global weather patterns shortly after World War I, they found high and low pressure systems migrating across the midlatitudes like writhing snakes (Fig. 18.19). A Norwegian scientist, Carl Rossby,

concluded that the single-cell model could not explain these migrating storms because it did not account for the Coriolis effect.

In the 1950s, global wind systems were modeled experimentally by climatologists at the University of

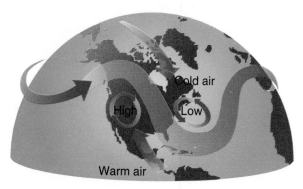

FIGURE 18.19 *High and low pressure systems that snake their way across the Northern Hemisphere cannot be explained by the Hadley one-cell model.*

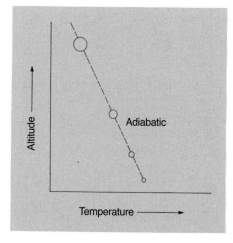

FIGURE 18.21 *Adiabatic cooling.*

Chicago. They mounted a circular pan on a variable-speed turntable and placed a heating element around the rim and a cooling coil at the center (Fig. 18.20). The pan represented the Earth, with the rim analogous to the Equator and the center analogous to the poles. They filled the pan with water and added dye to trace currents.

FIGURE 18.20 *An experimental model to demonstrate the Earth's winds.*

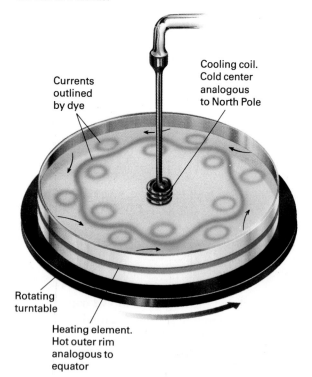

Currents outlined by dye

Cooling coil. Cold center analogous to North Pole

Rotating turntable

Heating element. Hot outer rim analogous to equator

When the pan was stationary, water rose at the heated edge, traveled across the surface, sank at the cooled center, and returned to the edge along the bottom, thus forming a Hadley cell. When the pan rotated slowly, this current was deflected by the Coriolis effect, but the single-cell pattern was retained. When the scientists increased the rotational speed, however, the cell broke apart. Midway between the edge and the center—the area representing the Earth's middle latitudes—the current diverged into whirls and eddies that looked like the real storms observed in the middle latitudes.

Our modern understanding of global climate is derived from this experiment combined with mathematical analysis and thousands of direct measurements of natural wind patterns. The modern model, called the **three-cell model,** depicts three convection cells in each hemisphere. As previously discussed, the Sun shines most directly at or near the Equator and warms the air near the Earth's surface. The warm air gathers moisture from the equatorial oceans. The warm, moist rising air forms a vast region of low pressure near the Equator, with little horizontal air flow. When a body of air rises, it ascends into a region of lower atmospheric pressure. As it moves into the zone of lower pressure, it expands, just as a balloon would expand if you placed it in a partial vacuum. As a result, the thermal energy spreads out over a larger volume, and the air cools. This effect is called **adiabatic cooling** (Fig. 18.21). (The word "adiabatic" means without loss or gain of heat.)

As the rising air cools adiabatically, the water vapor condenses and falls as rain. Therefore, local

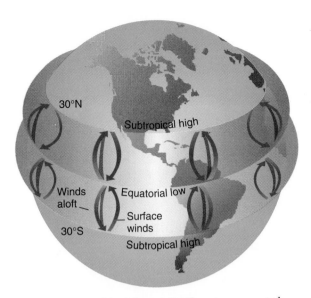

FIGURE 18.22   *Air rising at the Equator moves pole-ward at high elevations, falls at about 30° north and south latitudes, and returns to the Equator, forming trade winds. (Winds at higher latitudes are shown in Fig. 18.23.)*

squalls and thunderstorms are common in the equatorial region, but steady winds are rare. This hot, still region was a serious barrier in the age of sailing ships. Mariners called it the **doldrums,** and literature is alive with descriptions of their despair at being unable to move across the vast, windless seas. On land, the frequent rains near the equatorial low pressure zone nurture lush tropical rain-forests.

The air rising at the Equator splits to flow north and south at high altitudes. However, these high-altitude winds do not continue to flow due north and south as Hadley predicted because they are deflected by the Coriolis effect. Thus, their poleward movement is interrupted. In both the Northern and the Southern Hemispheres, this air veers until it flows due east, at about 30° north and south latitudes as shown in Figure 18.22. It then cools enough to sink to the surface, creating subtropical high pressure zones at 30° north and south latitudes. The sinking air warms adiabatically, absorbing water and forming clear blue skies. At the center of the high pressure area, the air moves vertically and not horizontally, and therefore few steady surface winds blow. This calm high pressure belt circling the globe is called the **horse latitudes.** The re-

gion was so named during the 1500s and 1600s because sailing ships were becalmed for long periods. Horses transported as cargo on the ships often died of thirst and hunger. The warm, dry descending air in this high pressure zone forms many of the world's great deserts, including the Sahara in North Africa, the Kalahari in South Africa, and the Australian interior desert.

Descending air at the horse latitudes splits and flows over the Earth's surface in two directions, toward the Equator and toward the poles. The surface winds moving toward the Equator are deflected by the Coriolis effect, so they blow from the northeast in the Northern Hemisphere and from the southeast in the Southern Hemisphere. In the days of sailing ships, sailors transporting goods for trade depended on these reliable winds and hence called them the **trade winds.** The winds moving toward the poles are also deflected by the Coriolis effect. The predominant winds in the midlatitudes are called the **prevailing westerlies.** They flow from the southwest in the Northern Hemisphere and from the northwest in the Southern Hemisphere as shown in Figure 18.23a.

The poles are cold year-round. The cold polar air sinks, creating yet another band of high pressure. The sinking air flows over the surface toward lower latitudes. In the Northern Hemisphere, these surface winds are deflected by the Coriolis effect to form the **polar easterlies.** The polar easterlies and prevailing westerlies converge at about 60° latitude. Air is forced upward at the convergence, forming a low pressure boundary zone called the **polar front.**

The three convection cells are bordered by alternating bands of high and low pressure as shown in Figure 18.23b. In the three-cell model, global winds are generated by heat-driven convection currents, and then their direction is altered by the Earth's rotation.

Figure 18.23 shows stationary linear boundaries between the cells. In reality, these boundaries migrate north and south with the seasons. They are also distorted by surface topography and local air movement. For example, in the Northern Hemisphere, storms develop along the polar front, as shown in Figure 18.24. These storms bring alternating rain and sunshine, which are favorable for agriculture. Consequently the great wheat belts of the United States, Canada, and Russia all lie between 30° and 60° north latitude.

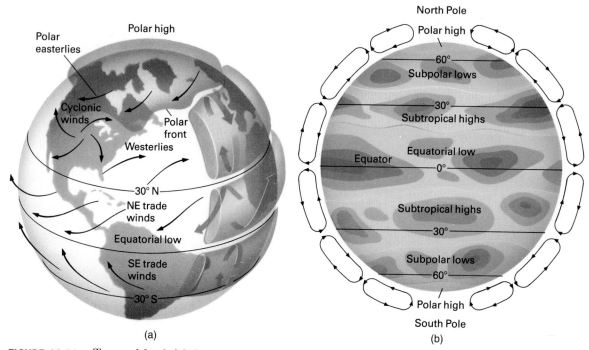

FIGURE 18.23 *Two models of global wind patterns as predicted by the three-cell theory.*
(a) An artist's rendition showing the three cells and the directions of surface winds.
(b) High and low pressure belts with the major convection currents shown on the edges of
the sphere.

A jet stream (as seen in Figure 18.24) is a narrow band of fast-moving high altitude air. Jet streams form at boundaries between the Earth's major climate cells as high altitude air is deflected by the rotation of the Earth. The **subtropical jet stream** flows between the trade winds and the westerlies, and the **polar jet stream** forms along the polar front. When you watch a weather forecast on TV, the meteorologist commonly shows the movement and direction of the polar jet stream as it snakes across North America. Storms commonly occur along this line. We can now understand this relationship. The jet stream marks the boundary between cold polar air and the warm, moist westerly flow that originates in the subtropics. Storms develop where the two converge.

FIGURE 18.24 *The polar front and jet stream. Cyclones and anticyclones that develop along this convergence distort the jet stream as shown in b. Weather forecasters plot the changes in the jet stream to predict the motion of storms.*

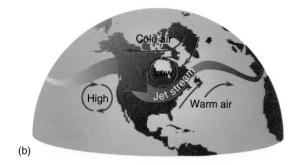

18.5 CONDENSATION AND RELATIVE HUMIDITY

If you boil water on a stove, you can see a steamy mist above the kettle, and then higher still the mist seems to disappear into the air. Of course, the water molecules have not been lost. In the pan, water is a liquid, and in the mist above the kettle the water exists as tiny droplets. These droplets then evaporate, and the water vapor mixes with air and becomes invisible. Air generally contains some water vapor. **Humidity** is the measure of the amount of water vapor in air. **Absolute humidity** is defined as the mass of water vapor contained in a given volume of air and is generally expressed in units such as grams per cubic meter (g/m³). But air cannot hold an unlimited quantity of water vapor. If you poured liquid water into a container of dry air, the water would start to evaporate; that is, some water molecules would leave the liquid and mix with the air molecules in the form of a gas. At the same time, some molecules would go the other way, from gas to liquid. When the two opposing rates become equal, net evaporation stops, and the air is said to be **saturated** with moisture.

At the saturation point, the air can hold no more moisture. The saturation quantity varies with temperature. Figure 18.25 shows that warm air can

FIGURE 18.25 *Maximum water content of air as a function of temperature.*

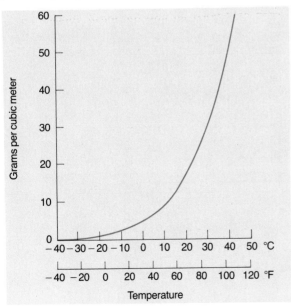

hold more water vapor than can cold air. For example, a cubic meter of air can hold 23 g of water vapor at 25°C, but if the air is cooled to 12°C, it can hold only half that quantity, 11.5 g/m³. The **relative humidity** is a measure of the amount of water vapor in the air compared with the saturation quantity at a given temperature.

Relative humidity (%) =

$$\frac{\text{actual quantity of water per unit of air}}{\text{saturation quantity at the same temperature}} \times 100\%$$

$$(18.1)$$

Suppose that there are 11.5 g/m³ of water vapor in a parcel of air at 25°C; that is, the absolute humidity is 11.5 g/m³. Because air at that temperature can hold 23 g/m³ when it is saturated, it is carrying half of the saturation quantity, and the relative humidity is

$$\text{Rel humidity} = \frac{11.5 \text{ g}}{23 \text{ g}} \, 100\% = 50\%$$

As an experiment, let us take some of this air and cool it without adding or removing any water vapor. Because cold air can hold less water vapor than warm air, the relative humidity rises even though the absolute humidity remains constant at 11.5 g/m³. If the air is cooled to about 12°C, the relative humidity reaches 100 percent because air at that temperature can hold only 11.5 g/m³ and that amount is already present. Any further cooling causes the water to condense and to form droplets.

You can observe how water condenses on cooling by performing a simple experiment. Heat some water on a stove until it is boiling rapidly. The clear air that lies just above the steamy mist will be hot and full of water vapor. Hold a drinking glass in this space and you will observe droplets of water condensing on the surface. Because the surface of the glass is cool, the warm, moist air that comes in contact with it is cooled. As the air cools, some of the water vapor condenses into liquid droplets. The same effect can be observed if you are inside a house on a cold day. If you breathe onto a window, droplets of water or crystals of ice appear as your moist breath cools on the glass.

CONCEPTUAL EXERCISE

Explain why frost forms on the inside of a refrigerator, assuming it is an old-fashioned one and not a modern frost-free unit. Would more frost tend to

FOCUS ON
RELATIVE HUMIDITY AND COMFORT

Many southwestern states claim that their high temperatures are more comfortable than the same temperature found, say, in a midwestern state. They base this statement on the fact that their arid climate lends itself to an atmosphere with a low relative humidity. Is this assertion true or false? The answer is that it is indeed true. One of the primary cooling mechanisms of the human body is perspiration, and relative humidity affects our comfort through its relationship to the rate of evaporation. As perspiration evaporates from the skin, the skin is cooled because, as we saw in Chapter 5, evaporation is a cooling process. This occurs because the heat required to change the droplets of sweat to a vapor is subtracted from the skin. Obviously, if the relative humidity is high, perspiration evaporates slowly from the skin, and you are left with a damp, sticky feeling. But just how

important is relative humidity to our comfort? The answer can best be demonstrated with reference to the table. To see how this table is used, first consider a state such as Arizona, which might have a temperature of 90°F but a relative humidity of only about 10 percent. Find 90°F along the horizontal and move down the column beneath it until you are directly across from the 10 percent relative humidity reading. The intersection of this row and column shows you that 90°F feels like 85°F when the relative humidity is only 10 percent. Consider a midwestern state on a day when the temperature is 90°F. On such a day, a relative humidity of 70 percent is not uncommon. Finding the value at the intersection of the row and column for these values shows that this particular 90°F day would actually feel more like 106°F. In this latter case, the air is already almost saturated, which makes it far more difficult for evaporation to take place.

Effective air temperatures as a function of relative humidity

| RELATIVE HUMIDITY | AIR TEMPERATURE (°F) | | | | | | |
| | 70 | 75 | 80 | 85 | 90 | 95 | 100 |
	EFFECTIVE TEMPERATURE						
0%	64	69	73	78	83	87	91
10%	65	70	75	80	85	90	95
20%	66	72	77	82	87	93	99
30%	67	73	78	84	90	96	104
40%	68	74	79	86	93	101	110
50%	69	75	81	88	96	107	120
60%	70	76	82	90	100	114	132
70%	70	77	85	93	106	124	144
80%	71	78	86	97	113	136	
90%	71	79	88	102	122		
100%	72	80	91	108			

form in (a) summer or winter? (b) In a dry desert region or a humid region? Explain.

Answer There is water vapor in the air inside the refrigerator. When this vapor-laden air comes in contact with the cooling coils, frost forms. More frost would tend to form in summer in a humid region where the air is warm and moist.

With this background, we can easily understand how **dew** is formed. On a typical summer evening in a moist temperate zone, the air is likely to be warm and laden with water vapor. After the Sun has set, the surfaces of various objects such as plants, houses, and windows begin to lose heat by radiation. In the early hours of the morning, when these surfaces are quite cool, water condenses from the warm moist air just as your breath condenses onto a window on a cold day. This condensation is called dew. The **dew point** is the temperature to which a sample of air must be cooled to become saturated with moisture.

If the temperature of the air is at or below freezing, the moisture in the air goes directly from gaseous vapor to solid ice on condensation. This is the process that occurs in the formation of **frost.**

18.6 CLOUDS

Clouds form when warm, moisture-laden air is cooled below its dew point. Condensation occurs most readily, however, when there are small particles present in the air. These particles, called **condensation nuclei,** serve as surfaces on which water molecules can collect. One type of cloud formation is **fog.** When warm, moist air from an ocean blows inland to a cold coast, the air cools, water condenses, and low clouds known as fog are formed. Thus, San Francisco, Seattle, and Vancouver all experience foggy winters accompanied by rain and drizzle.

Cloud formation can occur in other ways as well. Imagine that a moist mass of air in a given region is moving toward the Equator. If the temperature of the air could remain constant, the water would remain in a vapor form indefinitely. However, as this air moves into the tropics, it is heated, becomes less dense, and therefore starts to expand. It then rises, expands further, and begins to cool adiabatically. If conditions are favorable, the air

FIGURE 18.26 *Cirrus clouds are wisps of fine ice crystals.*

cools so much that water vapor condenses out and clouds form. Thus, cloud formation occurs most readily when moist air rises.

There are three basic types of clouds: **cirrus, stratus,** and **cumulus.** Subtypes exist under each of these three categories.

Cirrus

The name *cirrus* comes from the Latin meaning lock or wisp of hair. These clouds are the type often painted by artists because they can be applied to canvas as feathered patches of white. Figure 18.26 shows that this is an accurate rendering of this type of cloud. Cirrus clouds are usually found at high altitudes, normally between 6 and 11 km. Because of the low temperature at this altitude, these clouds are actually formed of ice crystals rather than water vapor.

Stratus

The name *stratus* is derived from the Latin for layer. As Figure 18.27 indicates, these clouds are characterized by their sheet-like or layered appearance. These clouds can form at any height ranging from near the surface of the Earth to about 6 km, and they consist of water droplets. A drab gray cloud

FIGURE 18.27 *Stratus are low clouds spread out across the sky like a continuous blanket.*

cover that blankets the entire sky consists of stratus clouds. These clouds have an almost constant thickness throughout their extent. Fog can be considered a low-lying stratus cloud.

FIGURE 18.28 *Cumulus clouds form over locations where there are rising currents of air. The base of these clouds is located at the height where condensation begins.* (Courtesy Kenneth R. Martin Collection)

Cumulus

The name *cumulus* comes from the Latin for heap or pile. In contrast to stratus clouds, cumulus clouds have a vertical rather than a horizontal development, as shown in Figure 18.28. They generally have a flat base with a towering, majestic pile mushrooming upward. These clouds can have their base at a height of anywhere from about 500 m to about 12 km. Cumulus clouds are formed by an upward movement of air. The base of the cloud forms at the height at which condensation begins. These clouds are often seen on bright summer days.

Nimbus

Nimbus is a word from the Latin that means precipitation and is used as a prefix or a suffix to describe clouds. For example, a nimbostratus cloud is a stratus cloud from which rain is falling. The familiar thunderhead seen during summer rains is an example of a cumulonimbus cloud. The dark appearance of this latter type occurs as condensation builds up in the cloud to the point at which sunlight is completely blocked. Figure 18.29 is an example of a cumulonimbus cloud.

18.7 PRECIPITATION

We can now understand a few simple relationships between atmospheric conditions and weather. When warm air rises, it expands and cools. The expansion makes it less dense, which means that it ex-

FIGURE 18.29 *A cumulonimbus cloud is a thunderhead that often brings thunder and lightning.*

erts less pressure and the barometer reading in that region is lower. As already noted, these same conditions also generally lead to cloud formation and, therefore, to the possibility of rain or snow. Thus, a falling barometer is a good indication that precipitation may soon follow. Alternatively, when cool air falls to the Earth, the barometric pressure rises. Air is compressed and heats up as it falls. Because warm air can hold more moisture than cold air, clouds generally do not form under high pressure conditions. Instead the warm air with low relative humidity tends to absorb moisture from the Earth's surface. Thus, a rising barometer generally predicts fair weather.

An understanding of this general relationship between barometric pressure and cloud formation does not instantly transform the casual observer into a seasoned weather forecaster. Often the barometric pressure drops but no rain falls, or, conversely, the pressure may rise amid cloudy skies. The temperature, rainfall, and wind patterns in a region are a result of many complex factors. Some of these are caused by local disturbances and others by global atmospheric patterns.

When clouds form, the droplets usually are light enough that they do not fall to the Earth. As a result, they remain suspended in near-equilibrium under the action of gravity (downward) and the movement of air (upward). However, if the mass of the droplets becomes too large, they will fall out of the sky as rain, snow, sleet, or hail.

Rain

The exact process that produces rain is not fully understood, but the method seems to depend on the type of cloud structure present. Raindrops falling from a cumulus cloud may have originated in a different way than those falling from a stratus cloud. Let us examine these two different processes.

The essential elements for the formation of rain within a cumulus cloud are (1) ice crystals, (2) a supersaturated vapor, and (3) a mixing process within the cloud, as shown in Figure 18.30. The first of these, the ice crystals, are present at the top of a cumulus cloud because the cloud extends high enough into the atmosphere for the crystallization process to occur. Supersaturated vapor is present at or near the base of the cloud, but we must first examine what supersaturated vapor really is. We have

FIGURE 18.30 *Ice crystals, a supersaturated vapor, and mixing are necessary to produce rain from a cumulus cloud.*

stated that when the temperature drops below the dew point, condensation occurs. However, it is possible for the temperature to drop below the dew point without condensation taking place, and when this occurs, the vapor is said to be supersaturated. In the presence of a mixing agent, supplied perhaps by air currents, the ice crystals and the supersaturated vapor can come into contact. The supersaturated vapor then condenses on the ice crystals, forming droplets, which, in turn, fuse with other droplets to form a drop of rain. The final result is a typical raindrop of about 2.5 to 6 mm in size and heavy enough to fall from the cloud as rain.

It can easily be demonstrated that the process just described is an important one in the formation of rain because scientists may hasten or help along the processes described. During a drought, the essential element often missing in the prescription for rain is the presence of ice crystals in the clouds. As a result, modern-day rainmakers seed the clouds, via airplane, with silver iodide or Dry Ice pellets. The structure of silver iodide crystals is similar enough to ice that they become a direct substitute for ice crystals. The Dry Ice crystals are at a temperature of about $-80°C$, and they cool the cloud enough to form ice crystals.

The rainfall from a stratus cloud is produced by a different, more direct method. In this case, the water droplets are formed by the coalescing of smaller droplets until a drop large enough to fall is formed. Because of this method of formation, the droplets formed are often small, of the order of 0.5 mm, and the rain may fall as a mist or drizzle.

Snow

After rain, the most common form of precipitation is snow. Snow is produced when the dew point is below 0°C, the freezing point of water. Under this condition, the water vapor passes directly from the vapor state to that of minute ice crystals. These crystals may fall as individual crystals, or, if the air temperature is high enough, several crystals may melt slightly and become stuck together to form a large snowflake.

Sleet and Freezing Rain

Sleet or freezing rain is formed when snow falls through a warm layer and melts to become raindrops. If these raindrops then refreeze when they fall through a lower-lying cold layer of air, they become pellets of ice called sleet. If instead the water droplets become supercooled in the cold layer, they freeze on contact with a solid surface on the Earth. This is known as freezing rain.

Hail

Occasionally, precipitation takes the form of large ice globules called **hail.** Hailstones vary from 5 mm in diameter to a record giant 14 cm in diameter, weighing 765 g (more than 1.5 pounds), that fell in Kansas. A 500 g (1 pound) hailstone crashing to Earth at 160 km (100 miles) per hour can shatter windows, dent car roofs, and kill people and livestock. Even small hailstones are destructive enough to damage crops. Hail always falls from a cumulonimbus cloud. Because cumulonimbus clouds form in columns with distinct boundaries, hailstorms occur in local, well-defined areas. Thus, one farmer may lose an entire crop while a neighbor is unaffected.

A hailstone consists of concentric shells of ice like the layers of an onion. Two mechanisms have been proposed for their formation. In one, ice crystals alternately fall and rise through the cloud. When the crystals first start falling, they grow in the lower portion of the cloud. If the rising air is strong enough, it blows the ice particles back upward. As they rise, more moisture collects, causing the particles to grow even larger. An individual particle may rise and fall several times until it is so large that it drops out of the cloud. In another proposed mechanism hailstones form in a single pass through

the cloud. During their descent, supercooled water freezes onto the ice crystals. The layered structure occurs because different temperatures and amounts of supercooled water exist in different portions of the cloud. Therefore, each layer is deposited in a different portion of the cloud.

18.8 OCEANS AND WORLD CLIMATE

The concepts discussed in the preceding sections enable us to predict wind and weather from season to season in a general way over vast areas of the Earth. But climates are influenced by local factors as well. Continents, oceans, prairies, and mountain ranges and the many dynamic interactions among them continuously influence winds, rains, and cloud formation. The ocean, in particular, affects climate, in large measure because water absorbs and stores heat at a rate different than does the solid earth.

Recall from our discussion of heat in Chapter 5 that if equal masses of water and ethyl alcohol are both heated on equal hot plates, the *temperature* of the alcohol rises nearly twice as fast as that of the water. Also, we saw that if water and sand absorb equal amounts of thermal energy, the temperature of the sand rises about five times as fast. Similarly, if equal masses of hot water and hot sand, both starting at the same temperature, are placed in a refrigerator, the sand cools faster. This means that more thermal energy is stored in 1 g of hot water than in 1 g of hot sand at the same temperature, which is to say that the specific heat of water is greater than that of sand. As a matter of fact, water has an unusually high specific heat—about five times that of sand, granite rock, or dry clay. Consider, then, what happens near the boundary between an ocean and a continental land mass. Suppose that both the water and the land are at the same temperature at some time in the spring. As summer approaches, both land and sea receive equal amounts of solar energy. The land warms up faster, however, just as the ethyl alcohol or the sand gets hotter than the water even though they both absorb equal amounts of thermal energy.

There is yet another factor that complements the difference in specific heats and makes the land warm up faster than the sea. The ocean is turbulent, and waves and currents carry warm surface water downward and bring the cold deep water up to

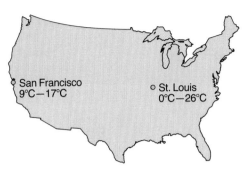

FIGURE 18.31 *Continental areas experience greater temperature extremes than do coastal areas. St. Louis: average January temperature = 0°C; average July temperature = 26°C. San Francisco: average January temperature = 9°C; average July temperature = 17°C.*

the surface. On land there is no comparable mixing process, and much of the thermal energy is concentrated in a shallow surface layer of soil or rock. Therefore, on land a thin surface layer becomes quite hot, so that you may burn your feet walking across dry sand or concrete. Yet a meter below the surface, the soil is 10° to 15°C cooler. In the sea, the temperature is much more uniform throughout. The same amount of energy that warms a large mass of water slightly causes a small layer of soil to become much hotter.

For these reasons, summertime temperatures are generally hotter in the central areas of large continents than near the seacoasts. In the winter, the opposite effect is observed, and the land tends to cool more than the sea. Thus, inland areas are generally colder in winter and warmer in summer than the coastal regions. For example, the coldest temperatures recorded in the Northern Hemisphere are found in central Siberia and not at the North Pole because Siberia is landlocked, whereas the North Pole is located in the middle of the Arctic Ocean. In the summer, however, Siberia is considerably warmer than the North Pole. Or, as another example, let us compare the climates of two large cities in the United States, San Francisco and St. Louis, that are both located at approximately 38° north latitude. As shown in Figure 18.31, St. Louis experiences much greater temperature extremes than San Francisco.

Coastal climate is also influenced by several other factors. The oceans are not like a huge bathtub full of water that sloshes about randomly. Water travels across oceans in well-defined, predictable

currents, much like the prevailing wind patterns in the atmosphere. These currents can be thought of as rivers that flow within the ocean.

Surface currents are caused partially by winds. When the trade winds blow steadily across the sea, they push along the upper layers of water, and form waves. But just as the air pushes the surface water, so the moving surface waves drag deeper waters along with them. Thus, the steady winds cause water to flow in predictable patterns across the oceans. There is considerable slipping between wind and water, and therefore the water moves much more slowly than the air; most midocean currents move at a rate of only 3 to 5 km per hour. Ocean currents, like wind systems, are deflected by the spin of the Earth. They are forced to veer clockwise in the northern oceans and counterclockwise south of the Equator. As the moving waters strike continental land masses, they are deflected even further, so that they rotate in circular motions called **gyres.**

Currents may carry either warm water toward the poles or cold water toward the Equator. Either way, the moving water profoundly affects climates on land. Portland, Maine, is at about the same latitude, 44° north, as the north coast of Spain. However, the Spanish coast is a warm resort area where people swim and sunbathe nearly year-round, whereas the Maine coast is noted for its hard, snowy winters and cool summers. Much of the reason for this climatic difference lies in the fact that the cold Labrador Current moves south from the Arctic Ocean to the coast of Maine, whereas the warm Gulf Stream tempers the climate of Spain (and the entire western coast of Europe). Figure 18.32 shows the global system of currents.

Sea Breezes

Anyone who has lived near an ocean or large lake has undoubtedly become acquainted with recurring winds that blow from water to land and from land to water. Local sea breezes are caused by uneven heating and cooling of the land and the ocean, as shown in Figure 18.33. We have already learned that the land heats up faster than the sea in summertime and cools more quickly in winter. The same effect occurs on a daily time scale. If land and sea exist at close to the same temperature on a summer morning, the land and the air above it will become hotter by noon. Hot air then rises over the land, a local convection current is established, and cooler air from the sea flows toward the land. Thus,

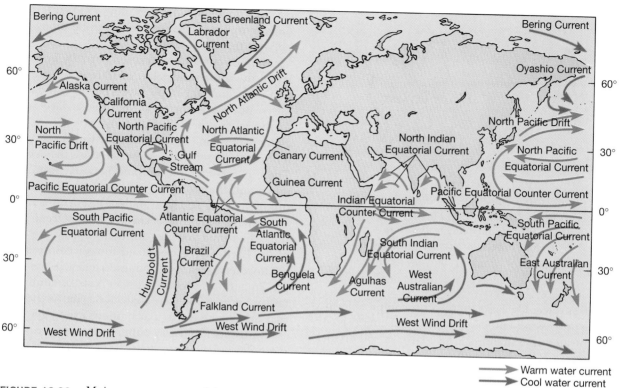

FIGURE 18.32 *Major ocean currents of the world.*

on a hot sunny day at the beach, winds generally blow from the sea inward. At night, the reverse process occurs. The land cools faster than the sea until it becomes colder by nighttime. Then the winds reverse, and breezes blow from the shore out toward the sea.

FIGURE 18.33 *Sea breezes blow inland during the day, and land breezes blow out to sea at night.*

(a)

(b)

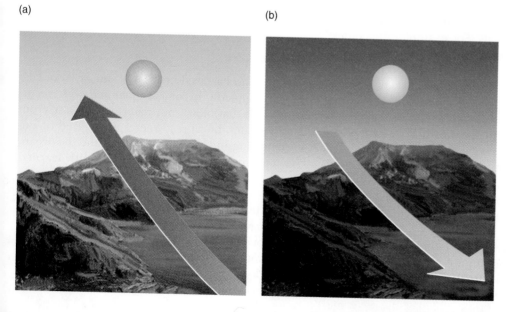

Monsoons

Sea breezes also occur on a larger, continental scale. Seasonally reversing winds caused by uneven heating of land and sea dominate the weather over most large land masses. The **monsoons** that are common in Asia and Africa are an example of such seasonal winds. In the summertime, the continents heat up faster and become warmer than the sea. The warm earth heats the air above it, and this warm air mass rises, creating a large low pressure area. This rising air draws moisture-laden sea breezes inland. When the wet air rises, clouds form and heavy rains fall. In the winter, the process is reversed. The land cools below the temperature of the sea, and air rises over the ocean to descend over the land, producing a continental high pressure zone. The surface winds blow from the land out to sea, and this dry high pressure air produces little rain. More than half of the inhabitants of the Earth depend on the monsoons for their survival because the predictable heavy summer rains bring water to the prairies and grain fields of Africa and Asia. If the monsoons fail to arrive, crops cannot grow, and people have nothing to eat.

18.9 WEATHER PATTERNS

In any location on Earth, seasonal changes are fairly predictable. Thus, ranchers in Colorado bring in the hay and round up scattered herds of cattle every fall in expectation of the winter snows they know will come. Skiers tune their skis in late November or early December. By the end of March, sports shops display tennis racquets, and the ranchers repair broken machinery and prepare for spring plowing. Yet no one can predict accurately when the first snow will fall or exactly when warm weather will return. **Weather** is a description of the day-to-day changes in temperature, wind, and precipitation.

Recall that precipitation is likely to occur when moist air rises. The rising air cools, causing the moisture to condense and fall as rain or snow. There are several ways in which moisture-laden air can rise. As mentioned previously, if the Sun heats the air in one region more than in another, the warmer air expands and rises. This type of pattern is responsible for the frequent rain squalls near the Equator. The same effect may occur on a local scale. Visualize a large area of plowed fields adjacent to a vast forest. Bare soil absorbs heat more efficiently than green leaves, and the air above the farms will be heated relative to adjacent air masses over the woods. This air may rise, and if conditions are favorable, local summer thunderstorms will occur.

Mountains also affect local weather. An air mass traveling across a mountain range must necessarily rise to flow over the mountains. If this rising air is laden with moisture, conditions will be favorable for condensation. Rain or snow will then be likely. Conversely, the air must fall on the downward side of the mountains. As it descends, it is compressed and thereby heated. Since this warm, high pressure air is unlikely to discharge moisture, a belt of dry climate, called a rain shadow, often exists on the downwind side of major mountain ranges (Fig. 18.34). Death Valley in California is a rain shadow desert (Fig. 18.35).

FIGURE 18.34 *Formation of a rain-shadow desert. Warm, moist air from the ocean rises when it encounters a mountain range. As it rises, it cools, and water vapor condenses to form rain. The dry, descending air on the downwind side absorbs moisture, forming a desert.*

Rising air generates low pressure, which leads to precipitation

Warm, moist air rises

Dry air descends, creating high pressure zone

Prevailing winds

Rain shadow desert

Klamath Mts
Coast Range
San Francisco
Santa Lucia Range
Salinas Valley
Mohave Desert
Sierra Nevada
Owens Valley
Inyo Mts
Death Valley
Great Valley

FIGURE 18.35 *Rainfall patterns in the state of California. Note how the rain shadow deserts lie east of the mountain ranges. Rainfall is reported in centimeters per year.*

18.10 FRONTAL WEATHER SYSTEMS

If you had an insulated container of cold, dry air and blew a breath of warm, moist air into it, the two parcels of air would eventually mix until the temperature and humidity in all parts of the container were the same. These changes are predicted by the second law of thermodynamics, which states that any undisturbed system will spontaneously tend toward maximum disorder, or sameness. Similar considerations apply whether the sample of air is a small volume or the entire Earth's atmosphere. If the atmosphere were insulated and undisturbed, it would drift toward sameness, and eventually its temperature would be identical at the poles, the Equator, and everywhere else. But the atmosphere is dynamic; it is warmed unevenly by the Sun, and it loses energy by radiation out into space.

Air that lies over the polar icecaps is frequently dry and cold, whereas air lying over tropical oceans is usually warm and moist. The term **air mass** refers to a large body of air that has approximately the same temperature and humidity throughout (Fig. 18.36).

As the Earth turns and winds blow, air masses move about and collide. When one air mass collides with another, at first the two behave as separate en-

tities, each with its own specific properties, like a drop of ink in a glass of water. If no other interactions occurred, they would, in time, mix to become one homogeneous parcel of air. But the Earth's atmosphere is not at all static. The Sun shines, winds blow, other air masses move in, and changes occur in a continuous and often unpredictable manner.

When two air masses collide, the zone where they meet is called a **front.** The movement of air along a weather front is fundamental to our understanding of weather. Whenever warm air comes in contact with cold air, the warm air, being less dense than the cold air, rises. This, in turn, leads to a zone of low barometric pressure, which frequently gives rise to cloud formation and precipitation. However, the character of the low pressure zone and the weather that results depend on the relative velocities of the two parcels of air.

Warm Fronts

If a warm air mass overtakes a mass of cold air, the warm air rises, as explained previously. But as it rises, it continues to move forward. The net result is that it flows up and over the cold air, pushing the cold air mass into a wedge. Figure 18.37 shows the shape of the two air masses at a range of elevations,

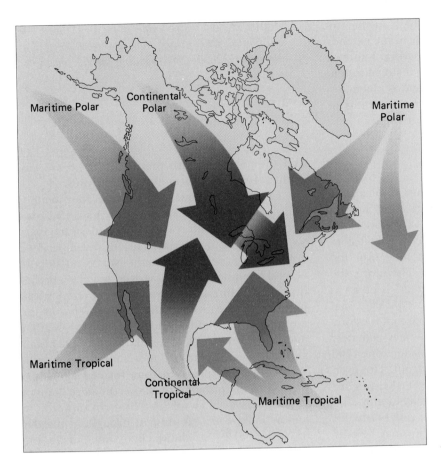

FIGURE 18.36 *Air masses classified by their source regions.*

FIGURE 18.37 *A profile of a warm front.*

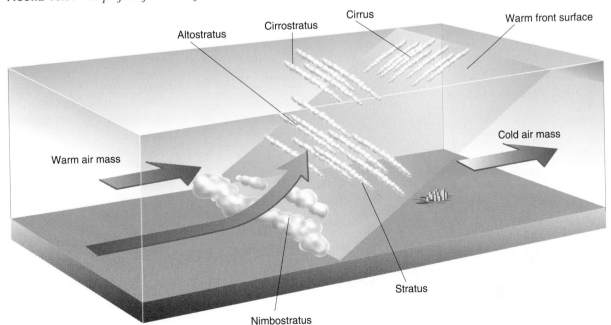

from ground level to an altitude high in the troposphere. You can see that the warm air extends over the cold air for a distance of several hundred kilometers. This rising air leads to a storm system that is often 500 to 600 km wide and is characterized by low barometric pressure, nimbostratus or stratus clouds, and persistent rain or snow.

Cold Fronts

A cold front, like a warm front, occurs along a zone where warm air and cold air masses come in contact. In this case, however, the cold air overtakes and displaces the warm air. During this collision, the faster moving cold air cannot slide over the warm air because it is more dense. Therefore, the cold air is distorted into a blunt wedge or bulge as it pushes under the warmer air. A typical frontal profile is shown in Figure 18.38. Note that the zone of rapidly rising air and precipitation is generally quite narrow, as opposed to the broader zone (500 to 600 km) observed for a warm front. However, the steep zone of contact between the two air masses causes the warm air to rise quite rapidly. The net result is a narrow band of rather violent squalls. In a typical situation, a line of storms may be only 50 to 150 km wide, but within this zone downpours

FIGURE 18.39   *Four weather maps for the western United States drawn from November 19 to November 22 with corresponding satellite photos.* ▶

and thunderstorms are prevalent. In extreme instances, tornadoes form along cold fronts.

Figure 18.39 shows a sequence of four weather maps of the western United States. A satellite photograph of the cloud cover accompanies each map. You can see from the map that on November 19 a cold front was developing in the Northwest, but the central mountain district was experiencing a period of high pressure. Extensive clouds covered Washington and Oregon, but the cloud cover over the mountains was lighter and broken. The weather map for November 20 shows that within the 24-hour period, the cold front had moved toward the south and east. A low pressure zone formed along the front, and the photograph shows the development of a thick layer of clouds that brought snow to the region. This storm continued throughout the day of November 21, as shown. By November 22, the cold front began to dissipate and sunny skies re-

FIGURE 18.38   *A profile of a cold front.*

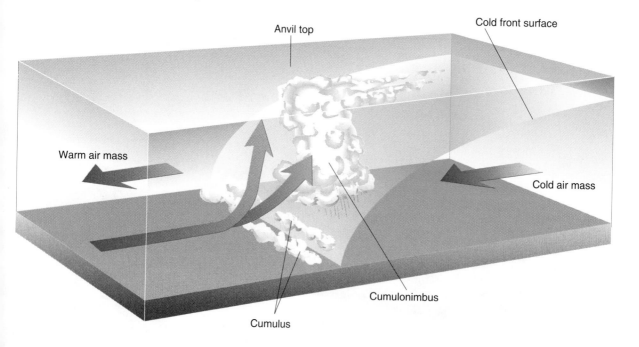

Anvil top

Cold front surface

Warm air mass

Cold air mass

Cumulonimbus

Cumulus

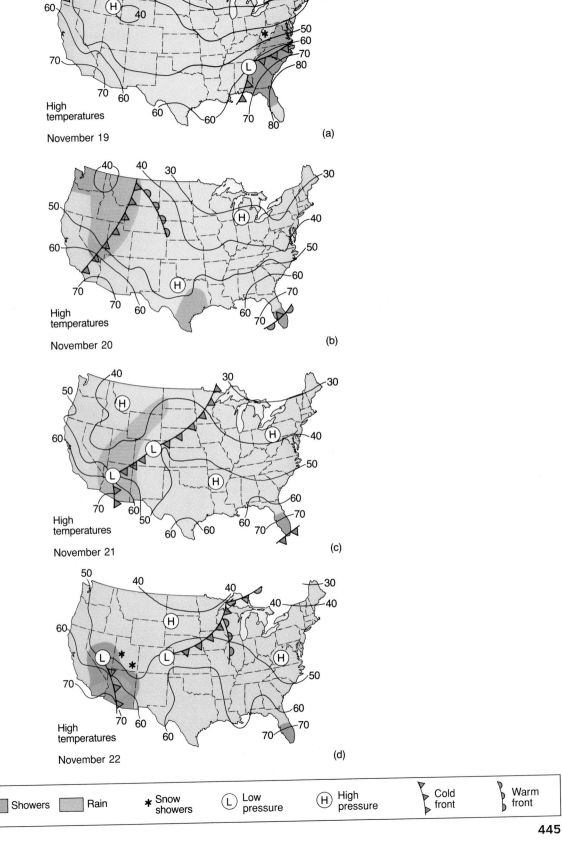

High
temperatures

November 19

(a)

High
temperatures

November 20

(b)

High
temperatures

November 21

(c)

High
temperatures

November 22

(d)

Showers   Rain   * Snow showers   L Low pressure   H High pressure   Cold front   Warm front

turned to most of the Rockies. The northern portion of the front continued into the Great Lakes, while the southwestern portion stalled and clouds hung over Arizona, Nevada, and southern California.

Occluded Fronts

A third type of system occurs when two fronts (and three air masses) collide. Figure 18.40 illustrates the case when a cold front overtakes a warm front. In this situation, a parcel of warm air is trapped between two parcels of cold air and is lifted completely off the ground. Such a system combines both the narrow zone of violent winds and precipitation of the cold front with the wider and more gentle pre-

cipitation zone of the warm front. The net result is a large zone of inclement weather.

Once again, it is important to understand that weather prediction is complex. An air parcel may rise and then, for any of a wide variety of reasons, precipitation may not occur. Storm fronts sometimes seem to build and then suddenly dissipate. Weather forecasters use sophisticated ground instruments, computer analysis, and extensive data from weather satellites. The accuracy of typical weather forecasting is about 80 percent for a 24 hour period. It is higher than this for a period of less than 12 hours but falls off for a 2 or 3 day interval. It falls off extremely rapidly for periods of more than 3 days.

FIGURE 18.40 *A profile of a storm caused by an occluded front.*

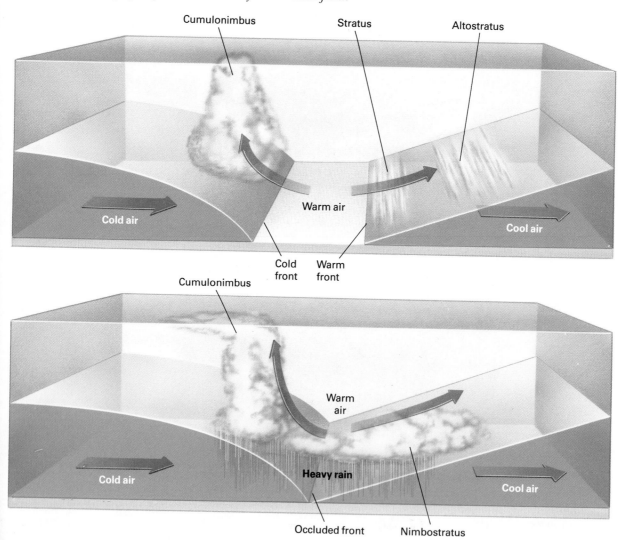

FOCUS ON
COMMON SYMBOLS USED IN WEATHER MAPS TO INDICATE
VARIOUS TYPES OF FRONTS

Note that the words "warm" and "cold" are relative only. Air that is situated over the central plains of Montana at a temperature of 0°C may be warm relative to polar air above northern Canada but cold relative to a 20°C air mass over the southeastern United States.

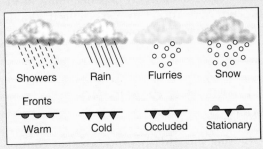

Common symbols used in weather maps to indicate various types of fronts.

CONCEPTUAL EXAMPLE 18.3   READING WEATHER MAPS

Study the weather map shown in the figure below and predict the weather in Salt Lake City, Chicago, and New York City 2 days after this map was drawn. Defend your prediction.

Solution Notice that the winds in the western part of the country were moving from the northwest, whereas there was no major motion of the low pressure area over New York City. Consequently, 2 days later, the low pressure that had been in Seattle reached Salt Lake City and it was snowing

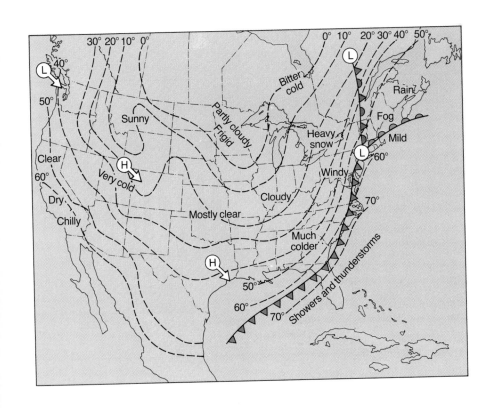

there. Cold, dry Canadian air swept into the northern Midwest, and the weather in Chicago was sunny and frigid. The storm that had been on the east coast had not passed out to sea, but without a new input of moisture was declining in intensity, and New York City experienced cloudy weather with occasional snow flurries.

18.11 EARTH'S CHANGING CLIMATE

After studying the orderly processes that control global wind systems, one might imagine that the average worldwide climate must be constant from year to year. But extreme changes occur over long spans of time. Thus, what is now the temperate zone in North America has experienced both tropical warmth and the cold of the Ice Ages in the distant past. One hundred fifty million years ago, giant dinosaurs wallowed in hot, humid swamps, whereas only 25,000 years ago, woolly mammoths roamed the edges of giant glaciers that covered much of the continent. In a series of experiments, scientists estimated global temperature during the past 40,000 years by studying geological records preserved in

glacial ice, ocean sediment, and ancient corals. All of these studies show that natural factors have caused temperature changes of 5° to 10°C in short periods of time, in some cases within 3 to 5 years (Fig. 18.41).

During the past 100 or so years, climate has continued to fluctuate on a smaller scale. Between 1880 and 1992, global temperature oscillated, with an overall upward trend. Although 1990 was the warmest year in the past century, the Earth cooled in 1991 and 1992 (Fig. 18.42). Most scientists attribute this cooling to the volcanic eruption of Mount Pinatubo in the Philippines (see the Focus On box, Volcanoes and Weather). However, even 1992 was warmer than the average temperature for the past 30 years. Notice, however, that the temperature fluctuations have been small, less than 1°C. You may ask: What difference does it make if the average temperature rises or falls 1°C? The answer is that small changes are not particularly important as far as human comfort is concerned, but they can be vital to agriculture. For example, if the temperature drops an average of 1°C in the northern United States or southern Canada, this usually means that the last killing frost occurs a week later in the spring and a week earlier in the fall than it did previously. Thus, a seemingly insignificant

FIGURE 18.41 *Temperature changes over the past 40,000 years in Greenland. The data were measured from ice cores as reported by the Greenland Ice Core Project.*

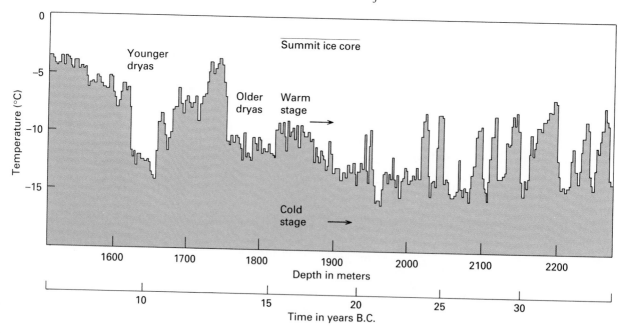

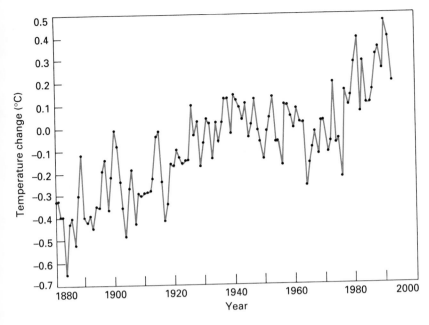

FIGURE 18.42   *Average global temperature from 1880 to 1992. The zero line represents the average from 1951 to 1980, and plus or minus values represent deviations from the average.*
(NASA/Goddard Space Center)

change in average temperature becomes quite significant when it reduces the growing season by 2 weeks. Even frost-resistant plants grow much more slowly if spring and fall temperatures are cold. In recent years, a 1°C cooling trend was blamed for a 25 percent decline in hay production in Iceland.

A warming trend could also have adverse effects. If the climate grows warmer, the zone of dry descending air, the horse latitudes, would tend to move toward higher latitudes. Approximately 5000 to 8000 years ago, the Northern Hemisphere was much warmer than it is now. During that time, grass grew in the Sahara, but little rain fell in parts of the Great Plains in North America and in the steppes of eastern Russia, as shown in Figure 18.43. Sand dunes blew across the now fertile fields and ranches of eastern Colorado and western Nebraska. Today the farms in North America and Eurasia produce a significant portion of the grain needed to feed a hungry world. No one can say for sure that a future warming trend would affect rainfall in the same manner, but many scientists believe that it could.

During the period between 1900 and the present, human development has accelerated rapidly. This era has witnessed an abnormally favorable global climate that has led to high agricultural productivity. Even with all the agricultural development that has occurred, millions of people are undernourished or starving to death. Food production and population have been pushed close to

the limit during an unusual period of particularly favorable climate. World population has increased exponentially since 1900. If the climate should change appreciably, farms in many regions of the world could fail. In fact, many climatologists believe that *either* a cooling or a warming trend might disrupt agriculture significantly.

18.12 NATURAL FACTORS AND CHANGE OF CLIMATE

Climate has fluctuated throughout the history of our planet, and therefore climatic change is a natural phenomenon. Some of the factors that may cause temperature and rainfall patterns to change are listed next.

Climate and Atmospheric Composition

If there were no atmosphere, the view from the Earth would be much like that which the astronauts saw from the Moon—a terrain where starkly bright surfaces contrast with deep shadows and a black sky from which the Sun glares and the stars shine but do not twinkle. The atmosphere protects us by serving as a light-scattering and heat-mediating blanket. As was shown in Figure 18.8, about half of the incident radiation from the Sun passes through the atmosphere to the Earth; the rest is reflected or ab-

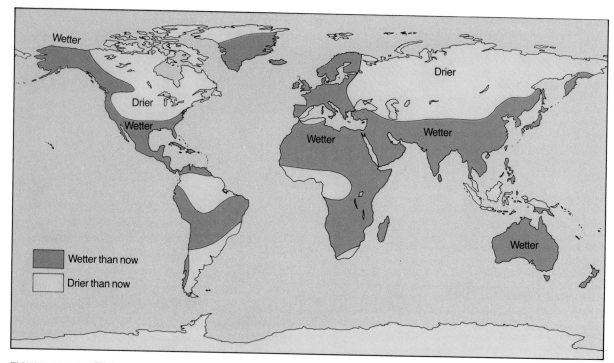

FIGURE 18.43 *This map reconstructs climate conditions 5000 to 8000 years ago, when the Earth was warmer than it is now.*

sorbed in the atmosphere. The energy absorbed by the Earth is eventually re-emitted to the atmosphere as infrared radiation. A large portion of this infrared energy is reabsorbed by the atmosphere and is conserved, with the result that the surface of the Earth is warmer than it would otherwise be.

Some molecules in the atmosphere absorb infrared radiation, and others do not. Oxygen and nitrogen, which together compose almost 99 percent of the total composition of dry air at ground level, do not absorb infrared. Molecules of water, carbon dioxide, methane, and ozone do absorb infrared and thereby warm the atmosphere. Water plays the major role in absorbing infrared because it is so abundant. Carbon dioxide is also important because it cycles in both natural and industrial systems, as shown in Figure 18.44. The natural cycles are quite varied. Only a small fraction of the total carbon near the Earth's surface is present in the atmosphere at any one time. In addition to the available pool of atmospheric carbon, large quantities of this element are (1) dissolved in seawater, (2) combined with other elements to form certain types of rocks, and (3) combined with other elements to form plant and animal tissue.

Thus:

$$\text{atmospheric } CO_2 \longleftrightarrow CO_2 \text{ dissolved in seawater}$$

$$\text{atmospheric } CO_2 \longleftrightarrow \text{carbonate rocks}$$

$$\text{atmospheric } CO_2 \longleftrightarrow \text{plant and animal tissue}$$

Each of the equations is written with arrows going both ways, for all of these reactions can proceed in either direction. Atmospheric carbon dioxide can dissolve in seawater; conversely, dissolved carbon dioxide can escape into the atmosphere. Similarly, gaseous carbon dioxide can react to form carbonate rocks, and carbonate rocks can react to form carbon dioxide gas. During photosynthesis, carbon dioxide gas is converted into plant tissue, and during respiration, plant and animal tissue react to form carbon dioxide and other products. Various physical factors such as pressure, temperature, and atmospheric composition affect all of these reactions. In addition, biological factors such as the vitality of ecosystems are important. The net result is that it is extremely difficult to predict how atmos-

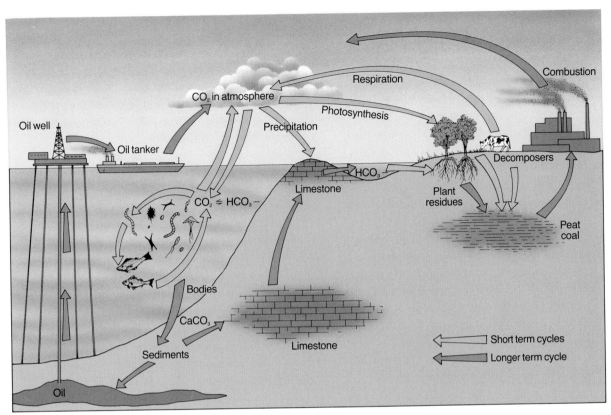

FIGURE 18.44   *A simplified carbon cycle. Carbon passes through the processes indicated by the lighter colored arrows much more rapidly than through those indicated by the darker colored arrows.*

pheric carbon dioxide levels are likely to change. Some scientists speculate that the Earth's systems are so well balanced that perturbations are compensated for. For example, if human activities release carbon dioxide into the atmosphere, most of the excess dissolves in the oceans or is converted to solid forms such as limestone. Other scientists disagree and argue that the entire balance of the Earth may be disrupted by a relatively small initial change.

One possible mechanism for such a change is outlined here: Suppose that for some reason the temperature of our planet were to rise by a few degrees. In the absence of other factors, this small increase in temperature would cause some of the carbon dioxide that is now dissolved in the oceans to be released to the atmosphere. But atmospheric carbon dioxide absorbs infrared and would warm the Earth further. Carbon dioxide is not the only compound that absorbs infrared radiation. Water is another important component of the climate cycle. If the Earth becomes warmer, there will be in-

creased evaporation from the oceans, thereby increasing the quantity of water vapor in the atmosphere. Because water vapor also absorbs infrared, this factor would add still further to a global warming. Thus the original temperature rise, which was caused by some totally unrelated factor, would be amplified by the increase in the concentrations of various atmospheric gases. A spiral could be envisioned whereby increased warming leads to more infrared-absorbing gases emitted into the atmosphere, which then leads to even a warmer climate. Scientists believe that a spiral of this sort has occurred on the planet Venus. Venus is similar to the Earth in many ways, but it is a little closer to the Sun. As a result, it would naturally receive more solar energy and be a little warmer than the Earth. This relatively small amount of additional heat has caused carbon dioxide and other compounds present on Venus to be released as gases. In turn, these gases absorbed infrared, causing the temperature to rise even more. Today the average tem-

FOCUS ON
VOLCANOES AND WEATHER

When Mount St Helens erupted on May 18, 1980, the ash darkened the sky over a vast area. In Yakima, Washington, 140 km from the mountain, drivers had to turn their car headlights on at noon. In Missoula, Montana, 620 km from Mount St. Helens, people observed an eerie darkness, or dry fog. Because clouds of volcanic ash and gas reflect light and heat from the Sun out into space, cooling and darkening occur on the Earth's surface. The immediate effects of a major volcanic eruption are easy to document. However, it is more difficult to assess the long-term climatological effects.

The largest volcanic eruption in recent history occurred in 1815 when Mount Tambora in the southwestern Pacific Ocean exploded, ejecting approximately 100 times as much magma and ash into the atmosphere as did Mount St. Helens. The following year was one of the coldest years in history and has been recorded as the "year without a summer" and "eighteen hundred and froze to death." Crop failures (compounded by the devastation caused by the Napoleonic Wars) led to widespread famine in Europe, a period that has been called "the last great subsistence crisis in the Western world." The question remains: Was this cold period caused by the Mount Tambora eruption, or was it merely a coincidence that the two events occurred together?

The figure shows a plot of global temperatures before and after eight major volcanic events that have occurred in recent times. This graph clearly shows a statistical correlation between global cooling and volcanic eruptions. Meteorological models also demonstrate that high altitude dust and gas act as an umbrella to prevent sunlight from reaching the Earth. Some scientists think that the eruptions of Mount Pinatubo in the Philippines in 1991 may have cooled the Earth by a few tenths of a degree Celsius.

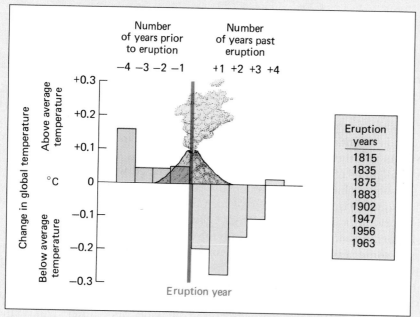

Temperature changes in the Northern Hemisphere in the 4 years immediately before and after eight large recent eruptions. (Michael Rampino, *Annual Review of Earth and Planetary Science* 16 (1988): 73–99)

perature on the surface of Venus is approximately 500°C, hot enough to melt lead.

Climate Change and Dust

About 65 million years ago, one-fourth of all known animal species and an uncounted number of plant species became extinct. Included in the list are the dinosaurs as well as many less conspicuous creatures such as species of clams, fish, single-celled plankton, and insects. Many theories proposed for this period of extinction focused on the demise of the dinosaurs alone and made no attempt to explain why numerous types of smaller creatures with vastly different food supplies, habitats, and reproductive mechanisms should accompany the larger reptiles into oblivion. In recent years, a plausible answer has been developed for the entire ecological catastrophe. This theory states that the extinctions were caused by a rapid and catastrophic climate change that was caused by the injection of large quantities of dust into the atmosphere. The dust blocked so much sunlight from reaching the Earth that plants died, rivers froze, and many entire species perished. There are two different theories about the original source of the dust. According to one, a giant meteorite or a rain of comets struck the Earth with such powerful impact that rock exploded, shooting dust and debris into the air. This dust cloud was carried by high altitude winds and dispersed throughout the upper atmosphere. Other scientists claim that the best evidence indicates that the dust was produced during a period of intense volcanic activity. A single volcano can eject large quantities of dust into the upper atmosphere. If many eruptions occurred in a geologically short period, the accumulated dust would be sufficient to cause the catastrophic cooling that is believed to have occurred.

Which of these theories is correct? Rocks that were formed just about the time of this extinction contain 100 times higher concentrations of the element iridium than do rocks that were formed just before or just after the extinction. Iridium is extremely rare on the surface of the Earth, but it is much more abundant in stony meteorites. Thus, the geochemical evidence indicates that a meteorite impact coincided with the extinction. Some geologists theorize that the meteorite impact blasted a hole or several holes in the crust, and that hot magma rose through these holes to form volcanic eruptions. Thus the meteorite theory and the volcano theory may be interrelated.

Although no cataclysmic volcanic eruptions have occurred in historical times, meteorologists believe that the largest ones have altered climate. This topic is discussed in the Focus On box.

Change of Climate and the Earth's Orbit

Many climatic variations occur in regular, periodic intervals, although the time period for these cycles differs widely. For example, the temperature of the Earth is known to oscillate on 23,000-, 41,000-, and 100,000-year cycles. During the cooler portions of these cycles, huge glaciers have advanced across the continents. Many different theories have been proposed to explain the advance and retreat of Ice-Age glaciers. Some scientists believe that there is a relationship between variations in the Earth's orbital path around the Sun and the Ice Ages. The graph in Figure 18.45 shows a possible correlation between the shape of the Earth's orbit and global temperature. If this theory is used to extrapolate future trends, a glacial period could begin in the next few thousand years.

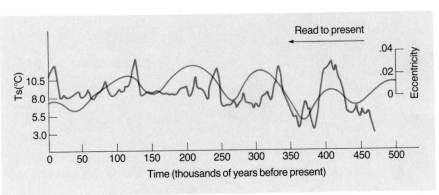

FIGURE 18.45 *Relationship between observed temperature (red line) and temperature changes predicted by calculating changes in the Earth's orbit (blue line).*

Change of Climate and the Sun

Astronomers have noted that the energy output of some stars varies with time, and there is increasing evidence that our Sun's output may vary as well. If this is true, some long-term climatic variations may be explained quite simply. When the Sun emits more energy, the Earth's climate becomes warmer; when the Sun cools, so does the planet. Unfortunately, not enough data are available to prove or to disprove this theory.

18.13 HUMAN ACTIVITIES AND GLOBAL CLIMATE CHANGE

Deforestation

A natural temperate forest system affects climate in many ways. The shade provided by the trees maintains a cool environment at ground level. In addition, a temperate forest floor is a thick bed of partially decayed leaves, needles, and rotting wood. This spongy mass retains moisture, absorbing excess water during the wet seasons and releasing it slowly through evaporation and runoff during drier months. Think of the difference between a parking lot at one extreme and a woodland at the other.

During even a mild rainstorm, puddles of water sit in depressions of the parking lot while tiny rivers and streams pour off high places. When the Sun comes out, this surface water disappears quickly, and usually the area is dry and hot within a few hours. Fresh rainwater penetrates into a forest floor and is retained in the soil and organic matter for long periods. Thus, a forest maintains a cool, moist environment. When a forest is cut and converted to a field of grain, many of these natural control mechanisms are disturbed. The shade, of course, is reduced. When soil is plowed and in some instances left bare for long periods, much of the organic matter is lost; poorly tended fields can be more similar to a parking lot than to the original forest system. Thus, as compared with a forest, a field tends to be hotter in the summertime and more vulnerable to cycles of flood and drought.

Greenhouse Effect

As previously discussed, some molecules in the atmosphere absorb infrared radiation and others do not (Fig. 18.46). Water absorbs infrared and, because it is so abundant, plays a major role in Earth's temperature balance. Four other gases—carbon dioxide, methane, chlorofluorocarbons, and nitro-

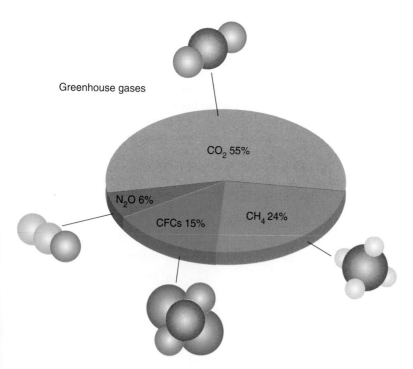

Greenhouse gases

CO_2 55%

N_2O 6%

CFCs 15%

CH_4 24%

FIGURE 18.46   *Greenhouse gases. Drawings of molecular models of the gases are included.*

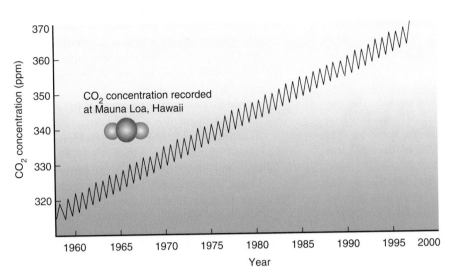

FIGURE 18.47   *Increase in atmospheric carbon dioxide concentration from 1955 to 1995. The small oscillations in the curve are caused by seasonal changes in plant respiration and photosynthesis.* (NOAA)

gen oxides—are called the **greenhouse gases** because they also absorb infrared.

Carbon dioxide is released into the air whenever any organic matter is burned or is consumed and digested by an animal. Four representative transformations are as follows:

1. Carbon (coal) $+ O_2 \xrightarrow{\text{burns}} CO_2$

2. Petroleum $+ O_2 \xrightarrow[\text{or rots}]{\text{burns}} CO_2 + H_2O$

3. Wood $+ O_2 \xrightarrow{\text{burns}} CO_2 + H_2O$

4. Sugar $+ O_2 \xrightarrow{\text{respiration}} CO_2 + H_2O$

Today, people are introducing large quantities of carbon dioxide into the atmosphere in two ways. Carbon dioxide is released whenever fossil fuels are burned. Global fossil fuel consumption is so large that the carbon dioxide emitted affects atmospheric composition. Deforestation also contributes. When a forest is cut, the logs are used for building or paper manufacture, but the branches and leaves are often burned. In addition, organic matter on the forest floor decomposes rapidly when it is disturbed by heavy machinery and exposed to air. Both burning and rotting release carbon dioxide.

The carbon dioxide concentration in the atmosphere has increased from about 290 ppm (parts per million) in 1870 to 350 ppm in 1990 (Fig. 18.47). Since carbon dioxide absorbs infrared, increased quantities of this gas could lead to global warming. Some scientists estimate that the carbon dioxide concentration could increase enough by the year 2040 to warm the Earth by as much as 2°

to 3°C. Growing seasons in the Northern Hemisphere may be prolonged by as much as 2 or 3 weeks, but, as mentioned previously, a warming trend may lead to drought conditions in many temperate zones. A significant global warming might also melt the polar icecaps sufficiently to raise the ocean level and flood coastal cities.

Methane, the second most influential greenhouse gas, is released whenever organic material rots. Both natural ecosystems and human activities produce large quantities of methane. Large quantities are released in Asia, where manure decomposes rapidly in flooded rice paddies. Additional quantities are produced by the bacteria in cows' digestive tracts and by termites that consume decaying wood. Incomplete combustion of fossil fuel also releases methane.

Dust

As mentioned previously, gases and small particles of dust injected into the stratosphere by volcanic activity tend to reflect solar radiation and cause a cooling effect. What, then, of the air pollutants injected into the atmosphere by human activity? These materials are derived from a variety of sources, including dust from agriculture and smoke from the incomplete combustion of coal, oil, wood, and garbage. Of course, much of the larger particulate matter falls to the ground or is washed down by rain. However, there is persistent introduction of small particles into the upper atmosphere. Recent research has indicated that industrial dust and sulfur aerosols may lead to global cooling. This cooling may oppose greenhouse warming, but the balance is delicate.

Dust that settles to the ground may also affect climate. Most of us have seen how snow in the city can become dirty after a few days. If you live in a rural area where snowfall is common, spread a thin layer of ashes on a 1 square meter section of snow on a warmish, sunny, winter day. By evening, you will notice that the snow under the ashes has melted faster than the snow nearby. The dark ash absorbs sunlight (lowers the snow's albedo) and causes the snow to melt. If increasingly larger quantities of dust from industry and agriculture were to settle on snow packs over wide areas, a change in global climate might possibly occur.

Chemical Pollutants and Ozone

In Chapter 17, we learned that various industrial pollutants serve as catalysts in the destruction of atmospheric ozone. In turn, if the ozone layer is destroyed, abnormal amounts of solar ultraviolet radiation penetrate the lower atmosphere. It is uncertain how this affects climate, although it is known that ultraviolet light can promote skin cancer and reduce the growth of certain crops.

18.14 NUCLEAR WARFARE AND CLIMATE CHANGE

Earlier in this chapter, it was pointed out that scientists believe that a cloud of dust ejected into the upper atmosphere by volcanic eruptions or an exploding meteorite led to the extinction of about two-thirds of all the species of animals alive at that time. What, then, about the dust that would be raised by multiple explosions in the event of a nuclear war? In October, 1983, an international conference of 100 atmospheric physicists and biologists from the United States, Western Europe, and the Soviet Union studied the problem of nuclear warfare. This group focused on the atmospheric and ecological consequences of such a war and not on the direct effects of the blasts and the ionizing radiation generated. The conclusions of the study are summarized:

1. Multiple ground-level nuclear blasts would lift phenomenal amounts of finely pulverized soil particles into the atmosphere. This soil would be accompanied by soot from fires initiated by the blasts. The soil and soot would be concentrated enough to block out 95 percent of the normal solar radiation. Temperatures in the Northern Hemisphere would plummet to $-25°C$ ($-13°F$) even if the war occurred during summertime. Crops and natural ecosystems would die, and billions of humans as well as an uncountable number of animals would starve to death.

2. The intense heat of the blasts would vaporize a large variety of different materials, which would then be carried aloft by dust clouds. Many of these would include industrial chemicals of all sorts. Some industrial chemicals such as pesticides, certain solvents, and a great many other compounds are already poisonous. Others are benign in their present composition but would be converted to poisons if heated. For example, polyvinyl chloride (PVC) is a clear plastic used for a variety of purposes, including the construction of shatterproof bottles for shampoo, cleansers, and other consumer products. PVC is an inert plastic, suitable as a general packaging material; but if it is heated or burned, toxic fumes containing hydrochloric acid and dioxin are released. Today, large quantities of different chemical compounds and products are integrated throughout our society. In the event of a nuclear war, many of these would be blasted into the atmosphere as vapors or aerosol particles and then dispersed over the entire globe in the form of deadly acid and chemical rains.

3. The heat of the blasts would convert large quantities of atmospheric nitrogen to nitrogen oxides. In turn, these compounds would destroy much of the ozone layer, exposing survivors and ecosystems to high levels of ultraviolet radiation. (See discussion of the ozone layer in Chapter 17.) In conclusion, an article published in *Science* in 1983 and co-authored by 20 prominent scientists states:

The extinction of a large fraction of the Earth's animals, plants, and microorganisms seems possible. The population size of Homo sapiens *conceivably could be reduced to prehistoric levels or below, and extinction of the human species itself cannot be excluded.* *

SUMMARY

One theory states that the composition of the Earth's atmosphere is maintained by living organisms. **Atmospheric pressure** is the weight of the atmosphere per unit area. Pressure decreases steadily with altitude, but temperature varies. The layers of the atmosphere, starting with the one closest to the Earth, are the **troposphere, the stratosphere, the mesophere,** and the **thermosphere.**

The energy balance of the Earth is achieved by an opposition of energy inflow and outflow to and from ground level and all layers of the atmosphere. The temperature gradient from the Equator to the poles and the change of seasons are caused by the spherical nature of the Earth and the tilt of the Earth on its axis.

Wind is a convection current powered by uneven heating of the Earth's surface. There are six major wind belts on the Earth. The movement of wind is affected primarily by convection and the **Coriolis effect,** which causes objects to be deflected toward the right as seen along the direction of motion when moving in the Northern Hemisphere. The basic wind systems are the **polar easterlies,** the **trade winds,** and the **westerlies.** The **doldrums** is a region around the Equator where heated air is rising. This air falls at the **horse latitudes.** Between the two, the **trade winds** blow in a steady, predictable pattern.

The three basic types of clouds are **cirrus, stratus, and cumulus. Condensation** occurs when moist air is cooled below its **dew point.** Condensation can lead to the formation of clouds, and when clouds form, there is a chance for precipitation to fall in the form of rain, snow, sleet, or hail.

Many wind systems arise because the temperature of the oceans changes slowly in response to a change in solar radiation, whereas the temperature of the land surfaces changes much more quickly.

Stormy weather generally occurs when warm and cold air masses collide. When two air masses collide, the warm air often rises, leading to low pressure, cloud formation, and often precipitation.

Climate change is affected by change in atmospheric composition, dust, changes in the Earth's orbit, and changes in radiation output of the Sun. Human activities such as burning fuels, cutting forests, or introducing dust and pollutants can alter climate. Nuclear war could potentially destroy life on Earth by altering climate catastrophically.

EQUATIONS TO KNOW

$$\text{Relative humidity } (\%) = \frac{\text{actual quantity of water per unit of air}}{\text{saturation quantity at the same temperature}}\ 100\%$$

KEY WORDS

Barometric pressure	Equinox	Humidity	Cumulus
Barometer	Coriolis effect	Absolute humidity	Nimbus
Troposphere	Polar easterlies	Relative humidity	Climate
Stratosphere	Trade winds	Dew	Weather
Mesosphere	Horse latitudes	Dew point	Sea breeze
Thermosphere	Doldrums	Adiabatic cooling	Monsoon
Albedo	Westerlies	Cirrus	Air mass
Solstice	Jet stream	Stratus	Front

*See R.P. Turco, O.B. Toon, T.P. Ackerman, J.B. Pollack, and C. Sagan: "Nuclear Winter: Global Consequences of Multiple Nuclear Explosions." *Science,* Vol. 222, Dec. 23, 1983, p. 1283ff. Also see P. Ehrlich, et al: "Long Term Biological Consequences of Nuclear War." *Science,* Vol. 222, Dec. 23, 1983, p. 1293ff.

CONCEPTUAL QUESTIONS

Evolution of the atmosphere

1. How did the primitive atmosphere differ from our atmosphere today?

2. Humans could not survive in the Earth's primitive atmosphere, yet life as we know it could not have evolved in the present one. Explain and discuss.

3. Explain why the compositions of the rocks on Earth, Venus, and Mars are all much more similar than the compositions of the atmospheres on the three different planets.

4. Many scientists are alarmed because industrial activities are altering the composition of the modern atmosphere. If people did not exist on Earth, would you expect the atmospheric composition to remain constant? Discuss the implications of your answer.

The atmosphere

5. Explain how plants help to maintain an atmosphere that can support animal life.

6. List the four primary layers of the atmosphere. Discuss the physical properties of each.

7. Imagine that enough matter vanished from the Earth's core that the mass of the Earth were reduced to half its present value. In what ways do you think the atmosphere would change? Would the normal pressure at sea level be affected? Would the thickness of the atmosphere change? Would more molecules be lost to outer space? Explain.

8. Climbers on high mountains must wear dark glasses to protect their eyes; also they get sunburns even when the temperature is below freezing. Explain.

9. Explain why mountaintops are generally colder than valleys, even though the mountains are closer to the Sun.

10. The temperature is quite high in the thermosphere, but little thermal energy is stored there. Explain.

Energy balance of the Earth

11. Refer to Figure 18.9a. What percent of the incident solar energy is received by the Earth? (b) Is the Earth growing warmer or colder, or is the global temperature fairly constant? Explain.

12. Assume that ocean circulation changed so that climate in North America became colder and wetter. Alpine glaciers would start to grow and flow farther into the valleys. Explain how the growing glaciers would themselves influence climate. Discuss a possible feedback loop between climate and glacial expansion.

13. If we lived on the surface of a flat Earth, would different regions experience different climates or similar ones? In answering, assume that the flat Earth is tilted 23.5° with respect to the plane of its orbit.

14. If the North Pole receives the same number of hours of sunlight per year as do the equatorial regions, why is it so much colder at the North Pole?

Wind

15. In what way are the winds analogous to a heat engine? What is the energy source that powers the wind?

16. Define pressure gradient, isobar, high pressure, and low pressure. Discuss the relationships among these terms.

17. If you were firing a long-range rocket and aimed it due north at a target due north of your launching pad, would you score a hit or a miss? What would happen if you fired due west at a target located due west? Explain.

18. Figure 18.18 refers to cyclones and anticyclones in the Northern Hemisphere. Draw a figure to illustrate cyclones and anticyclones in the Southern Hemisphere.

19. Discuss the scientific principle behind the Hadley model of global wind systems. What factor did Hadley ignore? How did that omission affect his conclusion?

20. What is a trade wind? Why are trade winds so predictable?

21. Why is the doldrum region relatively calm and rainy? Why are the horse latitudes calm and dry?

22. Would the exact location of the doldrum low pressure area be likely to change from month to month? From year to year? Explain.

23. Sailors traveling in the Northern Hemisphere expect to incur predictable winds from the northeast between about 5 and 30° north latitudes. Should airplane pilots expect northeast trade winds while flying at high altitudes in the same region? Explain.

Condensation

24. Discuss the difference between relative humidity and absolute humidity.

25. Which of the following conditions will produce frost? Which will produce dew? Explain. (a) A constant temperature throughout the day. (b) A warm summer day followed by a cool night. (c) A cool fall afternoon followed by a freezing temperature at night.

26. Explain why clouds frequently form when air rises. Discuss another mechanism that can lead to cloud formation.

27. Draw a picture of a cirrus, a stratus, and a cumulus cloud. If you were planning a picnic and looked out the window, which type of cloud would you prefer to see? Which type might cause you to alter your plans?

Climate

28. Discuss the effect of the oceans on world climate. Would it be correct to say that coastal regions are always warmer than inland areas? Explain.

29. Would a large inland lake be likely to affect the climate of the land surrounding it? Deep lakes seldom freeze completely in winter, whereas shallow ones do. Would a deep lake have a greater or a lesser effect on weather than a shallow one? Explain.

30. Would sea breezes be more likely to be strong on an overcast day or on a bright sunny one? Explain.

31. What is a monsoon? How is it formed? At what time of the year do monsoons generally bring rain?

32. Describe the difference between weather and climate. Which is more predictable?

Weather

33. What is an air mass? Describe what would be likely to happen if a polar air mass collided with a humid subtropical air mass.

34. Discuss four conditions that can lead to low pressure storms.

35. Explain why a wide band of rainy weather forms along a warm front, whereas a narrow band of more violent weather often develops along a cold front.

Climate change

36. Explain why a 2°C drop in global temperatures would be alarming.

37. Name four factors that may cause climate to change. How many of these factors are at least partially controlled by people? Discuss.

38. Discuss four ways in which human activities may be changing world climate. In each instance, explain what types of activities are causing the potential disruption and by what mechanism the climate may be changed.

39. Compare climate change in the past century with some extreme climate changes that have occurred in the past. Discuss the implications of your comparison.

CHAPTER 19

EARTH

A typical scene on Earth during the Jurassic Period.
(Courtesy The Smithsonian Institution)

19.1 INTRODUCTION

Imagine yourself on a rocky beach, walking toward the surf. You can see, hear, and feel the wind and the water. But you do not see the cliffs move, and the Earth does not shake under your feet. The solid Earth seems to be a firm base beneath the blowing winds and the breaking waves. This apparent rigidity is deceptive, however—Earth's crust is actually dynamic, not static. Continents move, mountains rise and fall, and rocks flow or are pushed from place to place. These movements escape most casual observations because they are generally slow, although every year volcanic eruptions, earthquakes, and other types of rapid movement occur somewhere on our planet.

There are two types of movement of solid material that affect our environment. Naturally occurring phenomena including mountain-building, continental migration, erosion, and other movements of large masses of materials are powered by energy sources of far greater magnitude than any that humans can harness. These phenomena will be discussed in this chapter. The second type of movement, to be discussed in Chapter 20, is initiated by human beings and involves comparatively tiny amounts of energy and relatively insignificant masses of material. We refer here mainly to mining and farming. Yet these activities, which are insignificant on a scale of global energy, are vitally important on a human scale. Farmers generally dig up less than 1 m of a planet whose radius is 6,400,000 m, and miners probe only a few kilometers downward, yet the impoverishment of soil or the depletion of mineral reserves has major technological, political, environmental, and economic consequences.

19.2 STRUCTURE OF EARTH

Earthquake Waves

In our study of the atom, we learned how scientists are able to understand the structures of objects that are too small to see. Now we ask how it is possible to study the interior of the Earth, thousands of kilometers below the deepest well. Again an analogy will be helpful. If you have ever gone to the store to buy a watermelon, you know that there is always a concern that you pick a juicy, ripe melon. The problem is that you have to make the choice without looking inside. One trick is to tap the melon gently with your knuckle. If you can hear a "sharp," "clean" reverberation traveling through the core of the melon, it is probably ripe; a dull thud indicates that it may be overripe and mushy. Of course, the words "sharp" and "clean" are not scientifically precise, but if you tap many different melons, you can hear differences. The point is that sound waves are affected by different types of liquid or solid media.

The same general technique is used to study the structure of the interior of the Earth. If you could somehow give the Earth a sharp tap, sound waves would travel through the Earth's interior. Geologists study waves in rock produced by conventional explosives, underground nuclear tests, and earthquakes. When an earthquake occurs near the surface of the Earth, built-up stress is released as huge segments of rock suddenly slip past each other. This movement initiates waves in the rock, just as a clanging bell initiates waves in water or air. By studying the properties of these waves scientists can deduce much about the structure of the Earth's interior.

The science of measuring and recording the shock waves of earthquakes is called **seismology.** Earthquakes are commonly referred to as seismic events or seismicity. The waves generated by earthquakes fall into two different categories, **surface waves** and **body waves.** Surface waves, as their name implies, travel along the surface of the Earth, and it is this type of wave that generally produces the most destruction associated with an earthquake. Body waves travel through the Earth, and it is these waves that scientists are most interested in because of what they can tell us about the interior of the Earth.

Three properties of waves provide information about the Earth's interior: speed, refraction, and behavior in liquid and solid media.

Recall from Chapter 7 that if a vibration is established in a medium, the speed of the resulting wave will be characteristic of the medium. Thus, sound travels in air with a speed of about 340 m/s. However, sound travels much faster through solids and liquids: It travels at 1500 m/s through water, 3810 m/s through marble, and 5200 m/s through iron. Thus, a measurement of the speed of a wave provides a clue to the composition of any medium through which it travels.

In 1909, seismologist Andrija Mohorovičić discovered that seismic wave velocities increase sharply at a depth varying from 7 to 70 km beneath the Earth's surface. This sudden velocity change indicates an abrupt transition in rock type at that depth. The upper layer is the Earth's crust, and the lower layer is the mantle. The boundary between the two is called the **Mohorovičić discontinuity** or the **Moho,** in honor of its discoverer.

Recall that light refracts as it travels from one medium to another. Earthquake waves also refract and, as discussed below, refraction patterns provide information about boundaries between layers within our planet.

Body waves are of two types, the P (for primary) wave and the S (for secondary) wave. A P wave is a longitudinal wave, which means that this wave causes the particles of the Earth to move back and forth along the direction of travel of the wave. You should recall from your study of physics that a sound wave is an example of a longitudinal wave. An S wave is an example of a transverse wave, meaning that this kind of wave causes the particles of the Earth to vibrate at right angles to the direction of travel of the wave. A wave on a string as shown in Figure 19.1 is another example of a transverse wave.

P waves move through solids or liquids, but S waves move *only through solids.* Because molecules in liquids and gases are only weakly bound together, they slip past each other and thus cannot transmit S waves. If the entire Earth were composed of one type of rock, P and S waves would travel everywhere and would be detected anywhere on the planet. However, Figure 19.2 shows that S waves are not detected beyond 105° from an earthquake. Because

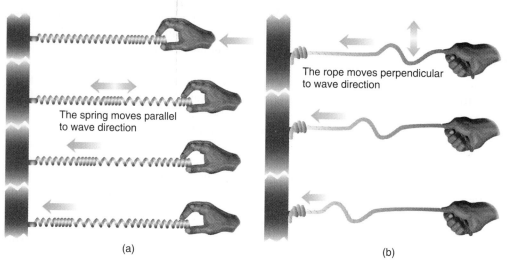

FIGURE 19.1   *Two different types of body waves travel through the Earth. Their characteristics are shown by a spring and a rope. (a) A P wave is a longitudinal wave illustrated by a stretched spring. The particles in the spring move back and forth along the direction in which the wave itself travels. (b) An S wave is a transverse wave illustrated by a rope. The particles in the rope move at right angles to the direction of travel of the wave.*

S waves travel through solids but not through liquids, these data tell us that a portion of the Earth's outer core is liquid.

Neither S nor P waves arrive in a "shadow zone" between 105° and 140° from an earthquake. Beyond 140°, direct P waves arrive, but direct S waves do

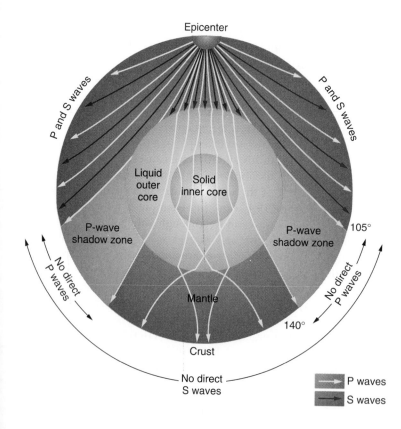

FIGURE 19.2   *Seismic waves curve gently as they pass through the Earth. They also bend sharply where they cross major layer boundaries in the Earth's interior. The blue S waves do not travel through the liquid outer core, and therefore direct S waves are observed only within 105° of an earthquake. The yellow P waves bend sharply at the core-mantle boundary to create a shadow zone of no direct P waves from 105° to 140°.*

not. The shadow zone, too, results from the change from solid rock to molten liquid at the core-mantle boundary. Earthquake waves curve gently as they refract through rocks of steadily increasing density in the mantle. P waves refract sharply as they pass across the abrupt transition from the mantle into the core. As a result, no P waves arrive in the shadow zone.

P waves refract sharply again when they pass from the outer core to the inner core, indicating another radical change in physical properties of the Earth's interior. In this case, the change in direction results from an abrupt transition from the molten outer core to the solid inner core. Thus, seismic data tell us that the core is composed of an inner solid sphere surrounded by an outer liquid shell.

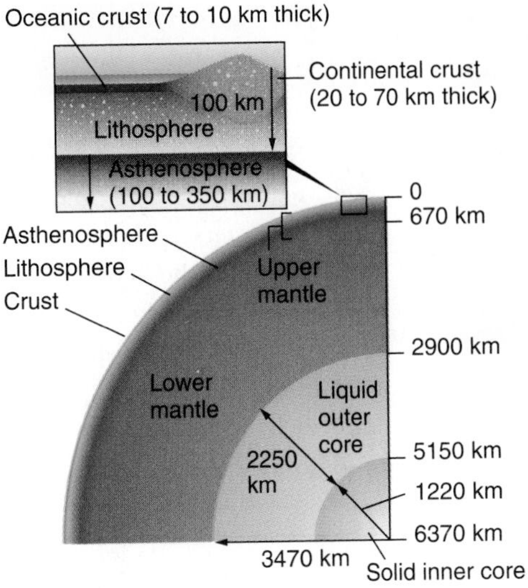

CONCEPTUAL EXAMPLE 19.1   SEISMOLOGY

Astronauts have established several permanent seismographic recording stations on the Moon. What can be learned about the structure of the Moon from these recorders? If no moonquakes occur, would you consider the experiment to be a failure? Explain.

Solution By studying moonquakes, we can learn whether or not the Moon is geologically active, and this information, in turn, will help us interpret whether the Moon's interior is hot or cold, how thick the crust is, and so forth. From these data we hope to learn something about the Moon's geological history. Few moonquakes have actually been recorded by the instruments left by Apollo astronauts; from these data we believe the Moon is internally cold. In a scientific experiment, a "no" answer can provide as much information as a "yes" answer. Thus the knowledge that the moon is not seismically active provides valuable insight into its structure and history.

The Earth's Interior

A cross-section of the Earth is shown in Figure 19.3. The **core** is predominantly iron with some nickel. The inner core is solid and is surrounded by an outer core of molten iron and nickel. A large solid layer called the **mantle** surrounds the core. The chemical composition of the entire mantle is fairly

FIGURE 19.3   *The Earth is a layered planet. The inset is drawn on an expanded scale to show near-surface layering.*

homogeneous. However, temperature and pressure generally increase with depth, and these factors affect the chemical properties of different portions of the mantle.

The thinnest Earth layer is the outer shell, called the **crust**. The crust under the ocean floor ranges from 7 to 10 km in thickness. In contrast, continental crust is much thicker than oceanic crust. The continents not only rise above the ocean floor, but also extend below it. Thus, continents have roots that protrude into the mantle. In mid-continent regions, the crust is about 20 to 40 km thick, whereas under major mountain ranges, its thickness increases to as much as 70 km.

The uppermost portion of the mantle and the crust above it are close enough to the surface that temperature and pressure are relatively low. Therefore, the rock in this region is rigid and brittle. Although the compositions of the crust and the mantle differ, the zone including crust and uppermost mantle is distinguished by its physical properties. Thus, the crust and the uppermost mantle together are called the **lithosphere** (Greek for "rock layer"). The lithosphere is about 100 km thick in most places. This lithosphere is broken into plates and rides on the semifluid layers beneath it, as we shall see in later sections.

At a depth of about 100 km, the cool, rigid, brittle rock of the lithosphere suddenly gives way

FOCUS ON
SEISMOMETERS

Earthquake waves are detected by devices called seismometers, designed as shown in the figure, and making use of the inertia of a suspended object to make their recordings. A frame and a rotating drum are supported directly by bedrock such that if the bedrock shakes, so do they. Suspended from the frame by a spring is a massive object that is capable of writing on the rotating drum either by an inked stylus or, in newer models, with a beam of light. The massive object is "suspended" in space by the spring and because of its inertia does not vibrate when an earthquake occurs. The amplitude of the vibration recorded on the drum is a measure of the strength of the earthquake.

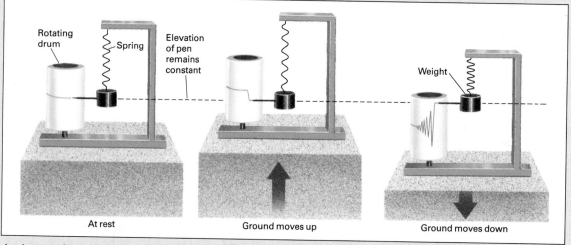

A seismograph records ground motion during an earthquake. When the ground is stationary, the pen draws a straight line across the rotating drum. When the ground rises abruptly during an earthquake, it carries the drum up with it. But the spring stretches so that the weight and pen hardly move. Therefore, the pen draws a line that drops on the scale. Conversely, when the ground sinks, the pen draws a line upward. During an earthquake, the pen traces a jagged line on the rotating drum, thus recording the up-and-down movement of the ground.

to hot, soft, plastic rock of the **asthenosphere.** The rock of this layer is so hot and plastic that it flows readily, even though it is solid. To visualize a solid that can flow, think of Silly Putty or road tar on a hot day. The asthenosphere extends from the base of the lithosphere to a depth of about 350 km.

The lithosphere is of lower density than the asthenosphere. Therefore, it floats on the asthenosphere, much as an iceberg or a block of wood floats on water. This concept of a floating lithosphere is called **isostasy** and is shown in Figure 19.4.

CONCEPTUAL EXAMPLE 19.2   THE EARTH'S CRUST

The oceanic crust is thinner than continental crust. In addition, oceanic crust is composed of basalt, which is denser than the granite that forms the con-

(a)

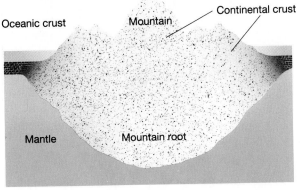

(b)

FIGURE 19.4 *The principle of isostasy. (a) The largest of the icebergs has the most material underwater and also the most above. (b) In an analogous manner, continental crust extends more deeply into the mantle beneath high mountains than it does under lower areas of the continents.*

tinents. Use this information to explain why the oceanic crust lies at a lower elevation than the continents.

Solution Even if both the ocean floor and the continents were composed of the same material, the thinner oceanic lithosphere would float lower in the asthenosphere, just as a thin iceberg sits lower in the water than a thick one does. The high density of basalt causes additional isostatic sinking of the ocean basins. Thus, to a geologist, the major difference between the ocean basins and the continents is not that one is wet and the other dry but that oceanic crust lies at a lower elevation because of its density and relative thinness.

Below the base of the asthenosphere, at a depth of about 350 km, the increasing pressure over-

whelms the effect of increasing temperature, and the mantle becomes wholly solid and less plastic. At a depth of about 670 km, pressure is great enough to cause some minerals to collapse and to form new minerals that are more dense. Thus, the chemical composition of the mantle doesn't change, but the mineral structure does. This change in mineral structure marks the boundary between the upper and lower mantle.

19.3 ROCK CYCLE

Pick up a rock that you find in your neighborhood and look at it carefully. Chances are that you will see particles of different kinds of materials. You may see a speck of pink matter, another of black, and a third that is white. A **rock** is a natural solid made up of one or more minerals or other natural solids (Fig. 19.5). A **mineral** is a naturally occurring inorganic solid that has a definite chemical composition and crystal structure. Quartz is a mineral; a piece of quartz can be a single crystal of silicon dioxide. Granite is a rock, made up of several different minerals, and there is no single chemical formula that can describe it.

Igneous Rocks

In the asthenosphere, there are pockets of molten rock, called **magma,** which have gases such as water vapor dissolved in them. When magma rises toward the Earth's surface and solidifies, it forms **igneous**

FIGURE 19.5 *Each of the colored grains in this granite is a different mineral.*

FIGURE 19.6 *A stream of molten lava and a fire fountain pour from Pu'u 'O'o vent, Hawaii, 1986.* (J.D. Griggs, U.S.G.S.)

FIGURE 19.7 *Obsidian is an igneous rock formed from lava that cools so quickly that crystals do not have a chance to form, so the rock becomes a type of glass. This material can be chipped relatively easily and formed into sharp objects such as knives and arrowheads. Obsidian was considered valuable to the Indians before Europeans brought metal tools and weapons to the North American continent.*

rock. Igneous rocks vary both in chemical composition and texture. Texture is determined by the rate of cooling and solidification, which, in turn, depends on the rock's history. When magma flows quickly to the surface of the Earth, a **volcano** is formed. The outpouring magma is called **lava** (Fig. 19.6). If lava cools within hours, the minerals do not have time to crystallize, and the resultant rock is a smooth volcanic glass (Fig. 19.7). If a magma solidifies over a few days to a few years, crystals begin to form, but they do not have time to grow to a large size. The result, as shown in Figure 19.8, is a fine-grained rock, one in which the crystals are too small to be seen with the naked eye. If magma protrudes into the crust but does not travel all the way to the surface, it will cool and crystallize over decades to thousands of years. During this time, crystals grow large enough to be seen with the naked eye (Fig. 19.9).

The most common igneous rocks are **granite** and **basalt.** The two have different chemical compositions. In addition, granite is coarse-grained rock that cooled and crystallized slowly deep in the crust, whereas basalt is a fine-grained volcanic rock that cooled quickly near the surface. Granite forms the continents, whereas the ocean floor is composed of basalt.

Sedimentary Rock

When igneous rocks reach the surface, they are exposed to surface forces. Wind, rain, freezing water, streams, glaciers, wave action, and biological processes chip off tiny pieces or dissolve the rocks. These small particles called **sediment** are carried downslope by streams and rivers, where they collect in valleys or where the river enters a lake or ocean. Sedimentary deposits are easy to find all around us. If there is a river or a creek near your home, notice that at a bend in the stream the water travels fastest on the outside of the turn and slowest on the inside, as shown in Figure 19.10. Go to where the water is slow and dig up a section of the river bottom. Most probably it contains numerous fine particles of different colors. Some are shiny, others dull, some reddish, some black, and so on. The sed-

FIGURE 19.8 *Basalt is a dark, fine-grained volcanic rock. Southeastern Idaho.*

FIGURE 19.10 *Sediment collects in places along a river where the current moves slowly. Aerial view of the Bitterroot River, southwest Montana.*

iment in your sample probably contains pieces of rock from different locations, mixed with organic debris. Thick layers of sediment are formed along deltas where rivers carry sediment to the sea or along the bottoms of the slopes of high mountains or plateaus.

If sediments are buried and compressed for long periods of time, the pieces cement together and coalesce to a solid known as **sedimentary rock,** such as that shown in Figure 19.11. Sedimentary rocks can be quite different from one another. **Limestone** is nearly pure calcium carbonate. **Sandstone** is a mixture of many small grains. Generally, most of the grains are quartz, but they may also include feldspar or, in rare instances, gold or any of a number of other types of rocks or minerals. **Shale**

is nearly pure clay, microscopic grains produced during the chemical weathering of other minerals.

Sediments can build up to impressive thicknesses. In the Grand Canyon, these layers, or strata, were exposed 1.5 km deep as the Colorado River gradually cut its way into a raised sedimentary plateau, and the strata are clearly visible today, as

FIGURE 19.11 *Most limestone is deposited as horizontal beds on the sea floor. As the rock is uplifted to form mountains, the beds are often folded or tilted. Rocky Mountains, Alberta, Canada.*

FIGURE 19.9 *Coarse-grained granite.* (Courtesy Geoffrey Sutton)

FIGURE 19.12   *Deer Creek Falls cascades into the Colorado River, Grand Canyon, Colorado.*

FIGURE 19.13   *Highly deformed gneiss, Baffin Island, Northwest Territories, Canada.*

shown in Figure 19.12. Other less visible sedimentary formations are even thicker yet. For example, a sedimentary layer that lies beneath parts of western Montana is 20 km thick.

Metamorphic Rock

Rock and sediment may also undergo geological change—**metamorphism**—if they are heated or compressed for long periods. During metamorphism, the texture of the mineral grains or their chemical composition changes. Thus, metamorphism differs from the formation of sedimentary rock in that metamorphism involves chemical and physical changes, whereas mineral grains are simply glued together to form a sedimentary rock. For example, clay, a naturally occurring component of soil, is soft and pliable. If clay is heated in a kiln, the minerals react chemically to form a hard, solid ceramic. A similar process may occur naturally. Suppose that hot magma came in close contact with a natural clay deposit. The heat from the magma would harden the clay. Many other types of heat-

hardening processes occur in the Earth. If limestone is subjected to heat and pressure, it is converted to **marble.** Granite can be metamorphosed as well. During metamorphism, the minerals in granite often react to form new minerals, and textures change. **Gneiss** forms during metamorphosis of either igneous or sedimentary rocks. Gneiss is characterized by parallel layers of different compositions, such as is shown in Figure 19.13. Note that in the examples given here, the minerals are not melted. Heat, pressure, and fluids alter the texture or chemical composition (or both) and form new minerals in the affected rock and generally increase its hardness. The resulting product is called **metamorphic rock.**

Mineral matter is slowly but continuously being changed from one form to another. Igneous rocks are broken apart to form sediments and then, depending on conditions, may be converted to sedimentary and then to metamorphic rock. Similarly, metamorphic rock may erode and be deposited as sediments. In addition, under certain conditions, surface rocks are slowly forced downward by geological action, where they are heated and melted. Once molten, the liquid magma mixes with other minerals and undergoes physical and chemical changes. This material may return to the surface millions or hundreds of millions of years later as newly formed igneous rocks. Thus, crustal material is formed, altered, and removed in a slow but continuous cycle called the **rock cycle** (Fig. 19.14).

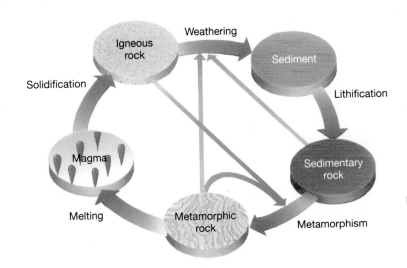

FIGURE 19.14   *The rock cycle shows that rocks change continuously over geological time. The arrows show paths that rocks can follow as they change.*

19.4 GEOLOGICAL TIME

Earth is some 4.6 billion years old, and parts of the solid crust today are close to 4 billion years old. Time spans of billions of years are difficult to comprehend. To grasp the concept of a billion, think first of a single coin, say a penny. A penny is about 2 mm thick. If a billion pennies were stacked in a pile, the pile would be 2000 km high.

How can scientists measure geological time? How do they know when certain events in the distant past occurred? The first studies of geological time were conducted on a relative basis. For example, imagine a series of layers of sedimentary rocks. Because the sediments were laid down in sequence one on top of another, one could presume that the deepest layer is the oldest and that the surface layer is the youngest. If an entire sedimentary formation lies on top of a bed of metamorphic material, it logically follows that the metamorphic rock is older than any of the sediments above it. Figure 19.15 shows a segment of the Grand Canyon in which we can clearly see the layers of sedimentary rock lying on the metamorphic rocks. The concept that rocks are deposited in sequential order is known as the **principle of superposition.** Although this principle is useful in a great many situations, the age of rocks is not *always* directly related to their position. For example, molten magma could push upward into the center of an older sedimentary layer, and in this situation, the youngest rocks would lie beneath older ones, as shown in Figure 19.16.

A relative time scale establishes a chronological sequence but does not specify absolute age. To measure the age of a rock in terms of years, one must search for some internal clock within the rock

FIGURE 19.15   *Sedimentary rocks lying over metamorphic rocks in the Grand Canyon. The sedimentary rocks are distinguished by the horizontal layers, called bedding. The metamorphic rocks near river level are considerably older than the sedimentary rocks above.*

FIGURE 19.16   *A basalt dike cutting across sedimentary rocks in the walls of Grand Canyon. The sedimentary rocks must be older than the dike.*

itself. One useful natural clock measures time by the process of radioactive decay. Consider a radioactive isotope such as uranium-238. As discussed in Chapter 13, a given radioactive isotope decomposes at a specific rate known as its half-life. Additionally the decay process produces a known set of products. For example, uranium-238 decays to form other radioactive isotopes, which decay relatively rapidly in turn to produce lead-206. The half-life for uranium-238 is 4.5 billion years. Therefore, if a sample of rock contains equal quantities of uranium-238 and lead-206, that rock is 4.5 billion years old. The absolute age of rocks is determined by studying the radioactive decay patterns of uranium-238 and a variety of other isotopes with long half-lives, especially potassium-40.

The geological time scale is shown in the figure on page 471. To put our own existence into perspective, imagine that the entire history of the Earth were recorded on a linear scale 1 m long. The early history, largely devoid of anything but one-celled life, would occupy 86 percent, or 86 cm, of that scale. The great coal-producing swamps and marshes would appear 5.5 cm from the end and the extinction of the dinosaurs at the 1.4 cm mark; the latest Ice Ages would be found in the last 0.03 cm; and recorded history would be squeezed into the

latest final 0.003 cm, a zone thinner than the diameter of the period at the end of this sentence.

19.5 HISTORICAL GEOLOGY

Let us pause in our discussion of the structure of the Earth and its rock formations to examine the various life forms that have inhabited this planet, when they arose, and, in a few cases, the causes for their demise. If we are to understand what may befall the inhabitants of Earth in the future, it is important to understand what has happened in the past.

Geological time is subdivided into time units called **eons,** which are divided further into **eras.** Eras are subdivided in turn into **periods** (the most commonly used time unit), which are divided even more finely into **epochs.** The subdivisions of geological time are based largely on fossils found in rocks that formed during each interval. Phanerozoic time is finely subdivided because fossils are abundant in rocks of that eon. Earlier eons are not subdivided because life was scarce, and, hence, fossils are uncommon.

Earliest Eons of Geological Time: Precambrian Time

The three earliest eons together are called **Precambrian time** because they preceded the Cambrian Period, when fossil remains first became abundant. The Earth formed about 4.6 billion years ago. For 100 to 900 million years, the surface was too hot for life to exist. The earliest fossils date to about 3.5 billion years ago, during the **Archean Eon** (Fig. 19.17). Thus, life formed within roughly 300 million to 1 billion years after the Earth's surface cooled. During the next 3 billion years, organisms remained predominantly single-celled and microscopic.

Rocks of the **Proterozoic Eon** (Greek for "earlier life"; 2.5 to 0.57 billion years ago) contain abundant fossil algae and other simple plants. Proterozoic animal fossils lacked hard parts, such as shells and skeletons. Consequently the fossils consist of imprints of their soft bodies in shale. Then within a very short time, at the end of the Proterozoic Eon and the start of the Phanerozoic Eon, large, multicellular plants and animals evolved abruptly, proliferated, and dominated the Earth.

Time Units of the Geologic Time Scale				Distinctive Plants and Animals	
Eon	**Era**	**Period**	**Epoch**		
Phanerozoic Eon (*Phaneros* = "evident"; *Zoon* = "life")	Cenozoic Era	Quaternary	Recent or Holocene	"Age of Mammals"	Humans
			Pleistocene —2—		
		Tertiary — Neogene	Pliocene —5—		Mammals develop and become dominant
			Miocene —24—		
		Tertiary — Paleogene	Oligocene —37—		
			Eocene —58—		
			Paleocene —66—		Extinction of dinosaurs and many other species
	Mesozoic Era	Cretaceous —144—		"Age of Reptiles"	First flowering plants, greatest development of dinosaurs
		Jurassic —208—			First birds and mammals, abundant dinosaurs
		Triassic —245—			First dinosaurs
	Paleozoic Era	Permian —286—		"Age of Amphibians"	Extinction of trilobites and many other marine animals
		Carboniferous — Pennsylvanian —320—			Great coal forests; abundant insects, first reptiles
		Carboniferous — Mississippian —360—			Large primitive trees
		Devonian —408—		"Age of Fishes"	First amphibians
		Silurian —438—			First land plant fossils
		Ordovician —505—		"Age of Marine Invertebrates"	First fish
		Cambrian —570—			First organisms with shells, trilobites dominant
Proterozoic		Sometimes collectively called Precambrian —2500—			First multicelled organisms
Archean		—3800—			First one-celled organisms
Hadean		—4600 ±—			Approximate age of oldest rocks / Origin of the Earth

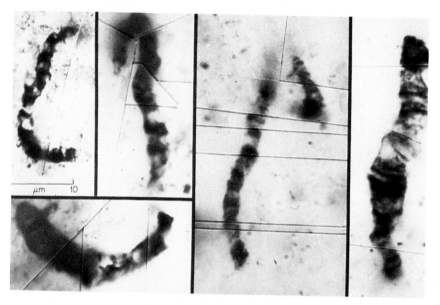

FIGURE 19.17   *Tiny, 3.5-billion-year-old, bacteria-like fossils from western Australia.* (J.W. Schopf and B.M. Packer)

This sudden explosion of life is a puzzle. Why did it take much longer for multicellular organisms to evolve from single-celled organisms than it did for life to evolve from nonliving molecules? Why have multicellular organisms existed for only the most recent 15 percent of the time that life has existed on the Earth? One theory is that the Earth's early atmosphere contained only a small amount of

FIGURE 19.18   *In Paleozoic time, the sea teemed with a wide variety of life.* (Smithsonian Institution)

oxygen. Although single-celled organisms could thrive in this environment, multicellular ones could not evolve until the oxygen concentration increased to suitable levels.

Phanerozoic Eon: The Last 570 Million Years

Four dramatic evolutionary changes occurred at the beginning of the Phanerozoic Eon:

1. Animals evolved shells and skeletons. Shells and skeletons are much more easily preserved than plant remains and soft body tissues. Thus, in rocks of earliest Phanerozoic time and younger, the most abundant fossils are the hard shells and skeletons.
2. The total number of individual organisms increased greatly.
3. The total number of species increased greatly.
4. The sizes of individual organisms increased from microscopic to macroscopic.

Paleozoic Era.　　The early Paleozoic oceans were dominated by snail-like gastropods, worms, brachiopods that looked like clams and oysters, colonies of corals, and horseshoe crab–like trilobites. Figure 19.18 is an artist's representation of these life forms. Algae and other simple plants shared the sea floor with these animals.

A few Paleozoic fossils exhibit a new, exciting adaptation. Worm-shaped animals called **chordates** had a primitive nerve cord along their backs. Muscles were arranged along the nerve cord in a series of V-shaped patterns. The muscles, controlled by the central nerve, propelled the animal through the water. By late Cambrian time, chordates had evolved into the first primitive fish.

During Cambrian time, algae spread into coastal swamps and eventually adapted to life in freshwater lakes and streams. Then, slowly, plants spread onto land during the Silurian or Ordovician Period.

Meanwhile, back in the ocean, simple chordates evolved into fishes with strong, bony skeletons. Although modern fish absorb oxygen from the water through gills, many Paleozoic fish had both gills and auxiliary lungs. The lungs enabled them to burrow into the mud and survive drought or periods of desiccation between high tides. During Devonian time, some of these bony lung fish

dragged themselves onto land and evolved into amphibians.

Amphibians prospered during the Carboniferous Period, between 360 and 286 million years ago. They were considerably more agile and better adapted to land than their lung fish predecessors, but they still needed to return to water to lay their unprotected eggs. The larvae hatched as fish-like organisms that metamorphosed into adults. Approximately 70 million years after the first amphibians, reptiles evolved that laid eggs enclosed by a waterproof exterior. The embryos passed through developmental stages within the protection of these eggs and hatched in a more adult-like form. Because the eggs were protected, the animals did not have to return to water to lay eggs. Reptiles, as shown in Figure 19.19, were the first large animals

FIGURE 19.19 *Paleozoic reptiles and amphibians in a landscape covered by ferns, ginkgoes, and conifers.* (Ward's Natural Science Establishment)

adapted fully to life on land. Alligators, snakes, and lizards are examples of modern reptiles. Paleozoic reptiles continued to evolve through the Permian Period, but their real dominance came during the Mesozoic Era.

During the Carboniferous Period, plants grew larger and woody trunks appeared. Many of these plants fell into warm swamps and partially rotted to form early coal deposits.

Ninety-five percent of all animal species and a large number of plant species suddenly disappeared at the end of the Paleozoic Era, 245 million years ago. Thus, the Paleozoic Era began with a sudden increase in types of organisms and ended with

a sudden **mass extinction.** Mass extinctions and their causes are discussed in the following section with reference to the more familiar demise of the dinosaurs at the end of the Mesozoic Era.

Mesozoic Era.   The extinction of most life at the end of the Paleozoic Era left a depopulated Earth. As the survivors crawled out from under the smoking wreckage left by whatever force caused the extinction, they found a world with great ecological imbalance and disrupted food chains. Prey, suddenly without predators, bred and multiplied astronomically, exploring the new Mesozoic world with rapid evolutionary experiments. Many new families of plants and animals quickly evolved to fill

FIGURE 19.20 *Evolution of the dinosaurs.* (From Colber, E.H. *Evolution of the Vertebrates.* New York, John Wiley & Sons, 1969.)

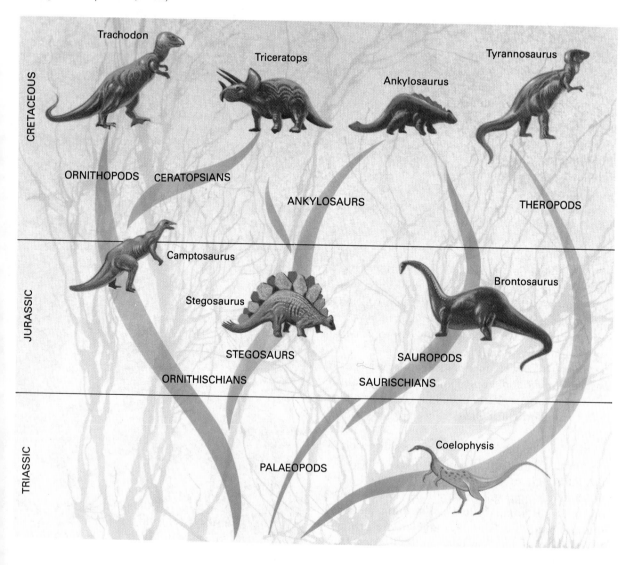

ecological niches so suddenly vacated by the extinction.

In early Mesozoic time, one surviving variety of reptile evolved into the dinosaurs, a group of animals that captures our fancy like no other prehistoric beast. Evolving from small, two-legged Triassic reptiles, dinosaurs developed rapidly into many species of all sizes and shapes and came to rule the Mesozoic landscape. Figure 19.20 shows the evolution of the dinosaurs. Dinosaurs have been traditionally portrayed as green, scaly, and cold-blooded. However, modern evidence suggests that many were warm-blooded, brightly colored, and either hairy or covered with feathers like modern birds. Evidence also shows that they cared for their young after they hatched.

Jack Horner, a paleontologist at the Museum of the Rockies in Bozeman, Montana, has reconstructed a scene of dinosaur life 100 million years ago in Montana. A shallow inland sea flooded much of North America east of the newly rising Rocky Mountains. Erosion of the mountains filled the seaway with mud and sand, creating a vast, swampy plain lying between the mountains and the sea. Horner envisions herds of thousands of duck-billed dinosaurs up to 30 feet long. Following warmth and blooming vegetation, the herds migrated with the seasons 1000 km or more north and south along the margins of the sea. They stopped along the way to build nests and raise their young, remaining in the nesting area until the new babies were able to travel with the herd (Fig. 19.21). Solitary carnivores followed the herds, preying on the young and infirm. Occasionally, great volcanic eruptions from the mountains buried an entire herd of dinosaurs in ash, preserving together eggs, baby dinosaurs in their nests, and adults.

At nearly the same time that one variety of reptile evolved into the dinosaurs, another gave rise to mammals. Mesozoic mammals were small, furtive, rodent-like animals. Mammals spent most of Mesozoic time improving their evolutionary abilities to adapt, survive, and reproduce. They developed an efficient system for maintaining body temperature that allowed them to thrive in both cold and warm climates. Near the end of Mesozoic time, the numbers of dinosaurs were declining even before the great extinction event. As dinosaurs became less numerous, mammals rapidly increased both in number of individuals and number of species.

FIGURE 19.21   *In this reconstruction, a mother duck-billed dinosaur nurtures her babies in their nest on a mudflat 100 million years ago in Montana.* (Museum of the Rockies)

Sixty-five million years ago, all species of dinosaurs suddenly became extinct. This event resembled the terminal Paleozoic extinction because it included not only dominant species, but also many other plants and animals. At least one-fourth of all animal species on Earth, both marine and terrestrial, vanished in this terminal Mesozoic extinction.

As discussed in Chapter 18, scientists have accumulated evidence that this mass extinction and perhaps earlier ones were caused when a giant meteorite struck the Earth. The impact vaporized both the meteorite and the Earth's crust at the point of impact, igniting massive fires. Soot from the fires and meteorite dust rose into the upper atmosphere, blocking out the Sun. Many plant and animal species froze or starved to death.

FIGURE 19.22   *Early Miocene mammals.* (National Museum of Natural History, J.H. Matternes)

Other researchers disagree with the meteorite theory of mass extinctions and have offered several other hypotheses. Most invoke sudden changes in global climate or in sea level. According to one hypothesis, increased volcanic activity 65 million years ago injected large quantities of ash into the atmosphere. In turn, this ash blocked out significant amounts of sunlight and led to rapid global cooling.

Cenozoic Era.   With the dinosaurs gone at the beginning of the **Cenozoic Era** (Greek for "recent life"; 66 million years ago to the present), mammals evolved rapidly to dominate the land. Grasses evolved in Miocene time, creating lush prairies and savannahs that supported large herds of newly evolving grazing animals, such as horses. Carnivores, as depicted in Figure 19.22, evolved to prey on the expanding herds. Only a few reptiles survived the extinction at the end of Mesozoic time. Birds, possibly the only surviving evolutionary descendants of dinosaurs, reproduced rapidly.

About 5 million years ago, near the end of Miocene time, creatures more resembling modern humans than apes appeared in India. Later, between 3 and 1 million years ago, several separate human-like lineages developed and are preserved as fossils in East Africa. The first *Homo sapiens* evolved between 140,000 and 100,000 years ago. By 30,000 years ago, early modern humans, called Cro-Magnon people, spread into Europe. They crafted well-made weapons and tools, developed religions, and created art. They are our immediate ancestors.

CONCEPTUAL EXERCISE

Lay a strip of cardboard along the edge of your classroom and draw in a geological time scale with time proportional to distance. Show the units of the geological time scale and indicate important events in the history of evolution. This exercise provides a visual image of geological time. You will see that multicellular life has only been on the planet for a small portion of Earth history. The dinosaurs dominated the planet for 180 million years, whereas all of human evolution fits into a thin line.

19.6 CONTINENTS AND CONTINENTAL MOVEMENT—THE IDEA IS BORN

Science usually creeps forward by innumerable little discoveries, each won by months or years of hard work in the field or laboratory. Occasionally, however, scientists gather all the little advances into a new idea or a new way of looking at old ideas to initiate a major scientific revolution. Such a revolu-

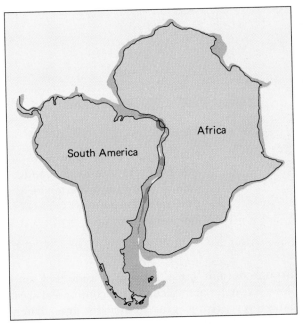

FIGURE 19.23 *The African and South American coastlines appear to fit together like adjacent pieces of a jigsaw puzzle.*

tion occurred in geology in the 1960s and 1970s. Its effects are as exciting and important to geologists as Einstein's theory of relativity was to physicists earlier in this century.

In the early twentieth century, a young German scientist named Alfred Wegener noticed that the African and South American coastlines on opposite sides of the Atlantic Ocean seemed to fit as if they were adjacent pieces of a jigsaw puzzle (Fig. 19.23). He realized that the jigsaw-like fit suggested that the continents had once been joined together and had later split and drifted apart. Studying world maps, Wegener noticed that not only did the continents on both sides of the Atlantic fit together

well, but that other continents, when moved properly, also fit like additional pieces of the same jigsaw puzzle. He constructed a map of the Earth based on the fit of continents. On his map, all the continents were joined together, forming one supercontinent (Fig. 19.24) that he called **Pangaea** from the Greek root word for "all lands." The northern part of Pangaea is called **Laurasia** and the southern part **Gondwanaland.**

Wegener then mapped the locations of fossil remains of several species of animals and plants that could neither swim well nor fly. Fossils of the same species are now found in Antarctica, Africa, Australia, South America, and India. Why would the same species be found on continents separated by thousands of kilometers of ocean? To solve this dilemma, Wegener plotted the same fossil localities on his Pangaea map. Figure 19.25 shows that all of them lie in the same region of Pangaea, suggesting that each species evolved and spread over a part of Pangaea, rather than mysteriously migrating across thousands of kilometers of open ocean.

Because certain types of sedimentary rocks form in certain climatic zones of the Earth (glaciers and glacial sediment, for example, concentrate in high latitudes), Wegener plotted sedimentary rocks that indicated climate and latitude on maps showing the modern distribution of continents. Figure 19.26a shows his map of 300-million-year-old glacial deposits. The white area shows how large the ice mass would have been if the continents had been in their present positions. Notice that the glacier would have crossed the Equator, and glacial deposits would have formed in tropical and subtropical zones. Figure 19.26b shows the same glacial deposits plotted on Wegener's Pangaea map. Here they are neatly clustered about the South Pole.

Wegener also noticed several instances in which an uncommon rock type or a distinctive sequence

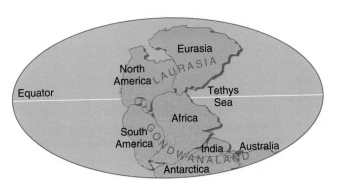

FIGURE 19.24 *Wegener's Pangaea, a supercontinent that existed about 200 million years ago. The northern part of Pangaea is called Laurasia, and the southern part is called Gondwanaland.*

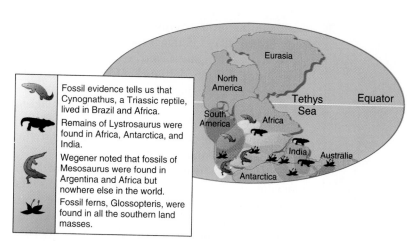

Fossil evidence tells us that Cynognathus, a Triassic reptile, lived in Brazil and Africa.

Remains of Lystrosaurus were found in Africa, Antarctica, and India.

Wegener noted that fossils of Mesosaurus were found in Argentina and Africa but nowhere else in the world.

Fossil ferns, Glossopteris, were found in all the southern land masses.

FIGURE 19.25 *Geographic distributions of plant and animal fossils indicate that a single supercontinent, called Pangaea, existed about 200 million years ago.*

of rocks on one side of the Atlantic Ocean was identical to rocks on the other side. When he plotted the rocks on a Pangaea map, those on the east side of the Atlantic were continuous with their counterparts on the west side (Fig. 19.27).

Wegener combined all of this evidence into a theory that he called **continental drift.** Skeptical scientists demanded an explanation of *how* continents could move. Wegener suggested two alternative possibilities: first, that continents plow their way through oceanic crust, shoving it aside as a ship plows through water; second, that continental crust slides over oceanic crust. Physicists immediately proved that both of Wegener's mechanisms were impossible. Oceanic crust is too strong for continents to plow through it. The attempt would be like trying to push a matchstick boat through heavy tar. The boat, analogous to the continents, would break apart. Furthermore, frictional resistance is too great for continents to slide over oceanic crust.

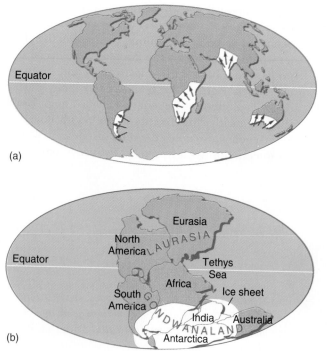

(a)

(b)

FIGURE 19.26 *(a) Three-hundred-million-year-old glacial deposits plotted on a map showing the modern distribution of continents. Arrows show directions of ice movement. (b) The same glacial deposits plotted on a map of Pangaea.*

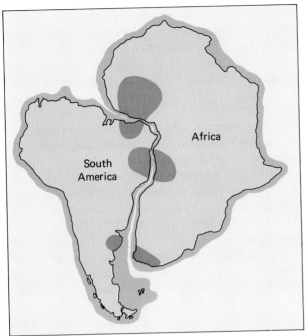

FIGURE 19.27 *Locations of distinctive rock types in South America and Africa, plotted on a portion of a Pangaea reconstruction.*

19.7 PLATE TECTONICS—THE MODERN THEORY

Wegener died in 1930 in a blizzard while traveling across Greenland on a scientific expedition, and for 35 years his ideas were largely forgotten. Shortly after World War II, oceanographers discovered a long submarine mountain range called the **Mid-Atlantic ridge** in the middle of the Atlantic Ocean (Fig. 19.28). Further studies showed that the Mid-Atlantic ridge is part of a continuous submarine mountain chain called the **mid-oceanic ridge,** which girdles the entire globe and is by far the Earth's largest and longest mountain chain.

One remarkable feature of the Mid-Atlantic ridge is that it lies directly in the middle of the Atlantic Ocean basin, halfway between Europe and Africa to the east and North and South America to the west. Sea floor rocks at the ridge axis are warm and young. The rocks become progressively older and cooler to the east and west.

In the mid-1960s, geologists suggested that new sea floor forms continuously from basaltic magma

FIGURE 19.28 *The mid-Atlantic ridge is a submarine mountain chain in the middle of the Atlantic Ocean. It is a segment of the mid-oceanic ridge, which circles the globe like the seam on a baseball.* (Marie Tharp)

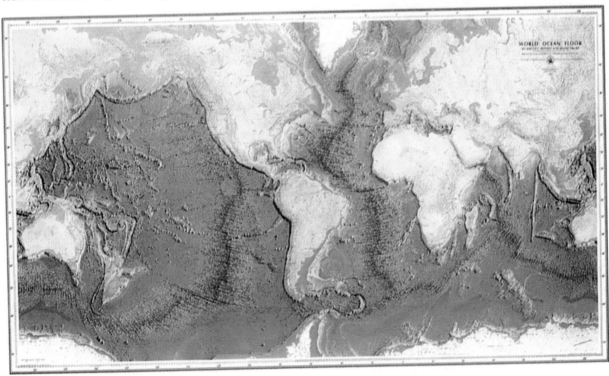

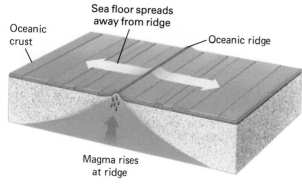

Sea floor spreads
away from ridge

Oceanic
crust

Oceanic ridge

Magma rises
at ridge

FIGURE 19.29 *New sea floor forms continuously from basaltic magma rising beneath the ridge axis. The new oceanic crust then spreads outward from the ridge.*

rising beneath the ridge axis. The new oceanic crust then spreads outward from the ridge as shown in Figure 19.29. This movement is analogous to two broad conveyor belts moving away from one another. As a result, the sea floor at the Mid-Atlantic ridge is young and becomes progressively older with greater distance from the ridge.

The plate tectonics theory is a model of the Earth in which the brittle lithosphere floats on the

hot, plastic asthenosphere. As shown in Figure 19.30, the lithosphere is broken into seven large plates and several small ones, resembling segments of a turtle's shell. The plates glide slowly over the weak, plastic asthenosphere at rates ranging from less than 1 to about 18 cm per year, almost as fast as a fingernail grows. Because the plates move in different directions, they bump and grind together at their boundaries. Mountain-building, volcanic eruptions, and earthquakes occur where two plates meet. Simple logic tells us that one plate can move relative to an adjacent plate in three different ways; they can separate, collide, or slide past one another. These three types of boundaries are shown in Figure 19.31.

A **divergent boundary,** also called a **spreading center,** occurs where two plates move apart horizontally (Fig. 19.32). As the two plates separate, hot, plastic asthenosphere rock flows upward to cool and form new lithosphere in the gap left by the diverging plates. The mid-oceanic ridge is a divergent boundary in oceanic crust. The basalt magma that oozes out onto the sea floor at the ridge forms oceanic crust on top of the new lithosphere. Note that it is not merely oceanic crust that spreads out-

FIGURE 19.30   *Lithospheric plate boundaries are shown in red. Gray arrows indicate directions of plate movement.* (Tom Van Sant, Geosphere Project)

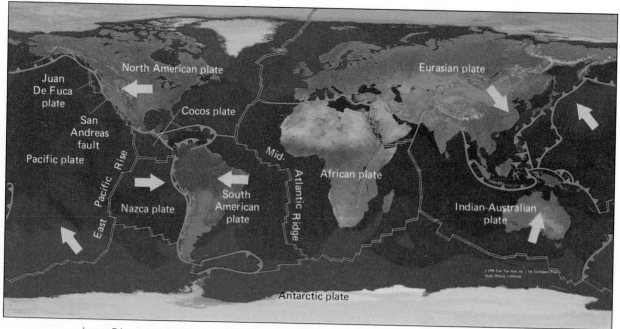

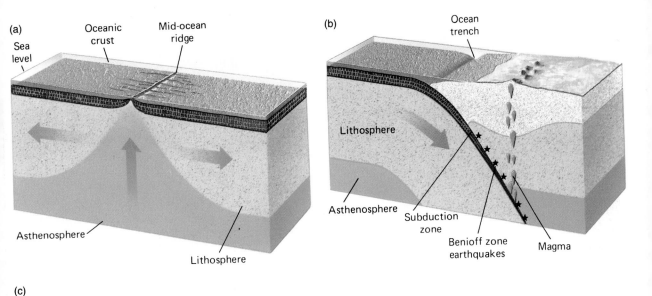

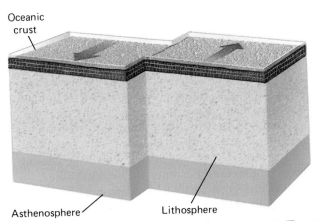

FIGURE 19.31 *Three types of plate boundaries: (a) Two plates separate at a divergent boundary. New lithosphere forms as hot asthenosphere rises to fill the gap where the two plates spread apart. (b) Two plates collide at a convergent boundary. Here an oceanic plate is sinking beneath a less-dense continental plate. Magma rises from the subduction zone, and a trench forms where the subducting plate sinks. (c) At a transform plate boundary, rocks on opposite sides of the fracture slide horizontally past each other.*

ward from the mid-oceanic ridge. Rather the entire lithosphere spreads, carrying the sea floor on top of it in piggyback fashion.

A convergent boundary develops where two plates collide (Fig. 19.32). Recall that oceanic crust is denser than continental crust. When a continental plate collides with a denser oceanic plate, the oceanic plate sinks beneath the continental plate and dives into the mantle. This process is called **subduction.** An **oceanic trench** is a long, narrow trough in the sea floor formed where a sub-

ducting plate bends downward to sink into the mantle. The deepest point on Earth is in the Mariana trench, in the southwestern Pacific Ocean where the sea floor is as much as 10.9 km below sea level, compared with the average sea floor depth of about 5 km. The oldest sea-floor rocks on Earth are only about 200 million years old because oceanic crust is continuously recycled back into the mantle at subduction zones. Far older rocks are found on continents because continental crust is not consumed by subduction.

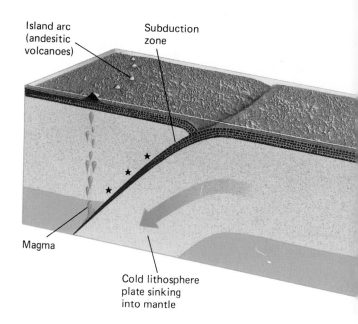

Island arc (andesitic volcanoes)

Subduction zone

Magma

Cold lithosphere plate sinking into mantle

FIGURE 19.32   *The outer few hundred kilometers of the Earth. In the center of the drawing, new lithosphere forms at a spreading center. At the sides of the drawing, old lithosphere sinks into the mantle at subduction zones.*

If two colliding plates are both covered with continental crust, subduction cannot occur because continental crust is too light to sink into the mantle. In this case, the two continents collide and crumple against each other, forming huge mountain chains in the collision zone. The Himalayas, the Alps, and the Appalachians all formed as results of continental collisions.

A **transform plate boundary** forms where two plates slide horizontally past one another (see Fig.

FIGURE 19.33   *According to one explanation, lithospheric plates are dragged along by mantle convection.*

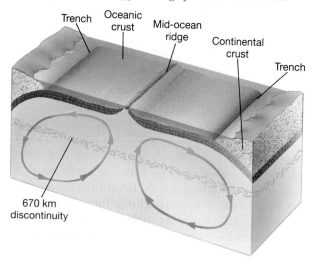

Trench

Oceanic crust

Mid-ocean ridge

Continental crust

Trench

670 km discontinuity

19.31c). California's San Andreas fault is a transform boundary between the North American Plate and the Pacific Plate.

What causes the plate to move? According to one theory, the asthenosphere, or perhaps even the entire mantle, may be heated unevenly, causing large convection currents to be established, as shown in Figure 19.33. The lithosphere may then ride along these convection currents much as icebergs are carried long distances as they float on ocean currents.

19.8 PLATE TECTONICS AND MOUNTAIN-BUILDING

Mountains form some of the most spectacular landforms on Earth and provide a constant reminder of the tremendous forces that exist within our dynamic planet. Several different types of mountain ranges can be distinguished.

Andes

The Andes Mountains lie in a long, thin line along the west coast of South America. This region marks a convergent plate boundary between the South American continental plate and an ocean plate to

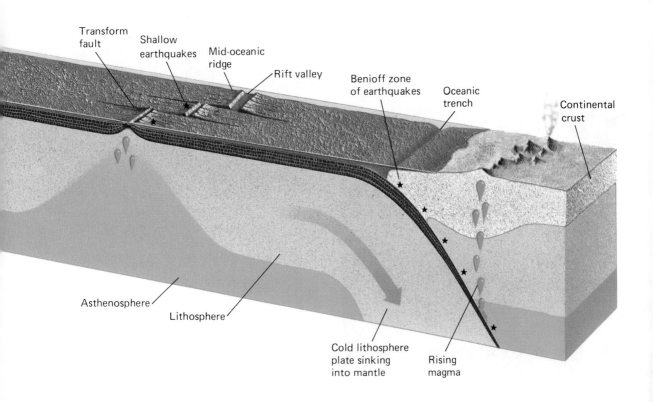

the west. A deep trench exists in the ocean floor along the collision zone. As explained earlier, the ocean plate is descending, forming the trench. As this ocean plate dives under the edge of the continent, the continental rock catches on the moving rock beneath it and is compressed, crumpled, and distorted. Think of trying to slip two pieces of sandpaper past each other. They will move, but one or the other will crumple, rise, and fall in the process. As the subsiding ocean plate reaches the hot asthenosphere, vast quantities of rock melt. Some of the molten material rises through the cracks in the stressed continental plate to form both volcanoes and large bodies of coarse-grained igneous rock. This combination of distorted, uplifted crust and rising magma has formed the Andes mountains, the second highest mountain range on Earth, as shown in Figure 19.34. The core of the range is composed mainly of volcanic and coarse-grained igneous rocks. The volcanic cones stand as visual reminders of the forces at work beneath the surface.

CONCEPTUAL EXAMPLE 19.3   MOUNTAIN-BUILDING
Explain why the cores of some of the mountains in the Andes are coarse-grained whereas others are volcanic cones.

Solution A diving subduction plate churns rock in the asthenosphere, forming a vast amount of magma. During the growth of the Andes, some of this magma rose and solidified in the crust to form coarse-grained rock. However, vast quantities of magma continued upward to erupt at the surface, forming numerous volcanoes.

Himalayas

Although the Andes were formed from the collision of a continental plate with an oceanic plate, the Himalayas were formed from the collision of

FIGURE 19.34   *(top) An unnamed mountain in the Peruvian Andes. (bottom) Volcanic cones in northern Chile were formed by frequent eruptions along the Andes subduction zone.*

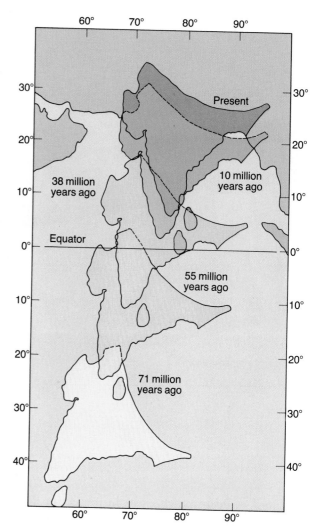

FIGURE 19.35   *India's northward movement. About 40 to 50 million years ago, the Indian subcontinent collided with the Asian continent, creating the Himalayan mountain range.*

two continents. Figures 19.35 and 19.36 show how the Indian subcontinent has collided with Eurasia. Measurements indicate that the two continents collided 50 to 40 million years ago, and since the initial impact, India has continued to push northward. Neither continental plate has been forced downward. Instead, they both have buckled upward and formed the Himalayan mountain range.

Rockies

One billion years ago, the western edge of North America did not appear as it does today. Parts of Alaska and western Canada and much of Washington, Oregon, western Idaho, and western Califor-

nia had not yet been joined to the continent (Fig. 19.37). Instead the area was ocean. As North America moved westward and the Pacific Ocean plate subducted beneath the western edge of the continent, islands were carried to the subduction zone. Because they were too light to follow the descending plate into the mantle, they jammed onto the western margin of the continent. As the islands crashed into the continent, they crushed rocks near the collision zone. This compression, coupled with volcanic activity, formed a mountain range only a

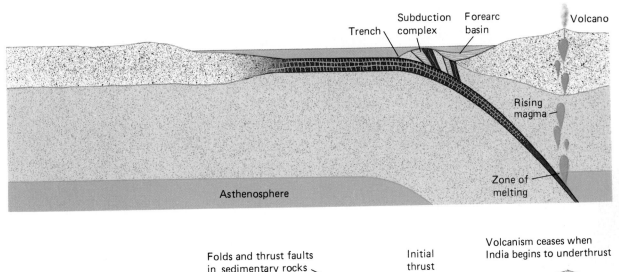

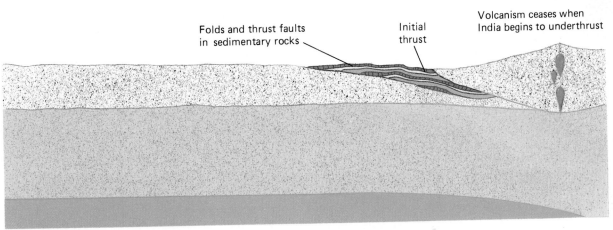

FIGURE 19.36   *(a) As India moved northward, subduction began at the southern margin of Asia. By 80 million years ago, an oceanic trench and subduction complex had formed. (b) By 40 to 50 million years ago, India had collided with Tibet. The leading edge of India was thrust under southern Tibet, forming the modern Himalayas.*

few hundred kilometers wide but extending from Mexico all the way into Alaska (Fig. 19.38). The last of the islands fused to North America about 80 million years ago, but mountain-building from the process continued until about 45 million years ago.

As North America moved westward, it drew closer to a portion of the mid-oceanic ridge called the **East Pacific rise** (Fig. 19.39). By 30 million years ago, most of the oceanic crust between North America and the East Pacific rise had been subducted, and the western edge of the continent reached the East Pacific rise near southern California. Today, this plate boundary runs through the Gulf of California where it is offset to the northwest across the state of California as the San Andreas Fault. It then proceeds into the North Pacific Ocean, as shown in Figure 19.39c. The relative motion along this fault initiated a lateral distortion of the rock. In some places, segments of the crust were stretched apart. At the same time, hot magma began to push upward. This combination of stretching and upward flow produced mountain formation to the east (Fig. 19.40).

We, as humans, live in one tiny period of geological history. Change is still occurring all around us. The continents continue to move. Many hot springs that exist throughout the Rocky Mountains remind us that hot magma remains near the surface, and geologists would not be surprised if a volcano erupted in Yellowstone National Park or the Pacific Northwest.

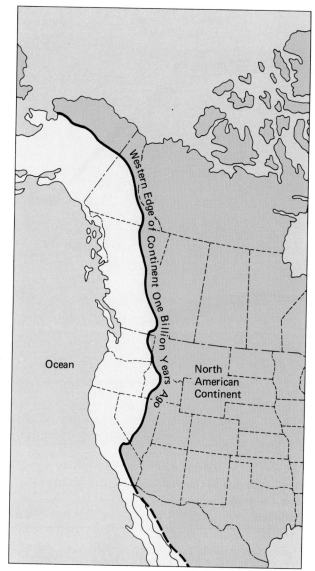

FIGURE 19.37   *Western North America from about 1 billion to 180 million years ago. The light blue area shows the modern outline of the continent.*

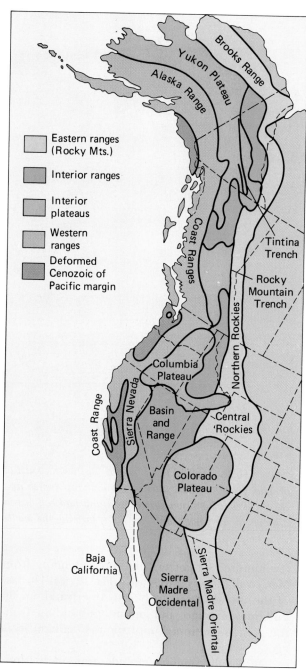

FIGURE 19.38   *The mountains of western North America are composed of many ranges that extend from Mexico into Alaska.*

19.9 EARTHQUAKES AND VOLCANOES

Try to imagine the forces that operate along tectonic plate boundaries. A large segment of rock—perhaps a whole continent—is being pushed in a given direction. Another continent or a lithospheric plate carrying sea floor may be pushing in the opposite direction. When this happens, tremendous stresses build up. Years or even centuries may elapse without motion, until the stress becomes so great that it suddenly overcomes the friction, and rocks on either side of the fault slip past each other. This slippage is called an **earthquake.** Figure 19.41 shows the damage left by the 1964 quake in Anchorage,

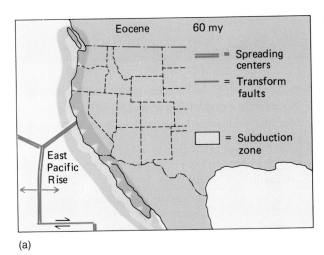

(a)

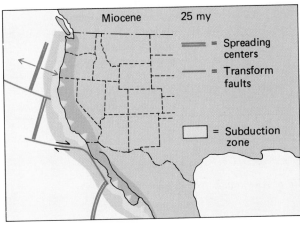

(b)

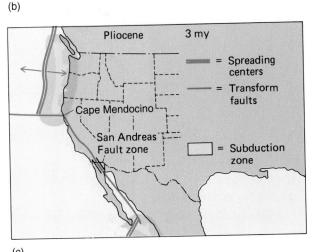

(c)

FIGURE 19.39   *The San Andreas fault developed where western North America overran the East Pacific rise, beginning about 30 million years ago. The fault has grown longer as more of California has hit the rise. Notice that subduction has ceased where California strikes the rise and the San Andreas fault forms.*

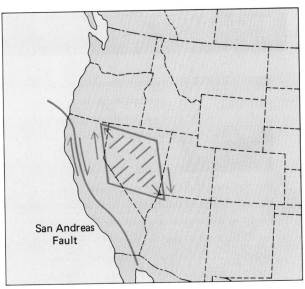

FIGURE 19.40   *Friction along the San Andreas fault pulls Nevada and nearby regions apart in a northwest-southeast direction (red arrows). At the same time, hot magma from below has uplifted the crust.*

Alaska. A place or zone in the rocks along which relative motion has occurred is known as a **fault.**

Earthquakes are some of the most devastating of all natural phenomena. They generally strike without forewarning, destroying homes, apartment buildings, and sometimes even entire cities. In many regions of the world, lumber is scarce, and people build their houses out of dried mud bricks called adobe. The roofs of these buildings are generally constructed of red clay tile. Adobe-tile buildings are easily toppled by earthquakes, and millions of people have died in recent years when the heavy mud blocks have collapsed.

The Earth's major earthquake zones coincide with tectonic plate boundaries (Fig. 19.42). The **San Andreas fault** in California is an example of an earthquake zone at a transform plate boundary (Fig. 19.43). This major fault system stretches from the Gulf of California northward beyond San Francisco and out to sea. The San Andreas fault forms the transform plate boundary between the Pacific Ocean plate and the North American plate.

Sometimes horizontal movement can be slow, gradual, and relatively nonviolent. Such motion is known as **fault creep.** In several locations in California, buildings have been unintentionally located

FOCUS ON
THE RICHTER SCALE

Perhaps the most widely recognized method used to measure the magnitude of an earthquake is the Richter scale, devised by C. F. Richter. This method specifies the magnitude of the earthquake based on the movement of the Earth at a distance of 100 km from the source of the quake. The scale is a logarithmic scale, like the decibel scale studied in Chapter 7. On this scale, an increase of one whole number corresponds to an earthquake wave with an amplitude ten times longer than the next lower number. Thus, an earthquake of magnitude 6 produces 10 times more ground motion than one of magnitude 5. Likewise, a magnitude 6 earthquake produces 100 times

the ground motion of a magnitude 4 earthquake. The table below specifies the severity of damage associated with several magnitudes on the Richter scale.

RICHTER MAGNITUDE	EFFECTS
less than 2.5	Generally goes unnoticed, but can be recorded
2.5–5.4	Occasionally felt. Produces only minor damage
5.5–6.0	Slight damage to structures
6.1–6.9	Destructive in populated regions
7.0–8.0	A major earthquake producing serious damage
greater than 8.0	A great earthquake, producing total destruction to adjacent regions

FIGURE 19.41 *The 1964 Alaska earthquake destroyed much of Anchorage.* (Ward's Natural Science Establishment)

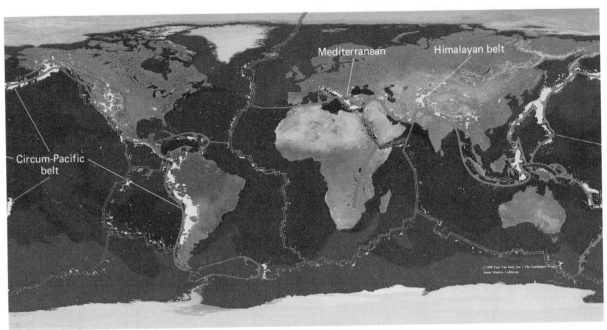

Divergent boundary   Convergent boundary   Transform boundary

FIGURE 19.42 *The Earth's major earthquake zones coincide with tectonic plate boundaries. Each yellow dot represents an earthquake that occurred between 1961 and 1967.*
(Tom Van Sant, Geosphere Project)

FIGURE 19.43 *The San Andreas fault slices the Earth's surface in San Luis Obispo County, California.*
(R.E. Wallace, U.S.G.S.)

so that they straddle the great San Andreas fault right where creep is occurring. Slowly the two sections of Earth slide past each other. Many tiny shakes and quakes occur, but none is severe enough to topple the buildings. However, as the Earth on each side of the fault moves, it cracks the foundation until walls fracture and break apart.

Such gradual slippage can destroy a few buildings along the fault, but because there is no serious sudden movement, nearby buildings are not harmed. However, if rock is stretched or compressed too severely without continual release by many small earthquakes, large sections of the Earth may move suddenly along a fault, resulting in a disastrous earthquake. For example, if part of a tectonic plate moves at a rate of 5 cm a year, and another part is held rigid by frictional forces, the rock may stretch like a giant rubber band. The slow stretch may accumulate for 100 years or more; then suddenly rock surfaces break loose and a large earthquake occurs. A 5 cm stretch per year for 100 years would result in a total displacement of 5 m.

In 1906, sections of rock near San Francisco jumped 4.5 to 6 m in a matter of seconds and then abruptly came to a halt, generating huge shock

FOCUS ON
THE LOS ANGELES EARTHQUAKE OF 1994: A CASE HISTORY

Los Angeles and its numerous suburbs straddle the San Andreas fault. Earthquakes have occurred repeatedly in the recent past, and geologists are certain that this activity will continue in the future. In 1857, an earthquake with an estimated Richter magnitude 8.0 struck just north of Los Angeles. Although this segment has not moved since 1857, the same fault to the north and south moved 4.6 m since 1857 by means of fault creep and frequent small quakes. Thus, the accumulated displacement of rock in the locked segment must be about 4.6 m. This

stretching is stored as elastic energy, waiting for a sudden fracture to be released.

In April, 1992, a magnitude 6.1 earthquake occurred in the Mojave desert, northeast of Palm Springs, California. Then, another earthquake with a magnitude of 7.3 struck in late June near Landers. Numerous additional quakes were recorded in the vicinity. The Landers earthquake occurred on a series of faults that intersect the San Andreas fault at an angle. Two geophysicists from Columbia University, Steven Jaume and Lynn Sykes, calculated that motion on the faults near Landers stretched rock along the San Andreas fault. Thus, they predicted an increased

Vehicles stranded by multiple failures of an overpass on Highway 14 near Los Angeles, resulting from the 1994 earthquake. (Douglas C. Pizac/Wide World Photo)

waves. Buildings were split and separated. In the city of San Francisco itself, disaster struck; some of the damage is shown in Figure 19.44. The rapidly

shifting rock caused the soil above it to move and settle. Many buildings whose foundations were anchored in soil toppled immediately. As buildings

probability of an earthquake near Los Angeles. In 1992, they wrote:

> *It had been estimated that the chance was about 60 percent that one or more great earthquakes would occur along the San Andreas Fault in Southern California from 1988 to 2018.[1] We conclude that the increase in earthquake activity since 1986 and the changes in stress along the San Andreas Fault associated with the Landers sequence indicate that the probability of a great shock is now higher than that estimated in 1988.[2]*

In January 1994, a magnitude 6.6 earthquake struck the San Fernando Valley just north of Los Angeles. Fifty-five people died, and property damage was estimated at $8 billion (total damage, which included lost work and business revenues, was much higher). This quake occurred on a buried fault west of the San Andreas fault. Geologists are now measuring earth motion in an effort to calculate forces and stresses in rock near the center of the 1944 quake. Ultimately, they are trying to forecast whether a great quake is imminent along the San Andreas fault itself. Although the 1994 quake was deadly and expensive, it was not a great earthquake. An 8.0 quake would release 125 times the energy of the San Fernando disaster. If such a shock were to occur during rush hour traffic, the death toll would be high.

[1] A great, or major, earthquake is defined as one with a magnitude of 7.0 or higher on the Richter Scale.

[2] S. C. Jaume and L. R. Sykes. "Changes in state of stress on the Southern San Andreas fault resulting from the California earthquake sequence of April to June 1992." *Science*, 258, Nov 20, 1992, p 1325.

fell and underground gas lines were cut by the moving Earth, great fires started. The fires spread throughout the city, causing widespread destruc-

tion that was more devastating than the effects of the quake itself.

Earthquake Prediction

In 1979, McCann and co-workers at the Lamont-Doherty Geological Observatory at Columbia University outlined the **seismic gap theory.** This theory starts with the simple premise that all segments of a fault such as the San Andreas eventually must move by the same amount because the fault is the boundary between two tectonic plates that are moving past each other. The theory continues with the assumption that when an earthquake occurs, the accumulated elastic energy in that region is released. However, strain continues to build in nearby regions where earthquakes have not occurred. Thus, after a quake, the probability of a second quake in the same place in the near future is low. Alternatively, regions along an active fault that have not experienced a recent major quake are at high risk of a large earthquake in the near future.

In 1991, Jackson and Kagan of the University of California at Los Angeles challenged the seismic gap theory. Using a catalogue of all the recorded earthquakes of magnitude 7 or higher along both coasts of the Pacific Ocean, they showed that no statistical correlation exists between earthquake activity and seismic gaps. Many earthquakes occur in the same region in quick succession. For example, in the 1800s, two significant *pairs* of earthquakes occurred in the San Francisco area. A quake on the

FIGURE 19.44 *California earthquake, 1906. The wrecked Hibernia Bank building in San Francisco.* (Photo by W.C. Mendenhall, U.S. Department of the Interior Geological survey)

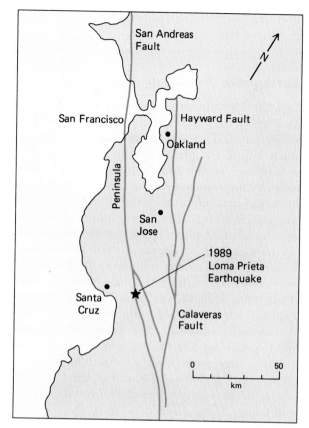

FIGURE 19.45 *Active faults in the San Francisco Bay area. The 1989 earthquake was centered in Loma Prieta, marked by a star.*

FIGURE 19.46 *Collapse of Interstate Highway 880 during the 1989 earthquake.* (Paul Scott/Sygma)

Hayward fault in the Oakland area in 1836 was followed by one on the San Andreas fault near San Francisco 2 years later. Then a second quake on the San Francisco peninsula in 1865 was followed 3 years later by a quake near Oakland. Thus, they concluded that an earthquake does not necessarily reduce the probability of another earthquake in the same vicinity in the near future. Alternatively, Jackson and Kagan did not record a significantly higher probability of earthquakes in seismic gaps that had been quake-free for a long time.

A magnitude 6.9 earthquake occurred near San Francisco in 1989; its location is indicated on the map in Figure 19.45, and the destruction it caused to one of the interstate highways is pictured in Figure 19.46. If the seismic gap theory is correct, stress was relieved, and the probability is low that another quake will occur in the region within the next few decades. However, if a pattern of pairs of quakes were to repeat itself, another major quake is overdue in the San Francisco area.

Short-term earthquake prediction depends on a reliable early warning system—a signal or group of signals that immediately precedes an earthquake. **Foreshocks** are small earthquakes that precede a large quake by an interval ranging from a few seconds to a few weeks. The cause of foreshocks can be explained by a simple analogy. If you try to break a stick by bending it slowly, you may hear a few small cracking sounds just before the final snap. If foreshocks consistently preceded major earthquakes, they would be a reliable tool for short-term prediction. Only about half of the major earthquakes in recent years, however, were preceded by a significant number of foreshocks. At other times, swarms of small shocks that could have been foreshocks were recorded, but a large quake did not follow.

Another approach to short-term earthquake prediction is to measure changes in the shape of the land surface. Seismologists monitor rising bulges and other unusual Earth movements with tiltmeters and laser surveying instruments. Some earthquakes have been successfully predicted with this method, but in other instances predicted quakes did not occur, or quakes occurred that had not been predicted.

One theory states that when rock is deformed to near its rupture point, microscopic cracks and pores open between the mineral grains. These tiny openings change several physical properties of the rock. For one, radon gas trapped in rocks and minerals is released. In some cases, an increased rate of radon release has preceded an earthquake. Also the formation of the pores changes the volume of the rock, which, in turn, alters the water table.

FIGURE 19.47 *Eruption of Mount St. Helens.*
(U.S.G.S., R.P. Hoblitt)

Thus, the water level in some wells has fluctuated just before an earthquake. In addition, air-filled holes do not conduct electricity as well as solid rock, so the electrical conductivity of rock decreases as pores open up.

In the early 1970s, Chinese geophysicists used seismic gap theory and historical data to issue a long-range warning for a major earthquake near the city of Haicheng. Seismic stations were established in and around the city. In January, 1975, scientists recorded swarms of foreshocks and unusual bulges in the land. When the foreshocks became especially intense on February 1, authorities ordered an evacuation of portions of the city. The evacuation was completed on the morning of February 4, and in the early evening of the same day, a large earthquake destroyed houses, apartments, and factories but caused few deaths.

Immediately after that success, geologists hoped that a new era of quake prediction had begun. A year later, however, Chinese scientists failed to predict an earthquake in the adjacent city of Tangshan. This major quake was *not* preceded by a swarm of foreshocks, no warning was given, and at least 250,000 people died.[*] Shortly after that fail-

ure, a quake was predicted in a third city and the city was evacuated, but the earthquake did not occur. At present, earthquake prediction remains unreliable.

Volcanoes

Perhaps the most spectacular and rapid geological event occurring on Earth is a volcanic eruption, which occurs when hot, molten magma moves upward through fissures in the rocks and escapes to the surface as fiery lava or hot ash (Fig. 19.47). The eruption is generally accompanied by the release of large quantities of steam and other gases. Sometimes magma is ejected in relatively calm lava flows, while in other instances violently explosive eruptions occur. Mauna Loa, a towering mountain that rises above the sea to form the second highest peak on the island of Hawaii, has erupted frequently in modern times, but the nonviolence of these events has enabled people to work and farm only kilometers from the volcanic crater. When the mountain becomes active, masses of molten rock flow smoothly down the sides of the mountain until they cool and solidify. These frequent gentle lava flows have gradually accumulated to form the mountain,

[*]Accurate reports of the death toll are unavailable. Published estimates range from 250,000 to 650,000.

which rises 4500 m from the ocean floor to the surface of the sea and another 4200 m above that. From ocean floor to summit, Mauna Loa is about as tall as Mt. Everest.

Not all volcanoes are so gentle, however. In the early 1800s, the island of Krakatoa in Indonesia was a landmark for clipper ships that carried tea and other freight from India. The mountain on the center of the island was conical, covered with trees, and rose nearly 800 m above sea level. On August 26, 1883, a huge volcanic explosion rocked the island. The crew of a ship sailing offshore witnessed an immense cloud of dust, ash, and steam that darkened the horizon. Lightning storms and intense squalls developed as the sailors headed out to sea to escape the violence. On the next day, four more great explosions rocked the island; when the dust had cleared away, the island of Krakatoa and its 800 m mountain had disappeared. A few tiny islets remained on what had been the rim of the former island, but the rest was gone. It is believed that approximately 20 km^3 of volcanic material shot skyward. As the exploding lava shot into the air, a huge hole appeared in the center of the island, and when the eruption subsided, the mountain had disappeared. A similar eruption about 7000 years ago formed Crater Lake, Oregon, by collapse of much of the mountain into the partially emptied magma chamber (Fig. 19.48).

All volcanic eruptions occur when molten magma forms in the upper mantle and rises toward the surface through fissures in the rock. The violence of the eruption is controlled by factors such as the chemical composition of the magma, its temperature, the shape and size of the fissures, and the quantity of gas in the fluid mixture. If the lava is thick and viscous and if large amounts of gas are trapped in the molten rock, eruptions are likely to be violent. The viscous lava does not flow easily and is prevented from moving upward until the gaseous pressure causes it to explode violently. On the other hand, the magma of a gentle volcano such as Mauna Loa is more fluid and contains comparatively less steam. The fluid lava rises through rock fissures easily, and, as there is little steam, extreme pressures and violent explosions never develop.

The boundaries between tectonic plates are particularly active volcanic zones. Thus, frequent eruptions occur along the mid-oceanic ridges where plates are separating. They also occur along convergent plate boundaries where ocean plates

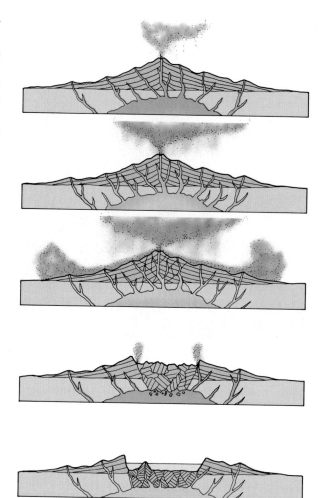

FIGURE 19.48   *The collapse of a mountain during a cataclysmic volcanic eruption.*

are colliding with continents. Portions of the subsiding ocean plate melt when they descend into the asthenosphere, and some of the magma rises to form volcanoes. Mount St. Helens was formed in this manner. Volcanoes are also observed in certain portions of the interior of plates where magma is rising.

Volcanoes have played a vital role in the evolution of planetary atmospheres and of life itself. When solid planets such as Earth, Venus, and Mars were originally formed, various gases were trapped in the rocky interiors. Many of these gases—compounds of hydrogen, carbon, nitrogen, oxygen, and sulfur—were released during volcanic eruptions. As discussed in the previous chapter, the volcanic gases accumulated to form an atmosphere on each planet; and on Earth some of these gases probably combined to form amino acids that served as the building blocks for proteins and finally for living organisms.

19.10 WEATHERING AND EROSION

Millions or tens of millions of years are required for rock to form, distort, and rise to become a mountain range. During this time, the rock is exposed to many different kinds of surface forces. These forces, acting collectively, remove small pieces of rock and carry them downslope. The wearing away and removal of material from the Earth's surface occurs in two stages. First, the rock is broken into small fragments by chemical and mechanical processes. The deterioration of rock into small pieces is called **weathering.** The small bits of rock and soil are then carried away by the action of running water, glacial ice, winds, or waves. This movement of material is called **erosion.**

Chemical Weathering

Air and water, especially when carrying impurities, may be corrosive and therefore can react with many types of rocks and minerals. For example, pure iron is a hard, strong metal. But there are few natural deposits of pure iron near the surface of the Earth. As Earth's crust formed, iron in contact with oxygen reacted to form iron oxides. The oxide known as rust (Fe_2O_3) usually becomes loose and flaky. The conversion of a hard, abrasion-resistant material (iron) to a softer, flaky one (rust) is an example of chemical weathering.

Water is another chemically active substance. Water never exists in its purest state in nature. Many minerals dissolve in pure water; a few, such as sodium chloride (used commonly as table salt), are highly soluble. Even distilled water contains dissolved air, which includes two important reactive components—oxygen (O_2) and carbon dioxide (CO_2). Dissolved oxygen is an oxidizing agent and contributes to the rusting of iron under water. Dissolved carbon dioxide reacts with water to form an acidic solution. Thus, rainwater is slightly acidic and therefore slightly corrosive. This solution is capable of dissolving many types of rocks and therefore carries dissolved mineral matter to the ocean. Other corrosive impurities may enter water systems from a large variety of sources. (Refer back to the discussion of acid rain in Chapter 17.) For example, sulfur compounds present in polluted air or in certain natural rock formations dissolve in water to form strongly acidic solutions. In many industrial regions, atmospheric pollutants have mixed with airborne water droplets to such an extent that the rainfall is acidic enough to kill fish and forests and to corrode statues and buildings. Salt from ocean spray, oxides of nitrogen from combustion or lightning, and many other substances may all dissolve in water and enhance its corrosiveness.

Mechanical Weathering

Temperature Changes. Rocks can also be broken apart by purely mechanical processes. For example, most liquids contract when cooled and shrink even more when they freeze. Water is anomalous: It expands when it freezes. Thus, if water drips into a crack in a rock and then freezes, the resultant expansion acts to push the rock apart. The ice holds the rock from falling, but when the ice melts, sections of rock crumble apart. If you ever climb in a high mountain range in the spring or early summer, when water freezes at night and thaws during the day, you will find that the mountains come alive with falling rocks. You can stand in a narrow valley and listen as the debris tumbles off the high cliffs.

Biological Processes. Plant roots also can crack rocks by expansion (Fig. 19.49). If a little bit of soil collects in a fissure in solid rock, a seed that falls there may start to grow. The roots then work their

FIGURE 19.49 *As this tree grew from a crack in bedrock, its roots widened the crack.*

FIGURE 19.50   *Abrasion rounded these rocks in a stream bed in Yellowstone National Park, Wyoming.*

way down into the rock. As the plant grows and the roots expand, they push the rock apart just as ice does.

FIGURE 19.52   *Windblown sand sculpts exotic figures in bedrock. Grand Canyon, Arizona.*

FIGURE 19.51   *The Grand Canyon formed as an uplifted plateau that was weathered and eroded.*

Moving Water.   Have you ever walked along a stream bed or ocean beach and looked at the rocks lying in or near the water? If you have, you may have noticed that many are rounded and smooth, as indicated in Figure 19.50. Pure water by itself has little abrasive power, but when water is moving rapidly, it picks up silt and sand. When these small particles are hurled against the rocks, the solid material is gradually ground away. During storms and floods, fist-sized stones or even large boulders are pushed by the violent water, and as they tumble along and rub against each other, small bits are broken off. Over long periods of time, the weathering and erosive action of streams and rivers can reshape huge land masses. In Utah and Arizona, the Colorado River has dug tremendous trenches below the level of uplifted plains to form the Grand Canyon and its tributaries, as shown in Figure 19.51. Ocean waves are also abrasive. They can carve away significant portions of a sandstone cliff in a single winter storm by rolling rocks against them in the surf.

Wind.   If purely gaseous air blows against a rocky mountainside, it has little effect. But if silt or sand is suspended in the air and is blown against the rocks by wind, it chips away at the solid mate-

(a) (b)

FIGURE 19.53 *(a) Aerial view of the Sherman Glacier, Alaska. (b) Granite Creek with an alpine glacier in the Eastern Chugach Mountains in Alaska.* (Courtesy of U.S.G.S.)

rial, as shown in Figure 19.52. Thus, the mechanism of weathering by wind is similar to the action by water.

Glaciers. In high mountains and near polar regions, the snow that falls in winter never melts completely during the summer and therefore accumulates from year to year. As the snow is melted and refrozen and compressed by the layers of snow above it, it gradually becomes dense enough to form glacial ice. When the ice builds to a thickness of about 40 m, it forms a **glacier,** as shown in Figure 19.53. When ice near the bottom of a glacier is subjected to the weight of thousands of tons of ice above, it flows slowly, like a semifluid plastic. Glaciers also slide downslope over the tilted bedrock beneath.

Huge glaciers exist in many parts of the Earth. In Greenland, the ice layer is 3000 m thick, whereas in Antarctica the ice cap is 4000 m deep in certain locations. Mountain glaciers are much smaller, sometimes being as thin as 40 to 90 m. As a glacier flows downhill, it picks up pieces of rock and soil, small stones, and even huge boulders weighing many tons. When the solid material is dragged sea-ward, grinding against bedrock, it carves huge valleys and shapes the topography of mountains or even continents.

Erosion

Once rock has been weathered, small pieces of it can be carried away by some moving substance. **Erosion** *is the process whereby weathered material is removed.* Running water, wind, and flowing ice, which are active weathering agents, cause erosion as well. Bare soil is particularly vulnerable to erosion. This movement of material with the subsequent loss of valuable farmland is a serious environmental problem that will be discussed in Chapter 20.

Weathering and erosion do not act equally on all types of rock. Some rocks are hard and resistant to weathering, whereas others are physically soft or chemically reactive. Therefore, some regions wear away faster than others. As a result, valleys may be cut between bastions of harder rock, and sharp cliffs are shaped or rounded in many different ways. Mountain-building, weathering, and erosion act to-

FIGURE 19.54   *Desert sandstone in Canyon Lands, Utah.*

gether to shape the Earth's crust. As subterranean forces push rock upward, surface forces scour the uplifting mass. Thus, mountains and other landforms that we see today were created by a series of opposing forces.

Sometimes it is relatively straightforward to read the history of a landform by studying its present structure. Consider the following two examples:

1. Figure 19.54 is a photograph of rocks in the Utah desert. Even a cursory examination reveals that there are horizontal beds of sedimentary rock. Sedimentary rock is formed in ocean basins or at the bottom of valleys, lakes, or rivers, not on the tops of mountains. Therefore, the structures shown here were once in a low region and have since been uplifted. If you draw imaginary lines from one pillar of rock to another, you see that the horizontal layers line up. From this evidence, we can deduce that all of the rock pillars were once part of a single deposit, which was uplifted and eroded.

2. Figure 19.55a is a cross-sectional map of a mountainous region, showing the different types of rock present in the cores of the mountains. Such a map tells us a great deal about the history of the region. To interpret the data, first draw imaginary lines to reconstruct the landform as if it were not sculptured by erosion, as shown in Figure 19.55b. The reconstructed structure shows a sequence of sedimentary beds overlying an older igneous granite. This sequence was folded upward. At some later time, a fracture developed, and hot magma pushed upward to cool and form a dike of hard, young rock cutting through the older sequence.

CONCEPTUAL EXAMPLE 19.4   WEATHERING AND EROSION

Baffin Island is located in the arctic region of Canada. The southern and eastern parts of the island are mountainous and are composed mostly of granite and other igneous rocks. The climate is ex-

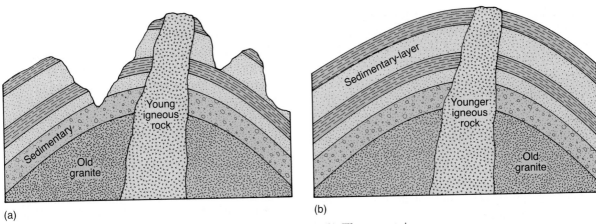

(a)   (b)

FIGURE 19.55 *(a) Cross-section of a hypothetical mountain. (b) The mountain reconstructed as if no weathering or erosion had occurred.*

tremely cold, with bitter winters, short summers, and consequently little plant growth. There is some precipitation year round, but the region is generally dry. Predict what types of erosion and weathering predominate.

Solution The rocks of southeastern Baffin are quite sharp and angular. These have been broken off of the parent formations by the action of ice and extreme differences in temperature. Because the land is frozen much of the year, there has been relatively little rounding action by running water and wind-blown soil.

19.11 FORMATION OF MINERAL DEPOSITS

If you collected a few samples of rock from your neighborhood and had them analyzed, chances are good that you would find traces of iron, copper, silver, gold, and a variety of other valuable metals. The concentration of these metals, however, would be so low that it would be impractical to start a mine in the area. There are a variety of natural processes that have concentrated certain minerals in distinct regions. One of the primary professional objectives of many geologists is to find new ore deposits.

Magmatic Processes

Cooling magma does not solidify and crystallize all at once. Instead, high temperature minerals crystallize first. Lower temperature minerals crystallize later, when the temperature drops. Solid materials are denser than liquid magma. Consequently, crystals that form first sink to the bottom of a magma chamber in a process called **crystal settling** (Fig. 19.56). These crystals form a layer that commonly consists of a single mineral or a mixture of minerals with similar melting points. If the minerals contain valuable metals, an ore deposit may form. The largest ore deposit formed in this way is the Bushveldt intrusion of South Africa. It is about 375 by 300 km in area—roughly the size of the state of

FIGURE 19.56 *Concentration of minerals by crystal settling in a magma chamber.*

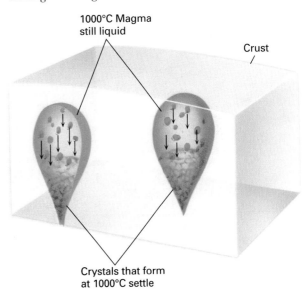

1000°C Magma
still liquid

Crust

Crystals that form
at 1000°C settle

Maine—and about 7 km thick. Large quantities of chromium and platinum are mined from the Bushveldt.

Hydrothermal Processes

Magma and underground rock contain large amounts of hot water with the same dissolved ions found in seawater. This mixture of hot water and dissolved ions is a **hydrothermal solution.** (The word hydrothermal is derived from the roots *hydro* for water and *thermal* for hot.) Hydrothermal solutions are corrosive and can dissolve metals such as copper, gold, lead, zinc, and silver from hot rock or magma. Most rocks contain low concentrations of many metals. For example, gold makes up 0.0000002 percent of average crustal rock, copper makes up 0.0058 percent, and lead makes up 0.0001 percent. Although the metals are present in the rock in low concentrations, hydrothermal solutions percolate slowly through vast volumes of rock, dissolving and accumulating large amounts of the metals. The metals are then deposited when the solutions encounter changes in temperature, pressure, or chemical environment. In this way, hydrothermal solutions scavenge metals from average crustal rocks and then deposit them locally to form ore.

If the metals precipitate in fractures in rock, a hydrothermal **vein** deposit forms as shown in Figure 19.57. Ore veins range from less than 1 mm to several meters in width. Single gold or silver veins have yielded several million dollars' worth of ore. The same hydrothermal solutions that flow rapidly through open fractures to form rich ore veins may also soak into large volumes of rock around the fractures. If metals precipitate in the rock, they may create large but much less concentrated **disseminated ore deposits.** Because they may form from the same solutions, vein deposits and disseminated deposits are often found together. The history of many mining districts is one in which early miners dug shafts and tunnels to follow the rich veins. After the veins were exhausted, miners used huge power shovels to extract low grade ore from disseminated deposits surrounding the veins. An open pit copper mine is shown in Figure 19.58.

Sedimentary Processes

Two types of sedimentary processes form ore deposits: sedimentary sorting and precipitation.

Sedimentary Sorting: Placer Deposits. Gold occurs naturally as a pure metal and is denser than any other mineral. Therefore, if a mixture of gold dust, sand, and gravel is swirled in a glass of water, the gold falls to the bottom first. Differential settling also occurs in nature. Many streams carry silt, sand, and gravel with an occasional small grain of gold. The gold settles first when the current slows

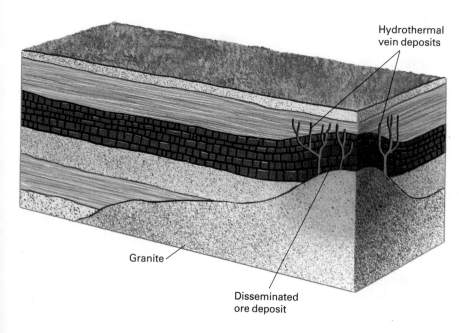

Hydrothermal vein deposits

Granite

Disseminated ore deposit

FIGURE 19.57 *Formation of hydrothermal deposits.*

FIGURE 19.58 *The Bingham open pit copper mine, a disseminated copper deposit.*

down. Thus, grains of gold concentrate near bedrock or in coarse gravel, forming a **placer deposit** (Fig. 19.59).

Precipitation.   As ground water percolates through rock, it dissolves minerals and carries off dissolved ions. In most environments, this water eventually flows into streams and then to the sea. Some of the dissolved ions, such as sodium and chloride, make seawater salty. In deserts, lakes develop with no outlet to the ocean. Water flows into

FIGURE 19.59 *Placer deposits occur in environments where water slows down and sediment is deposited.*

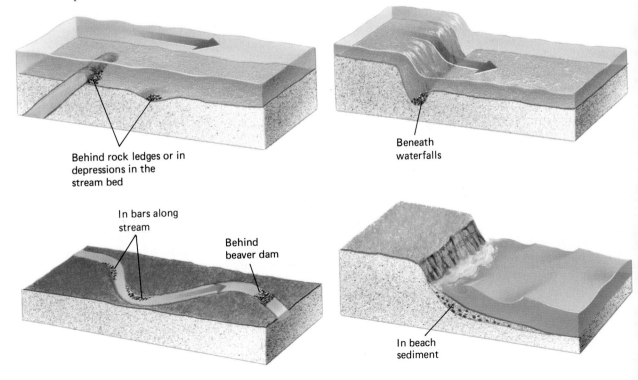

Behind rock ledges or in depressions in the stream bed

Beneath waterfalls

In bars along stream

Behind beaver dam

In beach sediment

(a) Litter falls to floor of stagnant swamp

(b) Debris accumulates, barrier forms, decay is incomplete

(c) Sediment accumulates, organic matter is converted to peat

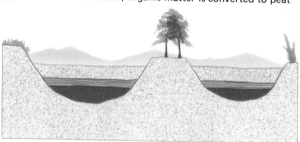

(d) Peat is lithified to coal

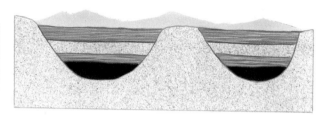

FIGURE 19.60   *Formation of coal deposits.*

the lakes but can escape only by evaporation. As the water evaporates, the dissolved ions concentrate until they precipitate.

You can perform a simple demonstration of evaporation and precipitation. Fill a bowl with warm water and add a few teaspoons of table salt. The salt dissolves, and you see only a clear liquid. Set the bowl aside for a day or two until the water evaporates. The salt precipitates and encrusts the sides and bottom of the bowl.

Salts that form by evaporation are called **evaporite deposits.** Evaporite minerals include table salt, borax, sodium sulfate, and sodium carbonate. These salts are used in the production of paper, soap, and medicines and for the tanning of leather.

Weathering Processes

In environments with high rainfall, the abundant water dissolves and removes most of the soluble ions from the soil and rock. The insoluble ions left behind form **residual deposits.** Both aluminum and iron have low solubility in water. **Bauxite,** our principal source of aluminum, forms in this manner, and in some instances iron also concentrates enough to become ore.

Fossil Fuel Deposits

Three organic deposits found in the Earth are coal, natural gas, and petroleum, the three fossil fuels. These materials are called fossil fuels because they were formed from the remains of decayed plants and animals that lived millions of years ago. Coal is composed primarily of carbon, and gas and petroleum are composed primarily of carbon and hydrogen. Living organisms are composed mainly of carbon, hydrogen, and oxygen. To convert organic tissue into fossil fuels, the tissue must be entrapped in an oxygen-poor environment. If tissue decays in an oxygen-rich environment, it burns or rots and is converted to carbon dioxide and water.

Coal was formed primarily from plant matter. In any forest or swamp, leaves, twigs, branches, and entire trees fall to the ground, where they slowly decompose. However, decomposition is not always complete. If newly fallen litter or sediment covers the old, partially rotted debris before it decomposes fully, some of the organic matter is preserved. If conditions are favorable, the surface layers prevent oxygen from penetrating the deeper sediments. Over time, heat and pressure gradually build up as this preserved matter is further compressed by accumulating debris and soil. The combined heat and pressure from accumulating layers initiate a series of transformations that change the buried plant tissue into **coal** (Fig. 19.60).

Go to a swamp or marsh and dig up a shovelful of the muck on the bottom. If you examine it closely, you will find that it is mostly decayed plant matter. If such a swamp bottom is covered with inorganic sediment and compressed for hundreds of thousands of years, a small coal deposit develops. However, most modern swamps are poor coal producers. It is estimated that a layer of compressed organic debris 12 m thick is required to produce a 1 m layer of coal. For a layer this large to accumulate, conditions must be favorable and stable in a region for many years. Coal deposits probably are being formed today in many areas, notably in the Ganges River delta in India, but the process is extremely slow—much slower than the exploitation of existing reserves. Therefore, because we cannot expect formation of new deposits to keep pace with use, there is compelling reason to conserve our present reserves.

Oil and natural gas formed from tiny marine microorganisms rather than from the debris of large plants. Organic matter eroded from land is carried to the oceans by streams and deposited with mud in shallow coastal waters. As plants and animals living in the sea die and settle to the bottom, they add more organic matter to the mud. This organic-rich mud is then buried by younger sediment. Burial increases temperature and pressure. These conditions convert the mud to shale and the organic material to liquid petroleum (Fig. 19.61). Oil cannot be recovered from shale because shale is relatively **impermeable;** that is, liquids do not flow through it rapidly. If conditions are favorable, petroleum is forced out of the shale and **migrates** to a nearby layer of sandstone or limestone with open spaces called pores between the grains. If liquids can flow readily through porous rock, the rock is said to be **permeable.**

Petroleum, being less dense than water and much less dense than rock, rises through permeable rock. An **oil trap** is any barrier to the upward migration of oil or gas. When oil or gas seeps into the rock below a trap, it accumulates there and forms a reservoir. Most reservoirs form in permeable sandstone or limestone. Many types of traps accumulate oil. In one common type, a dome of impermeable **cap rock,** such as shale, covers the permeable reservoir rock (Fig. 19.61d). The cap rock prevents the petroleum from rising farther.

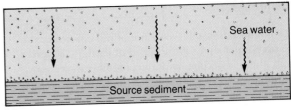

(a)

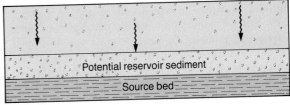

(b)

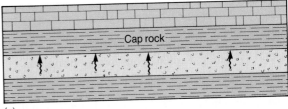

(c)

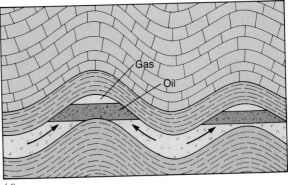

(d)

FIGURE 19.61 *Formation of petroleum. (a) Organic-rich marine sediments fall to the floor of the ocean. (b) This sedimentary layer accumulates. (c) The sediments are covered by a layer of impermeable rock. (d) The rock is deformed. Gas and oil slowly rise and are concentrated in the dome-shaped structures.*

It is important to emphasize that a petroleum reservoir is not an underground pool or lake of oil. Instead, it is permeable, porous rock saturated with oil, more like an oil-soaked sponge than a bottle of oil.

SUMMARY

Geologists deduce the structure of the interior of the Earth from a study of the speed of earthquake waves and their reflection and refraction. P waves move through solids or liquids, but S waves move only through solids. The inner core of the Earth is solid iron and nickel, surrounded by an outer core of molten iron and nickel. A large solid **mantle** surrounds the core. The **asthenosphere** is in an upper section of the mantle and is semifluid and plastic. The **lithosphere** includes the surface **crust** and the uppermost portion of the mantle. The lithosphere floats **isostatically** on the denser asthenosphere beneath.

Igneous rocks form directly from molten magma, or lava, and **sedimentary** rocks form from sediments that are compressed or cemented together. **Metamorphic** rocks form when other types of rock are heated (but not melted) and compressed for long periods of time.

Geological time is determined by **relative** and **absolute** age dating. Earliest fossils formed during the **Archean Eon.** Within a short time at the end of the **Proterozoic Eon** and the start of the **Phanerozoic Eon,** large multicellular plants and animals evolved and dominated the Earth. During the **Paleozoic Era,** organisms migrated from the sea to the land. Dinosaurs dominated the **Mesozoic** landscape; mammals and grasses proliferated during the **Cenozoic Era.**

Plate tectonic theory states that the Earth's surface is broken up into distinct plates that move relative to each other. Plates separate at **divergent boundaries.** The largest divergent boundaries lie in oceanic crust along the mid-oceanic ridges. Plates collide at **convergent boundaries.** If an ocean plate collides with a continental plate, the ocean plate subsides and the edge of the continental plate rises to form mountains. A **transform plate boundary** forms where two plates slide horizontally past one another.

The Andes formed along a subduction zone, the Himalayas formed when two continents collided, and the modern Rockies formed by compression followed by expansion. Earthquakes generally occur where plates slip past each other, where one is subducted, or at mid-oceanic ridges where plates are separating. Volcanoes form when magma rises rapidly to the surface.

The deterioration of rock into small pieces is called **weathering.** Weathering can occur by chemical action or mechanical processes, including temperature changes, biological processes, moving water, wind, and glaciers. **Erosion** is the movement of small bits of rock and soil. Mineral deposits are formed by separation by gravity, separation by differential solubility, formation of placer deposits in streams and lakes, and chemical precipitation of minerals. Fossil fuel deposits are partially decomposed organic debris.

KEY WORDS

Seismology
Surface wave
Body wave
Mohorovičić
 discontinuity
Crust
Asthenosphere
Isostasy
Mineral
Magma
Igneous rock
Sediment
Sedimentary rock
Limestone
Sandstone
Shale
Metamorphism
Marble

Gneiss
Metamorphic rock
Geological time
Principle of superposition
Era
Period
Epoch
Precambrian time
Archean Eon
Proterozoic Eon
Paleozoic
Chordates
Mass extinction
Mesozoic

Pangaea
Laurasia
Gondwanaland
Mid-Atlantic ridge
Mid-oceanic ridge
Divergent boundary
Spreading center
Subduction
Oceanic trench
Transform plate
 boundary
East Pacific rise
Earthquake
Fault creep
San Andreas fault
Seismic gap theory
Foreshock

Volcano
Weathering
Chemical weathering
Mechanical weathering
Glacier
Erosion
Magmatic processes
Crystal settling
Hydrothermal processes
Disseminated ore deposit
Sedimentary processes
Placer deposits
Precipitation
Evaporite deposit
Residual deposit

CONCEPTUAL QUESTIONS

Structure of the Earth

1. Explain how geologists use seismic waves to determine the composition of rock in the Earth's interior and changes in rock structure with depth.

2. Imagine you were studying seismic waves from three distant planets. Given the data below, describe how each planet differs from Earth. **Planet X:** Both P and S waves travel through the planet's core. **Planet Y:** Both P and S waves travel straight through the entire planet. **Planet Z:** Seismic wave velocity increases sharply at a depth of 250 km.

3. Briefly outline the interior structure of the Earth. What regions are solid and brittle? Plastic? Molten?

4. Explain the concept of isostasy. Give two consequences of isostatic behavior.

Rock cycle

5. What is magma? How does the rate of cooling of magma affect the texture of the resultant rocks? What do we call magma when it travels quickly and violently to the surface? What kind of rock is formed when magma flows slowly upward but cools and solidifies before it reaches the surface?

6. Trace possible geological processes for each of the following transformations: (a) metamorphic rock changing to sedimentary rock, (b) sedimentary rock changing to metamorphic rock, (c) sedimentary rock changing to igneous rock.

7. For each of the examples below, categorize the rock as igneous, sedimentary, or metamorphic. Discuss the reasoning behind your choice. (a) Obsidian is a volcanic glass formed when magma cools so rapidly that crystals do not have time to grow. (b) Schist is a rock with easily visible crystals that form when the clay grains in shale are altered by high temperature and pressure. (c) Conglomerate is composed of stones cemented together within a fine-grained matrix of sand or silt.

Geological time

8. Discuss the differences between relative and absolute age dating. What types of observations are required for each type of dating? What information is obtained?

9. The half-life of some common isotopes found in rocks are:

   potassium-40, 1.3 billion years

   uranium-238, 4.5 billion years

   rubidium-87, 47 billion years

   If you had a sample of rock and suspected, from relative age dating, that it was about 10 million years old, which isotope would you study to determine its absolute age? Defend your answer.

10. No traces of life are known in Hadean rocks, and abundant fossils are found in rocks of the Pro-

terozoic Eon. Therefore, life must have originated sometime during the Archean Eon. Discuss what the Archean environment must have been like for life to have evolved in that time.

11. According to one modern theory, periodic meteorite impacts have led to mass extinctions during the Phanerozoic Eon. Is it reasonable to assume that similar events occurred during the Archean? Discuss problems in searching for evidence for your conclusion.

12. Describe life in the Cambrian period.

13. When did organisms migrate onto land? Briefly describe early terrestrial organisms.

14. Describe life at the end of the Paleozoic era.

15. Explain why the Phanerozoic Eon contains so many subdivisions, whereas Precambrian time, which is about nine times longer, has so few.

Plate tectonics

16. List three types of reasoning that Wegener used to support his idea that the continents migrate across the globe.

17. When continental drift theory was first proposed, it was believed that the continents must somehow plow through mantle rock as a ship plows through the water. This analogy has since been discarded. Explain how the continents move and why this movement cannot be compared to a ship traveling across the sea.

18. The climate of many regions of the globe is changing. Could these climatic changes be caused by tectonic plate movement? Defend your answer.

19. The deepest ocean trenches in the world lie on the western edge of the Pacific Ocean. What types of geological activity would you expect to find in the region?

20. Describe each of the three types of plate boundaries and discuss the geological environments at each.

Mountain formation

21. Why are there deep ocean trenches adjacent to some large mountain ranges such as the Andes?

22. The Himalayas contain extensive regions of sedimentary rock, and marine fossils have been found at high elevations. Explain these facts in terms of the theory for the Himalayan formation as described in the text.

Earthquakes and volcanoes

23. Explain why some regions of the globe are more likely than others to experience earthquakes.

24. It has been suggested that engineers should inject large quantities of liquids into locked por-

tions of the San Andreas fault. Proponents of the plan believe that these liquids will reduce friction by lubricating the sides of the fault, and thus the rock will slide along slowly, pressure will be relieved, and a major earthquake will be averted. If you were the Mayor of San Francisco, would you encourage or discourage the injection of fluids into the fault structure? Defend your decision.

25. Explain why geologists can confidently state that earthquakes will occur in the Los Angeles area, but no one can forecast exactly when and where the next big one will strike.

26. Significant earthquakes have occurred in Parkfield, California, in 1857, 1881, 1901, 1922, 1944, and 1966. Draw a graph with the dates on the vertical (Y) axis and the numbers of the events (simply 1, 2, 3, and so on) spaced evenly on the X axis. Use your graph to forecast when the next earthquake might occur in Parkfield. What conclusion can you draw from your forecast?

27. Imagine that geologists predict a major earthquake in a densely populated region. The prediction may be right or it may be wrong. City planners may heed it and evacuate the city or ignore it. The possibilities lead to four combinations of predictions and responses, which can be set out in a grid as follows:

DOES THE PREDICTED
EARTHQUAKE REALLY OCCUR?

	YES	NO
IS THE CITY EVACUATED? YES		
NO		

For example, the space in the upper left corner of the grid represents the situation in which the predicted earthquake occurs and the city is evacuated. For each space in the square, outline the consequences of that sequence of events.

28. Are the Hawaiian Islands composed primarily of igneous, sedimentary, or metamorphic rock? Explain.

29. Two volcanoes that have erupted in recent time are Mt. St. Helens in Washington and Mauna Loa in Hawaii. Mt. St. Helens erupted violently, and Mauna Loa has erupted in gentle lava flows. What types of eruption would you predict for each mountain in the future? Defend your answer.

30. Do you think that volcanic activity might be likely to occur near earthquake zones, or would you guess that no correlation would be observed? Defend your answer.

Weathering and erosion

31. What process is responsible for each of the following observations and phenomena? Is the process mechanical or chemical? (a) A board is sawed in half. (b) A board is burned. (c) A cave forms when water seeps through limestone. (d) Calcite precipitates from a hot underground spring. (e) Meter-thick sheets of granite peel off a newly exposed rock face. (f) In mountains of the temperate region, rockfall is more common in the spring than in midsummer.

32. Arctic regions are cold most of the year, and summers are short. In the temperate regions, spring, fall, and summer are longer and winter is shorter. How do these differences affect weathering and erosion?

33. Explain how wildfires might contribute to erosion.

Mineral deposits

34. It is common for a single mine to contain fairly high concentrations of two or more minerals. Discuss how geological processes might favor the deposition of two similar minerals in a single location.

35. If one compound is to be separated from a complex mixture, it must somehow be transported away from the rest of the material. Explain how ores are moved out of a mixture in each of the following processes: (a) crystal settling, (b) hydrothermal process, (c) formation of placer deposits.

36. If you were searching for petroleum, would you search primarily for sedimentary rocks, metamorphic rocks, or igneous rocks? Which of these three would be the least likely to contain petroleum? Explain.

CHAPTER 20

ENVIRONMENTAL GEOLOGY

The Kennecott open pit copper mine near Salt Lake City is one of the largest in the world. (USDA)

Three thousand years ago, farmers in the desert valleys of the Middle East were forced to abandon their land when salt from irrigation water poisoned the soil. During the first century, B.C., Plato wrote emotionally about soil loss and ground water depletion after the Athenian hills were deforested.

> What now remains compared with what then existed is like the skeleton of a sick man, all the fat and soft earth having wasted away, and only the bare framework of the land being left. . . . [Before the deforestation, the land was] enriched by the yearly rains from Zeus, which were not lost to it, as now, by flowing from the bare land into the sea; but the soil it had was deep, and therein it received the water, storing it up in the retentive loamy soil to provide all the various districts with abundant supplies of spring-water and streams, whereof the shrines which still remain even now, at the spots where the fountains formerly existed.

Thus, since the beginning of civilization, people have understood that human activities can disrupt geological and biological systems. Today the study of interactions between humans and the environment has become a significant component of geology. Modern geologists must address environmental issues, not as advocates of any particular philosophy, but as scientists who ask questions, collect data, and seek answers based on their observations and experiments. In this chapter, we will study soil, water, and minerals, three of the basic necessities of civilization and of life itself. We will also see how human activities can endanger these resources.

20.1 SOIL

Rock exposed at the Earth's surface breaks into smaller fragments as it weathers, and much of it decomposes to clay and sand. Thus, a layer of loose

507

rock fragments mixed with clay and sand overlies bedrock. This material is called **regolith.** In engineering and construction, "regolith" and "soil" are interchangeable terms. Soil scientists, however, define **soil** as the upper layers of regolith that support plant growth. That is the definition we use here.

Components of Soil

Soil is a mixture of mineral grains, organic material, water, and gas. The mineral grains include clay, silt, sand, and rock fragments. Clay grains are so small and closely packed that water and gas do not flow through them readily. Pure clay is so impermeable that plants growing in clay soils often become waterlogged and suffer from lack of oxygen. In contrast, water flows easily through sandy soil. The most fertile soil is **loam,** a mixture of sand, clay, silt, and generous amounts of organic matter.

When plant or animal matter dies and falls, it retains its original shape until it begins to decay. Thus, if you walk through a forest or prairie, you can find bits of leaves, stems, and flowers on the surface. This material is called **litter** (Fig. 20.1). When litter decomposes so that you can no longer determine the origins of the individual pieces, it becomes **humus.** Humus is an essential component of most fertile soils. Scoop up some forest or garden soil with your hand. Soil rich in humus is light and spongy and readily absorbs water. It soaks up so much moisture that it swells after a rain and shrinks during a dry spell. This alternate shrinking and swelling keeps the soil loose, allowing roots to grow easily. A rich layer of humus also insulates deeper soil from heat and cold and reduces water loss by evaporation.

Soil nutrients are chemical elements necessary for plants to grow. Some examples are phosphorus, nitrogen, and potassium. Humus holds nutrients in soil and makes them available to plants. Intensive agriculture commonly destroys humus by exposing it to erosion and oxidation. Farmers then replace the lost nutrients with chemical fertilizers to maintain crop growth. However, loss of humus reduces the natural ability of soil to conserve water and nutrients. Water can then flow over the surface, eroding the soil. In addition, the muddy runoff carries fertilizer and pesticide residues, contaminating streams and ground water.

Soil Profiles

If you dig down through undisturbed soil, you can see several layers, or **soil horizons,** as shown in Figure 20.2. The uppermost layer of mature soil is the **O horizon,** named for its organic component. This layer is mostly litter and humus with a small proportion of minerals. The next layer down, called the **A horizon,** is a mixture of humus, sand, silt, and clay. The thicker layer including both O and A horizons is often called **topsoil.** A kilogram of average fertile topsoil contains about 30 percent by weight organic matter, including approximately 2 trillion bacteria, 400 million fungi, 50 million algae, 30 mil-

FIGURE 20.1 *Litter is organic matter that has fallen to the ground and started to decompose but still retains its original form.*

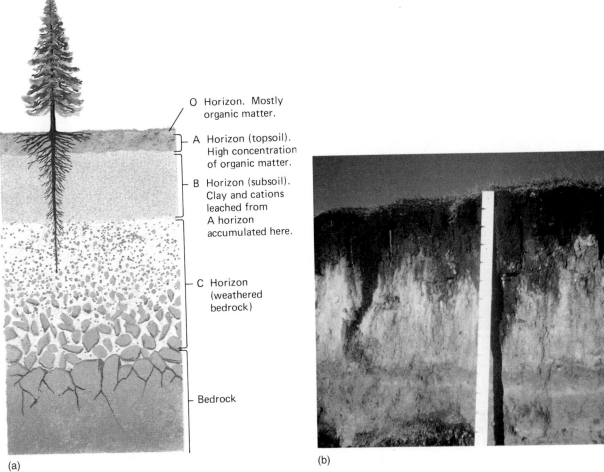

O Horizon. Mostly
organic matter.

A Horizon (topsoil).
High concentration
of organic matter.

B Horizon (subsoil).
Clay and cations
leached from
A horizon
accumulated here.

C Horizon
(weathered
bedrock)

Bedrock

(a)

(b)

FIGURE 20.2 *(a) A well-developed soil commonly shows several horizons. (b) Soil horizons are often distinguished by color and texture. The dark upper layer is the A horizon; the white lower layer is the B horizon.* (U.S. Department of Agriculture)

lion protozoa, and thousands of larger organisms such as insects, worms, and mites.

The third layer, the **B horizon,** or subsoil, is a transitional zone between topsoil and weathered parent rock below. Roots and other organic material occur in the B horizon, but the amount of organic matter is low. The lowest layer, called the **C horizon,** consists of partially weathered bedrock. It lies directly on unweathered parent rock. This zone contains little organic matter.

When rainwater falls on soil, it sinks into the O and A horizons. As it travels through the topsoil, it partially dissolves minerals and carries the dissolved ions to lower levels. This downward movement of dissolved material is called **leaching.** The A horizon

is sandy because water also carries clay downward but leaves the sand grains behind. Because materials are removed from the A horizon, it is called the **zone of leaching.**

Dissolved ions and clay carried downward from the A horizon accumulate in the B horizon, which is therefore called the **zone of accumulation.** This layer retains moisture because of its high clay content. Although moisture retention may be beneficial, if too much clay accumulates, a dense, waterlogged soil can develop. Rain seeps downward through soil, but other factors pull the water back upward. Roots suck soil water toward the surface, and water near the surface evaporates. In addition, water is electrically attracted to soil particles. If the

FIGURE 20.3 *Capillary action causes colored water to soak upward through the pores in a paper towel.*

pore size is small enough, water can be drawn upward by **capillary action.** Capillary action can be demonstrated by placing the corner of a paper towel in water and watching the water rise as in Figure 20.3.

During a rainstorm, water percolates down through the A horizon, dissolving soluble ions such as calcium, magnesium, potassium, and sodium. In arid and semiarid regions, when the water reaches the B horizon, capillary action and plant roots then draw it back up toward the surface, where it evaporates. As it evaporates or is taken up by plants, many of its dissolved ions precipitate in the B horizon, encrusting the soil with salts. A soil of this type is a **pedocal** (Fig. 20.4a).

In a wet climate, ground water leaches soluble ions from both the A and B horizons. The less soluble elements such as aluminum, iron, and some silicon remain behind, accumulating in the B horizon to form a soil type called a **pedalfer** (Fig. 20.4b). The subsoil in a pedalfer commonly is rich in clay, which is mostly aluminum and silicon, and has the reddish color of iron oxide.

In regions of high rainfall, such as a tropical rainforest, so much water seeps through the soil

that nearly all the cations are leached away. Only very insoluble aluminum and iron minerals remain. Soil of this type is called a **laterite** (Fig. 20.4c). Laterites are often colored rust-red by iron oxide. A highly aluminous laterite, called **bauxite,** is the world's main source of aluminum ore.

20.2 SOIL EROSION

Weathering decomposes bedrock, and plants add organic material to the regolith to create soil at the Earth's surface. However, soil does not continue to accumulate and thicken throughout geological time. If it did, the Earth would be covered by a mantle of soil hundreds or thousands of meters thick, and rocks would be unknown at the Earth's surface. Instead, flowing water, wind, and glaciers erode soil as it forms (Fig. 20.5). In addition, some weathered material simply slides downhill under the influence of gravity. In fact, all forms of erosion combine to remove soil about as fast as it forms. For this reason, soil is usually only a few meters thick or less in most parts of the world.

Once soil erodes, the sediment begins a long journey as it is carried downhill by the same agents that eroded it: streams, glaciers, wind, and gravity. During the journey, the sediment may come to rest in a stream bed, a sand dune, or a lake bed, but those environments are usually temporary stops. Sooner or later most sediment erodes again and is carried further downhill, until it is finally deposited where the land meets the sea. There it remains and is buried by younger sediment until it lithifies to form sedimentary rocks.

In nature, soil forms about as fast as it erodes. However, improper farming, livestock grazing, and logging can accelerate erosion. Plowing removes plant cover that protects soil. Logging often removes forest cover, and the machinery breaks up the protective litter layer. Similarly, intensive grazing can strip away protective plants. Rain, wind, and gravity then erode the exposed soil. Meanwhile, soil continues to form by weathering at its usual slow, natural pace. Thus, increased rates of erosion caused by farming and logging can lead to net soil loss.

Soil is eroding more rapidly than it is forming on about 35 percent of the world's croplands. About 23 trillion kg of soil are lost every year. The soil lost annually would fill a train of freight cars

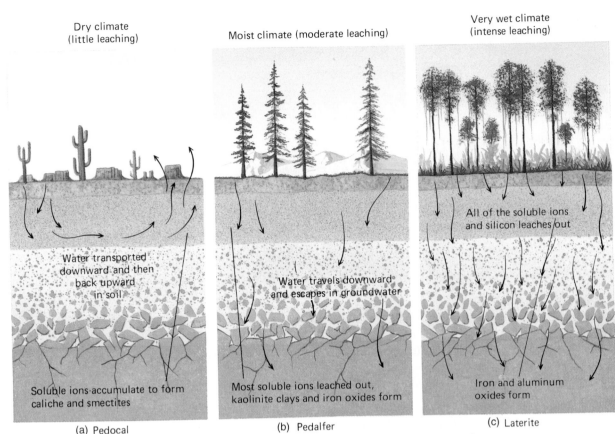

Dry climate
(little leaching)

Moist climate (moderate leaching)

Very wet climate
(intense leaching)

Water transported
downward and then
back upward
in soil

Water travels downward
and escapes in groundwater

All of the soluble ions
and silicon leaches out

Soluble ions accumulate to form
caliche and smectites

Most soluble ions leached out,
kaolinite clays and iron oxides form

Iron and aluminum
oxides form

(a) Pedocal

(b) Pedalfer

(c) Laterite

FIGURE 20.4 *Different climates allow for the formation of various soil types. (a) Pedocal soil is formed in a dry climate, (b) pedalfer soil in a moist climate, and (c) laterite soil in a very wet climate.*

FIGURE 20.5 *Flowing water has eroded gullies in this hillside.* (Don Hyndman)

long enough to encircle the Earth 150 times. In the United States, approximately one third of the topsoil that existed when the first European settlers arrived has been lost (Fig. 20.6). Erosion is continuing in the United States at an average rate of about 10.5 tons per hectare* per year. In some regions, however, the rate is considerably higher, and yearly losses of more than 25 tons per hectare are common. To put this number in perspective, a loss of 25 tons per hectare per year would lead to complete loss of the topsoil in about 150 years.

Soil erosion can be prevented by simple measures. For example, if farmers plow up and down a slope, each furrow acts as a tiny stream bed. Water accelerates downhill through these channels and

*One hectare equals 2.47 acres.

FIGURE 20.6   *Fertile topsoil being washed away in an untended field in eastern Missouri.* (From Levin, H., *Contemporary Physical Geology,* 3rd ed. Philadelphia, Saunders College Publishing)

erodes the soil. A better practice, called **contour plowing,** is to plow at right angles to the slope, so each furrow creates a small dam to interrupt the flow of water. On steeper hillsides, farmers build **terraces** to provide a series of flat surfaces along a slope. Terracing is especially popular in mountainous regions where food and arable land are scarce. Windbreaks check erosion by high wind. Figure 20.7 illustrates some of these soil conservation practices.

FIGURE 20.7   *(a) Farmers cut terraces into the steep hillsides surrounding Kangding in Western China. (b) The contour farming techniques used on this farm are an excellent example of conservation methods designed to preserve the soil resource.* (U.S. Department of Agriculture.)

(a)

(b)

Chemical Fertilizers and Soil Deterioration

Chemical fertilizers have increased global food production and have thus helped feed the expanding human population. For this reason, modern farmers all over the world use fertilizers in ever-increasing amounts, but the long-term effects of chemical fertilizers cause serious concern. In types of intensive agriculture in which chemical fertilizers are used, crop residues such as straw are removed and soil humus is thus continuously depleted. When humus is lost, soil retains less water, nutrients leach from the soil, and the agricultural system becomes dependent on irrigation and chemical fertilizers.

One solution to humus depletion is to use agricultural wastes or other organic matter as fertilizers. Natural fertilizers such as manure are usually more expensive than chemicals. However, many soil scientists calculate that if one were to view profit and loss over a period of decades, not just one or two growing seasons, it would be economically advantageous to use natural fertilizers.

Thus, soil preservation is largely an economic problem. An individual farmer can profit more in a single year by planting crops instead of rows of windbreak trees, by using inorganic rather than organic fertilizers, or even by leaving land without plant cover for a period. Because the economic loss from soil erosion may not be felt for several years or even until the next generation, it is tempting to respond to short-term economic needs rather than to conserve for the future. Although governments in many countries are beginning to respond to the problem of continued soil erosion, they face the difficult task of designing economic incentives that will encourage farmers to conserve soil.

CONCEPTUAL EXAMPLE 20.1 CHEMICAL AND ORGANIC FERTILIZERS

One of the advantages of chemical fertilizers is that they are so concentrated that only a small amount of material is needed to add the chemicals necessary for plant growth. Manure contains fewer nutrients per ton, so that fertilization with manure imposes much higher handling costs. What other arguments are important in deciding whether to use manure or chemical fertilizers on a farm?

Solution Although manure is more expensive to handle, it adds organic matter to the soil in addition to chemical nutrients. The resultant humus regulates flow of water and nutrients, improving the fertility and long-term health of the soil.

20.3 LOSS OF FARMLAND TO URBANIZATION

A person living in a rural community can raise enough vegetables on 0.07 hectare of land (0.17 acre) to support a family of four. In contrast, the average urban American requires almost twice as much land per person (0.13 hectare or 0.32 acre) for nonagricultural purposes—for homes, lawns, roadways, parking lots, shopping centers, and factories. Much of this urban development has occurred in regions that were once prime farming areas. The loss of farmland to urban sprawl has become a serious concern.

In the United States, a total of 1.2 million hectares of land (3 million acres) is paved with concrete and asphalt or replanted with ornamental lawns every year. This is equivalent to an area nearly the size of the state of Connecticut. Unfortunately the loss is concentrated in many of the most productive agricultural areas. If current trends continue, most of Florida's citrus groves, 16 percent of the vegetable-producing regions of Southern California, and 24 percent of the prime agricultural land in Virginia will be removed from production by the year 2000.

The same trend is occurring in the less developed countries, although at a somewhat slower rate. Although many people are urbanized in these countries, they use less space per person because many live in crowded slums, and elaborate roadways and shopping centers are practically nonexistent. Yet estimates show that if present trends continue, the cropland that will have been urbanized in these areas between 1980 and the year 2000 represents the agricultural capacity to feed 84 million people.

Is there any way this trend can be reversed? In a well-planned society, the rate of change could certainly be reduced. Two examples illustrate the point.

Suburban subdivisions are characterized by single story, one-family dwellings on small plots of land. Suburban shopping centers are often single story, sprawling stores and malls. If these were replaced with multistory apartments and department stores, a considerable area of land could be saved.

Transportation systems are similarly inefficient. A two-track local subway uses a roadbed 11 m wide and can carry 80,000 passengers per hour. On the other hand, an eight-lane superhighway is 38 m wide and carries only 20,000 people per hour under normal traffic conditions. A superhighway capable of carrying 80,000 people per hour would have to be 152 m wide (approximately 1.5 times as wide as the *length* of a football field). Therefore, a shift to mass transit would conserve land surfaces.

20.4 THE HYDROLOGIC CYCLE (WATER CYCLE)

If you travel widely on land, by sea, and in the air and look at the Earth's waters, three observations become apparent. First, much of the water is *stored* in places that look rather permanent. The largest quantities, of course, are in the oceans. But there are also the Greenland and Antarctic Ice Caps as well as many smaller glaciers and lakes. Second, much of the Earth's water is in motion: Snow and rain fall, clouds drift, and rivers flow toward the sea. Third, the water on land is unevenly distributed. As you wander through tropical jungles, everything is wet, and water often drips on you throughout the day. But you had better not try to trek across Australia, Libya, or even southern California without taking all your water with you. These regions get very little rainfall; most of the year they get none.

The movement of water on Earth is called the **hydrologic cycle.** Water transport occurs by evaporation, precipitation, and runoff, as shown in Figure 20.8.

Evaporation, or **vaporization,** is the transformation of liquid water to water vapor. Dissolved minerals remain behind when water evaporates. Most water vapor is produced by evaporation of liquid water from the surface of the oceans. Water can also vaporize *through* the tissues of plants, especially from leaf surfaces. This process is called **transpiration. Precipitation** means falling from a height. Referring to water, precipitation includes all forms in

FIGURE 20.8 *The hydrologic cycle shows that water is constantly recycled among the sea, the atmosphere, and land. Numbers are thousands of cubic kilometers of water transferred each year. Percentages are proportions of total global water in different portions of the Earth's surface.*

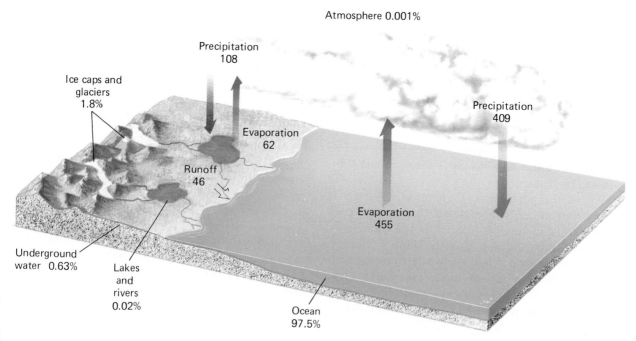

Atmosphere 0.001%

Precipitation
108

Precipitation
409

Ice caps and
glaciers
1.8%

Evaporation
62

Runoff
46

Evaporation
455

Underground
water 0.63%

Lakes
and
rivers
0.02%

Ocean
97.5%

which atmospheric moisture descends to Earth: rain, snow, hail, and sleet. The water that enters the atmosphere by vaporization must first condense into liquid (clouds and rain) or solid (snow, hail, and sleet) before it can fall. **Runoff** is the flow back to the oceans of the precipitation that falls on land. In this way, the land returns the water that was carried to it by clouds that drifted in from the ocean. Runoff occurs both from the land surface (rivers) and from underground water.

The numbers given in percentages in Figure 20.8 show the portion of the Earth's total water in different reservoirs. The total quantity of global water is about 1.35 billion km^3. Note that of this vast amount, 97.5 percent lies in the ocean and 1.8 percent is locked in glaciers. Only 0.8 percent is in the form of inland and underground waters and most of this quantity (0.6 percent) is underground. The remaining 0.2 percent is the inland surface water such as lakes and streams. The least amount of water is in the atmosphere (0.001 percent). Because all this water is in motion, it follows that all the waters of the Earth renew themselves; that is, they move from place to place.

An important question is: How long does it take for water in a given part of the Earth to renew itself? Consider, for example, two flows of pure water, one that goes into a small basin and the other into a large one, as shown in Figure 20.9. Both basins are well stirred, so that the water in each is always uniform throughout. If the water in both basins is polluted, the time it takes for the fresh water to rinse out the pollutant depends on the flow

TABLE 20.1	Average Residence Times of Water Resources	
LOCATION	**AVERAGE RESIDENCE TIME**
Atmosphere	9–10 days
Ocean	
Shallow layers	100–150 years
Deepest layers	30,000–40,000 years
World ocean average	3000 years
Continents	
Rivers	2–3 weeks
Lakes	10–100 years
Ice caps and glaciers	10,000–15,000 years
Shallow ground water	up to 100s of years
Deep ground water	up to 1000s of years

rate of the water and the volume of the basin. The greater the flow rate and the smaller the basin, the faster is the rinsing action. The *average* time that a water molecule spends in the basin is called the **residence time.** Table 20.1 gives average residence times for water in various parts of the hydrologic cycle.

Note that water spends the least time in the atmosphere and the longest time in the deepest ocean layers. Changes in global energy patterns can therefore readily affect atmospheric moisture and hence rainfall and agricultural productivity. The fresh water available for human use is the runoff from rivers and underground sources. Rivers renew themselves rapidly (in weeks), but ground water takes much longer (hundreds to thousands of years). Pollution of waters with long residence times is not easily reversed.

20.5 HUMAN USE OF WATER

Water is used in the home or office, in industry, in agriculture, and for recreation. Both the quantities used and the water quality needed vary widely, depending on the application. In the home, for example, the amount of water used in one toilet flush would satisfy the drinking requirements of an adult (2 L per day) for about 1½ weeks; the water used for one load of laundry in a clothes washer would be enough for drinking for almost 6 weeks. The amounts used in industry and agriculture are far greater than those needed for any personal use. For example, the water used in industry to refine a

FIGURE 20.9 *The same rate of water flow rinses pollutants out of a small basin faster than out of a large one.*

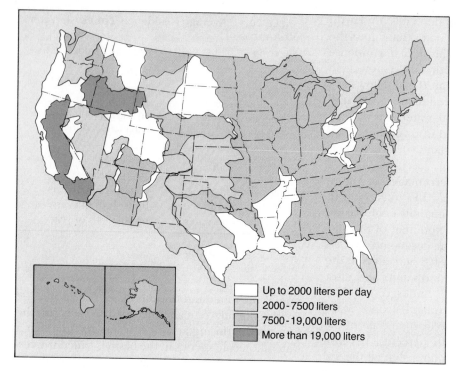

FIGURE 20.10 *Per capita fresh water consumption in the United States (1 gallon = 3.785 L). This includes all types of consumption—household, industrial, and agricultural.*

tonne of petroleum would be enough to do about 200 loads in a clothes washer. When crops are irrigated, it takes much more water to grow a tonne of grain than it does to manufacture a tonne of most industrial materials such as metals or plastics.

The strictest requirements for quality apply to drinking water for humans. The least strict requirements probably apply to water used for cooling, where the prime concern is its temperature. Seawater is therefore adequate. For some industrial applications, the most important consideration is whether the water will corrode the equipment; control of acidity is often the only requirement in such cases. For most human needs, however, including the large amounts used in agriculture and industry, water must be fresh, not salty. Figure 20.10 shows the pattern of consumption of fresh water in the United States. Note that the highest water consumption does not occur in the most densely populated areas. In general, rates of water consumption reflect the needs of agriculture much more than those of the home, of commerce, or of industry. One conclusion from this difference is that efforts to conserve water in the home, while locally helpful, cannot make a significant contribution to the demands of agriculture. A comparison of Fig-

ure 20.11 with Figure 20.10 shows that some of the areas where the surface water supply is often depleted are also the areas where the largest quantities of fresh water are needed. Even in years of average rainfall, much of the Midwest and Southwest of the United States depletes most of its surface waters. Therefore, people transport water large distances or use ground water.

20.6 WATER DIVERSION PROBLEMS

One obvious solution to the problem of local water shortages is to divert water from an abundant source to a dry region where the water is needed. Southern California offers examples of this type of water diversion. The All American Canal channels water from the Colorado River to the farms and cattle ranches of the Imperial Valley, east of San Diego. The great Los Angeles Aqueduct, completed in 1913, brings water south from Owens Valley to Los Angeles. More recently, several large diversion projects have been considered in other regions. Perhaps the most ambitious of these is a proposal to divert water from the southern portions of Hud-

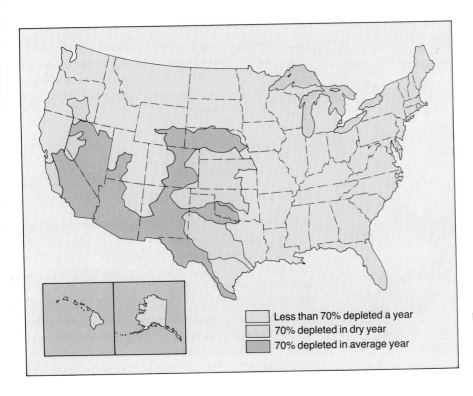

FIGURE 20.11 *Fresh water consumption compared with water supply in the United States.*

Legend:
Less than 70% depleted a year
70% depleted in dry year
70% depleted in average year

son's Bay, in northeast Canada, to the midwestern regions of the United States. Such large water diversion projects can create a number of problems at both ends—at the source and at the area to which the water is supplied.

1. **Encouragement of Waste.** Planners often underestimate the ability and willingness of people to conserve when it is necessary to do so. (The unexpected decline in energy use in the early 1980s is one such example.) Most people use water wastefully when they know that it is abundant, but when it is scarce, conservation is not seen as a serious burden. Thus, it is convenient to leave the water running while you are washing your hands or brushing your teeth. But if you must, you can use only about one-tenth as much water in a stoppered sink to accomplish the same purpose in about the same time. There are many other ways to conserve water in the home, on the farm, and in water-distribution systems. In many cases, these measures would be an adequate or at least a partial substitute for water diversion projects.

2. **Salinity.** Irrigation is almost as old as agriculture itself. The ancient Egyptians, Babylonians, Chinese, and Incas all brought water from nearby rivers to increase the yields of their crops. Today a large portion of the world's crops of vegetables and some grains depends on irrigation. With imported water, marginal farmland has become more productive, and even former deserts are being farmed. Despite these successes, irrigation leads to some environmental problems.

When rainwater falls on mountainsides, it collects in small streams and in ground water. As it flows downward, it filters over, under, and through rock formations. The water dissolves mineral salts present in the rock and soil. Therefore, river water is slightly salty. In most cases, you can't taste the salt, but it is there. If this water is used for irrigation, farmers are bringing slightly salty water to their fields. When water evaporates, the salt remains. Thus, over the years, the salt content of the soil increases slowly. Because most plants cannot grow in salty soil, the fertility of the land decreases. In Pak-

FIGURE 20.12　*Waters from the Nile even today are pumped up to farmland by human power, using a hand pump.*

istan, an increase in salinity decreased soil fertility alarmingly after 100 years of irrigation. In parts of what is now the Syrian desert, archaeologists have uncovered ruins of rich farming cultures. However, the land lying near the ancient irrigation canals is now too salty to support plant growth. California farms now produce about 40 percent of the vegetables consumed in the United States. Here, too, salinity is threatening productivity. As a result, elaborate and expensive drainage systems have been built to draw off salty water. These measures alleviate the problem but do not eliminate it entirely.

3. **Energy Consumption.** The great aqueducts of ancient Rome are sloped downward; there were no pumps. Ancient Egyptian farmers pumped the waters of the Nile a meter or so up to their farms by human or animal power; many do the same to this day, as shown in Figure 20.12. Modern water diversion projects, however, carrying water over hilly terrain, use electrically driven pumps. Diversion projects involve *large* quantities of water (hundreds of cubic me-

ters per second) and therefore need much energy. One estimate for an expanded California State Water Project foresees a use of 10 billion kWh of electricity in the year 2000—about as much energy as is used in 2 million homes.

20.7 POLLUTION OF INLAND WATER BY NUTRIENTS

All animals, even those that live under water, require oxygen to survive. On the surface of the Earth, oxygen is readily available. Oxygen is also dissolved in most bodies of water, but in general the supply under water is much less plentiful than it is on land.

In a natural system, the growth of all organisms is controlled by the quantity of the nutrients available. The entire system is delicately balanced. If the supply of any essential ingredients is increased or decreased, the entire system may be upset. Imagine that there is a clean, cool, flowing river that supports a healthy population of trout and salmon. The fish share the stream with populations of microorganisms, larger plants, small aquatic worms, and many other types of organisms. Now suppose that someone dumps some sewage into this stream. The sewage provides nutrients for plants and animals. Microorganisms and algae grow faster, as shown in Figure 20.13. The fish that eat these organisms are nourished as well. Yet the introduction of sewage may lead to severe ecological disruptions. In general, populations of smaller animals and plants grow and reproduce faster than populations of larger animals. All these animals consume large quantities of oxygen. If the growth is rapid enough, most of the oxygen in the water is used up, and fish may eventually suffocate and die.

It is important to understand that the sewage, by itself, does not kill fish. In fact, it nourishes them. But the sewage supports the growth of other forms of life that consume the oxygen. It is the lack of oxygen that kills fish. When so many nutrients have been added to a body of water that the fish die, the resulting condition is called **eutrophication.** Many lakes, rivers, and bays throughout the world are polluted in this manner.

Sewage is not the only material that can fertilize lakes and rivers and lead to eutrophication.

(a)   (b)

FIGURE 20.13   *(a) Crater Lake in Oregon is low in nutrients. The water is clear and blue.* (Rich Buzelli/Tom Stack and Associates) *(b) Eutrophic lakes and ponds, such as this one in western New York State, are often covered with slimy, smelly mats of algae and cyanobacteria.* (Visuals Unlimited/W.A. Banaszewski)

Sometimes fertilizers that have been spread on farmers' fields wash into waterways. Fertilizers promote plant growth on land and in the water. If they are applied to a cabbage field, the cabbage grows well. If they dissolve in rainwater and flow into aquatic systems, algae and other aquatic plants and the organisms that feed on them grow well, too. These growing organisms consume, and may eventually deplete, the available oxygen. Then, as mentioned, fish die.

There has been considerable public discussion of laundry detergents and their role in water pollution. Many modern detergents contain phosphates, which are an essential component of agricultural fertilizers. More importantly, phosphates are frequently in short supply in natural waters. Without this nutrient, aquatic plants cannot grow in abundance. When phosphate detergents are discharged into waterways, they supply a needed nutrient for the growth of plants. Sometimes, the results far exceed unsophisticated expectations. In many areas of the world, aquatic weeds have multiplied explosively. They have interfered with fishing, navigation, irrigation, and the production of hydroelectric power. They have brought disease and starvation to communities that depended on these bodies of water. Water hyacinth in the Congo, Nile, and Mississippi rivers; the water fern in southern Africa; and water lettuce in Ghana are a few ex-

amples of such catastrophic infestations. People have always loved the water's edge. To destroy the quality of these limited areas of the Earth is to detract from our humanity as well as from the resources that sustain us.

20.8   INDUSTRIAL WASTES IN WATER

In the early days of the Industrial Revolution, factories and sewage lines dumped untreated wastes into rivers. The first sewage treatment plant in the United States was built in Washington, D.C., in 1889, more than 100 years after the Revolutionary War. Soon other cities followed suit, but few laws regulated industrial waste discharge.

In November 1952, an oily film on the Cuyohoga River near Cleveland caught fire, spreading flame and smoke across the water (Fig. 20.14). Although the image of the burning river triggered public awareness of water pollution, Congress did not pass significant water pollution control legislation for nearly 20 years after the incident. In 1970, President Nixon declared that "the 1970s absolutely must be the years when America pays its debt to the past by reclaiming the purity of its air, its waters, and our living environment. It is literally now or

FIGURE 20.14   *Fire on the Cuyohoga River.* (*The Plain Dealer, Cleveland, Ohio*)

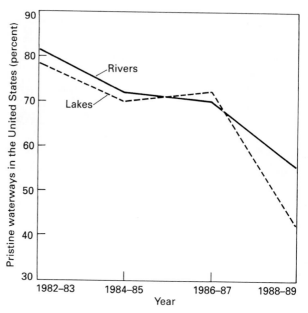

FIGURE 20.15   *The percent of pristine rivers and lakes in the United States between 1982 and 1989. A pristine waterway is defined by the Clean Water Act as one that fully supports the uses and aquatic ecosystems that it supported in its natural state.* (From R. Adler, J. Landman, and D. Cameron, *The Clean Water Act 20 Years Later.* Washington, D.C., Island Press, 1993. 320 pp.)

never." The **Clean Water Act,** passed in 1972 (over Nixon's veto) stated that:

> The objective of this Act is to restore and maintain the chemical, physical, and biological integrity of the Nation's waters. In order to achieve this objective it is hereby declared that . . . .
>
> (1) it is the national goal that the discharge of pollutants into the navigable waters be eliminated by 1985;
>
> (2) it is the national goal that wherever attainable, an interim goal of water quality which provides for the protection and propagation of fish, shellfish, and wildlife and provides for recreation in and on the water be achieved by July 1, 1983;
>
> (3) it is the national policy that the discharge of toxic pollutants in toxic amounts be prohibited.

Analyzing the progress under the Clean Water Act is akin to determining whether a glass of water is half empty or half full. Between 1972 and 1992, the proportion of the United States population served by sewage treatment plants jumped from 32 to 74 percent. In the same period, emissions of industrial toxic pollutants declined by 99 percent, and discharge of metal ions from mining and metal refining declined by 98 percent. At the same time, the number of pristine waterways has declined, and the number of polluted waterways has increased. Figure 20.15 shows trends in water quality for rivers and lakes in the United States between 1982 and 1989.

How could the number of polluted waterways increase at the same time that the quantity of pol-

lutants discharged decreased? One possible answer is that pollution is not necessarily getting worse, but our measurement of the problem is improving. Over the past two decades, the criteria for defining a waterway "fully supporting original aquatic ecosystems" have become stricter, so a level of pollution deemed insignificant in 1970 would be judged more serious today. In addition, the number of testing sites has increased, and thus we have a heightened awareness of the problem.

Certainly, goods can be manufactured more cheaply if industrial wastes are discharged directly into the environment. But then everyone suffers from the effects of unhealthy water or unclean air. Therefore, it is a legitimate role of government to regulate waste disposal practices; the arguments arise when trying to decide *how much* regulation is desirable. Pollution control is not a yes or no, on or off affair. Imagine that a chemical factory with no pollution control devices releases a certain quantity of wastes into the water every month. Equipment can be designed to remove any portion of the pollutant, from a minor amount to practically all of it. In general, the more pollution that is removed, the more expensive the process becomes. Limited pollution control can be relatively inexpensive, but an essentially pollution-free environment is costly.

How much pollution control should we pay for? Some people suggest that pollution control measures should be applied only when it can be shown that there is a positive economic return on the investment. Opponents of this argument claim that money is not the only measurement of the quality of life. How can you place a dollar value on the annoyance of a vile odor or of unclean water? What about recreational opportunities? How much is it worth to be able to float quietly down a river and fly-fish for trout? Going beyond annoyance, how can you measure the dollar value of human suffering and misery caused by illness or the value of a human life that ends too soon? Many people believe that such costs are beyond our right to judge.

20.9 GROUND WATER

When rain falls on dry soil, the first raindrops simply wet the soil; they do not flow down or away. As the rain continues after all the land surface is wet, gravity pulls the excess water down through the pores in soil or the underlying rock, sand, or gravel. Eventually the downward flow is stopped when the water meets rock that is impermeable. Because the water can go no farther, it backs up, filling all the pores in the rock above the barrier. This completely wet section is called the **zone of saturation,** as shown in Figure 20.16. The upper boundary of the zone of saturation is called the **water table.** Below the water table, the ground is saturated; above it, the ground may be moist but is not saturated.

Once underground, the water moves at widely varying rates, depending on a variety of geological conditions. At one extreme, moisture can flow rapidly through subsurface voids or caverns, much like an underground river. However, underground caverns are rare. By far the largest proportion of underground water moves slowly through pores in rock. The total quantity of slow-moving ground water is large. Note, however, from Table 20.1 that these waters are replaced *very* slowly (in up to thousands of years). Much of the water in some underground reservoirs was accumulated many centuries ago in wetter climates than the present one. Under such conditions, deep ground water may be considered to be, for all practical purposes, nonrenewable. Just as coal and petroleum are called fos-

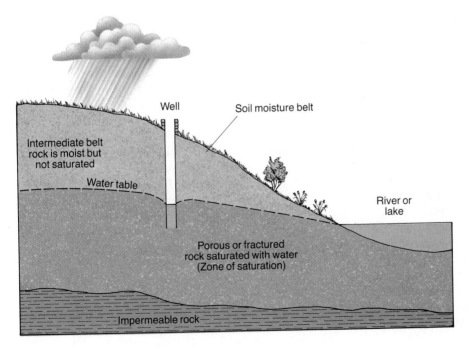

FIGURE 20.16 *Distribution of ground water.*

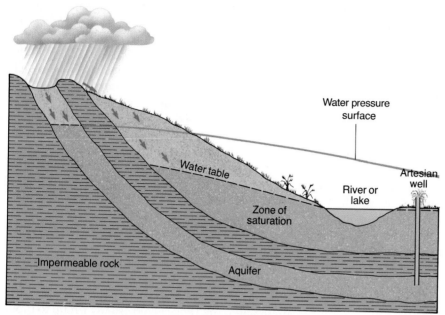

FIGURE 20.17   *An aquifer is a layer of water-bearing rock. An artesian aquifer forms when permeable rock is sandwiched between two layers of impermeable rock and the rock beds are sloped as shown in the drawing. Water will flow to the surface of the well without being pumped. This is called an artesian well. Water that is trapped for long periods in an aquifer is called "fossil" water.*

sil fuels, so is deep ground water sometimes called "fossil" water. The removal of deep ground water is therefore analogous to mining.

An **aquifer** is a body of rock that is porous and permeable enough to yield economically significant quantities of water. If an aquifer is trapped between two layers of impermeable rock, as shown in Figure 20.17, an artesian aquifer forms. Water in an aquifer can move horizontally, but its vertical movement is limited.

The Ogallala aquifer of the American Midwest is one of the world's largest reservoirs of fresh ground water. Farmers in the Midwest have found that it is profitable to "mine" this water and use it to irrigate field crops. As a result, more water is taken out of the Ogallala aquifer than is replaced by rainfall. It is estimated that some of the Ogallala ground water levels are being lowered at rates of several centimeters to about half a meter a year. At such rates, serious depletion of the aquifers can occur early in the next century, as shown in Figure 20.18. It is important to understand that the via-

bility of an agricultural-industrial-urbanized society that depends on fossil water can be at risk when that source is "seriously depleted."

Two other problems besides depletion can arise as a result of the excessive removal of ground water. One of these problems is **subsidence,** or settling, of the ground as deep ground water is removed. This removal allows the rock particles to shift somewhat closer to each other, filling some of the space left by the departed water. As a result, the volume of the entire rock layer decreases, and the surface of the ground subsides, as shown in Figure 20.19. (Removal of oil from oil wells has the same effect.) Subsidence rates can reach 5 to 10 cm per year, depending on the rate of water removal. These effects have been observed in such areas as the San Joaquin Valley of California, Houston (Texas), and Mexico City. Unfortunately, subsidence is not a readily reversible process. In many cases, the pores in the underground rock are squeezed shut by the weight of surface rock and soil. As a result, the water-holding capacity of a depleted aquifer may be

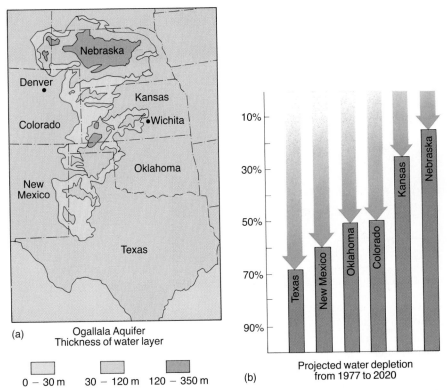

Ogallala Aquifer
Thickness of water layer

(a)

0 – 30 m   30 – 120 m   120 – 350 m

(b) Projected water depletion
from 1977 to 2020

FIGURE 20.18 *(a) Location of the Ogallala aquifer. (b) Projected water depletion in various sections of the Ogallala aquifer, from 1977 to 2020. (Numbers are given in percent of original supply.)*

permanently reduced so that it cannot be completely recharged even when water becomes abundant again.

The other problem is **saltwater intrusion,** as shown in Figure 20.20. As ground water is removed from a coastal area, the zone of fresh water saturation is reduced both from above and below. From above, the water table declines. From below, saltwater seeps in. As a result, saltwater may be drawn into wells, making the water unfit for drinking.

20.10 POLLUTION OF GROUND WATER

Whenever wastes are buried or spread over the ground, they may migrate downward to pollute ground water, as is shown in Figure 20.21. More than 50 percent of the people in the United States obtain their drinking water from ground water. Studies have shown that 45 percent of municipal ground water supplies in the United States are con-

FIGURE 20.19 *An extreme example of subsidence. In some regions of Florida, large quantities of water were withdrawn from local aquifers during the late 1970s and early 1980s. Certain underground rock structures collapsed, drawing houses, cars, and commercial buildings into the gaping holes. (Courtesy of Wide World Photos)*

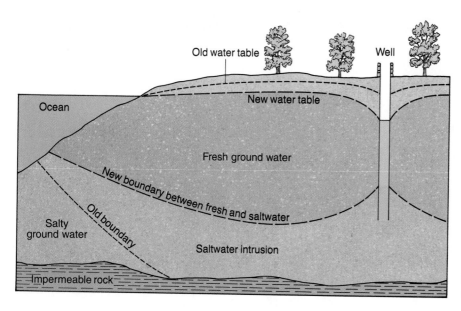

FIGURE 20.20 *Saltwater intrusion.*

taminated with synthetic organic chemicals, and wells in 38 states contain pesticide levels high enough to pose a threat to health. Every major aquifer in New Jersey is contaminated. In Florida, the water table is only 3 m below the surface in places and yet 92 percent of the population relies on ground water for drinking. More than 1000 wells have been closed as a result of excess contamina-

tion, and over 90 percent of the remaining wells have detectable levels of industrial organic compounds.

CONCEPTUAL EXAMPLE 20.2 SEPTIC SYSTEMS

In a septic system, domestic sewage flows into a large tank. The solids settle to the bottom of the

FIGURE 20.21 *Sources of ground water pollution.*

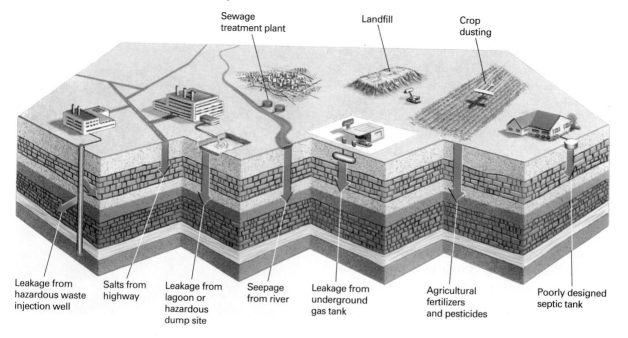

FOCUS ON
THE LIPARI LANDFILL: A CASE HISTORY

 The Lipari Landfill was opened in 1958, before modern environmental laws were enacted. It was situated in an abandoned sand and gravel pit near Pitman, New Jersey. Three million gallons of chemical wastes were legally dumped into the site before it was closed and covered in 1971. But engineers only sealed the surface, and wastes seeped into the ground water. By the mid-1970s, nearby Alcyon Lake turned orange and purple. Local residents said that fumes rising from the soil brought tears to their eyes and sometimes even a bittersweet taste on their tongues. In 1983, the Environmental Protection Agency (EPA) called the Lipari the worst hazardous waste site in the United States. New Jersey Senator Frank Lautenberg proclaimed that "Lipari is a symbol" and promised rapid action.

EPA engineers reasoned that although the sand and gravel in the dump site are permeable a lower clay layer that lies beneath the porous material is not. Therefore they could isolate the landfill by digging a trench around its perimeter to the clay and then filling the trench with concrete. However, the year after the wall was finished, approximately 2600 gallons of polluted water leaked outside the perimeter. Perhaps the wall had cracked, the bond between the wall and the clay was permeable, or the clay layer itself was fractured.

Next the EPA sunk numerous wells into the landfill and into the ground outside the wall. They pumped clean water into some of the wells and removed polluted water from others. The engineers then pumped the polluted water to a purification plant. This process, begun in 1989, is expected to continue for 7 to 10 years.

Local citizens are not happy. They contend that the plan is analogous to pouring water into a leaky bathtub and that the process will *increase* ground water contamination. In addition, some of the wastes in the landfill were sealed in metal drums. What will happen if the drums rust through after the flushing project is completed? These critics argue that the only solution is to dig up the landfill, remove the polluted material, and recover the sealed drums. In addition, they recommend dredging Alcyon Lake. But where should the EPA put the thousands of cubic meters of contaminated sediment from the landfill and the lake? In another landfill?

The Lipari Landfill is only one of tens or hundreds of thousands of hazardous waste sites in the United States. There are no easy solutions. Even partial clean-up would cost tens of millions of dollars. People cannot agree on how much they are willing to spend and how much pollution they are willing to live with.

---

tank and the waste water flows out of the tank into the soil. Bacteria and other organisms in the soil decompose the organic matter, purifying the waste water. Would a septic system be more effective in deep sandy soil or in thin clay soils underlaid by granite rock?

Solution If waste water percolates slowly through the soil, decay organisms have enough time to consume the organic wastes and the septic system is effective. If the waste water travels rapidly, it may contaminate streams or other wells before the decay organisms purify it. Water flows slowly and is readily purified in sandy soil. Clay is impermeable, however, and water does not flow through it. In this example, waste water would likely collect on the granite rock and flow downslope rapidly along the rock surface, thus contaminating nearby regions.

FOCUS ON
OIL IN THE OCEAN: THE WRECK OF THE EXXON VALDEZ: A CASE HISTORY

Petroleum from Alaska's north coast is pumped through the Alaska pipeline to Valdez on the south coast of Alaska (see figure). From there the petroleum is loaded into tankers that carry it to refineries. In March 1989, as the supertanker *Exxon Valdez* steamed out of port, the captain, who had been drinking, left the bridge in the hands of an inexperienced third mate. As the mate attempted to steer around floating ice, he ran the ship aground on a rock, spilling 42 million L (11 million gallons) of petroleum into Prince William Sound (see figure).

In preparation for such an emergency, shipping companies had stored booms and pumps at the dock in Valdez. If they had been deployed quickly, they could have contained and retrieved much of the spilled oil, but days passed before the effort was coordinated. During the following

few weeks, oil fouled more than 5000 km of coastline, killing 30,000 birds, 3500 to 5500 sea otters, 30 seals, 22 whales, and an unknown number of fish. The Exxon oil company spent $2 billion to doctor sea mammals, wash oil off birds, and scour beaches. Some scientists argued that the clean-up effort in many cases did more harm than good. The detergents used to scour the rocks had their own negative impact on the beaches. The crews trampled shoreline organisms such as barnacles, clams, mussels, eelgrass, and rockweed, killing them and destroying their habitats. As a result, many untreated beaches regained productivity faster than those that had been cleaned.

While the short-term effects of the oil spill were devastating, the long-term ecosystem damage is difficult to ascertain. Petroleum is a complex mixture of many different compounds. The lightest ones are volatile enough to evaporate within days or weeks. The heaviest ones glob to-

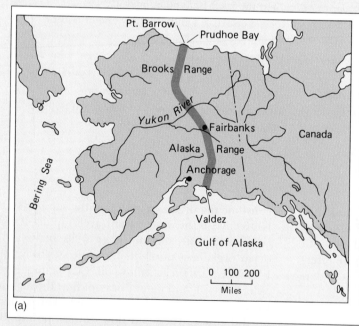

(a)

(b)

(a) The route of the Alaska pipeline. (b) The Alaska pipeline is built in a zigzag fashion so it will not rupture when the permafrost soil expands and contracts. (Alyeska Pipeline Company)

gether into tar balls that persist for decades but are relatively harmless. Compounds of medium molecular weight remain in the environment for years until bacteria slowly degrade them. In 1993, 4 years after the accident, scientists from Exxon and NOAA (National Oceanic and Atmospheric Administration) debated the long-term effects of the spill. Both groups of scientists agreed that a wide variety of plants and animals in Prince William Sound contained high concentrations of hydrocarbons in their tissue. Government scientists claimed that the tissue analysis proved that contamination from the spill still lingered in the ecosystem. However, Exxon scientists disagreed. They argued that every sample of crude oil has a specific ratio of hydrocarbons that distinguishes it from all other samples, much as a fingerprint identifies an individual. Because the petroleum fingerprints from tissue samples did not match the fingerprint from the oil spilled by the *Exxon Valdez*, the Exxon scientists argued that the tissue contamination originated from other sources such as natural seeps, discharges from fishing boats, and lingering contamination from older

spills. Government scientists argued, though, that hydrocarbon ratios change when the oil is ingested by living organisms, so the fingerprint changes.

More is at stake in this argument than a scientific debate. Exxon is currently facing lawsuits for spill-related damages. If they can prove that the contamination from the spill no longer threatens the ecosystem, their liability will decrease substantially. In June 1994, a federal jury found Exxon liable for the spill, and deliberations began on how large the fine will be. Plaintiffs had asked for $1.5 billion in actual damages and $15 billion in punitive damages.

The *Exxon Valdez* accident was not an isolated incident; shipwrecks, offshore drilling, and war have all led to severe spills in recent years. In 1979, an offshore oil well in the Gulf of Mexico spilled nearly 700 million L; in 1991, Sadaam Hussein deliberately poured 1 billion L of crude oil into the Persian Gulf; in 1993, the tanker *Braer* lost power off the Shetland Islands and crashed into the rocks, spilling most of its 100 million L of light crude oil. Each event was caused by a sequence of seemingly avoidable circumstances: a negligent captain, a broken drill pipe, an angry dictator, and an engine malfunction complicated by a delayed rescue tug. But accidents occur; between 1980 and 1988, tankers in the United States were involved in 468 groundings, 371 collisions, 97 rammings, and 55 fires or explosions. In 1990, in the wake of the *Exxon Valdez* disaster, Congress passed a bill designed to reduce tanker accidents. The law increases shipowner liability, establishes Coast Guard spill response teams, and requires double hulls for all oil transport vessels. (If the *Exxon Valdez* had been equipped with a double hull, it would not have spilled oil after it went aground.) This legislation should reduce oil pollution, but ships have sunk as long as people have sailed the seas, so even the strictest precautions cannot guarantee an end to the problem.

The Exxon Valdez after it had run aground and begun to spill oil. The slick appears off the ship's bow. (Wide World Photos)

20.11 OCEAN POLLUTION

Nearly all rivers and most ground water carry their pollutants into the sea. In addition, people dump sewage, industrial waste, and municipal trash directly into the ocean. Although the ocean is large enough to dilute these pollutants eventually, dispersal is slow and pollutants concentrate along shallow coastlines, especially in protected bays and estuaries with limited access to the deep sea. Worldwide, coastal pollution is roughly proportional to the population density along drainage basins (Fig. 20.22).

Sewage and Municipal Waste in Coastal Waters

In the United States, 35 percent of municipal sewage from coastal cities is dumped into the ocean without treatment. Before 1990, many cities also dumped municipal trash offshore. During the summer of 1988, used hypodermic syringes, intravenous tubing, and blood sample vials washed up on beaches along the Eastern seaboard and health inspectors recorded high levels of disease-causing bacteria in coastal waters (Fig. 20.23). Nearly 500 beaches from Maine to Florida were closed for health reasons. In response to public outcry, Con-

FIGURE 20.23   *An Aquasphere Project volunteer collects debris washed up on a beach. Some beaches have been contaminated with medical and other hazardous wastes.* (Mark Elias/ Wide World Photos)

gress passed the Ocean Dumping Ban Act, banning ocean disposal of municipal trash. However, the Act has had little effect on sewage treatment plants, and in 1992, the number of beach closings increased to 2600. The fivefold increase in beach closures reflects both continued pollution and increased public awareness and testing.

Toxic Chemicals in Coastal Waters

When commercial whaling was banned along the mouth of the Saint Lawrence River in the 1950s, biologists estimated the beluga whale population at about 1200. By 1988, the population had declined to 450. In that year, marine biologists from the University of Western Ontario performed autopsies on 72 dead whales that had washed up on the beaches. They detected 30 chemical pollutants, including DDT, PCBs (polychlorinated biphenyls), Mirex (a pesticide), mercury, and cadmium. A majority of the whales had died of septicemia, a form of blood poisoning that attacks the immune system. Other diseases included pneumonia, hepatitis, perforated gastric ulcers, and bladder cancer. Dr. Joseph Cummings, a researcher on the project, believes that a strong link exists between the toxic substances and the fatal diseases in whales.

FIGURE 20.22   *The relationship between pollution of coastal water (as measured by the mean annual nitrate concentration) and the population density along adjacent river drainages.*

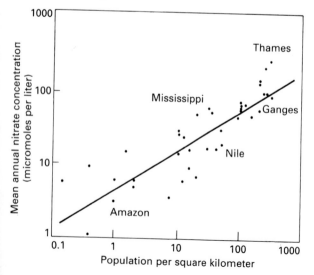

FIGURE 20.24 *A fisherman pulls in nets on Chesapeake Bay.* (Chesapeake Bay Foundation)

Toxins also affect the people who eat contaminated seafood. According to the EPA, fish from 4000 test sites in the United States were contaminated with high enough concentrations of toxins to pose a threat to human health.

Habitat Destruction and Coastal Ecosystems

Although coastal seas account for only 10 percent of the ocean surface area and 0.5 percent of its volume, approximately 90 percent of marine animals rely on river mouths and coastal wetlands for some portion of their life cycles. Worldwide, over half of the coastal wetlands have been destroyed by commercial development. The world fish catch rose from 1950 to 1989 then fell from 1989 to 1992 (Fig. 20.24). The combined effects of overfishing, pollution, and habitat destruction caused the decline.

20.12 NONRENEWABLE MINERAL RESOURCES

Living organisms use a source of energy (ultimately the Sun) to convert nutrients from the environment into body tissues. The time scale of these conversions is the time scale of the spans of life—months for plant fibers such as cotton, years for animal material such as bone and hide, and decades for wood. On the time scale of human lives, therefore, these materials are **renewable;** if they are not consumed faster than they are produced, they need never be exhausted. Geological processes, like those of life, can also organize and concentrate materials, but here the time spans extend to millions and billions of years. Since humans cannot wait that long, mineral resources are said to be **nonrenewable.**

An **ore** is considered to be a rock mixture that contains enough valuable minerals to be mined profitably with currently available technology. The **mineral reserves** of a region are defined as the estimated supply of ore in the ground. Reserves are depleted when they are dug up, but our reserve supply may be increased by either of two circumstances. First, new reserves may be discovered. Second, the value of a known deposit may change. For example, many known deposits are not being mined because it would not be profitable to do so under the current economic climate. If technology improves so that the materials can be refined cheaply or if the market price of the metal increases, the deposit will suddenly become an ore reserve.

Many of the high grade, concentrated, and easily accessible ores, such as the 50 percent iron deposit of the Mesabi Range in Minnesota, are being used up rapidly and either have been or will be depleted in the near future. These mines are essentially nonrenewable. Once they are gone, our civilization will have suffered an irreplaceable loss. But our technological life will not end with the exhaustion of these rich reserves because less concentrated deposits are still available. In some situations, the less concentrated ores are more plentiful than the concentrated ones are. Returning to our example of iron, in 1966, it was estimated[*] that the global resource reserve of iron was about 5 billion tonnes. At that time, the global annual consumption rate was about 280 million tonnes. If these figures were correct, and if consumption continued at a constant rate, the iron reserve of the Earth would have been consumed in 18 years (5 billion/280 million), bringing the end of iron reserves to 1966 + 18 years, or 1984. There must be something wrong with such calculations. Thus, we see

---

[*]B. Mason: *Principles of Geochemistry.* 3rd ed. New York, John Wiley & Sons, 1966, Appendix III.

again that reserves are not a constant factor but can change markedly with exploration and with the development of methods suitable for processing ores of lower grade. In the case of iron, the big change has been the development of improved methods of processing taconite rock for its iron content.

Would it be reasonable to expect similar dramatic improvements in the technology of mining other materials? If so, continued progress would make it profitable to mine less and less concentrated sources, and then we need never run out of anything. Such an approach ignores some serious difficulties. As discussed in more detail in the following paragraphs, as lower grade ores are mined, more energy is needed to produce a tonne of product. Imagine for a moment that you were to mine iron, for example, in your back yard. Iron could undoubtedly be found in the rock and soil around your house. The problem is one of concentration. For many metals, the concentrations in ordinary rock may be as low as one ten-thousandth of those in commercial ores. The energy consumption re-

quired to mine this rock and the pollution resulting from such a mine would be prohibitive.

The question remains: Are we, or are we not, in danger of running out of mineral resources? There is no clear-cut answer to this question; instead, there are two opposing opinions. One view holds that as rich deposits are depleted and people must mine increasingly lower grade reserves, the cost and the total environmental consequences of mining will become so great that many minerals will become prohibitively expensive.

Others disagree. They claim that as the richest mines become depleted, three factors will act together to ensure that acute shortages do not develop. (1) Mining technology will continue to improve, and new reserves will be found. (2) As minerals become more scarce, recycling will automatically become attractive, thereby extending the life of present reserves. (3) When certain minerals do become expensive, satisfactory substitutes will be found. Therefore, our technological existence will continue uninterrupted.

FIGURE 20.25   *The depletion of the Comstock silver lode in Nevada. The first discoveries were rich in silver. Later periods of mining followed, but the quantity of silver produced in each period was less than in the previous one. This pattern is followed in many other mining districts. The graph reminds us that mineral depletion is a problem and that once rich mines are exhausted, a significant loss has occurred.*

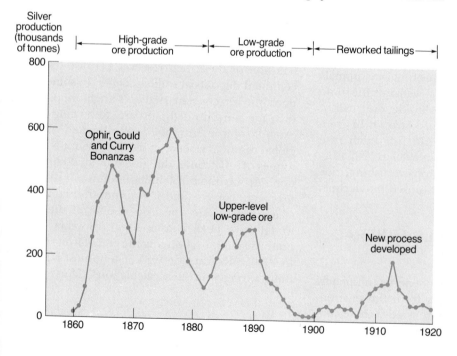

FIGURE 20.26 *Copper deposit being mined in Butte, Montana. Butte was once the copper capital of the West, but now the mines have shut down because the quality of the remaining ore is too low to mine profitably.*

Argument: Mineral Reserves Will Be Depleted in the Near Future

The ancient Greeks and Romans found copper ore under 10 cm or so of soil. Today the depths of mines are measured in kilometers. It is impossible to go back to the days of picks and shovels. As increasingly lower grade ores are sought, the technological problems inherent in all aspects of mining and refining rise sharply (Fig. 20.25). Dependence on technology to solve all problems may lead to disappointment. In addition, the future availability of metals depends on many factors besides the quantity of ore in the ground and the state of refining technology. Some of these are discussed in the following paragraphs.

Availability of Energy.   To extract metal from ore, the dirt and rock must be dug up and crushed, the ore itself must be separated and chemically reduced to the metal, and the metal must finally be refined to purify it. Each step, especially the chemical reduction, requires energy. Low grade ores require much more energy to process than do high grade ores. Some low grade ores differ from high grade ores not only in concentration, but also in chemical composition. Some chemicals are easier to purify than others. Chemists can purify these low grade ores, but in many cases more energy is needed. Thus the price and availability of many ores are linked to the price and availability of energy.

Pollution and Land Use.   Most mining processes cause pollution of land, water, and air (Fig. 20.26). For example, sulfur is found in large quantities in many ore deposits. This sulfur, chemically bound to metals in the Earth's crust, is brought to the surface when minerals are mined. Sulfur reacts with water in the presence of air to produce sulfuric acid, which runs off into the streams below the mine. This pollution, known as **acid mine drainage,** kills fish and disrupts normal life cycles in streams, rivers, and lakes. When sulfur combines with other chemicals during the refining processes, it is often converted to gaseous air pollutants such as hydrogen sulfide and sulfur dioxide. These compounds, in turn, react in the atmosphere and fall to the ground as acid rain. Sulfur, of course, is not the only polluting chemical from mining operations. Many other mine pollutants cause serious air and water pollution.

Just as more energy is required to handle low grade ores than high grade ones, more pollution generally results from processing these impure materials. The pollution can be controlled, but such measures are expensive and add to the total cost of refining ore.

The world is running short of food, energy, and recreational areas as well as high grade mineral deposits. What should our policy be if a valuable ore or fuel lies under fertile farmland or a beautiful mountain? Which resource takes precedence? At present, this question is being raised principally with respect to exploitation of fuel reserves, for vast coal seams lie under the fertile wheat fields of North America. If large areas of low grade ore must be exploited, the problem will extend to metal reserves as well.

Argument: Mineral Reserves Will Last for Generations to Come

Those who believe that our mineral reserves are not likely to be depleted in the foreseeable future point to past successes. They argue that pessimists predicted iron shortages by the early 1980s when, in fact, no such shortages occurred at that time. Moreover, this line of reasoning continues, there are several options in addition to mining conventional, land-based reserves.

Nonconventional Reserves. As conventional mines become depleted, new ones can be explored. What about mining the sea floor? Various explorations have indicated that the mineral deposits on the sea floor are vast. Much of this material is concentrated in the form of round, flat, or odd-shaped pieces, typically weighing about a kilogram or so, that are rich in manganese. They are called **manganese nodules,** but they also contain copper, iron, nickel, aluminum, cobalt, and about 30 or 40 other metals. It is estimated that there are a trillion or more tonnes of these nodules on the sea floor. The sea floor does not have to be drilled or blasted, and explorations can be done with undersea television cameras. Furthermore, no one "owns" the sea floor, but the question of just how the mining rights are to be allocated is still unclear. Nonetheless, the technology of collection is complicated, and costs are expected to be high. The sea is not the easiest

FIGURE 20.27   *Mining the sea floor for manganese nodules.*

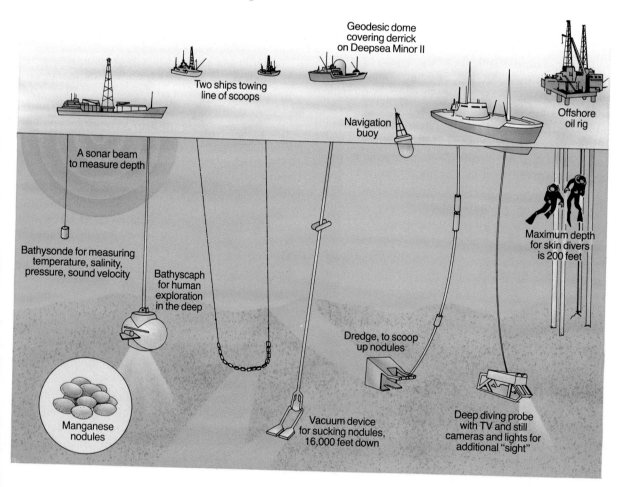

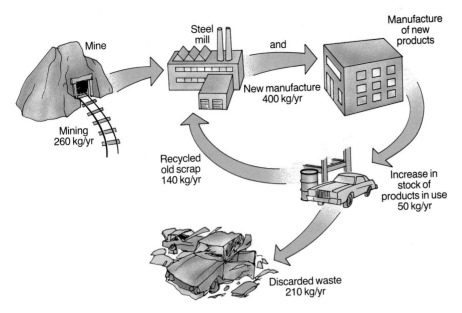

FIGURE 20.28 *Estimated annual manufacture and flow of iron and steel in the United States, in kilograms per capita.*

environment in which to operate complex machinery. Possible methods of collection include scoops, dredges, and vacuum devices, as shown in Figure 20.27. Various groups of corporations are already involved in exploration and in planning undersea mining operations.

Conservation and Recycling. An alternative approach is to use less rather than to mine more. Consumption can be reduced by either conservation or recycling. Figure 20.28 shows the extent to which recycling in the United States could increase the life of iron reserves. The depletion of reserves is accounted for by the increase in stock plus waste, but the waste accounts for over 80 percent (210 kg/260 kg per year per person). This means that if all the discarded iron and steel were recycled, reserves could last four times as long.

Such objectives could also be realized by manufacturing products that are smaller and last longer or by taking better care of them. (It is much easier to keep on using a device that still works than it is to recycle it after it stops working.) Perhaps best of all is an improvement in technology that results in conservation of materials. A good example is the modern pocket calculator replacing the old slow, mechanical, office calculating machine, which weighed about 20 or 25 kg.

Substitutes. Finally, what about substitutes for at least some nonrenewable mineral resources? Why not go back to the simpler styles of using wood, stone, and fiber, or go forward to using plastics that

can be synthesized from coal, air, and water? To some extent, both paths are being followed. The use of sand and gravel (for concrete) and of stone is increasing, as shown in Figure 20.29. Wood, too, is still a desirable construction material. The use of many varieties of plastics is increasing much more rapidly than the use of metals. Improvements in the chemistry as well as in the mechanical makeup of plastics have made them competitive with metals in many applications, even where strength is an important factor.

FIGURE 20.29 *Concrete is often used in place of steel in the construction of highway bridges.*

OVERVIEW

Seven different factors must be estimated to predict the future availability of metals. These are (1) the quantities, concentrations, and locations of minerals in the Earth's crust; (2) the availability of energy; (3) the effects of pollution from mining and refining; (4) land-use conflicts between mining and agriculture or urbanization; (5) future demands for metals; (6) future population levels; and (7) future rates of recycling. Because there are large differences in predictions of any *one* of these factors, it becomes nearly hopeless to predict how all seven will operate in concert. However, we can be sure of two things: The reserves of many important minerals are not inexhaustible, and once they are dispersed and discarded in old dumps where they are mixed with many other waste materials, they will be expensive to recover.

SUMMARY

Soil is an intimate mixture of pulverized rock and organic debris. Soil erosion and loss of soil to urbanization are significant environmental problems.

Water leaves the atmosphere as rain or snow, moves on the Earth in the form of ice or liquid water, and returns to the atmosphere by evaporation. Only a small fraction of the Earth's water is in the atmosphere; most of it is in the oceans, which serve as the ultimate sink for water-borne impurities. Humans use the least amounts of water for drinking, but this water must be of the highest quality. The largest amounts are used for industry and agriculture. **Water-diversion projects** are used to bring water to regions where it is needed. These systems are expensive. In addition, they encourage waste, they may cause salinity, and some require large consumption of energy for pumping. Nutrients pollute water by promoting the rapid growth of organisms that then deplete the dissolved oxygen. Many water pollutants, such as heavy metals and organic chemicals, come from industrial sources. Some **ground water** reserves are "fossil" and are not replaced quickly. Ground water can be polluted by contamination from a variety of sources. Natural purification of ground water is slow because it is not readily diluted and does not have access to air. A major source of pollution in the oceans is petroleum from tankers. The ultimate effect of pollution on aquatic life is uncertain, although the immediate and local effects can be devastating.

An **ore** is a rock mixture that contains enough valuable mineral matter to be mined profitably. **Mineral reserves** are the estimated supplies of ore in the ground. This estimate can change with the discovery of new reserves, with improvements in extraction and refining, with new prices for minerals and for energy, and with increased pollution associated with the mining and processing of ores. Future prospects for metals cannot be predicted accurately because they depend on changes in reserves, on conservation and recycling, on changes in technology, and on the use of substitute materials. However, once the prime ores are depleted, the poorer sources will be more expensive to mine, will require much more energy, and will cause much more pollution. One possible source of new mineral resources is the store of manganese nodules on the sea floor.

KEY WORDS

Soil	Transpiration	Capillary action	Ore
Humus	Precipitation	Zone of saturation	Mineral reserve
Chelates	Runoff	Water table	Manganese nodules
Hydrologic cycle	Residence time	Aquifer	Acid mine drainage
Evaporation	Eutrophication	Subsidence	
Vaporization	Ground water	Salt water intrusion	

CONCEPTUAL QUESTIONS

Soil and soil erosion

1. What are the components of healthy soil? What is the function of each component?

2. Characterize the four major horizons of a mature soil.

3. How does fertile soil differ from beach sand?

4. Would humus help prevent excess soil salinity caused by irrigation? Why or why not?

5. Discuss the soil characteristics of: (a) A tropical rainforest; (b) a temperate desert; (c) land close to your home or school.

6. Explain how farming and logging accelerate erosion. What measures can reduce the impact?

7. In a typical suburban development, each family owns a small yard. Many land use planners suggest that people live, instead, in multistory apartments and the remaining land be set aside as parks and open spaces. Discuss the relative merits of each system. Which would you prefer? Why?

The hydrologic cycle

8. In which physical state (solid, liquid, or vapor) does most of the Earth's free water exist? Which physical state accounts for the least?

9. Many urban areas draw their drinking water from a nearby river. Much of this water is then returned to the river in the form of treated sewage. How does this practice affect the hydrologic cycle of the region? How does it affect the water quality?

10. Assume that you had waste available from the following five sources: (1) rainwater drained from your roof; (2) good well water or tap water; (3) water from the wash cycle of your dishwasher or clothes washer; (4) water from the rinse cycle of your dishwasher or clothes washer; (5) water drained from your bath or shower. List all the applications in your home, in your garden, for your pets, or for other purposes for which each of these water supplies could be used.

11. List as many uses as you can think of for water in industry, in agriculture, and for recreation.

12. Imagine that you live in an area with abundant water, and it is proposed that some of the excess water be diverted to another region that needs it. List the questions that people on both sides of the pipeline should consider before construction is started.

13. Explain how irrigation can, in some instances, be economically favorable in the short term but harmful over the long range.

Water pollution

14. In its article on "Sewerage," the eleventh edition of the *Encyclopedia Brittanica*, published in 1910, states, "Nearly every town upon the coast turns its sewage into the sea. That the sea has a purifying effect is obvious. . . . It has been urged by competent authorities that this system is not wasteful, since the organic matter forms the food of lower organisms, which in turn are devoured by fish. Thus the sea is richer, if the land is the poorer, by the adoption of this cleanly method of disposal." Was this statement wrong when it was made? Defend your answer. Comment on its appropriateness today.

15. Explain how a nontoxic organic substance, such as chicken soup, can be a water pollutant.

16. What is eutrophication? Explain how it occurs and why it is hastened by the addition of inorganic matter such as phosphates.

17. List an activity that would pollute (a) inland surface waters, (b) ground water, (c) the oceans. Discuss the effects of this pollution on humans and on plants and other animals.

18. Pesticides are poisonous to many species of plants and animals, whereas phosphate is a form of fertilizer. Yet both are considered to be water pollutants. Explain the differences.

19. List the three major goals outlined by the Clean Water Act of 1972. Do you think that these goals were realistic at the time the law was passed? Compare current water quality in the United States with the objectives of the Act.

Ground water

20. (a) Describe what happens to rainwater that starts to fall onto a dry area and then continues heavily for several days. (b) Describe what happens to this water when there is no more rain for a month.

21. Why is pollution of ground water potentially more serious than pollution of surface waters?

22. A team of engineers proposed to dig a deep well to supply irrigation water for a village in the desert in Egypt. The town elders asked whether the water from the well would be like water from a river or like oil from an oil well. What do you think they meant by the question?

23. Explain why land subsides when ground water is depleted. If the removal of ground water is stopped, will the land necessarily rise again to its original level? Defend your answer.

Ocean pollution

24. Between 1990 and 1993, harvests in Prince William Sound were below the average of the previous decade. Fishermen contend that the *Exxon Valdez* oil spill caused the decline in fish populations. *Exxon Valdez* lawyers contend that fish populations have fluctuated for centuries, and the current decline could be caused by other unrelated factors such as natural population cycles, deep ocean fishing, or pollution of inland spawning grounds. Discuss the merits of

each argument. What data would help provide an answer to the argument?

Mineral reserves

25. Explain why energy cost and availability are important factors in estimating future mineral reserves.

26. Petroleum is generally burned as a fuel, but in many applications the chemical compounds in the oil are used for the manufacture of plastics and other materials. Explain why petroleum cannot be economically recycled after it is burned, but if it is used for the synthesis of plastics, recycling may be possible.

27. A noted environmental scientist reported that the world tin reserves may be depleted in the year 2000. In making this prediction, he assumed that (a) mining technology and world economic activity will remain constant, (b) consumption levels and population will remain constant, and (c) no new deposits will be discovered. Do you think that these assumptions are reasonable? If so, defend your position. If not, suggest more likely assumptions.

28. Economist Julian Simon argues that we are not facing shortages of minerals because human ingenuity will extend our reserves. Explain how human ingenuity can extend the life of a geological resource.

29. Sand and bauxite, which are the raw materials for glass and aluminum, respectively, are plentiful in the Earth's crust. If we are in no danger of depleting these resources in the near future, why should we concern ourselves with recycling glass bottles and aluminum cans?

30. Referring to a discussion of the difficulty of recovering ores from the soil and rock in a back yard, Professor Peter Frank of the Department of Biology of the University of Oregon wrote, "The second law of thermodynamics comes in with a vengeance." Explain what he means.

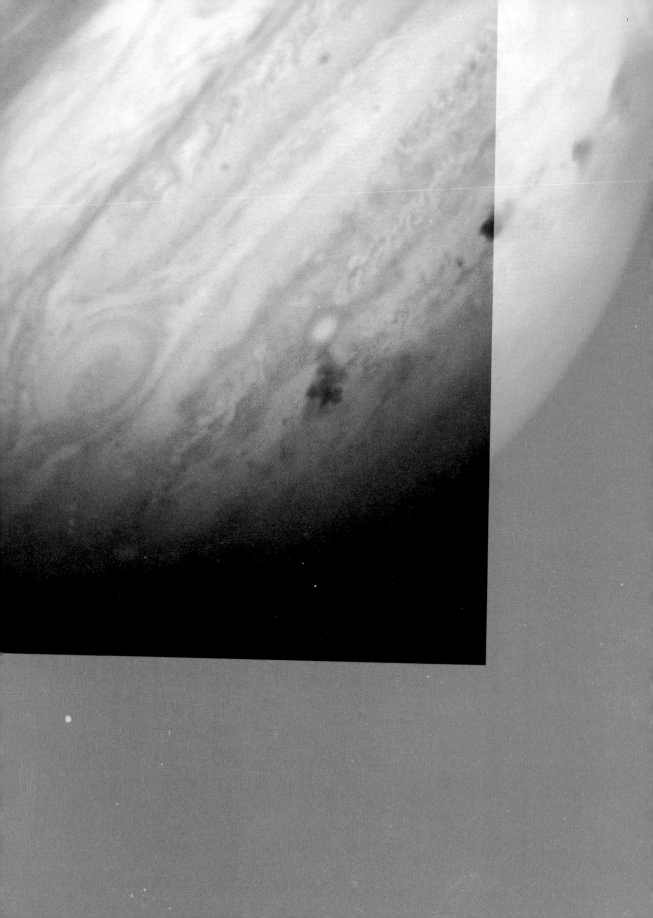

PART FOUR

ASTRONOMY

When Newton discovered that the same law that dictates why an apple falls to Earth also explains why the Moon orbits the Earth, he ended our perception of living in a duoverse, where different laws rule the heavens than the Earth, to explain that we truly live in a universe. The dynamo that powers quasars, the most distant objects known in the heavens, can be explained, at least in principle, by the same laws that are at work in our everyday lives. Likewise, no astronomer has ever found an atom, a molecule, or any chemical combination in the heavens that has not also been found on Earth.

Astronomy is the oldest of the sciences, and in many ways our yearning to understand motion in the heavens provided the impetus for scientific investigations here on Earth. Also, one of the most dynamic of sciences, astronomy is ever-changing with new developments. Because of the rapid advances in technology, we are continually able to see and learn more about the Universe in which we live.

Where did the Universe come from, and where is it going? A glimpse at the answers to these questions seems an appropriate way to end our investigations of the physical sciences.

In July 1994, fragments of comet Shoemaker-Levy collided with Jupiter over a period of a few days. This image of Jupiter made with NASA's Hubble Space Telescope's Planetary Camera shows eight impact sites, the features of which changed on timescales of a few days. The smallest features in this image are less than 200 km across. (Courtesy NASA)

BEN PEERY
INTERVIEW

Ben Peery is Professor of Astronomy and Chairman of the Department of Astronomy at Howard University in Washington, D.C. He grew up in Minnesota and attended the University of Minnesota. As a physics student he became interested in astronomy during his senior year in college and went on to graduate school in astronomy at the University of Michigan. After he finished his thesis, he accepted a faculty appointment at the University of Indiana, where he rose to the rank of professor. In 1977, he was lured to Howard University to build an astronomy program there.

Professor Peery has long been especially interested not only in his research fields but also in problems of education. In 1990, he served at a workshop that met at the National Academy of Sciences to formulate recommendations for improving federal efforts in astronomical education. Professor Peery is well acquainted with the ways of Washington, not only from his permanent position there but also from the two years he spent in a visiting staff position in the Astronomical Sciences Division of the National Science Foundation as one of the professional astronomers who "rotate" into such positions for fixed periods and then return to their home institutions.

WHAT IS YOUR FAVORITE PART OF ASTRONOMY?

It has been primarily stars—stellar interiors from an observational point of view and stellar evolution. The goal has been to see if we could find observational support for theories of the deep interior of stars. This is what I have spent most of my research time in astronomy pursuing. I call myself a "star person."

HOW DID YOU GET INVOLVED IN RESEARCH ON STARS?

I suppose that interest came from Lawrence Aller. Neither of us knew it at the time, but he was probably the man most responsible for influencing me to work on stars when I was a graduate student back in Michigan. Lawrence and I were pretty close, and stellar research is where his interests lay. It just seemed to me that this was what a person does. [Aller, then on the faculty at the University of Michigan, was later on the faculty at UCLA.] Different things come into vogue in any field of research and then give way to some other aspect of the science. And I think that during that time—the 1950's—when I was a graduate student, matters of stellar interiors and stellar evolution were just at the place where we knew enough of the appropriate physics to do something about these problems, both from a theoretical side and from an observational side. At Michigan this was just a preoccupation of mine, and now that I think about it, it wasn't everybody's preoccupation, but at that time stellar evolution and stellar structure were really having their hour.

AND HOW DID YOUR CAREER PROGRESS FROM THERE?

It was a different style in those days, as far as a grad student finding something he could get paid for doing. Even before I had finished my thesis, people had decided that probably I should go to Indiana University. Quite literally, in those days when the field was so much smaller than it is today—five or six people throughout the nation might get their Ph.D.'s in one year—some of the elders in the field would actually discuss a student who was doing his thesis in stellar interiors from an observational point of view: "Where do you think he ought to go?" "Do you need somebody like this?" I didn't apply for a job. The phone rang two or three times. People had decided that I was the kind of guy they would be interested in.

One of these people was Frank Edmondson at Indiana University. They needed somebody to give some new strength to their observational programs down there. I liked it very much down in Indiana. People were warm and encouraging, and I thought that was where I would like to go. And I spent 18 years there before coming to Howard University. Those were great times at Indiana; I can't imagine a better place to cut my teeth than that place was. I carried on some research in stellar evolution from an observational point of view—testing theories and interpreting what I was observing in terms of current theories of stellar evolution and stellar structure—and it was great.

AND THEN WHAT LED YOU TO MOVE ON?

I just couldn't imagine leaving Indiana University. One of my colleagues left and I thought the man had gone perfectly bonkers to want to leave, but people at Howard University were trying to get me to come. I thought the whole thing was preposterous, but they showed me that it might be the time for me to come and

I THINK THAT ASTRONOMY IS A BEAUTIFUL AVENUE TO SCIENTIFIC THOUGHT—NOT JUST ASTRONOMICAL THOUGHT— AND TO SCIENTIFIC PRACTICE.

build up something at Howard. That was a time of great agony for me.

I happened to be on sabbatical leave in 1975–76, and I was agonizing over this thing. The more I thought about it, the more ridiculous it seemed to go to Howard where practically nothing existed in astronomy. Still, the idea kept coming on me that maybe it was the time to come out and meet a different kind of challenge. My friend Leo Goldberg, who had been the chairman of the department back at the University of Michigan when I was a graduate student, had left Michigan and gone to Harvard and on to the directorship of the Kitt Peak National Observatory, where I was on sabbatical. I was just pondering that thing, and I guess the agony showed as I walked past Leo's office. He said, "Ben, what's the matter? You've been looking so glum." When I told him, he said, "Go, go right away. Take the chance to build something and do it your way. Go, this is a great opportunity for you." What a relief that was, and I came to Howard in 1977.

DO YOU SEE A HOPE OF SPREADING SCIENTIFIC LITERACY OR ASTRONOMICAL LITERACY?

Scientific literacy is the way we ought to be thinking. There has never been as much excitement

in this mission as there is right now, both on the part of scientists and on the part of those who think it is a good thing and want to support us. What we are really after is not simply spreading the gospel about astronomy but that astronomy is just so accessible as a science. We can't all get into high-energy elementary particle labs, for example, but all of us are at some level familiar with astronomy because the sky is accessible to us.

Unfortunately, this is not always the case since you can't even really see stars from Washington. So many of the students have not really noticed that there were stars up there because the sky is so polluted, not only with light but also with little fine particles as well.

I think that astronomy is a beautiful avenue to scientific thought—not just astronomical thought—and to scientific practice. I persist in believing that a person somehow is better off with some degree of familiarity with reason and being reasonable— not in a social sense but in a logical sense, being able to reason through problems, rather than respond emotionally to problems that come along. Not that I'm down on emotional responses at all—they have their obvious place—but they are a pretty poor way of solving problems. So it is this taste of the rational, a taste of reason, that we really should think about as the kind of desirable experience that we want our students to have.

I just think that astronomy represents the best style, the best spirit, that we can possibly set upon to give this kind of rational experience. And, of course, the students are really wild about it.

My students certainly come in with a certain curiosity. There are other things they could have chosen, and the fact that they have chosen astronomy shows that there is some curiosity. But they often have zero ideas about what astronomers are after. So they get very, very turned on about it. All the while I am stressing this business of trying to understand the difference between a good theory and a bad theory, understanding the criteria for the validity of statements that one makes in science, and this process is really revealing since students often have thought only very little in these terms.

HOW DID YOU GET INTERESTED IN ASTRONOMY IN THE FIRST PLACE?

As a child I used to look at stars and mildly wonder what they were, but it wasn't until I was a senior at the University of Minnesota that it struck me how pitiable it was that I could stand out there in the darkness and look at the points of light in the sky and not have the vaguest idea what they were. And it bugged me. I didn't know any astronomers and I didn't have any friends who knew astronomy. I started rummaging around in the library, and I found a book by Eddington, *The Expanding Universe,* and it was certainly the most influential book I ever read because it blew my mind. I had no idea about this great laboratory in the sky; I had no way to find out about it. It was that book and the next book and so forth. When it came time to go to graduate school, I knew I would have to go into astrophysics. (Somewhere along the line I ran into that

YOU INEVITABLY LEARN YOUR CONNECTIONS WITH THE BROADER UNIVERSE. IT CAN'T HELP BUT CHANGE YOUR CONCEPT OF THE MEANING OF YOUR OWN TIME ON THIS LITTLE SPECK OF SPACE.

word.) So I became interested in astrophysics through sheer curiosity plus the advantage of being able to see the stars from my backyard.

I ended up at the University of Michigan without ever having seen an astronomer before, without ever having seen a telescope. The first telescope I saw was an old relic telescope built in 1911, literally with rivets, but having never seen another telescope at that time, I thought that this had to be one of the wonders of the world.

What did I expect about my reception at the University of Michigan? I did not know. Why weren't there any black astronomers? Why was it that no one thought that this was a sensible way to make a living? Was there any sort of bias that was keeping them out? Were there people who were interested and curious who had been discouraged from getting into the profession? I didn't know any of these things; I absolutely didn't know. I did approach Michigan with a certain degree of trepidation and in a gingerly style but my reception there was fine. I was just one of the guys. It was a small group and we were very intimately con-

nected with our professors. I just got completely overwhelmed by the whole experience.

I remember as a student that first meeting of the American Astronomical Society I went to. In those days, meetings were small. If you had 200 astronomers at a meeting, it was a fairly successfully attended meeting. Naturally, I was looking around to see if there were any African-Americans there besides me. It didn't make all that much difference to me in that I was going to hang around even if there weren't any—and I was awfully curious—and still am. If you think I am going to give you an answer to the question, "Where are all the blacks in American astronomy?" I can't give you that answer. It is one of my driving questions, but I can't give you that answer.

I am glad to report that at Howard we are beginning to graduate undergraduates with a great interest in astrophysics. Our first astrophysics majors are graduating this spring, and we are very excited about this. It has taken a long time to get to this stage. It might seem we should have had some immediately, but it is very difficult, and there is a barrier. Part of it is something that I must sympathize with. I have not tried to talk physics students into astrophysics. In past years, there has been a distinct feeling, not only among blacks but also among other minorities, that you are not doing the right thing unless you are doing something practical that can make an impact and meet the challenge to improve one's own community. So engineering is very much in vogue, very much in order, an obvious contribution with practical consequences, while astronomy is at the end of everyone's list.

"Whoever changed the world with astronomy?" students ask. I try to help them understand that some of the most profound changes have occurred through astronomy, but it is not obvious to most students, and this argument doesn't fly very well unless the student has given it a try and had his or her eyes opened about his or her own existence. You inevitably learn your connections with the broader Universe. It can't help but change your concept of the meaning of your own time on this little speck of space. And many students after taking astronomy courses couldn't help but comment on this sort of thing.

YOU CAME TO ASTRONOMY ONLY AT THE END OF YOUR UNDERGRADUATE YEARS, BUT YOU HAD NOT BEEN TURNED OFF FROM PHYSICS THE WAY MOST 12-YEAR-OLDS ARE.

I didn't grow up in metropolitan circumstances. I lived in two or three small towns in southern Minnesota on the banks of the Mississippi, low-population areas. Peer group pressures were far less devastating than they are in high-population-density areas, where to be different really takes a pound of your flesh. You were free to go your own way and do your own thing, and I was always interested and curious about science. My interests tended toward physical science.

We moved to Minneapolis when I was in my early teens, and in junior-high school I really began to get a conscious exposure to the things of science. In 9th grade, I remember, I had this magnificent science teacher. She was just so good and she took a great deal of interest in me. It was there that for the first time it occurred to me, "I'm going to be a scientist," and for the first time I began to focus. I am still so grateful. I wanted to dedicate my Ph.D. thesis to her, only to find that she had died a few years earlier. She was one of the people to help me realize what direction I had been travelling in for so long.

So I got into astronomy late, but the general direction was there for me. I'm sure that nearly every person who is a scientist would say the same thing.

This interview was conducted by Jay M. Pasachoff and originally appeared in his text *Astronomy: From Earth to the Universe*, Fourth Edition, Saunders College Publishing, 1991. The interview appears here in an abbreviated form.

CHAPTER 21

MOTION IN THE HEAVENS

When first launched, defects in the optics of the Hubble Space Telescope caused it to produce poor-quality images. In 1994, astronauts used the Space Shuttle to capture the HST so they could make some much-needed repairs. (Courtesy NASA)

Astronomy, the study of the heavens, holds the unique position of being both the oldest science and the youngest. It is the oldest in the sense that probably the first intelligent creatures to walk this planet looked toward the heavens and asked: Why? Every civilization has had its legends concerning the stars, the Milky Way, and the wonders of its known universe. Astronomy was also the first of the sciences to be investigated on a systematic basis. Many of the reasons for these studies were based on practical concerns such as using the stars for navigational purposes. Other reasons were far from practical, even though they were thought to be practical at the time because they arose from a belief in astrology. Astrologers believed that the characteristics of an individual's personality were influenced by the position of the stars and planets at the time of birth. The detailed observations made by astrologers were futile in predicting the course of human history, but they did have some value in that many of the observations of the heavens made by them helped eventually to form that portion of physics referred to as mechanics.

Astronomy is also a new science because items from the popular press frequently discuss new discoveries made by astronomers that cause us to alter our perception of our place in the scheme of the heavens. Thus, astronomy is evolving rapidly, even today, which is the characteristic of a new science. Whatever the reasons, the human fascination with astronomy was strong in the past and continues to be just as compelling today.

Our study of astronomy will be from the ground up. We will look first, in this chapter, at motion in the heavens, as seen from Earth, and at some of the central figures from the past who have contributed to our present knowledge of astronomy. Chapter 22 will deal with the Solar System, and then in Chapter 23 we shall focus our attention on the stars. Finally, in Chapter 24, we will look at galaxies and other unusual objects that populate our Universe, such as quasars, black holes, and pulsars.

21.1 CONSTELLATIONS

The fixed stars that dot the night sky form the backdrop for observing motion in the heavens. These stars are called fixed stars because day in and day out, year in and year out, these stars seem to maintain their same position relative to one another. Everyone at some time has gazed at clouds drifting overhead and observed familiar shapes formed by them. Likewise, ancient observers gazed heavenward and found that their imagination could associate an image with the arrangement of a particular grouping of stars. The Chinese, the Egyptians, and finally the Greeks did this, and these groupings of stars are called **constellations.** Today, however, a constellation refers to a particular area of the sky and not to a specific grouping of stars. The region of sky referred to as the constellation Leo obviously includes the grouping of stars commonly identified as the lion, Leo, but it now also includes every star or spot of space that is within the area designated for Leo. Modern-day astronomers still make use of these constellations because they provide a road map in the sky. For example, if you want to tell someone the general location of an object in the sky, to say it is in or near the constellation Orion provides a simple method for doing so. There are 88 of these recognized constellations today, and most of them were first identified by the Greeks. Many of these have names that are Latin translations of the names given originally by the Greeks. Do not be concerned if you look at a constellation and see no apparent relationship at all with its namesake. For example, the best-known of all the constellations is Ursa Major, the Great Bear. Finding a bear in the shape of this grouping of stars, however, is not easy. In fact, part of this constellation is more commonly known in the United States as the Big Dipper, a shape that is more easily discernible. Be aware that the stars that make up a constellation are seldom related to one another in any way except for the fact that they are seen from Earth in roughly the same part of the sky. One star

in the constellation may be relatively close to Earth, whereas another may be trillions of miles farther out in space.

The star charts following this chapter show the primary constellations as seen at midlatitudes from the Northern Hemisphere during the four seasons. Because there is something satisfying about being able to find your way around the sky, we encourage you to spend a few evenings under the stars seeing if you can find these constellations. We do not intend to give you an exhaustive discussion of these star groupings, but let us take a brief look at a few of them.

Turn to the star chart for the month of December and locate the constellation Orion, the Hunter. An easy way to locate Orion in the sky is to find the three stars that are aligned to form the belt of the hunter. This constellation is one of the most easily found and prettiest in the sky. It contains several bright stars, including Rigel and Betelgeuse. Rigel is the brightest star in Orion and has a brilliant blue color when observed through a telescope. It locates the left kneecap of Orion. The star Betelgeuse (from the Arabic for armpit) is the second brightest star in Orion and has a distinctive red color. It locates the right shoulder of Orion. Now refer to the star charts and note how the constellations appear to move in the sky. In December, Orion has just risen in the east at 9 P.M. Turning to the star chart for March, we find that Orion has moved across the sky toward the western horizon at the same time of night. Thus, the general motion of the constellations through the seasons is from east to west. Finally, in June, Orion has moved completely out of the sky at 9 P.M.

Let us look at another constellation on the charts, Ursa Major, the Big Bear. In June, Ursa Major is found in the northern part of the sky and slightly toward the western horizon. Moving to the sky chart for September, we see that Ursa Major has moved more toward the western horizon and has also moved lower down on the northern horizon. In December, the constellation is still low on the northern horizon, but it has now shifted slightly toward the eastern portion of the sky. Finally, in March, it has moved higher in the northern sky and is still located slightly toward the east. The geographical latitude of these charts is 34°N, and at this latitude, the constellation Ursa Major is never

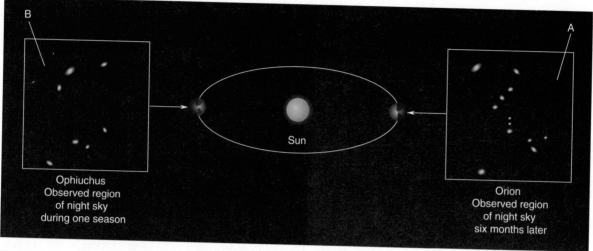

FIGURE 21.1 *The constellations seen at any particular time of year are those that are in the direction away from the Sun. During the winter season, all the stars in the region of the sky labeled as A are seen, and 6 months later, the stars in region B make their appearance during the summer season.*

out of the sky. Any constellation that is always above the horizon at a particular location on Earth is called a **circumpolar constellation.** Before we leave the constellation Ursa Major, look at the star chart for March. Note that the two stars forming the outer edge of the bowl of the dipper can be used as pointer stars to find the location of the star Polaris, the North Star. (The dashed lines in the figure point the way.) Polaris is not a particularly bright star but is famous because of its particular location in the northern sky. If one were to extend a line outward into space through the axis of rotation of Earth, Polaris would be almost on this line.

CONCEPTUAL EXAMPLE 21.1 STAYING OUT ALL NIGHT

We have noted that the motion of the constellations during the course of a few weeks is such that they tend to move toward the west. But what about their movement during the course of a single night? We will provide the answer, but a question like this can be answered a little more satisfactorily if you go outside and observe for yourself. On a clear night, find a particular constellation and periodically check its position over the course of a few hours. You can gauge the movement of the stars by comparing their position to an object on the Earth's horizon. How do they move?

Solution This kind of motion is called daily motion because it can be observed during the course of a single day. You should have noted that the daily motion of the stars also appears to be from east to west.

21.2 MOTION IN THE HEAVENS

In the last section, we noted two different types of motion that the stars seem to undergo—one during the course of several months and another during the course of a single day. Let us now investigate why these motions occur.

Figure 21.1 shows why the stars shift position during the course of a year. This shift occurs because as the Earth orbits the Sun, different portions of the heavens are visible in the night sky at different times of the year. For example, during the winter season, the constellation Orion and all other constellations on the right side of Figure 21.1 are visible at, say, 9 P.M. Those stars to the left of the position of the Earth at this time are lost in the glow of the Sun and cannot be seen. Six months later, an entirely different vista is visible at the same hour of the night. At this time, an observer will see Ophichius and all other constellations toward the left on the diagram, but now Orion will be lost in the Sun's glow.

FIGURE 21.2 *A time-exposure showing circular arcs of stars around the North Pole. Photo was taken with the McMath Pierce Telescope on Kitt Peak.* (Courtesy NOAO).

Before we continue with our discussion, let us pause briefly to examine an important vocabulary item. The words "revolve" and "rotate" are frequently used interchangeably in everyday life, but scientists distinguish between the two. The word "**revolve**" is used for those situations in which *one object moves around another.* Thus, the Earth revolves around the Sun. The word "**rotate**" is reserved to designate an object *turning on its axis.* Thus, the Earth rotates on its axis once each day.

The daily motion of the stars, as discussed in Example 21.1, takes place because of the rotation of the Earth on its axis. The stars appear to rise in the east and set in the west, but this is only an apparent motion caused by the rotation of the Earth from west to east. The axis of rotation of the Earth points toward the star Polaris, and the motion of the Earth on its axis can be seen by a time-exposure photograph (Fig. 21.2). This time-exposure shows star trails as they circle about Polaris, which remains essentially motionless because it is within 1° of the axis of rotation of the Earth. If you were to live at the North Pole directly below the star Polaris, you would observe that the stars would always circle above you in the sky with the North Star as the center of their circular path. As a result, you would never see any stars sink below the horizon. This means that if you lived at the North Pole, all the stars you could see would be circumpolar. An observer at the Earth's Equator would see Polaris on the northern horizon, and all the stars circling

about it would rise and set. In fact, all the stars seen would rise straight up on the eastern horizon and set straight down on the western horizon. Thus, at the Equator there are no circumpolar stars.

So far in this section we have readily spoken of the motion of the Earth in its orbit about the Sun and of the rotation of the Earth on its axis. As we shall see before the end of this chapter, however, such notions were not easily arrived at by early astronomers. The Sun appears to rise each morning in the east and set each evening in the west. But such an apparent motion could result either if the Earth rotated once every 24 hours on its axis or if the Sun were orbiting around our planet. Until the sixteenth century, most astronomers believed that the Earth was the center of the Universe and the Sun and the stars revolved around it. It is necessary to recognize that scientists had not yet discovered the other planets in our Solar System. If there were only two objects in the Universe, the Earth and the Sun, there would be no way to tell which one was moving around which, and it would be meaningful to say only that they were orbiting relative to each other.

To understand how astronomers eventually proved that the Earth revolves around the Sun, consider the following situation. Suppose that you were on a raft drifting on an ocean. If there were no landmarks anywhere, it would be impossible to tell whether you were moving or stationary. But now suppose that there were two islands in sight, one

(a)   (b)

FIGURE 21.3   *The concept of parallax is illustrated by these two photographs. (a) The photographer is in line with the row of columns. (b) When the photographer moved, the relative positions of the columns with respect to each other appear to have shifted. The same effect has been observed in astronomical studies. As the Earth revolves about the Sun, the relative positions of the stars with respect to each other appear to shift.*

close to your boat and the other farther away. If you were stationary, the two islands would remain in the same position relative to each other. If you moved, however, they would appear to shift positions. This apparent shift in position of objects against a background, caused by the motion of the observer, is known as **parallax.** This shift is illustrated by the two photographs of the columns of a building shown in Figure 21.3.

The ancients understood the concept of parallax, and they correctly reasoned that if the Earth did move around the Sun, the stars should change position relative to one another, as shown in Figure 21.4. Since no such shift was observed, they concluded that the Earth must be stationary. This was the only logical conclusion possible from the data at hand. The mistake arose not out of faulty reasoning but out of inaccurate measurements. The diameter of the Earth's orbit is small in comparison with the distance to even the nearest stars. Therefore, the parallax angle changes little in 6 months' time and was not measurable until more accurate instruments became available. Only then could astronomers prove that stellar positions change relative to one another and therefore that our planet is in motion.

We now know that the Sun is the center of our Solar System and that nine planets revolve around the Sun. In addition to revolving, the planets simultaneously rotate. The Earth rotates approximately 365 times for each complete orbit around the Sun. Each complete rotation of the Earth represents one day.

FIGURE 21.4   *A nearby star appears to change position with respect to the distant stars as the Earth orbits the Sun. When the Earth is on the right side of the Sun in the drawing, the star is seen at position A against the background stars. Six months later, when the Earth has moved to the left side of the diagram, the star is seen at position B against the background stars. This drawing is greatly exaggerated; in reality, the distance to the nearest star is so much greater than the diameter of the Earth's orbit that the parallax angle, p, is only a small fraction of a degree.*

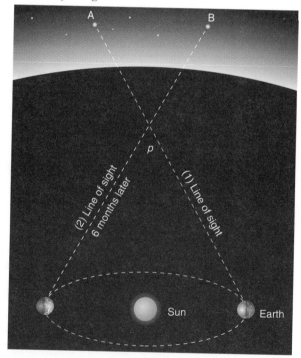

FIGURE 21.5   *The phases of the Moon.* (Courtesy Kevin Reardon, Williams College, Hopkins Observatory. From Pasachoff, *Astronomy: From Earth to the Universe,* 4th edition, 1993.)

The story of how scientists determined that objects move in the heavens reads at times more like a detective story with a strange cast of characters than like a study of pure science. Later in this chapter, we will look at the contributions made to our understanding of motion in the heavens by these early observers of the sky.

21.3 MOTION OF THE MOON

Phases of the Moon

Even the most casual observer of the Moon has seen that it appears to change shape on a regular basis. If on one evening the Moon is circular and bright, it is said to be full. A few evenings later, part of the disk is darkened. As the days progress, the dark portion grows until only a tiny sliver of the Moon is visible. This phase of the Moon is called a crescent. Finally, about 15 days after the Moon was full, it becomes invisible. This day is called the time of the new Moon. Shortly after the new Moon, the thin crescent reappears. As the nights go by, the visible portion of the Moon grows larger and larger until the Moon is full again after a total cycle of about 29.5 Earth days. Further observation reveals that

when the Moon is full, it rises approximately at sunset, and on each successive evening it rises, on the average, about 53 minutes later, so that in 7 days it rises in the middle of the night. In about a month, the cycle is complete and the next full Moon rises on schedule in the early evening. The phases of the Moon are shown in Figures 21.5 and 21.6.

To understand the phases of the Moon, we must first realize that the Moon does not emit its own light but simply reflects light from the Sun. Thus, the amount of moonlight received depends on the relative positions of the Sun, the Moon, and an observer on the Earth. The Earth revolves around the Sun in a flat orbit, as if the two were positioned on an imaginary table top. As shown in Figure 21.7, the Moon's orbit around the Earth is tilted with respect to this plane, so that generally the Moon, the Sun, and the Earth do not lie in a straight line. As the Moon orbits, the half of the sphere facing the Sun is continuously illuminated (see Fig. 21.6). If the Moon is positioned directly behind the Earth away from the Sun, the entire sunlight area of the Moon is visible from Earth, and the Moon appears to be full. If the Moon is between the Earth and the Sun, however, the sunlit side is facing away from the Earth, and the unlit surface that faces the Earth cannot easily be seen. The

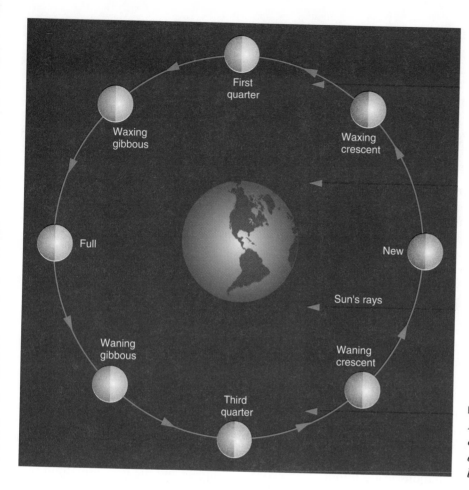

FIGURE 21.6 *Phases of the Moon. The Sun is so far away that the light from it is assumed to travel in parallel lines.*

Moon is then virtually invisible to us, and it is said to be new. (The word "virtually" is used in the last sentence because there is a slight illumination of the Moon caused by light reflected from the Earth to the Moon.) In Figure 21.6, the position of the Moon is shown in its different phases. Midway be-

tween a new Moon and a full Moon, half of the side of the Moon facing Earth is illuminated. Because the Moon at this time is one-quarter of the way through its complete cycle, starting with the new Moon as the beginning of the cycle, this Moon is called a first-quarter Moon. Note that during the

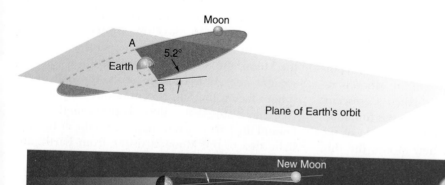

FIGURE 21.7 *The Sun and the Earth lie in one plane, while the Moon's orbit around the Earth lies in another. (Scales are exaggerated for emphasis.)*

interval between a new Moon and a full Moon, the Moon appears to grow as seen from the Earth. Those phases of the Moon in which we see a little more of the Moon on one night than we did the night before are said to be **waxing phases.** Thus, a few days after a new Moon, we have a waxing crescent phase, and a few days after a first-quarter Moon, we have a waxing gibbous Moon. As the Moon moves from the full phase back to the new phase, we see a little less of the Moon each night than we did the night before, and these phases are referred to as the **waning phases.** Thus, a few nights after a full Moon, we have a waning gibbous phase, which eventually becomes a waning crescent phase and then a new Moon once again. Midway between a full Moon and a new Moon, half of the face of the Moon visible from Earth is again illuminated, and we have a third-quarter Moon.

Eclipses of the Sun and Moon

Recall that the Moon's orbit is tilted with respect to the plane of the Earth's orbit about the Sun. But two planes tilted with respect to each other must necessarily intersect, as shown in Figure 21.8. Therefore, as the Moon orbits, it must pass through the Earth-Sun plane twice during each revolution. These points are called the **nodes** of the orbit. Figure 21.9a shows the planes of the Earth's and Moon's orbits, and the nodes are along the line where these planes intersect. If a line is drawn between the Sun and the Earth, the nodes of the Moon's orbit do not normally fall on that line. Instead the alignment

is as shown in Figure 21.9b or c. But the nodes of the Moon's orbit do line up with the Sun on occasion.

What happens if there is a new Moon at the time when the lunar orbital nodes line up with the Sun? The Moon's shadow falls onto the Earth, thereby blocking out, or eclipsing, the Sun, as shown in Figure 21.9e. When a total solar eclipse occurs, an unnatural darkness descends, and the Earth becomes still and quiet. Birds seem to become confused, return to their nests, and cease their singing. While the eclipse is total, the Sun itself is hidden; but the outer solar atmosphere (called the corona), normally invisible because of the Sun's brilliance, appears as a halo around the dark Moon.

Now suppose that there is a full Moon when the lunar orbital nodes line up with the Sun, as shown in Figure 21.9f. The Moon is positioned behind the Earth, but now instead of lying above or below its orbital plane, it is directly behind the Earth, in its shadow. When this happens, the full Moon becomes temporarily invisible from Earth. This phenomenon is called a **lunar eclipse.** Lunar eclipses last a few hours from beginning to end and occur at periodic intervals.

The solar corona, or atmosphere, is seen during this 1980 total eclipse of the Sun. (Courtesy of Serge Koutchmy, Institut d'Astrophysique, Paris. From Pasachoff: *Astronomy: From Earth to the Universe,* 4th edition, 1993.)

FIGURE 21.8 *Two planes intersect in a straight line.*

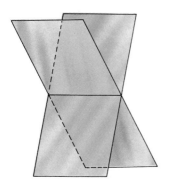

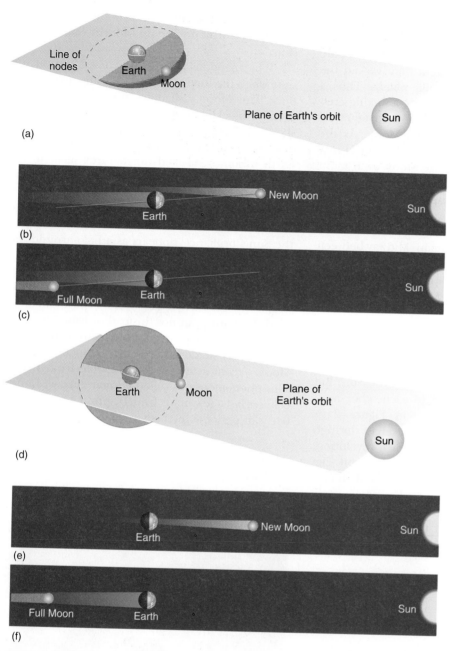

FIGURE 21.9 *Eclipses of the Sun and Moon. The Sun and the Earth lie in one plane, while the Moon's orbit around the Earth lies in another. (Scales are exaggerated for emphasis.) (a) Normally the Moon lies out of the plane of the Earth-Sun orbit, (b) so at new Moon the Moon's shadow misses the Earth, (c) and at full Moon the Earth's shadow misses the Moon. (d) However, if the Moon passes through the Earth-Sun plane when the three bodies are aligned properly, an eclipse will occur. (e) An eclipse of the Sun occurs when the Moon is directly between the Sun and the Earth, and the Moon's shadow is cast on the Earth. (f) An eclipse of the Moon occurs when the Earth's shadow is cast on the Moon.*

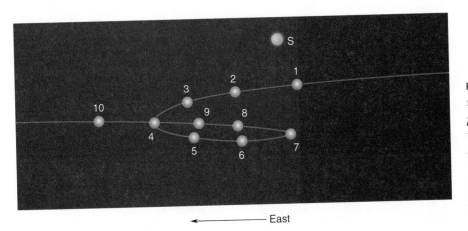

← ——————— East

FIGURE 21.10 *Retrograde motion of a planet. The planet drifts eastward with respect to the background star S until position 4. It then moves westward with respect to S until it reaches position 7. It then resumes its eastward march.*

21.4 MOTION OF THE PLANETS AND THE SUN

The word "planets" is derived from the Greek word "*planetes*," which is translated as *wanderers.* They were so named because the planets change their position relative to the fixed, or background, stars. Early observers of the heavens noted that the Sun and Moon did not stay fixed in position relative to the backdrop of the stars. Instead, these two objects drift eastward with respect to the stars behind them. Most of the time, the planets also obey this eastward drift. This means that if you observe a planet on a particular night and pay attention to a particular star close to it in the sky, in a couple of weeks the position of the planet has changed such that it is farther east than is the background star.

A problem that confronted early astronomers was that the Sun and Moon always drift eastward, but at times the planets do not. Sometimes the planets move westward. Why do they wander around in the way that they do? To see why, consider Figure 21.10. On the first day of observation, the planet is at position 1 on the figure, and on day 2 it has moved to position 2. Note that the planet is drifting eastward in the sky relative to the background star S. This gradual eastward drift continues until we reach position 4. It then begins to back up in the sky, drifting westward with respect to the star until it reaches position 7. The planet then reverts to its original eastward drift as shown by positions 8 through 10. When a planet is backing up in the sky, drifting westward, the planet is said to be un-dergoing **retrograde motion.** What causes this? In a later section, we shall see the incorrect explanation of early astronomers and then discuss what actually is happening.

To study the motion of the Sun, astronomers have devised a figure to depict its path across the sky. This large sphere with its center concentric with the Earth, as shown in Figure 21.11, is called the **celestial sphere.** The projection of the Earth's Equator on this sphere is called the **celestial equator.** The Sun appears to move around the sky, changing its position with respect to the background stars by about 1° each day. This path of the Sun is called the **ecliptic.** Because of the tilt of the Earth's axis of rotation, the celestial equator and the ecliptic are also tilted with respect to one another by 23.5°. Note that the celestial equator and the ecliptic intersect at two points called the equinoxes (from the Latin for equal nights). At these times, there are 12 hours of daylight in both the Northern and Southern Hemispheres. When the Sun passes the point indicated as the **vernal equinox,** spring begins in the Northern Hemisphere. This occurs each year about March 21. The Sun continues its northward movement in the sky until it reaches a point called the **summer solstice** on about June 21. At this time, the Sun stands as high in the sky as it is going to get. As the Sun continues its motion, it reaches a point called the **autumnal equinox** on about September 22, and fall begins for the Northern Hemisphere. Finally, on December 21, the winter season begins when the Sun reaches its lowest point in the sky at the **winter solstice.**

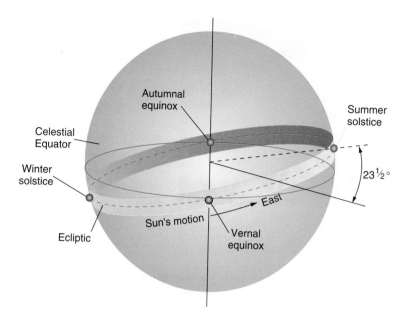

FIGURE 21.11 *The celestial sphere. The celestial equator is the extension of the Earth's equator into space. The ecliptic is the apparent path of the Sun around the celestial sphere during the course of a year. The Moon and planets are always close to the ecliptic, and those constellations close to the ecliptic are called the constellations of the zodiac. The ecliptic and the celestial equator intersect at the vernal equinox and the autumnal equinox. The Sun is highest in the sky, in the Northern Hemisphere, at the summer solstice, and lowest at the winter solstice.*

As the Sun makes its yearly journey around the heavens, it passes through 12 constellations that are called the constellations of the **zodiac.** These constellations lie within a belt that extends about 9° on each side of the ecliptic. Some of these constellations are so dim that they are hardly visible to the naked eye, yet more people have heard of these than of any other constellations in the sky because they form the basis of astrological horoscope tables. Virtually everyone knows his or her "sign." But what does it mean if you are a Libra, for example? Astrologers would have you believe that a person born under this sign has a tendency toward certain behaviors that are governed by the stars. Nothing could be further from the truth. There is absolutely no scientific way in which objects as remote as the stars could affect a person's life simply because they were born under a particular grouping of them. If you are a Libra (October), this means that if your father or mother had taken you outside on the day you were born and let you look at the sky, the constellation Libra would have been in the sky directly behind the Sun. You would, obviously, not have been able to see it because it would be hidden in the Sun's glow, but you would have been a Libra nonetheless. To cast one more aspersion on the beliefs of astrology before we move forward, you actually would not be a Libra in today's world. When the beliefs in astrology were first being formed, the constellation Libra *was* behind the Sun in October.

Since that time, however, the position of the Sun as seen from Earth has shifted slightly, and now the constellation that is behind the Sun is Virgo. Libra has moved to a position such that it would be known as a November sign if astrologers were developing their charts in the present day.

Finally, let us consider one more feature of the motion of the planets. If you were to go to the North Pole of the Sun and go directly upward in space, after having traveled far out into space, you would see a view of the Solar System as shown in Figure 21.12. All the planets would be seen to be revolving counterclockwise about the Sun, and most of them would be seen to be rotating counterclockwise. As we shall see, there are two notable exceptions to this: Venus and Pluto revolve counterclockwise but rotate clockwise. Any theory of the formation of the Solar System must account for these features of planetary motion.

CONCEPTUAL EXERCISE

In Chapter 18 we saw that the seasons are caused by the 23.5° tilt of the Earth's axis. In Figure 21.11, we see that in the summer season in the Northern Hemisphere the Sun stands 23.5° above the horizon. How does the Earth's tilt explain this?

Answer If the Earth were not tilted, we would see the Sun always directly above the horizon and above the Equator. In the summer, we tilt toward the Sun, so we see it gradually rising above the horizon as we tip toward it. Finally, at maximum tilt, it is 23.5° above the horizon. For the same reason, it is 23.5° below the horizon at maximum tilt during the winter season in the Northern Hemisphere.

21.5 ARISTOTLE

We encountered Aristotle when we examined some of the early ideas about motion in our study of physics. Aristotle wrote on a wide variety of subjects ranging from politics to medicine, and among his fields of philosophical thought was astronomy. Let us examine a few of his viewpoints concerning the heavens. His philosophy held that the Earth was the center of the Universe about which all the planets, the Sun, the Moon, and the stars revolved in perfectly circular orbits. This idea came, in time, to be called the **geocentric** (Earth-centered) model of the Universe. The circle was selected as the shape of the orbits of these objects because the ancients tended to fit their concepts in with a spiritual view of the heavens. Heavenly objects seemed to behave differently than earthbound objects. A moving object on Earth always stops, but the planets and other

objects in the sky never stop their motion. There was just something different about the heavens and the Earth. The circle seemed to fit in with this view of heavenly perfection. The circle is symmetrical; it has no beginning and no end. It seems a shape destined for the heavens and for the presumed perfection that resides there.

Another concept that Aristotle held that fit in with this perfection of the heavens was that the Moon was perfectly spherical and polished like a looking glass so that it could reflect the light from the Sun and reveal Man in all his glory. Granted, the Moon can be observed to have some blemishes. Even with the naked eye, light and dark regions are obvious. However, a few imperfections should be expected because the Moon is the heavenly body that is closest to the corrupt and imperfect Earth.

We have already discussed the fact that most of these Aristotelian ideas of the way the Universe works are wrong, and we shall see shortly that, in fact, *all* of the ideas mentioned are incorrect. What is so important about the fact that Aristotle was wrong in his view of the Universe? Many scientists have been wrong about many topics throughout the course of human history, and scientists are often incorrect or incomplete about the results of their research even today. Yet these scientists are not mentioned in a textbook. The reason that Aristotle is so important in the progress of astronomy lies with the fact that his beliefs became inculcated into religious belief. In fact, this reached a climax in the thirteenth century owing to the urging of Saint Thomas Aquinas. Aquinas believed that there should not be a conflict between religious thought and beliefs concerning how the Universe works and how it is laid out. The end result of this expression was that Aristotle's concept of the heavens came to hold second place only to the revelations of the Bible. Aristotle's teachings became so strongly ingrained in Christian thought that it became difficult to question the ideas of the ancients in any way. Thus, an idea that the Earth might not be at the center of the Universe was not simply viewed with skepticism; the originator of such an heretical idea could be subject to severe religious persecution.

It should be noted here that not all of Aristotle's ideas were incorrect. For example, he correctly concluded that the Earth was round. He did so by observing that the shadow cast on the Moon by the Earth during an eclipse is always round. The only shape that the Earth could have and still al-

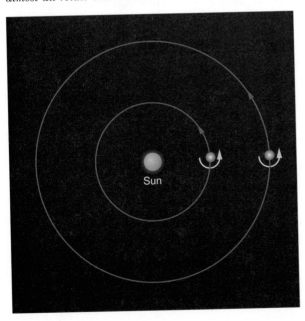

FIGURE 21.12 *From above the north of the Earth, all the planets revolve counterclockwise about the Sun and almost all rotate counterclockwise.*

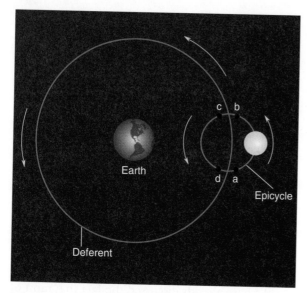

FIGURE 21.13   *The deferent and an epicycle for a planet revolving about the Earth according to the geocentric theory of the Universe.*

ways produce a round shadow is a sphere. If the Earth were disk-shaped, it would occasionally cast a shadow that was a line. Additionally, he noted that a traveler moving north was able to observe stars that had not been visible before. This bore witness to the curvature of the Earth.

The geocentric model was brought to its greatest fruition by Claudius Ptolemy about A.D. 140. He developed a model of the Universe that was able to predict the positions and motions of the planets with reasonable accuracy. Figure 21.13 shows the basic approach that he used. He assumed that the motion of a planet about the Earth followed a circular path that he called the **deferent.** Thus, if the planets always moved eastward against the background stars as they followed their deferents, all would have been well. As we have seen, however, the planets often back up in their orbits and drift westward in retrograde motion. To explain this aspect of the motion of a planet, Ptolemy added **epicycles,** which were small orbits that the planet followed as it moved on the deferent, as shown in Figure 21.13. Thus, we have circles rolling on circles. When the planet is moving from a to b on the epicycle of Figure 21.13, it is moving eastward, but when the planet moves from c to d, it is moving against the direction of its basic motion along the deferent, so it drifts westward. This scheme solves

the problem of retrograde motion, and by adding epicycles on epicycles, and by choosing the distances and speeds of the planets with care, it is possible to explain most of the basics of planetary motion. The scheme is cumbersome, however, and the way was open for a better idea.

21.6 COPERNICUS

The idea that the Sun and not the Earth is the center of the Universe was proposed by Nicolaus Copernicus in the early sixteenth century. This basic idea of a Sun-centered, or **heliocentric,** theory of the Universe was not in accord with the religious beliefs of the time. As a result, Copernicus did not publish his ideas until 1543 when he was on his death bed. Copernicus was not able to detect the parallax shift of the stars, which would have verified that the Earth was in motion. Instead, he based his reasoning on the fact that the motions of the planets could be explained much more simply if the Sun were the center of the Solar System. He retained many of the traditional ideas, such as the belief that the orbits of the planets were circular. In his book, *De Revolutionibus,* he set forth his postulates, but the book was not widely read and caused little controversy at the time.

We have seen that much of the work done by Ptolemy with his deferents and epicycles was an attempt to explain features of planetary motion such as retrograde motion. He was able to do so, but the scheme relied on complex geometrical manipulations. The explanation based on the heliocentric idea is considerably easier to comprehend. Figure 21.14 shows how retrograde motion can be explained for the planet Mars in a Sun-centered Solar System. We first find the Earth in the figure at position 1 while Mars is in its orbit more distant from the Sun at its position 1. The line of sight toward Mars of an observer on Earth is indicated by the dashed line, which passes through both points 1 in the figure. This observer will see certain stars behind the planet at this time. As the weeks pass, the Earth moves to position 2 in its orbit, and Mars also moves to a new position 2. Note, however, that the Earth is catching up with Mars because it is moving faster in its smaller orbit. The line of sight of the observer on Earth is shown by the dashed line passing through these points labeled as 2. Different stars are now seen behind Mars by the observer,

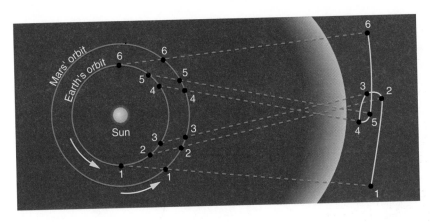

FIGURE 21.14 *Retrograde motion from the point of view of the helio-centric theory.*

but note that the motion of the two planets has been such that Mars has appeared to drift eastward against the backdrop of stars. Note now what happens between those positions labeled as 3 and 4 for Earth and Mars. During this interval, the Earth has caught up with Mars and passed it. The dashed lines in the figure indicate that during this interval Mars has appeared to drift westward. Thus, retrograde motion has been occurring. You can continue this procedure through points 5 and 6 to see that this retrograde motion soon stops, and the normal eastward drift of the planet resumes.

Thus, Mars never really moves backward. It just moves forward more slowly than Earth and seems to move backward when the Earth catches up to it and passes it.

21.7 TYCHO

Tycho Brahe (1546–1601) was a Danish astronomer who was born into a family of nobility. At the age of 14, he witnessed a partial eclipse of the Sun. He was so amazed at the fact that astronomers could predict such an event that he determined at that time to devote his career to observing the heavens. As a result of his reputation and connections, he gained the patronage of Frederick II and was able to establish an excellent observatory on the island of Hveen. Tycho's work was done before the discovery of the telescope, but he employed all of the best available instruments of the time, including some that he developed himself, in an effort to chart the course of the Sun, Moon, and planets. He is regarded by many as the most systematic observer of the heavens in the history of astronomy. He compiled enormous amounts of data on the position of all the known bodies in the Solar System. This in-

formation enabled him to note that the positions of the planets varied from predictions that were based on the work of Ptolemy.

If you had lived at that time and for some reason wanted to know the position of the planet Venus in the sky on the night of June 15, 1585, Tycho would be the one you would ask. Thus, Tycho was an observer of considerable note, but other predilections of his may have been responsible for his never developing any laws or theories based on these observations. For example, Tycho had his nose cut off in a duel as a young man. To cover this deformity, he had false noses fashioned out of gold and silver. Tycho also had a strong inclination for late-night drinking parties. In fact, this led to his death because at a command appearance at a noble's function, he drank to such excess that his bladder ruptured.

When Frederick II died in 1597, Tycho was forced to leave Denmark by the enemies he had accumulated in abundance because of his argumentative nature and large ego. He moved to Prague where he undertook the same duties for Emperor Rudolph of Bohemia. While there, Tycho made a great contribution to astronomy when he hired a young assistant named Johannes Kepler.

21.8 KEPLER

Johannes Kepler (1571–1630) fled his Catholic homeland of Germany as a Protestant refugee to work with Tycho. The relationship was not a pleasant one, however, because Tycho was jealous of Kepler's brilliance. As a result, Tycho kept much of his observational data from Kepler. When Tycho died, however, this wealth of information came into Kepler's hands, and he studied it for approximately

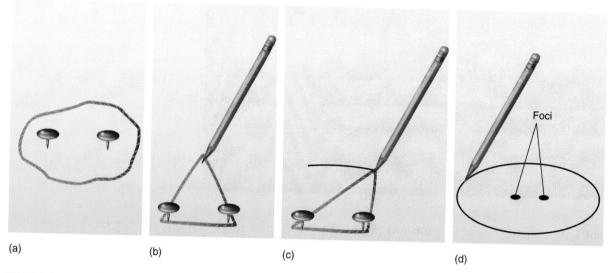

(a)   (b)   (c)   (d)

FIGURE 21.15 *(a) Drape a loop of string over two thumbtacks, which will represent the foci. (b) Draw the string taut with a pencil. (c) Keep the string taut and trace out a curve, and (d) you end up with an ellipse. To really understand this procedure, you should do it. As an exercise, collect the necessary items and draw an ellipse.*

20 years. The first of Kepler's works was published in his book *The New Astronomy* in 1609. Today the work of Kepler is summarized in his three laws of planetary motion. It should be emphasized here that Kepler's laws are representative of a class of laws called empirical laws. This means that there was at the time no method by which they could be derived based on the known laws of physics. He simply observed that they were true and did not attempt to show that they should be valid based on other, more basic, principles of science. The status of these laws as empirical laws came to an end when Newton formulated his law of universal gravitation. Based on his law, Newton was able to show that Kepler's laws were a necessary consequence of any two objects bound to one another by an inverse-square attractive force. Let us take a look at each of Kepler's laws in turn.

Kepler's First Law

To understand Kepler's first law of planetary motion, we must first understand a geometrical curve called an ellipse. Figure 21.15 shows how to draw an ellipse. First, two thumbtacks are pressed into a table, as in part (a); the location of the tacks will be referred to as the *foci* (singular *focus*) of the el-

lipse. Note that a loop of string is placed over the tacks. In part (b) of the figure, the string has been drawn taut by a pencil pressed against it. A curve is then drawn by tracing out those points that the pencil can reach while the string is kept taut. This is indicated in part (c). The result is a curve called an ellipse, like that shown in part (d). The distance from the center of the ellipse, through one of the foci, and out to the ellipse is called the **semimajor axis.** This distance is d in Figure 21.16.

Kepler's first law states that:

The orbit of a planet is an ellipse with the Sun at one focus. The other focus is empty.

FIGURE 21.16 *The distance d from the center of an ellipse through a focus is called the semimajor axis of the ellipse.*

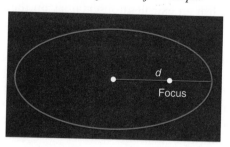

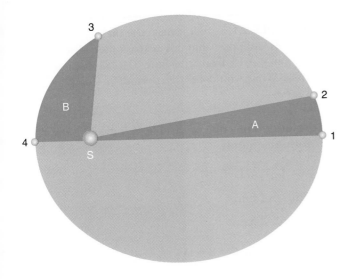

FIGURE 21.17 *Kepler's second law, the law of equal areas.*

It should be pointed out here that although Kepler formulated his law based on observations of the planets, the basic form of the law is found to hold for any object that orbits another. Thus, the Moon moves about the Earth in obedience to Kepler's first law, as do double star systems orbiting one another and as does a human-made satellite orbiting the Earth.

Kepler's Second Law

Kepler's second law is often referred to as the law of equal areas. To see why it is given this designation, consider Figure 21.17, which shows the orbit of a planet moving about the Sun. First, we draw a line from the Sun to the planet when the planet is at position 1 in its orbit; then we wait a certain period of time, say 1 month. In that month's time, the planet will have moved to a new position in its orbit, position 2 in the figure. After this time interval, we draw a second line from the Sun to the planet. These two lines form a pie-shaped segment of the ellipse, as indicated in the figure by area A. Now, let us wait a while until the planet is at some new position in its orbit, say position 3 in the figure. We now draw a line from the Sun out to position 3 and wait for the *same* time interval as when we were sketching out our previous area, 1 month. The planet has again moved in its orbit during this month and is now located at position 4. Again a pie-

shaped area, B, has been swept out by the line during this interval of time. Kepler found that segments A and B had the same area. He stated his second law as:

The line from the Sun to a planet sweeps out equal areas in equal intervals of time.

As an outgrowth of this law, it can be seen that the planets do not move in their orbits with a constant speed. To see this, consider Figure 21.17 once again. If the areas A and B are to be equal, the planet would have to move faster in its orbit when close to the Sun as it swept out area B than it would when farther from the Sun sweeping out area A. Thus, the Earth moves faster in its orbit during the winter season when it is close to the Sun than it does in the summer when it is farther from the Sun.

Kepler's Third Law

Kepler's third law is a mathematical relationship between the period of a planet and its average distance from the Sun. The period is the time it takes a planet to make one complete revolution in the orbit, and in Kepler's formulation of the law, this period is expressed in Earth-years. *The average distance from the Sun is the length of the semimajor axis of its elliptical path, and this distance must be expressed in* **astronomical units (A.U.).** The average distance of the Earth from the Sun is about 93 million miles, and this distance is defined to be the astronomical unit. Table 21.1 gives the period of the planets that were known in Kepler's time and their average distance from the Sun in A.U., the units appropriate to Kepler's third law.

Kepler's third law is expressed in terms of a mathematical relationship between the period P and the average distance from the Sun d. The equation is

$$P^2 = d^3$$

In words, the law is stated as:

The squares of the periods (in years) of the planets are equal to the cubes of their average distance (in A.U.) from the Sun.

EXAMPLE 21.2   EXAMINING THE THIRD LAW

Consider the information give in Table 21.1 and show that Kepler's third law is satisfied by the motion of Venus.

Solution From the table, we find that the period of Venus is 0.61 year. Let us square this number to find

$$P^2 = (0.61)^2 = 0.37$$

Likewise, we find from the table that the average distance of Venus from the Sun is 0.72 A.U. Let us cube this number.

$$d^3 = (0.72)^3 = 0.37$$

Thus, we find that $P^2 = d^3$. Try this for a few of the other planets. Does the law work as well for Saturn, a distant planet, as it does for Mercury, which is close to the Sun?

TABLE 21.1

PLANET	PERIOD (YEARS)	DISTANCE (A.U.)
Mercury	0.24	0.39
Venus	0.61	0.72
Earth	1.00	1.00
Mars	1.88	1.52
Jupiter	11.86	5.20
Saturn	29.46	9.54

EXAMPLE 21.3   AN INVISIBLE PLANET

Let us suppose that an astronomer discovers a planet 4 A.U. from the Sun. (No such planet exists.) What would be the period of this planet in years?

Solution The cube of this planet's average distance from the Sun is

$$d^3 = (4)^3 = 64$$

Thus, we find

$$P^2 = 64$$

from which, $P = 8$ years.

21.9 GALILEO

We have discussed in our study of physics some of the discoveries and methods used by Galileo in his investigation of mechanics, and we found that his adherence to experimental observations became a basic part of science. Let us now turn our attention to his contribution to astronomy, along with the trials and tribulations that he encountered along the way.

Galileo was born about a century after Copernicus. During his life, the teachings of Copernicus were still banned by religious authorities. However, Galileo came to support the idea of a heliocentric solar system based on his observations with a telescope. Galileo did not invent the refracting telescope, but he has gained distinction as the first person to use it in a systematic way for observing the heavens. Let us consider a few of the things that Galileo saw through his telescope that caused him to accept the Copernican viewpoint.

1. Galileo turned his telescope to the hazy pathway across the sky that we call the Milky Way and saw that it was not simply a misty blur, but instead was a vast collection of individual stars too faint to be distinguished individually by the naked eye. This drew him into controversy with contemporary thought because the fundamental outlook on science at the time was that if it were true that the Milky Way was a vast collection of stars, the ancients would have already mentioned it.

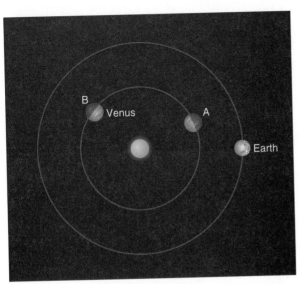

FIGURE 21.18 *At position A, Venus is in a crescent phase; at B, it is in a gibbous phase.*

There was no room for new information about the heavens. This observation had nothing to say about the validity of the heliocentric theory, but it caused Galileo to ponder the idea that if the ancients could be wrong about this, perhaps they could also be wrong about the geocentric theory.

2. When he turned his telescope toward Jupiter, he found that it was circled by four moons. (We now know that Jupiter has many more moons than this, but the best telescope that Galileo ever owned had a magnification of only about 30, and as a result, he was not able to see these smaller, dimmer moons.) This was an important observation for Galileo and a blow to his detractors because they had held that *everything* orbited the Earth. Now Jupiter was found to have moons orbiting it. The idea that the Earth was the center of all motion was in serious trouble.

3. Galileo's telescope revealed that the Moon had craters and tall mountains. This was in contradiction to the idea of the time that heavenly objects were more nearly perfect than is the Earth. That is, they were smoother and more symmetrical. Yet, here was an object that had about the same kind of blemishes on it as could be found on the "corruptible" Earth. Thus, the Earth could

also find its place among the heavenly objects. This observation did not prove the Copernican system to be correct, but it once more raised the issue that if one idea can be wrong, others can be incorrect also.

4. The deciding factor that convinced Galileo of the truth of the Copernican concept was his observations of the planet Venus. He observed that Venus goes through phases just like the Moon. At some observation times, Galileo would find Venus in a crescent phase; at others, he would find it to be in a gibbous phase, and so forth. The only way that Venus could ever be seen in a gibbous phase is if it is circling the Sun. Figure 21.18 shows why Venus undergoes these phase changes.

Based on his telescopic observations, Galileo became a staunch defender of the heliocentric theory of the Solar System. But old ideas die hard. In 1633, Galileo was taken before a high court of the Church and asked to recant his belief that the Sun and not the Earth was the center of the Solar System. Galileo bowed to the pressure of the Church rather than face imprisonment and torture. The main issue of the controversy concerning Galileo centered on his book *Dialogue Concerning Two Chief World Systems.* The book was written in Italian, a language understandable by the common man. It consisted of a series of discussions, or dialogues, among three characters, Salviati, Sagredo, and Simplicio. The name Salviati implies Savior, in the sense that he is an adherent of the heliocentric theory. Sagredo is neutral, but his name implies wisdom or intelligence. The third character is a defender of the geocentric theory, and his name Simplicio, or simpleton, needs no explanation. The end result of the dialogue is that Salviati convinces Sagredo of the wisdom of the heliocentric system.

Following his inquisition by the Catholic Church, Galileo's book was published in Protestant Holland, where his ideas met with widespread acceptance. Thus, the death knell had been struck for the geocentric view of the Solar System.

CONCEPTUAL EXERCISE

To gain some practice in finding phases, consider Figure 21.18. **(a)** What phase would Venus be in if it were at a position in its orbit such that it was on a straight line between the Earth and the Sun? (Ignore any problems you might have with being able to see the planet.)

Answer The side of Venus pointing toward the Earth would not receive any rays of light from the Sun. Thus it would be dark, and you would see a "new" Venus.

(b) Repeat if Venus were on a straight line with the Earth and Sun but on the opposite side of the Sun.

Answer In this case, all of the side of Venus that is facing the Earth is lit up. Thus, you would see a "full" Venus.

(c) If you were on Venus at positions A and B in Figure 21.18, in what phase(s) would you see the Earth?

Answer In both positions most of the side of the Earth facing you would be lit up. Thus, you would see a gibbous Earth.

21.10 NEWTON

Galileo was to spend the rest of his life under house arrest, but his work was to serve as the inspiration for a sickly English lad named Isaac Newton. They were close to being contemporaries, but the worlds in which they worked were considerably different. Galileo suffered from persecutions because of his scientific teachings, but Newton was to gain great fame and wealth during his lifetime because of his contributions to the endeavor of science.

The first great contribution to astronomy that Newton made was his invention of the reflecting telescope, discussed in Chapter 11. It was his three laws of motion and the law of universal gravitation, however, that united the heavens with the Earth. It had been held that the heavens operated under a different set of laws and rules than did the Earth. For example, it was Aristotle's belief that the natural tendency of moving objects on the Earth was that they should slow down. This did not, of course, apply to the heavens because they obeyed a different set of rules. The motion of the stars, the Sun, the Moon, and the planets never ceased. Newton showed that it was the natural tendency of objects to continue to move endlessly unless a force, such as friction, acted to retard their motion. This brought the heavens closer to Earth. With this law of gravitation, Newton was able to show that Kepler's laws of planetary motion were natural consequences of the fact that they were held in their orbits by a force of mutual attraction described by his expression for the gravitation force. So, Newton moved Kepler's laws from the realm of empirical equations to equations that could be derived from more fundamental principles. Thus, it may be said that Newton's laws unified the Universe.

SUMMARY

Constellations are groupings of stars that are imagined to have certain shapes from Earth. There are 88 of these now recognized. Those constellations that are visible but never set at a particular location on Earth are said to be **circumpolar constellations** for that location.

Astronomers proved that the Earth revolves around the Sun by showing that the stars change position relative to each other during the course of a year. The phases of the Moon are caused by the revolution of the Moon around the Earth in a plane slightly tilted with respect to the plane of Earth's orbit. **Eclipses** occur when the Sun, Moon, and Earth all lie in a straight line and the Moon is in the plane of the Earth's orbit.

The general motion of the planets is toward the east against the background of fixed stars. At certain times, however, the motion shifts to a westward drift. This backing up of a planet in its orbit is called **retrograde motion** and can be explained from both the geocentric and heliocentric theories of the Universe.

The path followed by the Sun as seen from Earth is called the **ecliptic**. The time when the Sun is highest in the Northern Hemisphere is called the **summer solstice**, and the Sun is lowest in the sky at the **winter solstice**. The ecliptic and the **celestial equator** intersect at the **autumnal equinox** and the **vernal equinox**.

Aristotle and Ptolemy were primarily responsible for the geocentric theory of the Universe. Copernicus is primarily responsible for the heliocentric theory. Tycho is famed for his careful observations of the heavens. Kepler is known for his three laws of planetary motion. Galileo is sometimes called the father of modern science, primarily because he introduced into science the method of experimentation as the basis for deciding between conflicting theories. Newton, by means of his law of universal gravitation, showed that motion in the heavens obeys the same physical laws as on Earth.

EQUATIONS TO KNOW

$P^2 = d^3$ (Kepler's third law)

KEY WORDS

Constellations
Circumpolar
  constellations
Parallax
Revolve
Rotate
Phases of the Moon

Waxing
Waning
Eclipses
Planets
Retrograde motion
Celestial sphere
Celestial equator

Ecliptic
Vernal equinox
Summer solstice
Autumnal equinox
Winter solstice
Zodiac
Geocentric theory

Deferent
Epicycle
Heliocentric theory
Kepler's laws
Astronomical Unit

CONCEPTUAL QUESTIONS

Constellations and motion of the Earth

1. Using the star charts at the end of the chapter, go outside and find several constellations now visible and several of the brighter stars now in the sky.

2. Discuss why the constellations are still important today.

3. Use the star charts at the end of the chapter to identify several constellations that are circumpolar and several that are not.

4. Explain why all constellations are circumpolar at the North Pole but none are at the Equator.

5. Are the same stars visible in the sky on a winter night as a summer night? If you were in a balloon high in the atmosphere, could you see the summer stars in winter? Explain.

6. Suppose that you are driving a car and the speedometer reads 80 km per hour. A person sitting next to you and looking at the speedometer at the same time may read a different value, perhaps 76 km per hour. Explain how two people can look at the same instrument at the same time and read different values.

7. With rigidly fixed modern telescopes, astronomers can determine how far away some stars are by observing the parallactic shift as the Earth travels around the Sun. Using this technique, would it be easier to estimate distance to nearby stars or to distant stars, or would it be equally difficult for all stars? Defend your answer.

Motion of the Moon

8. How is the Moon positioned with respect to the Earth and the Sun when it is (a) full, (b) new, (c) crescent?

9. What time does a new Moon rise? A full Moon?

10. Imagine yourself to be riding on the Moon watching the Earth as the Moon goes through its complete cycle of phases. When the Moon is in each of its phases, what is the corresponding phase of the Earth? That is, is there a full Earth at the same time there is a full Moon?

11. In Figure 21.7, assume the sunlight is coming in from the left of the page and that the Moon circles counterclockwise. Locate the position of each phase of the Moon.

12. Suppose the Sun, Earth, and Moon lay in a single plane, but all other motions were the same as now. Describe the lunar cycle in this hypothetical situation.

13. Many societies use a lunar calendar instead of a solar one. In a lunar calendar, each month represents one full cycle of the Moon. How many days are there in a lunar month? If there are 12 months to a lunar year, will a lunar year be longer or shorter than a solar year? Explain.

14. Why does the Moon rise approximately 50 minutes later each day than it did the previous day?

15. Draw a picture of the Sun, Earth, and Moon as they will be during an eclipse of the Sun. Draw a picture of the Sun, Earth, and Moon as they

will appear during an eclipse of the Moon. Explain how these positions produce eclipses.

Motion of the planets and the Sun

16. If the tilt of the Earth's axis of rotation with respect to its orbital plane were (a) greater than it is now, how would Earth's seasons be affected? (b) Repeat assuming the tilt is less than.

17. Describe the motion of the stars about Polaris as seen from the Equator.

18. Explain how the tilt of the Earth's axis of rotation produces the seasons. (You may want to refer to the discussion in Chapter 18.)

19. (a) Is the Sun farther from the ecliptic in the summer or winter? (Be wary of trick questions.) (b) In what season is the Sun farthest from the celestial equator? (c) When is the Sun closest to the celestial equator? (d) Do the ecliptic and the celestial equator ever intersect?

20. Explain why it is cold during the winter in the Northern Hemisphere while it is warm at the same time in the Southern Hemisphere.

21. How would our view of the heavens be changed (all other features being the same) if (a) Earth revolved clockwise rather than counterclockwise and (b) rotated clockwise rather than counterclockwise?

Ancient astronomers

22. Present the main arguments for and against the geocentric theory and the heliocentric theory.

23. It is noted that often when a planet is in retrograde motion, it appears brighter than average. Why should this occur?

24. List some of the beliefs that Aristotle held about astronomy that are now known to be incorrect.

25. Explain retrograde motion by means of both the geocentric theory and the heliocentric theory.

26. In the heliocentric theory, we state the order of the planets from the Sun outward. Thus, the order is Mercury, Venus, Earth, Mars, Jupiter, and Saturn. How would this ordering change from the point of view of the geocentric theory?

27. It was once hypothesized that a planet similar to the Earth circled the Sun such that it was always behind the Sun as seen from Earth. What distance would this planet have to be from the Sun? Why?

28. Prove that Kepler's third law is valid for Jupiter.

29. If the Earth moved in a perfectly circular orbit rather than its elliptical orbit, would our seasons be affected?

30. Use Kepler's second law to verify that the Earth moves fastest in its orbit during the winter season.

31. List the observations made by Galileo through his telescope and show how they tended to disprove the geocentric ideas of Aristotle.

32. Show the position of Venus when it is in a full phase. Why could this position never be seen from Earth?

33. Show the position of Venus for its various phases from the point of view of the (a) geocentric theory and (b) the heliocentric theory.

34. Explain the statement that Newton changed our outlook on the heavens from that of a duoverse to that of a universe.

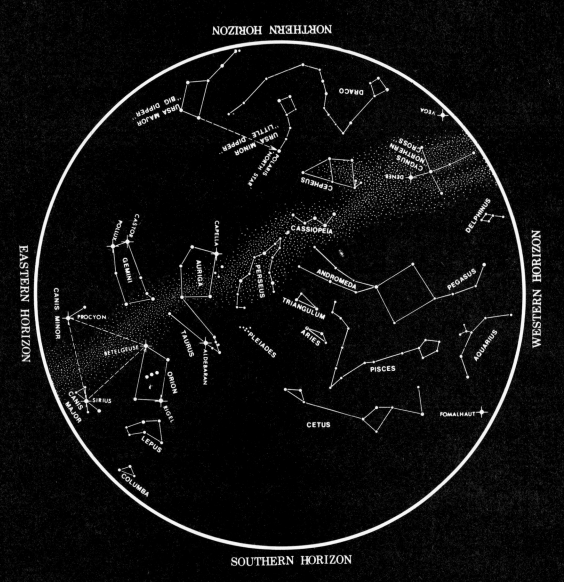

THE NIGHT SKY IN DECEMBER

Latitude of chart is 34°N, but it is
practical throughout the continental
United States.

Chart time (Local Standard):

10 p.m. First of month

9 p.m. Middle of month

8 p.m. Last of month

Star Chart from *GRIFFITH OBSERVER*, Griffith Observatory, Los Angeles

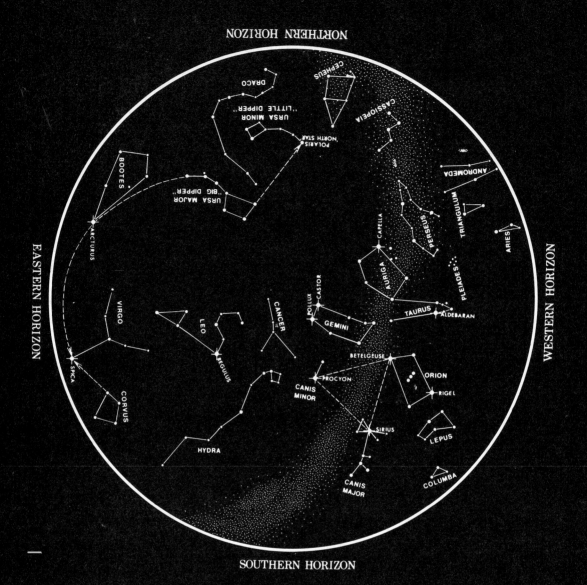

THE NIGHT SKY IN MARCH

Latitude of chart is 34°N, but it is practical throughout the continental United States.

Chart time (Local Standard):
10 p.m. First of month
9 p.m. Middle of month
8 p.m. Last of month

Star Chart from *GRIFFITH OBSERVER*, Griffith Observatory, Los Angeles

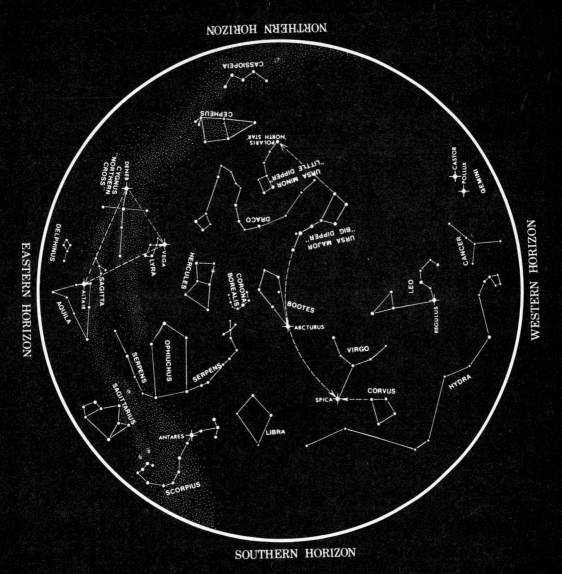

THE NIGHT SKY IN JUNE

Latitude of chart is 34°N, but it is practical throughout the continental United States.

Chart time (Local Standard):
10 p.m. First of month
9 p.m. Middle of month
8 p.m. Last of month

Star Chart from *GRIFFITH OBSERVER*, Griffith Observatory, Los Angeles

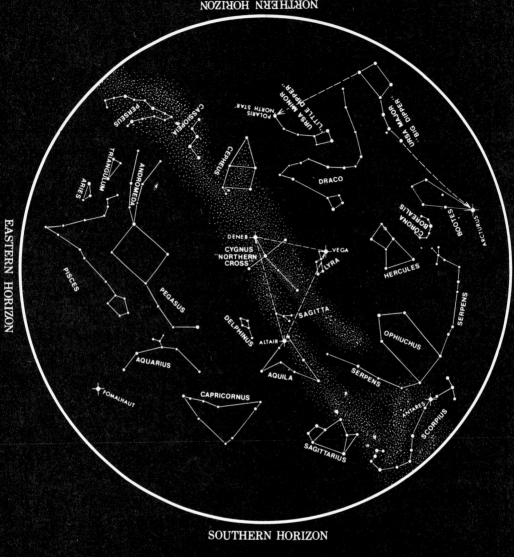

THE NIGHT SKY IN SEPTEMBER

Latitude of chart is 34°N, but it is
practical throughout the continental
United States.

Chart time (Local Standard):
10 p.m. First of month
9 p.m. Middle of month
8 p.m. Last of month

Star Chart from *GRIFFITH OBSERVER*, Griffith Observatory, Los Angeles

CHAPTER 22

THE SOLAR SYSTEM

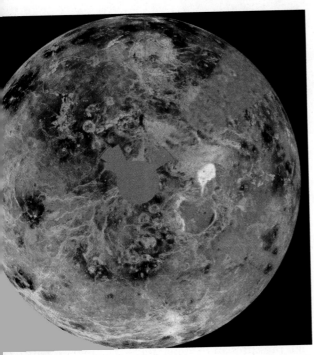

The Northern Hemisphere is seen in this global view of the surface of Venus. Radar mosaics from the Magellan *orbiter were mapped onto a computer-simulated globe to create this image.* (Courtesy NASA)

Now that we have examined the principles of motion that affect the Earth, Sun, and other heavenly bodies, our journey *through* the Universe actually starts with this chapter. Here we focus our attention on our nearest neighbors in space: the planets and their moons and our Sun. Our progression will be from the Sun outward. Much is known about these objects, but there is still much to be learned: In 1979, for the first time, a spacecraft passed close enough to some of the outermost planets to provide us with surprising details about them and their moons. As a result of voyages such as these, many statements reported as facts in astronomy textbooks a few years ago are now known to be incorrect. Perhaps no other science has changed as rapidly in the last fifteen years as has astronomy.

22.1 FORMATION AND STRUCTURE OF THE SOLAR SYSTEM

There are several conflicting scientific theories that attempt to explain the origin of the Solar System and the planet Earth. According to the most widely accepted of these ideas, the **nebular hypothesis,** the Solar System evolved from a cloud of gas and frozen dust. This cloud was nothing like the clouds you can see in the atmosphere. For one thing, it was extremely diffuse and would be considered to be a good vacuum by terrestrial standards. Second, it was composed mainly of light elements and is believed to have been approximately 79 percent hydrogen, 20 percent helium, and 1 percent other elements. (Percentages are by mass.)

Approximately 5 billion years ago, billions of years after the galaxy itself had taken form, the cloud that was to become the Solar System began to contract. This diffuse mass was originally rotating quite slowly. The material in the cloud was quite cold, with temperatures perhaps as low as 3 K. At

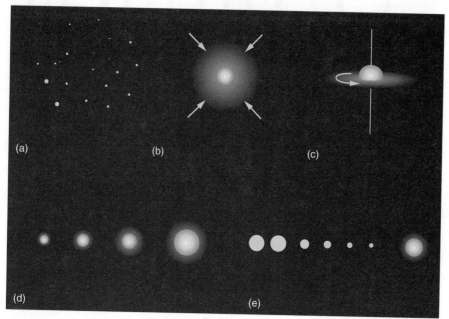

FIGURE 22.1   *Formation of the Solar System. (a) The Solar System was originally a diffuse cloud of dust and gas. (b) This dust and gas began to coalesce under its internal gravitation. (c) The shrinking mass began to rotate and was distorted. (d) The mass broke up into a discrete protosun orbited by large planets composed primarily of hydrogen and helium. (e) The Sun heated up until fusion temperatures were reached. The heat from the Sun then drove most of the hydrogen and helium away from the closest planets, leaving small, solid cores behind. The massive outer planets remain mostly composed of hydrogen and helium.*

these extremely low temperatures, particles move about so slowly that even a slight force will be able to affect them appreciably. Some scientists believe that the slight gravitational attractions among the dust and gas particles themselves caused the cloud to condense slowly into a spherical ball, as shown in Figure 22.1a and b. Alternatively, perhaps a star exploded in a nearby region of space, and the shock wave triggered the condensation. As the condensation continued, more matter gravitated toward the center of the newly formed sphere. Therefore, the density and pressure began to increase within the cloud. As atoms were pulled inward, they accelerated under the influence of the gravitational field. Some of the energy from the collapse was converted into heat, and the temperature of the gas and dust began to rise. This shrinking ball began to rotate more rapidly as it contracted.

Rotational motion distorted the ball to form a body shaped something like that shown in Figure 22.1c. The center of this cloud then coalesced into a large mass that is called the **protosun,** meaning

the "earliest form of the Sun." Within the protosun, the gravitational attraction was great, and the gases were pulled inward, compressing them to high pressures and temperatures. The highest temperature was found in the center of the protosun, where atoms were accelerating inward at the fastest rate. As the contraction progressed, the temperature continued to rise. Eventually, when the critical temperature of 10 million K was reached, collisions between hydrogen nuclei were so forceful that nuclear fusion reactions were started. Hydrogen fused into helium, great quantities of energy were released, and the Sun was born. Our Sun and all the other stars in the sky live out their lives under a delicate balance between forces tending to collapse them and forces tending to make them expand, as shown in Figure 22.2. The force of gravity tends to make the star fall inward on itself, while thermal pressure created by the intense heat and radiation present at the core of the star tries to make it expand. At some radius, these two forces are equal, and an equilibrium is achieved that is maintained

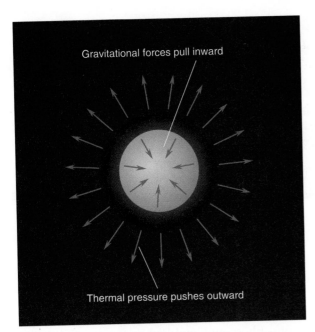

Gravitational forces pull inward

Thermal pressure pushes outward

FIGURE 22.2 *Equilibrium in a star. What are the forces that keep a blown-up balloon in equilibrium?*

until the star begins to die. We will see what happens then in the next chapter.

Not all of the material present in the original cloud was drawn inward to form the Sun, and some of the dust and gas remained in the orbiting disk. As fusion reactions were initiated in the core of the Sun, the temperature of the Solar System began to rise. In the inner regions of the system, light elements such as oxygen and nitrogen existed in the gaseous state, while the heavier elements remained in the solid state as particles of dust. In time, many of these dust particles came together to form small chunks of rock, which, in turn, coalesced to form miniplanets, called **planetesimals.** This aggregation continued, and the planetesimals collided to form larger masses that eventually became the four inner planets, Mercury, Venus, Earth, and Mars (see Fig. 22.1d and e). As these planets grew in size, many changes occurred. Heat was generated when particles and planetesimals fell together. This heat melted certain solids, vaporized some of the liquids, and boiled many of the gases off into space. Later, the planets cooled, and over the course of time additional solid particles and gases were drawn inward until the present structure and composition were achieved.

There is some disagreement about the evolution of the giant outer planets—Jupiter, Saturn,

Uranus, and Neptune. These four have a composition similar to that of the Sun and consist mainly of hydrogen and helium. According to one idea, the outer giants evolved directly from the condensation of the original cloud of dust and gas, much as the Sun evolved. The major difference was that the masses of the outer planets were not great enough to pull the elements inward with sufficient speed to raise their temperature to the point that fusion reactions could occur. Another theory states that the formation of the outer planets was similar to that of the inner ones except that at the extremely cold temperatures of the outer Solar System, more of the lighter elements were retained.

Most of the mass in the Solar System is concentrated in the Sun. Nine major planets orbit the Sun: Mercury, Venus, Earth, Mars, Jupiter, Saturn, Uranus, Neptune, and Pluto. Most of the planets have one or more satellites of their own orbiting about them. For example, one moon orbits Earth, there are two moons orbiting Mars, and 16 revolve around Jupiter. In addition to the planetary system, other bodies such as asteroids, meteoroids, and comets orbit about the Sun. Figure 22.3 is a pictorial representation of the entire system. The four planets closest to the Sun—Mercury, Venus, Earth, and Mars—are all relatively small, dense, and rocky. They are called the **terrestrial,** or Earth-like, planets. The next four—Jupiter, Saturn, Uranus, and Neptune—are much larger, more massive, and less dense, being composed mainly of hydrogen and helium. They are called the **Jovian** (for Jupiter) planets. Pluto, the farthest known planet, has a density of 2 g/cm^3, approximately equal to that of the four gas giants, but it is much smaller, about the size of the Moon.

Several clouds exist in outer space that are believed to be condensing to form new stars. But most of these are in too early a stage of evolution to determine whether they will condense into a single star or break up into smaller fragments that will become planets. In 1984, however, a satellite recorded unusual infrared emissions from a star, Beta Pictoris, about 50 light-years from Earth. Further observations from ground-based instruments showed a halo of light around the star (Fig. 22.4). If this light is being reflected off many small particles, as astronomers believe, the photograph might be a record of a solar system in formation. Another solar system candidate was found by NASA's *Infrared Astronomical Satellite (IRAS),* which discovered a ring

FIGURE 22.3 *A schematic view of the Solar System (not to scale). Try to identify all of these planets.*

of large particles surrounding Vega, the third brightest star in the sky (see Fig. 22.4b). Vega is a relatively young star (less than 1 billion years old), so the material around it cannot have reached the same stage of evolution as our Solar System (which is believed to be 4.6 billion years old). The odds seem tremendously in favor of there being many other such solar systems circling the myriad of stars that dot our galaxy and other galaxies in space.

22.2 THE SUN

Ancient Greek scientists and philosophers believed that the Sun was a perfect, symmetrical, homogeneous sphere, unblemished in any way. This belief was first questioned by Galileo in the seventeenth century. While studying the Sun with the aid of a telescope, he noted occasional dark spots appearing on its surface. Many critics disagreed with

FIGURE 22.4 *(a) This photograph may be an image of another Solar System in the process of formation. The thin disk around the central star, Beta Pictoris, is composed of bits of dust, which are believed to be similar to the material that condensed to form the planets of our own Solar System.* (Courtesy University of Arizona and Jet Propulsion Laboratories)
(b) NASA's Infrared Astronomical Satellite (IRAS) has discovered a shell or ring of large particles surrounding Vega, the third brightest star in the sky. This material could be a solar system in formation. (Courtesy NASA)

(a)

(b)

Galileo, although they refused to look through his telescope. They argued instead on philosophical grounds, claiming that if Galileo questioned the perfect symmetry of the Sun, he was simultaneously questioning the perfection of God, and this was heresy of the highest order. We now have a great deal of evidence to show that the Sun is, in fact, a complex object. Some of the evidence has been obtained visually, from frequent observations of dark spots and granular structures on the surface of the Sun and from huge flares of hot gas that occasionally shoot outward from its surface. Yet our visual information represents only a small part of the solar data that have been collected. Astronomers also study many other forms of electromagnetic emissions: radio, infrared, ultraviolet, X-rays, and gamma rays. Additionally, much of what is now deduced about the unseen internal structure of the Sun has been inferred by calculating how matter must behave under the conditions of the solar environment. Of course, such inferences are uncertain, and many solar phenomena remain poorly understood.

Although approximately 70 elements have been detected in the Sun's atmosphere, most exist only in trace quantities. The main ingredient is hydrogen, which alone accounts for nearly 75 percent of the total matter by mass. Helium, second in abundance, makes up close to 25 percent of the total, and the sum of all the other elements accounts for about 1 percent.

The central **core** of the Sun is extremely hot and dense, reaching temperatures of over 15,000,000 K at a pressure 1 billion times that of atmospheric pressure on Earth. What happens to the hydrogen and helium under these conditions? Individual gaseous atoms cannot remain electrically neutral at these extreme conditions. Interatomic collisions are so intense and so frequent that the electrons are simply knocked out of their atoms. The result is a homogeneous mixture of hydrogen and helium nuclei surrounded by a rapidly moving sea of electrons. This is a fourth state of matter called a **plasma.** Suppose that, within this plasma, hydrogen nuclei approach each other on a collision course. If their kinetic energies are great enough to overcome their mutual electrical repulsions, they fuse together to form helium. This nuclear fusion releases energy. Eventually, most of the Sun's hydrogen will be converted to helium; when that happens, some 5 billion years from now, the Sun will change dramatically, as we will see in the next chapter.

A great deal of energy is released from the core during these fusion reactions. This energy is carried radially outward by energetic photons. However, radiation cannot pass from the core directly through the body of the Sun to the surface. The photons are absorbed by many particles along their route, re-emitted, and finally absorbed again in a region just under the surface. Energy is carried outward from there by convection. The structure of the Sun is shown in Figure 22.5.

At the visible surface of the Sun, called the **photosphere,** energy is radiated out into space in the form of photons. Typically the temperature of the photosphere is 6000 K with a pressure only one-tenth that of air at the Earth's surface. The part of the Sun that we see is an extremely diffuse region of glowing, gaseous hydrogen and helium. Close examination reveals that it has a granular structure (Fig. 22.6). The bright spots are regions where hot gases are rising upward, and the dark spots are cooler areas of descending gases. Thus, there is direct visual evidence that convection carries energy to the surface of the Sun.

Large dark spots, called **sunspots,** also appear regularly on the surface of the Sun (Fig. 22.7). These are the same phenomena that were observed by Galileo over 350 years ago. A single sunspot may be small and last for only a few days, or it may be as large as 150,000 km in diameter and remain visible for several months. From spectral analysis, astronomers now know that sunspots are simply regions of the Sun that are 1000 K cooler than the gases around them. Because they are cooler, they radiate less energy and hence appear dark by comparison with the rest of the photosphere. No one is able to predict exactly when or where a particular sunspot will appear or how long it will last, but it is known that sunspot activity becomes more frequent on a 22-year cycle.

Because heat normally flows from a hot body to a cooler one, it would seem logical to suppose that a large, cool region of gas would be heated quickly and disappear. Yet relatively long-lived sunspots have been observed. Sunspot formation is associated with the presence of intense magnetic fields on the Sun. Astronomers believe that these strong magnetic fields restrict the solar turbulence and somehow inhibit the transfer of heat from hot regions into the nearby sunspots.

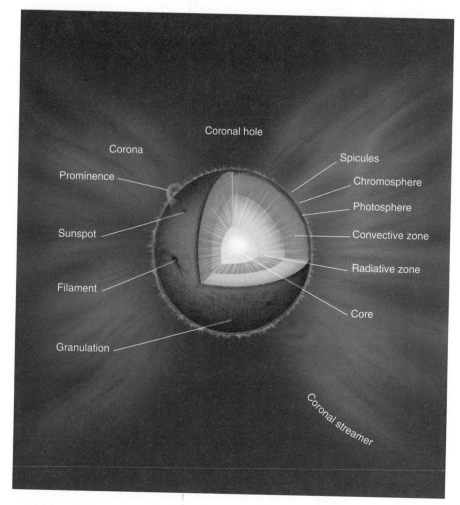

Coronal hole

Corona

Prominence

Sunspot

Filament

Granulation

Spicules

Chromosphere

Photosphere

Convective zone

Radiative zone

Core

Coronal streamer

FIGURE 22.5 *The structure of the Sun.* (From Pasachoff: *Astronomy: From Earth to the Universe,* 4th edition, 1993)

Large flares of hot gas sometimes explode from amid a group of sunspots. Gases accelerate upward and outward, attaining speeds often in excess of 900 km per hour, and shock waves smash through the solar atmosphere. Truly the Sun is not a static, homogeneous, symmetrical sphere such as the Greek philosophers envisioned. Sometimes these flares are powerful enough to have direct effects here on Earth; they initiate reactions that can interrupt radio communication and are associated with auroral displays in our polar regions. Also, such solar activity may affect terrestrial climate, but no definite relationship has been proven.

Sun's Outer Layer

The Sun is not bounded by a sharply defined surface. Rather, its gaseous regions extend far out into

FIGURE 22.6 *A view of the Sun showing its granulated structure.* (Courtesy NASA)

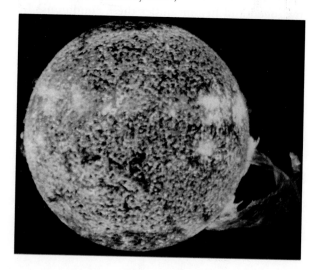

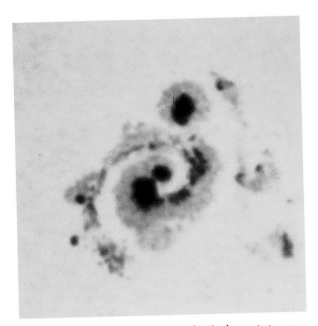

FIGURE 22.7   *This unprecedented spiral sunspot was photographed on February 19, 1982. Sunspots are normally seen as irregularly shaped dark holes. This spiral sunspot had a diameter of approximately 50,000 miles—about six times the diameter of the Earth—and held its shape for about 2 days before it broke up and changed its form.*

Solar Wind and Auroras

Matter behaves differently at one-billionth of an atmosphere and 2 million degrees than it does in any terrestrial system. Electrons are stripped off their atoms. Hydrogen and helium are reduced to bare nuclei in a sea of electrons, and many of these particles fly off into space. This stream of nuclei and electrons is called the **solar wind.** It surrounds the Earth and extends outward toward the far reaches of the Solar System. When you think about the solar wind, don't think of a gentle breeze blowing against your face. The atmosphere on Earth contains approximately 10^{19} particles/cm^3, and you can feel the effect of these particles striking your cheek. The solar wind contains only 5 particles/cm^3 and certainly cannot be felt.

FIGURE 22.8   *Numerous spicules in the chromosphere of the Sun.* (Courtesy NOAO)

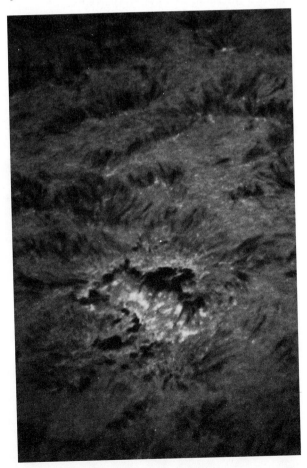

space. Above the photosphere lies a turbulent, diffuse, gaseous layer called the **chromosphere.** The chromosphere consists of a series of spikes, or spicules, that look something like flames from a burning log. A representative spike is about 700 km across and 7000 km high and lasts about 5 to 15 minutes. These spikes are composed of hot gases that are shot upward from the turbulent photosphere below (Fig. 22.8).

Farther out, beyond the chromosphere, there exists an even more diffuse region called the **corona.** The corona can be observed as a beautiful halo around the Sun during a total solar eclipse. The corona is extremely hot, about 2 million K. Even though the temperature is extremely high, there is comparatively little thermal energy in the region. Recall that the thermal energy in a sample of matter is related both to the temperature and the mass of the substance. The corona has a density equal to one-billionth of the density of the atmosphere at the surface of the Earth. In a physics laboratory, such a density would be considered an almost perfect vacuum.

Aurora borealis, the Northern Lights, occurs when charged particles originating mainly from the Sun become trapped in the Earth's magnetic field. When they are dumped out over the poles, they collide with other atoms, producing spectacular patterns of visible light. (Courtesy Jack Finch/Science Photo Library)

Despite this incredibly low density, the solar wind affects the Earth in many ways. These effects arise out of interactions between the charged particles of the solar wind and the magnetic field of the Earth. Recall from the discussion of magnetism in Chapter 9 that two different types of interactions can occur when a charged particle moves into a magnetic field. One possibility is that the particle will be deflected. The other possibility is that the charged particle will be trapped by the magnetic field, as we will discuss later. Both types of interactions occur between the solar wind and the magnetic field of our planet. When high-speed nuclei and electrons from the Sun intersect the Earth's magnetic field, most of them are deflected and move around the Earth, leaving a comet-shaped area around our planet called the **magnetosphere,** as shown in Figure 22.9. Thus, the magnetic field of the Earth is a protective shield that prevents most of these high-energy particles from reaching the surface.

Other particles, those that come in almost parallel to the magnetic field lines of the Earth, may become trapped in this field and spiral from pole to pole circling about a field line. The number of particles trapped in this way builds up to the point that some of them are dumped out. These smash into the upper atmosphere, ionizing some of the atoms and molecules in the region and exciting others to higher energy states. When these molecules or atoms return to their original condition by acquiring an electron or by losing energy, they emit

FIGURE 22.9   *The Earth and its magnetosphere, showing how auroras are produced.*

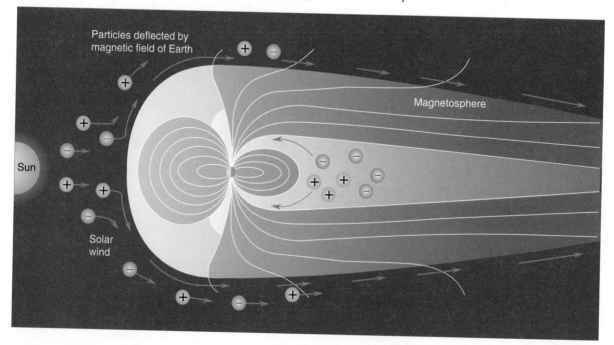

radiation of characteristic wavelengths. The visible wavelengths produce the auroras that are commonly seen from high latitudes on Earth.

22.3 STUDY OF THE MOON AND THE PLANETS

Before the exploration of space in the 1960s, the study of the Moon and nearby planets had been a great deal more difficult than the study of the Sun. Scientists are able to study the Sun and other stars through the interpretation of spectral data. The Moon and the planets, however, do not emit their own light; they reflect the rays of the Sun. Therefore, the spectrum of moonlight is nearly identical to that of sunlight, and it teaches us little about the

Moon. The nature of some regions of planetary atmospheres can be deduced by noting which frequencies of sunlight are absorbed, but this information tells us nothing about the nature of the surfaces of the planets. Unfortunately, all earthbound studies of the surface detail of planets and moons must always be limited. Let us suppose, for example, that it were possible to build an optically perfect telescope as large as we wanted. No matter how efficient the telescope, it would not enable us to see the details of the surface of another planet. The problem arises because the Earth's atmosphere is turbulent and heterogeneous. Light entering this atmosphere is refracted unevenly. This effect produces an inherent blurring in any observation of a distant object, thus limiting the resolution of any terrestrial telescope. This problem was unsolvable

FOCUS ON
HOW IS THE GEOLOGICAL HISTORY OF THE MOON AND THE PLANETS DEDUCED?

 Considerable information about the geological history of a moon or planet can be deduced by studying the number and type of meteorite craters. If there are two regions, and one is marked by a series of meteorite craters while the other is smooth and level, we can deduce that both regions were once covered with craters, for external bombardment would affect the entire planet equally. At a later date, some of these craters must have been obliterated by volcanic or tectonic activity or through extensive erosion. Thus, the cratered region is geologically older, and the smooth region would have been formed by geological processes. If a few scattered craters appear in the smooth areas, and if smaller craters lie inside larger ones, it seems reasonable to believe that a second, later era of meteorite bombardment followed the first.

This chronology serves as a rough guide to the sequence of events but does not date the various time periods. To establish lunar chronology,

astronauts have collected samples of rock from various representative areas on the Moon and brought them back to laboratories on Earth. These have been dated by studying patterns of radioactive decay, and the oldest date back 3.9 billion years.

No one has yet retrieved samples of rock and soil from any other celestial bodies for careful analysis here on Earth. The lunar chronology, however, may be used to establish a general geological time scale for the Solar System. First, we assume that any significant period of meteorite bombardment must have affected the Moon and nearby planets at the same time. Then it is a simple matter to compare crater patterns on planets with those on the Moon. If the densities, sizes, and general shapes of the craters on a planet are similar to those on the Moon, we may deduce the age of the planetary rock from our knowledge of the Moon. This type of deductive reasoning is not foolproof, but at the moment it is the best we have.

This outstanding view of a full Moon was photographed from the Apollo 11 spacecraft from a distance of 10,000 nautical miles. Note the cratered regions and the flat plains, which are erroneously called seas or maria. Go outside and look at the moon with your naked eye. Can you spot a "sea"?

until it became possible to build rockets to carry telescopes, cameras, and other scientific instruments above the atmosphere.

These instruments have provided pictures and recorded data about planetary temperatures, the strengths of magnetic and gravitational fields, the chemical composition of planetary atmospheres, the nature of atomic particles in interplanetary space, and much more. Astronauts have landed on the Moon, carried out a variety of experiments, and returned to Earth with valuable samples of lunar rock and soil. As a result of all this exploration, our knowledge of the Solar System has expanded at an extremely rapid rate.

22.4 THE MOON

Until the Apollo astronauts flew by the back side of the Moon, observers on Earth had never seen any of the features of that side of its surface. Let us begin our study of the Moon with a discussion of why it and many other objects in the Universe always keeps the same face pointed toward the object

about which they revolve. This kind of motion occurs when the period of revolution and rotation is the same, and in such circumstances, the orbiting, smaller object is said to be in **synchronous rotation** with the larger. Synchronous rotation occurs when the orbiting body is slightly more massive on one side than it is on the other. In this case, the gravitational force exerted on it by the object about which it revolves is slightly off-center, and this uneven gravitational force will cause the smaller one to move into a synchronous rotation after several million revolutions. The orbiting object must also be small and close to the central object. To see why the same face of the orbiter always faces the central object, consider Figure 22.10. At the beginning of the revolution, an observer at A is facing the Sun, and as the revolution progresses to B, one-fourth of the way through, the rotation carries the observer also one-fourth of the way through a complete turn. As we follow the motion through C and D, we see that when the rotation and revolution rates are the same, the observer is always facing the central body.

The Moon is close enough that its gross surface features are detectable by telescope. Its mountain ranges, smooth flat plains (called **maria** from the Latin for seas), and thousands of craters have been well known since the time Galileo first observed them in the 1600s. Before the Apollo space program, however, astronomers knew little about the geology of the Moon. The two most significant questions asked during the space exploration program were: First, how was the Moon formed, and how did it begin to orbit the Earth? Second, what is the geological history of the Moon: Was it once hot and molten like the Earth, and if so, does it still have a molten core?

The answer to the question of how the Moon was formed has been and continues to be the subject of considerable debate. There are several theories.

Theory: The Moon split off from the Earth soon after the formation of the planet. According to this theory, the Earth was spinning so rapidly that a large chunk of molten matter spun off, much as water spins out of wet clothes on the spin cycle of a washing machine.

This concept is largely rejected because no one can explain how the proto-Earth came to spin so rapidly, and if it did rotate at this rate, no one can

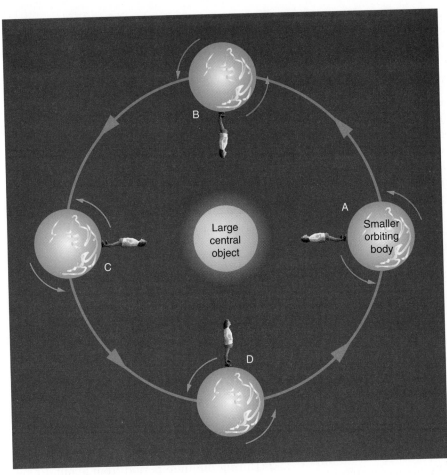

FIGURE 22.10 *The smaller orbiting body is in synchronous rotation about the larger central object. Note that the same face of the orbiting body is always seen from the central object.*

explain how the rate of rotation of the Earth-Moon system has decreased to its present value.

The surface of the Moon showing numerous craters. (Courtesy NASA)

Theory: The Moon was formed in some remote region of the Solar System and then was captured by the gravitational field of the Earth at some later time.

This event is highly improbable. Imagine that a Moon-sized object were flying randomly through space. The probability of passing close enough to a planet to be captured is small. Even if it did pass closely enough to a planet, its trajectory and energy would have to be within precise limits for capture to occur. Such an event might happen once, but most of the planets in our Solar System have moons, and some have many. It is difficult to believe that captures would have occurred repeatedly.

Theory: The Earth and Moon were formed simultaneously from a single cloud of dust and gas. As this cloud condensed, a high concentration of heavy elements gravitated toward the center of this sphere to form the Earth, whereas the outer ring, which became the Moon, was made up of minerals of lower density.

This theory has been popular for many years but has one major drawback: The chemical compositions of the Earth and Moon appear to be more different from one another than can be satisfactorily explained by this theory.

Theory: Shortly after the Earth was formed, a giant object, perhaps a miniplanet, smashed into the Earth. Parts of this object and segments of the Earth's crust were ejected into space by the force of the collisions, and these fragments began to orbit the central planet. Eventually the fragments coalesced into a single satellite, the Moon.

Both calculations and comparative observations of the other planet-satellite systems support this theory as the most plausible hypothesis. However, questions still remain, and any theory, by its nature, is open to question and dispute.

The second question, posed earlier, is: What is the geological history of the Moon, and how did it evolve in the eons after its original formation? One of the most significant discoveries of the Apollo program was that the entire surface of the Moon is covered with different types of igneous rock. Since igneous rock is formed only when lava cools, it is clear that at least the surface was once hot and liquid. The next logical question arose: How was it heated? The Earth was heated both by radioactive decay from within its interior and by intense meteorite bombardment from outer space. But what about the Moon? Theoretical geologists have calculated that it would take roughly 1 billion years for radioactive decay inside the Moon to build up enough heat to melt the rock. Analysis of lunar rock showed that the oldest igneous rocks were crystallized from molten lava when the Moon was a mere 400 million years old. Thus, because there was insufficient time for the rock to have been melted by the heat from radioactive decay, at least some of the melting of the Moon was caused by meteorite bombardment. Further study of lunar geology has given us a partial picture of the rest of its history, as outlined:

1. Astronomers believe that the Moon was formed shortly after the original condensation of the Earth, some 5 billion years ago. A few hundred million years after its formation, much or all of the lunar material grew hot enough to melt. Possibly some of the heat required to melt the Moon was gen-

Earthrise, as seen from the Moon. (Courtesy NASA)

erated during the condensation of the protomoon, whereas most of the rest was supplied by intense meteorite bombardment. Within the molten interior, the denser elements such as iron and nickel gravitated toward the center, while the less dense minerals floated upward toward the surface. Thus, the Moon, like the Earth, has a small metallic core surrounded by a shell of rocks of lower density. Later, the Moon cooled and solidified so that this structure was maintained.

2. During the time period from about 4.2 billion years to 3.9 billion years ago, a second gigantic series of meteor showers rained down on the Moon. Billions of meteors, some small and others as large in diameter as the state of Rhode Island, smashed into the surface and gouged out most of the craters that can be seen today.

3. At the same time that the meteor shower was marking the surface, the lunar interior was being heated by radioactive decay. Huge lava beds were formed deep beneath the surface. About 3.8 billion years ago, some of this molten material flowed upward through the surface crust to form many active volcanoes. Smooth lava flows covered vast regions of the Moon, forming what is now called the lunar maria, or seas. (The use of the word "seas" is a misnomer; on the contrary, these regions are dry, barren, flat expanses of rock. Early observers of the Moon believed that these dark regions actually were seas, and thus the term arose.) This volcanic activity lasted ap-

Close-up of lunar plain with mountain range in background. The vehicle on the plain is the Lunar Rover. (Courtesy NASA)

proximately 700 million years and ended about 3 billion years ago.

4. Because the Moon is small, the heat produced by radioactive decay was quickly dissipated into space, so that the Moon soon cooled considerably and now lies geologically quiet and inactive. Seismographs left on the lunar surface by Apollo astronauts indicate that the energy released by moonquakes is only one-billionth to one-trillionth as much as is released by earthquakes on Earth. Seismic data indicate that the core of the Moon is probably molten, or if it is not, it is at least hot enough to be soft and plastic. The cool, solid upper mantle and crustal layers, however, are thick enough to inhibit appreciable seismic activity. Meteor bombardment of the Moon has continued throughout this history of geological dormancy, but there has never again been an intense rain of meteors such as the one that occurred about 4 billion years ago.

The lunar experiments tell us a great deal about the history of the Earth. If the Moon was subject to intense meteor bombardment 4 billion years ago, the Earth, being larger and exerting a greater gravitational force, must have attracted more debris and thus must also have been melted by the same process at the same time. We can imagine that the Earth also cooled at the end of this epoch, only to be reheated by its own internal radioactivity. Because the Earth is so much larger than the Moon, it has not cooled as rapidly; as a result, part of the core continues to be molten, and the mantle is hot and geologically active to this day. Most of the original rocks of the Earth's crust have long since been pushed down into the mantle by tectonic activity, to be remelted and re-formed in a continuous and dynamic process, and other ancient landforms have been altered or destroyed by erosion. Thus, evidence of the first 1.5 billion years of the Earth's history has been largely lost. But the Moon—cold, lifeless, and not subject to erosion from wind and water—has preserved a record of its history and has allowed us to probe more deeply into the origins of the Earth.

22.5 MERCURY

Studies of lunar geology have taught us much about the state of the Solar System several billion years ago. To complete the picture and to understand what conditions were like throughout our local region of space, we must also consider the planets.

Mercury has a radius less than four-tenths that of the Earth. It is also the closest planet to the Sun, and therefore it orbits the Sun faster than any other planet, in obedience to Kepler's third law. Each Mercurial year is only 88 days. Mercury rotates on its axis rather slowly, so that it rotates only three times for each two complete revolutions around the Sun.

Mercury is close to the Sun, and because it keeps one face pointing toward the Sun for a long period of time, its sunny side becomes extremely hot. Studies of infrared and radio emissions indicate a daily high of about 450°C (hot enough to melt tin or lead). The temperature of the side of the planet facing away from the Sun drops to lows of −175°C (nearly cold enough to liquefy oxygen). These extremes of temperature are enhanced by the fact that there is virtually no atmosphere on Mercury, so there can be no wind to carry heat from one region to another and no cloud cover to retain heat on the dark side. The absence of an atmos-

FIGURE 22.11   *A close-up view of Mercury. The photograph shows an area 550 km from side to side.* (Courtesy NASA)

phere can be explained by the fact that the temperatures on the planet are so high that air molecules would achieve a high enough speed to escape the small gravitational pull of a planet as low in mass as Mercury.

Little was known about the surface of Mercury before the spring of 1974. At that time, a sophisticated spacecraft called *Mariner 10* passed close to Mercury and began relaying information back to Earth. The first *Mariner 10* photographs revealed a cratered surface, similar in many respects to that of the Moon (Fig. 22.11). The similarities in surface contours indicate that the geological histories of the two bodies must be similar. Thus, although many questions remain unanswered, most scientists agree that, after its initial formation, Mercury was subject to a period of intense meteor bombardment, similar to the bombardment that occurred on the Moon. Perhaps the interior of the planet was once hot, but no one is sure, for there are no vast lava plains comparable to the lunar maria. The ancient meteor craters stand out sharply, unmarked by extensive erosion or tectonic leveling, for there has been little geological activity during the past few billion years.

Perhaps the most striking discovery of the Mariner probe was that Mercury has a small but distinctly measurable magnetic field. Scientists believe that the Earth's magnetic field results from the effect of the relatively rapid rotation of its iron core. The discovery of a magnetic field on Mercury was

a surprise because this planet rotates slowly. This anomaly is best explained by assuming that the metallic core of Mercury is relatively larger than the core of the Earth. In turn, this assumption can also be justified. Mercury is closer to the Sun than the Earth, so during its formation a greater percentage of the lighter elements would have vaporized into space, leaving behind a proportionally larger core of dense metals.

Because the surface of Mercury is so hot, and the gravitational field of the planet is so small, we would expect that any atmospheric gases would have vaporized into space long ago. An atmosphere has been detected on Mercury, however, which is about one-billionth as dense as that on Earth.

22.6 VENUS

Of all the planets in the Solar System, Venus most closely resembles Earth with respect to size, density, and mean distance from the Sun. From this information alone, astronomers once believed that environmental conditions on Venus might be expected to be similar to those on Earth and thus that some form of life might be found there. It was impossible to study the surface of Venus, however, for it is wrapped in a thick, dense atmosphere with an opaque cloud cover (Fig. 22.12). But now it can be said with certainty that no life exists on Venus. Temperatures on the planet are extremely high; the sur-

FOCUS ON
HOW A LOT OF INFORMATION CAN BE DEDUCED FROM
A SMALL AMOUNT OF DATA

In 1975, a Soviet spacecraft landed on the surface of Venus. The intense heat and pressure rapidly destroyed the instruments aboard, but before the radio transmitter failed, a single photograph was sent back to Earth. This picture shows sharply angular rock, as can be seen in Figure 22.14a. A second Soviet craft soon landed 2000 km away and returned a photograph of a smooth landscape interspersed with sections of cooled lava or highly weathered rock (Figure 22.14b). A wealth of information is contained in just two photographs. The angular rocks must be geologically young, for they would be expected to erode rapidly in the harsh conditions of the Venusian atmosphere. The smooth landscape and weathered rock of the second landing site would necessarily be older. Thus, scientists were able to deduce from these data that there has been recent tectonic activity on Venus.

FIGURE 22.12   *Venus. Note that the solid surface of the planet is obscured by a turbulent cloud cover.* (Courtesy NASA)

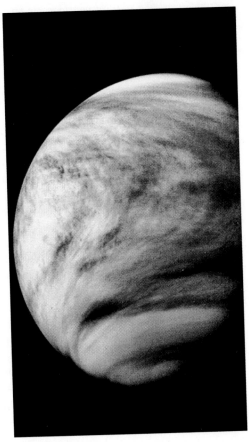

face actually has a slightly higher average temperature, about 460°C, than that of Mercury. As noted in Chapter 21, most of the planets and moons in the Solar System revolve and rotate counterclockwise as seen from north of the Solar System. Venus revolves in this counterclockwise fashion, but its rotation is retrograde—it rotates clockwise as shown in Figure 22.13.

Atmosphere of Venus

Venus is slightly smaller than the Earth with a radius equal to 0.95 that of Earth, and its gravitational force is also less. Nonetheless, the atmospheric density at the surface of Venus is 90 times greater than that of our planet. Thus, the pressure acting on an object on the surface of Venus is equal to the pressure on an object that is 1000 m beneath the surface of the ocean on Earth. The Venusian atmosphere is composed of 96 percent carbon dioxide and 3.5 percent nitrogen, with the remainder being helium, neon, and other gases. The clouds we see contain a high percentage of sulfuric acid. Thus, a rainfall on the planet would bring down the same acidic solution that is in a car battery.

One is immediately made to wonder why conditions on Venus should be so different from those on Earth. The most widely accepted answer to this question centers on the greenhouse effect, discussed briefly in Chapter 5. Recall that both carbon

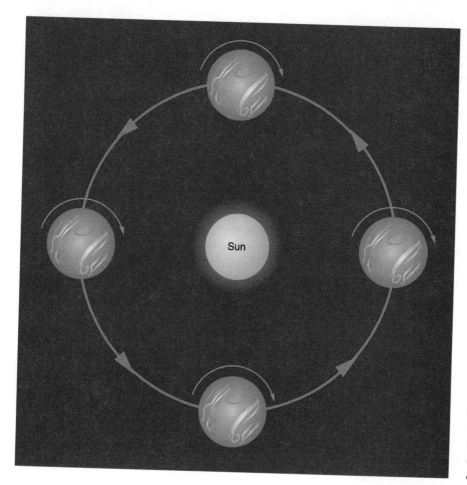

FIGURE 22.13 *Venus has a counterclockwise revolution about the Sun but a clockwise rotation on its axis.*

dioxide gas and water vapor absorb infrared radiation and warm the environment. Carbon dioxide can exist as a gas in the atmosphere, can be dissolved in water, or can be chemically bonded with other substances to form rock. Water commonly exists in its vapor form, or as a liquid, or as ice, a solid. According to the most widely accepted theory, at one time Venus was much cooler than it is today. If this premise is correct, rivers would have flowed over its surface, and seas would have filled the central basins. Because Venus is slightly closer to the Sun than is the Earth, Venus naturally receives more sunlight if all other conditions are equal. Therefore, the rate of evaporation on Venus must have been just a little greater than it was on Earth. But water vapor absorbs infrared radiation, so a greenhouse warming occurred. In turn, as the temperature increased, more water evaporated. Conditions spiraled. As more water evaporated, the carbon dioxide that was dissolved in the seas was released

as well. Thus, Venus became hotter and hotter until eventually the liquid water all boiled away, and then the carbonate rocks, if they existed, decomposed, releasing even more carbon dioxide into the atmosphere. Eventually an equilibrium temperature of about 460°C was reached.

Despite the incredible harshness of the Venusian environment, the atmosphere at the surface is not as turbulent and changeable as that on Earth. Venus rotates quite slowly on its axis, so slowly in fact that 1 day on this planet is actually longer than a year. Astronomers believe that the slow rotation is one fact that is responsible for the relative stability of the lower atmosphere. In contrast, the upper atmosphere is quite windy, with speeds up to 300 km per hour. An ongoing problem in planetary astronomy is to explain these weather conditions on Venus. Such knowledge might help us to better understand wind circulation patterns on Earth

(a)

(b)

FIGURE 22.14 *(a) The surface of Venus photographed from* Venera 9. *Notice the sharply angled rocks. (b) The surface of Venus 2000 km from the* Venera 9 *landing site, photographed by* Venera 10.

Another important difference between Venus and the Earth is that Venus has no magnetic field. In the absence of a protective magnetic shield, the particles of the solar wind penetrate easily into the upper atmosphere. In turn, these particles ionize many of the atmospheric gases, and the electrical turbulence generates continual lightning storms that travel back and forth across the upper layers of the clouds.

Surface of Venus

In contrast to the surfaces of the Moon and Mercury, the surface of Venus shows many signs of relatively recent tectonic activity, as shown in Figure 22.14. The opaque cloud cover of Venus makes it impossible to see the surface of Venus from Earth, but the surface has been studied extensively by radar mapping. In 1978, the spacecraft *Pioneer Venus 1* was placed in orbit about Venus and bounced radar waves off the planet. Radar at airport control towers can tell the distance an airplane is away from the airport, and in the same fashion radar waves

could be used to detect how far away the spacecraft was from the surface of the planet. The results are shown in Figure 22.15. Most of the terrain of Venus

FIGURE 22.15 *The surface of Venus as based on the* Pioneer Venus *radar map. Two continents exist on Venus: Aphrodite Terra, which is comparable in scale with Africa, and Ishtar Terra, which is comparable in scale with the continental United States or Australia.* (Courtesy NASA)

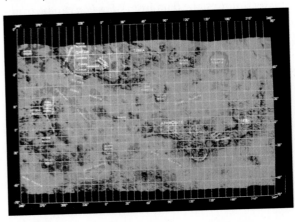

FIGURE 22.16   *A view of Venus as seen by the* Magellan *orbiter.* (Courtesy NASA)

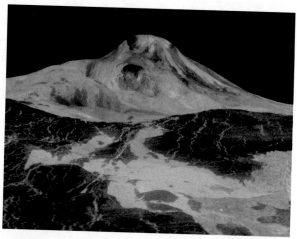

FIGURE 22.17   *A simulated view of Maat Mons using the* Magellan *data. The mountain is 8 km tall. Lava flows from it spread out from its base.* (Courtesy NASA)

consists of plains, with only about 10 percent of the surface considered highland regions comparable to continents on Earth. A large highland region, Aphrodite, is along the equator of the planet, stretching almost halfway around the globe. The northern highland region, Ishtar, is about the same size as Australia. This "continent" contains the highest mountains on the planet, the Maxwell Mountains, which reach a height of about 11 km above the lowland regions. In 1990, the *Magellan* orbiter returned pictures with a much higher resolution; features as small as 100 m across could be detected. Figure 22.16 shows a view of Venus as seen by *Magellan*. Using computer enhancement and processing techniques, the data from *Magellan* can be transformed to show views of the surface in perspective, as shown in Figure 22.17. There are not many craters seen on the images, so the surface of the planet must have been renewed by lava flows.

22.7 MARS

Mars has captured the imagination of scientists and lay people alike, since early observations seemed to indicate that if extraterrestrial life were to be found in the Solar System, it would be found on that planet. A part of the reason for this began with

some observations of Mars made in 1863. An Italian astronomer, Giovanni Schiaparelli, drew a map of the planet that contained some lines he called *canali*. The best translation of this word to English would be channels; however, the translation to English became canals. This left the impression in the mind of most lay people that these were canals laid out by intelligent beings on Mars. In fact, in 1894, a noted American astronomer, Percival Lowell, drew a map of the planet based on his own observations that showed details of 500 canals. We now know that these canals were products of an active imagination.

An additional observation of Mars that led to the idea that it might have a hospitable environment for life was the existence of polar caps. In contrast to Venus, Mars is covered by a thin, nearly cloudless atmosphere that enables astronomers to view its surface easily with telescopes. Figure 22.18a is a mosaic of the planet as seen from the *Viking* orbiter. Several hundred years ago, it was noted that the Martian polar regions are white and that the white ground cover shrinks in summer and expands in winter, as shown in Figure 22.18b. These changes strongly suggest that the white regions are ice caps. If this speculation were correct, and if the ice caps melt significantly in summer, water must be available. This led to the idea that the purpose of the canals was to bring water from the polar ice caps to irrigate desert regions near the equator. Astronomers also observed that, each spring, large re-

(a)

(b)

FIGURE 22.18 *(a) A mosaic of Mars as seen from the* Viking *orbiter. The great canyon* Valles Marineris *is seen along the center. To the left are three giant volcanoes.* (Courtesy NASA) *(b) A view of the south polar cap of Mars.* (Courtesy NASA)

gions of the globe near the equatorial region darken, only to become light in winter. Many people thought that these color changes were caused by annual blooms of vast areas of vegetation. The axis of rotation of Mars is tipped at an angle of approximately 24° with respect to the plane of its orbit. As you recall, it is such a tilt that causes the Earth, and hence Mars, to have seasons. Thus, seasonal activity in the form of the growth of plant life might be expected, and this idea helped to solidify the theory of life on Mars.

In the 11 years between 1965 and 1976, a total of 12 United States and Soviet spacecraft were sent to Mars, including two spectacularly successful *Viking* vehicles that landed on the surface of the planet, where they collected and analyzed samples of Martian soil. The data from the spacecraft have drastically changed our picture of the Martian environment and have led to the conclusion that probably no life, not even that of microorganisms, now exists, or ever has existed on the planet. We can be sure that the postulated vast forests and canal systems do not exist. The seasonal changes of color are actually caused by great, dry dust storms powered by seasonal winds. When the large global winds subside, bright particles of dust are thought to settle in certain areas, causing these regions to be light in color. In the spring, local winds stir up the dust into suspension in the atmosphere. This sweeping action reveals the darker underlying surface of these regions.

You should not think of these dust storms as being of the type one might observe in a desert sandstorm. The atmosphere of Mars is extremely rarefied, with the total amount of atmospheric gases being about 1 percent of that on Earth and consisting largely of carbon dioxide. Thus, little material is moved about in one of these storms. In fact, cameras located on the surface of the planet have barely detected any loss of visibility at all, even during an intense storm.

The winter ice caps, once thought to be the source of spring floods, are largely composed of frozen carbon dioxide, commonly called Dry Ice.

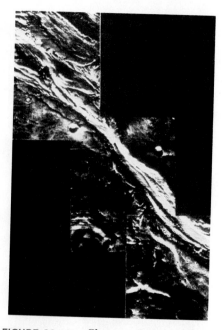

FIGURE 22.19   *Close-up of Mars taken by* Mariner 9 *spacecraft, showing a deep canyon. This channel is thought to have been formed by running water in Mars' geological past.* (Courtesy NASA)

FIGURE 22.20   *The largest volcano on Mars, and probably in the Solar System, is Olympus Mons.* (Courtesy NASA)

Considerable quantities of water ice are present as well, but polar temperatures remain below the freezing point of water throughout the entire year, and this water ice never melts. Thus, the shrinking of the polar caps during the summer months is caused by sublimation of the Dry Ice, leaving behind a permanently frozen water ice cap. At some time in the distant past, however, liquid water probably did exist on Mars. Rivers must have flowed across the surface, gouging out stream beds and deep canyons, for these features can be observed today (Fig. 22.19). However, no rain has fallen for millions or hundreds of millions of years, and now there is no liquid water anywhere on the surface of the planet. This barren, rocky land is also quite cold. Near the equator, midday temperatures may reach as high as 20°C during the summer season, but at night these same locations experience a temperature drop to as low as −140°C. These extremes of temperature during a single day are explained by the thin atmosphere that is too rarefied to hold in the heat during the cloudless nights.

When the spacecraft *Mariner 9* was sent to Mars, one of the dust storms mentioned earlier arose, and there was considerable fear that the batteries on board would be drained before the storm subsided to the point where any surface details could be seen. For a while, only three features could be seen near the equatorial belt of the planet. They were dark spots that could be explained only by assuming that they were mountains high enough to be seen above the enveloping dust storm. When the storm subsided, the spots were, indeed, found to be gigantic extinct volcanoes. The largest of these has been named Olympus Mons (Fig. 22.20). Its height is about 24 km, and its base would reach across the state of California. Thus, a getaway vacation to Mars would have to take you to this natural wonder. While there, you should also visit a series of huge canyons, named *Valles Marineris* in honor of the *Mariner* spacecraft, that stretches away from the base of Olympus Mons. This canyon makes the Grand Canyon look like a small gully. On Earth, *Marineris* would stretch from New York City to Los Angeles and in some places would be as much as 4 miles deep.

Moons of Mars

When seen, even by the naked eye, Mars has a distinct reddish coloration. This color is due to a complex series of interactions that arise basically because of the thin atmosphere of the planet. On Earth, little ultraviolet radiation reaches the surface of the planet because it is absorbed by the relatively thick atmosphere. On Mars, however, this protective blanket is not present, and the ultraviolet radiation penetrates easily. One of the effects that this has on the planet is that any water vapor that was present in the atmosphere has been broken down into hydrogen and oxygen. The oxygen forms compounds with other elements present, notably iron. In general, iron oxides are distinguished by their red color. The chemical origin of the color of Mars was not known by ancient observers, but its color was known. As a result, the distinctive color of the red planet may have brought to mind the color of blood and had some influence on its having been named for Mars, the god of war. Students of mythology may recall that the Mars of myth had two colleagues who helped him with his dastardly deeds. They were aptly named Phobos (fear) and Deimos (panic). Thus, it seems appropriate that the two Martian moons should also bear these names.

Both Phobos and Deimos are irregularly shaped moons, as shown in Figure 22.21. Phobos is about 29 km at its longest point and about 16 km across at its narrowest point. Deimos is even smaller, with an approximate size of about 14 by 11 km. Phobos has the distinction of being the moon that is the closest to its planet of any in the Solar System. Its revolution rate is about 7 hours and 10 minutes. Both moons are in synchronous rotation about Mars; the surfaces of both are heavily cratered; they are dark in color, and they also have densities of about twice that of water. The dark coloration and the low densities are characteristics of asteroids. Thus, there is much speculation that these moons are asteroids that have been captured by the gravitational pull of Mars.

CONCEPTUAL EXERCISE

You were told that both Phobos and Deimos are in synchronous rotation about Mars. What features of these moons should have led you to know this before you were told?

Answer Synchronous rotation occurs when the orbiting object is small, close to the central body, and has its mass slightly off center. All of these facts are true for the martian moons.

22.8 JUPITER

Mercury, Venus, Earth, and Mars constitute a foursome called the terrestrial planets. They are all relatively small, have a solid mineral crust, orbit close to the Sun, and have rotation rates that are relatively slow. The giant outer planets—Jupiter, Saturn, Uranus, and Neptune—are considerably different from the terrestrial group. Visualize once again the primordial dust cloud that was eventually to condense and become the Solar System. As the cloud shrank and broke apart, the protosun and all the protoplanets were originally composed mainly of hydrogen and helium. The protosun's gravitation was so great that it pulled its gases inward with enough force to initiate fusion reactions. The gravitational fields of the terrestrial planets were so weak that most of their light gases escaped and boiled off into space or were blown away by the solar wind. Jupiter is much larger than any of the terrestrial planets, yet much smaller than the Sun.

FIGURE 22.21   *(a) The highly irregular shape of Phobos is easily seen with Mars in the background as in this photo. Phobos appears quite dark relative to Mars.* (Courtesy NASA) *(b) This is a computer-generated color photo of Deimos. Resolution in this photo shows objects as small as 200 m. Deimos is a uniform gray color.* (Courtesy NASA)

(a)

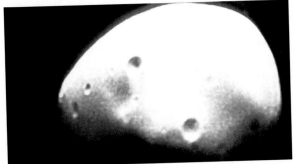

(b)

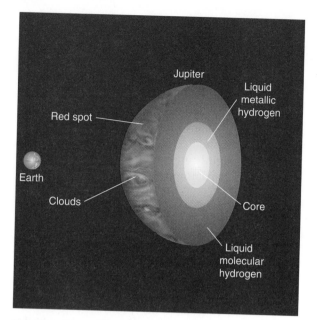

FIGURE 22.22 *The current model of the interior of Jupiter.*

rocky sphere about 20 times as massive as the Earth and probably composed of iron, nickel, and other metals and minerals.

The Jovian atmosphere contains a mixture of gases, liquid droplets, and crystalline particles consisting of hydrogen, helium, ammonia, methane, water, hydrogen sulfide, and other substances. This atmosphere is indeed a turbulent region, as even a casual glance at Figure 22.23 reveals. It is heated from above by the Sun and from below, to an even greater extent, by the interior of the planet. Moreover, the giant planet spins quite rapidly, rotating once approximately every 10 hours. (A rapid rotation rate is a characteristic of all the gas-giants.) All of these effects combine to generate turbulent wind systems, great storms, and changing weather patterns on the surface. Most of the recognizable storm systems appear to form, distort, and move on within a few hours or days, but some of them are surprisingly stable over long periods.

Over 300 years ago, two European astronomers reported seeing a **Great Red Spot** on the surface

Therefore, it is physically similar to neither. Jupiter is the largest of the planets. In fact, its size is such that approximately 1300 Earths could fit inside it. The axis of rotation of Jupiter is tipped only about 3° with respect to the plane of its orbit, so it does not have seasons like the Earth.

As Jupiter was being formed, the inward condensation provided enough energy to heat the dust appreciably, but fusion temperatures were never reached. Yet the internal mass was sufficient for the gravitational forces to retain most of the original hydrogen and helium. Therefore, the chemical composition of Jupiter is much like that of the Sun, but its internal temperature and structure are different. There is no hard, solid, rocky surface where an astronaut could land or walk about. The surface of the planet and more than half of the volume of its interior is a vast sea of cold, liquid hydrogen (see Fig. 22.22). An inner layer, composed of hydrogen in a different form, lies beneath this hydrogen ocean. At the extreme pressures and temperatures found there, the electrons become separated from the nucleus. In this state, the substance is referred to as **liquid metallic hydrogen.** The term "metallic" is used to indicate that it is a good conductor of electricity. Jupiter's core is believed to be a solid,

FIGURE 22.23 *This is the first true-color photograph of Jupiter as taken by NASA's Hubble Space Telescope. Notice the different colors in the clouds, which form a band structure around the planet. Also, note the Great Red Spot near the equator.* (Courtesy NASA)

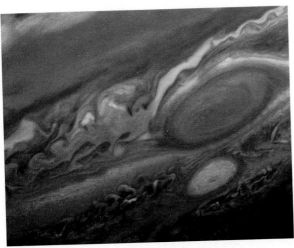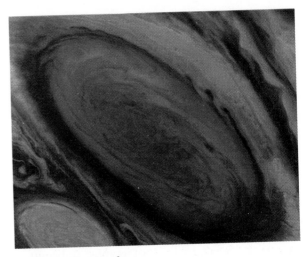

FIGURE 22.24   *Two close-up views of the Great Red Spot, a hurricane that has been swirling for over 300 years.* (Courtesy NASA)

of Jupiter; although its shape and color have changed noticeably from year to year, the spot remains intact to this day, as shown in Figure 22.24. Data from space probes of the planet indicate that the spot is a giant hurricane-like storm. If the entire Earth's crust were peeled off like a giant orange rind and laid flat, it would fit entirely within the Great Red Spot.

Another obvious feature of Jupiter that can be noted from Figure 22.23 is the dark-colored and light-colored bands that circle the planet. The light areas are regions where hot gas is rising from the interior of the planet, and the parallel darker regions are locations where cooler gases are descending toward the interior.

Apparently, another characteristic of the Jovian planets is that they have rings around them. The rings of Saturn are well known by elementary schoolchildren, and until recently it was thought that this was a feature only of Saturn. The *Voyager 1* spacecraft, however, returned photographs of a thin ring circling Jupiter as well. The particles in this ring are extremely small, about the size of those found in cigarette smoke. As a result, these rings could not have been around since the formation of the Solar System; the solar wind would have driven them off into space. This means that they are constantly being replenished by some source, perhaps by particles from Jupiter itself.

Jupiter's Moons

Recall from Chapter 21 that the discovery of Jupiter's moons played an important role in the development of our present understanding of the Solar System. In 1610, Galileo discovered four tiny specks of light close to Jupiter. He noted that they distinctly orbited Jupiter and correctly reasoned that they were satellites of the giant planet. This direct visual evidence that at least some objects did not orbit the Earth was the first concrete proof that the Earth was not the center of all motion in the heavens.

A total of 16 moons revolve around Jupiter. Of these, the four that were originally discovered by Galileo have been the most widely studied. The innermost of the so-called Galilean moons, Io, is small, dense, and rocky, whereas the outermost, Callisto, is significantly less dense and is believed to consist largely of water ice. Thus, the Jovian moon system is reminiscent of the Solar System itself, for, as we have seen, the inner four planets are much denser than the five outer ones.

These similarities imply that perhaps Jupiter and its moons were formed simultaneously, much as the Solar System itself was formed. Just as the Sun and planets condensed out of a single nebula, it is possible that Jupiter and its moons condensed as sort of a mini–solar system of their own. As

Jupiter coalesced, huge quantities of energy must have been generated by gravitational forces. This energy in the form of heat was sufficient to boil most of the lighter elements off the surface of Io, leaving behind a relatively dense sphere. The outer two Galilean moons, Ganymede and Callisto, retained more of their lighter elements and are less dense than Io.

Io   Io is a small satellite, only two-thirds the size of the Earth's Moon and just slightly more dense. Because it is too small to have retained the energy released by radioactive decay in its interior, some observers expected to see a cold, lifeless, cratered, lunar-like surface. Nothing could be further from the truth. Spectacular photographs transmitted from *Voyager* spacecraft showed clear images of volcanoes erupting on the surface of Io, as shown in Figure 22.25. The pictures provided the first evidence of currently active extraterrestrial volcanism in the Solar System. In Figure 22.25 the eruption carries material to about 160 km, but other photos show huge masses of gas and rock ejected higher than 200 km above the surface. This material is not lava, steam, or carbon dioxide, which are the normal components of the material ejected from terrestrial volcanoes. Instead, the material ejected from the volcanoes of Io consists of sulfur and sulfur dioxide. The plumes from the volcanoes cool as they rise, condense, and drift downward as "snowfall" that covers wide areas of the surface. Sulfur and sulfur compounds generally have colors that

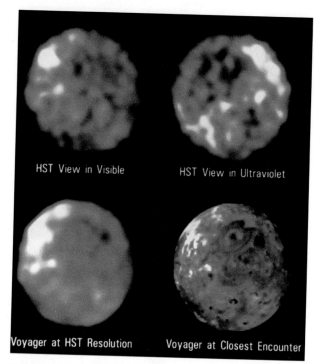

HST View in Visible   HST View in Ultraviolet

Voyager at HST Resolution   Voyager at Closest Encounter

FIGURE 22.26   *Various views of Io. Upper left: A visible light image of Io from the Hubble Space Telescope (HST) taken when Io was 414 million miles from Earth. HST resolves features as small as 150 miles. Upper right: An ultraviolet (UV) light image of Io. Some regions that look bright in visible light are dark in UV. This probably is caused by absorption of the UV by a sulfur dioxide frost. Sulfur dioxide is a strong absorber of UV. Bottom right: An image of Io taken in 1979 by the* Voyager *spacecraft at a distance of 250,000 miles. Bottom left: A "synthetic"* Voyager *image modified to match HST's resolution. (HST was 1000 times farther away than* Voyager *when the photos were taken.) No change in the large-scale distribution of surface material is seen in the 13 years that elapsed between the two observations.* (Courtesy NASA)

FIGURE 22.25   *A volcanic eruption, on the horizon, as photographed by* Voyager 1. *Solid material is being thrown up to an altitude of about 160 km.* (Courtesy NASA)

range from white through all the hues and shades of orange and red. As a result, a photograph of Io looks like a huge pizza because of its unusual coloration (Fig. 22.26). The average daytime temperature on Io is about 130 K, and the ejected gases give it a rarefied atmosphere.

A question that scientists had to grapple with was: Why is Io geologically active? The answer seems to be that Io is affected by a complex interplay of gravitational forces between Jupiter and two other moons of Jupiter, Europa and Ganymede. Jupiter is about 300 times more massive than the Earth, yet Io is about the same distance from that planet as is our own Moon from Earth. Thus, Jupiter exerts enormous gravitational forces on Io. In addition, Europa and Ganymede also exert gravitational forces, but in the direction opposite to that of Jupiter. The opposition of these forces causes Io to exhibit tidal flexing, and this flexing causes enough friction within the interior of Io to generate the heat required for nearly continuous volcanic activity. The frequent lava flows have obliterated all ancient land forms, giving Io a smooth and nearly craterless surface.

Europa   The next moon out from Jupiter is Europa (Fig. 22.27). Because of its greater distance from Jupiter, Europa is not subjected to enormous tidal forces, and as a result, it has no active volcanoes. The surface, however, is smooth and relatively craterless. Because meteor bombardment probably occurred sometime in Europa's history, the smooth surface seen today indicates that some kind of geologic activity has produced a degree of self-renewal on its surface.

Europa basically has a rocky composition, but its surface is covered with ice. Large streaks were observed in this ice surface, which might be caused by cracks in the ice that have opened and then been refrozen.

FIGURE 22.28   *A close-up view of the surface of Jupiter's largest satellite, Ganymede.* (Courtesy NASA)

FIGURE 22.27   *Europa has a surface of water ice, crossed by complex cracks.* (Courtesy NASA)

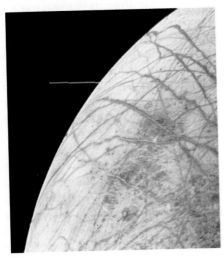

Ganymede   Ganymede is the largest moon in the Solar System, with a diameter of 5270 km, larger than the planet Mercury. It is a large sphere composed of a mixture of rock and water ice (Fig. 22.28). Two distinctly different types of terrain are observed on this moon; one is heavily cratered, whereas the other is grooved and contains fewer craters. Astronomers believe that the heavily cratered regions are more than 4 billion years old, whereas the smooth regions are much newer and have been formed by recent internal activity.

Callisto   Callisto, the outermost Galilean moon, is heavily cratered and shows no grooved or smooth terrain (Fig. 22.29). These data indicate that its surface crust is old. This moon has a diameter of 4820 km, about the same as the planet Mercury. The density of Callisto, however, is about one-third that of Mercury, which tells us that it is an icy body. The temperature of Callisto is about 120 K at the equator. Its most significant surface feature is a huge bull's-eye impact crater. The meteor that

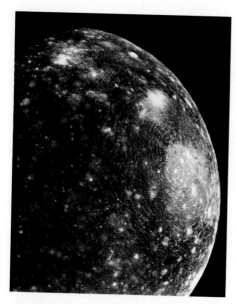

FIGURE 22.29   *Callisto photographed from* Voyager 1 *is seen to be heavily cratered.* (Courtesy NASA)

FIGURE 22.31   Voyager *view of the cloud structure of Saturn, which is much less spectacular than that on Jupiter.*

caused the bull's-eye did such extensive damage to the moon that it almost shattered it.

22.9 SATURN

Saturn, the second largest planet, is similar in many respects to Jupiter. It has the lowest density of all the planets, so low in fact that the entire planet could float on water if there were a bathtub large enough to hold it (Fig. 22.30). This low density implies that it, too, must be composed primarily of hy-

FIGURE 22.30   *Saturn's density is lower than that of water, so it would float in water.*

drogen and helium. The atmosphere of the planet is similar to that of Jupiter. Dense clouds cover the planet, and several distinct storm systems have been photographed, although none as notable or persistent as the Great Red Spot of Jupiter. Additionally, the atmosphere of Saturn has the same banded structure as does Jupiter, as shown in Figure 22.31. Like Jupiter, and all the gas giants, Saturn has a rapid rotation rate, turning on its axis in about 10 hours 40 minutes. Because of its high rate of rotation and low density, Saturn is the most oblate of all the planets. Its diameter at the equator is about ten percent greater than its diameter as measured pole to pole. Its axis of rotation is tipped about 27° with respect to a line drawn perpendicular to its orbital plane, so it does have seasons. The mean temperature of Saturn is lower than that of her sister planet, Jupiter; thus some substances that are gases on Jupiter are frozen solid on Saturn. In addition, the colder temperatures have favored the formation of certain organic molecules such as ethane, C_2H_6.

Certainly the most distinctive feature of Saturn is its spectacular rings that are readily visible from Earth, even through a small telescope, as shown in Figure 22.32. In addition to the rings, there are at least 17 satellites orbiting the planet. Before the space program, astronomers had observed three

FIGURE 22.32   *The ring system of Saturn as it can be seen by a good telescope on Earth.* (Courtesy NASA)

distinct rings with dark gaps separating them. Observations from the *Voyager* space probe, however, showed that the rings are incredibly more complex than was previously expected. A total of seven major rings was recorded, and each of these is further differentiated into thousands of smaller ringlets, as shown in Figure 22.33. This ring system is extremely thin. Estimates vary, but observations indicate that the rings are only 10 to 25 m thick, considerably thinner than the length of a football field. They have a large diameter, however, and cover a distance of some 425,000 km from the inner edge to the outer edge. If you were to make a scale model of the ring system and you chose a plastic disk the

thickness of a phonograph record, your model would have to be 30 km in diameter.

The rings are not rigid but are composed of many separate particles of dust, rock, and ice. The particles in the outer rings are only a few ten-thousandths of a centimeter in diameter, but the innermost ones are made up of larger chunks a few meters across. Each piece orbits the planet independently, and some are moving faster, some slower, in a continuous jumbled parade. Two major questions have been asked about the rings: (1) How were they formed? (2) Why are they so intricate and complex?

According to current theory, these rings are believed to be the scattered remnants of a moon that was never formed or that was formed and then ripped apart by the gravitational field of Saturn. Small objects, such as a rock or an ice cube, are held together by electrical attractions between the atoms and molecules. Large objects, such as the Sun, the planets, and their satellites, are held together mainly by gravitational forces. Imagine what happens to a small satellite orbiting a larger central planet. The surface of the moon closest to the planet is attracted more strongly than the region farther away, as shown in Figure 22.34. A strong force on one side of the moon and a weaker force on the other tend to elongate the moon and break it into pieces. Such forces are called **tidal forces.** Tidal forces between the Earth and Moon cause ocean movements on the Earth and seismic rumblings on the Moon, whereas forces between Io and Jupiter are comparatively greater, causing the internal layers of rock in Io to move and heat up. If a satellite is too close to its planet, the tidal forces can be greater than the internal forces that hold

FIGURE 22.33   *Close-up of view of Saturn's rings.* (Courtesy NASA)

FIGURE 22.34   *The force on side A is greater than on side B. This tends to stretch the moon and to create internal stress.*

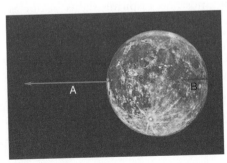

the moon together, and the moon is pulled apart or never formed in the first place.

The explanation of the fine structure of the rings involves three separate mechanisms. Each of these mechanisms originates independently of the others, yet the three combine to create a single system.

1. From fundamental laws of physics, any rotating, flattened disk that is made up of independent particles and is orbiting a central massive object will generate a ringed structure. This type of structure is clearly seen when billions of stars orbit around a massive galactic core. The same effect occurs with the small pieces of debris that orbit Saturn.

2. The particles in the rings of Saturn are attracted not only by the planet, but also by the 17 moons that orbit it. Over the years, the gravitational pull of the satellites has altered the orbits of the individual particles, spreading out the ring system and creating gaps within it. In some instances, two satellites work together to "shepherd" particles into precise orbital zones, much as two sheep dogs work together to channel a flock of sheep into a thin column that can pass through a narrow gate.

3. The third mechanism is not gravitational but electromagnetic. Saturn has a strong magnetic field. It follows that the field would trap charged particles in belts, as does the Earth. Scientists believe that the solid particles of the rings become charged, and complex attractions and repulsions between these particles then alter their orbital paths.

Although it is difficult to analyze each of these independent factors mathematically and then integrate them into a single, unified picture, the overall effect of these factors has produced the enormous, beautiful ring pattern that makes Saturn one of the most awe-inspiring sights in our Solar System.

Titan, Saturn's Largest Moon

Of the 17 moons of Saturn, Titan, the largest, is also one of the most unusual of the moons in the Solar System because it is the only satellite with an appreciable atmosphere. This atmosphere is largely methane, CH_4, a material that is commonly used on Earth as a fuel and is the major component of natural gas. The mean temperature on the surface of Titan is 90 K ($-183°C$), and the atmospheric pressure is 50 percent greater than the atmospheric pressure on the surface of the Earth. These conditions are close to the point where methane can exist in the solid, liquid, or vapor form. Therefore, small changes in atmospheric conditions on Titan cause methane to freeze, melt, vaporize, or condense. This situation is analogous to the environment on Earth, where water exists in each of its three phases and transfers back and forth readily between them. At present, astronomers believe that the surface of Titan is covered by a methane ocean.

At one time, astronomers held out the hope that conditions on Titan might be suitable for life. It was thought that the thick atmosphere might, via the greenhouse effect, have produced high enough surface temperatures so that some microbial life might have formed. However, the conditions as presented here by way of data from the Voyager probes are so harsh that we cannot visualize any such life forms.

22.10 URANUS AND NEPTUNE

Uranus and Neptune, both invisible to the naked eye from Earth, were unknown to the ancients. They are so distant that even today ground-based observation of them is limited. They are both quite large in size and low in density. The best evidence indicates that they are similar in structure to Jupiter and Saturn, being composed of a dense atmosphere, a liquid surface, and a solid mineral core. In 1977, scientists discovered rings around Uranus. In 1989, astronomical observations from the *Voyager 2* spacecraft found five rings around Neptune.

Data from the *Voyager 2* satellite indicate that, as expected, Uranus is enveloped in a thick atmosphere, composed primarily of hydrogen and helium with smaller amounts of compounds of carbon, nitrogen, and oxygen. The surface is believed to be a sea of methane, ammonia, and possibly water, and the core is probably rocky and about the

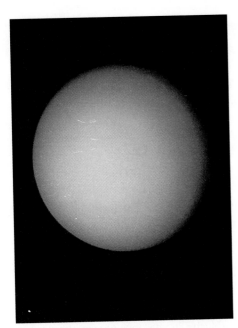

FIGURE 22.35 *The cloud structure of Uranus is much less spectacular than that on either Jupiter or Saturn. Wavelengths in the orange and red are absorbed out of the sunlight by the methane in the atmosphere of Uranus. The reflected light is, thus, rich in blues and greens, giving Uranus its characteristic color.* (Courtesy NASA)

size of the Earth. Figure 22.35 is a view of Uranus 1 week before the closest approach of *Voyager 2*.

Uranus is "tipped over." This means that its axis of rotation is in the same plane as its plane of revolution (Fig. 22.36). Thus, at times the polar regions are aimed toward the Sun. To make matters even more confusing, the magnetic pole is tilted

55° off the planet's rotational axis and offset from it. (Earth's magnetic North Pole is tilted only 11.7° from its geographic pole.) Scientists speculate that maybe a collision with a giant object knocked the planet over, giving it its strange tilt. The anomaly of the magnetic field, however, has not yet been explained.

A complex ring system and a total of at least 15 moons have been discovered orbiting Uranus. Several of these moons are small and irregular, indicating that they may be debris from the postulated collision that tipped the planet. The largest moons show complex surfaces indicative of a varied geological history. For example, ten different kinds of terrain have been identified on the moon Miranda.

In contrast to the symmetrical rings of Saturn, those around Uranus are warped, tilted, bizarrely elliptical, and varied in width. Some of these features can be explained by the proximity of "shepherd moons," but further analysis of the data is needed before a full explanation is available.

Neptune is often referred to as a triumph of gravitational theory. This description occurs because its presence and position were predicted before it was found by telescope. The story began as astronomers attempted to explain some unusual variations in the orbit of Uranus. Specifically, it was not following the exact orbital path predicted by gravitational theory. The only way to explain these deviations was to postulate the presence of another planet that was producing them. Thus, two mathematicians, Adams and Leverrier, predicted the location of the planet and are given credit for its discovery even though it was first actually seen by the astronomer Galle. Photographs taken by *Voyager 2* during its 1989 flyby revealed a Great Dark Spot near the equatorial region that seems to be a storm

FIGURE 22.36 *The axis of rotation of Uranus is tipped over.*

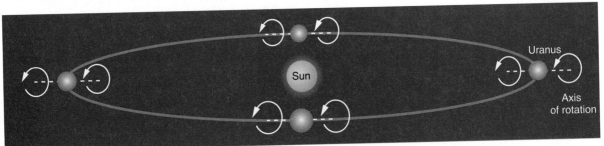

FOCUS ON
EXTRATERRESTRIAL LIFE

For many years, people have asked whether or not the Earth is unique in its ability to support life. In the beginning, the search for extraterrestrial life was concentrated on the nearby planets, Venus and Mars. But, as we have seen, Venus is too hot to support life, and Mars, although potentially more hospitable, has been found, thus far, to be completely devoid of any living organisms. Today, astronomers are expanding their search for life toward other regions of the galaxy.

As for life in other solar systems, one can only make some sort of guess. As we will learn in Chapter 23, there are roughly 100 billion stars in an average galaxy and billions of galaxies in the Universe. Thus, there are so many stars in existence that it is reasonable to believe that solar systems have formed around other stars. Distances are so large in the Universe that the only reasonable way to search for life is via radiotelescopes. Such a telescope, pictured in the figure, detects radio waves from space instead of visible light. In 1960, a radiotelescope followed two nearby stars looking for any tell-tale signs of radio emissions different from what would be expected normally from a heavenly object. Later in 1970, about 600 nearby stars were scanned in the same manner, but thus far there is no indication of extraterrestrials. In 1971, a joint project between NASA and the American Society of Engineering Education suggested that a giant array of 100 radiotelescopes be constructed for the sole purpose of searching for signals from extraterrestrial life. Thus far, this project has not been funded.

The acronym used to describe searches such as those described is SETI, for **s**earch for **e**xtra **t**errestrial **i**ntelligence. There is another side to the coin, called CETI, for **c**ommunication with **e**xtraterrestrial **i**ntelligence. By this, it is meant that we should also use the radiotelescopes mentioned above as transmitters of information in an effort to let others "find us." You should be aware that if such an effort ever meets with success, the dialogue will be considerably different from that normally found in radio transmissions here on Earth. The distance to even the nearest stars is such that if we say "hello," the message would not reach a nearby star for about 4 years, and it would take another 4 years minimum to receive a reply. For more distant stars, the answer to a question could require several lifetimes.

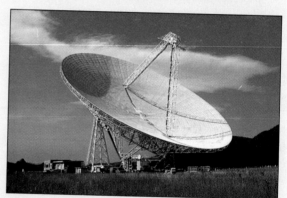

The 91 m radio telescope at the National Radio Astronomy Observatory site in Green Bank, West Virginia, before its collapse in 1988.

like that seen in the Great Red Spot of Jupiter (Fig. 22.37). Six new moons have been found, to give a total of eight. It has also been observed that the planet has bands around the polar caps.

Two of Neptune's moons, Triton and Nereid, are among the most unusual in the Solar System. Triton, only slightly smaller than our own Moon, has the unusual feature of a retrograde revolution in that it revolves about Neptune clockwise. Triton is the only moon in the Solar System to exhibit this behavior. The moon Nereid is also unusual in that it has the most eccentric orbit of any moon in the

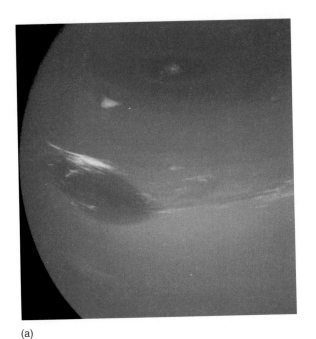

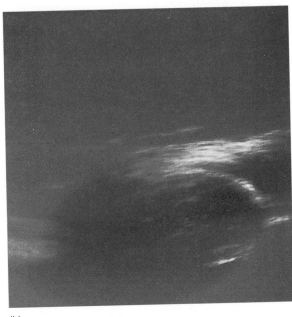

(a)   (b)

FIGURE 22.37   *(a) A view of Neptune as seen from* Voyager 2 *and (b) a close-up of the Great Dark Spot.*

Solar System. (The more eccentric the orbit of an object, the more squashed and elongated is the orbit. For example, a perfectly circular orbit is said to have zero eccentricity.) These unusual features of Triton and Nereid have led many astronomers to believe that some catastrophe happened early in the history of the formation of the Solar System. One of the most common speculations is that a near-collision occurred between these moons and some larger object. The near-miss sent Triton into its retrograde orbit and caused Nereid to deviate into its strange orbit.

22.11 PLUTO

Pluto, the outermost of the known planets, is quite small, roughly about the size of Earth's Moon. Spectroscopic studies of Pluto's surface indicate that methane ice exists on the surface. Therefore, temperatures must be extremely low, about 40 K ($-230°$C). For many years, astronomers believed that Pluto was dense and rocky, similar to the terrestrial planets. Our understanding was altered drastically in 1978, however, when a satellite, now named **Charon,** was discovered orbiting this distant planet, as shown in Figure 22.38. This moon has

the distinction of being the largest moon relative to the size of its planet of any in the Solar System.

FIGURE 22.38   *The Hubble Space Telescope's Faint Object Camera has obtained the clearest image ever of Pluto and its moon Charon. Pluto is the bright object at the center of the frame; Charon is the fainter object in the lower left. Charon's orbit around Pluto is a circle seen nearly edge-on from Earth.* (Courtesy NASA)

Using Kepler's third law, it is possible to calculate the relative masses of the planet and satellite when the radius of the satellite's orbit and its period are known. Using the best available data, the density of Pluto has been estimated to be only slightly greater than that of Saturn, indicating that it is composed of light elements.

22.12 VAGABONDS OF THE SOLAR SYSTEM

The imposing parts of our Solar System include the Sun, the planets, and the moons, but now we shall turn our attention to some objects of lesser significance—meteoroids, asteroids, and comets. As we shall see, some of these objects produce brief flickers in our night sky, others have caused tremendous devastation on our planet in the past, and others can lead to dramatic displays sufficient to cause people to ask the age-old question: Why? Let us look at these vagabonds of the Solar System in turn.

Meteoroids

The terrestrial planets are no longer subject to the intense bombardment from outer space that once generated enough heat to melt large volumes of crustal rock. But if you sit outside on almost any clear night, watching the sky for a few hours, you may see a fiery streak called a **meteor,** or colloquially a **shooting star,** descend toward Earth (Fig. 22.39). Shooting stars appear when small bits of interplanetary solid matter called **meteoroids** are caught by Earth's gravity and accelerated through the atmosphere. Friction between the meteoroid and the atmosphere produces enough heat to raise the temperature of the meteoroid to a high level. The fiery streak across the sky, seen by an observer on Earth, is heated air and vapor from the meteor. Most meteoroids are barely larger than a grain of sand when they enter the atmosphere and are completely vaporized before they reach the Earth. If they are larger, say the size of a basketball, some of the original material falls to the ground. A fallen meteoroid is called a **meteorite.** Examination of fallen debris indicates that most meteorites are roughly as old as the Solar System, approximately

FIGURE 22.39 *A meteor, or shooting star.* (Courtesy Smithsonian Astrophysical Observatory)

4.6 billion years old. Some meteorites have never been subjected to planetary heating and remelting, and scientists believe that these fragments represent the kind of primordial material that originally condensed out of the interspacial dust to form the terrestrial planets.

One fascinating observation is that some meteorites contain fairly complex organic compounds, including amino acids and other molecules that are vital components of living organisms. Thus, com-

A meteorite about the size of a basketball. (Courtesy NASA)

FOCUS ON
THE TITIUS-BODE RULE

In 1766, Johann Titius found a series of numerical steps that could be used to find the approximate distance of each planet from the Sun. The procedure was first published in 1772 by Johann Bode and is now known as the **Titius-Bode rule.** The rule uses a sequence of numbers starting with 0, with the next number equaling 3. From then on, the next number in the sequence is found by doubling the preceding one. Thus, the basic sequence is 0, 3, 6, 12, 24, and so on. Then the number 4 is added to each number in the sequence and the result is divided by 10. The resulting answer gives the distance of the planets from the Sun in astronomical units (AU). The procedure is outlined in the table for all the planets. The distances are predicted quite well for the innermost planets, including the asteroid belt, but the match for the outermost planets is not as good.

PLANET	TITIUS-BODE RULE (AU)	OBSERVED (AU)
Mercury	$(0 + 4)/10 = 0.4$	0.387
Venus	$(3 + 4)/10 = 0.7$	0.723
Earth	$(6 + 4)/10 = 1.0$	1
Mars	$(12 + 4)/10 = 1.6$	1.524
Asteroids	$(24 + 4)/10 = 2.8$	2.75
Jupiter	$(48 + 4)/10 = 5.2$	5.203
Saturn	$(96 + 4)/10 = 10.0$	9.539
Uranus	$(192 + 4)/10 = 19.6$	19.18
Neptune	$(384 + 4)/10 = 38.8$	30.06
Pluto	$(768 + 4)/10 = 77.2$	39.44

Some astronomers believe that the procedure is a result of pure chance and that regardless of what the orbital distances might be, some such numerical procedure would reproduce the orbits. Others believe that the law rests on some underlying physical reason that simply has not yet been found.

plex organic molecules are not unique to the Earth but have been formed in other regions of the Solar System. This fact by itself does not mean that life exists elsewhere in our Solar System, but it does tell us that the molecules of life have been synthesized by inorganic (nonbiological) reactions occurring in outer space.

Asteroids

Eighteenth-century astronomers noted that the dimensions of the planetary orbits increased in a regular pattern, starting with Mercury's orbit, the smallest, and going on to those of Venus, Earth, and Mars. There seemed to be a gap before Jupiter. Based on this interruption in the pattern, it was predicted that a planet might be found in the "open

space" between Mars and Jupiter. Instead of a full-sized planet, however, observers have found tens of thousands of smaller bodies orbiting in a wide ring. These bodies are called **asteroids.** The largest asteroid, Ceres, has a diameter of 770 km. Three others are about half that size, and most are far smaller. The orbits of some asteroids are not permanently fixed like that of a planet. If one of these passes too close to a nearby planet, it will be pulled toward it and fall onto the planet's surface. If an asteroid passes by a planet without getting too close, however, the gravitational force of the planet can pull the asteroid out of its current orbit and deflect it into a new orbit about the Sun. Thus, a given asteroid may change its orbit frequently in an erratic manner. Fortunately, most of the orbits in the asteroid belt are relatively fixed. Frequently, small fragments and pieces of dust are deflected helter-

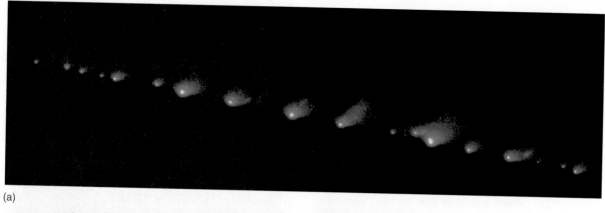

(a)

(b)

(c)

(a) A Hubble Space Telescope image of Comet Shoemaker-Levy 9 taken in May 1994. When the comet was observed, its train of icy fragments stretched across 710,000 miles. (b) A composite photo showing the comet's approach to Jupiter. (c) Eight impact sites are visible from the collision of the comet on Jupiter's surface.

skelter in widely divergent directions. Some of these fragments cross the orbit of the Earth and are attracted by our gravitational field. These then fall through the atmosphere and are visible as meteors. Asteroids are not the only source of meteors. When a comet passes too near the Sun, it can be broken up by tidal forces. When this occurs, the debris drifts through space, covering a large volume of the Solar System. When the Earth passes through this trail, intense meteor showers are visible. In fact, most meteors are remnants of comets, not asteroids.

In July 1994, a series of approximately 20 fragments from the comet Shoemaker-Levy collided with Jupiter over a six-day period starting on July 16th. The comet had been fragmented because of a close encounter with Jupiter. This was the first opportunity that astronomers have had to know in advance about a comet-Jupiter collision and to observe its effects on Jupiter. The collisions occurred on the side of Jupiter facing away from Earth, but the rapid rotation of the planet brought the impact sites into view in less than an hour where they could be seen by the Hubble Space Telescope. Images returned to Earth showed vast disturbances in the Jovian atmosphere with huge plumes of gas thrown hundreds of kilometers into space.

What is the probability that a large fragment, or even an entire asteroid, will strike the Earth some day? Geological evidence indicates that such events

FOCUS ON
HALLEY'S COMET

"When beggars die, there are no comets seen;

The heavens themselves blaze forth the death of princes."

Shakespeare, Julius Caesar

Actually, no comet "blazed forth" the death of Julius Caesar, but one was seen suspended over Rome just before the general and statesman Marcus Agrippa died in 11 B.C. This was the fourth recorded return of what is now called Halley's Comet.

In 1705, the British scientist Edmund Halley noted that the orbits of bright comets that had appeared in 1531, 1607, and 1682 were about the same. This similarity, coupled with the regularity of return, suggested that the three appearances were of a single orbiting comet. Halley predicted that it would return again in 1758, which turned out to be 16 years after Halley's death. Since then Halley's Comet, as it is now called, has reappeared three times, in 1835, 1910, and most recently in late 1985 and early 1986. The 1910 sighting was particularly spectacular. The comet was so close that it was observed to stretch across a 100° to 120° arc in the night sky. Some people predicted that poisonous gases within the comet's tail would pass through the Earth's atmosphere and destroy civilization, but responsible scientists understood that a comet's tail is too diffuse to cause any significant impact on terrestrial life.

In 1986, Whipple's model of a comet was largely confirmed when a spacecraft, the *Giotto*, sent out by the European Space Agency, flew to within 350 miles of Halley's comet. (The *Giotto* spacecraft got its name from an Italian painter who had painted an image of the comet in a fresco in the fourteenth century showing it as the star of Bethlehem.) The figure shows a photo of

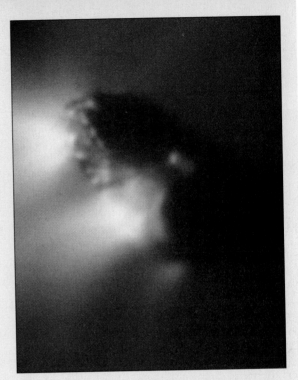

The potato-shaped nucleus of Halley's comet is about 16 km by 8 km across. Two jets spray gas and dust toward the Sun. (Courtesy Max Planck Institut für Aeronomie)

the comet taken by the spaceship. It is shown to be a potato-shaped object about 16 km by 8 km in size that is as dark as velvet because the comet seems to be covered by a black, tarry crust. Two bright jets of material are spewing out of the side of the nucleus that is facing the Sun. These jets are produced as the Sun vaporizes the interior of the comet and the gases escape through fractures in the tarry surface. The earthbound view of the comet was disappointing compared with earlier encounters. The comet was considerably fainter than anticipated, and it was not an impressive naked-eye object.

FIGURE 22.40   *Aerial view of the meteor crater near Flagstaff, Arizona.* (Courtesy Meteor Crater, Inc.)

have occurred in the past. For example, there is a large crater near Flagstaff, Arizona, approximately 1.5 km in diameter, that was formed by a falling meteorite about the size of a large semi-truck (Fig. 22.40). This meteorite is believed to have landed in recent geological history, and the crater is perhaps no more than 50,000 years old. Other curious circular basins exist on the surface of the Earth that may well be eroded remnants of large meteorite craters.

Comets

Occasionally a fuzzy object appears in the sky, travels slowly around the Sun, and then disappears again out into space. These objects are called comets after the Greek word for "long-haired," and a typical one is shown in Figure 22.41. Despite their fiery appearance, comets are quite cold, and most of the light that we see is reflected sunlight. In 1950, Harvard astronomer Fred L. Whipple proposed that a comet was similar to a dirty snowball composed of a tenuous collection of rock and dirt embedded in water ice and frozen carbon dioxide. As the comet approaches the Sun, solar radiation vaporizes its surface, and the force of the solar wind blows some of the lighter particles away from the head to form a long tail, the "long hair." As the comet orbits the Sun, the solar wind constantly blows the tail so that it always points away from the Sun, as shown in Figure 22.42. Comet tails have been observed to be over 90 million miles long, almost as long as the distance from the Earth to the Sun. There is little matter in a comet tail. By terrestrial standards, this region would represent a good, cold laboratory vacuum, yet viewed from a celestial perspective, such an object looks like a hot, dense, fiery arrow.

FIGURE 22.41   *Halley's comet as photographed on March 12, 1986.* (Courtesy Royal Observatory, Edinburgh)

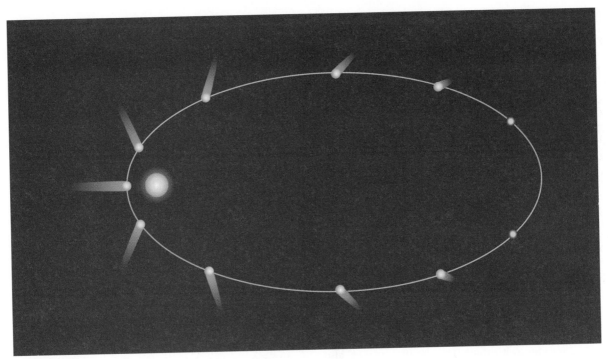

FIGURE 22.42   *A comet's tail always points away from the Sun.*

SUMMARY

The Solar System was formed from a mass of dust and gas that was rotating slowly in space. Within the center, the gravitational attraction was so great that the gases were pulled inward with enough speed that fusion temperatures were reached. The planets were also formed from coalescing clouds of matter, but fusion temperatures were not achieved.

The **central core** of the Sun is hot and dense. Hydrogen fusion occurs in this region. The visible surface of the Sun, called the **photosphere,** is appreciably cooler and has a pressure of one-tenth of an atmosphere. **Sunspots** are magnetic storms on the surface of the Sun. The outer layers, called the **chromosphere** and the **corona,** are turbulent and diffuse. The **solar wind** is a stream of particles coming from the Sun. **Auroras** occur when particles from the solar wind interact with the magnetic field of the Earth.

One theory states that the Moon was formed from the debris of a collision between a giant object and the Earth. The Moon was heated by the energy released during condensation, by radioactive decay, and by meteorite bombardment, but it is cold and inactive today.

Mercury is hot on the sunny side and cold on the shaded portion. Its topography is similar to that of the Moon. Venus has a hot, dense atmosphere and a surface that shows signs of tectonic activity. Mars is a dry, cold planet with a thin atmosphere, but the surface bears signs of ancient periods of erosion and tectonic activity.

Jupiter, Saturn, Uranus, and Neptune are all large planets with a low density. All are believed to have an appreciable atmosphere, a surface region composed largely of hydrogen, an inner zone of liquid metallic hydrogen, and a solid inner mineral core. The largest moons of Jupiter are Io, which is heated by gravitational forces; Europa, which is smooth and ice covered; and Ganymede and Callisto, which are large spheres made up of rock and ice. The rings of Saturn are made up of many small particles of dust, rock, and ice. They were formed from a moon that was pulled apart (or not allowed to form) by gravitational forces. Titan, the largest moon of Saturn, has an atmosphere and may be tectonically active. Pluto has a low density and is quite small.

A **meteorite** is a fallen **meteoroid,** a piece of matter from interplanetary space. **Asteroids** are small planet-like bodies. **Comets** are diffuse collections of solid mineral particles coated with frozen films of various compounds.

KEY WORDS

Protosun	Maria	The Great Red Spot	Pluto
Planetesimal	Mercury	Europa	Charon
Plasma	Synchronous rotation	Ganymede	Meteor
Photosphere	Venus	Callisto	Shooting star
Sunspot	Mars	Io	Meteoroid
Solar flare	Phobos	Saturn	Meteorite
Chromosphere	Deimos	Titan	Asteroid
Corona	Jupiter	Uranus	Comet
Solar wind	Terrestrial planets	Neptune	
Magnetosphere	Liquid metallic hydrogen	The Great Dark Spot	

PROBLEMS AND CONCEPTUAL QUESTIONS

Problems requiring numerical work are identified with a blue number.

Formation and structure of the Solar System

1. Explain why the protosun gradually became warmer as it shrank. Did the protoplanets also become warmer as they coalesced? If so, why are they so much cooler today?

2. Astronomers have observed that all the planets of the Solar System revolve around the Sun in the same direction and in nearly circular orbits. Is the theory of the origin of the Solar System offered in this text consistent with this ordered revolution? Defend your answer.

3. Explain why a star would appear to wobble as seen from Earth if it is being circled by a planet.

4. Venus rotates slowly clockwise, while all the other planets rotate counterclockwise. What does this suggest about the history of Venus?

The Sun

5. What are the most abundant elements present on the Sun? How does this composition differ from that of the Earth? Explain how this difference evolved.

6. Draw a diagram of the Sun, labeling the core, the photosphere, and the corona. Label the temperatures and the relative densities of each region.

7. What is the fundamental source of energy within the Sun? Is the Sun's chemical composition constant, or is it continuously changing? Explain.

8. How does energy travel from the core of the Sun to the surface? How does it travel from the surface of the Sun to the Earth?

9. Compare and contrast the core of the Sun with its outer surface.

10. Describe the formation of sunspots. Why do sunspots appear black to an observer here on Earth?

11. What did the first observations of sunspots have to say about earlier theories of the heavens?

12. Auroras are observed on Earth. On which of the following would you expect to observe auroras: the Moon, Mercury, Venus, Jupiter, Saturn? Defend your answer in each case.

13. Briefly outline some changes that would occur if the Earth's magnetic poles shifted so that they lay on the Equator.

14. Compare and contrast the solar wind with a breeze that blows on the surface of the Earth.

The Moon

15. What leads us to believe that the Moon was hot at one time in its history? According to modern theory, how was the Moon heated?

16. It is best to observe the Moon through a telescope when it is between crescent and gibbous phases, with the full Moon being a bad time for observation. Explain why this is true.

17. Suppose the oldest igneous rocks on the Moon had been formed when the Moon was 1 billion years old. What conclusions would we then draw about the geological history of the Moon? Could we positively answer the question: Was the Moon heated by internal radioactivity, or by external bombardment? Defend your answer.

18. Explain how we can learn about Earth's geology by studying the Moon.

Mercury

19. Give a brief description of the planet Mercury. Include its atmosphere, magnetic field, surface

temperature, type of terrain, and speed of revolution around the Sun.

20. Why is the presence of a magnetic field around Mercury a curious phenomenon?

21. If Mercury rotated once every 24 hours as the Earth does, would you expect daytime temperatures on that planet to be higher or lower than they are today? Defend your answer.

22. Why was it a surprise to astronomers to discover an atmosphere (even a thin one) on Mercury?

Venus

23. At one time, Venus and Earth probably had similar environments except that Venus was about 20° warmer. If you could somehow cool the surface of Venus by 20°, would conditions on that planet be likely to become similar to those on Earth? Explain.

24. Why are there few meteorite craters visible on Venus?

25. Venus is often referred to as the sister planet to Earth. Why?

26. If you were to take a vacation on Venus, list several features of the planet that you would like to see.

Mars

27. Discuss the evidence that indicates that the Martian atmosphere was once considerably different than it is today. Explain why this atmosphere could not have been similar to that of the Earth for long periods of time.

28. Refer to Figure 22.19. Imagine that you knew nothing about Mars except that it is a planet and this photograph was taken of its surface. What information could you deduce from this picture? Defend your conclusions.

29. In the late 1980s, President George Bush called for a manned mission to Mars. Do you believe this should be undertaken? Defend your answer.

30. Discuss the early arguments that led people to believe that there was life on Mars.

31. At the surface of Phobos, the acceleration due to gravity is 0.001 that of Earth. If a baseball is released from the top of a tall building on that moon, what speed will it have after falling for 3 seconds? (*Hint*: Refer to the equations for a freely falling object given in Chapter 2.)

32. It is said that a baseball player standing on Phobos could throw a baseball such that it would go into orbit around that moon. Based on the acceleration due to gravity on Phobos as given in Question 31, do you believe this is likely?

33. Compare and contrast Earth with the other terrestrial planets.

34. Why does water still exist on Earth but is absent on Mercury, Venus, and Mars?

35. Explain how the relative masses and distance from the Sun of Venus, Earth, and Mars led to markedly different environments on each of these three planets.

Jupiter

36. Describe the composition of the planet Jupiter. How does it differ from that of Earth?

37. Why is it unlikely that a planet like Jupiter would be formed near the Sun?

38. Compare and contrast the satellite system of Jupiter with the Solar System.

39. Explain why the mass of Jupiter was an important factor in determining its present composition and structure.

40. About 4 billion years from now, the Sun will probably grow significantly larger and hotter. How will this change affect the composition and structure of Jupiter?

41. Compare and contrast the four Galilean moons of Jupiter. Describe some similarities between the Galilean moon system and the Solar System as a whole.

Saturn

42. Compare and contrast Saturn with Jupiter.

43. Compare and contrast Titan with the Earth.

44. Would you expect to find gases in the ring system of Saturn? Why or why not?

45. A science fiction story describes a visit by a spaceship to a planet with an enormous gravitational field. Before the ship could land, however, it was ripped apart by tidal forces. Explain how tidal forces could have done this.

46. Give arguments for and against the possibility of life on Titan.

Uranus and Neptune

47. Why is the discovery of Neptune often called a triumph of gravitational theory?

48. Draw a sketch of Uranus at several positions in its orbit showing its axis of rotation with respect to the plane of its orbit.

49. List several characteristics common to the Jovian planets.

Pluto

50. At one time, it was thought that Pluto might once have been a moon of Neptune. However, the discovery of the moon Charon has caused some astronomers to doubt this scenario. Why should Charon play a role in this debate?

51. Sometimes Pluto is the eighth planet and Neptune is the ninth. Sketch possible orbits for these two planets that would allow for this possibility.

Meteoroids and comets

52. Astronomers once thought that perhaps meteoroids were fragments from a planet that formed and then exploded or was destroyed in a collision with another planet. Do you think that such an origin is likely? Defend your answer.

53. Is a comet really hot, dense, and fiery? If so, what is the energy source? If not, why do comets look as though they consist of burning masses of gas?

54. How are comets and meteor showers related?

55. When meteor showers occur, they are most easily visible after midnight. Why should the time of day be a factor?

ANSWERS TO SELECTED NUMERICAL PROBLEMS
31. 0.03 m/s

CHAPTER 23

THE LIFE AND DEATH OF STARS

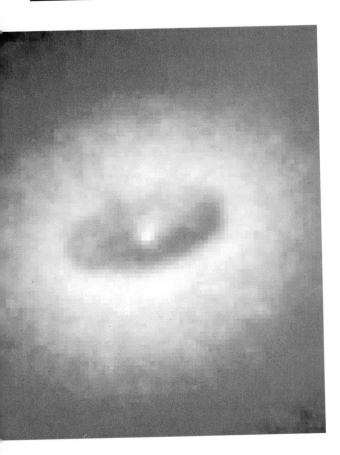

A Hubble Space Telescope image of a giant disk of gas and dust—about 300 light-years across—fueling a suspected black hole. The bright hub presumably harbors the black hole. The disk is at the core of a galaxy in the Virgo cluster, located 45 million light-years away. (Courtesy NASA)

Throughout most of this book we have tried to explain physical phenomena in terms of concepts that can be related to our everyday experiences. But in some instances this is difficult to do. For example, in our discussion of relativity, such effects as the increase in mass with speed, the decrease in length, and the slowing down of high-speed clocks do not fall into the realm of common observation. Likewise in this chapter, we will find some happenings in space that stagger the imagination. Large objects and great distances are difficult to comprehend, and when the mysteries of distant galaxies and the vast regions of intergalactic space are considered, we simply cannot use common "house and garden" analogies. Such comparisons are simply inadequate. For example, some distant galaxies are over 8 billion light-years from Earth. That means that light moving at 3.00×10^8 m/s must travel for 8 billion years to reach us here. Think about it: The light we see now left that galaxy 3.5 billion years before our Solar System was even formed. It has been traveling all this time at 3.00×10^8 m every second (186,000 miles/s).

Not only are distances in space incomprehensibly large, but also forces are similarly unimaginable. For example, some dead and dying stars become compressed so severely that protons and electrons are squeezed together to form neutrons. Other stars, compressed even more vigorously, become so dense that nothing, not even light, can escape from their gravitational tug. If we are to comprehend these and other related concepts, we must release our imagination and let our thoughts fly beyond terrestrial standards of force, size, mass, distance, and time.

23.1 STUDYING THE STARS

Other than the sun, the nearest star to Earth (Proxima Centauri) is about 4.2 light-years distant from

FOCUS ON
ASTRONOMICAL DISTANCES

As we have seen, astronomical distances are incredibly large, even within the neighborhood of our own Solar System. For example, in Chapter 22, we found it convenient to speak of distances in terms of astronomical units, where 1 AU is equal to the average distance from the Earth to the Sun, 1 AU = 93,000,000 miles. Likewise, when speaking of distances beyond our Solar System, it is convenient to measure distances in terms of light-years,

where, as we have seen, 1 light-year is the distance that light travels in 1 year, 1 light-year = 9.5×10^{12} km. In these units, the closest star to our Solar System is about 4.2 light-years distant. A light ray leaving that star for Earth must travel for over 4 years before it reaches us. If our focus of attention is expanded farther out into space, even a distance of 4 light-years can seem small. The Andromeda Galaxy is 2.25 million light-years from Earth, and the most distant galaxies are about 10 billion light-years.

us, and most are much farther. Yet, as distant as these stars are, astronomers have been able to piece together much information about them. But how do we know these facts? How do we know their temperatures, their masses, their chemical compositions, their velocities through space, and many other pieces of information? The answer to how we found out most of the information lies with the study of the spectra of the stars. We have discussed the various kinds of spectra previously, but in this section we shall review them briefly and then make use of our findings to see how information can be gleaned from starlight.

Spectra and Chemical Composition of Stars

As we have seen, one of the most useful methods available to astronomers is the study of atomic and molecular spectra. As you recall from Chapter 12, an excited atom emits light when an electron returns to a lower energy state from a higher state. The light, however, is not emitted in a random pattern. Instead, only certain frequencies are observed. Each element has a characteristic **emission spectrum** and can be identified by it. If the spectrum of an unknown element matches that of neon, for example, the unknown element must be neon.

Light is emitted deep inside a star as a **continuous spectrum** over a wide range of frequencies. As

this light passes through the outer layers, some of it is selectively absorbed by the various atoms in the cooler gas layers surrounding the star. Therefore, an observer on Earth sees a spectrum showing lines of darkness crossing the continuous band of colors. This is called an **absorption spectrum.** To see how an absorption spectrum can be used to determine the chemical composition of a star, consider Figure 23.1a. There we see an imaginary spectrum from a similarly imaginary star. There are only a few absorption lines missing from the spectra, and the task is to find out what element did the absorbing. Figure 23.1b shows an absorption spectrum produced here on Earth by allowing a continuous spectrum to pass through hydrogen gas. Note that the lines in (a) and (b) match up perfectly. Thus, we can conclude correctly that this star contains hydrogen. In actual practice, one would never find an absorption spectrum from a star looking quite as clean as the one in Figure 23.1. Because a star contains many different gases, a stellar spectrum contains the superimposed images of many individual absorption spectra. As an example, the absorption spectrum of the Sun is shown in Figure 23.2. (The complete spectrum ranging from red to blue has been folded over so that it fits in the space for the diagram.) By a careful matching up of lines with spectra taken in earthbound laboratories, astronomers can identify which of these lines belong to what elements, and thus they can determine the chemical composition of the star.

Blue   Red

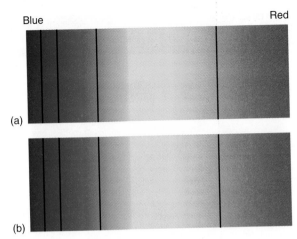

(a)

(b)

FIGURE 23.1 *(a) Absorption spectrum from a star. (b) Absorption spectrum of hydrogen made from a light source on Earth.*

Temperature of a Star

There is a simple method for determining rough information about a star's temperature that you are probably aware of from your common experience. Consider what happens to a piece of metal when you heat it. As the temperature of the metal increases, it eventually begins to glow with a dull red color. Then, as the temperature continues to increase, the color changes to orange and then to yellow. If you could continue this process, you would find the color would go to white and then to blue. Similarly the overall color of starlight changes with temperature, so by studying the color of a star, astronomers can obtain an idea of stellar tempera-

FIGURE 23.2 *The absorption spectrum of the Sun; similar spectra are observed from certain stars.* (Courtesy NOAO)

tures. The color of some bright stars can be detected with the naked eye. Examples are the reddish star Betelgeuse on the right shoulder of Orion the Hunter and the bluish star Rigel at his knee. A telescope provides even better visual observation. If you have access to a telescope, slowly scan across the sky and pay attention to the various hues and shades that you find. Each color has its own story to tell about the temperature of that star.

Spectra and the Speed of Stars

Stars are all traveling through space, and scientists can determine their speed relative to Earth by studying stellar spectra. To recall how this is done, return to Chapter 7 and review the discussion of the Doppler effect. We found in our discussion of the Doppler effect that objects moving toward us have their sound frequencies shifted toward a higher value than when the object is stationary. Similarly, the sound frequency is shifted toward smaller values if the object is in motion away from us. The frequency of light waves also changes with relative motion. Spectral lines reaching us from a star that is flying rapidly away from Earth appear at a lower frequency (closer to the red end of the spectrum) than would be expected if the star were stationary with respect to Earth. For example, Figure 23.3 shows a spectrum of hydrogen taken in a laboratory on Earth and a spectrum from a star that contains hydrogen. Note that the spectral lines from the star are not quite where they should be. Instead,

FIGURE 23.3 *(a) Hydrogen spectrum on Earth and (b) hydrogen spectrum from a star. Note the shift of the lines from the star toward the red.*

Blue   Red

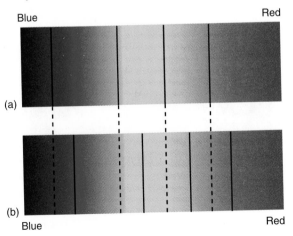

(a)

(b)

Blue   Red

FOCUS ON
MEASURING THE DISTANCE TO NEARBY STARS BY PARALLAX

The distance to the nearest stars can be measured by use of a technique called the method of parallax. To understand how this works, let's take an example from ordinary life. Imagine that you and a friend are driving down a straight highway in two separate cars. You spot a tree off in the distance, and you decide that you would like to know how far the tree is away from the highway. The technique of parallax can provide the answer, as shown in the figure. The two of you stop your cars along the side of the road and measure the distance between the two. This distance indicated is called the baseline. Each of you then measures the angles shown as angle A and B. We will not go through the details of how to find the distance to

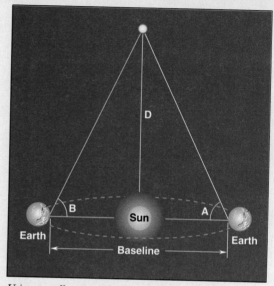

Using parallax to measure the distance to a star.

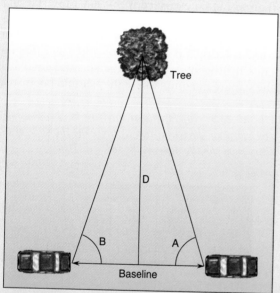

The distance between two cars is the baseline for measuring the distance to the tree by parallax.

the tree, indicated by the distance D in the figure. It suffices here to say that it is a simple technique from trigonometry to find D if you know the length of the baseline and the angles A and B. Now, let's move into space and see how this same procedure can be used to find the distance to some stars.

First, we must set up a baseline, and the next figure shows how this is done. We observe a nearby star and measure the angle A to it at a particular time of year. We then wait until the Earth has moved to the other side of its orbit and again measure the angle to the star, indicated as B. Our baseline is now equal to twice the distance from the Earth to the Sun, 2 × 93,000,000 miles, and we know the angles A and B. As before, simple trigonometry enables us to find the distance to the star D.

To see why this technique works only for nearby stars, consider the third figure. Note what

they are shifted slightly toward the red end of the spectrum. We say that this light has been **red-shifted.** Such a red shift is indicative of the fact that the frequency of the light from the star has been decreased because of its motion away from Earth. Similarly, light from stars traveling toward us appears at higher frequencies and is said to be **blue-shifted.**

Distance to a Star

As discussed in the Focus box on measuring distance, the distance to nearby stars can be found by use of parallax. This method breaks down, however, for more distant objects because it is impossible to measure such small angles precisely enough. Yet we have been talking about finding the distance to galaxies as far away as 8 billion light-years, and the distance to the farthest objects in our Universe is now thought to be between 18 and 20 billion light-years. How do we know these distances?

In the 1920s, the astronomer Vesto M. Slipher of the Lowell Observatory noted that light from almost every galaxy he studied was red-shifted. According to this observation, all the galaxies are moving away from us. In 1929, Edwin Hubble confirmed Slipher's observations and took the additional step of showing that there is a pattern in the way in which the galaxies move apart. Specifically, he found that the distance to a galaxy is proportional to its red shift. Thus, these galaxies are moving outward at great speeds, whereas the closer ones are receding more slowly. This relationship has been quantified and is known as Hubble's law, as shown in Figure 23.4. To plot the lowest point on the diagram, a representative galaxy from a cluster of galaxies in the direction of the constellation Virgo is selected. The remainder of the points are for representative galaxies in the direction of Ursa Major, Corona Borealis, Bootes, and Hydra.

Plotted is the speed in kilometers per second away from Earth versus the distance to these galaxies in light-years. The straight line indicates that there is a direct relationship between these two quantities. The exact relationship is given by **Hubble's law,** which is stated as

$$v = Hd \tag{23.1}$$

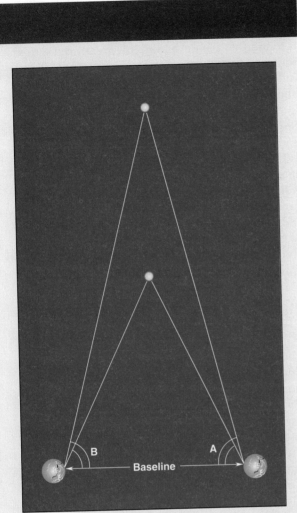

The angles A and B become impossible to measure accurately for distant stars.

is happening to the angles A and B as the distance to a star becomes larger and larger. Even in the best of circumstances, the angles A and B are close to 90°. (They are shown here as being considerably smaller than 90° for convenience.) As the distance increases, however, the angles become so close to 90° that the measuring instruments simply cannot find the exact value of the angle. Thus, the distance to only about 700 nearby stars can be found by this method.

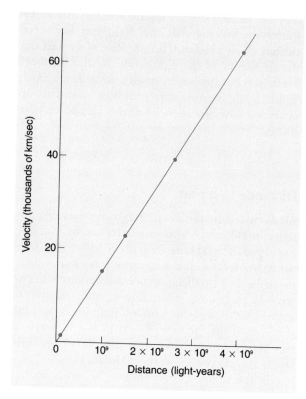

FIGURE 23.4   *The Hubble diagram for several different galaxies. Note that the relationship between the velocity of recession and the distance to the galaxies is a straight line.*

where v is the speed of the galaxy or the star in km/s, d is the distance of the galaxy or star in millions of light-years (Mly), and H is Hubble's constant given by

$$H = \frac{15 \text{ km/s}}{\text{Mly}}$$

EXAMPLE 23.1   HOW FAR AWAY IS THAT STAR?
The red shift in the light from a distant object indicates that it is receding from the Earth at a speed of 30,000 km/s. What is the distance to this object?

Solution This is a direct application of Hubble's law, which we can solve for the distance d as

$$d = \frac{v}{H} = \frac{30000 \text{ km/s}}{15 \dfrac{\text{km/s}}{\text{Mly}}} = \boxed{2000 \text{ Mly}}$$

CONCEPTUAL EXAMPLE 23.2   RED SHIFTS
When the spectrum of a blue star is investigated, it is found that the light from it is greatly red-shifted, while the light from a nearby red star is only slightly red-shifted. What do these observations tell you about these stars?

Solution The color of the stars provides a relative gauge of the temperature of the stars. The blue star is considerably hotter than the red star. The fact that the blue star has a large red shift compared with the red one means that the blue is moving more rapidly away from Earth than the red. Finally, the two stars are not really nearby; they only appear to be close together in space. The greater the red shift, the farther away is the star. So, the blue star is really the more distant of the two.

22.2 BRIGHTNESS OF STARS

About 2000 years ago, Hipparchus, a Greek astronomer, developed a system of classifying stars according to their relative brightness as seen on Earth. On his scale, the brightest stars were said to have an **apparent brightness** of 1, and those stars just barely visible to the naked eye have an apparent brightness of 6. Such a rating scale has limited usefulness because it tells us nothing about some intrinsic property of that star. For example, a star might appear to be bright as seen from Earth and yet be pouring out only a relatively small amount of energy into space. The reason that it appears bright is simply because it is close to Earth. A star that appears relatively dim from Earth could actually be a gigantic powerhouse, pumping enormous amounts of energy out into space. The reason for it appearing dim is that it is far from Earth.

What is important to know about a star is a quantity called its **intrinsic brightness.** This is a number that truly rates a star according to how much energy it is pouring out into space. That is, if a star has an intrinsic brightness of 1, it is honestly pouring out far more energy than is a star that has an intrinsic brightness of 6.

We shall not do the calculations here, but a three-way relationship exists between apparent brightness, intrinsic brightness, and the distance to a star. If any two of these three are known, the third can be found. This means that for a typical star, at least in principle, it is a relatively straightforward task to find its intrinsic brightness, and as we shall see in the next section, knowing the intrinsic brightness of a group of stars and their temperature led to an enormous jump in our understanding of the evolution of stars.

23.3 HERTZSPRUNG-RUSSELL DIAGRAM

Shortly after the beginning of the twentieth century, two astronomers, Hertzsprung and Russell,

plotted a graph that was to have enormous consequences in our understanding of the life history of stars. This graph, called an H-R diagram after its originators, plots intrinsic brightness along the vertical axis and the temperature of the stars along the horizontal (Fig. 23.5). The vertical axis follows the general pattern of plotting graphs in that as one moves upward along the axis, one is moving toward brighter stars. The horizontal axis, though, is a little unusual in that as one moves out along this axis, one moves toward stars having a lower temperature. When such a graph is drawn for a group of stars, it is found that approximately 90 percent of the stars

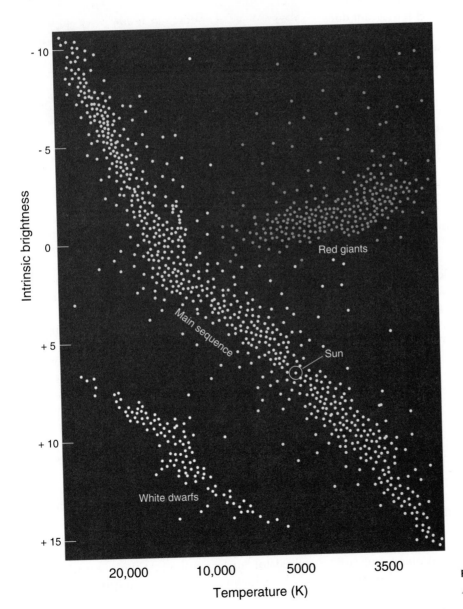

FIGURE 23.5 *The Hertzsprung-Russell diagram.*

fall along a diagonal line called the **main sequence.** There are two other distinct groupings of stars, however, that always appear, one in the upper right corner of the graph and one in the lower left corner. The stars in the upper right corner are bright, large stars, yet cool enough to be red in color. These are called **red giant** stars. The stars at the lower left corner of the graph are dim, small stars, hot enough to be white in color; these are called **white dwarf** stars. When this graph was first published, the question of why the stars grouped together in distinct regions of the graph immediately arose. Does the position on the graph tell us something about the stage of evolution that a star is in? The answer is that the position of a star on an H-R diagram does indeed tell us something about the life history of a star. As we shall see, those stars on the main sequence are stars that are in the prime of their life. Eventually, however, all things must die, including stars, and when a star begins to die, it moves off the main sequence to become perhaps a red giant, a white dwarf, or, in some instances, something truly exotic. In the next section, we shall begin our exploration of the life sequence of stars.

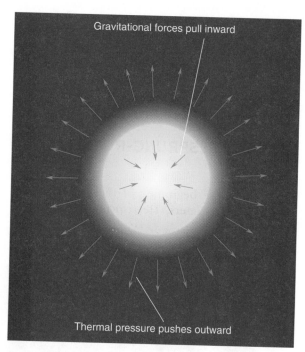

FIGURE 23.6 *Equilibrium in a star is a balance between the gravitational forces pulling matter inward and the radiation pressure pushing outward.*

CONCEPTUAL EXAMPLE 23.3   FIND THE SUN

Locate our star, the Sun, on the Hertzsprung-Russell diagram.

Solution The Sun would be close to the center of the main sequence, but slightly closer to the red dwarf end of the series. Once again, there seems to be nothing special or spectacular about our position of importance in the Universe.

23.4 LIFE OF A STAR

A star is a large, spherical mass of gases that is so hot that nuclear fusion reactions are occurring within the central regions. As described in the last chapter, tremendous gravitational pressures pull these gases inward, yet the star does not collapse. In fact, the outer layers of most stars are so diffuse and tenuous that they are less dense than Earth's atmosphere at sea level. Because stars do not collapse, some pressure must be pushing outward against the force of gravity. This outward pressure

is created by energetic photons and fast-moving subatomic particles released by nuclear reactions at the core. These particles and photons push against the outer layers of gas and prevent the gases from falling inward. The resultant density of a star is determined by two opposing forces: the gravitational force pulling in and thermal forces pushing out (Fig. 23.6). These forces leave a star like the Sun in a condition of equilibrium that can last for billions of years.

Over a period of many millions or billions of years, stars evolve, change, and die, and new ones are born. Throughout the life of a star, the gravitational and thermal forces oppose each other until, as we shall see, the star dies and undergoes radical change.

When a large, diffuse cloud of cold dust and gas is pulled together by gravitation, it forms the early stages of a star, known as a **protostar.** As the particles are pulled inward even faster, they eventually reach the extremely high temperatures required for nuclear fusion. At the onset of fusion, additional energy is released, and the gases move

Gigantic clouds of dust and gas like the Orion Nebula are regions where star formation occurs. (Courtesy NASA)

even more rapidly in all directions. These extremely energetic particles push outward against the gravitational force. Therefore, the gravitational force pulling inward is balanced by the thermal and radiation forces pushing outward. At equilibrium, an average-sized star has a dense core surrounded by a less dense envelope.

The outer region of a star is considerably cooler and less dense than the core. The surface temperature of the Sun, for example, is only about 6000 K, whereas the temperature in the core soars to 15,000,000 K. At the relatively cool outer temperatures, hydrogen nuclei do not collide forcefully enough to fuse, so no energy is produced in these regions. Within the dense core, however, temperatures are so high that hydrogen nuclei fuse to form helium, with the release of large amounts of energy (Fig. 23.7a).

The process described here happens to all stars regardless of their mass while they are living out their normal lives. Take particular note of the result once again. The fuel being consumed by the nuclear reactions is hydrogen, whereas the resultant ash is helium. Eventually, as in all normal processes, the fuel will all be consumed at the core. At this point, conditions change drastically, and the star enters the first stage of its death throes.

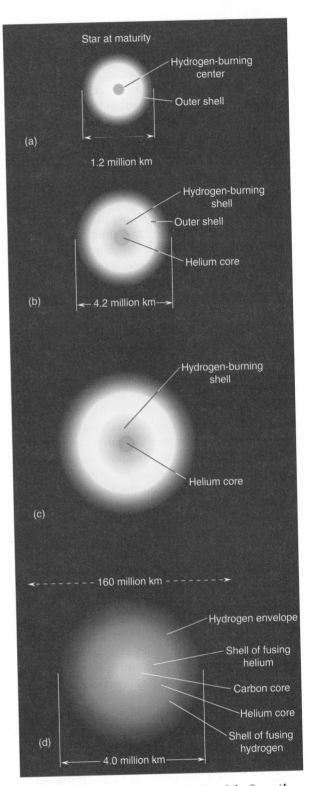

FIGURE 23.7 *Aging of a star the size of the Sun. (In all cases, the core region has been drawn larger than scale to show detail.)*

23.5 DEATH OF A STAR LIKE THE SUN

As a star grows older, increasing quantities of the ash, helium, accumulate at the core of the star, and the fuel, hydrogen, is depleted in the core. The helium nuclei do not fuse with each other at the temperature of an average mature star like the Sun. Recall that for fusion to occur, two nuclei must be pushed close together. A hydrogen nucleus contains only one proton, so to fuse two of them, the electrostatic repulsion of one proton for another must be overcome. Each helium nucleus, however, contains two protons, so for helium fusion to occur, a much greater repulsion must be overcome. As a result, the helium nuclei cannot fuse at this time. A star the size of the Sun has enough hydrogen fuel to last for about 10 billion years. After that time, the outer shell still contains large quantities of hydrogen, but the central core is mostly hot helium (Fig. 23.7b). The star now begins to behave quite differently than it did earlier in its life. Because little nuclear energy is produced in the central core, the core cools. When the core temperature decreases, the outward pressure that kept the star from falling inward also decreases, and the central regions start to shrink under gravitational forces. This gravitational contraction causes the core to stop cooling and begin to grow hotter. It seems a paradox that when the nuclear fire starts to diminish, the core should get hotter, but that is exactly what happens. To repeat,

when hydrogen fusion ends within the core, the equilibrium of the star is upset so that the central core is compressed by gravity and its temperature rises.

The hot shell of hydrogen around the helium core then fuses more rapidly. At this point, the star is releasing hundreds of times as much energy as it did when it was mature. The situation is analogous to a flash fire of hydrogen that is rapidly burning just outside the helium-rich core. The intense energy output now causes the outer parts of the star to expand (Fig. 23.7c). As this outer shell grows in size, it cools. In fact, its temperature rapidly reaches the stage at which the star appears red. When the Sun goes through this phase of its death throes, its outer shell will expand so that its surface will be somewhere between Earth and Mars (Fig. 23.8). To recap, the star has grown cool as it has expanded,

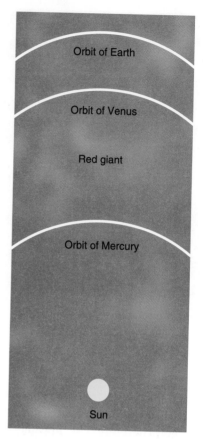

FIGURE 23.8   *The Sun, shown in its original position, will expand in the red giant stage, indicated by the red haze, to a size that it will engulf the Earth.*

and it has become enormous compared with its original size. From these two facts, a star in this stage of its life is given its name: a **red giant.**

Meanwhile, the inner core continues to contract under the influence of its gravitation. Because of this contraction, the core gets hotter and hotter until the critical temperature of 100 million degrees is reached. At this temperature, a new nuclear fusion reaction becomes possible. Helium nuclei begin to fuse together to form carbon, according to the following reaction sequence:

$$^4_2\text{He} + {}^4_2\text{He} + \text{energy} \longrightarrow {}^8_4\text{Be}$$

This first reaction indicates that two helium nuclei have fused to form beryllium. (Note that an input of energy is required to make this reaction occur.) The beryllium nuclei produced then undergo a fusion reaction with another helium nucleus as

$$\frac{4}{2}He + \frac{8}{4}Be \longrightarrow \frac{12}{6}C + energy$$

Thus, the end result of this chain of events is that helium is being burned and the ash that is now accumulating at the core of the star is carbon, $\frac{12}{6}C$. The onset of the helium fusion reactions initiates several drastic, rapid changes. After a few hundred thousand years of instability, the star shrinks and then enters a new, stable phase. The newly structured core is now composed of fusing helium. Gradually, as increasing quantities of helium fuse to form carbon, the carbon starts to accumulate in the core just as helium had done during the early life of the star (Fig. 23.7d). As the helium fuel is consumed, the nuclear fire diminishes once again and the carbon core contracts. This gravitational contraction causes the core to become hotter once again. At this point, the history of a star depends on its initial mass. If it is about the mass of the Sun, the temperature will not rise enough to initiate fusion of the carbon nuclei. The gravitational collapse provides enough thermal energy to blow a shell of gas away from the star. This expanding shell is called a planetary nebula (Fig. 23.9). Meanwhile, the core continues to contract. A star as massive as the Sun will eventually shrink so that its diameter will be approximately that of Earth. Such a shrunken star no longer produces energy of its own and glows solely from the residual heat produced during past eras. The star is no longer kept from collapsing by ther-

FIGURE 23.10 *The size of a white dwarf is not different from the size of the Earth. A white dwarf, however, contains about 300,000 times more mass than does the Earth.*

FIGURE 23.9 *The Ring Nebula.* (Courtesy NOAO)

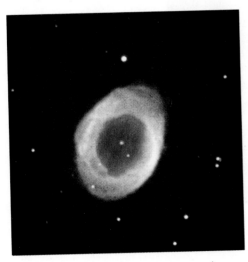

mal pressure. Rather, it has become so small that particles are squeezed together. Finally, a point is reached at which the particles resist further squeezing, and further contraction stops. At this stage, the star is still hot; in fact, it is white hot, and it is also small. Hence, such a star is called a **white dwarf** star. The size of a white dwarf is not very different from the size of Earth (Fig. 23.10). A white dwarf, however, contains about 300,000 times more mass than does Earth. Thus, the star is extremely dense; in fact, if you were to pick up a teaspoon of it, you would find that it would weigh about a ton.

White dwarf stars are extremely common in the Universe. Our own galaxy, the Milky Way, is estimated to contain several billion. Such a star will cool slowly over the course of tens of billions of years, but it will never change size again. No nuclear fire will cause it to expand, and the gravitational force is not strong enough to compress it further. Such a star will cool until it reaches its final stage as a **black dwarf** star. Eventually the Universe will be populated by trillions of black dwarf corpses, but none are seen now. Many astronomers believe that even if the sequence of events for the death of a star had occurred shortly after the formation of the Universe, there has not been sufficient time for a white dwarf to have cooled to the black dwarf stage.

23.6 DEATH OF MASSIVE STARS

Some stars do not die as gently as will the Sun. If a star has a mass greater than about four times the mass of the Sun, the slowdown of helium fusion and the contraction of the resulting carbon core does not terminate in a white dwarf. Instead, as the core contracts slightly, the fusion reactions accelerate, and the outer regions expand into a second red giant phase.

After some time, the gravitational contraction of the carbon core will be strong enough to heat the core to such extreme temperatures that new fu-

sion reactions start to occur. As an indication of what can happen, consider the following processes: (1) When the temperature of the core reaches 600 million K, carbon burning begins, and the ash produced consists primarily of magnesium and neon. (2) As the core continues to collapse, the temperature can soar to 1 billion K. At this temperature, neon burning begins, and the ash is largely oxygen and magnesium. (3) At a temperature of 1.5 billion K, oxygen burning begins, and the principal by-products of this reaction are sulfur, silicon, and phosphorus. (4) For a massive star, a final gravitational collapse can cause the temperature to reach 3 billion K. Here, silicon can burn, and the primary ash is a crucial one in the life of the star, iron. Iron is important to the future life of the star because when fusion occurs with iron, energy is *absorbed*. Thus, the fusion of iron nuclei *cools* the core. When

(a)

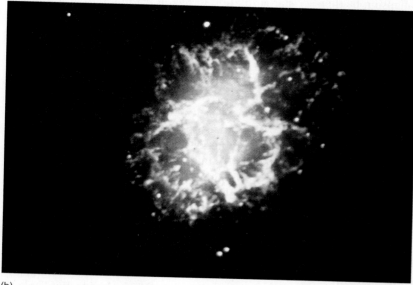

(b)

FIGURE 23.11   *(a) Direct photographs of a region of space during and after a supernova explosion. (b) The Crab Nebula in Taurus is the remains of a supernova of* A.D. *1054.*

[(a) Courtesy Lick Observatory; (b) Courtesy NOAO]

this happens, the thermal pressure that pushed the stellar gases outward is reduced, and the star begins to collapse under its own gravitation. This collapse releases large amounts of heat. In turn, the intense heat initiates a complex series of nuclear reactions that quickly lead to cataclysmic changes. Within a few seconds—a fantastically short period of time in the life of a star that is measured in terms of billions of years—the temperature reaches trillions of degrees, and the star explodes, hurling matter outward into space. This exploding star is called a **supernova** (Fig. 23.11). The word *nova* is translated to the English equivalent of *new*. Thus, the word implies that a super, or very bright, new star has appeared in the sky. This is an accurate way of describing such an explosion. Imagine the following scenario. You look at a particular region of the sky, and you see no star at all. There *is* a star there, but it is so dim that you cannot distinguish it with your naked eye. However, the next night when you come out and look at that particular region of space, there is a star there that is brighter than any near

it. What has happened, of course, is that the star at that location in space has undergone a supernova explosion, and you now are able to see it. To the ancients, however, it was as though a new star had miraculously appeared. This new star will not retain its brilliance for long. It will fade and disappear to the naked eye within a few weeks or months.

A supernova explosion is truly fantastic. For a brief period, a single star shines as brightly as hundreds of billions of stars and can emit as much energy as an entire galaxy. Four of these events have been seen in our galaxy in the last 1000 years, and one occurred in 1987 in a nearby galaxy, the Large Magellanic Cloud. A supernova explosion is violent enough to fragment many atomic nuclei, thereby shooting subatomic particles about in all directions. Shock waves—giant sonic booms—race through the atmosphere of the star. Under these conditions, many of the nuclear particles collide with sufficient energy both to fuse and to split apart. These processes form all of the known elements heavier than iron.

Thus, in studying the evolution of stars, scientists have learned how the natural elements were formed. We now believe that in the beginning of time, when the Universe was first formed, hydrogen was the predominant element. There was some helium, but there were no elements heavier than boron present. Within the cores of stars, hydrogen was converted to helium, helium to carbon, carbon to magnesium, and so forth, until many elements, including iron, became abundant. Then, in giant supernova explosions, the rest of the elements were formed by fission and fusion of atomic nuclei and nuclear fragments.

The death of a giant star is, in a way, only a beginning. A great number of nuclei are shot out into space. As countless supernova explosions occur in the course of time, the concentration of heavy elements slowly accumulates. Then, when conditions are favorable, the heavy elements, mixed with quantities of hydrogen and helium, condense into new stars and new solar systems. The Sun and the Earth are condensates of the remnants of supernova explosions that occurred billions of years ago. In fact, every object in the Universe that is composed of elements heavier than the lightest ones originated in some supernova explosion. Since you are made of such heavy elements, you too are the remnant of

Supernova in 1987A in the Large Magellanic Cloud appears as a bright object near the lower right area of this photograph. (Courtesy NOAO)

some long-dead stars. Life itself rises out of the debris of dying stars.

23.7 NEUTRON STARS, PULSARS, AND BLACK HOLES

In a supernova explosion, not all the original matter is shot out into space. Some of it (perhaps half) is left behind with the core, compressed into a tight sphere. In the 1930s, scientists tried to imagine the physical characteristics of the residual matter. If the sphere were more than 1.4 times as massive as the Sun, the gravitational forces would be extremely intense, so intense, in fact, that the star could not resist further compression in the same manner as a white dwarf. Instead, the theory said that in every atom the electrons would be squeezed into the nucleus where they would join together with protons to form neutrons as

electrons + protons $\longrightarrow$ neutrons

or

$$-_{-1}^{0}e + {}_{1}^{1}H \longrightarrow {}_{0}^{1}n$$

The neutrons would then resist further compression and remain tightly packed. This ball of compressed neutrons is called a **neutron star.** A neutron star would be so dense that a typical star would become about the size of a medium-sized city. It would have a radius of about 6 km. If one could pick up a tablespoon of this star, it would have about the same weight as Mount Everest (Fig. 23.12). Theory predicts that such stars should exist, but because of their small size, they would be very dim. Thus, astronomers had never seen one and had little hope of finding one in space until an accidental discovery in 1967.

In 1967, Jocelyn Bell (now Burnell) was a graduate student in astronomy at Cambridge University in England. Her doctoral dissertation was to be concerned with radio emissions from distant galaxies, and as a result, she constructed a telescope to detect frequencies within the radio range. In one part of the sky, she detected a radio signal that pulsed

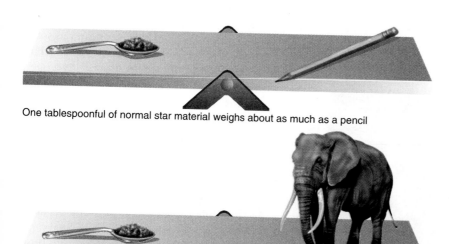

One tablespoonful of normal star material weighs about as much as a pencil

One tablespoonful of white dwarf material weighs about as much as an elephant

One tablespoonful of neutron star material weighs about as much as Mt. Everest

FIGURE 23.12 *Comparison of the densities of some stellar bodies.*

with a frequency of about one pulse every 1.33 seconds. When such a signal is amplified and fed into the speaker of a conventional radio, you hear a beep, beep, beep evenly spaced at one beep every 1.33 seconds. There are many radio emissions arriving at Earth from outer space, but what made these particularly unusual was that they were (1) sharp, (2) regular, and (3) spaced only a little over one second apart. At first, astronomers seriously considered that they might represent a signal from intelligent life, so for a short while they called the pulse signals LGM, for "little green men." But when Burnell found a second similar pulsating source in a different region of the sky, scientists ruled out the possibility of two widely divergent civilizations, each sending out bizarre signals in almost identical fashion and in a way that was quite wasteful of energy. Once it was established that the signals were of natural origin, the unknown pulsing sources were called **pulsars.** But naming the objects didn't help to explain them. How is it possible to study an ob-

ject that is many trillions of kilometers away and that emits no visible light, only radio signals?

The first step was to estimate the size of the mysterious object. Not all parts of an object in space are equidistant from Earth (Fig. 23.13). If a large sphere emits a sharp burst of radio signals from over its entire surface, some of the photons start off on their journey significantly closer to Earth than others and therefore arrive sooner. A person on Earth listening to the radio noise hears not a sharp beep but a more pronounced beeeeep because it takes a while for all the photons to arrive. Alternatively, a signal from a small sphere is much sharper, for the difference in distance is not nearly so great. The pulsar signals were unusually sharp, indicating that the source must be extremely small, perhaps 30 km in diameter. Was the source a starlike object only 30 km across? If so, the pulsar might be the long-searched-for neutron star.

The question remains: Why does a neutron star beep periodically? The following theory, sometimes

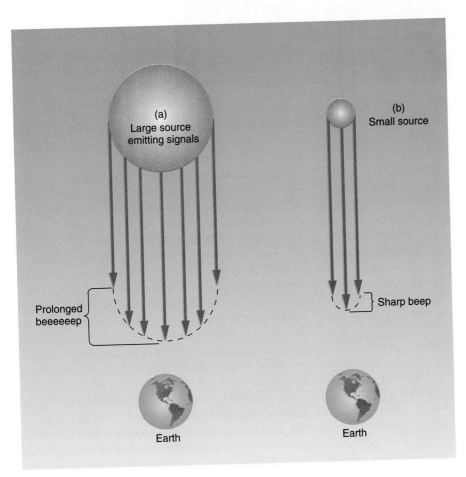

FIGURE 23.13 *A sharp signal from a larger sphere (a) arrives over a longer time interval than a sharp signal from a smaller sphere (b).*

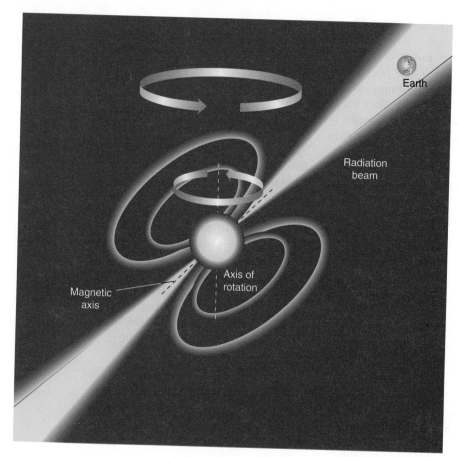

Earth

Radiation
beam

Axis of
rotation

Magnetic
axis

FIGURE 23.14   *The radiation beam of the pulsar is observed only when it sweeps across the Earth.*

called the **lighthouse theory,** explains how this occurs. Imagine yourself lost at sea. Safety comes when you see a lighthouse beam blinking on and off in the distance. In the presence of a fog, you could actually see the lighthouse beam sweeping around and coming toward you, but in the absence of fog, you will see the light only when it is pointing directly toward you. Thus, the conclusion that astronomers drew is that *a pulsar is a rotating neutron star.* Let's see how this works.

Stars, including the Sun, have magnetic fields, and if the Sun were shrunk to the size of a neutron star, the magnetic field would become quite intense. As we have seen, the magnetic field of the Earth traps charged particles in our atmosphere, and they make their presence known when they spill out over the poles to produce auroras. A somewhat similar occurrence takes place on a pulsar except the radiation is produced by accelerating and decelerating charges. Recall that X-rays are produced when electrons are decelerated on collision with a block of metal. Any charged particle radiates

energy when it is accelerated or decelerated, and such changes in velocity occur for particles falling into a pulsar. Additionally the beam from the star is concentrated in a narrow beam away from the star's poles. The north and south magnetic poles of the Earth are not located directly at the geographic north and south axis of rotation of the Earth, and neither are these locations the same on a pulsar. As shown in Figure 23.14, we see the radiation from the rotating neutron star only when the beam is directed toward us on Earth. Thus, there may be many pulsars in space that we will never see because their lighthouse beam never sweeps across our planet.

The crucial test for the pulsar as rotating-neutron-star theory came when radio astronomers looked at the Crab Nebula. The Crab is the remnant of a supernova explosion that occurred in the constellation of Taurus (the bull) in A.D. 1054 (see Fig. 23.11). The event was recorded by Chinese astronomers, who called it a "guest star." According to their records, the star was visible in the daytime

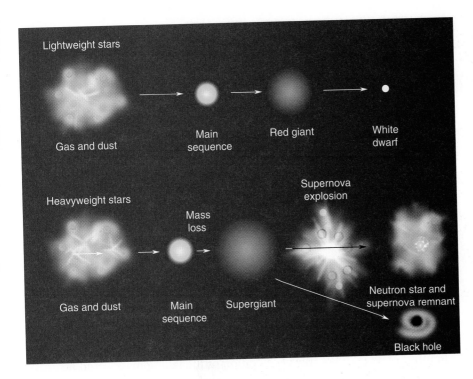

Lightweight stars

Gas and dust

Main sequence

Red giant

White dwarf

Heavyweight stars

Gas and dust

Main sequence

Mass loss

Supergiant

Supernova explosion

Neutron star and supernova remnant

Black hole

FIGURE 23.15   *A summary of the stages of evolution for stars of different masses.*

for 23 days and was bright enough to read by at night. The radio astronomers found a pulsar in the center of this debris. Thus, a pulsar was found precisely where a neutron star should be. This made it even more convincing that pulsars must be neutron stars that are emitting radiofrequency energy.

Figure 23.15 summarizes the death cycles of lightweight stars like the Sun and of heavyweight stars. The end result for the light star is a white dwarf and for the heavier, it is a supernova remnant with a neutron star at the center. Also notice, however, that there is an alternative pathway for even heavier stars, leading to a remnant that borders on science fiction, the **black hole.** Let us investigate these fantastic creations of the Universe.

What happens when a *very* large star dies? Astronomers believe that if the central core remaining after a supernova explosion is greater than three to five times the mass of the Sun, the neutrons are not able to resist the inward gravitational force. Then the star shrinks to a size much smaller than a neutron star and becomes a black hole. Such a collapse is impossible to imagine in earthly terms. A tremendous mass, perhaps a trillion, trillion, trillion kilograms of matter, shrinks smaller and smaller. If known laws of physics are obeyed, this faint star will contract to the size of a pinhead and then continue to shrink to the size of an atom, and

then even smaller. Eventually, it will collapse to a point of infinite density, called a singularity.

A black hole is so small and massive that it creates an extremely intense gravitational field. According to Einstein's theory of general relativity, a photon is affected by a gravitational field as though it had mass. As proof of this, light from nearby stars has been observed to be bent by the gravitational field of the Sun. If a star becomes dense enough, its gravitational field becomes so intense that photons cannot escape. If you throw a baseball up into the air here on Earth, it will return to you, as you well know. If you throw it such that when it leaves your hand, it is traveling at a speed of 7 miles per second, however, it will have achieved escape velocity, and it will travel upward never to return. On a black hole, the escape velocity exceeds 3×10^8 m/s, the speed of light. Since no light can escape such an object, it must always be invisible; hence the name "black hole." If you were to shine a flashlight beam, or a radar beam, or any kind of radiation at a black hole, the photons would simply be absorbed. The beam could never be reflected back to your eyes; therefore, you would never see the light again. It would be as if the beam just vanished into space. Similarly, if a spaceship flew too close to a black hole, it would be sucked in forever. No rocket engine could possibly be powerful enough

FIGURE 23.16 *An artist's conception of the disk of swirling gas that would develop around a black hole (right) as its gravity pulled matter off the companion supergiant star (left). The X-ray radiation would arise in the disk.*

to accelerate the ship back out, for no object can travel faster than the speed of light.

The search for a black hole has been even more difficult than the search for a neutron star. How do you find an object that is invisible, emitting no energy in any form? In short, how do you find a hole in space? Although it is theoretically impossible ever to see a black hole itself, it is possible to observe the effects of its gravitational field. Many of the stars in the Universe exist in pairs or small clusters. If two stars are close together, they will orbit about each other. What if one of the pair were a black hole? The visible one would appear to be orbiting around an invisible partner. Astronomers have studied several stars that appear to vary in this unusual manner. In at least one case, the invisible member of the pair has been shown to have more than five times the mass of the Sun. Since a normal star this massive would be visible at that distance, the invisible partner might be a black hole. Observation of such movement, however, is not complete proof that a black hole exists. The unseen partner could be some other object, such as a neutron star, although theory indicates that a neutron star could not be so massive.

Astronomers have calculated that if a supergiant star were mutually orbiting with a black hole, great masses of gas from the supergiant would be sucked into the black hole, to disappear forever from view (Fig. 23.16). As this matter started to fall into the hole, it would naturally accelerate, just as a meteor accelerates as it falls toward Earth. The gravitational field of a black hole is so intense that the acceleration of the falling mass would be great. As we saw in our discussion of X-rays in Chapter 12, when charged matter, such as a stellar plasma, is accelerated, radiation is produced. For accelerations as large as those that would be produced by a black hole, the photons produced would be in the X-ray range. These X-rays might then be detected here on Earth because they are produced far enough from the black hole that they can escape its gravitational tug. Just such a scenario seems to be being played out in the constellation of Cygnus (the swan). There an intense X-ray source, called Cygnus X-1, is pouring X-rays out into space. At present, this seems to be the best candidate for a black hole, but several others have also been nominated.

CONCEPTUAL EXAMPLE 23.4   ANGULAR MOMENTUM AND PULSARS

Use the law of conservation of angular momentum from the physics section to explain why a pulsar can rotate as rapidly as it does.

Solution The conservation of angular momentum predicts that an object rotates faster as its mass moves closer to the axis of rotation. Thus, if a star's core is initially very large and has only a very slow rotation, the rotational speed will increase greatly because it shrinks to such a small size after the supernova explosion.

SUMMARY

Atomic spectra are used to determine various physical characteristics of an object in space. The relative velocity of unknown objects can be determined by measuring the **Doppler shift.** The distance to a nearby star can be found by parallax measurements and to more distant ones by Hubble's law.

Stars are formed from condensing clouds of dust and gas. During maturity, hydrogen nuclei in the core of a star fuse to form helium, with the release of large amounts of energy. When the hydrogen fuel in the core is exhausted, and hydrogen fusion ends within the core,

the central mass is compressed by gravity, and the temperature rises. Fusion in the outer shell accelerates and the star expands to become a **red giant.** In the following stage, helium nuclei fuse to form carbon. In an average-size star, after helium fusion ends, the star sends off a **planetary nebula** and then shrinks to become a **white dwarf.** A larger star continues to undergo a sequence of fusion steps, then explodes to become a **supernova.** The remnant can contract to become a **neutron star** or, if the mass is great enough, a **black hole.**

EQUATIONS TO KNOW

$v = Hd$ (Hubble's law)

KEY WORDS

Spectrum
Absorption spectrum
Emission spectrum
Doppler effect
Apparent brightness

Intrinsic brightness
H-R diagram
Hubble's law
Protostar
Red giant

Planetary nebula
White dwarf
Black dwarf
Supernova

Neutron star
Pulsar
Lighthouse theory
Black hole

PROBLEMS AND CONCEPTUAL QUESTIONS

Problems requiring numerical work are identified with a blue number.

Studying the stars and the distance to a star

1. Explain how the chemical composition, the temperature, the velocity of a distant star, and the distance to a star can be determined.

2. Explain why the method of parallax would work only for measuring the distance to nearby stars.

3. The following facts are noted about two stars A and B. What do they tell us about these stars? A has a much greater red-shift than B. A is blue, and B is red. The same absorption spectrum is observed for both stars.

4. In recent years, astronomers have learned a great deal about our Universe by studying radio, X-ray, and other invisible radiation. Do you think that more information could be gained by building large microphones to detect the sounds of giant explosions in space? Defend your answer.

5. What happens to the (a) wavelength of light from a star moving away from us, (b) the frequency of the light from that star, and (c) the speed of the light from that star?

6. A particular star is observed to be light blue, and on careful observations, the hydrogen and helium lines predominate and these lines are red-shifted. Based on these pieces of information, what can you determine about that star?

7. As a review of the Doppler shift, consider dipping one tine of a vibrating tuning fork into a pool of water and then moving the fork through the water. If you were in front of the moving tuning fork, would you observe more waves, fewer waves, or the same number of waves per unit time than if you were behind the fork? Explain.

8. Many of the stars in our galaxy exhibit spectra that are blue-shifted. Does this invalidate Hubble's law? Explain.

9. In Example 23.1, we calculated the distance to a star based on the value of Hubble's constant of $15\frac{km/s}{Mly}$. In the last few years, disagreements have arisen over the exact value of this constant. Some astronomers believe that a more accurate value is $25\frac{km/s}{Mly}$. If this number should prove to be correct, what is the distance to this star?

10. Some galaxies and other strange objects have been detected as far away as 15 billion light-years. According to Hubble's law, what is the speed of recession of these objects?

Life and death of stars

11. Hydrogen burns in air according to the following equation:

$$2H_2 + O_2 \longrightarrow 2H_2O$$

Is this chemical combustion of hydrogen an important process within a star? Why or why not? Explain.

12. Why does nuclear burning always begin at the center of a star and not on its surface?

13. Almost 90 percent of the stars seen in the sky are on the main sequence. Why is the percentage so high?

14. What is the primary determining factor as to whether helium, carbon, and so forth can be burned in the interior of a star?

15. Explain why the density of the gases near the surface of the Sun is less than the density of the gases near the surface of the Earth, even though the gravitational force of the Sun is much greater.

16. Explain why the core of a star becomes hotter after the nuclear fusion reactions diminish.

17. What could you tell about the past history of a star if you knew that its core was composed primarily of carbon? Explain.

18. Compare and contrast a white dwarf with a red giant. Can a single star ever be both a red giant and a white dwarf during its lifetime? Explain.

19. Would you be likely to find life on planets orbiting around a star in which the primary fusion process is that of the "burning" of carbon? Explain.

20. Certain stars that lie above and below the plane of the Milky Way contain fewer heavy elements than our own Sun has. From this information alone, what can you tell about the history of these stars?

21. What is a supernova? Do all stars eventually explode as supernovae? Explain. What is a neutron star? Do all supernovae lead to the formation of neutron stars?

Pulsars and black holes

22. Explain why the pulsar signals detected by Jocelyn Bell Burnell could not have originated from (a) a star, (b) an unknown planet in our Solar System, (c) a distant galaxy, or (d) a large magnetic storm on a nearby star.

23. The first name given to pulsars was LGM (for little green men). Why was the idea discarded that the signal from pulsars originated from other civilizations trying to contact us?

24. If the Sun should suddenly become a black hole, how would the orbit of the Earth be affected?

25. Compare the blackness of a black hole to the color of the black ink used in a textbook.

26. Do you think that a black hole could be hidden in our Solar System? Explain.

27. Some astronomers believe that there are tiny black holes flying through space. Would such objects represent a hazard to a rocket ship traveling to distant stars? Could the crew of such a rocket detect a black hole well in advance and avoid an encounter? Explain.

28. Do you think that a heavy concentration of black holes in intergalactic space could be detected? What about a low concentration? Discuss.

29. Why would parallax measurements work better for a planet than a star?

30. Trace the steps in the death of a star in which the remnant is (a) a white dwarf, (b) a neutron star, (c) a black hole.

31. Use the lighthouse theory of pulsars to explain why there are probably many more of these in our Galaxy than we have detected or can detect.

32. Theory predicts that neutron stars are very smooth, with "mountains" about an inch high. It is also estimated that to climb such a mountain would require the energy expenditure of a full lifetime. Explain why so much energy would be required.

33. Why is the Doppler effect of no value for determining the speed of a star moving perpendicular to the line of sight from the Earth to that star?

ANSWERS TO SELECTED NUMERICAL PROBLEMS

9. 1200 MH
10. 225,000 km/s

CHAPTER 24

GALAXIES AND TIME

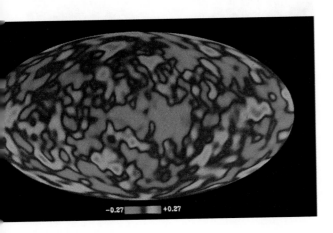

The temperature of the background radiation reaching the Earth from all directions in space gives us information about the early history of the Universe. (Courtesy Astronomical Society of the Pacific)

This chapter begins with a look at collections of stars, called galaxies, and moves from there to objects, called quasars, that are the most distant objects in the Universe. Finally, we conclude with an attempt to answer some questions that have been asked by all civilization: Where did we come from? and Where are we going?

24.1 MILKY WAY GALAXY

Structure of Our Galaxy

Because of light pollution in our environment caused by street lights, lighted parking lots, and so forth, modern people do not have the appreciation of the sky that their ancestors had. You should make an effort to get away from the city lights sometime to take a good look at the sky from a dark country meadow. From such a vantage point, you will note that most of space contains a diffuse scattering of stars that are spaced so far apart that you can see large expanses of black space between these points of light. In one region of the sky, a nearly continuous band of light, called the **Milky Way,** stretches across the sky, as shown in Figure 24.1. In 1610,

FIGURE 24.1 *The Milky Way stretches across the sky like a sheet of gauze.*

Galileo turned his telescope toward the Milky Way and found that this haze of light is produced by multitudes of stars. Approximately 150 years later, the astronomer Thomas Wright hypothesized (correctly) that the Earth is located in a disk-shaped group of stars. When we look toward the band of light, we are looking into the plane of the disk. What you are seeing are the billions of billions of stars, some relatively close to Earth and some farther away, that make up the disk of the Milky Way. A view perpendicular to this disk shows only a few scattered stars. This grouping of stars is now known to be only one of many such groupings throughout the Universe, and these bundles of stars are referred to as **galaxies.**

A galaxy is a large volume of space containing many billions of stars, dust, and gas all held together by their mutual gravitational attraction.

In 1785, the idea of the geocentric universe returned in a slightly different form, and again it was incorrect. In that year, the noted astronomer William Herschel and his sister Caroline did a telescopic star count in all directions from the Earth. As you might expect, they found that there were many more stars in the plane of the Milky Way than

FIGURE 24.3 *A globular cluster in the constellation Hercules.* (Palomar Observatory, California Institute of Technology)

FIGURE 24.2 *According to the star-counting procedure of the Herschels, our galaxy resembles this shape, with the Sun at the position indicated. This picture was later found to be incorrect.*

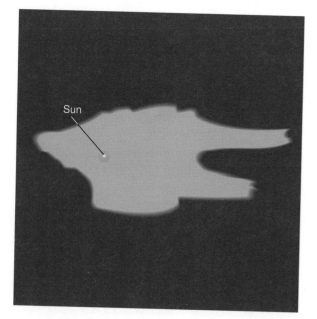

Sun

there were in directions away from this plane. The count revealed, however, that the number of stars were about the same in all directions when looking in the plane of the galaxy. They concluded that the shape of the galaxy is as shown in Figure 24.2, with the Sun at the position indicated. Thus, the old idea that there is something special about the Earth that puts it at the center of the Universe reared its head again. Much later it was discovered that the star count was inaccurate because of large clouds of interstellar dust that block our view of space. Herschel was actually counting only stars that were within about 2000 light-years of Earth because the light from stars more distant than this is blocked, in some directions, by obscuring clouds of dust and gas.

The first accurate estimate of the size of our galaxy and of our position within it came in 1917 from work done by Harlow Shapley, who at that time was studying **globular clusters.** Globular clusters are collections of hundreds of thousands of stars bundled together by gravity. Figure 24.3 shows a typical globular cluster. At the center of one of these clusters, the stars are packed together such that their average distance of separation is only about 0.5 light-year.

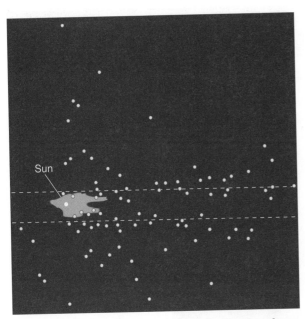

FIGURE 24.4 *A copy of the diagram by Shapley, showing the distribution of globular clusters in a plane perpendicular to the Milky Way. Also shown is the galaxy according to the star-counting procedure of the Herschels.*

Shapley's primary interest in these clusters was to find the distance to them and to plot their relative distribution in space. There were 93 known globular clusters at the time of Shapley's work, and they were easy to study because most of them did not lie in the plane of the galaxy where they would be obscured by the intervening dust clouds. He found that the clusters were grouped into a roughly spherical shape whose center was approximately 25,000 to 30,000 light-years away and in the direction of the constellation Sagittarius. Figure 24.4 shows a two-dimensional plot of his observations, along with the location of the galaxy according to the star-counting procedure of the Herschels. Based on these observations, Shapley proposed that the center of these globular clusters was actually the center of our galaxy. Many pieces of information since that time have confirmed Shapley's theory.

Additional information about the shape of our galaxy came about unintentionally from telescopic observations made by Charles Messier in 1781. Messier was an inveterate comet hunter who did not want the distractions of any other heavenly objects to veer him from his single-minded path. As he scanned the night sky in his relentless search for new comets, he would occasionally happen upon a nebulous-looking object in space that was so hazy and indistinct that he could not tell whether or not it was a comet. After a few nights of observation, he would find that the object did not move and thus could not be a comet. Because there were others interested in comet hunting, he made a detailed catalog of these so-called nebulae in which he listed 109 of them under such names as M41 (for Messier object number 41). His goal with this catalog was to help others by telling them, "Don't waste time looking here; this is not a comet." Fortunately, other astronomers in later years did not take Messier's advice because these objects he rejected as not being worth the time for observation turned out to be of extreme astronomical importance.

Many of these objects were later found to be star clusters or clouds of dust and gas in our own galaxy. However, some stood out as having a distinct spiral structure like that shown in Figure 24.5. The idea gradually grew that some of these groupings of stars might be "island universes" far removed from our own Milky Way galaxy. Others held that this idea was ridiculous. Shapley was one of these. He believed that they were simply unusually shaped groupings of stars that were a part of our galaxy, just as were the globular clusters that he had been

FIGURE 24.5 *The Whirlpool galaxy.* (Courtesy Naval Observatory)

FIGURE 24.6   *The Andromeda galaxy.* (Courtesy NOAO)

studying. The controversy was finally resolved by Edwin Hubble, who abandoned the study of law in favor of astronomy. He was able to find the distance
to one of these Messier objects, M31, in 1924 and
found it to be about 2 million light-years from
Earth. It is now known to be 2.25 million light-years
away. Thus, for the first time, we began to appreciate the enormous size and complexity of the Universe. M31 is a large spiral-shaped galaxy now called
the Andromeda galaxy (Fig. 24.6).

The observation that many of these island universes are spiral in shape gave astronomers the clue
to what shape our own Milky Way might have. We
now know that the Milky Way is a spiral galaxy

shaped as shown in Figure 24.7. The disk of our
galaxy is approximately 5000 light-years thick and
nearly 100,000 light-years in diameter. As seen from
the edge, the Milky Way looks like two fried eggs
placed back to back. The central bulge is called the
nucleus of the galaxy, and it is about 10,000 light-
years thick. Extending outward from this central
bulge are relatively diffuse spiral arms. The Sun lies
in one of the spiral arms of the Milky Way, some
30,000 light-years from the center. The concentration of stars in the galactic core is perhaps 1 million times greater than is found in the outer disk.
This core is like the globular clusters in the sense
that if you could visit a planet orbiting one of these
stars, you would never experience night, for the accumulated starlight would provide ample light all
the time. Stable solar systems, however, are not
likely to exist at all in this region, for collisions or
near collisions between stars would rip planets out
of their orbits fairly frequently, as judged by the expanses of astronomical time.

FIGURE 24.7   *An edgewise view of the Milky Way
galaxy.*

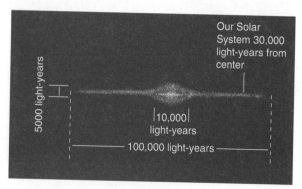

CONCEPTUAL EXAMPLE 24.1   THE CENTER OF OUR GALAXY

As we shall see in the next section, the center of
the Milky Way galaxy is a site of chaotic, turbulent
motion. Assuming that something really dramatic,
such as a large-scale explosion, is occurring there,
should we be concerned about our welfare?

Solution At our position on the outskirts of the galaxy, we would have to wait 30,000 years for the first wave of the explosion to reach us. (This assumes that the explosion travels outward at the speed of light.) Thus, we are in no immediate danger.

Core of the Milky Way

The core of the Milky Way lies in the direction of the constellation Sagittarius and is shrouded by dust, so it must be observed by nonoptical telescopes. It is now known that the core is much denser than the disk and is the site of much greater agitation and turmoil than the relatively serene spiral arms. Hydrogen clouds in the core are undergoing turbulent, chaotic motion at extremely high speeds. Also, two jets of hydrogen have been detected hurtling out from the center toward us and away from us at speeds of the order of 100 km/s. It is not as easy to determine the motion of a jet moving perpendicular to our line of sight from Earth, but it is probable that the hydrogen outflow is part of a shell of gas blown outward from the center. Finally, energetic gamma rays are streaming out from a small, unseen object in the center of all this activity.

The most widely accepted explanation of these findings is both fantastic and conjectural. It is thought that the only source capable of producing the tremendous amounts of energy detected is gravity. The present theory states that a gigantic black hole is located at the center of the galaxy. The energy in the form of gamma rays is released as matter spirals into the black hole in a process similar to that discussed in Chapter 23. It has been found that there are stars orbiting the galactic center at high speeds, which would be the case if they were orbiting about a massive black hole. In fact, their motion indicates that the black hole may have a mass of about 5 million times the mass of the Sun.

24.2 GALAXIES AND CLUSTERS OF GALAXIES

As mentioned earlier, one of the pioneers who provided us with much of our present information about the galaxies was Edwin Hubble. He found that galaxies occur in four basic shapes: spiral, barred spiral, elliptical, and irregular.

The Milky Way is an example of a **spiral galaxy,** as is the Andromeda galaxy, which is the closest large galaxy to us. Spirals all have the basic shapes shown in Figure 24.8a and b, but they differ in the size of the central bulge and the winding of the spiral arms. Spirals range between the extremes of those having a large nucleus with tightly wound spiral arms to those with a small nucleus and loosely wound spiral arms.

FIGURE 24.8 *(a) An "edge" view of a spiral galaxy.* (Courtesy NASA) *(b) A "head-on" view of a spiral galaxy.* (Courtesy Anglo-Australian Telescope Board)

(a)

(b)

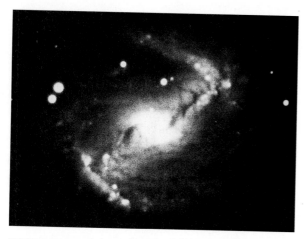

FIGURE 24.9   *A barred spiral in the constellation Canes Venatici.* (Courtesy NOAO)

An example of a **barred spiral** is shown in Figure 24.9. These are characterized by a bar that runs through the nucleus and from the ends of which the spiral arms emerge. These galaxies range over the same gamut as do the spirals, from those with a large nucleus and closely wound arms to a small nucleus with loosely wound arms.

An **elliptical galaxy** is shown in Figure 24.10. These are named for the shape, and as the figure shows, they have no spiral arms. These ellipticals are found in shapes ranging from the almost spher-

FIGURE 24.10   *A giant elliptical galaxy in the constellation Virgo.* (Courtesy NOAO)

ical to the highly flattened elliptical. These galaxies have almost no interstellar dust and gas, and perhaps because dust clouds are the maternity wards in space for star formation, there are no young stars found in ellipticals. The largest galaxies found in the Universe are elliptical in shape, as are the smallest. Giant ellipticals are relatively rare, but the smaller, dwarf ellipticals, containing only a few million stars, are numerous.

Irregular galaxies, like those shown in Figure 24.11, have no obvious geometrical shape. The two galaxies closest to us are irregulars. They are the Large Magellanic Cloud and the Small Magellanic Cloud. As their name implies, they were discovered by the crew of Magellan on his around-the-world voyage from 1519 to 1522. These galaxies cannot be seen from the Northern Hemisphere.

Some obvious questions always arise at this point concerning the evolution of galaxies. Do galaxies begin as irregulars, evolve into spirals, and finally, after many revolutions, wind their arms in on themselves to become ellipticals? Or does the progression move in some other direction, such as from irregular to elliptical, which then throw off spiral arms? Let us examine some features of these possible scenarios. First, elliptical galaxies do not evolve into spirals because it is well known that spiral galaxies contain large amounts of gas and dust, and that elliptical galaxies are virtually devoid of both. Any dust and gas that may have been present initially in an elliptical has already been used up in the process of star formation. But what about the other direction—from spiral to elliptical? This evolutionary direction seems to be precluded by the laws of physics. Via rotation, an elliptical or spherical body could evolve into a spiral disk, but the laws do not allow a disk to coalesce into a symmetrical elliptical shape.

One theory of the formation of galaxies states that ellipticals formed early in the history of the Universe from nebulae that were densely packed with dust and gas. In such a dense cloud, stars would have formed rapidly, and there would not have been enough time for the rotation of the cloud to produce a disk-shaped structure. The end result would have been an elliptical collection of stars that do not interact strongly with one another. If the cloud had been less dense, star formation would have taken place at a more leisurely pace, and there would have been sufficient time for the rotation of the cloud to produce a flattened disk shape.

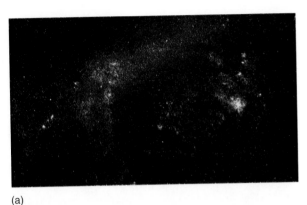 

(a)   (b)

FIGURE 24.11   *(a) The Large Magellanic Cloud and (b) the Small Magellanic Cloud.*
(Courtesy NOAO)

A more recent idea in competition with this theory states that early in the history of the Universe spiral galaxies were first to form, and then ellipticals formed as spirals coalesced or merged. The lack of dust and gas in ellipticals could be explained because the merger of spirals would produce significant turbulence and distortion in the colliding galaxies, and this would enhance star production, leading to the cleaning out of the dust and gas. Another feature of the Universe that supports this idea is that in regions of space where galaxies are packed closely together, most of them are elliptical, a fact that seems reasonable because collisions between galaxies would have been frequent. In regions where the spacing between galaxies is greater than average, most of the galaxies are spirals.

Approximately 1 billion galaxies can be observed from Earth. When the distribution of galaxies was mapped, astronomers learned that they exist in **clusters.** Thus, just as there are galaxies of stars, there are also galaxies of galaxies. For example, the Milky Way, Andromeda, the Magellanic Clouds, and about another two dozen small ellipticals are bound into a cluster that is called the **Local Group.** These clusters range in size and number of galaxies from relatively sparse ones, like ours, to giant clusters that contain more than 10,000 galaxies. This is not the end of the story, however. Clusters also bind together to form **superclusters.** Our local group is part of a supercluster that also contains the Virgo cluster and a few other smaller clusters. Between these superclusters exist huge empty regions completely devoid of galaxies.

Clusters of galaxies have been in the forefront of astronomical investigations because of what they may have to say about the future of the Universe.

The question is, why do galaxies stay bound together in clusters? It is well known that galaxies have random velocities through space and thus should be capable of freeing themselves from the tenuous hold of their neighbors. Yet, they do not. They stay bound together as stable configurations. Why? Since it is known that they do stay bound together, one can calculate how much mass the cluster would have to contain to keep all of its constituents bound together by gravity. When this calculation is performed for the Virgo cluster, it is found that there should be 50 times more mass present than there actually is. This deficiency is called the **missing-mass problem,** and because we cannot see this mass, it is referred to as **dark-matter.** This matter is in a form too cool to emit enough radiation for us to detect. As we shall see in Section 24.4, the presence or absence of this dark-matter has profound implications for the fate of our Universe.

24.3 ENERGETIC GALAXIES AND QUASARS

Recall that the cores of many spiral galaxies exhibit evidence of explosions or unusually powerful sources of energy. Even more energetic reactions have been recorded in other galaxy-sized objects. In 1951, astronomers discovered a distant object that glowed only dimly in the visible region of the spectrum but was an intense emitter of energy at radio frequencies. More than a million times as much radio-frequency energy was radiated from this source as is emitted by the entire Milky Way galaxy. This unusual object appeared to be a galaxy-sized collection of stars and thus was called a radio galaxy. Since that time, many other similar objects have been detected. Pictures of some radio galaxies show large lobes of gas that seem to be flying away from the central core. Do these lobes represent material ejected from the center of the galaxy by a giant explosion? Perhaps. If so, the intense radio signals are being emitted by electrons that were accelerated by shock waves or by intense magnetic and electric fields generated by the explosion.

As the study of the Universe has progressed during the past few decades, the list of unusual and highly energetic objects has grown. The most energetic objects discovered so far are called **quasars,** short for quasi-stellar radio sources. Quasars exhibit tremendous red shifts, far greater than any other objects yet observed in the Universe. This indicates that they are moving away from us at great speeds. Indeed, the fastest quasar is traveling at 92 percent of the speed of light. According to Hubble's law, these tremendous speeds mean that they must be far away, on the order of several billion light-years from Earth. Quasars emit tremendous quantities of energy, perhaps 100 times as much as the largest galaxies known. Despite their energy output, quasars are quite small; most are less than 1 light-year in diameter (recall that the Milky Way galaxy is 100,000 light-years in diameter), and one has been found that is as small as 1 light-week in diameter.

Often when astronomers look at objects in space, they say that they are looking backward in time. We have touched on the reason for this earlier, but let us reconsider it for a moment and see what it has to say about quasars. Light from the Sun takes about 8 minutes to reach the Earth. Thus, when we look at the Sun, we see it not as it is right at this moment, but as it was 8 minutes ago. Now, consider the Andromeda Galaxy, which is 2.25 million light-years from us. When we focus a telescope on this galaxy, we see it not as it is now, but as it was 2.25 million years ago. Thus, when you think about quasars, it is important to remember that the light we see from them is billions of years old. In fact, the most distant quasar is about 14 billion light-years from Earth. This distance is one piece of evidence that is used by astronomers to determine the age of the Universe. As a result, the Universe is now determined to be between 14 and 20 billion years old.

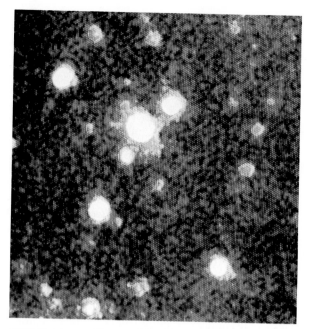

FIGURE 24.12 *Quasar 3C 275.1, the first to be found at the center of a galactic cluster, appears as the brightest object near the center of this pseudocolor image. The quasar nucleus is surrounded by a rotating elliptical gas cloud. The quasar is believed to be some 7 billion light-years away, and the light that formed this image left the quasar more than 2 billion years before our Solar System was formed.* (Courtesy NOAO)

Quasars are a window to the past, a view of what the Universe was like long ago. Extending this line of reasoning, quasars might be young galaxies in formation and might provide a picture of the Milky Way as it was long, long ago. Many astronomers believe that a quasar is the core of an evolving galaxy. To support this hypothesis, the best photographs of quasars indicate that at least a few of them actually lie in the center of a galaxy, as shown in Figure 24.12. If stars surround other quasars, they are hard to see because of their great distance.

One theory states that the core of a quasar is a massive black hole, and the emissions are coming from material that is accelerated by this intense source of gravitational energy. If this idea is correct, it could provide both an explanation for the energy in a quasar and a model for the formation of our own galaxy, but at present the theories are still highly speculative and our view of the Universe may change as more data become available.

24.4 COSMOLOGY—A STUDY OF THE BEGINNING AND THE END OF TIME

During our brief look at astronomy, we have studied many of the celestial bodies and some of the physics of the events that occur within our Solar System and the regions beyond. The knowledge that we have gained does not detract from the mystery and wonder, for one can hardly contemplate the objects in space and the distances between them without being filled with awe. Perhaps the ultimate mystery, the great question that never ceases to boggle the mind, is: When did time and space begin, and when and how will it end? **Cosmology** is the study of the origin, structure, and fate of the Universe. To contemplate cosmological questions, you must stretch your imagination even farther than you have done previously in this chapter and attempt to contemplate some amazing concepts of space, time, and nothingness.

Big Bang Theory

Almost all astronomers now accept an explanation of the origin of Universe called the **Big Bang theory.** According to this theory, all the matter and energy in the Universe were once compressed into an infinite density. Then a cataclysmic event, the Big Bang, occurred about 15 to 20 billion years ago that exploded this apart and marked the beginning of the Universe and the start of time. From that time down to the present, the Universe has been expanding, and as matter was formed and grouped together to form galaxies, it too shared in the expansion.

An observer on Earth sees everything moving outward away from us, and because of this, it is tempting to think of ourselves as being at the center of the Universe. But there is no center. Think of a baker making a loaf of raisin bread, as shown in Figure 24.13. Imagine that each raisin is a galaxy and that the dough represents the empty space between them. As the dough rises, each raisin moves farther away from every other one. Any observer on any raisin always sees all the other raisins moving away, and thus no one can truly be considered to be at the center. This conclusion is somehow disquieting, and you may say, "If everything, all matter, was once in one spot, then that spot had to be the center." But, to repeat, that type of reasoning is not correct; there is no center. The formation of the Universe was the creation of space itself; before the beginning, there was nothing at all—not even emptiness, not even time.

Let us return once again to the original cosmic egg of infinite density that is believed to have con-

FIGURE 24.13 *From every raisin in a raisin cake, every other raisin seems to be moving away from you at a speed that depends on its distance from you. This leads to a relation like the Hubble law between the speed and the distance. Note also that each raisin would be at the center of the expansion measured from its own position, yet the cake is expanding uniformly. For a better analogy with the Universe, consider an infinite cake; clearly, there is no center to its expansion.*

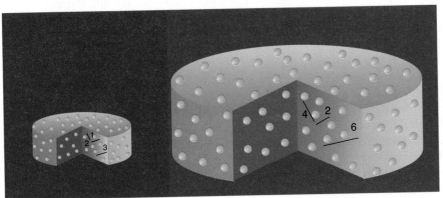

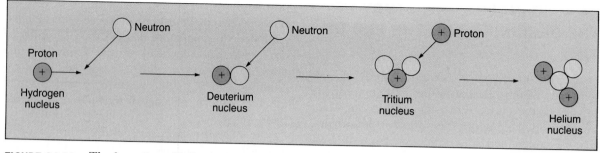

FIGURE 24.14   *The formation of helium nuclei in the primordial Universe. This process predominated when the Universe was of the order of several minutes old. Approximately 25 percent (by mass) of the protons and neutrons of the Universe combined to form helium.*

tained all the matter of the Universe. It is meaningless to try to describe this collection of all matter compressed into no space because there is no frame of reference to start with. If you think about this egg, you cannot help but feel its mystery. Such feelings may be expressed in religious or other personal ways, but they are beyond science. Science can deal only with events that started after the initial explosion of the Universe.

Matter and energy, in the sense in which we think of them, came into existence with this explosion. The evolution of the Universe from this initial fireball down to the present time is believed to have taken place in the following steps. The first 10^{-43} s of the Universe is called the ultra-hot epoch with temperatures of the order of 10^{32} K. By 10^{-35} s after the bang, the temperature had dropped to 10^{29} K; this is the hot epoch. At about 0.1 s after the start of time, the temperature of the Universe was about 100 billion K. Certainly, no molecules or even atomic nuclei could exist at these temperatures, and the Universe consisted of a uniform sea of photons and subatomic particles, such as electrons, positrons, and neutrinos. Approximately 1 s after the Big Bang, the Universe cooled to 10 billion K, and particles joined together to form protons and neutrons. Within a few minutes, temperatures dropped sufficiently so that fusion reactions could occur, and the simplest atomic nuclei were formed. About 25 percent of the mass of the Universe combined to form helium, as shown in Figure 24.14. Nearly all the rest remained as hydrogen, which even to this day composes about 75 percent by mass of all known matter. After about 1 million years, the temperature was low enough that

electrons and simple nuclei could condense to form atoms. At this point, photons left over from the Big Bang itself were no longer absorbed efficiently, for remember that atoms can absorb light only at certain specific frequencies. These primordial photons dispersed unhindered into the void, and as we shall soon see are still around to be detected.

As matter expanded farther, it separated into galaxy-sized agglomerations held together by mutual gravitation. These galaxies continued to fly apart, as part of the expanding Universe. Meanwhile, within each galaxy, matter collected into stars, and the stars slowly aged. Supernova explosions ripped through dying stars to produce heavier elements, and, in at least one case, a solar system was formed and life evolved. This evolution of the Universe is pictured in Figure 24.15.

Cosmic Background Radiation

About 1 million years after the Big Bang, the temperature of the Universe had fallen to about 3000 K, low enough for electrons and simple nuclei to join and form atoms. At this point, photons left over from the Big Bang could no longer be absorbed because atoms can absorb light only at certain specific frequencies. Thus, this leftover radiation from the Big Bang should still be in existence and should be detectable. The existence of this background radiation was first postulated in the 1940s, and many of the details of its characteristics were predicted at that time. One of its features was that it should be coming toward the Earth uniformly from all direc-

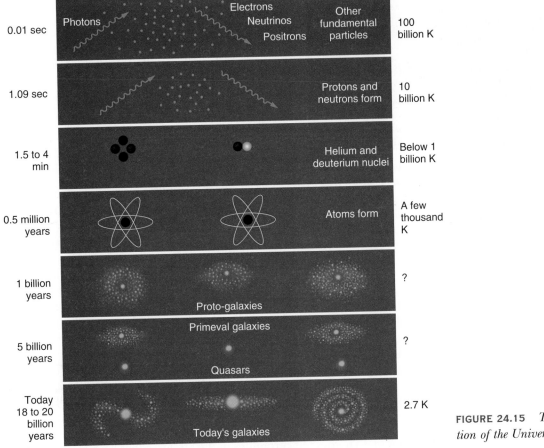

0.01 sec	Photons Electrons Neutrinos Positrons Other fundamental particles	100 billion K
1.09 sec	Protons and neutrons form	10 billion K
1.5 to 4 min	Helium and deuterium nuclei	Below 1 billion K
0.5 million years	Atoms form	A few thousand K
1 billion years	Proto-galaxies	?
5 billion years	Primeval galaxies Quasars	?
Today 18 to 20 billion years	Today's galaxies	2.7 K

FIGURE 24.15 *The evolution of the Universe.*

tions. A second prediction was that the radiation should be characteristic of the radiation emitted by an object that is at a temperature of 3 K. This occurs because this background radiation was formed early in the history of the Universe and is now coming in to the Earth from great distances. As a result, it has been red-shifted to a great extent, down from a frequency characteristic of a 3000 K object to that characteristic of a 3 K object.

These predictions were first made in 1948, but at that time, there was no way to detect the weak signal that this radiation would produce. In 1965, however, two engineers at Bell Laboratories, Penzias and Wilson, found their way into history by accidentally discovering it. These two scientists were interested in improving radio communication by way of microwave transmissions, but they were frustrated by noisy interference that seemed to be coming in uniformly from everywhere in space. In 1978,

they were awarded a Nobel Prize for identifying this interference as being caused by the cosmic background radiation.

The discovery of the cosmic background radiation provides strong confirmation of the Big Bang theory of the formation of the Universe and indeed is largely responsible for the wide acceptance of this theory. It also produces another problem, however. Scientists pondered the problem of the uniformity of the radiation. It was held that there would have to be a slight fluctuation in this background for objects such as galaxies and quasars to form. In 1989, NASA launched a satellite called COBE (pronounced KOH-bee), for **Co**smic **B**ackground **Ex**plorer, to study this radiation in great detail. In 1992, George Smoot of the Lawrence Berkeley Laboratory, based on the data collected, found that the background was not perfectly uniform. Instead, there were irregularities of only 0.00003 K in the

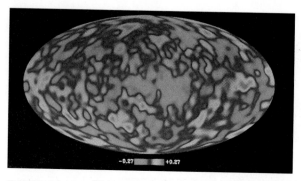

FIGURE 24.16   *This map of the sky shows the temperature of the incoming background radiation. If the temperature were truly the same in all directions, the sky would be shown as a uniform color. The mottled appearance is evidence of different temperatures of the radiation in certain directions.* (Courtesy NASA)

background, as shown by the mottled appearance of the temperature map of the sky shown in Figure 24.16. It is these small temperature variations that provided nucleation sites for the formation of the galaxies and other objects we now see in the sky.

The End of the Universe

Although scientists feel fairly confident that they know something of what happened in the beginning of time, they feel less sure about what will happen in the future. One possibility is that the Universe will expand forever. Galaxies will continue to disperse, and the space between them will gradually grow larger and emptier, as shown in Figure 24.17. Simultaneously, fuel for stellar fusion reactions will be consumed, so the energy output within each galaxy will slowly diminish. Eventually, space will become emptier and emptier, and only small bits of cold matter in a frozen void will remain.

Another possibility is that the expansion will eventually come to a halt. Although the galaxies are flying away from each other, they are also being pulled inward by mutual gravitation. If the gravitational attraction is sufficiently large, it will cause the galaxies to decelerate gradually to a standstill, then accelerate back inward, falling closer and closer until they all join together. All the galaxies, all the stars, all matter, and all space will collapse again into the original cosmic egg. This giant agglomeration of matter containing all the mass in the Uni-

verse would be exactly identical to the mass that was present at the beginning of our current Universe. This dense matter might then "explode" again and begin expanding to form a new Universe. This scenario is called the **oscillating Universe,** which alternately expands and contracts indefinitely. In this version, the death of one universe would simultaneously represent the birth of a new one. Thus, the oscillating universe cosmology predicts innumerable beginnings and endings in an infinite continuum, as shown in Figure 24.17b.

Will our Universe slowly expand into space and gradually fade into oblivion, or will it be pulled back together by its own gravitation so that all matter becomes reunified into a dense volume? Astronomers trying to answer this question have attempted to measure the density of the entire Universe. They reason that if they knew the density of the system, they could calculate its internal gravitation. Then, knowing the momentum of the receding galaxies, it would be possible to calculate whether they would be slowed down and drawn in again or whether they would continue to move outward forever. If space contains approximately one hydrogen atom in every cubic meter, that is enough mass to stop the expansion and cause the Universe to collapse.

It is difficult to calculate the amount of mass in the Universe, and there are a great many uncertainties in the attempts that have been made. To understand the nature of the problem, let us start by focusing on something that we can observe today rather than on an event that may or may not occur sometime in the distant future. Throughout the Universe, there are many large groups, or clusters, of galaxies. Each cluster is held together by the mutual gravitation of the component galaxies. However, if astronomers calculate the mass of each galaxy as measured by the estimated number of stars in it, they come to a disturbing conclusion. The galactic masses calculated in this fashion account for only about 10 percent of the mass required to hold the galaxies together. Ninety percent of the mass is unaccounted for. The same conclusion is reached from the study of the motion of stars within a single galaxy. Stars revolve around the galactic core, and this motion obeys Newton's laws. The galactic revolutions, however, do not follow the known laws of physics unless the galaxies are much more massive than we measure them to be. In observation after observation, the same conclusion has been reached: More than 90 percent of

(a) Forever-expanding cosmology

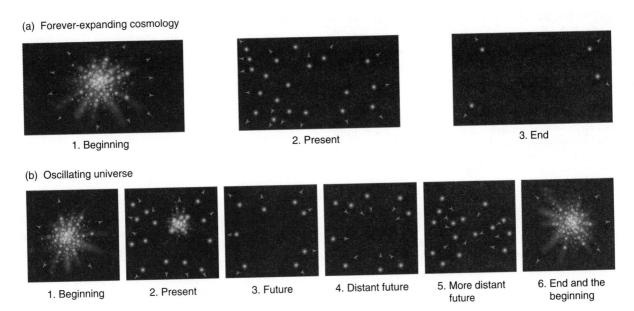

(b) Oscillating universe

FIGURE 24.17   *The formation and ultimate fate of the Universe. (a) A forever-expanding cosmology. (b) An oscillating Universe.*

the Universe is not visibly accounted for. Scientists have added the masses of interstellar and intergalactic dust and stray hydrogen atoms, but still there is not nearly enough mass to cause the Universe to behave the way it does.

What about black holes? Could they provide the missing mass? The answer appears to be no. To understand this conclusion, let us return to our study of the origin of the Universe. Recall that when the Universe was young—on the order of several minutes old—there were large quantities of deuterium nuclei floating about in a sea of protons and neutrons. If deuterium is struck first by a neutron and then by a proton, it will react to form helium. The rate of the reaction is partially dependent on the density of the various particles. If the density of deuterium, protons, and neutrons were high, all the deuterium would have reacted to form helium; if the density were low, little helium would have been formed. By measuring the ratio of helium to deuterium observed today, it is possible to calculate the density of the nuclear particles in the primordial Universe. This calculation indicates that there was not enough of these particles to form enough black holes to provide the missing mass.

Another consideration is neutrinos. Physicists estimate that throughout the Universe there are 10 billion neutrinos for every atom. If each neutrino had even a tiny mass, all of them together would add up to quite a lot of mass. But recall from Chapter 13 that scientists are not sure whether or not neutrinos have any mass at all.

The search for the missing mass, often called the "dark-matter," continues. Some physicists postulate that there are other subatomic particles that may exist but have never been detected that may make up some of the missing mass. But this line of reasoning is highly conjectural, to say the least.

In short, no one knows how massive the Universe is and whether it will expand forever or oscillate.

EPILOGUE

In a search for the origin and conclusion of the Universe, we have thought about events that must have occurred many billions of years ago or that will occur many billions of years from now. We have tried to extend our vision across expanses of space, and in our imagination we have traveled to the edge of what is known and beyond.

Let us ask the final question: Have we gone beyond the bounds of the merely unknown and perhaps encroached on the unknowable? The Big Bang cosmologies assume a beginning—a small, dense, hot proto-universe. Can science address itself to the question of how that Universe originally appeared or where it came from? If there were nothingness before, what was nothingness like or even what is meant by "before"? At the present, there is no experiment, not even a thought experiment, that begins to bear on this question. It is inherently unanswerable. If you think about it long, it is difficult not to wonder: If our Universe is a unique "one-shot" affair, why are we so special? If we live in an oscillating Universe, the search for a beginning may be meaningless, for perhaps there was none; perhaps matter has been expanding and contracting for all of past time and will continue to oscillate forever.

SUMMARY

A **galaxy** is a large volume of space containing many billions of stars, all held together by their mutual gravitational attraction. Our Sun lies in the plane of the disk of the **Milky Way galaxy.** There are large, diffuse clouds of dust and gas between the stars in our galaxy. There is evidence that there has been a gigantic explosion at the core of the galaxy. One theory is that the central core exploded in a colossal supernova-type explosion, and a giant black hole was left behind. There are four basic types of galaxies. They are **spiral** galaxies like our Milky Way, **elliptical** galaxies, **barred spirals,** and **irregulars.** Galaxies also form collections held together by gravitational attraction. The Milky Way is part of the **Local Group,** an assemblage of about 28 galaxies.

Scientists believe that **quasars** are objects that are far away, emit perhaps 100 times as much energy as an entire galaxy, and are quite small compared with the size of a normal galaxy. Quasars may be young galaxies in the process of evolution. (Remember, the light we see from them is several billion years old.)

The **Big Bang theory** states that about 18 to 20 billion years ago, the Universe was confined to a small globule of pure energy. This exploded to form the Universe itself. The original cloud of matter eventually cooled and separated into galaxies, stars, and planets, and in at least one place, living creatures came into being. Our Universe is expanding, and it is not known whether this expansion will continue or whether the galaxies will eventually fall back together. If so, another Big Bang could occur, creating another universe. This latter concept is called the **oscillating Universe** theory.

KEY WORDS

Galaxy	Elliptical galaxy	Local Group	Big Bang theory
Globular clusters	Barred spiral galaxies	Radio galaxy	Oscillating Universe theory
Milky Way	Irregular galaxies	Quasar	Dark-matter
Spiral galaxy	Clusters of galaxies	Cosmology	

CONCEPTUAL QUESTIONS

Milky Way galaxy

1. In what way were our early perceptions of our place in the galaxy similar to those of the geocentric theory of the Solar System?
2. The view of the Milky Way in the Northern Hemisphere is far more spectacular in July than in December. Why is this true?
3. How would our view of the Milky Way change if the Sun were (a) on the extreme edge of the galaxy, (b) at the center of the galaxy?
4. If a massive black hole has evolved at the center of our galaxy, are we in danger of being pulled into it?

5. What evidence indicates that the center of our galaxy was once the scene of a violent explosion?
6. Explain how globular clusters were important in determining the position of the Sun in the galaxy.
7. If a gigantic explosion should occur at the center of our galaxy at this moment, should we be concerned about the effects that this might have on us during our lifetime? How long would an explosion with its front traveling at the speed of light take to reach us?
8. It is often said that if you lived on a planet circling a star embedded in a globular cluster, your

knowledge of astronomy would be limited. Explain this statement.

Galaxies and clusters of galaxies

9. What is the closest spiral galaxy to the Milky Way? What is the closest irregular galaxy?

10. Why did the text not give a more comprehensive treatment of the evolution of galaxies?

Energetic galaxies and quasars

11. Would you consider a quasar to be similar to a star, a galaxy, or neither? Discuss.

12. How can a quasar be distinguished from a dim star?

13. It is often said that the first way in which matter condensed on a large scale was in the form of quasars. Defend this statement.

14. Quasars are all very far away. Using the current theory of quasars, explain why they are not found (a) in our Solar System, (b) in our galaxy, (c) approximately the same distance from Earth as the closest galaxies.

15. Based on your knowledge of physics, devise an explanation for how the enormous energy of a quasar could be produced.

16. Why is it often said that quasars are "windows on time"?

Cosmology

17. Arrange the following different environments in the order of increasing densities: (a) intergalactic space, (b) the core of the Sun, (c) the corona of the Sun, (d) the region of space between planets in our Solar System, (e) galactic space, (f) the Earth's atmosphere at sea level.

18. What factor(s) will determine whether the Universe will expand forever or eventually begin to contract?

19. According to the Big Bang theory, the Universe was once a homogeneous mass of photons and atomic particles. Trace the evolution of hydrogen, helium, and the heavier elements.

20. What evidence supports the statement that our Universe is very old?

21. Some Big Bang cosmologists predict ours is an ever-expanding Universe, whereas others predict an oscillating Universe. Discuss the similarities and differences between the two ideas. What information is needed to show which one is correct?

22. If we live in an oscillating Universe, would there be any way to find out about what happened in previous universes?

23. What evidence supports the contention that the Universe is evolving?

24. What evidence supports the Big Bang theory?

25. The contention is often made that space and time began at the instant of the Big Bang. Defend this statement.

Appendices

A.
SIGNIFICANT FIGURES

The discussions in this text have not dealt with significant figures in measurement, but if the reader wishes to do so, this section will serve as a guide.

Suppose you weigh an object four times on a balance that provides readings to the hundredth of a gram, and you get the following values:

> 5.14 g
> 5.13 g
> 5.12 g
> 5.13 g

You assume that the mass is actually constant; why, then, are there different values? Perhaps the balance is influenced by uneven air currents, or maybe you don't eye the scale the same way each time. These variations may change the reading over a range of one or two hundredths of a gram, so it would be reasonable to state that the mass of the object is between 5.12 and 5.14 g, or 5.13 ± 0.01 g. More commonly, the value would be written simply as 5.13 g, with the understanding that there may be some uncertainty in the information provided by the last figure.

Now suppose that the same object is weighed on a more sensitive balance, with the following results:

> 5.12904 g
> 5.12903 g
> 5.12904 g

The mass would now be expressed as 5.12904 ± 0.00001 g, or simply as 5.12904 g, again with the understanding that there may be some uncertainty in the last figure. With the more sensitive balance you are now sure of the 5.1290, but the last place is still uncertain.

In the value of 5.13 g there are three digits that provide significant information; in the value 5.12904 g there are six such digits. A **significant figure,** therefore, is defined as a digit that is believed to be correct, or nearly so. The value 5.13 g has three significant figures, and 5.12904 g has six. Note that even though the last number is uncertain, it is still counted as a significant figure, because its uncertainty is limited, usually to ± 1 or 2.

Decimal points have nothing to do with the significant figures. If the value of 5.12904 g is expressed in mg, it becomes 5129.04 mg, which still has six significant figures.

The concept of significant figures does not apply to all kinds of numbers; it applies only to numbers that express measurements or to computations derived from measurements. Thus, the concept does not apply to the numbers in the relationships 1 m = 100 cm, or 1 ft = 12 in, because there are exact definitions, and the numbers are not measured ones.

The following rules govern the use of significant figures:

Rule 1

To count the significant figures in a measured number, read the number from left to right starting with the *first digit that is not zero.* Count that first number and all the numbers that follow, including all the later zeros. The position of the decimal point, if any, is irrelevant. Thus, the value of 0.10 mg has two significant figures. If the same mass is expressed as 0.00010 g, it still has two significant figures. There are instances, however, where this rule is not followed, and your common sense will allow you to recognize such exceptions. For example, if you read a statement that Mt. Everest is "about 9000 meters high," you would probably not conclude that the statement means 9000 ± 1 m, implying 4 significant figures. First, you may doubt that the measurement could be so precise; second, you may think it unlikely that the height would be a whole number of thousands of meters; and third, the word "about" implies a rough approximation. The three zeros are in fact "spacers," or place-holders, to define the magnitude of the first digit. The ambiguity could be avoided by using exponential notation, such as 9×10^3 m, which shows only 1 significant figure, or 9.0×10^3 m, which shows 2 significant figures, as explained in Rule 2 below.

EXAMPLE 1

How many significant figures are there in each of the following measured quantities? **(a)** 0.00406 mm; **(b)** 31.020 L; **(c)** 0.020 s; **(d)** $6.00 \cdot 10^8$ kg; **(e)** 50,000,000 years.

Answer: (a) 3; **(b)** 5; **(c)** 2; **(d)** 3; **(e)** ambiguous, but probably only 1.

Rule 2

If the number of digits needed to express the magnitude of a measurement exceeds the permissible number of significant figures, exponential notation should be used. For example, assume that a length is measured to be 5.2 m. This value contains two significant figures. What is that length expressed in mm? Multiplying 5.2 mm by 1000 gives 5200 mm, but that appears to have four significant figures, which are too many. The answer should therefore be expressed at 5.2×10^3 mm, which expresses the correct magnitude and retains the correct number of significant figures.

Rule 3

In addition or subtraction, the value with the *fewest decimal places* will determine how many significant figures are used in the answer:

308.7810 g	(4 decimal places; 7 significant figures)
0.00034 g	(5 decimal places; 2 significant figures)
10.31 g	(2 decimal places; the fewest; 4 significant figures)
Sum: 319.09 g	(2 decimal places; 5 significant figures)

Rule 4

In multiplication and division, the value with the *fewest significant figures* will determine how far the significant figures should be carried in the answer:

$$\frac{3.0 \times 4297}{0.0721} = 1.8 \times 10^5$$

Note that the number with the fewest significant figures is 3.0 (two significant figures), and therefore the answer must have also two significant figures.

Rule 5

When a number is "rounded off" (that is, non-significant figures are discarded), the last significant figure is increased by 1 if the next figure is 5 or more and is unchanged if the next figure is less than 5:

4.6349, rounded off to four significant figures → 4.635

4.6349, rounded off to three significant figures → 4.63

2.815, rounded off to three significant figures → 2.82

B.
LOGARITHMS TO THE BASE 10 (THREE PLACES)

N	0	1	2	3	4	5	6	7	8	9
1	000	041	079	114	146	176	204	230	255	279
2	301	322	342	362	380	398	415	431	447	462
3	477	491	505	519	532	544	556	568	580	591
4	602	613	623	634	644	653	663	672	681	690
5	699	708	716	724	732	740	748	756	763	771
6	778	785	792	799	806	813	820	826	833	839
7	845	851	857	863	869	875	881	887	892	898
8	903	909	914	919	924	929	935	941	945	949
9	954	959	964	969	973	978	982	987	991	996

C.
DIRECT AND INVERSE PROPORTIONALITY

Two variable properties are said to be directly proportional to each other when a change in one of the variables produces a proportionate change in the other. If one variable is doubled, the other doubles; if one triples, the other triples. An example is Charles' law (Chapter 15) which states, "The volume of a gas is directly proportional to its absolute temperature." The expression is

$$V \propto T$$

(The symbol $\propto$ means "is proportional to.") For example, if a sample of gas at 100 K occupies 2 L, if we double the absolute temperature to 200 K, the gas volume also doubles, to 4 L.

In an inverse proportionality, when one variable changes by a given factor, the other variable changes by the *reciprocal* of that factor. For example, Boyle's law (Chapter 15) states that "the volume of a gas is inversely proportional to its pressure." That is,

$$V \propto \frac{1}{P}$$

Thus, if the pressure of a sample of gas is doubled, multiplied by 2, the volume of a gas is halved, multiplied by 1/2.

D.
ALGEBRA

When algebraic operations are performed, the laws of arithmetic apply. Symbols such as x, y, and z are usually used to represent quantities that are not specified or have unknown values.

First, consider the equation

$$4x = 12$$

If we wish to solve for x, we can divide (or multiply) each side of the equation by the same factor without affecting the validity of the equation. In this case, let us divide both sides by 4. We have

$$\frac{4x}{4} = \frac{12}{4}$$

or

$$x = 3.$$

Next, consider the equation

$$x + 5 = 9$$

In this type of expression, we can add or subtract the same quantity from each side without affecting the validity of the equation. If we subtract 5 from each side, we get

$$x + 5 - 5 = 9 - 5$$

or

$$x = 4$$

Now, let us put these facts together to solve a slightly more complex equation. Consider the expression

$$\frac{2x}{5} + 3 = 13$$

The overall objective in solving any algebraic expression such as this is to isolate x by itself on one side of the equation. We can begin this task by getting rid of the 3 on the left side of the equation. We do so by subtracting 3 from each side of the equation, as

$$\frac{2x}{5} + 3 - 3 = 13 - 3$$

which reduces to

$$\frac{2x}{5} = 10$$

Now, let us get rid of the 5 on the left side of the equation. This is done by multiplying both sides of the equation by 5.

$$\frac{2x}{5} \times 5 = 10 \times 5$$

or

$$2x = 50$$

Finally, we get rid of the 2 on the left side of the equation by dividing both sides of the equation by 2. We find

$$\frac{2x}{2} = \frac{50}{2}$$

We are left with

$$x = 25$$

Regardless of how complicated an equation in this text may look, it can be solved by carefully following the rules specified above.

TABLE OF ATOMIC WEIGHTS

NAME	SYMBOL	ATOMIC NUMBER	ATOMIC WEIGHT	NAME	SYMBOL	ATOMIC NUMBER	ATOMIC WEIGHT
Actinium	Ac	89	227.028	Neodymium	Nd	60	114.24
Aluminum	Al	13	26.98154	Neon	Ne	10	20.179
Americium	Am	95	(243)	Neptunium	Np	93	237.0482
Antimony	Sb	51	121.75	Nickel	Ni	28	58.70
Argon	Ar	18	39.948	Niobium	Nb	41	92.9064
Arsenic (grey)	As	33	74.9216	Nitrogen	N	7	14.0067
Astatine	At	85	(210)	Nobelium	No	102	(259)
Barium	Ba	56	137.33	Osmium	Os	76	190.2
Berkelium	Bk	97	(247)	Oxygen	O	8	15.9994
Beryllium	Be	4	9.01218	Palladium	Pd	46	106.4
Bismuth	Bi	83	208.9804	Phosphorus	P	15	30.97376
Boron	B	5	10.81	Platinum	Pt	78	195.09
Bromine	Br	35	79.904	Plutonium	Pu	94	(244)
Cadmium	Cd	48	112.41	Polonium	Po	84	(209)
Calcium	Ca	20	40.08	Potassium	K	19	39.0983
Californium	Cf	98	(251)	Praeseodymium	Pr	59	140.9077
Carbon	C	6	12.011	Promethium	Pm	61	(145)
Cerium	Ce	58	140.12	Protactinium	Pa	91	231.0359
Cesium	Cs	55	132.9054	Radium	Ra	88	226.0254
Chlorine	Cl	17	35.453	Radon	Rn	86	(222)
Chromium	Cr	24	51.996	Rhenium	Re	75	186.2
Cobalt	Co	27	58.9332	Rhodium	Rh	45	102.9055
Copper	Cu	29	63.546	Rubidium	Rb	37	85.4678
Curium	Cm	96	(247)	Ruthenium	Ru	44	101.07
Dysprosium	Dy	66	162.50	Samarium	Sm	62	150.4
Einsteinium	Es	99	(254)	Scandium	Sc	21	44.9559
Erbium	Er	68	167.26	Selenium	Se	34	78.96
Europium	Eu	63	151.96	Silicon	Si	14	28.0855
Fermium	Fm	100	(257)	Silver	Ag	47	107.868
Fluorine	F	9	18.998403	Sodium	Na	11	22.98977
Francium	Fr	87	(223)	Strontium	Sr	38	87.62
Gadolinium	Gd	64	157.25	Sulfur	S	16	32.06
Gallium	Ga	31	69.72	Tantalum	Ta	73	180.9479
Germanium	Ge	32	72.59	Technetium	Tc	43	(97)
Gold	Au	79	196.9665	Tellurium	Te	52	127.60
Hafnium	Hf	72	178.49	Terbium	Tb	65	158.9254
Helium	He	2	4.00260	Thallium	Tl	81	204.37
Holmium	Ho	67	164.9304	Thorium	Th	90	232.0381
Hydrogen	H	1	1.0079	Thulium	Tm	69	168.9342
Indium	In	49	114.82	Tin	Sn	50	118.69
Iodine	I	53	126.9045	Titanium	Ti	22	47.90
Iridium	Ir	77	192.22	Tungsten	W	74	183.85
Iron	Fe	26	55.847	Unnilennium	Une	109	266
Krypton	Kr	36	83.80	Unniloctium	Uno	108	265
Lanthanum	La	57	138.9055	Unnilhexium	Unh	106	263.12
Lawrencium	Lr	103	(260)	Unnilquadium	Rf	104	261.11
Lead	Pb	82	207.2	Unnilpentium	Ha	105	262.11
Lithium	Li	3	6.941	Unnilseptium	Uns	107	261
Lutetium	Lu	71	174.967	Uranium	U	92	238.029
Magnesium	Mg	12	24.305	Vanadium	V	23	50.9415
Manganese	Mn	25	54.9380	Xenon	Xe	54	131.30
Medelevium	Md	101	(257)	Ytterbium	Yb	70	173.04
Mercury	Hg	80	200.59	Yttrium	Y	39	88.9059
Molybdenum	Mo	42	95.94	Zinc	ZN	30	65.38
				Zirconium	Zr	40	91.22

Glossary

absolute humidity   See *humidity*.

absolute zero   The zero point on the Kelvin scale of absolute temperature. No substance can be cooled below zero kelvin, or 0 K. The equivalent temperature on the Celsius scale is $-273.15°$

absorption spectrum   See *spectrum*.

acceleration   The change in velocity of an object per unit time. Acceleration is a vector quantity. A body is accelerating when it is speeding up, slowing down, or changing direction.

acceleration due to gravity   The acceleration of a freely falling body under the influence of the Earth's gravitation at sea level. This term is symbolized by the letter g and is equal to approximately 9.8 m/s^2.

acid   A substance that can supply hydrogen ions (protons) to another substance, known as a base.

acid mine drainage   Runoff from sulfur-rich mining wastes that has reacted with water to form sulfuric acid, which pollutes natural waterways.

adiabatic   Referring to a process that occurs without loss or gain of heat.

adiabatic cooling   A cooling process that occurs without loss or gain of heat. Under certain conditions, a rising air mass may cool adiabatically. If no heat is transferred, a rising air mass will expand and, as it expands, it performs work and therefore cools.

adsorption   The process by which molecules from a liquid or gaseous phase become concentrated on the surface of a solid.

air mass   A large body of air that has approximately the same temperature and humidity throughout.

air pollution   The deterioration of the quality of air that results from the addition of impurities.

albedo   A measure of the reflectivity of a surface, measured as the ratio of light reflected to light received. A mirror or bright snowy surface has a high albedo, whereas a rough flat road surface has a low albedo.

allotropes   Different forms of the same element.

alpha particle   The nucleus of a helium atom.

alternating current (ac)   An electric current that oscillates in a wire. When an alternating current is established in a wire, electrical energy is transmitted, but the electrons themselves do not move in a concerted direction. See also *direct current*.

ampere   A measure of electrical current equal to the movement of one coulomb of charge past a given point in a wire in one second.

amplitude (of a wave)   The magnitude or height of a wave, measured as the distance between the zero point of the wave to the point of maximum displacement. The amplitude is one half of the vertical distance between the crest and the trough.

angle of incidence   When a light ray is beamed onto a surface, the angle of incidence is defined as the angle between the light ray and a line drawn perpendicular to the surface.

angle of reflection   The angle between a reflected light beam and a line drawn perpendicular to the surface.

angular momentum   The product of the mass of an object, m, the radius of its orbit, r, and its speed, v. That is, mrv.

anode   The positive terminal in a vacuum tube. The electrons emitted by the cathode are collected at the anode. In an electrochemical cell, the anode is the electrode where oxidation (loss of electrons) occurs.

apparent brightness   A rating scale developed by Hipparchus such that the brightest stars in the sky are said to have an apparent brightness of 1 while those just visible to the naked eye have a rating of 6.

aquifer   An underground layer of rock that is porous and permeable enough to store significant quantities of water.

Archean Eon   A division of geological time 3.8 to 2.5 billion years ago. The oldest known rocks formed at the beginning of or just prior to the start of the Archean Eon.

asteroid   One of the small planetary bodies that orbits the Sun. Asteroids range in size from less than 1 km to 1000 km in diameter.

asthenosphere   The plastic part of the Earth's mantle just below the lithosphere.

atmosphere   The predominantly gaseous envelope that surrounds the Earth.

atmosphere (standard of pressure)   The pressure exerted at sea level by a column of mercury 76 cm high. This corresponds to the normal pressure exerted by the Earth's atmosphere at sea level.

atom   The fundamental unit of the element.

atomic nucleus   The small positive central portion of the atom that contains its protons and neutrons.

atomic number   The number of protons in an atomic nucleus.

atomic orbital   See *orbital*.

aurora   A luminous atmospheric display appearing mainly in the high latitudes that is created by interactions between the particles of the solar wind and the magnetic field of the Earth.

Avogadro's law   Equal volumes of all gases (at the same temperature and pressure) contain the same number of molecules.

background radiation   The level of radiation on Earth from natural sources.

barometer   A device used to measure atmospheric pressure.

barometric pressure   The pressure (force/area) exerted by the atmosphere.

basalt   A dark-colored, fine-grained volcanic rock that forms most of the oceanic crust.

base   Chemistry: A substance that accepts protons from an acid. The reaction is said to neutralize the acid.

beta particle   An electron emitted from an atomic nucleus.

Big Bang   An event that is thought to mark the beginning of our Universe. The theory assumes that, some 10 to 20 billion years ago, all matter that was to form the Universe exploded into space from an infinitely compressed state. See also *oscillating universe cosmology*.

black hole   A small region of space that contains matter packed so densely that an intense gravitational field is created, from which light cannot escape.

body wave   A seismic wave that travels through the interior of the Earth.

Boyle's law   The volume of a gas (at constant temperature) is inversely proportional to its pressure.

branching chain reaction   A chain reaction in which each step produces more than one succeeding step.

breeder reactor   A nuclear reactor that produces more fissionable material than it consumes.

bit   A single unit of binary information such as is stored electronically in a computer.

calorie   A unit used to express quantities of thermal energy. When "calorie" is spelled with a small c, it refers to the quantity of heat required to heat 1 g of water 1°C. When "Calorie" is spelled with a capital C, it means 1000 small calories, or one kilocalorie, the quantity of heat required to heat 1000 g (1 kg) of water 1°C.

capillary action   The movement of water upward against the force of gravity through the action of electrical attractions to the surfaces of small openings.

catalyst   A substance that influences (usually speeds up) the rate of a chemical reaction and that is not consumed in the reaction.

cathode   The negative terminal in a vacuum tube that emits electrons. These electrons then travel across free space toward the anode. In an electrochemical cell, the cathode is the electrode where reduction (gain of electrons) occurs.

cathode ray   A beam of electrons emerging from the cathode in a cathode ray tube.

Celsius scale   The temperature scale used in the metric system. On this scale, water freezes at 0°C and boils at 100°C at sea level.

centrifugal force   An outward force observed by a person from within an accelerating frame of reference.

centripetal force   The inward force that is necessary to keep a body in circular motion.

chain reaction   A reaction that proceeds in a series of steps, each step being made possible by the preceding one. See also *branching chain reaction*.

Charles' law   The volume of a gas (at constant pressure) is directly proportional to its absolute temperature.

chemical bond   The force that holds atoms together to form molecules. See also *covalent bond* and *ionic bond*.

chemical change   A transformation that results from making or breaking of chemical bonds.

chemical energy   The energy that is absorbed when chemical bonds are broken or that is released when chemical bonds are formed. A substance that can release energy by undergoing chemical reactions is said to have chemical energy.

chemical formula   A combination of symbols of elements that shows the composition of a molecule or a substance.

chemical weathering   The chemical decomposition of rocks and minerals by exposure to air, water, and other chemicals in the environment.

China syndrome   A facetious expression referring to a nuclear meltdown in which the hot radioactive mass melts its way into the ground toward China.

chip   A tiny circuit board containing many different electronic switching and/or amplifying devices.

chromosphere   A turbulent diffuse gaseous layer of the Sun that lies above the photosphere.

cirrus   A wispy, high-level cloud.

climate   The composite pattern of weather conditions that can be expected in a given region. Climate refers to yearly cycles of temperature, wind, rainfall, etc., and not to daily variations. See also *weather*.

closed system   A system that is isolated so that neither mass nor energy can enter or leave.

coherent light   A beam of light in which all the component waves are traveling in phase, at the same frequency, and in exactly the same direction. Coherent light is produced by a laser.

comet   A celestial body moving about the Sun, usually in a highly elliptical path or orbit. Comets appear to have a fairly dense core surrounded by a "fuzzy," "fiery" halo, but in actuality comets are quite cold. When a comet approaches the Sun, the force of the solar wind blows matter outward from the comet, forming a long "tail."

compound (compound substance)   A substance that consists of a fixed composition of elements and has a fixed set of properties.

condensation   Conversion of vapor to liquid.

conduction (of thermal energy)   The process by which thermal energy is transmitted directly through materials. Conduction occurs because energetic atoms or molecules move rapidly and collide with neighboring atoms or molecules. Kinetic energy is transferred during the collision process, and the neighboring molecules accelerate and become energetic, or "hot."

conductor (electrical)   A material that offers little resistance to the movement of electric current.

continental drift   The theory stating that the continent-sized masses of the Earth's crust are moving slowly relative to one another.

control rod   A neutron-absorbing medium that controls the reaction rate in a nuclear reactor.

convection   The process by which thermal energy is transmitted through gases and liquids by the action of currents that circulate in the fluid.

core (of the Earth)   The central portion of the Earth, believed to be composed mainly of iron and nickel.

Coriolis effect   The deflection of air or water flow caused by the rotation of the Earth.

corona   The luminous irregular envelope of highly ionized gas outside the chromosphere of the Sun.

cosmic ray   A form of high-energy radiation consisting mainly of high-speed atomic nuclei and other atomic particles that move through space and frequently strike the Earth's atmosphere.

cosmology   The study of the origin and the end of the Universe.

coulomb   A unit quantity of electricity transported in one second by a current of 1 ampere.

Coulomb's law   A relationship that states that the force of attraction (or repulsion) between two charges is directly proportional to the product of the charges and inversely proportional to the square of the distance between them.

covalent bond   A chemical bond between atoms that is characterized by shared electrons.

crest   The highest point in a wave.

critical condition   A condition under which a chain reaction continues at a steady rate, neither accelerating nor slowing down.

critical mass (in a nuclear reaction)   The quantity of fissionable material just sufficient to maintain a nuclear chain reaction.

crust (of the Earth) The solid outer layer of the Earth; the portion on which we live.

crystal settling   A process in which the crystals that solidify first from a cooling magma settle to the bottom of a magma chamber because the solid minerals are more dense than liquid magma.

crystalline solid   See *solid*.

cumulus   A column-like cloud with a flat bottom and a billowy top.

decibel (dB)   A unit of sound intensity. The decibel scale is a logarithmic scale used in measuring sound intensities relative to the intensity of the faintest audible sound.

degeneracy pressure   The strength of the atomic particles that holds a white dwarf star from futher collapse.

density   Mass per unit volume.

detergent   A synthetic cleaning compound.

deuterium   Isotope of hydrogen with mass number of 2. Also called "heavy hydrogen."

dew   Moisture condensed from the atmosphere, usually during the night, when the ground and leaf surfaces become significantly cooler than the surrounding air.

dew point   The temperature to which a sample of air must be cooled to become saturated with moisture.

diffraction   The ability of a wave to pass around an obstacle or to bend past an opening.

direct current (dc)   An electric current moving in one direction only. When a direct current moves through a wire, electrons travel progressively through the wire. See also *alternating current*.

distillation   A process in which a liquid is vaporized and the vapor is condensed to a liquid again.

divergent boundary   The boundary or zone where lithospheric plates separate from each other. (syn: *spreading center*)

doldrums   A region of the Earth near the Equator in which hot, humid air is moving vertically upward, forming a vast low-pressure region. Local squalls and rainstorms are common, and steady winds are rare.

domain (magnetic)   Microscopic bundles of magnetic atoms that are held together by electrical forces. The atoms in a domain are aligned so that the entire bundle produces a net magnetic field.

Doppler effect   The observed change in frequency of light or sound that occurs when the source of the wave is moving relative to the observer.

dust   An airborne substance that consists of solid particles typically having diameters greater than about 1 micrometer.

earthquake   A sudden traumatic movement of part of the Earth's crust.

eclipse   A phenomenon that occurs when a heavenly body is shadowed by another and therefore rendered invisible. When the Moon lies directly between the Earth and the Sun, we observe a *solar eclipse;* when the Earth lies directly between the Sun and the Moon, we observe a *lunar eclipse.*

electric charge   The net quantity of electricity or the electric energy possessed by a substance. It is a measure of the excess or deficiency of electrons.

electric circuit   A complete path of conducting materials that allows electric current to flow.

electric current   A concerted and continuous movement of charged particles in response to a potential gradient.

electric field   A region of space in which electric forces can be detected.

electric force   The force that results from the interaction of charged bodies. Electrical force is repulsive if the bodies carry like charges($+ +$ or $- -$) and is attractive if the bodies are oppositely charged ($+ -$ or $- +$).

electric potential   See *volt.*

electrode   A terminal in an electric circuit where electrons enter or leave a gas, a vacuum, or an ionic liquid. See also *anode* and *cathode.*

electromagnet   A device consisting of an iron core wrapped with wire that is magnetized when a direct current is passed through the wire.

electromagnetic field   The combined electric and magnetic field produced by an oscillating charged particle or particles.

electromagnetic induction   The induction of an electric current in a wire when a magnetic field changes near the wire or when the wire moves across a magnetic field.

electromagnetic spectrum   The entire range of electromagnetic radiation.

electromagnetic wave   A periodically oscillating electromagnetic field.

electromagnetism   The force generated by magnets or by charged objects.

electron   The fundamental atomic unit of negative electricity.

electron shell   An energy level in an atom that can accommodate a specific number of electrons and that is designated by a principal quantum number.

element   A substance all of whose atoms have the same atomic number.

emission spectrum   See *spectrum.*

energy   The capacity to perform work. See also specific types of energy such as *chemical, kinetic, nuclear,* and *potential*

engine   A device that converts heat to work.

entropy   A thermodynamic measure of disorder. It has been observed that the entropy of an isolated system always increases during any spontaneous process; that is, the degree of disorder always increases.

epoch   The smallest unit of geological time. Each period is divided into epochs.

era   A geological time unit. Eons are divided into eras, and eras, in turn, are subdivided into periods.

equinox   Either of two times during a year when the Sun shines directly overhead at the Equator. During the equinox, every portion of the Earth receives 12 hours of daylight and 12 hours of darkness.

erosion   The process by which parts of the Earth's surface are transported to new locations by water, wind, waves, ice, or other natural agents.

eutrophication   The pollution of a body of water by enrichment with nutrients, with a consequent increase in the growth of organisms and a resultant depletion of dissolved oxygen.

evaporation   Conversion of liquid to vapor.

evaporite deposit   A chemically precipitated sedimentary rock that formed when dissolved ions were concentrated by evaporation of water.

excited electron   An electron in an atom or molecule that has been promoted to a higher energy level and therefore exists in an excited state.

excited state   A state of an atom or molecule in which one or more electrons have absorbed energy and exist in higher energy levels. See also *ground state.*

fault   A crack or weak point in the Earth's surface; a potential site of earthquake activity.

fault creep   The gradual non-traumatic slippage of land surfaces past each other.

fiber optics   The use of glass fibers to transmit information by means of a pulsed laser beam.

field of force   See *electric field.*

First law of thermodynamics   See *thermodynamics.*

fission (of atomic nuclei)   The splitting of atomic nuclei into approximately equal fragments.

fog   A low cloud formation usually formed when warm, moisture-laden air is cooled on contact with land or water.

force   Any influence that can cause a body to accelerate. Force is commonly measured in newtons in the SI and pounds in the British system.

foreshocks   Small earthquakes that precede a large quake by an interval ranging from a few seconds to a few weeks.

frequency   The number of wave disturbances (can be measured as the number of crests) that pass a given point in a specific amount of time. Frequency is usually expressed in cycles/s, or hertz. 1 Hz = 1 cycle/s.

friction   A type of force that opposes the motion of one body past another when the two are in contact.

frontal weather system   A weather system that develops when air masses collide

fulcrum   The support or point of rest of a lever.

fundamental frequency   The lowest frequency of a musical sound.

fusion (of atomic nuclei)   The combination of nuclei of light elements to form heavier nuclei.

galaxy   A large volume of space containing many billions of stars, all held together by mutual gravitation.

gas   A state of matter that consists of molecules that are moving independently of each other in random patterns.

generator   A device that produces electrical energy when a coil of wire is rotated in a magnetic field. Generators must be driven by some external source of energy.

glacier   A large mass of flowing ice. A glacier usually takes the form of a river of ice.

glass (or glassy solid)   A rigid state of matter in which the atoms or molecules are randomly arranged.

gneiss   A foliated rock with banded appearance formed by regional metamorphism.

granite   A medium- to coarse-grained igneous rock that forms the base of the continents.

graphite   A soft, black form of the element carbon that consists of crystalline layers, which can slide past one another.

gravitation   A universal force of mutual attraction between all bodies

gravitation, acceleration due to   See *acceleration due to gravity.*

Great Red Spot   A large atmospheric storm on the surface of Jupiter that was first observed over 300 years ago.

greenhouse effect   The effect produced by certain gases, such as carbon dioxide or water vapor, that causes a warming of the atmosphere by absorption of infrared radiation.

ground state   A state of an atom or molecule in which all the electrons are in their lowest allowed energy levels. See also *excited state.*

ground water   Water contained in soil and bedrock. All subsurface water.

gyre   A curved or circular ocean current formed when currents traveling northwards or southwards are deflected by the spin of the Earth and the shape of the continents.

half-life (of a radioactive substance)   The time required for half of a sample of radioactive matter to decompose.

hard water   Water containing Ca^{2+} and/or Mg^{2+} ions.

heat   The energy that is transferred from one system to another when two systems at different temperatures are in contact.

heat engine   A mechanical device that converts heat to work.

heat of fusion   The heat required to melt 1 g of a solid at constant temperature.

heat of vaporization   The heat required to vaporize 1 g of a liquid at constant temperature.

heavy oil   Petroleum deposits that are too viscous to be extracted and pumped using conventional oil well technology.

heliocentric theory   The theory, now known to be true, that the Sun, and not the Earth, is the center of the Solar System.

hertz (Hz)   A unit that expresses the frequency of a wave form. When one crest of a wave passes a given point every second, that wave is said to have a frequency of 1 hertz. One hertz is therefore one cycle per second.

Hertzsprung-Russell diagram   A plot of intrinsic brightness along the vertical axis versus temperature of a star along the horizontal axis.

heterogeneous   Referring to a nonuniform substance that has different properties and compositions in different regions.

hologram   A three-dimensional photograph produced using laser light.

homogeneous   Referring to a uniform substance having the same properties throughout the sample.

horse latitudes   A region of the Earth, lying at about 30 degrees north and south latitudes, in which air is moving vertically downward, forming a vast high-pressure region. Generally dry conditions prevail, and steady winds are rare.

Hubble's law   A law that relates the red shift of an object outside our galaxy to its distance from Earth. See *red shift*.

humidity   A measure of the amount of moisture in the air. *Absolute humidity* is defined as the amount of water vapor in a given volume of air. *Relative humidity* is defined as the ratio of the amount of moisture in a given volume of air divided by the amount of moisture that can be held by that volume at a given temperature when the air is saturated

humus   The complex mixture of decayed organic matter that is an essential part of healthy natural soil.

hydrocarbon   A compound of hydrogen and carbon.

hydrogen bond   A bond formed between a hydrogen atom bonded to a very electronegative element (F, N, O) and another atom of one of the three electronegative elements.

hydrologic cycle (water cycle)   The cycling of water, in all its forms, on the Earth.

hydrophilic   Water-liking.

hydrophobic   Water-fearing.

hydrothermal processes   Changes in rock that are caused primarily by migrating hot water and by ions dissolved in the hot water.

igneous rock   Rock formed directly from cooling magma.

inclined plane   A ramp or tilted surface used as a machine element.

inertia   That property of a body that compels it to remain at rest or at constant velocity unless it is forced to change.

insulator   A material that offers substantial resistance to the movement of an electric current.

interference   A process whereby two or more waves combine when they reach a single point in space at the same time. The new wave is formed by the addition of the wave components.

intrinsic brightness   A number that rates a star according to how much energy it pours out into space.

ion   An electrically charged atom or group of atoms.

ionic bond   A chemical bond formed by attraction between oppositely charged ions.

isoelectronic   Having the same electronic structure.

isoelectronic structures   Atoms or ions that have the same electronic composition, such as F^-, Ne, and Na^+.

isomers   Different substances that have the same molecular formula, such as ethyl alcohol and dimethyl ether, both C_2H_6O.

isostasy   A principle that states that the Earth's crust is floating on denser, fluid layers beneath it.

isotopes   Atoms of the same element that have different mass numbers.

jet stream   A high-altitude, fast-moving air current.

joule   The fundamental unit of work in the SI. One joule equals 1 newton-meter.

kelvin   The SI unit of temperature. One kelvin is the same as a difference in temperature of $1°C$. The Kelvin temperature scale starts at 0 K, which equals $-273.15°C$.

kinetic energy   The energy possessed by a moving object, equal to $1/2\ mv^2$.

kinetic theory   A theory that accounts for the nature of gases by assuming that they consist of independently moving molecules.

laser   Acronym for **l**ight **a**mplification by **s**timulated **e**mission of **r**adiation. A device that produces a short, intense flash of coherent light in which all the component waves are traveling in phase, at the same frequency, and in exactly the same direction.

laser fusion   A nuclear fusion reaction triggered by a laser beam.

lava   The material produced when magma pours rapidly onto the surface of the Earth through fissures in the crust. A site where lava appears is called a volcano.

lens   A curved piece of transparent material that refracts light in such a way that the apparent sizes of objects are altered.

lever   A rigid bar positioned over a fulcrum that is used as a machine element.

light-year   The distance traveled by an electromagnetic wave in one year, approximately 9.5×10^{15} m.

lighthouse theory   The theory that states that a rapidly rotating neutron star may be the source of pulsar signals.

limestone   A sedimentary rock consisting chiefly of calcium carbonate.

liquid   A non-rigid state of matter in which the molecules are arranged rather randomly and adhere to each other just strongly enough to form a well-defined boundary. Liquids therefore flow readily and take the shape of their containers.

liquid crystal   A state of matter of substances that are crystalline when undisturbed but that can easily be made to flow like liquids.

liquid metallic hydrogen   Hydrogen that is exposed to such extremes of high pressure that it becomes liquid and exhibits metallic properties.

lithosphere   The outer shell of the Earth, including the crust and the uppermost portion of the mantle.

longitudinal wave   See *wave*.

mach number   A unit of speed related to the speed of sound. The mach number is the speed of the object divided by the speed of sound.

machine   A type of tool that alters the magnitude or direction of an applied force.

magma   A fluid material, lying in the upper layers of the Earth's asthenosphere, consisting of melted rock mixed with various gases such as steam and hydrogen sulfide.

magnetosphere   A region of magnetic field around a planet.

manganese nodules   Manganese-rich, potato-shaped masses found on the ocean floor.

mantle   The solid but partly semiplastic portion of the Earth that surrounds the central core and lies under the crust.

maria   Smooth, flat plains on the surface of the Moon. (Originally these were erroneously thought to be seas—hence the "maria.")

mass   The quantity of matter in an object. Mass is fundamentally defined in terms of the inertia of a body, its resistance to a change in velocity, by the equation: $F = ma$, or $m = F/a$, where m is mass, F is force, and a is acceleration.

mass number   The sum of the number of protons and neutrons in an atomic nucleus.

mechanical weathering   The disintegration of rock into smaller pieces by physical processes.

mesosphere   The layer of air that lies above the stratosphere and extends from about 33 km upward to 80 km above the surface of the Earth.

Mesozoic Era   The part of geological time roughly 245 to 65 million years ago. Dinosaurs rose to prominence and became extinct during this era.

metal   An element characterized by a great ability to conduct heat and electricity, to reflect light, and to maintain its crystal structure even when its shape is deformed.

metamorphic rock   Rock formed when sedimentary or igneous material is altered by heat and pressure within the Earth's crust.

metamorphism   The process by which rocks and minerals change in response to changes in temperature, pressure, and chemical conditions.

meteor   A meteoroid that glows in the Earth's atmosphere when it is accelerated downward by gravity. (Called also *shooting star*.)

meteorite   A meteoroid that has fallen to the surface of the Earth.

meteroid   A small bit of solid matter traveling through space.

metric system   See *Système International d'Unités*.

Mid-Atlantic Ridge   A large ridge running under the surface of the Atlantic Ocean that is split in the middle by a sharp rift or valley. Igneous material rises out of the Mid-Atlantic Ridge, forming new crustal rock as the tectonic plates move apart.

Mid-Oceanic Ridge   A continuous submarine mountain chain that forms at the boundary between divergent tectonic plates within oceanic crust.

Milky Way   The large spiral galaxy that contains our Solar System.

mineral   A naturally occurring inorganic solid with a definite chemical composition and a crystalline structure.

mineral reserve   The known supply of ore in the ground.

moderator   A medium used in a nuclear reactor to slow down neutrons.

Mohorovicic discontinuity (MOHO)   The boundary between the crust and the mantle identified by a change in the velocity of seismic waves.

mole   The amount of substance that contains as many elementary particles (atoms or molecules) as there are carbon atoms in exactly 12 g of the carbon-12 isotope. Also, the amount of substance that contains 6.02205×10^{23} elementary particles. Also, the atomic weight of an element or the molecular weight of a compound expressed in grams.

molecule   An aggregate of at least two atoms in a definite arrangement held together by chemical bonds.

momentum   The product of the mass of a body times its velocity.

monsoon   A continental wind system caused by uneven heating of land and ocean surfaces. Monsoons generally blow from the sea to the land in the summer, when the continents are warmer than the ocean, and bring predictable rainstorms. In winter, when the ocean is warmer than the land surfaces, the winds reverse.

net force   The vector sum of all the forces acting on a body.

neutralization   The reaction of an acid with a base to produce a salt and water.

neutrino   A subatomic particle that bears no charge and has very little, if any, mass.

neutron   A subatomic particle that is electrically neutral and has a mass approximately equal to that of a proton.

neutron star A small (by stellar standards) dense core of neutrons that remains intact after a supernova explosion. See also *pulsar.*

newton The SI unit of force. One newton is equal to the force needed to give a mass of 1 kg an acceleration of 1 m/s^2.

nimbus A prefix or suffix that denotes precipitation in a cloud.

noble (or "inert") gases The elements of Group 8 of the periodic system, starting with helium.

nodes (of an orbital system) The points where the planes of two orbits intersect. An eclipse of the Sun or the Moon can occur only when the Moon passes through the nodes of the planes of the Earth-Sun and Earth-Moon orbits.

noise Unwanted sound or an unwanted signal.

nuclear energy The energy transferred in nuclear reactions. This energy can be released by natural radioactivity or in the transformations in nuclear reactors or atomic bombs.

nuclear reactor A device that utilizes nuclear reactions to produce useful energy.

nucleus See *atomic nucleus.*

oceanic trench A long, narrow trough in the sea floor formed where a subducting plate bends downward to sink into the mantle.

octet rule States that an atom other than hydrogen tends to form bonds until it is surrounded by 8 electrons.

ohm A measure of the electrical resistance of a material.

Ohm's law A relationship that states that the voltage between two points of an electric circuit is equal to the current flowing through the circuit multiplied by the resistance between the two points.

orbital The shape that defines the space occupied by an electron at a given energy level.

ore A rock mixture that contains enough valuable minerals to be mined profitably with currently available technology.

oscillating universe cosmology A version of the Big Bang theory that assumes that the Universe expands and contracts in an endless sequence.

overtone An acoustical frequency that is an even multiple greater than the fundamental frequency.

oxidation The chemical reaction of a substance with oxygen. More generally, oxidation is a loss of electrons.

Paleozoic Era The part of geological time 570 to 245 million years ago. During this era invertebrates, fishes, amphibians, reptiles, ferns, and cone-bearing trees were dominant.

Pangaea A supercontinent identified and named by Alfred Wegener that existed from about 300 to 200 million years ago and included most of the continental crust of the Earth.

parallax The apparent displacement of an object caused by the movement of the observer.

parallel circuit An electric circuit with two or more resistors arranged so that any one resistor completes the circuit independently of the other.

pascal The SI unit of pressure. One pascal is one newton per square meter.

period A geological time unit longer than an epoch and shorter than an era.

periodic table (of the elements) An arrangement of the symbols of the elements that shows that their properties are periodic functions of their atomic numbers.

pH A measure of acidity. pH = −log (hydrogen ion concentration).

photochemical reaction A chemical reaction initiated by a photon.

photoelectric effect A process whereby a ray of light can produce an electric current. The photoelectric effect can be realized if high-frequency light is shined on a cathode constructed of a metal such as potassium or cesium.

photon The smallest burst, or packet, of electromagnetic energy.

photosphere The surface of the Sun visible to us here on Earth.

photosynthesis The process by which chlorophyll-bearing plants use energy from the Sun to convert carbon dioxide and water to sugars.

placer deposits Surface mineral deposits formed by the mechanical concentration of mineral particles (usually by water) from weathered debris.

Planck's constant The constant that relates the frequency of a photon to its energy. h = 6.63 × 10^{-34} J · s.

plasma A gas at such a high temperature that the electrons have been stripped from their atoms, resulting in a mixture of nuclei surrounded by rapidly moving electrons.

plate tectonics The theory stating that tectonic plates move about and collide with one another. See *continental drift.*

polar ice cap The permanent layer of ice that covers parts of the Arctic Ocean.

pollution The impairment of the quality of some portion of the environment by the addition of harmful substances.

positron A positive electron.

potential energy   The energy posessed by an object that can be released sometime in the future. Gravitational potential energy is available when an object at some height has the potential to fall down to a lower level. Other forms of potential energy include chemical and nuclear energy.

power   The amount of energy delivered in a given time interval.

ppm   Abbreviation for parts per million.

Precambrian time   All of geological time before the Paleozoic Era, encompassing approximately the first 4 billion years of Earth's history.

precipitation (of water)   All forms in which atmospheric moisture descends to Earth.

pressure   Force per unit area.

principal quantum number   A whole number, 1, 2, 3, . . . ,that characterizes an electron shell.

principle of superposition   The principle that states that in any undisturbed sequence of sediments or sedimentaty rocks, the age becomes progressively younger from bottom to top.

proton   A fundamental particle of the atom that bears a unit of positive charge.

protoplanets   The planets in their earliest, incipient stage of formation.

protosun   The Sun in its earliest, incipient state of formation. The protosun was a cold condensing agglomeration of dust and gas.

Proterozoic Eon   The portion of geological time from 2.5 billion to 570 million years ago when most life was single-celled or simple multicelled.

pulsar   A neutron star that emits a pulsating radio signal.

quantum   A small, discrete quantity. Specifically, a discrete quantity of energy; a photon.

quark   A fundamental subatomic particle that, in various combinations, makes up other particles such as protons and neutrons.

quasar   A region of space, less than one light-year in diameter and very distant from Earth, that emits extremely large quantities of energy.

radiation   The process by which energy is emitted and transmitted as electromagnetic waves.

radio galaxy   A galactic-sized collection of stars that emit large quantities of radio-frequency energy.

radioactive dating   Measuring the ages of objects by calculations from rates of radioactive decay.

radioactivity   The spontaneous emission of radiation by atomic nuclei.

radioisotope   A radioactive isotope.

red giant   A stage in the life of a star when the core is composed of helium that is not undergoing fusion. A hot shell of hydrogen around this core is fusing at a rapid rate, producing enough energy to cause the star to expand greatly and glow brightly.

red shift   The frequency shift of light waves observed in the spectrum of an object traveling away from an observer. This shift is caused by the Doppler effect, which states that the observed frequency of a wave will decrease if the object is traveling away from an observer and increase if the object is approaching the observer.

reduction   A chemical change in which oxygen is removed from a substance. More generally, reduction is a gain of electrons.

redundancy   In the context of safety systems, redundancy refers to the provision of a series of devices that duplicate each other's functions and that are programmed to go into operation in sequence if a preceding device in the series fails.

reflection   The phenomenon that occurs when waves bounce back from an object in their path.

refraction   The change in direction of a wave motion as it moves from one transparent medium to another and strikes the second medium at an angle.

relative humidity   See *humidity*.

relativistic speeds   Speeds close to the speed of light.

residence time (for water)   The average time that a water molecule spends in a particular region such as a lake or underground aquifer.

resonance   A process whereby a periodic disturbance acts upon a second medium, thereby causing it to vibrate with the same frequency as the original disturbance.

respiration   The process by which plants and animals combine oxygen with sugar and other organic matter to produce energy and maintain body functions. Carbon dioxide and water are released as by-products.

revolve   To orbit a central point. A satellite revolves around the Earth. See also *rotate*.

rift   A deep and very narrow split or crack in the Earth's crust.

rotate   To turn or spin on an axis. A top rotates. See also *revolve*.

rotational motion   The movement of an object around an axis of rotation.

runoff   The flow of water toward the ocean through surface and underground pathways.

salt water intrusion   The movement of salt water from the ocean to terrestrial groundwater supplies that occurs when the water table in coastal areas is reduced.

sandstone   Clastic sedimentary rock composed predominantly of clay minerals.

saturation point   In meteorology, the maximum concentration of water vapor that can ordinarily exist in air at a given temperature.

scientific notation   A shorthand notation in which all numbers are written as a number between one and ten along with a power of ten that locates the decimal.

sea breeze   A local wind caused by uneven heating of land and ocean surfaces.

Second law of thermodynamics   See *thermodynamics*.

sedimentary rock   Rock formed from compressed sediment.

sediments   Small particles of mineral and organic matter deposited by erosion.

seismograph   A device used to detect earthquakes.

seismology   The science of measuring and recording the shock waves of earthquakes.

semiconductor   A material that is a moderately effective conductor of electricity but whose conductivity can be sensitively controlled by regulating various factors. Semiconductors are used in the construction of transistors.

semimetal   A substance that combines metallic and non-metallic properties.

series circuit   An electrical circuit with two or more resistors arranged so that the electric current travels through each one of them in turn.

shale   A fine-grained sedimentary rock composed predominantly of clay minerals.

shooting star   See *meteor*.

short circuit   A phenomenon that occurs when a circuit is completed with materials of low resistance only. Because the resistance is low, a great deal of current is allowed to flow, and large quantities of heat are generated.

SI   See *Système International d'Unités*.

sine wave   A smoothly oscillating symmetrical wave form. Mathematically it is defined as a wave described by the following equation: $y = \text{sine } x$.

sinusoidal   Having the character of a sine wave.

soap   Sodium salts of organic molecules that have a long hydrocarbon chain at one end and a polar group at the other.

soft water   Water that is mostly free of Ca^{2+} and/or Mg^{2+} ions.

soil   A mixture of mineral grains, organic material, water, and gas that lies on the surface of the Earth and supports plant growth.

solar cell   A device that converts sunlight directly into electricity.

solar collector   A device used to collect solar energy and concentrate it for useful purposes such as space or water heating.

solar design (active)   A solar heating system in which a working substance (usually water or air) is actively pumped from a solar collector to some type of radiator within the building.

solar design (passive)   A series of design features used in building construction to capture solar heat without the use of mechanical collection or pumping.

solar eclipse   See *eclipse*.

solar wind   A stream of atomic particles shot out into space by violent storms occurring in the outer regions of the Sun's atmosphere.

solid (crystalline)   A rigid state of matter in which the atoms or molecules are arranged in an orderly pattern.

solstice   Either of two times per year when the Sun shines directly overhead farthest from the Equator. One solstice occurs on or about June 21 and marks the longest day of the year in the Northern Hemisphere and the shortest day in the Southern Hemisphere; the other solstice occurs on or about December 22, marking the longest day in the Southern Hemisphere and the shortest day in the Northern Hemisphere.

sonic boom   The sharp disturbance of air pressure caused by the reinforcing waves that trail an object moving at supersonic speed.

sonic speed   The speed of sound.

specific heat   The quantity of energy required to raise the temperature of 1 g of a substance 1°C.

spectrum (electromagnetic)   A pattern of wavelengths into which a beam of light or other electromagnetic radiation is separated. The spectrum is seen as colors, or is photographed, or is detected by an electronic device. An *emission spectrum* is obtained from radiation emitted from a source. An *absorption spectrum* is obtained after radiation from a source has passed through a substance that absorbs some of the wavelengths.

speed   The distance traveled by an object in a given time interval.

speed of light   The speed traveled by an electromagnetic wave in a vacuum. The speed of light is a universal constant: 2.998×10^8 m/s.

spreading center   The boundary or zone where lithospheric plates rift or separate from each other. (syn: *divergent plate boundary*)

standard atmosphere   See *atmosphere (standard of pressure)*.

standing wave   A wave formed by constructive interference of the original wave whose properties and position do not change with time.

steady state   A condition in which the inflow of material or energy is equal to the outflow.

stimulated emission Emission of a photon caused by an electron transition that is stimulated by another photon of the proper frequency. Stimulated emission is the fundamental process used to create a laser beam.

stratosphere A layer of air of fairly constant temperature that lies just above the troposphere.

stratus A horizontally layered, sheet-like cloud.

strong force The force that holds atomic nuclei together.

structural formula A chemical formula that shows the sequences of atomic linkages.

subduction A process whereby one continental plate is forced downward during a collision with another.

sublimation Direct conversion of solid to vapor.

subsidence The settling of the surface of the ground as an ore, oil, or deep groundwater is removed.

subsonic speed Less than the speed of sound.

sunspot A cool region of the Sun formed by an intense magnetic disturbance. Sunspots are observed as dark blotches on the surface of the Sun.

superconductor A material that offers almost zero resistance to the flow of electric current. Superconductivity is shown by certain materials when they are cooled to temperatures close to absolute zero.

supernova A star that has collapsed under intense gravitation and then exploded, hurling matter into space and sometimes emitting as much energy as an entire galaxy.

supersonic speed Greater than the speed of sound.

surface wave An earthquake that travels along the surface of the Earth.

synfuels An abbreviation for synthetic fuels. Any fuel that is manufactured by a chemical conversion from one type of fuel to another. The gasoline produced by conversion of coal or extraction of oil shale is a synfuel.

Système International d'Unités (SI) An outgrowth of the metric system of measurement used in all scientific circles and by lay people in most nations of the world. The base units in the SI are: length—meter; mass—kilogram; time—second; electric current—ampere; temperature—kelvin; luminous intensity—candela; and amount of substance—mole.

tectonic plate A large, continent-sized piece of the Earth's crust that may move and collide with other tectonic plates.

temperature A measure of the warmth or coldness of an object with reference to some standard. Temperature should not be confused with thermal energy. Thermal energy is the quantity of energy possessed by a body; the temperature is just a measure of how hot or cold it is.

terrestrial planets Mercury, Venus, Earth, and Mars; the four innermost planets of our Solar System that are all relatively small, dense, and rocky.

thermal energy The combined energy of motion of all the particles in a sample.

thermal pollution A change in the quality of an environment (usually an aquatic environment) caused by raising its temperature.

thermodynamics The science concerned with thermal energy and work and the relationships between them.

First law of thermodynamics Energy cannot be created or destroyed.

Second law of thermodynamics It is impossible to derive mechanical work from any portion of matter by cooling it below the temperature of the coldest surrounding object.

thermonuclear reaction A nuclear reaction, specifically fusion, initiated by a very high temperature.

thermosphere An extremely high and diffuse region of the atmosphere lying above the mesosphere.

tides The cyclic rise and fall of ocean water caused by the gravitational force of the Moon and, to a lesser extent, by the gravititational force of the Sun.

trade winds The winds that blow steadily from the northeast in the Northern Hemisphere and southeast in the Southern Hemisphere between 5 and 30 degrees north and south latitudes.

transformer A device that changes the magnitude of the voltage and current of an electric signal but does not by itself produce electric power.

transform plate boundary A boundary between two lithospheric plates where the plates are sliding horizontally past one another.

translational motion The movement of an object such that the entire object travels from one place to another.

transpiration The vaporization of water through the tissues of plants, especially through leaf surfaces.

transmutation (of elements) The conversion of one element to another.

transverse wave See *wave*.

tritium Radioactive isotope of hydrogen with mass number of 3.

troposphere The layer of air that lies closest to the surface of the Earth and extends upward to about 12 km.

trough The lowest point in a wave.

uncertainty principle The theory that tells us that we cannot know both the position and the velocity of a particle with infinite accuracy.

valence The chemical combining capacity of an element.

valence shell   The highest-energy electron shell of an atom, which houses the electrons usually involved in chemical changes.

vaporization   The transformation of liquid water to water vapor.

vector quantity   A quantity that has both magnitude and direction. Velocity, force, and acceleration are all vector quantities.

vein (of a rock)   A thin layer of one type of rock embedded in a dominant rock formation.

velocity   A description of the speed of a body and its direction of motion. Velocity is a vector quantity.

vibrational motion   The movement of an object, or a portion of the object, back and forth without any permanent displacement away from a fixed position. The motion of a struck tuning fork is vibrational.

volcano   A fissure in the Earth's crust through which lava, steam, and other substances are expelled.

volt   A measure of the electric potential energy per unit charge. Voltage is a potential and must always be measured with respect to some other point.

wake   A high-energy wave produced when an object such as a boat or an airplane moves through some medium such as water or air at a rate faster than the speed of the wave.

water table   The upper level of water in the zone of saturated subsurface soil and rock.

watt   The SI unit of power. A watt is a joule per second.

wave   A periodic disturbance in some medium. A wave carries energy from one point to another, but there is no net movement of materials. A *longitudinal* or *elastic wave* is a wave that manifests itself as a series of compressions and expansions of an elastic medium. Sound waves are elastic waves. A *transverse wave* moves at right angles to the motion of the medium along which it travels. *Electromagnetic waves* are qualititively different from all other waves. They can be propagated in a vacuum.

wavelength   The distance between successive disturbances of the same type in a wave, such as between neighboring crests.

weather   The temperature, wind, and precipitation conditions that prevail in a given region on a particular day. See also *climate*.

weathering   The sum of processes that fracture and decompose surface rock.

weight   The force of gravity acting on a body.

white dwarf   A stage in the life of a star when fusion has halted and the star glows solely from the residual heat produced during past eras. White dwarfs are very small stars.

work   The energy expended when something is forced to move. Work is defined as the force exerted on an object multiplied by the distance that the object is forced to travel.

zone of saturation   The subsurface zone of soil and rock that is completely saturated with water.

Index

ff = following pages   fb = focus box

A

A horizon, 508
Absolute time scale, 469
Absolute zero, 96, 348
Absorption spectrum, 252, 610 ff
Acceleration, 27
    acceleration due to gravity, 28
    constant acceleration, 29
Acid, 390 ff
Acid mine drainage, 531
Acid rain, 394 ff
Acidic solutions, 390
Action (reaction), 36
Action-at-a-distance forces, 32
Adiabatic cooling, 430
Air mass, 442
Albedo, 421
Alchemy, 294
Alcohol, 378
Alkali metals, 365
Allotropes, 339
Alloys, 343
Alpha decay, 294
Alpha particle, 273, 294
Alternating current, 206
Amines, 378
Ampere, 171
Amphibians (evolution of), 473
Amplitude of wave, 135
Anaerobic, 416
Anderson, Carl, 315
Andes, 482
Angle of incidence, 216
Angle of reflection, 216
Anion, 370
Angular momentum, 68
Annihilation, 316
Anode, 271
Antimatter, 315
Antiparticle, 315
Antiproton, 316
Apparent brightness, 614

Aquifer, 522
Aquinas, St. Thomas, 555
Archean Eon, 470
Aristotle, 23, 75, 555 ff
Artificially produced radioisotope,
    299
Asteroids, 601
    Ceres, 601
Asthenosphere, 464
Astronomical unit (AU), 559,
    610 fb
Atmosphere, 415 ff
    composition and climate change,
    449
    composition of, 417
    evolution of 415 ff
Atmospheric circulation, 428
Atmospheric pressure, 418, 425
Atom, 269 ff
Atomic bomb, 310 fb
Atomic mass, 389
Atomic nucleus, 274
Atomic number, 275
Aurora, 199
Autumnal equinox, 553
Avogadro, Amedeo, 349
    Avogadro's law, 349 ff, 351

B

B horizon, 509
Bacteria, magnetic, 192 fb
Balanced equation, 366
Bar, 419
Barometer, 419
Barometric pressure, 418, 437
Basalt, 466
Base, 390 ff
Base units, 8
Basic, 390
Barton, Jacqueline (interview), 334
Battery, 45

Bauxite, 502, 510
Beats, 146
Becquerel, Henri, 291
Bell, Jocelyn, 622
Beta decay, 295
Beta rays, 294
Big Bang Theory, 637
Binary number system, 284
Black dwarf, 619
Black hole, 622 ff
Blue-shift, 613
Body waves, 461
Bohr, Niels, 247, 276
Bohr atom, 276 ff
Boiling, 99 fb
Bombardment reaction, 299
Boyle, Robert, 346
    Boyle's law, 346, 351
Brahe, Tycho, 557
Brightness
    apparent, 614
    intrinsic, 614
British thermal unit (BTU), 98
Broglie, Louis de, 278
Buckminsterfullerene, 339
Buckyballs 339, fb
Bunsen, Robert, 252
Burglar alarm, 244 fb
Byte, 284

C

C horizon, 509
Callisto, 593
Calorie, 98
Caloric, 97
Cap rock, 503
Capillary action, 510
Carbon dating, 300 ff
Carbon dioxide (as a greenhouse
    gas), 450, 454
Carbon-oxygen cycle, 450

Carboniferous period, 474
Carboxylic acids, 378
Catalysis, 397
Catalyst, 397
Catalytic converter, 403
Cathode, 271
Cathode ray tube, 271
Cathode rays, 271
Cation, 370
CD player, 284
Celestial equator, 553
Celestial sphere, 553
Celsius, 96
Cenozoic era, 476
Centrifugal force, 89
Centripetal acceleration, 84 ff
Centripetal force, 85
Ceres, 601
Chadwick, James, 299
Chain reaction, 317 ff
Change of phase, 103 ff
Charles, Jacques, 347
    Charles' law, 347, 351
Charon, 599
Chemical changes, 270, 337
Chemical equation, 366 ff
Chemical forces, 343
Chemical formula, 342
Chemical reaction, 337
Chemistry, 333 ff.
Chlorofluorocarbons, 406, 454
Chordates, 473
Chu, Steven (interview), 18
Circular motion, 83 ff
Circumpolar constellation, 546
Cirrus clouds, 435
Clean Water Act, 520
Climate, 423, 438, 448 ff
Clouds, 435 ff
    cirrus, 435
    cumulus, 436
    nimbus, 436
    stratus, 435
Clusters, 633
Coal, 502
Cogeneration, 126
Coherent, 281
Cold front, 444
Color, 256 fb
    primary additive, 256
    subtractive primaries, 257
Combustion, 392
Comets, 604
    Halley's Comet, 603 spt
Compact disk, 284
Compound, 270, 342 ff

Concave mirror, 219
Condensation, 433 ff
Condensation nuclei, 435
Conduction, 105
Conduction band, 383
Conductor, 105, 172
Conservation of angular momen-
    tum, 68
Conservation of energy, 61
Conservation of matter, 337
Conservation of momentum, 53
Constellations, 545
Contact forces, 32
Continental drift, 476 ff
Continuous spectra, 250, 610 ff
Contour plowing, 512
Control rods, 321
Convection, 107
Converging lens, 225
Convergent boundary, 481
Conversion factors, 10
Convex mirror, 220
Coolant, 321
Cooling ponds, 124
Copernicus, 230, 556
Core
    of Earth, 463
    reactor, 319
Coriolis effect, 427
Cosmic background radiation, 259
Cosmic rays, 198
Cosmology, 637
Coulomb, 158, 289
Coulomb, Charles, 158
    Coulomb's law, 158
Covalent bond, 373
Covalent substances, 373 ff
Critical mass, 310 fb
Crust of earth, 463
Crystal, 353
Crystal lattice, 353
Crystal settling, 499
Crystalline solid, 353
Cumulus clouds, 436
Curie Marie and Pierre, 292
Currents
    ocean, 439
    electrical, 169
Cyclic structures, 377

D

Dalton, John, 270 fb
Dark matter, 635, 640
Davisson, C. J., 278
de Broglie, Louis, 278

Decibel, 139
Deferent, 580
Deforestation (and climate
    change), 454
Deimos, 590
Democritus, 270
Density, 44, 383
Derived units, 8
Detergent, 403
Dew, 435
Dew point, 435
Diffraction grating, 249
Dinosaurs, 475
Dirac, P. A. M., 315
Direct current, 169 ff
Displacement, 25
Disseminated ore deposits, 500
Divergent boundary, 480
Diverging lens, 225
DNA, 379 fb
Doldrums, 431
Domains, magnetic, 198
Doppler, Christian, 150
Doppler effect, 150 ff
    and light, 254
    and stars, 611
Dot symbols, Lewis, 371
Double bonds, 374
Ductile, 383
Dust, 453, 455
Dynamics, 22

E

Earth
    history of, 469
    interior of, 463
    structure of, 461
Earthquake waves, 461
Earthquakes, 486 ff
East Pacific rise, 485
Eclipses
    of Sun and Moon, 551
Ecliptic, 553
Einstein, Albert, 304 ff.
Electric charge, 158, 289
Electric circuit, 168 ff
    parallel circuit, 176
    series circuit, 174
Electric field, 160
Electric generator, 204
Electric motor, 203
Electrical conductivity, 383
Electrical potential, 165
Electrical potential energy, 165
Electric power, 179

Electricity, 155 ff
  conductors, 172
  electric current, 169
  electric field, 160
  insulators, 172
  Ohm's law, 173 ff
  positive/negative poles, 156
  resistance, 171
  static electricity, 155
  superconductor, 172 fb
  voltage (potential difference), 165
Electrodes, 369
Electromagnet(s), 194
Electromagnetic spectrum, 255 ff.
Electromagnetic wave, 108, 261 ff.
Electromagnetism, 184, 343
Electron, 156, 272 ff.
Electron shell, 360
Electronegative, 375
Elements, 270, 338 ff.
Emission spectra, 251, 610 ff
Energy, 57
  gravitational potential, 60
  kinetic, 57, 70
  law of conservation of, 61
  mass equivalence of, 314
  thermal, 65
Energy balance of the Earth, 420
Entropy, 120
Eon (geologic), 470
Enzymes, 398
Epicycles, 556
Epoch (geologic), 470
Equinox, 424, 553
Erosion, 497, 510
Ether, 305 ff
Europa, 593
Eutrophication, 518
Evaporation (or vaporization), 99, 574
Evaporite deposits, 502
Excited state, 247
Exponential notation, 11 ff
Extensive properties, 338
Extraterrestrial life, 598 fb
Exxon Valdez, 527 fb
Eye, 228

F

Faraday, Michael, 200, 159
Farenheit, 96
Farsighted eye, 229
Fault, 487
Fault creep, 487

Fermi, Enrico, 297
Fertilizer, 513
Fiber optics, 232 fb
Fields
  electric, 160 ff
  magnetic, 189 ff
Fish (evolution of), 473
Fission, see nuclear fission
Focal length, 220, 226
Fog, 435
Force, 25 ff.
  action-at-a-distance force, 32
  contact force, 32
  resultant (net) force, 31
Foreshocks, 492
Fossil fuels, 502 ff
Franklin, Benjamin, 271
Fraunhofer, Joseph von, 253
Free fall, 28
Freezing rain, 438
Frequency of wave, 133
Friction, 42
  Kinetic friction, 43
  static friction, 42
Fronts, see Weather fronts
Frost, 435
Fulcrum, 68
Fusion, see Nuclear fusion

G

Galaxies, 629 ff
  Andromeda, 633
  barred spiral galaxy, 634
  elliptical galaxy, 634
  irregular galaxies, 634
  Milky Way, 629
  spiral galaxy, 633
Galileo, 23, 213, 230, 560, 572
Galle, 597
Gamma decay, 295
Gamma rays, 261, 294
Ganymede, 593
Gas, 344 ff
  gas pressure, 349 fb
  kinetic theory of gases, 350
Geiger, 273
Geiger counter, 292 fb
Generator, electric, 204
Geocentric model, 230, 555
Geologic time, 469
Geologic time scale, 470
Germer, L., 278
Giant molecule, 381
Glacier, 497
Glass (glassy solid), 354

Globular clusters, 630
Gneiss, 468
Gondwanaland, 477
Grand Canyon, 467
Granite, 466
Gravitational field, 159
Gravitational potential energy (PE), 60
Gravity, 74 ff.
  artificial, 86
  law of, 76
  universal gravitational constant (G), 76
Greenhouse effect, 109
Ground state, 246
Groundwater, 521 ff
  pollution of, 523 ff
Guitars, electric, 201
Gulf stream, 439
Gyres, 439

H

Haagen-Smit, A. J., 401
Hadley, George, 428
Hahn, Otto, 316
Hail, 438
Half-life, 296 ff, 470
Halley's Comet, 603 fb
Halogens, 365
Hard water, 404
Heat, 98
  of evaporation, 103
  of fusion, 103
Heat engine, 115 ff
Heat radiation, see Infrared radiation
Heliocentric model, 230, 556
Henry, Joseph, 199
Herschel, William, 630
Hertzsprung-Russell diagram, 615
Heterogeneous, 342
High pressure, 425, 429
Himalayas, 483
Hologram, 283 fb, 284
Homogeneous, 342
Hooke, Robert, 239
Horner, John, 475
Horse latitudes, 431
Hubble, Edwin, 613, 631
Hubble's law, 613
Humans (evolution of), 476
Humidity, relative 433, 434 fb
Humus, 508
Hydrocarbons, 376
Hydrogen bond, 375
Hydrologic cycle, 541

Hydrophilic, 404
Hydrophobic, 404
Huygens, Chriatian, 239

I

Igneous rock, 465
Images, 217
    real images, 217
    virtual image, 217
Impermeable, 503
Inclined plane, 66
Indicators, 390
Induced voltage, 200
Inertia, 35
Infrared radiation, 108, 255, 260
Insulators, 105
Intensive properties (specific properties), 338
Interference(waves), 140 ff
    constructive, 142
    destructive, 142
Intermolecular forces, 344
Internal combustion engine, 117
Internal energy, 114 ff
International System of Units, 8
Intrinsic brightness, 614
Io, 592
Ion, 370
Ionic substances, 368 ff
Isobars, 425
Isoelectronic, 371
Isomers, 377
Isostacy 464
Isotopes, 275, 338

J

Jet stream, 432
    polar jet stream, 432
    subtropical jet stream, 432
Joliot-Curie, Irene and Frederick, 299
Joule (J), 56
Jovian planets, 571
Jupiter, 589 ff
    Callisto, 593
    Europa, 593
    Ganymede, 593
    Great Red Spot, 590
    Io, 592
    Moons, 591 ff

K

Kelvin (K), 96, 348
Kepler, Johannes, 557 ff
    First law, 558
    Second law, 559
    Third law, 559
Kilogram, 9
Kinematics, 22
Kinetic energy, 57
    derivation of equation, 70
Kinetic friction (force of), 43
Kinetic theory of gases, 350
Kirchhoff, Gustav, 252
Krakatoa, Mount, 494

L

Labrador current, 439
Laser, 280 ff
Laterite, 510
Laurasia, 477
Lava, 466
Leaching, 509
Lenses, 225 ff
    converging, 225
    diverging, 225
Leucippus, 270
Lever, 68
Leverrier, 597
Lewis dot symbols, 371
Lewis, Gilbert, 371, 373
Light-year, 214
Limestone, 467
Lipari landfill, 525 fb
Litter, 508
Liquid, 344, 351 ff
Liquid crystal, 345, 354
Liquid metallic hydrogen, 590
Lithosphere, 463
Litmus, 390
Loam, 508
Local group, 635
Longitudinal wave, 131
Los Angeles earthquake, 490 fb
Low pressure, 425, 429
Lowell, Percival, 586
Luster, 383

M

Mach number, 149
Machines, simple, 66 ff
Magellanic cloud,
    large, 621, 634
    small, 634
Magma, 465
Magmatic processes, 499
Magnetic domain, 198
Magnetic field, 189 ff
Magnetosphere, 576
Magnets, 188

Magnifier, 226
Main group elements, 363
Malleable, 383
Mammals (evolution of), 476
Manganese nodules, 532
Mantle, 463
Marble, 468
Maria, 578
Mars, 586 ff
    Moons, 588
Marsden, 273
Mass, 36, 314
    and energy, 314
Mass extinction, 474
Mass number, 275
Mechanics, 22
Meitner, Lise, 316
Mendeleev, Dmitrii Ivanovich, 363
Mercury, 581 ff
Mesosphere, 420
Mesozoic era, 474
Messier, Charles, 631
Metal, 340, 382 ff
Metallic bond, 383
Metalloids, 341
Metamorphic rock, 468
Metamorphism, 468
Methane (as a greenhouse gas), 454
Meteorite bombardment, 577 fb
Meteoroids, 600
    meteorite, 600
Meteorology, 415 ff.
Meter, 9
Meyer, Julias Lothar, 363
Michelson, Albert, 213
    Michelson-Morley experiment 306 ff
Microscope, 229
Microwaves, 258
Mid-Atlantic ridge, 479
Mid-Oceanic ridge, 479
Migration of petroleum, 503
Milky Way, see Galaxies
Miller-Urey experiment, 378
Mineral, 465
Mineral reserves, 499 ff, 529 ff
Mirage, 223
Mirrors
    Concave, 219
    Convex, 220
    rearview, 218
Mixture, 342
Moderator, 319
Moho, 461
Mohorovicic, Andrija, 461
Mohorovicic discontinuity, 461
Mole, 388 ff.

Molecular mass, 390
Molecule, 270, 342 ff, 373
Momentum, 53
    law of conservation of, 53 ff.
    angular, 68
Monsoons, 441
Moon, 549 ff,
    crescent, 549
    full, 549
    lunar eclipse, 551
    Maria, 578
    new, 549
    waning phases, 551
    waxing phases, 551
Motors, electric, 203
Muons, 313

N

Nearsighted eye, 229 f b
Nebular hypothesis, 569
Neptune, 596
    Great Dark Spot, 597
    Triton and Nereid, 598
Network covalent substances, 380 ff
Neutral, 391
Neutralization, 390, 391
Neutrino, 297, 621
Neutron, 156, 275
Neutron moderator, 319
Neutron star, 622 ff
Newlands, John, 363
Newton, (unit of force), 40
Newton, Isaac, 30, 562
    first law of motion, 34 ff.
    light, 239
    second law of motion, 39 ff.
    third law of motion, 36 ff.
Nimbus clouds, 436
Nitrogen oxides (as greenhouse
    gases), 454
Noble gases, 363, 380
Noise, 140 f b
Nuclear fission, 316 ff
Nuclear fuel, 318
Nuclear fusion, 322 ff
Nuclear power plant, 319
Nuclear warfare, 456
Nucleus, 156

O

O horizon, 508
Obsidian, 466
Occluded front, 446
Oceanic trench, 481
Oceans, 438 ff, 526

Octet rule, 373
Oersted, Hans Christian, 192
Ogallala aquifer, 522
Ohm, George Simon, 173
Ohm's law, 173 ff
Oil, see petroleum
Oil trap, 503
Optically active, 264
Orbit (Earth) and climate change,
    453
Ore, 529
Organic chemistry, 376 ff
Oscillating universe, 640
Oxidation, 392
Oxygen
    formation of in atmosphere, 416
Oxygen-carbon cycle, 449
Ozone, 399 ff, 420, 456

P

P waves, 461
Pair annihilation, 315
Pair production, 315
Paleozoic era, 473
Pangaea, 477
Parallax, 548, 612 f b
Parallel (circuit), 176
Pauli, W., 297
Pedalfer, 510
Pedocal, 510
Peery, Ben (interview), PO4.3
Periods (geologic), 470
Periodic table, 362 ff
    groups, 363
    periods, 363
Permeable, 503
Petroleum, 503
Permian period, 474
pH, 391 ff
Phanerozoic eon, 473
Phobos, 589
Photochemical smog, 402
Photoelectric effect, 242 ff
Photon, 243
Photosynthesis, 417
Physical change, 337
Pinatubo, Mount, 448, 452 f b
Placer deposit, 500
Planet, 553 ff
Planetesimals, 571
Plank's constant (h), 243, 278
Plasma, 345, 573
Plate tectonics, 479
Plum pudding model, 272
Pluto, 599
    Charon, 599

Polar easterlies, 431
Polar front, 431
Polarized light, 263 ff
    polarized wave, 263
    polarizer, 264
    Polaroid, 263
    unpolarized light, 263
Pollution (automobile), 401 if
    water, see Water pollution
Positron, 315
Potential difference, 167
Potential energy, see Energy
Power, 65, 179
Powers of telescope, 231 ff
Precambrian time, 470
Precipitation
    meteorological 436, 514
    ore formation, 501
Pressure, 46
    atmospheric, 418 , 425
Pressure gradients, 425
Prevailing westerlies, 431
Primary colors, 256
Principal quantum number, 360
Principle of superposition, 469
Projectile motion, 80 ff
Proterozoic eon, 470
Proton, 156, 274,
Protostar, 616
Protosun, 570
Proxima Centauri, 609
Ptolemy, Claudius, 556
Pulsars, 622 ff
    lighthouse theory, 624
Pythagorean theorem, 32

Q

Quantum, 243
Quark, 158, 326
Quasar, 636

R

Radiation, 108
Radio waves, 255 ff
    AM, 255
    FM, 255
Radioactivity, 291 ff
Radon, 291 f b
Rain, 437
Rain shadow desert, 441
Rainbow, 226 f b
Ray, 215
Rearview mirror, 218 f b
Red giant, 616, 618
Red-shift, 611

Reduction, 392
Reference frame, 304
Reflecting telescope, 234
Reflection, 216 ff
    angle of incidence, 216
    angle of reflection, 216
    Law of reflection, 216
    total internal reflection, 235
Refracting telescope, 230
Refraction, 221 ff
Relative time scale, 469
Relativity, 304 ff.
    energy, 314
    length, 312
    mass, 314
    special relativity, 308
    time, 309,
Renewable (or nonrenewable) min-
    eral resources, 529 ff
Reptiles (evolution of), 473
Residence time, 515
Residual deposits, 502
Resistance, 171
Resonance (waves), 144 fb
Resultant force, 31
Retrograde motion, 553
Revolve, 547
Richter scale, 488 fb
Rock, 465
Rock cycle, 465
Rockies, 484
Rossby, Carl, 429
Rotate, 547
Runoff, 515
Rutherford, Lord Ernest, 273, 294,
    298

S
S wave, 461
SI, see International System of
    Units
Saint Helens, Mount, 452 fb, 494
Salinization, 517
Saltwater intrusion, 523
San Andreas fault, 482, 485, 487
San Francisco earthquake, 489
Sandstone, 467
Saturn, 594
    rings, 594
    shepherd moons, 596
    tidal forces, 595
    Titan, 596
Scalar, 24
Schneider, Stephen, (interview)
    410

Schrödinger, Erwin, 279
Scrubbers, 397
Sea breezes, 439
Seatbelts, 37
Sediment, 466
Sedimentary rock, 466
Sedimentary sorting, 500
Seismic gap theory, 491
Seismology, 461
Seismometers fb, 464
Series (circuit), 174 ff
Shadow zone, 462
Shale, 467
Shapley, Harlow, 630
Shoemaker-Levy, 602
Significant figures, 10
Simple machines, 66 ff
Sleet, 438
Smog, 402 ff
Smoke detector, 293 fb
Snow, 438
Soap, 403
Soddy, Frederick, 294
Soft water, 404
Soil, 507 ff
Soil composition, 508
Soil erosion, 510 ff
Soil horizons, 508
Soil nutrients, 508
Solar System, 569 ff
Solar wind, 575
Solid, 345, 353 ff
Solstice, 423, 553
Solute, 343
Solution, 343
Solvent, 343
Sonic boom, 148 fb
Sound waves, 135 ff
    beats, 146
    constructive interference, 142
    decibel, 139
    destructive interference, 142
    Doppler effect, 150
    elasticity, 137
    fundamental frequency, 143
    interference, 140
    overtone, 144
    resonance, 144 fb
    sonic booms, 148 fb
    speed of sound, 137
    standing waves, 143
    threshold of hearing, 139
    threshold of pain, 139
    ultrasound, 138
Special relativity, see Relativity
Specific heat, 100 ff, 438

Spectroscope, 250
Spectrum, 250
    absorption spectra, 252, 610
    continuous spectrum, 250, 610
    emission spectra, 251, 610
    red-shift or blue-shift, 611
Speed, 23
    average speed, 24 ff
    instantaneous speed, 24 ff.
Speed of light, 213
Spreading center, 480
Stable state, 281
Standing wave, 143 ff
Stars, 609 ff
Static electricity, 155
Static friction (force of), 42
Steam, 345
Stimulated emission, 280
Stratosphere, 420
Stratus clouds, 435
Strong nuclear force, 289
Structural formula, 376
Subduction, 481
Subsidence, 522
Substance, 338 ff, 342
Subtractive colors, 257
Summer solstice, 423, 553
Sun, 572 ff
    auroras, 199, 576
    chromosphere, 575
    core, 573
    corona, 575
    magnetosphere, 576
    photosphere, 573
    solar wind, 575
    sunspots, 573
Superclusters, 635
Superconductor, 172 fb
Supernova, 621
Superposition principle, 469
Surface waves, 461

T
Tambora, Mount, 452 fb
Tape recorders, 202 fb
Tectonic plates, 479 ff
Telephone frequencies, 133 fb
Telescopes, 230 ff
    eyepiece, 230
    light gathering power, 231
    magnifying power, 231
    objective lens, 230
    reflecting telescope, 234
    refracting telescope, 230
    resolving power, 231

Television, 166 fb
Temperature, 95 ff, 98
    Celsius, 96
    Farenheit, 96
    Kelvin (K), 96, 348
Terraces, 512
Theory of continental drift, 476 ff
Theory of sea floor spreading, 479
Thermal energy, 65
Thermal pollution, 124
Thermodynamics, 113
    first law of thermodynamics, 115
    second law of thermodynamics,
        118 ff
Thermonuclear reaction, 323
Thermosphere, 420
Thompson, B., 98
Thomson, J.J., 271
Tides, 76 fb
Titan, 596
Topsoil, 508
Total internal reflection, 235
Trade winds, 431
Transform plate boundary, 482,
    487
Transformer, 207
    primary coil, 207
    secondary coil, 207
    step-up/step-down transformers,
        207
Transmutation, 294
Transpiration, 514
Transverse wave, 130
Triangle method, 25
Triple bonds, 374
Triton, 598
Troposphere, 420
Turbine, 123

U

Ultraviolet radiation, 260
Upper air winds, 432

Upper mantle, 463
Uranus, 596 ff
Urbanization of farmland, 513

V

Valence, 361
Valence electron, 366
Valence shells, 360
Van Allen, James, 199
Van de Graaff, Robert J., 163
Van de Graaff generator, 163 ff
Vaporization, 514
Vector, 25
Vein, 500
Velocity, 26
Venus, 582 ff
Vernal equinox, 424, 553
Viscosity, 352
Volcanic glass, 466
Volcanoes, 452 fb, 466, 483 ff
Voltage, 165

W

Warm front, 442
Water cycle, see hydrologic cycle
Water diversion, 516 ff
Water pollution, 518 ff
    chemical wastes, 519 ff
    nutrients, 518
    oil pollution, 526 ff
Water table, 521
Watt (W), 65
Wave, 130 ff
    amplitude, 135
    compression, 136
    constructive interference, 142
    destructive interference, 142
    frequency, 133
    Hertz, 134
    interference, 142
    longitudinal wave, 130

    rarefaction, 136
    sound waves, 136 ff
    standing wave, 143
    transverse wave, 130
    wavelength, 134
Wave function, 279
Wavelength, 134
Weather, 441 ff
Weather fronts, 442 ff
    cold front, 444
    occluded front, 446
    warm, 442
Weather maps, 444, 447
Weathering, 495 ff
    chemical, 495
    mechanical, 495
    ore deposits from, 502
Wegner, Alfred, 477 ff.
Weight, 40
Weightlessness, 79
Westerlies, 431
White dwarf, 616, 619
Wind, 425 ff, 496
Wind chill index, 426 fb
Winter solstice, 424, 553
Work, 56, 114

X

X-rays, 261

Y

Young, Thomas, 240
    Double-slit experiment, 240

Z

Zodiac, 554
Zone of accumulation, 509
Zone of leaching, 509
Zone of saturation, 521

is

A Harcourt Higher Learning Company

Now you will find Saunders College Publishing's distinguished innovation, leadership, and support under a different name . . . a new brand that continues our unsurpassed quality, service, and commitment to education.

We are combining the strengths of our college imprints into one worldwide brand: Harcourt

Our mission is to make learning accessible to anyone, anywhere, anytime—reinforcing our commitment to lifelong learning.

We are now Harcourt College Publishers. Ask for us by name.

One Company
"Where Learning Comes to Life."

www.harcourtcollege.com
www.harcourt.com